Encyclopaedia of

LINEAR ALGEBRA

Encyclopaedia of
LINEAR ALGEBRA

Volume 2

R.K. Pandey

ANMOL PUBLICATIONS PVT. LTD.
NEW DELHI - 110 002 (INDIA)

ANMOL PUBLICATIONS PVT. LTD.
H.O.: 4374/4B, Ansari Road, Daryaganj,
New Delhi-110 002 (India)
Ph.: 23278000, 23261597
B.O.: No. 1015, Ist Main Road, BSK IIIrd Stage
IIIrd Phase, IIIrd Block,
Bangalore - 560 085 (India)
Visit us at: www.anmolpublications.com

Encyclopaedia of Linear Algebra

First Edition, 2009

ISBN 978-81-261-3914-9 (Set)

PRINTED IN INDIA

Printed at Mehra Offset Press, Delhi.

Contents

Preface

Linear Algebra is one of the foundations of modern Natural Science: wherever we are interested in calculations, from Engineering to String Theory, we use Linear Equations, Quadratic Forms, Matrices, Linear maps and their Cohomologies. There is a widespread feeling that the non-linear world is very different, and it is usually studied as a sophisticated phenomenon of Interpolation between different approximately-Linear Regimes.

Our aim in this book is basically a very simple one: to show how the familiar methods and ideas used in the analysis of the system of "Numbers" have been and are being extended and generalized to enable more diverse mathematical systems to be analyzed. In particular the concept of Factorization or Decomposition into "simpler" pieces underlies essentially all of our efforts. However, before we can proceed with this programme of study we must try to establish a common starting a point and path of approach. Even before doing this we must agree upon some basic assumptions, terms and notations, and these are set down in this book.

The book is written for a first-year graduate student. Very few undergraduates would have the necessary background and mathematical sophistication to appreciate this book. Not only can the book be used as a textbook but engineers and scientists will frequently consult it as a reference.

Author

List of Symbols

±	⊄	⊂	⊆	∈	∉	ℜ	∂	≠	≈	∠	∇
Σ	Π	Ω	ϕ	α	β	χ	ε	γ	η	λ	μ
ν	⊥	π	θ	ρ	σ	τ	ℜ	ω	ξ	ψ	ζ
ϒ		∃	∫	∬	∭	×	√	∵	≥	∮	⇒

Chapter 1

Determinant

GEOMETRY OF DETERMINANT

DETERMINANT OF 2 BY 2 MATRIX

The determinant of a 1 by 1 matrix is $det(a) = a$.

The determinant of a 2 by 2 matrix is

$$\det\begin{bmatrix} a & b \\ c & d \end{bmatrix} = ad - bc.$$

The expression appeared in the criterion for the invertibility of a 2 by 2 matrix: A 2 by 2 matrix A is invertible $\Leftrightarrow$ the determinant is nonzero. The criterion clearly also holds for a 1 by 1 matrix, and will be shown to be true for general n by n matrices.

Let $u = (a, c)$, $v = (b, d)$ be the two columns of the matrix. Then the determinant $det[u\ v]$ is a function of two vectors.

The absolute value of $det[u\ v]$ is the area of the parallelogram spanned by u and v.

The claim can be seen through the following picture.

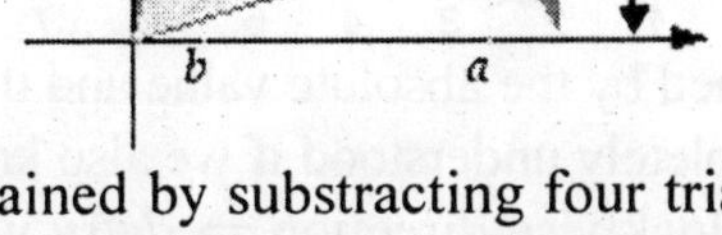

The parellelogram is obtained by substracting four triangles from a rectangle. Thus area of the parallelogram = $(a + b)(c + d) - (a + b)c - b(c + d) = ad - bc$.

Note that the argument is done for the vectors in the given positions, and the determinant turns out to be positive for the given situation. One may also carry out similar arguments for the other situations.

Example: The area of the triangle with the vertices (1, 2), (7, 1), (3, 5) is half of the area of the parallelogram spanned by $u = (7, 1) - (1, 2) = (6, -1)$ and $v = (3, 5) - (1, 2) = (2, 3)$. Thus

$$\text{area} = (1/2)\left| det\begin{bmatrix} 6 & 2 \\ -1 & 3 \end{bmatrix}\right| = 10.$$

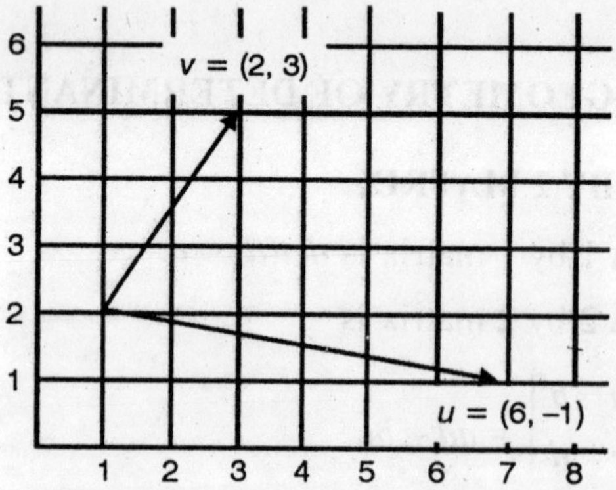

The area of the 4–gon with the vertices (1, 2), (5, 1), (7, 6), (3, 5) is the sum of two triangles. Thus

$$\text{area} = (1/2)\left| det\begin{bmatrix} 4 & 6 \\ -1 & 5 \end{bmatrix}\right| + (1/2)\left| det\begin{bmatrix} 6 & 2 \\ 5 & 3 \end{bmatrix}\right| = 17.$$

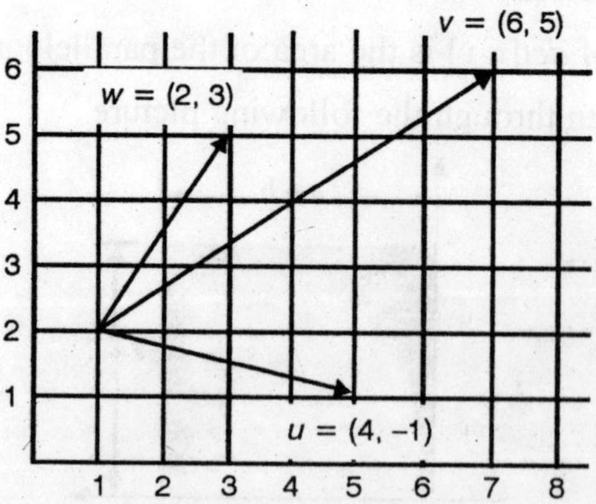

A real number is determined by the absolute value and the sign. Thus the determinant of a 2 by 2 matrix would be completely understood if we also know the meaning of the sign.

u moves to v in counterclockwise direction $\Rightarrow det[u\ v] \geq 0$.

u moves to v in clockwise direction $\Rightarrow det[u\ v] \leq 0$.

In other words, the sign is given by the orientation from the first colunm vector to the second.

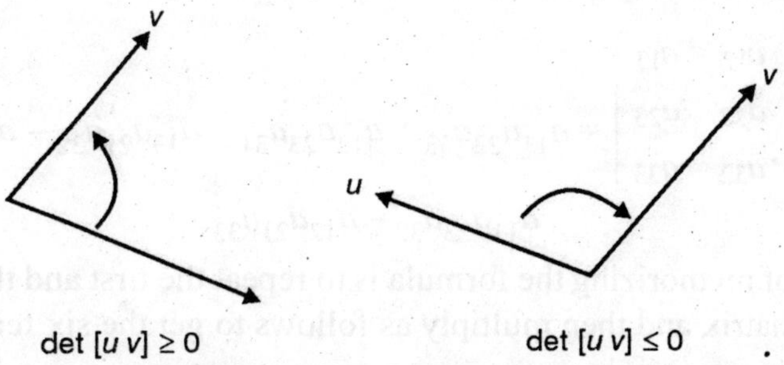

To see why the orientation determines the sign, we consider the following movie for the counterclockwise situation.

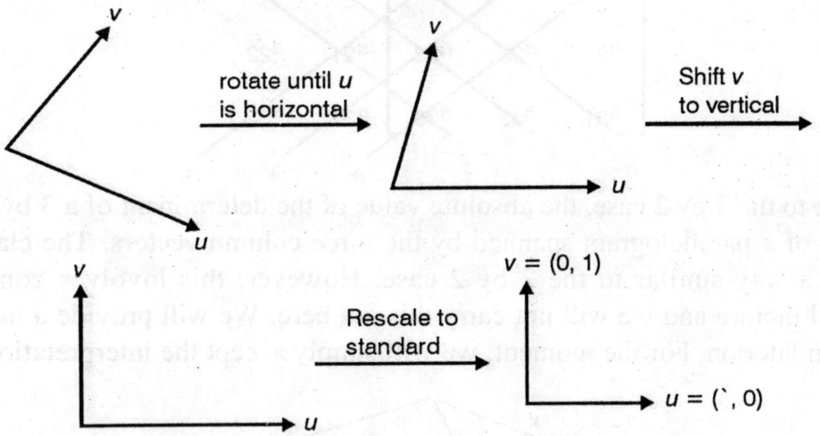

Since the vectors are changed continuously, the determinant is also changed continuously. Moreover, the determinant never becomes zero since the absolute value, being the area of the parallelogram, is obviously never zero.

Since a nonzero continuous function never changes sign, we conclude that $det[u\ v]$ and $det[e_1\ e_2] = 1$ have the same sign. In other words, $det[u\ v] > 0$.

For the clockwise situation, one may construct a similar movie and conclude that $det[u\ v]$ and $det[e_1\ -e_2] = -1$ have the same sign.

What about the case $det[u\ v] = 0$? In this case the area of the parallelogram is zero. This can happen only if *u* and *v* are *parallel.*

By an earlier discussion, this is equivalent to the vectors are linearly dependent. Then by this result, this is also equivalent to that the matrix $[u\ v]$ is not invertible.

Thus we have a geometrical proof of the earlier criterion for the invertibility of a 2 by 2 matrix.

DETERMINANT OF 3 BY 3 MATRIX

The determinant of a 3 by 3 matrix is

$$det\begin{bmatrix} a_{11} & a_{12} & a_{13} \\ a_{21} & a_{22} & a_{23} \\ a_{31} & a_{32} & a_{33} \end{bmatrix} = a_{11}a_{22}a_{33} + a_{12}a_{23}a_{31} + a_{13}a_{21}a_{32} - a_{13}a_{22}a_{31} - a_{11}a_{23}a_{32} - a_{12}a_{21}a_{33}.$$

A simple way of memorizing the formula is to repeat the first and the second columns to the right of the matrix and then multiply as follows to get the six terms.

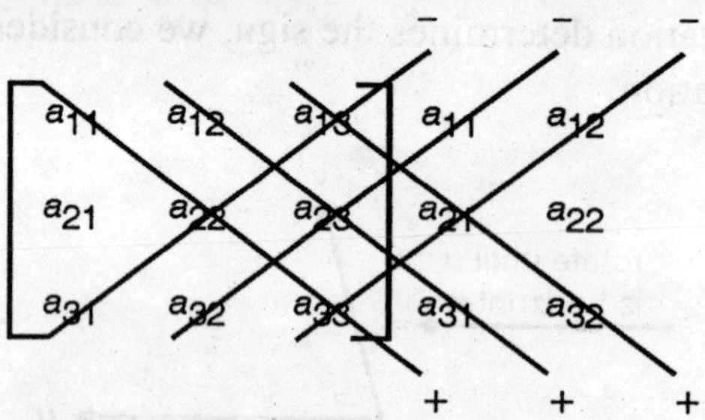

Similar to the 2 by 2 case, the absolute value of the determinant of a 3 by 3 matrix is the volume of a parallelogram spanned by the three column vectors. The claim may be verified in a way similar to the 2 by 2 case. However, this involves complicated 3 dimensional picture and we will not carry this out here. We will provide a more general explaination later on. For the moment, we will simply accept the interpretation.

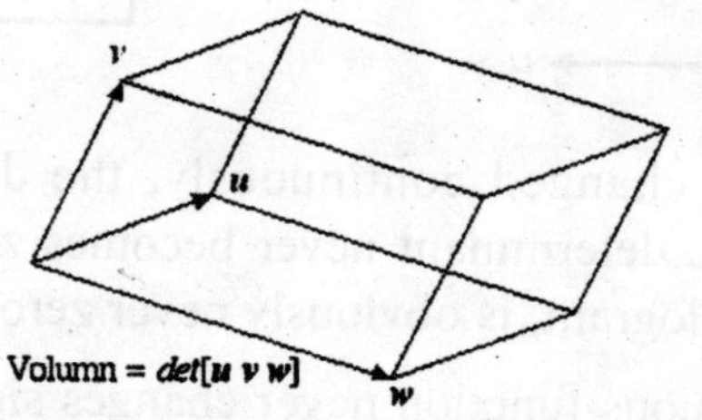

A consequence of the interpretation is that

$det[u\ v\ w] = 0$

$\Leftrightarrow$ $u\ v\ w$ lie on a plane

$\Leftrightarrow$ $u\ v\ w$ are linearly dependent

$\Leftrightarrow$ $[u\ v\ w]$ is not invertible.

This clearly generalizes the criterion for the invertibility of a 2 by 2 matrix. The criterion will still be true for general n by n matrices, as long as the absolute value of the general determinant still measures the volume of the parallelogram.

Example: The volume of the tetrahedron with the vertices (1, –2, 1), (2, 1, 2), (0, –1, 4), (2, –3, 1) is one sixth of the volume of the parallelogram spanned by $u = (2, 1, 2) - (1, -2, 1) = (1, 3, 1)$, $v = (0, -1, 4) - (1, -2, 1) = (-1, 1, 3)$, $w = (2, -3, 1) - (1, -2, 1) = (1, -1, 0)$, which is

$$(1/6)\left| det\begin{bmatrix} 1 & -1 & 1 \\ 3 & 1 & -1 \\ 1 & 3 & 0 \end{bmatrix}\right| = (1/6)|1 \times 1 \times 0 + (-1) \times (-1) \times 1 + 1 \times 3 \times 3 - 1 \times 1 \times 1$$

$$- 1 \times (-1) \times 3 - (-1) \times 3 \times 0 \,| = 2.$$

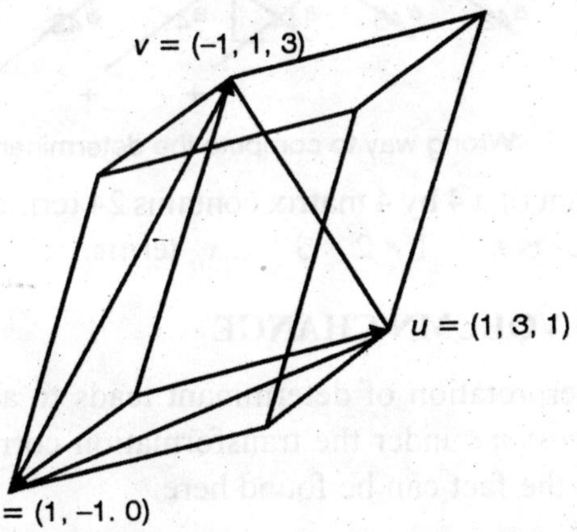

Similar to the 2 by 2 case, the sign of the determinant of a 3 by 3 matrix is also determined by the orientation of the three column vectors. Specifically,

u, *v*, *w* follows the right hand rule $\Rightarrow det[u\ v\ w] \geq 0$

u, *v*, *w* follows the left hand rule $\Rightarrow det[u\ v\ w] \leq 0$

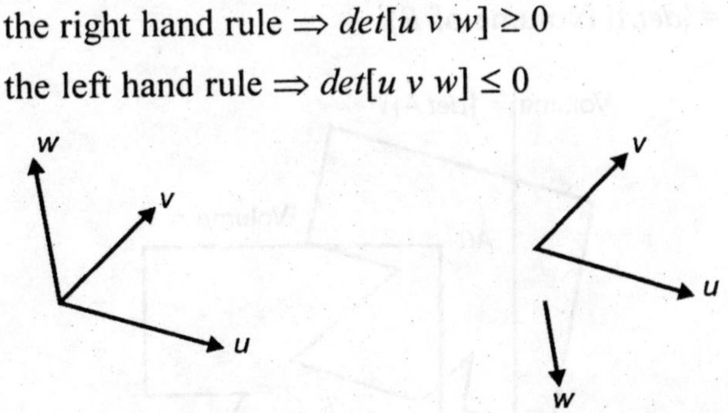

The reason for the orientation to determines the sign is similar to the 2 by 2 case. We may construct a (somewhat more complicated) movie to deform the given arrangement into one of the two standard arrangements.

The extension of the determinant to matrices of bigger size should reflect the requirements that the absolute value is the volume of a parallelogram and the sign should be the orientation of the arrangement of the column vectors. We emphasis that the following naive extension to 4 by 4 matrices does not satisfy the requirements.

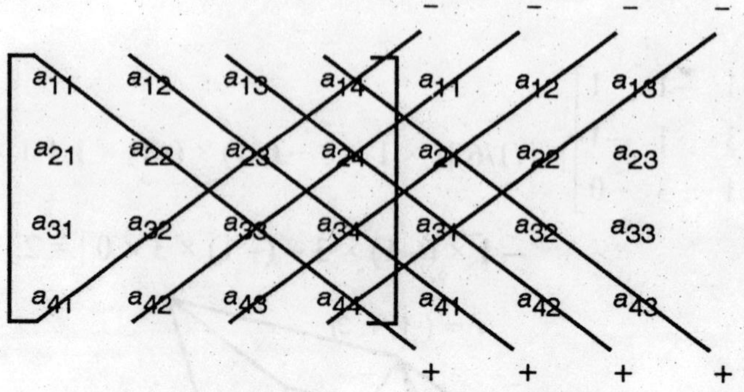

Wrong way to compute the determinant

In fact, the determinant of a 4 by 4 matrix contains 24 terms. In general, the determinant of an n by n matrix contains $n! = 1 \times 2 \times 3 \times \ldots x_n$ terms.

DETERMINANT AND VOLUMN CHANGE

The geometrical interpretation of determinant leads to another interpretation as the *ratio for the change of volume* under the transformation corresponding to the matrix. A detailed explaination for the fact can be found here.

Let A be an n by n matrix. Let R be a region in R^n, and AR be the image under the transformation given by A. Then

(volume of AR) = $|detA|$ (volume of R).

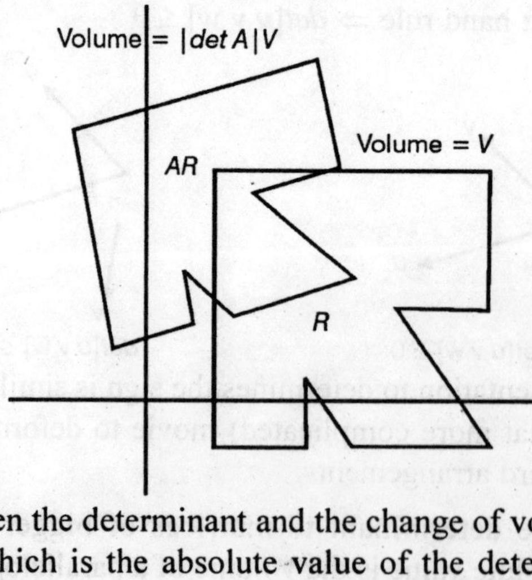

The relation between the determinant and the change of volume is the reason behind the use of Jacobian (which is the absolute value of the determinant of the first order derivative matrix) in the change of variable formula in multivariable integration.

Example: In an earlier example, we computed the area of the triangle R with vertices $(1, 2)$, $(7, 1)$, $(3, 5)$. The conclusion was area(R) = 10. If we apply the linear transformation by the matrix

$$A = \begin{bmatrix} 2 & -1 \\ 1 & 1 \end{bmatrix},$$

then the image AR is a triangle with vertices $(0, 3)$, $(13, 8)$, $(1, 8)$.

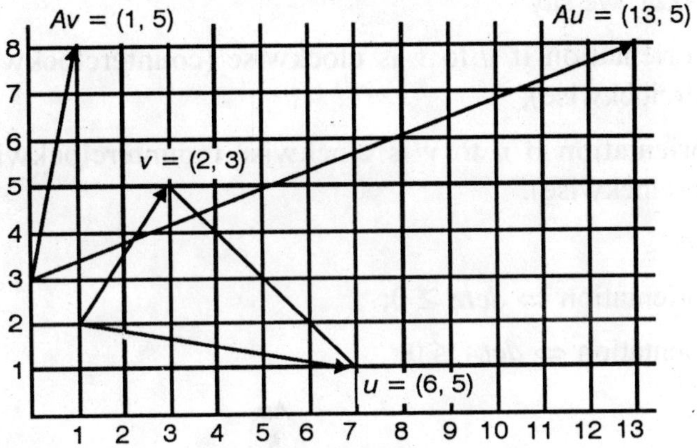

As in the earlier example, we have

$$\text{area}(AR) = (1/2)\left|\, det\begin{bmatrix} 13-0 & 1-0 \\ 8-3 & 8-3 \end{bmatrix}\right| = 30.$$

On the other hand, we have $detA = 3$, and the equality area(AR) = $|detA|$ area(R) indeed holds.

Example: We would like to compute the area of the ellipse $9x^2 + 6xy + 5y^2 \leq 36$. The inequality is the same as $(3x + y)^2 + (2y)^2 \leq 6^2$. Thus the ellipse becomes the unit disk $w^2 + z^2 \leq 1$ under the transformation $(x, y) \leftrightarrow (w, z)$ given by $w = (3x + y)/6$, $z = 2y/6$.

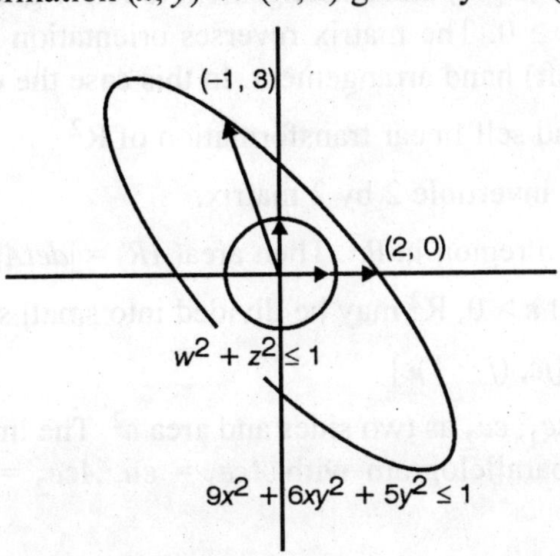

Alternatively, we may consider the ellipse as the image of the unit disk under the transformation $(w, z) \to (x, y) = (2w - z, 3z)$. Thus

$$\text{area(ellipse)} = \left|det\begin{bmatrix} 2 & -1 \\ 0 & 3 \end{bmatrix}\right| \text{area(unit disk)} = 6\pi.$$

The sign of the determinant also has interpretation in terms of linear transformations. For a 2 by 2 matrix A, we say

A preserves orientation if u to v is clockwise (counterclockwise) $\Rightarrow$ Au to Av is clockwise (counterclockwise);

A reverses orientation if u to v is clockwise (counterclockwise) $\Rightarrow$ Au to Av is counterclockwise (clockwise).

Then we have

A preserves orientation $\Leftrightarrow$ $detA \geq 0$;

A reverses orientation $\Leftrightarrow$ $detA \leq 0$.

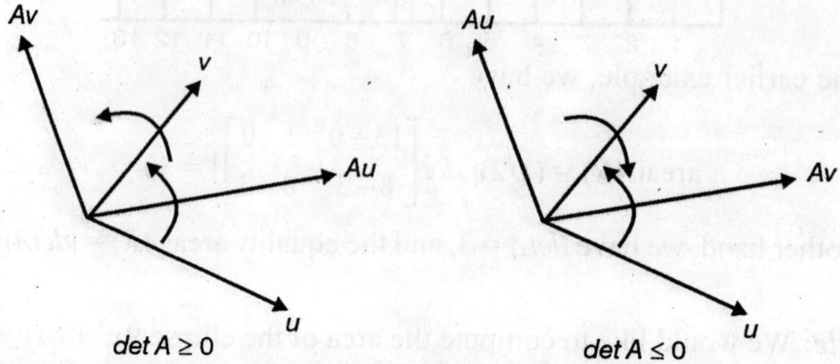

The similar discussion applies to 3 dimensional case. A 3 by 3 matrix preserves orientation if it takes left (right) hand arrangement to left (right) hand arrangement. In this case the determinant is ≥ 0. The matrix reverses orientation if it takes left (right) hand arrangement to right (left) hand arrangement. In this case the determinant is ≤ 0.

The determinant and self linear transformation of R^2

Let $A = [u\ v]$ be an invertible 2 by 2 matrix.

Theorem: Let R be a region in R^2. Then area(AR) = $|detA|$area(R).

Proof For any small $\varepsilon > 0$, R^2 may be divided into small squares

$S_{ij} = [i\varepsilon, (i + 1)\varepsilon]\text{X}[j\varepsilon, (j + 1)\varepsilon]$

which has vectors εe_1, εe_2 as two sides and area ε^2. The images AS_{ij} is also a division of R^2. AS_{ij} is a small parallelogram with $A\varepsilon e_1 = \varepsilon u$, $A\varepsilon e_2 = \varepsilon u$ as two sides and area $|det[u\ v]| = \varepsilon^2|detA|$.

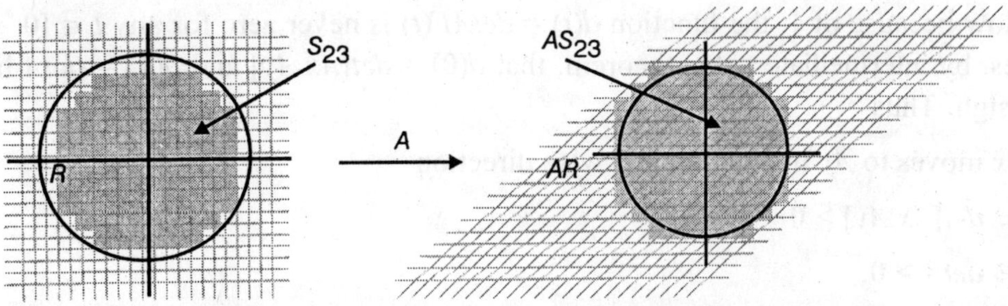

Let $S(1), S(2),..., S(k)$ be the small squares contained in R and $R_\varepsilon = S(1) \cup S(2) \cup ... \cup S(k)$. Then

area(R_ε) = area($S(1)$) + area($S(2)$) +... + area($S(k)$), and area(R) = $\lim_{\varepsilon\to 0}$ area(R_ε).

On the other hand, among the small parallelogram AS_{ij}, $AS(1), AS(2),..., AS(k)$ are the ones contained in AR, and $AR_\varepsilon = AS(1) \cup AS(2) \cup ... \cup AS(k)$. Moreover,

area(AR_ε) = area($AS(1)$) + area($AS(2)$) +... + area($AS(k)$), and area(AR) = $\lim_{\varepsilon\to 0}$ area(AR_ε).

Then

area (AS_{ij}) = $\varepsilon^2|detA|$, area(S_{ij}) = ε^2

$\Rightarrow$ area($AS(i)$) = area($S(i)$)$|detA|$

$\Rightarrow$ area(AR_ε) = area(R_ε)$|detA|$

$\Rightarrow$ area(AR) = area(R)$|detA|$.

Remark: The key point in the argument is that if S is a square with $\varepsilon e_1, \varepsilon e_2$ as two sides, then AS is a parallelogram with $\varepsilon Ae_1, \varepsilon Ae_2$ as two sides and has area $\varepsilon^2|detA|$. The follows from the geometrical meaning of the determinant. Since the meaning is still valid in higher dimension, similar argument also implies volume(AR) = $|detA|$ volume(R) in higher dimension.

Theorem: Let x moves to y in counterclockwise direction. Then Ax moves to Ay in counterclockwise direction $\Leftrightarrow detA \geq 0$.

Proof. As explained here, it is possible to construct a movie of linearly independent vectors, starting with (x, y) and ending with (e_1, e_2). Strictly speaking, we have a 2 by 2 matrix $F(t) = [f_1(t)\, f_2(t)]$, such that

- all four entries of $F(t)$ are continuous functions of $t \in [0, 1]$;
- $F(0) = [x\ y]$, $F(1) = [e_1\ e_2]$;
- $f_1(t), f_2(t)$ are linearly independent (i.e., $F(t)$ is invertible) for any t.

Then $AF(t)$ is also a matrix–valued function satisfying 1 (linear combinations of continuous functions are continuous) and 3 (product of invertible matrices is invertible). Moreover, $AF(0) = [Ax\ Ay]$, $AF(1) = A$.

Now because of 1, the function $d(t) = detAF(t)$ is never zero for any $t \in [0, 1]$. This implies, by intermediate value theorem, that $d(0) = det[Ax\ Ay]$ and $d(1) = detA$ have the same sign. Thus

Ax moves to Ay in counterclockwise direction

$\Leftrightarrow det[Ax\ Ay] > 0$

$\Leftrightarrow detA > 0.$

Exercise: For the following arrangements of vectors, show that that area of the parallelogram spanned by u and v is $|det[u\ v]|$.

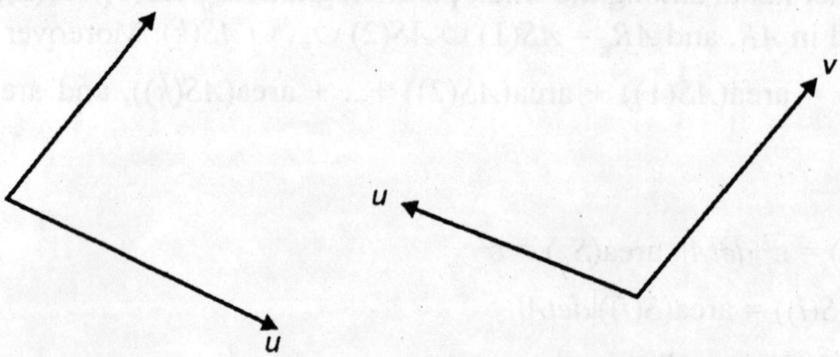

Answer: From the following picture, we have

area of the parallelogram = $(a + b)(-c + d) - a(-c) - bd = ad - bc$

where we note that $c < 0$, so that the length is $-c$.

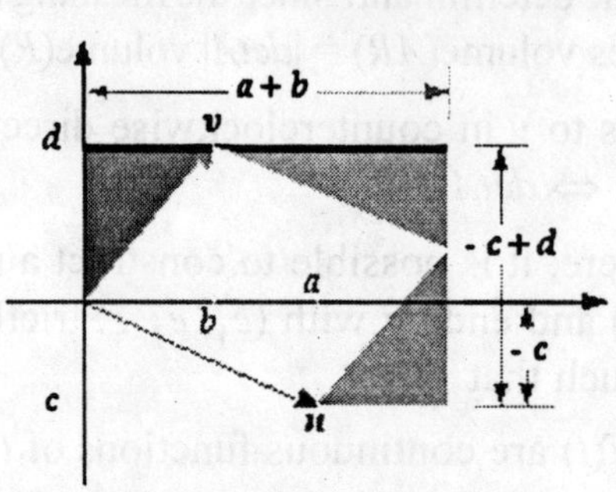

From the following picture, we have

area of the parallelogram = $(-a + b)(c + d) - (-a)c - bd = -ad + bc$

where we note that $a < 0$, so that the length is $-a$. Moreover, the determinant is negative, and the area is the negative (i.e., the absolute value) of the determinant.

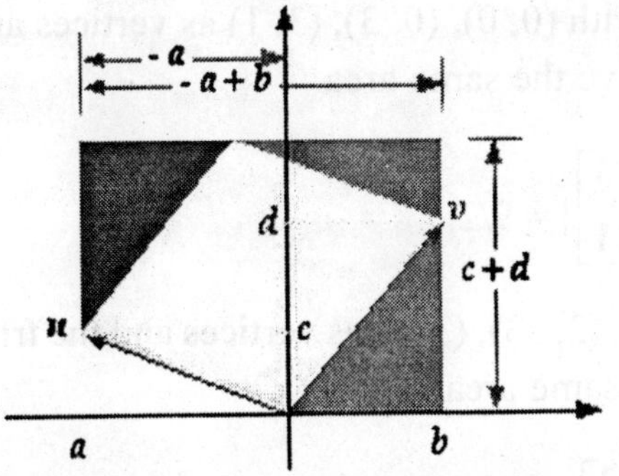

Exercise: Find the areas of the polygons with the following points as vertices.

1. (1, 1), (3, 6), (4, 2), (8, 1)

Answer: The area of the triangle with (1, 1), (3, 6), (4, 2) as vertices is

$$(1/2)\left| \det\begin{bmatrix} 3-1 & 4-1 \\ 6-1 & 2-1 \end{bmatrix} \right| = (1/2)|2\times 1 - 5\times 3| = 13/2.$$

The area of the triangle with (1, 1), (4, 2), (8, 1) as vertices is

$$(1/2)\left| \det\begin{bmatrix} 4-1 & 8-1 \\ 2-1 & 1-1 \end{bmatrix} \right| = (1/2)|3\times 0 - 1\times 7| = 7/2.$$

Thus the area of the 4–gon is 13/2 + 7/2 = 10.

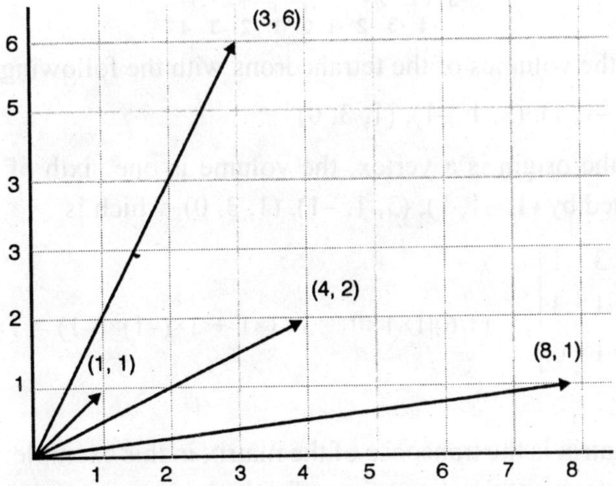

2. (0, 3), (2, –3), (–2, –3), (3, 1), (–3, 1)

Answer: The triangle with (0, 0), (0, 3), (3, 1) as vertices and the triangle with (0, 0), (0, 3), (–3, 1) as vertices have the same area

$$(1/2)|\ det\begin{bmatrix} 0 & 3 \\ 3 & 1 \end{bmatrix}| = 9/2.$$

The triangle with (0, 0), (2, –3), (3, 1) as vertices and the triangle with (0, 0), (–2, –3), (–3, 1) as vertices have the same area

$$(1/2)|\ det\begin{bmatrix} 2 & 3 \\ -3 & 1 \end{bmatrix}| = 11/2.$$

The area of the triangle with (0, 0), (2, –3), (–2, –3) as vertices is

$$(1/2)|\ det\begin{bmatrix} 2 & -2 \\ 3 & 3 \end{bmatrix}| = 6.$$

Thus the area of the 5–gon is 2×(9/2) + 2×(11/2) + 6 = 26.

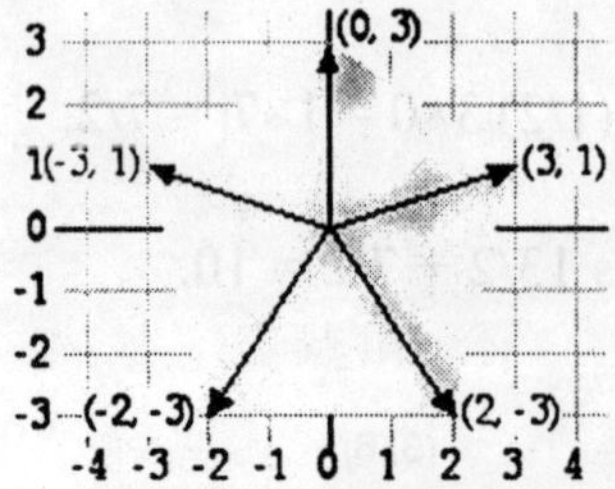

Exercise: Find the volumes of the tetrahedrons with the following points as vertices.

1. (0, 0, 0), (1, –1, 1), (3, 1, –1), (1, 3, 0)

Answer: Since the origin is a vertex, the volume is one sixth of the volume of the parallelogram spanned by (1, –1, 1), (3, 1, –1), (1, 3, 0), which is

$$(1/6)|\ det\begin{bmatrix} 1 & 3 & 1 \\ -1 & 1 & 3 \\ 1 & -1 & 0 \end{bmatrix}| = (1/6)|1\times1\times0 + 3\times3\times1 + 1\times(-1)\times(-1) - 1\times1\times1 - 1\times3\times(-1) -$$

$3\times(-1)\times0| = 2.$

Remark: The matrix is the transpose of the matrix in this example. The results are the same. In fact, the six terms of the determinant for the two matrices clearly corresponds, so that the determinants are equal.

Please verify, for general 2 by 2 and 3 by 3 matrices, that $detA = detA^T$.

2. (–1, –1, 0), (0, –2, 1), (2, 0, –1), (0, 2, 0)

Answer: The volume is one sixth of the volume of the parallelogram spanned by (0, –2, 1) – (–1, –1, 0) = (1, –1, 1), (2, 0, –1) – (–1, –1, 0) = (3, 1, –1), (0, 2, 0) – (–1, –1, 0) = (1, 3, 0). The parallelogram is the same as the one in the previous problem. Therefore the volumn of the tetrahedron is still 2.

3. (1, 1, 1), (1, 2, 3), (2, 3, 4), (3, 4, 5)

Answer: The volume is one sixth of the volume of the parallelogram spanned by (1, 2, 3) – (1, 1, 1) = (0, 1, 2), (2, 3, 4) – (1, 1, 1) = (1, 2, 3), (3, 4, 5) – (1, 1, 1) = (2, 3, 4). Since

$$det\begin{bmatrix} 0 & 1 & 2 \\ 1 & 2 & 3 \\ 2 & 3 & 4 \end{bmatrix} = 0\times2\times4 + 1\times3\times2 + 2\times1\times3 - 2\times2\times2 - 0\times3\times3 - 1\times1\times4 = 0,$$

the volumn is zero.

In fact, the determinant is zero because the columns are linearly dependent: (0, 1, 2) – 2(1, 2, 3) + (2, 3, 4) = (0, 0, 0)

Exercise: Find the determinant of the following matrices from their geometric meanings and from the formulae. Compare the results.

1. $\begin{bmatrix} 2 & 2 \\ -3 & 5 \end{bmatrix}, \begin{bmatrix} -6 & 1 \\ 3 & 3 \end{bmatrix}, \begin{bmatrix} -3 & -3 \\ -3 & 1 \end{bmatrix}$

Answer: The parallelograms spanned by the column vectors are illustrated below. The areas of the parallelogram are twice of the areas of the indicated triangles.

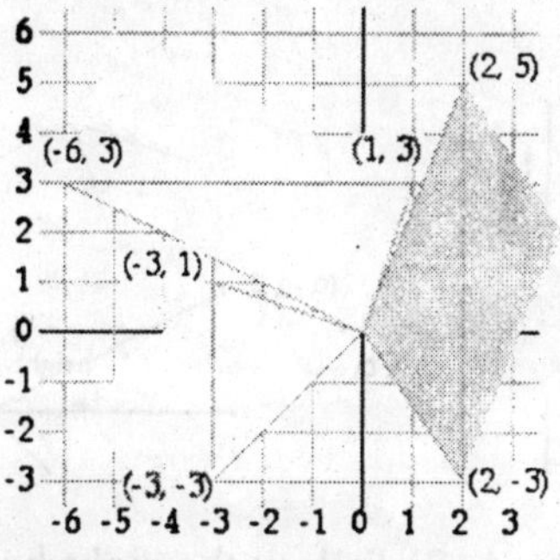

For the pink one, the triangle has base length 5 – (–3) = 8. The height is 2. Therefore the area of the parallelogram is 8 × 2 = 16. Since the vector (2, –3) moves to (2, 5) in counterclockwise direction, the determinant of the first matrix is positive and is equal to16.

For the yellow one, the triangle has base 7 and height 3. Since (–6, 3) moves to (1, 3) in clockwise direction, the determinant of the second matrix is nagative and is equal to –7 × 3 = –21.

For the green one, the triangle has base 4 and height 3. Since (–3, 3) moves to (–3, 1) in clockwise direction, the determinant of the third matrix is nagative and is equal to –4 × 3 = –12.

Now we compare the results with direct computations.

$$det\begin{bmatrix} 2 & 2 \\ -3 & 5 \end{bmatrix} = 2 \times 5 - (-3) \times 2 = 16,$$

$$det\begin{bmatrix} -6 & 1 \\ 3 & 3 \end{bmatrix} = (-6) \times 3 - 3 \times 1 = -21,$$

$$det\begin{bmatrix} -3 & -3 \\ -3 & 1 \end{bmatrix} = (-3) \times 1 - (-3) \times (-3) = -12.$$

2. $\begin{bmatrix} 2 & 2 & 1 \\ -3 & 5 & -1 \\ 0 & 0 & 2 \end{bmatrix}, \begin{bmatrix} -2 & 0 & 0 \\ 1 & -6 & 1 \\ -1 & 3 & 3 \end{bmatrix}, \begin{bmatrix} 1 & 0 & 0 \\ 1 & 1 & 0 \\ 1 & 1 & 1 \end{bmatrix}$

Answer: For the first two matrices, the parallelograms spanned by the column vectors are illustrated below. We take the pink and yellow 2–dimensional parallelograms as the bases. Note that their areas have been computed in the previous problem.

Moreover, both have height 2. Thus the absolute values of the determinants are respectively 2 × 16 = 32 and 2 × 21 = 42.

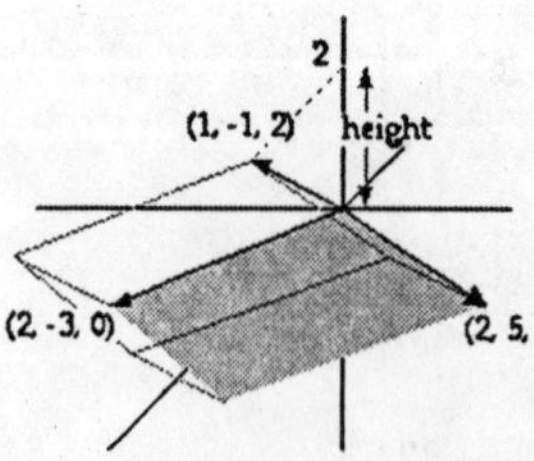

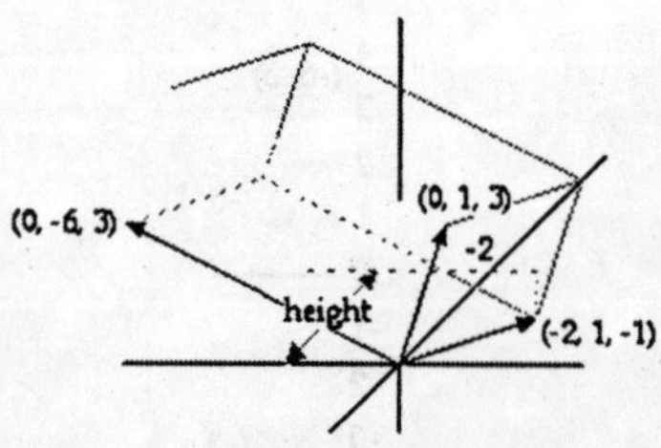

Since (2, –3, 0), (2, 5, 0), (1, –1, 2) follow the right hand rule, the determinant of the first matrix is positive and is 32. Alternatively, the direct computation gives us

$$det\begin{bmatrix} 2 & 2 & 1 \\ -3 & 5 & -1 \\ 0 & 0 & 2 \end{bmatrix} = 2 \times 5 \times 2 + 2 \times (-1) \times 0 + 1 \times (-3) \times 0 - 1 \times 5 \times 0 - 2 \times (-1) \times 0$$

Since (–2, 1, –1), (0, –6, 3), (0, 1, 3) also follow the right hand rule, the determinant of the second matrix is also positive and is 42. Alternatively, the direct computation gives us

$$det\begin{bmatrix} -2 & 0 & 0 \\ 1 & -6 & 1 \\ -1 & 3 & 3 \end{bmatrix} = (-2) \times (-6) \times 3 + 0 \times 1 \times (-1) + 0 \times 1 \times 3 - 0 \times (-6) \times (-1) - (-2) \times 1 \times 3 - 0 \times 1 \times 3 = 42.$$

The following picture illustrates the parallelogram spanned by the columns of the third matrix. We use the 2-dimensional green parallelogram as the base, which has area 1 × 1 = 1, and the height is 1. Thus the volumne is 1. Moreover, since (1, 1, 1), (0, 1, 1), (0, 0, 1) follow the right hand rule, we conclude that the determinant of the third matrix is 1.

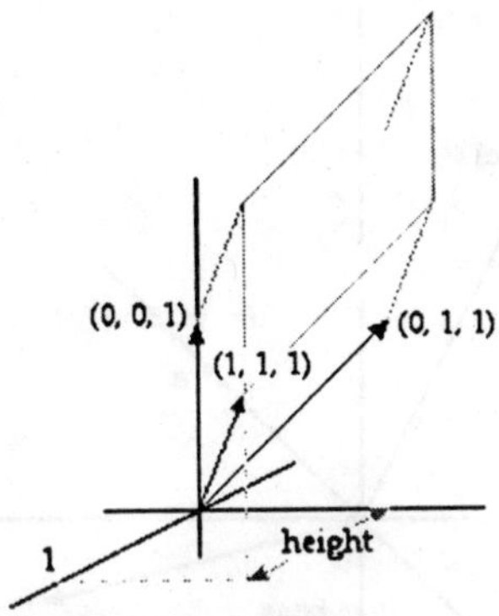

Alternatively, the direct computation gives us

$$det\begin{bmatrix} 1 & 0 & 0 \\ 1 & 1 & 0 \\ 1 & 1 & 1 \end{bmatrix} = 1 \times 1 \times 1 + 0 \times 0 \times 1 + 0 \times 1 \times 1 - 0 \times 1 \times 1 - 1 \times 0 \times 1 - 0 \times 1 \times 1 = 1.$$

Exercise: Give a geometrical explaination that the absolute value of the determinant of

$$\begin{bmatrix} a & d & f \\ 0 & b & e \\ 0 & 0 & c \end{bmatrix}$$

is the volume of the parallelogram spanned by $(a, 0, 0)$, $(d, b, 0)$, (f, e, c).

Answer: The case $a, b, c > 0$ is illustrated by the picture, the volumn = (area of the base)X(height). The height is c. The base is a 2-dimensional parallelogram and, by the picture on the right, has area ab. Thus the absolute value of the determinant is abc. Moreover, the vectors u, v, w follow the right hand rule. Therefore the determinant is positive and is equal to abc.

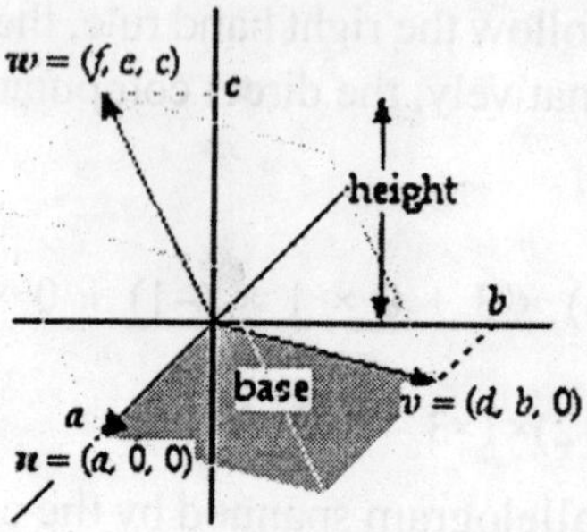

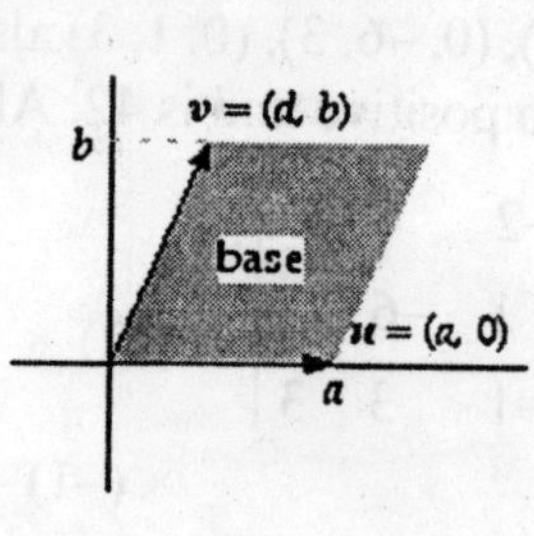

The following illustrates the case $a < 0$ and $b, c > 0$. The volumn is still $|abc| = -abc$. However, the vectors u, v, w follow the left hand rule. Therefore the determinant is negative and is equal to $-(-abc) = abc$.

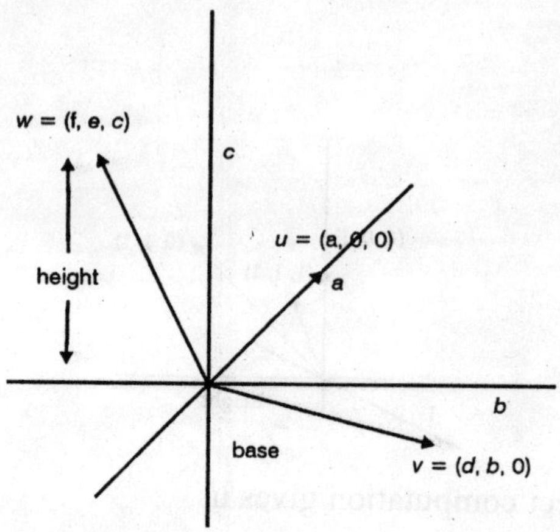

One may also check out the other possibilities and find out that the determinant is always abc.

Exercise: Find the area of region.

1. $4x^2 - 12xy + 13y^2 + 4x - 2y \leq 2$

Answer: We rewrite the left side of the inequality

$4x^2 - 12xy + 13y^2 + 4x - 2y$

$= (4x^2 - 12xy + 4x + 9y^2 - 6y + 1) + 4y^2 + 4y - 1$

$= (2x - 3y + 1)^2 + [(2y)^2 + 4y + 1] - 2$

$= (2x - 3y + 1)^2 + (2y + 1)^2 - 2.$

Thus the region is $(2x-3y+1)^2+(2y+1)^2 \le 2^2$, which becomes the disk $(w+1)^2+(z+1)^2 \le 2^2$ of radius 2 under the linear transformation $w = 2x - 3y$, $z = 2y$. Then

$$\text{area}((w, z)\text{–disk}) = \left|det\begin{bmatrix} 2 & -3 \\ 0 & 2 \end{bmatrix}\right| \text{area}((x, y)\text{–ellipse}) = 4\ \text{area}((x, y)\text{–ellipse}).$$

Since the (w, z)–disk has radius 2 and area 4π, we conclude that the (x, y)–ellipse has area π.

2. $x^2 + 5y^2 + 11z^2 + 2xy - 2xz + 2yz \le 9$

Answer: We rewrite the left side of the inequality

$$\begin{aligned} & x^2 + 5y^2 + 11z^2 + 2xy - 2xz + 2yz \\ &= (x^2 \quad + 2xy - 2xz + y^2 + z^2 - 2yz) + 4y^2 + 10z^2 + 4yz \\ &= (x + y - z)^2 + (4y^2 + 4yz + z^2) + 9z^2 \\ &= (x + y - z)^2 + (2y + z)^2 + (3z)^2. \end{aligned}$$

Thus the region becomes the ball $x'^2 + y'^2 + z'^2 \le 3^2$ of radius 3 under the linear transformation $x' = x + y - z$, $y' = 2y + z$, $z' = 3z$. Then

$$\text{volume}((x', y', z')\text{–ball}) = \left|det\begin{bmatrix} 1 & 1 & -1 \\ 0 & 2 & 1 \\ 0 & 0 & 3 \end{bmatrix}\right| \text{volume}((x, y, z)\text{–ellipsoid}) = 6\ \text{volume}((x, y, z)\text{-ellipsoid}).$$

Since the (x', y', z')–ball has radius 3 and volume $(4/3)3^3\pi = 36\pi$, we conclude that the (x, y, z)-ellipsoid has volume 6π.

Remark: The way we rewrite the quadratic expressions is the very useful technique of completing the square. Generally, it is done step by step. In the first step, we put all the terms involving the first variable together (the red terms in the two computations above) and try to add some additional terms (the green terms) so that we have a square (the blue term). The whole expression is then a square involving the first variable plus the remaining quadratic expression that does not involve the first variable. Then in the second step, we try to complete a square around the second variable, and so on.

Chapter 2

Computation of Determinant

ROW/COLUMN OPERATIONS

In general, determinants are defined for square matrices. Nonsquare matrices do not have determinants. Except for 2 by 2 or perhaps 3 by 3 matrices, the determinant is never computed by the formula in the definition. The computation is usually done by combining two methods.

- Row/column operations to simply matrices (i.e., creating zeros)
- Cofactor expansion to reduce the size of matrices

We discuss the first method in this section and the second method in the next section.

The row operations change the determinant as follows. The change by the column operations is similar.

A changed to *B* bydeterminant changed by

r[row i] + [row j]$detB = detA$

[row i] $\leftrightarrow$ [row j]$detB = -detA$

d[row i], $d \neq 0$ $detB = d\ detA$

The proof of the property will be given here, after the rigorous definition of determinants.

For the column operations on 2 by 2 matrices, the changes can be explained by the geometric meaning of the determinant. We will see later that $detA = detA^{\mathrm{T}}$, which means that there is no difference between rows and columns as long as the determinant is concerned. Then the geometric explaination also applies to the row operations.

The following pictures compare $det[u\ v]$ (the red part) with $det[u\ ru + v]$ (the blue part). The left one has $r > 0$, and the right one has $r < 0$. In either case, the red and the blue parallelograms have the same base and height. Moreover, the orientation from u to v is the same as the orientation from u to $ru + v$ (you are encouraged to draw the pictures for the case *u* moves to *v* in clockwise direction). With the same volume and orientation, we conclude $det[u\ ru + v] = det[u\ v]$.

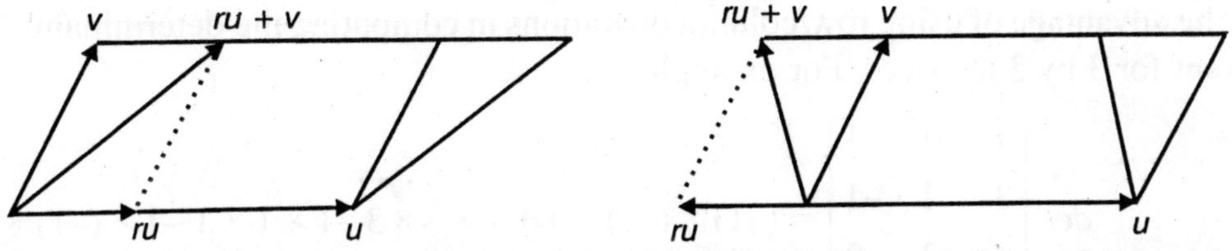

The following picture compares $det[u\ v]$ with $det[v\ u]$. The switching of the two vectors do not change the parallelogram. However, the orientation from *v* to *u* (the blue arrow) is the opposite of the orientation from *u* to *v* (the red arrow). With the same volume and different orientation, we conclude $det[v\ u] = -\ det[u\ v]$.

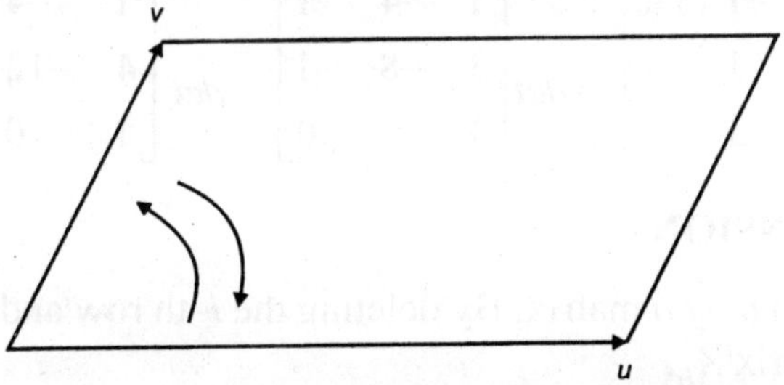

· The following pictures compare $det[u\ v]$ (the red part) with $det[du\ v]$ (the blue part). The left one has $d > 0$, and the area of the blue parallelogram is *d* times of the red one. Moreover, the orientation from *u* to *v* is the same as the orientation from *du* to *v*. Thus we conclude $det[du\ v] = d\ det[u\ v]$ for the case $d > 0$. The right picture shows the case $d < 0$. In this case, the area of the blue parallelogram is $|d| = -d$ times of the red one. On the other hand, the orientation from *u* to *v* is different from the orientation from *du* to *v*. Therefore $det[du\ v] = -\ |d| det[u\ v] = d\ det[u\ v]$.

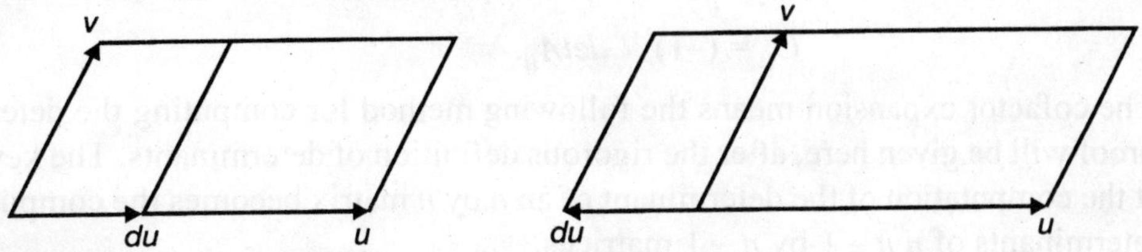

Example: The determinant

$$det\begin{bmatrix} 6 & 2 \\ -1 & 3 \end{bmatrix} = 6 \times 3 - (-1) \times 2 = 20$$

consists of 2 terms. If we do 6[row 2] + [row 1], then

$$det\begin{bmatrix} 6 & 2 \\ -1 & 3 \end{bmatrix} = det = 0 \times 3 - (-1) \times 20 = -(-1) \times 20 = 20$$

consists of only one term. We see from the example that the more zeros we create, the fewer terms we need to compute.

The advantage of using row/column operations in computing the determinants is more apparent for 3 by 3 matrices. For example,

$$det\begin{bmatrix}1 & -1 & 1\\3 & 1 & -1\\1 & 3 & 0\end{bmatrix} = (-1)\times(-1)\times 1 + 1\times 3\times 3 - 1\times 1\times 1 - 1\times(-1)\times 3 = 12$$

consists of essentially 4 terms (the terms containing zero are omitted). If we use (–3)[col 1] + [col 2] and [row 1] + [row 2], then we get

$$det\begin{bmatrix}1 & -1 & 1\\3 & 1 & -1\\1 & 3 & 0\end{bmatrix} = det\begin{bmatrix}1 & -4 & 1\\3 & -8 & -1\\1 & 0 & 0\end{bmatrix} = det\begin{bmatrix}1 & -4 & 1\\4 & -12 & 0\\1 & 0 & 0\end{bmatrix} = -1\times(-12)\times 1 = 12.$$

COFACTOR EXPANSION

Let $A = (a_{ij})$ be an n by n matrix. By deleting the i–th row and the j–th column, we get an $n - 1$ by $n - 1$ matrix A_{ij}.

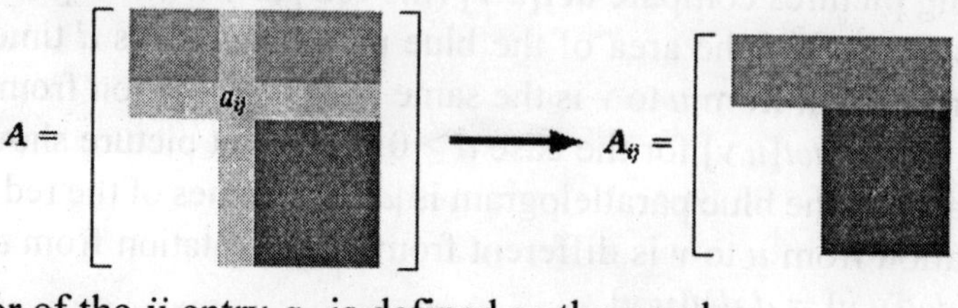

The cofactor of the ij-entry a_{ij} is defined as the number

$$C_{ij} = (-1)^{i+j}\, detA_{ij}.$$

The cofactor expansion means the following method for computing the determinant. The proof will be given here, after the rigorous definition of determinants. The key feature is that the computation of the determinant of an n by n matrix becomes the computation of the determinants of n $n - 1$ by $n - 1$ matrices.

For any i–th row, $detA = a_{i1}C_{i1} + a_{i2}C_{i2} + ... + a_{in}C_{in}$.

For any j–th column, $detA = a_{1j}C_{1j} + a_{2j}C_{2j} + ... + a_{nj}C_{nj}$.

For a 3 by 3 matrix, the cofactor expansion along the first row is

$$det\begin{bmatrix}a_{11} & a_{12} & a_{13}\\a_{21} & a_{22} & a_{23}\\a_{31} & a_{32} & a_{33}\end{bmatrix} = a_{11}det\begin{bmatrix}a_{22} & a_{23}\\a_{32} & a_{33}\end{bmatrix} - a_{12}det\begin{bmatrix}a_{21} & a_{23}\\a_{31} & a_{33}\end{bmatrix} + a_{13}det\begin{bmatrix}a_{21} & a_{22}\\a_{31} & a_{32}\end{bmatrix}$$

where the darker row indicates the row along which the cofactors are expanded, and the negative sign for the second 2 by 2 determinant is due to $(-1)^{1+2} = -1$. You are encouraged to compare both sides by using the usual formulae for 2 by 2 and 3 by 3 determinants.

Example: In an earlier example, we computed the determinant of

$$A = \begin{bmatrix} 1 & -1 & 1 \\ 3 & 1 & -1 \\ 1 & 3 & 0 \end{bmatrix}$$

by direct computation and row/column operations. Now we apply the cofactor expansion along the third row

$$det \begin{bmatrix} 1 & -1 & 1 \\ 3 & 1 & -1 \\ 1 & 3 & 0 \end{bmatrix} = 1 \times (-1)^{3+1} det \begin{bmatrix} -1 & 1 \\ 1 & -1 \end{bmatrix} + 3 \times (-1)^{3+2} det \begin{bmatrix} 1 & 1 \\ 3 & -1 \end{bmatrix} = 1 \times 0 - 3 \times (-4) = 12,$$

where the term $a_{33}C_{33}$ is omitted because $a_{33} = 0$. We see that cofactor expansion along the rows or columns with maximal number of zeros is often more efficient.

We may also compute the determinant by column expansion. Again we prefer expanding along the third column because it contains zero.

$$det \begin{bmatrix} 1 & -1 & 1 \\ 3 & 1 & -1 \\ 1 & 3 & 0 \end{bmatrix} = 1 \times (-1)^{1+3} det \begin{bmatrix} 3 & 1 \\ 1 & 3 \end{bmatrix} + (-1) \times (-1)^{2+3} det \begin{bmatrix} 1 & -1 \\ 1 & 3 \end{bmatrix} = 1 \times 8 - (-1) \times 4 = 12.$$

Example: To compute a 4 by 4 determinant, we use the cofactor expansion to reduce the computation to several 3 by 3 determinants.

$$det \begin{bmatrix} 2 & 1 & 4 & 2 \\ 0 & 2 & 1 & 2 \\ 0 & 2 & 1 & 1 \\ 2 & 0 & 1 & 0 \end{bmatrix} = 2det \begin{bmatrix} 2 & 1 & 2 \\ 2 & 1 & 1 \\ 0 & 1 & 0 \end{bmatrix} - 2det \begin{bmatrix} 1 & 4 & 2 \\ 2 & 1 & 2 \\ 2 & 1 & 1 \end{bmatrix} = 2 \times 2 - 2 \times 7 = -10.$$

COMBINATION OF TWO METHODS

The row/column operations can simplify determinants. The cofactor expansions can reduce size. By combining the two methods, we can efficiently compute determinants.

Example: In an earlier example, we computed a 4 by 4 determinant by cofactor expansion only. The computation involes two 3 by 3 determinants. Now we mix two methods and get a more efficient computation.

$$det\begin{bmatrix} 2 & 1 & 4 & 2 \\ 0 & 2 & 1 & 2 \\ 0 & 2 & 1 & 1 \\ 2 & 0 & 1 & 0 \end{bmatrix} = det\begin{bmatrix} 0 & 1 & 3 & 2 \\ 0 & 2 & 1 & 2 \\ 0 & 2 & 1 & 1 \\ 2 & 0 & 1 & 0 \end{bmatrix} = -2det\begin{bmatrix} 1 & 3 & 2 \\ 2 & 1 & 2 \\ 2 & 1 & 1 \end{bmatrix}$$

$$= -2det\begin{bmatrix} -5 & 3 & -1 \\ 0 & 1 & 1 \\ 0 & 1 & 0 \end{bmatrix} = -2det = \begin{bmatrix} -5 & -1 \\ 0 & 1 \end{bmatrix} -10.$$

In the computation, the red equalities = mean row/columns operations. The first red equality means (–1)[row 4] + [row 1], and the second one means (–2)[col 2] + [col 1] and (–1)[col 2] + [col 3]. Moreover, the green equalities = mean cofactor expansions, along the darkened row or column.

Example: By [col 3] + [col 2] and (–1)[row 2] + [row 3], we have

$$det\begin{bmatrix} 1-\lambda & 3 & -3 \\ -3 & 7-\lambda & -3 \\ -6 & 6 & -2-\lambda \end{bmatrix} = det\begin{bmatrix} 1-\lambda & 0 & -3 \\ -3 & 4-\lambda & -3 \\ -6 & 4-\lambda & -2-\lambda \end{bmatrix} = det\begin{bmatrix} 1-\lambda & 0 & -3 \\ -3 & 4-\lambda & -3 \\ -6 & 0 & 1-\lambda \end{bmatrix}.$$

Expanding along [col 2], the determinant is

$$(4-\lambda)det\begin{bmatrix} 1-\lambda & -3 \\ -3 & 1-\lambda \end{bmatrix} = (4-\lambda)[(1-\lambda)^2-(-3)^2] = (4-\lambda)(1-\lambda-(-3))(1-\lambda+(-3)).$$

Thus we conclude

$$det\begin{bmatrix} 1-\lambda & 3 & -3 \\ -3 & 7-\lambda & -3 \\ -6 & 6 & -2-\lambda \end{bmatrix} = (4-\lambda)^2(-2-\lambda).$$

Example: The 4 by 4 Vandermonde matrix is

$$V_4(a, b, c, d) = \begin{bmatrix} 1 & a & a^2 & a^3 \\ 1 & b & b^2 & b^3 \\ 1 & c & c^2 & c^2 \\ 1 & d & d^2 & d^3 \end{bmatrix}.$$

To find its determinant, we apply (–*a*)[col 3] + [col 4], (–*a*)[col 2] + [col 3], (–*a*)[col 1] + [col 2], and then expand along the first row

$$detV_4(a, b, c, d) = det \begin{bmatrix} 1 & 0 & 0 & 0 \\ 1 & b-a & b(b-a) & b^2(b-a) \\ 1 & c-a & (c-a) & c^2(c-a) \\ 1 & d-a & d(d-a) & d^2(d-a) \end{bmatrix}$$

$$= det \begin{bmatrix} b-a & b(b-a) & b^2(b-a) \\ c-a & c(c-a) & c^2(c-) \\ d-a & d(d-a) & d^2(d-a) \end{bmatrix}$$

Note that in the 3 by 3 matrix, [row 1] is $(b-a)$ multiplied to $(1, b, b^2)$, and similarly for [row 2] and [row 3]. By the third operation, we get

$$detV_4(a, b, c, d) = (b-a)(c-a)(d-a)det \begin{bmatrix} 1 & b & b^2 \\ 1 & c & c^2 \\ 1 & d & d^2 \end{bmatrix}$$

$$= (b-a)(c-a)(d-a)detV_3(b, c, d).$$

This suggests a pattern for reducing the determinant of a Vandermonde matrix to another Vandermonde matrix of smaller size. Following the pattern, we get

$$\begin{aligned} detV_4(a, b, c, d) &= (b-a)(c-a)(d-a)detV_3(b, c, d) \\ &= (b-a)(c-a)(d-a)(c-b)(d-b)detV_2(c, d) \\ &= (b-a)(c-a)(d-a)(c-b)(d-b)(d-c)detV_1(d) \\ &= (b-a)(c-a)(d-a)(c-b)(d-b)(d-c). \end{aligned}$$

In general, the n by n Vandermonde matrix is

$$V_n(a_1, a_2, a_3, \dots, a_n) = \begin{bmatrix} 1 & a_1 & a_1^2 & .. & a_1^{n-1} \\ 1 & a_2 & a_2^2 & & a_2^{n-1} \\ 1 & a_3 & a_3^2 & & a_3^{n-1} \\ : & : & : & & : \\ 1 & a_n & a_n^2 & .. & a_n^{n-1} \end{bmatrix}$$

From the relation

$$detV_n(a_1, a_2, a_3, \dots, a_n) = (a_n - a_1)\dots(a_3 - a_1)(a_2 - a_1)detV_{n-1}(a_2, a_3, \dots, a_n),$$

we get

$$detV_n(a_1, a_2, a_3, \dots, a_n) = \Pi_{1 \le i < j \le n}(a_j - a_i).$$

PROPERTIES OF DETERMINANT

PROPERTIES

Let us begin with the determinant of product of matrices. The proof of the property will be given here, after the rigorous definition of determinants.

Let A and B be square matrices of the same size. Then $detAB = detA\ detB$.

Taking the absolute value, the property means that the ratio for the change of volume under the composition AB is the product of the ratios under the transformations A and B. Intuitively, this is quite obvious: enlarging the volume twice and then enlarging again three times is the same as enlarging six (two times three) times.

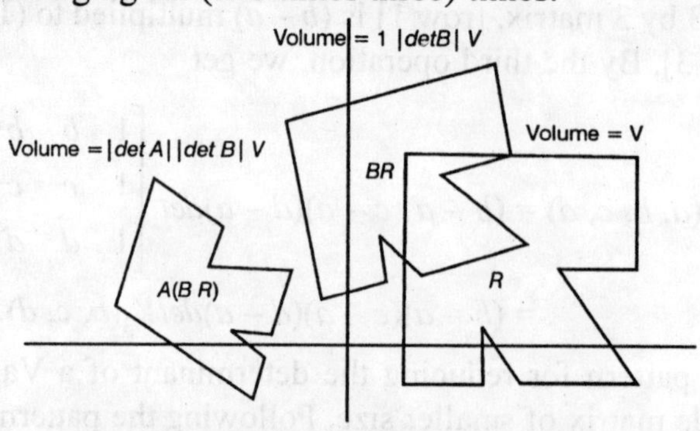

The total ratio of change under AB is $|detA|\,|det\ B|$

Taking the signs, we may consider various combinations. For example, if 2 by 2 matrices A and B preserve orientations (i.e., $detA > 0$, $detB > 0$), then

(u, v) clockwise $\Rightarrow$ (Au, Bv) clockwise $\Rightarrow$ (ABu, ABv) clockwise,

so that AB also preserves orientation (i.e., $detAB > 0$). We may also verify signs of other combinations in a similar way.

We also note that the product property implies

$det(A^k) = (detA)^k$, for $k > 0$.

The next property indicates that, as far as the determinant is concerned, there is no difference between rows and columns. In particular, one may use row operations as well as column operations. The proof of the property will be given here, after the rigorous definition of determinants.

$detA^{\mathrm{T}} = detA$.

The next property is quite useful for computation. The proof of the property will be given here, after the rigorous definition of determinants.

Let A and B be square matrices. Then $det\begin{bmatrix} A & \# \\ O & B \end{bmatrix} = detA\ detB$.

Geometrically, the absolute value of the determinant represents the volumn of a high dimensional parallelogram with B–paralleogram as "base" and A–parallelogram (skewed by # term) as "height". The size of the base is $|detB|$, and the size of the height is $|detA|$. Therefore the size of the whole parallelogram is $|detA\ detB|$.

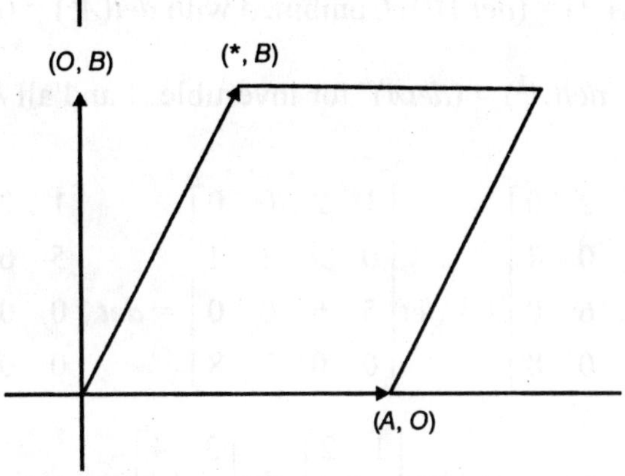

The triangle property can be generalized (by induction and transpose, for example). For square matrices $A_1, A_2,..., A_k$, we have

$$\det\begin{bmatrix} A_1 & \# & .. & \# \\ O & A_2 & .. & \# \\ : & : & & : \\ O & O & .. & A_k \end{bmatrix} = \det\begin{bmatrix} A_1 & O & .. & O \\ \# & A_2 & .. & O \\ : & : & & : \\ \# & \# & .. & A_k \end{bmatrix} = detA_1\ detA_2 ...\ detA_k,$$

In particular, for upper and lower triangular matrices, we have

$$\det\begin{bmatrix} a_1 & \# & .. & \# \\ 0 & a_2 & .. & \# \\ : & : & & : \\ 0 & 0 & .. & a_n \end{bmatrix} = a_1a_2...a_n,$$

$$\det\begin{bmatrix} a_1 & 0 & .. & 0 \\ \# & a_2 & .. & 0 \\ : & : & & : \\ \# & \# & .. & a_n \end{bmatrix} = a_1a_2...a_n.$$

Further specializing to the identity matrix, we have

$$detI = 1.$$

Geometrically, this simply means that the cube with unit side length has volume 1, and the standard basis has the positive orientation. Moreover, from $AA^{-1} = I$, we have $detA\ detA^{-1} = 1$, so that $det(A^{-1}) = (detA)^{-1}$. Combined with $det(A^k) = (detA)^k$ for $k > 0$, it is not hard to conclude that

$$det(A^k) = (detA)^k \text{ for invertible } A \text{ and all } k.$$

Example:

$$det\begin{bmatrix} 1 & 0 & 2 & 0 \\ 0 & 3 & 0 & 4 \\ 5 & 0 & 6 & 0 \\ 0 & 7 & 0 & 8 \end{bmatrix} = -det\begin{bmatrix} 1 & 2 & 0 & 0 \\ 0 & 0 & 3 & 4 \\ 5 & 6 & 0 & 0 \\ 0 & 0 & 7 & 8 \end{bmatrix} = det\begin{bmatrix} 1 & 2 & 0 & 0 \\ 5 & 6 & 0 & 0 \\ 0 & 0 & 3 & 4 \\ 0 & 0 & 7 & 8 \end{bmatrix}$$

$$= det\begin{bmatrix} 1 & 2 \\ 5 & 6 \end{bmatrix} det\begin{bmatrix} 3 & 4 \\ 7 & 8 \end{bmatrix} = (-4)(-4) = 16.$$

Example: Let A and B be n by n matrices. Suppose $detA = 7$, $detB = 2$. Then

$det(AB) = detA\ detB = 14$,

$det(A^T AB^2 A^{-3} B) = detA^T\ detA\ (detB)^2\ detA^{-3}\ detB = (detA)^{-1}\ (detB)^3 = 8/7.$

Finally, we remark that

$$det(A + B) \neq detA + detB.$$

However, the determinant is multilinear with respect to rows and columns.

In dA, the number d is multiplied to each row of A. By the second of this property, for an n by n matrix A, we have

$$det(dA) = d^n\ detA \neq d\ detA.$$

INVERTIBILITY AND DETERMINANT

If A is an invertible matrix. Let B be the inverse of A. Then by the properties in this section, we have

$detA\ detB = detAB = detI = 1.$

In particular, we just proved the $\Rightarrow$ direction of the following property.

A square matrix A invertible $\Leftrightarrow detA \neq 0$.

To prove the $\Leftarrow$ direction, we need to use the assumption $detA \neq 0$ to construct an inverse of A. We illustrate the idea by a 4 by 4 matrix.

$$A = \begin{bmatrix} a_{11} & a_{12} & a_{13} & a_{14} \\ a_{21} & a_{22} & a_{23} & a_{24} \\ a_{31} & a_{32} & a_{33} & a_{34} \\ a_{41} & a_{42} & a_{43} & a_{44} \end{bmatrix}$$

Recall the cofactor expansion along the first row

$$a_{11}C_{11} + a_{12}C_{12} + a_{13}C_{13} + a_{14}C_{14} = det \begin{bmatrix} a_{11} & a_{12} & a_{13} & a_{14} \\ a_{21} & a_{22} & a_{23} & a_{24} \\ a_{31} & a_{32} & a_{33} & a_{34} \\ a_{41} & a_{42} & a_{43} & a_{44} \end{bmatrix}$$

If we replace the terms $a_{11}, a_{12}, a_{13}, a_{14}$ in the formula by $a_{21}, a_{22}, a_{23}, a_{24}$, then we have

$$a_{21}C_{11} + a_{22}C_{12} + a_{23}C_{13} + a_{24}C_{14} = det \begin{bmatrix} a_{21} & a_{22} & a_{23} & a_{24} \\ a_{21} & a_{22} & a_{23} & a_{24} \\ a_{31} & a_{32} & a_{33} & a_{34} \\ a_{41} & a_{42} & a_{43} & a_{44} \end{bmatrix}$$

The determinant vanishes by the following result.

A has a zero row (column) $\Rightarrow detA = 0$.

A row (column) in A is a scalar multiple of another row (column) $\Rightarrow detA = 0$.

Proof If A has a zero row, then by cofactor expansion along the zero row, we get $detA = 0$. If [row i] is r multiple of [row j], then apply $(-r)$[row i] + [row j] to A to get B.

By this property, we have $detA = detB$. Moreover, the j–th row of B is zero, so that $detB = 0$.

The argument we had clearly generalizes to the following result.

$$a_{i1}C_{j1} + a_{i2}C_{j2} + a_{i3}C_{j3} + a_{i4}C_{j4} = \begin{cases} \det A & i = j \\ 0 & i \neq j \end{cases}$$

In other words, if we define the adjugate

$$adjA = \begin{bmatrix} C_{11} & C_{21} & C_{31} & C_{41} \\ C_{12} & C_{22} & C_{32} & C_{42} \\ C_{13} & C_{23} & C_{33} & C_{43} \\ C_{14} & C_{24} & C_{34} & C_{44} \end{bmatrix}$$

then we have

$$A\ adjA = (detA)I.$$

Since A is a square matrix, by this property, we conclude

$detA \neq 0 \Rightarrow A$ is invertible, with $A^{-1} = (detA)^{-1}\ adjA$.

We had the 2 by 2 case of this result before.

Example: In an earlier example, we used row operations to find the inverse of

$$A = \begin{bmatrix} 3 & 1 & -1 \\ 1 & -1 & 1 \\ 2 & 2 & 1 \end{bmatrix}$$

Now we use our new formula. First,

$$detA = det\begin{bmatrix} 3 & 4 & -4 \\ 1 & 0 & 0 \\ 2 & 4 & -1 \end{bmatrix} = -det\begin{bmatrix} 4 & -4 \\ 4 & -1 \end{bmatrix} = -12.$$

Then

$$C_{11} = (-1)^{1+1}det\begin{bmatrix} -1 & 1 \\ 2 & 1 \end{bmatrix} = -3,\ C_{21} = (-1)^{2+1}det\begin{bmatrix} 1 & -1 \\ 2 & 1 \end{bmatrix} = -3,\ C_{31}$$

$$= (-1)^{3+1}det\begin{bmatrix} 1 & -1 \\ -1 & 1 \end{bmatrix} = 0,$$

$$C_{12} = (-1)^{1+2}det\begin{bmatrix} 1 & 1 \\ 2 & 1 \end{bmatrix} = 1,\ C_{22} = (-1)^{2+2}det\begin{bmatrix} 3 & -1 \\ 2 & 1 \end{bmatrix} = 5,$$

$$C_{32} = (-1)^{3+2}det\begin{bmatrix} 3 & -1 \\ 1 & -1 \end{bmatrix} = -4,$$

$$C_{13} = (-1)^{1+3}det\begin{bmatrix} 1 & -1 \\ 2 & 2 \end{bmatrix} = 4,\ C_{23} = (-1)^{2+3}det\begin{bmatrix} 3 & 1 \\ 2 & 2 \end{bmatrix} = -4,$$

$$C_{33} = (-1)^{3+3}det\begin{bmatrix} 3 & -1 \\ 1 & -1 \end{bmatrix} = -4,$$

and

$$A^{-1} = (-1/12)\begin{bmatrix} -3 & -3 & 0 \\ 1 & 5 & -4 \\ 4 & -4 & -4 \end{bmatrix} = \begin{bmatrix} 1/4 & 1/4 & 0 \\ -1/12 & -5/12 & 1/3 \\ -1/3 & 1/3 & 1/3 \end{bmatrix}$$

One find the result to be the same as the earlier computation. However, the computation here is much more complicated. It is even more complicated if you try the inverse of matrices of bigger size. Therefore row operation is still the practical way of computing the inverse.

CRAMER'S RULE

With the help of explicit formula for the inverse matrix, we may get explicit formula for the solution of the system $Ax = b$.

Again let us consider the 4 by 4 case as an example.

$$A = [a_1\ a_2\ a_3\ a_4] = \begin{bmatrix} a_{11} & a_{12} & a_{13} & a_{14} \\ a_{21} & a_{22} & a_{23} & a_{24} \\ a_{31} & a_{32} & a_{33} & a_{34} \\ a_{41} & a_{42} & a_{43} & a_{44} \end{bmatrix},\ x = \begin{bmatrix} x_1 \\ x_2 \\ x_3 \\ x_4 \end{bmatrix},\ b = \begin{bmatrix} b_1 \\ b_2 \\ b_3 \\ b_4 \end{bmatrix}.$$

If $detA \neq 0$, then

$$x = A^{-1}b = (detA)^{-1}\begin{bmatrix} C_{11} & C_{21} & C_{31} & C_{41} \\ C_{12} & C_{22} & C_{32} & C_{42} \\ C_{13} & C_{23} & C_{33} & C_{43} \\ C_{14} & C_{24} & C_{34} & C_{44} \end{bmatrix}\begin{bmatrix} b_1 \\ b_2 \\ b_3 \\ b_4 \end{bmatrix}$$

$$= (detA)^{-1}\begin{bmatrix} C_{11}b_1 + C_{21}b_2 + C_{31}b_3 + C_{41}b_4 \\ C_{12}b_1 + C_{22}b_2 + C_{32}b_3 + C_{42}b_4 \\ C_{14}b_1 + C_{24}b_2 + C_{34}b_3 + C_{44}b_4 \end{bmatrix}$$

is the unique solution. Now recall the cofactor expansion along the first column

$$a_{11}C_{11} + a_{21}C_{21} + a_{31}C_{31} + a_{41}C_{41} = det\begin{bmatrix} a_{11} & a_{12} & a_{13} & a_{14} \\ a_{21} & a_{22} & a_{23} & a_{24} \\ a_{31} & a_{32} & a_{33} & a_{34} \\ a_{41} & a_{42} & a_{43} & a_{44} \end{bmatrix}$$

If we replace the terms $a_{11}, a_{21}, a_{31}, a_{41}$ in the formula by b_1, b_2, b_3, b_4, then we have

$$x_1 = (detA)^{-1}(C_{11}b_1 + C_{21}b_2 + C_{31}b_3 + C_{41}b_4) = (detA)^{-1}\, det\begin{bmatrix} b_1 & a_{22} & a_{23} & a_{24} \\ b_2 & a_{22} & a_{23} & a_{24} \\ b_3 & a_{32} & a_{33} & a_{34} \\ b_4 & a_{42} & a_{43} & a_{44} \end{bmatrix}$$

Similarly, we get the formula for other coordinates of x:

$$x_1 = det[b\ a_2\ a_3\ a_4]/det[a_1\ a_2\ a_3\ a_4]$$

$$x_2 = det[a_1\ b\ a_3\ a_4]/det[a_1\ a_2\ a_3\ a_4]$$

$$x_3 = det[a_1\ a_2\ b\ a_4]/det[a_1\ a_2\ a_3\ a_4]$$

$$x_4 = det[a_1\ a_2\ a_3\ b]/det[a_1\ a_2\ a_3\ a_4]$$

In general, we have the Cramer's rule.

If $A = [a_1\ a_2 ...\ a_n]$ is an invertible matrix, then the unique solution of $Ax = b$ is given by

$$x_i = det[a_1\ a_2 ...\ b ...\ a_n]/det[a_1\ a_2 ...\ a_i ...\ a_n].$$

Example: In an earlier example, we found the solution of the system

$$3x_1 + x_2 - x_3 = 2$$

$$x_1 - x_2 + x_3 = 2$$

$$2x_1 + 2x_2 + x_3 = 6$$

Now we use Cramer's rule. The augmented matrix is

$$[a_1\ a_2\ a_3\ b] = \begin{bmatrix} 3 & 1 & -1 & 2 \\ 1 & -1 & 1 & 2 \\ 2 & 2 & 1 & 6 \end{bmatrix}$$

In another earlier example, we computed the determinant

$$det[a_1\ a_2\ a_3] = -12.$$

We also have

$$det[b\ a_2\ a_3] = det\begin{bmatrix} 2 & 1 & -1 \\ 2 & -1 & 1 \\ 6 & 2 & 1 \end{bmatrix} = det\begin{bmatrix} 2 & 1 & -1 \\ 4 & 0 & 0 \\ 6 & 2 & 1 \end{bmatrix}$$

$$= -4det\begin{bmatrix} 1 & -1 \\ 2 & 1 \end{bmatrix} = -12,$$

$$det[a_1\ b\ a_3] = det\begin{bmatrix} 3 & 2 & -1 \\ 1 & 2 & 1 \\ 2 & 6 & 1 \end{bmatrix} = det\begin{bmatrix} 3 & 2 & -1 \\ 4 & 4 & 0 \\ 5 & 8 & 0 \end{bmatrix}$$

$$= -det\begin{bmatrix} 4 & 4 \\ 5 & 8 \end{bmatrix} = -12.,$$

$$det[a_1\ a_2\ b] = det\begin{bmatrix} 3 & 1 & 2 \\ 1 & -1 & 2 \\ 2 & 2 & 6 \end{bmatrix} = det\begin{bmatrix} 3 & 1 & 2 \\ 4 & 0 & 4 \\ -4 & 0 & 2 \end{bmatrix}$$

$$= -det\begin{bmatrix} 4 & 4 \\ -4 & 2 \end{bmatrix} = -24.$$

Thus

$$x_1 = det[b\ a_2\ a_3]/det[a_1\ a_2\ a_3] = (-12)/(-12) = 1,$$
$$x_2 = det[a_1\ b\ a_3]/det[a_1\ a_2\ a_3] = (-12)/(-12) = 1,$$
$$x_3 = det[a_1\ a_2\ b]/det[a_1\ a_2\ a_3] = (-24)/(-12) = 2.$$

The computation is much more complicated than the row operation method. In fact, Cramer's rule is never a practical way of computing the solution. Moreover, the rule cannot be applied to the system in which the coefficient matrix is not invertile (including not square).

Exercise: Compute the determinant.

1. $\begin{bmatrix} 8 & 88 & 888 \\ 8888 & 88888 & 888888 \\ 88888888 & 88888888 & 888888888 \end{bmatrix}$

Answer: In the first equality, we use (–1)[row 1] + [row 2], (–1)[row 1] + [row 3]. In the second equality, we note that [row 3] is a scalar (=1001) multiple of [row 2] and use this result to conclude that the determinant must be zero.

$$det\begin{bmatrix} 8 & 88 & 888 \\ 8888 & 88888 & 888888 \\ 88888888 & 88888888 & 888888888 \end{bmatrix}$$

$$det\begin{bmatrix} 8 & 88 & 888 \\ 8880 & 88800 & 888000 \\ 88888880 & 88888800 & 888888000 \end{bmatrix} = 0.$$

2. $\begin{bmatrix} 0 & 0 & 1 & 3 \\ 0 & 0 & 2 & 4 \\ 1 & 1 & \sqrt{5} & \sqrt{7} \\ 1 & -1 & \sqrt{6} & \sqrt{8} \end{bmatrix}$

Answer: In the first equality, we move [col 3] to [col 1] (by two column exchanges) and the move [col 4] to [col 2] (by another two column exchanges). In the second equality, we use this result.

$$det\begin{bmatrix} 0 & 0 & 1 & 3 \\ 0 & 0 & 2 & 4 \\ 1 & 1 & \sqrt{5} & \sqrt{7} \\ 1 & -1 & \sqrt{6} & \sqrt{8} \end{bmatrix} = det\begin{bmatrix} 1 & 3 & 0 & 0 \\ 2 & 4 & 0 & 0 \\ \sqrt{5} & \sqrt{7} & 1 & 1 \\ \sqrt{6} & \sqrt{8} & 1 & -1 \end{bmatrix}$$

$$= det\begin{bmatrix} 1 & 3 \\ 2 & 4 \end{bmatrix} det\begin{bmatrix} 1 & 1 \\ 1 & -1 \end{bmatrix} = (-2) \times (-2) = 4.$$

3. $\begin{bmatrix} 15 & 0.2 & 0.1 & 4 \\ \pi & 2\pi & \pi & -\pi \\ 3 & \sqrt{12} & \sqrt{3} & -\sqrt{2} \\ \sqrt{2} & 2 & 1 & -\sqrt{2} \end{bmatrix}$

Answer: Since [col 2] is twice of [col 3], by this result we conclude that the determinant must be zero.

4. $\begin{bmatrix} 2-\lambda & 0 & 1 & 0 \\ 0 & 2-\lambda & 0 & 1 \\ 1 & 0 & 2-\lambda & 0 \\ 0 & 1 & 0 & 2-\lambda \end{bmatrix}$

Answer: In the first equality, we use [col 2] "! [col 3]. In the second equality, we use [row 2] "! [row 3]. In the third equality, we use this result.

$$det\begin{bmatrix} 2-\lambda & 0 & 1 & 0 \\ 0 & 2-\lambda & 0 & 1 \\ 1 & 0 & 2-\lambda & 0 \\ 0 & 1 & 0 & 2-\lambda \end{bmatrix} = -det\begin{bmatrix} 2-\lambda & 1 & 0 & 0 \\ 0 & 0 & 2-\lambda & 1 \\ 1 & 2-\lambda & 0 & 0 \\ 0 & 0 & 1 & 2-\lambda \end{bmatrix}$$

$$= \det\begin{bmatrix} 2-\lambda & 1 & 0 & 0 \\ 1 & 2-\lambda & 0 & 0 \\ 0 & 0 & 1 & 2-\lambda \\ 0 & 0 & 1 & 2-\lambda \end{bmatrix} = \det\begin{bmatrix} 2-\lambda & 1 \\ 1 & 2-\lambda \end{bmatrix}^2$$

$$= [(2-\lambda)^2 - 1^2]^2 = (3-\lambda)^2(1-\lambda)^2.$$

5. $\begin{bmatrix} 1 & 2 & 3 & 4 & 5 \\ 2 & 3 & 4 & 5 & 6 \\ 3 & 4 & 5 & 6 & 7 \\ 4 & 5 & 6 & 7 & 8 \\ 5 & 6 & 7 & 8 & 9 \end{bmatrix}$

Answer: In the first equality, we use (–1)[row 1] + [row 2], (–1)[row 1] + [row 3]. In the second equality, we note that [row 3] is twice of [row 2] and use this result to conclude that the determinant must be zero.

$$\det\begin{bmatrix} 1 & 2 & 3 & 4 & 5 \\ 2 & 3 & 4 & 5 & 6 \\ 3 & 4 & 5 & 6 & 7 \\ 4 & 5 & 6 & 7 & 8 \\ 5 & 6 & 7 & 8 & 9 \end{bmatrix} = \det\begin{bmatrix} 1 & 2 & 3 & 4 & 5 \\ 1 & 1 & 1 & 1 & 1 \\ 2 & 2 & 2 & 2 & 2 \\ 4 & 5 & 6 & 7 & 8 \\ 4 & 5 & 6 & 7 & 8 \end{bmatrix} = 0.$$

Exercise: Compute the determinant and find the condition on the parameter so that the matrix is invertible.

1. $\begin{bmatrix} 1 & 1 & 1 & 1 \\ 1 & 0 & 0 & 0 \\ 2 & 2 & 0 & a \\ 0 & 1 & 2 & a \end{bmatrix}$

Answer: The condition for invertibility is $a \neq 3/2$.

$$\det\begin{bmatrix} 1 & 1 & 1 & 1 \\ 1 & 0 & 0 & 0 \\ 2 & 2 & 0 & a \\ 0 & 1 & 2 & a \end{bmatrix} = -\det\begin{bmatrix} 1 & 1 & 1 \\ 2 & 0 & a \\ 1 & 3 & a \end{bmatrix} = -\det\begin{bmatrix} 1 & 1 & 1 \\ 2 & 0 & a \\ -2 & 0 & a-3 \end{bmatrix} = 2(2a-3).$$

2. $\begin{bmatrix} 0 & 1 & 1 & a \\ 1 & 0 & a & 1 \\ 1 & 1 & 0 & 0 \\ 0 & a & -1 & 1 \end{bmatrix}$

Answer: In the first equality, we use (–1)[col 1] + [col 2]. In the third equality, we use [row 1] + [row 2]. In the fourth equality, we use (–1)[col 2] + [col 3]. The condition for invertibility is $a \neq -1$ or 2.

$$det\begin{bmatrix} 0 & 1 & 1 & a \\ 1 & 0 & a & 1 \\ 1 & 1 & 0 & 0 \\ 0 & a & -1 & 1 \end{bmatrix} = det\begin{bmatrix} 0 & 1 & 1 & a \\ 1 & -1 & a & 1 \\ 1 & 0 & 0 & 0 \\ 0 & a & -1 & 1 \end{bmatrix}$$

$$= det\begin{bmatrix} 1 & 1 & a \\ -1 & a & 1 \\ a & -1 & 1 \end{bmatrix} = -det\begin{bmatrix} 1 & 1 & a \\ 0 & a+1 & a+1 \\ a & -1 & 1 \end{bmatrix} = -det\begin{bmatrix} 1 & 1 & a-1 \\ 0 & a+1 & 0 \\ a & -1 & 2 \end{bmatrix}$$

$$= (a+1)\ det\begin{bmatrix} 1 & a-1 \\ a & 2 \end{bmatrix} = -(a+1)(a^2-a-2) = -(a+1)^2(a-2).$$

3. $\begin{bmatrix} 1 & 0 & 0 & 0 \\ a & 1 & 0 & 0 \\ 0 & a & 1 & 0 \\ 0 & 0 & a & 1 \end{bmatrix}$

Answer: The determinant = 1 is never zero. The matrix is always invertible.

4. $\begin{bmatrix} 1 & 1 & 1 & 1 \\ a & 1 & 1 & 1 \\ a & a & 1 & 1 \\ a & a & a & 1 \end{bmatrix}$

Answer: In the first equality, we use (–1)[row 1] + [row 2], (–1)[row 1] + [row 3], (–1)[row 1] + [row 4]. In the third equality, we use this result. The condition for invertibility is $a \neq 1$.

$$det\begin{bmatrix} 1 & 1 & 1 & 1 \\ a & 1 & 1 & 1 \\ a & a & 1 & 1 \\ a & a & a & 1 \end{bmatrix} = det\begin{bmatrix} 1 & 1 & 1 & 1 \\ a-1 & 0 & 0 & 0 \\ a-1 & a-1 & 0 & 0 \\ a-1 & a-1 & a-1 & 0 \end{bmatrix} = det\begin{bmatrix} a-1 & 0 & 0 \\ a-1 & a-1 & 0 \\ a-1 & a-1 & a-1 \end{bmatrix} = (a-1)^3.$$

5. $\begin{bmatrix} a & b & b & b \\ a & a & b & b \\ a & a & a & b \\ a & a & a & a \end{bmatrix}$

Answer: In the first equality, we use (–1)[row 3] + [row 4], (–1)[row 2] + [row 3], (–1)[row 1] + [row 2]. In the second equality, we use this result. The condition for invertibility is $a \neq 0$ or b.

$$det\begin{bmatrix} a & b & b & b \\ a & a & b & b \\ a & a & a & b \\ a & a & a & a \end{bmatrix} = det\begin{bmatrix} a & b & b & b \\ 0 & a-b & 0 & 0 \\ 0 & 0 & a-b & 0 \\ 0 & 0 & 0 & a-b \end{bmatrix} = a(a-b)^3$$

Exercise: Let A be an invertible matrix. Prove that

$$det\begin{bmatrix} A & B \\ C & D \end{bmatrix} = detA\ det(D - CA^{-1}B).$$

Answer: We have

$$\begin{bmatrix} A & B \\ C & D \end{bmatrix} = \begin{bmatrix} I & O \\ CA^{-1} & I \end{bmatrix}\begin{bmatrix} A & B \\ O & D-CA^{-1}B \end{bmatrix}$$

Thus

$$det\begin{bmatrix} A & B \\ C & D \end{bmatrix} = det\begin{bmatrix} I & O \\ CA^{-1} & I \end{bmatrix} det\begin{bmatrix} A & B \\ O & D-CA^{-1}B \end{bmatrix}$$
$$= detI\ detI\ detA\ det(D - CA^{-1}B) = detA\ det(D - CA^{-1}B).$$

Exercise: Prove that if A is a skew–symmetric matrix of odd size, then $detA = 0$.

Answer: A is skew–symmetric means $A^T = -A$. Thus

$$\begin{aligned} detA = detA^T &= det(-A) \\ &= (-1)^n detA \\ &= -detA \end{aligned}$$

Thus we have $\quad detA = 0.$

Exercise: Show that the concepts of trace and determinant can be extended to linear transformations from a finite dimensional vector space to itself. Then compute the trace and determinants of the following linear transformations.

Answer: Given any linear transformation $T: V \to V$ and a basis B of V, we have a matrix $A = [T(B)]_B$ for T with respect to B. Moreover, if B' is another basis of V, then the matrix $B = [T(B')]_{B'}$ for T with respect to B' is similar to A

$$B = PAP^{-1},$$

where $P = [B]_{B'}$ is the matrix for changing the B-coordinates to B'-coordinates. The upshot is that if $F(X)$ is a function of n by n matrices satisfying $F(XY) = F(YX)$, then $F(B) = F(AP^{-1}P) = F(A)$. Therefore we may define

$$F(T) = F([T(B)]_B)$$

which is independent of the choice of the basis B.

Since the trace and the determinant satify the property $F(XY) = F(YX)$, the concepts can be extended to any linear transformation $T: V \to V$.

Remark: Other functions satisfying $F(XY) = F(YX)$ include $tr(X^2)$, $tr(X^3)$,...

Next we compute some concrete examples.

1. $T(p(t)) = p'(t)$: $P_n \to P_n$

Answer: With respect to the standard basis $\{1, t, t^2,..., t^n\}$, the matrix of the transformation is

$$A = \begin{bmatrix} 0 & 1 & 0 & 0 & .. & 0 & 0 \\ 0 & 0 & 2 & 0 & .. & 0 & 0 \\ 0 & 0 & 0 & 3 & .. & 0 & 0 \\ \vdots & \vdots & \vdots & \vdots & & \vdots & \vdots \\ 0 & 0 & 0 & 0 & .. & 0 & n \\ 0 & 0 & 0 & 0 & .. & 0 & 0 \end{bmatrix}$$

Both the trace and the determinant are zero.

2. $T(p(t)) = (1 - t)p'(t) - 2p(t) + p(1)$: $P_n \to P_n$

Answer: With respect to the standard basis $\{1, t, t^2,..., t^n\}$, the matrix of the transformation is

$$A = \begin{bmatrix} -1 & 2 & 2 & 2 & .. & 1 & 1 \\ 0 & -3 & 2 & 0 & .. & 0 & 0 \\ 0 & 0 & -4 & 3 & .. & 0 & 0 \\ 0 & 0 & 0 & -5 & .. & 0 & 0 \\ \vdots & \vdots & \vdots & \vdots & & \vdots & \vdots \\ 0 & 0 & 0 & 0 & .. & -n-1 & n \\ 0 & 0 & 0 & 0 & .. & 0 & -n-2 \end{bmatrix}$$

The trace is

$tr = (-1) + (-3) + (-4) + ... + (-n-2) = 2 - (n+2)(n+3)/2 = -(n^2 + 5n + 2)/2.$

The determinant is

$det = (-1)(-3)(-4)...(-n-2) = (-1)^{n+1}(n+2)!/2.$

3. $T(X) = X^T: M(n, n) \to M(n, n)$

Answer: Let E_{ij} be the matrix with 1 as ij–entry and 0 as all the other entries. Then all the E_{ij} form a basis of $M(n, n)$. We arrange the basis in the order $\{E_{11}, E_{22}, ..., E_{nn}, E_{12}, E_{21}, E_{13}, E_{31}, ..., E_{(n-1)n}, E_{n(n-1)}\}$, in which we begin with all E_{ij} satisfying $i = j$, followed by pairs E_{ij}, E_{ji} satisfying i “ j. Then $T(E_{ij}) = E_{ji}$ means that, with respect to this basis, the matrix for T is of the form

$$A = \begin{bmatrix} I & O \\ O & B \end{bmatrix},$$

where I is the n by n identity matrix, and

$$B = \begin{bmatrix} 0 & 1 & 0 & 0 & .. & 0 & 0 \\ 1 & 0 & 0 & 0 & .. & 0 & 0 \\ 0 & 0 & 0 & 1 & .. & 0 & 0 \\ 0 & 0 & 1 & 0 & .. & 0 & 0 \\ : & : & : & : & & : & : \\ 0 & 0 & 0 & 0 & .. & 0 & 1 \\ 0 & 0 & 0 & 0 & .. & 1 & 0 \end{bmatrix}$$

is an $n^2 - n$ by $n^2 - n$ matrix. Thus

$$tr = trI + trB = n,$$

$$det = detI\ detB = (-1)^{(n2-n)/2}.$$

4. $T(X) = X + X^T: M(n, n) \to M(n, n)$

Answer: Let B be a basis of the subspace of symmetric matrices. Let B' be a a basis of the subspace of skew–symmetric matrices. Then this exercise shows that B □B' is a basis of $M(n, n)$. For any b □ B, we have $T(b) = 2b$. For any b □ B', we have $T(b) = 0$. Therefore the matrix of T with respect to the basis is

$$A = \begin{bmatrix} 2I & O \\ O & O \end{bmatrix},$$

where the size of I is the dimension $n(n+1)/2$ of the space of symmetric matrices. Thus

$tr = n(n + 1),$

$det = 2^{n(n+1)/2}.$

5. $T(X) = AX$: $M(m, n) \to M(m, n)$, where A is a fixed m by m matrix

Answer: By looking at individual columns of X, T is really n copies of $R^m \to R^m$, $x \to Ax$, which has matrix A with respect to the standard basis. Thus the matrix for T is an n by n block matrix

$$\begin{bmatrix} a & o & .. & o \\ o & A & .. & O \\ : & : & & : \\ O & O & .. & A \end{bmatrix}.$$

Thus

$$tr = n\, trA,$$

$$det = detA^n.$$

6. $T = id$: $V \to V$, where V is any finite dimensional vector space

Answer: Let $dimV = n$. The the matrix of T with respect to any basis of V is the n by n identity matrix. Thus

$tr = trI_n = n,$

$det = detI_n = 1.$

THEORY OF DETERMINANT

DEFINITION OF DETERMINANT

Let V be a vector space. Let $D(v_1, v_2,..., v_n)$ be a function with n vectors of V as variables.

D is multilinear if it is linear in each vector variable

$D(..., u + v,...) = D(..., u,...) + D(..., v,...),$

$D(..., cu,...) = c\, D(..., u,...).$

D is alternating if switching two variables introduces a negative sign

$D(..., u,..., v,...) = - D(..., v,..., u,...).$

Note that taking $u = v$ in the alternating property gives $D(..., u,..., u,...) = - D(..., u,..., u,...)$, so that $D(..., u,..., u,...) = 0$. Conversely, if $D(..., u,..., u,...) = 0$ for all u, then by the multilinearity of D, we have

$0 = D(..., u + v,..., u + v,...)$

$= D(..., u,..., u + v,...) + D(..., v,..., u + v,...)$

$$= D(\ldots, u, \ldots, u, \ldots) + D(\ldots, u, \ldots, v, \ldots) + D(\ldots, v, \ldots, u, \ldots) + D(\ldots, v, \ldots, v, \ldots)$$
$$= D(\ldots, u, \ldots, v, \ldots) + D(\ldots, v, \ldots, u, \ldots),$$

which is the alternating property. Thus we proved the following.

If D is multilinear, then
$D(\ldots, u, \ldots, v, \ldots) = -D(\ldots, v, \ldots, u, \ldots) \Leftrightarrow D(\ldots, u, \ldots, u, \ldots) = 0.$

A function $D(A)$ of n by n matrices may be considered as a function with n column vectors of A as variables. Assuming D is multilinear and alternating, we will derive a formula for D.

First for a 2 by 2 matrix

$$A = \begin{bmatrix} a & b \\ c & d \end{bmatrix} = [ae_1 + ce_2, be_1 + de_2],$$

the multilinearity and the alternating property give us

$$\begin{aligned} D(A) &= D(ae_1 + ce_2, be_1 + de_2) \\ &= aD(e_1, be_1 + de_2) + cD(e_2, be_1 + de_2) \\ &= abD(e_1, e_1) + adD(e_1, e_2) + cbD(e_2, e_1) + cdD(e_2, e_2) \\ &= adD(e_1, e_2) + cbD(e_2, e_1) \\ &= D(e_1, e_2)\,(ad - cb) \\ &= D(I_2)\, detA, \end{aligned}$$

which is a constant $D(I_2)$ multiplied to the determinant.

For the 3 by 3 case

$$A = \begin{bmatrix} a_{11} & a_{12} & a_{13} \\ a_{21} & a_{22} & a_{23} \\ a_{31} & a_{32} & a_{33} \end{bmatrix} = [a_{11}e_1 + a_{21}e_2 + a_{31}e_3, a_{12}e_1 + a_{22}e_2 + a_{32}e_3, a_{13}e_1 + a_{23}e_2 + a_{33}e_3],$$

the multilinearity gives us

$$\begin{aligned} D(A) &= D(a_{11}e_1 + a_{21}e_2 + a_{31}e_3, a_{12}e_1 + a_{22}e_2 + a_{32}e_3, a_{13}e_1 + a_{23}e_2 + a_{33}e_3) \\ &= a_{11}a_{12}a_{13}D(e_1, e_1, e_1) + a_{11}a_{12}a_{23}D(e_1, e_1, e_2) + a_{11}a_{12}a_{33}D(e_1, e_1, e_3) + \ldots \\ &= \Sigma_{1\le i,j,k\le 3}\, a_{i1}a_{j2}a_{k3}D(e_i, e_j, e_k) \end{aligned}$$

where i, j, k can be any of 1, 2, 3. By the alternating property, if any two of i, j, k are the same, then $D(e_i, e_j, e_k) = 0$. This leaves only the terms with distinct i, j, k

$$D(A) = a_{11}a_{22}a_{33}D(e_1, e_2, e_3) + a_{11}a_{32}a_{23}D(e_1, e_3, e_2) + a_{21}a_{12}a_{33}D(e_2, e_1, e_3) + \ldots$$
$$= \Sigma_{i,j,k \text{ distinct}}\, a_{i1}a_{j2}a_{k3}D(e_i, e_j, e_k)$$

Since i, j, k are distinct numbers chosen from 1, 2, 3, they must be a permutation (rearrangement of positions) of (1, 2, 3). A permutation (i, j, k) can be changed to (1, 2, 3) by a sequence of transpositions (exchange of *two* positions). For example, (2, 3, 1) is changed to (1, 2, 3) by

$$(2, 3, 1) \to (2, 1, 3) \to (1, 2, 3),$$

where the blue color highlights the transpositions. By the alternating property, the sequence of transpositions tells us

$$D(e_2, e_3, e_1) = -\,D(e_2, e_1, e_3) = D(e_1, e_2, e_3).$$

In general, we always have $D(e_i, e_j, e_k) = \pm\, D(e_1, e_2, e_3)$, where the sign is positive if the number of transpositions is even, and is negative if the number is odd.

Now we can list all the terms with distinct (i, j, k) and how the signs are changed.

Permutation	***Transpositions to* (1, 2, 3)**	***Term in D(A)***	***Equal to***
(1, 2, 3)	= (1, 2, 3)	$a_{11}a_{22}a_{33}D(e_1, e_2, e_3)$	$a_{11}a_{22}a_{33}D(e_1, e_2, e_3)$
(2, 3, 1)	→ (2, 1, 3) → (1, 2, 3)	$a_{21}a_{32}a_{13}D(e_2, e_3, e_1)$	$a_{21}a_{32}a_{13}D(e_1, e_2, e_3)$
(3, 1, 2)	→ (1, 3, 2) → (1, 2, 3)	$a_{31}a_{12}a_{23}D(e_3, e_1, e_2)$	$a_{31}a_{12}a_{23}D(e_1, e_2, e_3)$
(3, 2, 1)	→ (1, 2, 3)	$a_{31}a_{22}a_{13}D(e_3, e_2, e_1)$	$-a_{31}a_{22}a_{13}D(e_1, e_{2}, {}_3)$
(2, 1, 3)	→ (1, 2, 3)	$a_{21}a_{12}a_{33}D(e_2, e_1, e_3)$	$-a_{21}a_{12}a_{33}D(e_1, e_{2}, {}_3)$
(1, 3, 2)	→ (1, 2, 3)	$a_{11}a_{32}a_{23}D(e_1, e_3, e_2)$	$-a_{11}a_{32}a_{23}D(e_1, e_{2}, {}_3)$

Originally, $D(A)$ is the sum of all the terms in the third column. By using the alternating property, this is equal to the sum of all the terms in the fourth column. Comparing with the definition of determinant for 3 by 3 matrices, we get

$$D(A) = D(e_1, e_2, e_3)\, detA = D(I_3)\, detA,$$

which is a constant $D(I_3)$ multiplied to the determinant.

The pattern we saw in the computation of $D(A)$ leads us to the following definition.

The determinant is a function *det* of n by n matrices, such that

- *detA* is multilinear in each column of A.
- *detA* is alternating in columns of A.
- $detI_n = 1$.

Several issues need to be settled in order to rigrously justify the definition.

- The function as described in the definition really exists;

- The function is also unique
- The function has all the properties we gave to the determinant early on.

GENERAL FORMULA

Consider a general n by n matrix

$$A = \begin{bmatrix} a_{11} & a_{12} & \dots & a_{1n} \\ a_{21} & a_{22} & \dots & a_{2n} \\ \vdots & \vdots & & \vdots \\ a_{n1} & a_{n2} & \dots & a_{nm} \end{bmatrix} = [a_{11}e_1 + a_{21}e_2 + \dots + a_{n1}e_n, a_{12}e_1 + a_{22}e_2 + \dots + a_{n2}e_n, \dots].$$

Our detailed computation for multilinear and alternating functions of 2 by 2 and 3 by 3 matrices give all the clues on the general formula for *detA*. The multilinearity implies

$$detA = \Sigma_{1 \le i1, i2, \dots, in \le n}\, a_{i11}a_{i22}\dots a_{inn}\, det(e_{i1}, e_{i2}, \dots, e_{in}).$$

By the alternating property, if any two of the indices i_* are the same, then $det(e_{i1}, e_{i2}, \dots, e_{in}) = 0$. Thus only the terms with distinct $i_1, i_2, \dots, i_n$ remain

$$detA = \Sigma_{i1, i2, \dots, in \text{ distinct}}\, a_{i11}a_{i22}\dots a_{inn}\, det(e_{i1}, e_{i2}, \dots, e_{in}).$$

The following facts

- The total number of indices is n,
- Indices are integers between 1 and n,
- Indices are distinct,

imply that $(i_1, i_2, \dots, i_n)$ fits into the following definition.

A permutation $(i_1, i_2, \dots, i_n)$ of $(1, 2, \dots, n)$ is a rearrangement of positions of the numbers. Equivalently, the map $k \to i_k$ is an invertible map from $\{1, 2, \dots, n\}$ to itself. The sign of the permutation is

- $Sign(i_1, i_2, \dots, i_n) = 1$ if $(i_1, i_2, \dots, i_n) \to (1, 2, \dots, n)$ by even number of transpositions
- $Sign(i_1, i_2, \dots, i_n) = -1$ if $(i_1, i_2, \dots, i_n) \to (1, 2, \dots, n)$ by odd number of transpositions

For example, the permutation (3, 1, 4, 2) represents the following invertible map from {1, 2, 3, 4} to itself:

$$1 \to 3,\ 2 \to 1,\ 3 \to 4,\ 4 \to 2.$$

Moreover, the following sequence of transpositions

$$(3, 1, 4, 2) \to (1, 3, 4, 2) \to (1, 3, 2, 4) \to (1, 2, 3, 4)$$

tells us $sign(3, 1, 4, 2) = -1$. Correspondingly,

$det(e_3, e_1, e_4, e_2) = -\,det(e_1, e_3, e_4, e_2) = det(e_1, e_3, e_2, e_4) = -\,det(e_1, e_2, e_3, e_4) = -1 = sign(3, 1, 4, 2)$.

where the red equality follows from the third condition in the definition of the determinant.

The discussion works in general. In particular, even (odd) number of transpositions will change sign even (odd) number of times and give us positive (negative) sign at the end. Thus we have

$det(e_{i1}, e_{i2},..., e_{in}) = sign(i_1, i_2,..., i_n)$.

Now we are able to write down the explicit formula for the determinant.

$detA = \Sigma_{\text{all permutations } (i1,i2,...,in) \text{ of } (1,2,...,n)}\, sign(i_1, i_2,..., i_n)\, a_{i11}a_{i22}...a_{inn}$.

We remark that, if we did not use the condition $detI_n = 1$ in the red equality above, then we get the following conclusion.

D is a multilinear and alternating function of n by n matricies $\Rightarrow D = d\ det$ for a constant $d = D(I)$.

EXISTENCE AND UNIQUENESS

Based on the three properties in the definition, we derived the formula

$detA = \Sigma_{\text{all permutations } (i1,i2,...,in) \text{ of } (1,2,...,n)}\, sign(i_1, i_2,..., i_n)\, a_{i11}a_{i22}...a_{inn}$.

We have two remarks.

- Since the functions satisfying the three properties must be given by the explicit formula, the determinant (if it exists) must be *unique*;
- We may prove the *existence* of *det* by verifying that the formula indeed satsify the three properties.

In the subsequent verification that the formula satisfies the three properties, the formula will be temporarily denoted $D(A)$ (we are not supposed to use the *det* until the properties are verified).

Multilinear Since linear combinations of linear functions are linear, we only need to verify that for each fixed permutation $(i_1, i_2,..., i_n)$, the function $a_{i11}a_{i22}...a_{inn}$ is linear in columns of A. For the linearity in [col 1], we fix entries in [col 2],..., [col n] and consider entries in [col 1] as changing. In our case, $a_{i22}...a_{inn}$ is a fixed number and a_{i11} changes. Then the composition

$$a_1 = \begin{bmatrix} a_{11} \\ a_{21} \\ : \\ a_{n1} \end{bmatrix} \to a_{i11} \to a_{i11}a_{i22}...a_{inn}$$

is linear because the first projection is linear and the second scalar multiplication by a fixed number is also linear. The reason for the linearity in the other columns is similar.

Alternating Let B be obtained from A by exchanging [col 1] and [col 2]. Then a_{i2} is an entry in [col 1] of B, a_{i1} is an entry in [col 2] of B, and the others are not changed. The change leads to

$$D(B) = \Sigma_{\text{all permutations } (i1,i2,\ldots,in) \text{ of } (1,2,\ldots,n)}\, sign(i_1, i_2,\ldots, i_n)\, a_{i12}a_{i21}a_{i33}\ldots a_{inn}$$
$$= \Sigma_{\text{all permutations } (i1,i2,\ldots,in) \text{ of } (1,2,\ldots,n)}\, sign(i_1, i_2,\ldots, i_n)\, a_{i21}a_{i12}a_{i33}\ldots a_{inn}.$$

Now we claim

$$sign(i_2, i_1,\ldots, i_n) = -\, sign(i_1, i_2,\ldots, i_n).$$

To see why, let

$$(i_1, i_2,\ldots, i_n) \to \ldots \to (1, 2,\ldots, n)$$

be a sequence of transpositions changing $(i_1, i_2,\ldots, i_n)$ to $(1, 2,\ldots, n)$. Then by adding the transposition $(i_2, i_1,\ldots, i_n) \to (i_1, i_2,\ldots, i_n)$ to the front, we get a sequence

$$(i_2, i_1,\ldots, i_n) \to (i_1, i_2,\ldots, i_n) \to \ldots \to (1, 2,\ldots, n)$$

of transpositions changing $(i_2, i_1,\ldots, i_n)$ to $(1, 2,\ldots, n)$. The upshort is that the number of transpositions in the two sequences have different parity. Thus $sign(i_2, i_1,\ldots, i_n)$ and $sign(i_1, i_2,\ldots, i_n)$ have different signs. With the help of the discussion on sign, we conclude (in the second equality, we relabeled i_2 as i_1 and i_1 as i_2)

$$D(B) = \Sigma_{\text{all permutations}} - sign(i_2, i_1,\ldots, i_n)\, a_{i21}a_{i12}a_{i33}\ldots a_{inn}$$
$$= -\,\Sigma_{\text{all permutations}}\, sign(i_1, i_2,\ldots, i_n)\, a_{i11}a_{i22}\ldots a_{inn}$$
$$= -\,D(A).$$

Identity In the identity matrix I we have $a_{ij} = 0$ for $i \neq j$ and $a_{ii} = 1$. Therefore the term $a_{i11}a_{i22}\ldots a_{inn}$ is nonzero only if the permutation $(i_1, i_2,\ldots, i_n) = (1, 2,\ldots, n)$. Thus

$$D(I) = Sign(1, 2,\ldots, n)\; 1 \times 1 \times \ldots \times 1 = 1.$$

GENERALIZATION

The determinant can be defined through multilinear and alternating properties. By extending the properties to nonsquare matrices, the concept of determinant can be extended.

For the 2 by 3 case

$$A = \begin{bmatrix} a_{11} & a_{12} & a_{13} \\ a_{21} & a_{22} & a_{23} \end{bmatrix} = [a_{11}e_1 + a_{21}e_2, a_{12}e_1 + a_{22}e_2, a_{13}e_1 + a_{23}e_2],$$

the multilinearity implies

$$D(A) = D(a_{11}e_1 + a_{21}e_2, a_{12}e_1 + a_{22}e_2, a_{13}e_1 + a_{23}e_2)$$
$$= a_{11}D(e_1, a_{12}e_1 + a_{22}e_2, a_{13}e_1 + a_{23}e_2) + a_{21}D(e_2, a_{12}e_1 + a_{22}e_2, a_{13}e_1 + a_{23}e_2)$$

$$\begin{aligned}
&= a_{11}a_{12}D(e_1, e_1, a_{13}e_1 + a_{23}e_2) + a_{11}a_{22}D(e_1, e_2, a_{13}e_1 + a_{23}e_2) \\
&+ a_{21}a_{12}D(e_2, e_1, a_{13}e_1 + a_{23}e_2) + a_{21}a_{22}D(e_2, e_2, a_{13}e_1 + a_{23}e_2) \\
&= a_{11}a_{12}a_{13}D(e_1, e_1, e_1) + a_{11}a_{12}a_{23}D(e_1, e_1, e_2) \\
&+ a_{11}a_{22}a_{13}D(e_1, e_2, e_1) + a_{11}a_{22}a_{23}D(e_1, e_2, e_2) \\
&+ a_{21}a_{12}a_{13}D(e_2, e_1, e_1) + a_{21}a_{12}a_{23}D(e_2, e_1, e_2) \\
&+ a_{21}a_{22}a_{13}D(e_2, e_2, e_1) + a_{21}a_{22}a_{23}D(e_2, e_2, e_2).
\end{aligned}$$

Note that in each $D(e_i, e_j, e_k)$, at least two from i, j, k are equal. Therefore the alternating property would further imply $D(e_i, e_j, e_k) = 0$. We conclude that any multilienar and alternating function of 2 by 3 matrices must be zero.

Similarly, for the 3 by 2 case

$$A = \begin{bmatrix} a_{11} & a_{12} \\ a_{21} & a_{22} \\ a_{31} & a_{32} \end{bmatrix} = [a_{11}e_1 + a_{21}e_2 + a_{31}e_3, a_{12}e_1 + a_{22}e_2 + a_{32}e_3],$$

the multilinearity implies

$$\begin{aligned}
D(A) &= D(a_{11}e_1 + a_{21}e_2 + a_{31}e_3, a_{12}e_1 + a_{22}e_2 + a_{32}e_3) \\
&= a_{11}D(e_1, a_{12}e_1 + a_{22}e_2 + a_{32}e_3) \\
&+ a_{21}D(e_2, a_{12}e_1 + a_{22}e_2 + a_{32}e_3) \\
&+ a_{31}D(e_3, a_{12}e_1 + a_{22}e_2 + a_{32}e_3) \\
&= a_{11}a_{12}D(e_1, e_1) + a_{11}a_{22}D(e_1, e_2) + a_{11}a_{32}D(e_1, e_3) \\
&+ a_{21}a_{12}D(e_2, e_1) + a_{21}a_{22}D(e_2, e_2) + a_{21}a_{32}D(e_2, e_3) \\
&+ a_{31}a_{12}D(e_3, e_1) + a_{31}a_{22}D(e_3, e_2) + a_{31}a_{32}D(e_3, e_3).
\end{aligned}$$

The alternating property further implies

$$\begin{aligned}
D(A) &= (a_{11}a_{22} - a_{21}a_{12})D(e_1, e_2) + (a_{11}a_{32} - a_{31}a_{12})D(e_1, e_3) + \\
&\quad (a_{21}a_{32} - a_{31}a_{22})D(e_2, e_3) \\
&= D(e_1, e_2)\, det\begin{bmatrix} a_{11} & a_{12} \\ a_{21} & a_{22} \end{bmatrix} + D(e_1, e_3)\, det\begin{bmatrix} a_{11} & a_{12} \\ a_{31} & a_{32} \end{bmatrix} + D(e_2, e_3)\, det\begin{bmatrix} a_{21} & a_{22} \\ a_{31} & a_{32} \end{bmatrix}.
\end{aligned}$$

Thus the function $D(A)$ is determined by three numbers $D(e_1, e_2)$, $D(e_1, e_3)$, $D(e_2, e_3)$. The three numbers can actually be arbitrarily chosen, because for any numbers D_{12}, D_{13}, D_{23}, the function

$$D(A) = D_{12}det\begin{bmatrix} a_{11} & a_{12} \\ a_{21} & a_{22} \end{bmatrix} + D_{13}det\begin{bmatrix} a_{11} & a_{12} \\ a_{31} & a_{32} \end{bmatrix} + D_{23}det\begin{bmatrix} a_{21} & a_{22} \\ a_{31} & a_{32} \end{bmatrix}$$

is multilinear and alternating. A fancy way of saying this is that all the multilinear and alternating functions on 3 by 2 matrices form a vector space of dimension 3, with the three 2 by 2 determinants as a basis.

In general, for m by n matrix A, the multilinearity gives us

$$D(A) = \Sigma_{1\le i1,i2,...,in\le m}\, a_{i11}a_{i22}...a_{inn}\, D(e_{i1}, e_{i2},..., e_{in}).$$

By the alternating property, $D(e_{i1}, e_{i2},..., e_{in}) \neq 0$ only if $i_1, i_2,..., i_n$ are distinct. Since the n integers $i_1, i_2,..., i_n$ are chosen from $1, 2,..., m$, this can happen only if $m \geq n$. In other words, if $m < n$, then $D(A) = 0$ and this generalization of determinant is trivial.

Now assume $m \geq n$. Then $D(A)$ will be nontrivial. Moreover, we may rearrange each choice $(i_1, i_2,..., i_n)$ into ascending order $J = (j_1, j_2,..., j_n)$, $1 \leq j_1 < j_2 <... < j_n \leq m$. In other words, we think of $(i_1, i_2,..., i_n)$ as a permutation of J and rewrite $D(A)$ as follows (Σ_J is the sum over all subsets of $\{1, 2,..., n\}$ that contain m numbers)

$$D(A) = \Sigma_J\, \Sigma_{\text{all permutations } (i1,i2,...,in) \text{ of } J}\, a_{i11}a_{i22}...a_{inn}\, D(e_{i1}, e_{i2},..., e_{in}).$$

Then by alternating property, we have

$$D(e_{i1}, e_{i2},..., e_{in}) = sign(i_1, i_2,..., i_n)D(e_{j1}, e_{j2},..., e_{jn})$$

where $sign(i_1, i_2,..., i_n)$ is computed by a sequence of transpositions that changes $(i_1, i_2,..., i_n)$ to ascending order. For example, the following sequence of transpositions

$$(6, 3, 8, 5) \to (3, 6, 8, 5) \to (3, 6, 5, 8) \to (3, 5, 6, 8)$$

tells us $sign(6, 3, 8, 5) = -1$. Thus

$$D(A) = \Sigma_J\, D(e_{j1}, e_{j2},..., e_{jn})\, \Sigma_{\text{all permutations } (i1,i2,...,in) \text{ of } J}\, sign(i_1, i_2,..., i_n)\, a_{i11}a_{i22}...a_{inn}.$$

Now $\Sigma_{\text{all permutations } (i1,i2,...,in) \text{ of } J}\, sign(i_1, i_2,..., i_n)\, a_{i11}a_{i22}...a_{inn}$ is the same as the explicit formula for the determinant, except $(1, 2,..., n)$ is now replaced by $J = (j_1, j_2,..., j_n)$. In other words, if we denote

$$A(J) = A(j_1, j_2,..., j_n) = n \text{ by } n \text{ matrix formed by [row } j_1\text{], [row } j_2\text{],..., [row } j_n\text{] of } A,$$

then

$$\Sigma_{\text{all permutations } (i1,i2,...,in) \text{ of } J}\, sign(i_1, i_2,..., i_n)\, a_{i11}a_{i22}...a_{inn} = detA(J)$$

and we conclude the following.

Suppose $D(A)$ is a multilinear and alternating function on columns of m by n matrices A. Then $D(A) = 0$ for $m < n$, and

$$D(A) = \Sigma_{1\le j1<j2<...<jn\le m}\, D(e_{j1}, e_{j2},..., e_{jn})\, detA(j_1, j_2,..., j_n)$$

for $m \geq n$.

Conversely, for arbitrarily chosen numbers D_J (one for each subset J of $\{1, 2,..., n\}$ that contain m numbers), the function $D(A) = \Sigma_J D_J\, detA(J)$ is a multilinear and alternating function. The total number of J's is $n!/m!(n-m)!$. In the special case $m = n$, the number is 1, and we recover this result.

Finally, the functions described in this section can be reformulated as linear functions over a suitable vector space $\Lambda^n R^m$, called exterior algebra, which is a fundamental tool in modern mathematics.

Exercise: Compute the signs of the permutations.

(1, 2, 3, 4, 5)

(5, 4, 3, 2, 1)

(3, 2, 5, 1, 4)

(3, 1, 2, 4, 5)

(3, 2, 1, 6, 5, 4)

ANSWER *sign*(1, 2, 3, 4, 5) = 1. The transpositions

(5, 4, 3, 2, 1) → (1, 4, 3, 2, 5) → (1, 2, 3, 4, 5)

(3, 2, 5, 1, 4) → (1, 2, 5, 3, 4) → (1, 2, 3, 5, 4) → (1, 2, 3, 4, 5)

(3, 1, 2, 4, 5) → (1, 3, 2, 4, 5) → (1, 2, 3, 4, 5)

(3, 2, 1, 6, 5, 4) → (1, 2, 3, 6, 5, 4) → (1, 2, 3, 4, 5, 6)

imply *sign*(5, 4, 3, 2, 1) = 1, *sign*(3, 2, 5, 1, 4) = –1, *sign*(3, 1, 2, 4, 5) = 1, *sign*(3, 2, 1, 6, 5, 4) = 1.

Exercise: Write down the detailed formula for the determinant of a general 4 by 4 matrix.

Answer: The following is the list (in alphabetical order) of all the permutations of (1, 2, 3, 4) and their signs.

Permutation	***Transpositions to (1, 2, 3)***	***Sign***
(1, 2, 3, 4)	= (1, 2, 3, 4)	1
(1, 2, 4,	→ (1, 2, 3, 4)	–1
(1, 3, 2, 4)	→ (1, 2, 3, 4)	–1
(1, 3, 4, 2)	→ (1, 2, 4, 3) → (1, 2, 3, 4)	1
(1, 4, 2, 3)	→ (1, 2, 4, 3) → (1, 2, 3, 4)	1
(1, 4, 3, 2)	→ (1, 2, 3, 4)	–1
(2, 1, 3, 4)	→ (1, 2, 3, 4)	–1
(2, 1, 4, 3)	→ (1, 2, 4, 3) → (1, 2, 3, 4)	1

Permutation	*Transpositions to (1, 2, 3)*	*Sign*
(2, 3, 1, 4)	→ (1, 3, 2, 4) → (1, 2, 3, 4)	1
(2, 3, 4, 1)	→ (1, 3, 4, 2) → (1, 2, 4, 3) → (1, 2, 3, 4)	–1
(2, 4, 1, 3)	→ (1, 4, 2, 3) → (1, 2, 4, 3) → (1, 2, 3, 4)	–1
(2, 4, 3, 1)	→ (1, 4, 3, 2) → (1, 2, 3, 4)	1
(3, 1, 2, 4)	→ (1, 3, 2, 4) → (1, 2, 3, 4)	1
(3, 1, 4, 2)	→ (1, 3, 4, 2) → (1, 2, 4, 3) → (1, 2, 3, 4)	–1
(3, 2, 1, 4)	→ (1, 2, 3, 4)	–1
(3, 2, 4, 1)	→ (1, 2, 4, 3) → (1, 2, 3, 4)	1
(3, 4, 1, 2)	→ (1, 4, 3, 2) → (1, 2, 3, 4)	1
(3, 4, 2, 1)	→ (1, 4, 2, 3) → (1, 2, 4, 3) → (1, 2, 3, 4)	–1
(4, 1, 2, 3)	→ (1, 4, 2, 3) → (1, 2, 4, 3) → (1, 2, 3, 4)	–1
(4, 1, 3, 2)	→ (1, 4, 3, 2) → (1, 2, 3, 4)	1
(4, 2, 1, 3)	→ (1, 2, 4, 3) → (1, 2, 3, 4)	1
(4, 2, 3, 1)	→ (1, 2, 3, 4)	–1
(4, 3, 1, 2)	→ (1, 3, 4, 2) → (1, 2, 4, 3) → (1, 2, 3, 4)	–1
(4, 3, 2, 1)	→ (1, 3, 2, 4) → (1, 2, 3, 4)	1

The formula for the determinant of a general 4 by 4 matrix is

$$\begin{aligned}
&a_{11}a_{22}a_{33}a_{44} - a_{11}a_{22}a_{43}a_{34} - a_{11}a_{32}a_{23}a_{44} + a_{11}a_{32}a_{43}a_{24}\\
&+ a_{11}a_{42}a_{23}a_{34} - a_{11}a_{42}a_{33}a_{24} - a_{21}a_{12}a_{33}a_{44} + a_{21}a_{12}a_{43}a_{34}\\
&+ a_{21}a_{32}a_{13}a_{44} - a_{21}a_{32}a_{43}a_{14} - a_{21}a_{42}a_{13}a_{34} + a_{21}a_{42}a_{33}a_{14}\\
&+ a_{31}a_{12}a_{23}a_{44} - a_{31}a_{12}a_{43}a_{24} - a_{31}a_{22}a_{13}a_{44} + a_{31}a_{22}a_{43}a_{14}\\
&+ a_{31}a_{42}a_{13}a_{24} - a_{31}a_{42}a_{23}a_{14} - a_{41}a_{12}a_{23}a_{34} + a_{41}a_{12}a_{33}a_{24}\\
&+ a_{41}a_{22}a_{13}a_{34} - a_{41}a_{22}a_{33}a_{14} - a_{41}a_{32}a_{13}a_{24} + a_{41}a_{32}a_{23}a_{14}
\end{aligned}$$

Exercise: How many terms are in the formula for the determinant of a general n by n matrix?

Answer: We construct a permutation by choosing the numbers one at a time.

For the first position, we choose i_1, which can be any integer between 1 and n. Therefore the number of choices for i_1 is n.

After i_1 is fixed, we choose i_2 for the second position. i_2 can be any integer between 1 and n and not equal to i_1. Therefore the number of choices for i_2 is $n - 1$.

Similarly, after i_1 and i_2 are fixed, we can choose i_3 to be any integer between 1 and n and not equal to i_1, i_2. The number of choices for i_3 is $n - 2$.

The argument goes on. The total number of permutations is the product of the total number of choices, which is $n! = n\,(n-1)\,(n-2)...\,3\,2\,1$.

Exercise: Prove that the number of permutations with positive sign is equal to the number of permutations with negative sign.

Answer: Let S_n be the set of all permutations of (1, 2,..., n). Let S_n^+ and S_n^- be the subsets of permutations with positive and negative signs. Consider the map

$\sigma(i_1, i_2, i_3, i_4,..., i_n) = (i_2, i_1, i_3, i_4,..., i_n): S_n \to S_n$

that switches the first two positions of a permutation. If

$(i_1, i_2, i_3, i_4,..., i_n) \to ... \to (1, 2, 3, 4,..., n)$

is a sequence of transpositions that changes $(i_1, i_2, i_3, i_4,..., i_n)$ to the standard one. Then by adding one more transposition,

$(i_2, i_1, i_3, i_4,..., i_n) \to (i_1, i_2, i_3, i_4,..., i_n) \to ... \to (1, 2, 3, 4,..., n)$

is a sequence of transpositions that changes $\sigma(i_1, i_2, i_3, i_4,..., i_n)$ to the standard one. Since the number of transpositions in the two sequence differ by 1, we have

$sign(\sigma(i_1, i_2, i_3, i_4,..., i_n)) = -sign(i_1, i_2, i_3, i_4,..., i_n)$.

Thus σ takes S_n^+ to S_n^- and S_n^- back to S_n^+. Moreover, σ is clearly invertible, with σ to be its own inverse. Therefore σ is a one–to–one correspoindence between S_n^+ and S_n^-. In particular, this implies that S_n^+ and S_n^- have the same number of elements.

PROOF OF PROPERTIES

TRANSPOSE

We prove that the determinant is not changed by the transpose.

If we denote the ij–entry of A by a_{ij}, then the ij–entry of A^T is a_{ji}. Consequently, if we change a_{ij} to a_{ji} in the explicit formula

$detA = \Sigma_{\text{all permutations}}\, sign(i_1, i_2,..., i_n)\, a_{i11} a_{i22} ... a_{inn}$,

then we get

$detA^T = \Sigma_{\text{all permutations}}\, sign(i_1, i_2,..., i_n)\, a_{1i1} a_{2i2} ... a_{nin}$.

To see why $detA^T$ can be equal to $detA$, let us consider the 4 by 4 case. Consider a term $sign(3, 1, 4, 2) a_{13} a_{21} a_{34} a_{42}$ in $detA^T$. By looking at the product of entries, we have

$a_{13} a_{21} a_{34} a_{42} = a_{21} a_{42} a_{13} a_{34}$.

Correspondingly, we have the term $sign(2, 4, 1, 3) a_{21} a_{42} a_{13} a_{34}$ in $detA$. It turns out that $sign(3, 1, 4, 2) = sign(2, 4, 1, 3)$, so that the corresponding terms in $detA^{\mathrm{T}}$ and $detA$ are equal (together with the sign).

In general, a similar argument can be carried out as follows. Any term $sign(i_1, i_2,..., i_n)$ $a_{1i1}a_{2i2}...a_{nin}$ in $detA^{\mathrm{T}}$ can be rewritten as $sign(i_1, i_2,..., i_n)\ a_{j11}a_{j22}...a_{jnn}$, which is compared with the corresponding term $sign(j_1, j_2,..., j_n)\ a_{j11}a_{j22}...a_{jnn}$ in $detA$. The equality $detA^{\mathrm{T}} = detA$ will then be the consequence of the equality $sign(j_1, j_2,..., j_n) = sign(i_1, i_2,..., i_n)$.

To prove $sign(j_1, j_2,..., j_n) = sign(i_1, i_2,..., i_n)$, we first need to understand the rewriting process. In more detail, the process is the following. We start with a 2 by n matrix

$$\begin{matrix}\text{first indices in } detA^{\mathrm{T}} \Rightarrow \\ \text{second indices in } detA^{\mathrm{T}} \Rightarrow\end{matrix} \begin{bmatrix} 1 & 2 & \dots & n \\ i_1 & i_2 & \dots & i_n \end{bmatrix}$$

A sequence of transpositions that changes $(i_1, i_2,..., i_n)$ to $(1, 2,..., n)$ can be applied to the columns of the matrix, giving us a sequence of transpositions on 2 by n matrices

$$\begin{bmatrix} 1 & 2 & \dots & n \\ i_1 & i_2 & \dots & i_n \end{bmatrix} \to \dots \to \begin{bmatrix} j_1 & j_2 & \dots & j_n \\ 1 & 2 & \dots & n \end{bmatrix} \begin{matrix}\Leftarrow \text{first indices in } detA \\ \Leftarrow \text{second indices in } detA\end{matrix}.$$

For the 4 by 4 example $a_{13}a_{21}a_{34}a_{42}$, this means the following

$$\begin{bmatrix} 1 & 2 & \dots & n \\ i_1 & i_2 & \dots & i_n \end{bmatrix} \to \begin{bmatrix} 2 & 1 & 3 & 4 \\ 1 & 3 & 4 & 2 \end{bmatrix} \to \begin{bmatrix} 2 & 1 & 4 & 3 \\ 1 & 3 & 2 & 4 \end{bmatrix} \to \begin{bmatrix} 2 & 4 & 1 & 3 \\ 1 & 2 & 3 & 4 \end{bmatrix}$$

where we note that the result corresponds to $a_{21}a_{42}a_{13}a_{34}$. The upshot is that, the sequence of transpositions that changed (3, 1, 4, 2) to (1, 2, 3, 4) also changes (1, 2, 3, 4) to (2, 4, 1, 3). Equivalently, the transpositions in the sequence

$(2, 4, 1, 3) \to (2, 1, 4, 3) \to (2, 1, 3, 4) \to (1, 2, 3, 4)$

is the reverse of the transpositions in the sequence

$(3, 1, 4, 2) \to (1, 3, 4, 2) \to (1, 3, 2, 4) \to (1, 2, 3, 4)$.

Since the reversing does not change the number of transpositions, we conclude that $sign(2, 4, 1, 3) = sign(3, 1, 4, 2)$.

In general, given a sequence of transpositions

$(i_1, i_2,..., i_n) \to ... \to (1, 2,..., n)$,

the reverse sequence of transpositions gives us

$(j_1, j_2,..., j_n) \to ... \to (1, 2,..., n)$.

Since the two sequences consist of the same number of transpositions, we have $sign(j_1, j_2,..., j_n) = sign(i_1, i_2,..., i_n)$.

JUSTIFICATION OF COMPUTATION

We prove the two computational tools for the determinant: row/column operations and cofactor expansions.

Column Operation The properties, stated for columns, are

A changed to B by	determinant changed by
r[col i] + [col j]	$\det B = \det A$
[col i] $\leftrightarrow$ [col j]	$\det B = -\det A$
d[col i], $d \neq 0$	$\det B = d \det A$

For the first operation, we have

$$\begin{aligned} det(..., u,..., ru + v,...) &= r\, det(..., u,..., u,...) + det(..., u,..., v,...) \text{ (det is multilinear)} \\ &= det(..., u,..., v,...) \text{ (this consequence of alternating property)} \end{aligned}$$

For the second operation, the property is exactly the alternating property, which is one of the defining properties of *det*.

For the third operation, the property is the (scalar part of) multilinearity property, which is also built into the definition of *det*.

Row Operation This follows from the column operation case and the property $\det A = \det A^T$ proved before.

Cofactor Expansion Along Column Let us consider the cofactor expansion along the first column of a 4 by 4 matrix

$$A = [a_1\ a_2\ a_3\ a_4] = \begin{bmatrix} a_{11} & a_{12} & a_{13} & a_{14} \\ a_{21} & a_{22} & a_{23} & a_{24} \\ a_{31} & a_{32} & a_{33} & a_{34} \\ a_{41} & a_{42} & a_{43} & a_{44} \end{bmatrix}$$

By the linearity in the first column, we have

$$\begin{aligned} \det A &= det(a_{11}e_1 + a_{21}e_2 + a_{31}e_3 + a_{41}e_4, a_2, a_3, a_4) \\ &= a_{11}det(e_1, a_2, a_3, a_4) + a_{21}det(e_2, a_2, a_3, a_4) \\ &\quad + a_{31}det(e_3, a_2, a_3, a_4) + a_{41}det(e_4, a_2, a_3, a_4). \end{aligned}$$

The first of the four determinants is

$$det(e_1, a_2, a_3, a_4) = det \begin{bmatrix} 1 & a_{12} & a_{13} & a_{14} \\ 0 & a_{22} & a_{23} & a_{24} \\ 0 & a_{32} & a_{33} & a_{34} \\ 0 & a_{42} & a_{43} & a_{44} \end{bmatrix} = det \begin{bmatrix} 1 & a_{22} & a_{23} & a_{24} \\ 0 & a_{12} & a_{13} & a_{14} \\ 0 & a_{32} & a_{33} & a_{34} \\ 0 & a_{42} & a_{43} & a_{44} \end{bmatrix} = D(A_{11}).$$

where we used the (just proved) column operations $(-a_{12})$[col 1] + [col 2], $(-a_{13})$[col 1] + [col 3], $(-a_{14})$[col 1] + [col 4] for the second equality, and

$$A_{11} = \begin{bmatrix} a_{22} & a_{23} & a_{24} \\ a_{32} & a_{33} & a_{34} \\ a_{42} & a_{43} & a_{44} \end{bmatrix}$$

Since $D(A_{11})$ is clearly multilinear and alternating in columns of A_{11}, by this result, $D(A_{11}) = d\, detA_{11}$, where $d = D(I_3) = det(e_1, e_2, e_3, e_4) = detI_4 = 1$. Thus we conclude $det(e_1, a_2, a_3, a_4) = D(A_{11}) = detA_{11} = C_{11}$.

The other determinants can be computed by first using the (just proved) row operations to change to a determinant of the form $det(e_1, a_2, a_3, a_4)$ and then use the computation above. For example,

$$det(e_2, a_2, a_3, a_4) = det\begin{bmatrix} 0 & a_{12} & a_{13} & a_{14} \\ 1 & a_{22} & a_{23} & a_{24} \\ 0 & a_{32} & a_{33} & a_{34} \\ 0 & a_{42} & a_{43} & a_{44} \end{bmatrix} = -\,det\begin{bmatrix} 1 & a_{22} & a_{23} & a_{24} \\ 0 & a_{12} & a_{13} & a_{14} \\ 0 & a_{32} & a_{33} & a_{34} \\ 0 & a_{42} & a_{43} & a_{44} \end{bmatrix}$$

$$= -\,detA_{21} = C_{21}.$$

At then end, we get

$$detA = a_{11}C_{11} + a_{21}C_{21} + a_{31}C_{31} + a_{41}C_{41}.$$

For the expansions along some other columns, we may first use the (just proved) column operations to move the specific column to the first column and then use the result just proved. Finally, the idea of the argument for the 4 by 4 case clearly works for the general case.

Cofactor Expansion Along Row This follows from the column expansion case and the property $detA = detA^T$ proved before.

JUSTIFICATION OF GEOMETRY

In this section, several properties of the determinant are given. Moreover, all except the transpose property were given geometrical meaning. The transpose property has been proved here. Below we will prove the other properties.

Product Property Let A and B be square matrices of the same size, then $detAB = detA\; detB$.

We fix A and consider $D(B) = detAB$ as a function of B. The following shows that D preserves addition in columns of B.

$$\begin{aligned} D(..., u+v,...) &= det(A[..., u+v,...]) \\ &= det[..., A(u+v),...] \\ &= det[..., Au+Av,...] \\ &= det[..., Au,...] + det[..., Av,...] \\ &= det(A[..., u,...]) + det(A[..., u,...]) \\ &= D(..., u,...) + D(..., v,...). \end{aligned}$$

Similarly, we may verify that D preserves scalar multiplication in columns of B. Moreover, the following shows D is alternating in columns of B.

$$\begin{aligned} D(..., v,..., u,...) &= det(A[..., v,..., u,...]) \\ &= det[..., Av,..., Au,...] \\ &= -\, det[..., Au,..., Av,...] \\ &= -det(A[..., u,..., v,...]) \\ &= D(..., u,..., v,...). \end{aligned}$$

Thus by this result, we have

$detAB = D(B) = D(I)\ detB = detAI\ detB = detA\ detB.$

Triangle Property Let A and B be square matrices. Then $det\begin{bmatrix} A & O \\ \# & B \end{bmatrix} = detA\ detB.$

We may use the idea similar to the proof of the product property. For fixed A, it is easy to verify that the function

$$D(B) = det\begin{bmatrix} A & O \\ \# & B \end{bmatrix}$$

is multilinear and alternating in columns of B. Then by this result, we have $D(B) = D(I)\ detB$. Now

$$D(I) = det\begin{bmatrix} A & O \\ \# & I \end{bmatrix}$$

may be computed by repeated use of cofactor expansion (already proved here) along the last column. The result is $D(I) = detA$. This completes the proof that $D(B) = detA\ detB$.

Chapter 3

Vector Algebra

A study of motion will involve the introduction of a variety of quantities which are used to describe the physical world. Examples of such quantities include distance, displacement, speed, velocity, acceleration, force, mass, momentum, energy, work, power, etc. All these quantities can by divided into two categories-vectors and scalars. In this chapter, we will study about vectors and operations on them.

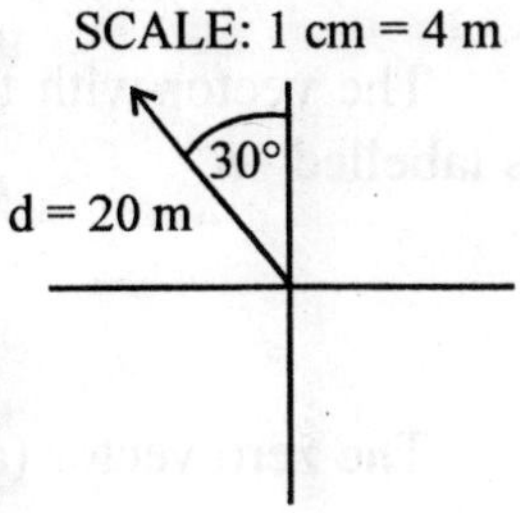

Scalars are quantities that only have a magnitude like mass and speed. Many times it is often useful to have a quantity that has not only a magnitude but also a direction; such a quantity is called a vector.

As an example of when vectors are necessary, suppose a plane traveling at 300 mph to the north with no wind present encounters a westerly crosswind of 50 mph. The resultant velocity of the plane is the sum of the velocities of the wind and the plane. To find this resultant velocity, we must use vectors.

Vectors alone have limited use other than providing a simple, yet effective, means of displaying quantities possessing both a magnitude and direction. The real power in vectors resides in the ability to perform mathematical operations on them.

SCALARS AND VECTORS

Scalars: A scalar is a real number.

Examples: Mass, volume, temperature, work, etc.

Vectors: A vector is a quantity with magnitude and a direction. The magnitude of the vector is also called its length.

Examples: Velocity, acceleration, momentum, force, electric and magnetic field intensities.

Note that all of these quantities not only have a magnitude (such as speed - the magnitude of the velocity vector) but also occur or act in a given direction. A vector is represented by a directed line segment. The direction of the vector is indicated by an arrow pointing from the tail to the head.

If the tail is at point A and the head is at point B, the vector from A to B is written $\overrightarrow{AB}$.

The length (magnitude) of a vector $\vec{v}$ is written $|\vec{v}|$. The length is always a non-negative real number.

VECTORS AS DIRECTED LINE SEGMENTS

Vectors are represented in a natural way by directed line segments, the length of the segment measuring the magnitude of the vector and the arrowhead indicating the direction of the vector. The arrowhead is usually placed at one end (or near the end) of the line segment; this end is called the head of the vector and the other end is called the tail of the vector.

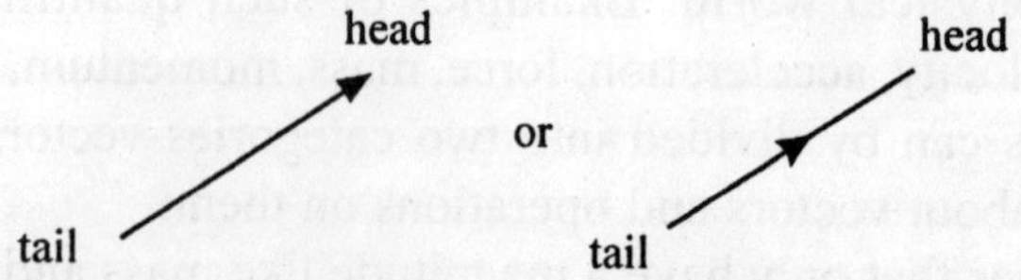

The vector with the same magnitude as u but opposite direction, the negative of u, is labelled $-u$.

The zero vector (a vector with zero length and no direction) is written 0.

REPRESENTATION OF A VECTOR

Given points A and B, we may think of the position vector $v = \overrightarrow{AB}$ as the vector that acts on the point A to get B.

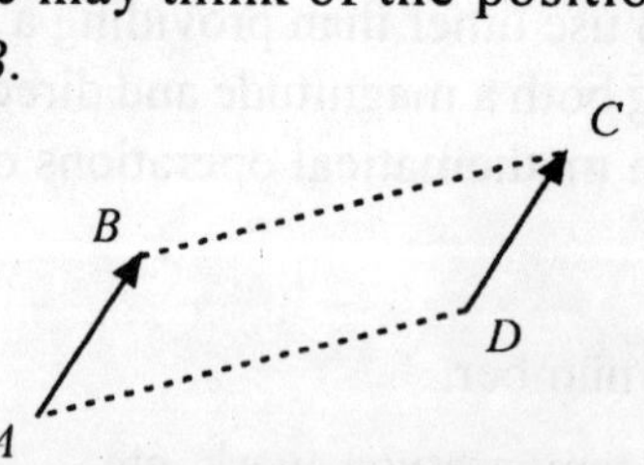

If the line segments AB and DC have the same direction and the same length, then $ABCD$ is a parallelogram and the position vectors $\overrightarrow{AB}$ and $\overrightarrow{DC}$ are equal; we write this as $\overrightarrow{AB} = \overrightarrow{DC}$.

The vector v is a free vector because although it has a definite direction and length it does not have any particular position in space.

When using vectors to model physical phenomena it is not always appropriate to use free vectors. For example, in order to completely describe the effect of a force we need to give a vector representing the magnitude and direction of the force as well as a point of application of the force.

A geometric vector is a quantity that can be represented by the directed line segment with initial point *A* and terminal point *B* as shown in the figure.

Its magnitude is the length of the directed line segment proportional to the magnitude of the vector. The magnitude is sometimes called the norm.

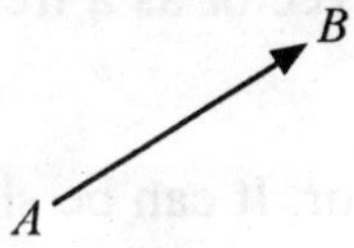

Fig.

Analytically a vector is represented by a symbol $\overrightarrow{AB}$ or a bold face letter *U*. The magnitude of $\overrightarrow{AB}$ AB is written as $|\overrightarrow{AB}|$ or $|AB|$ or simply *AB*. The magnitude of *U* is written as $|U|$ or simply *U*.

ZERO OR NULL VECTOR

Even though we have said that all vectors have length and direction, there is one exception, the zero vector.

The zero vector $\hat{0}$ is the vector from a point A to the same point A. We can write $0 = \overrightarrow{AA}$ no matter where the point *A* is.

The zero vector does not have a direction but it does have a length; not surprisingly, its length is 0. Thus we can write $|0| = 0$, i.e., it is a vector of magnitude zero. It is written as $\hat{0}$.

Thus zero vector is a vector, having no specific direction. Graphically it is a point. If $\overrightarrow{AB}= 0$, then $|\overrightarrow{AB}| = 0$ and A and B coincides.

EQUALITY OF VECTORSS

Two vectors *a* and *b* whose magnitudes are equal i.e. $a = b$ and which are in the same direction regardless of their initial points, are called equal vectors.

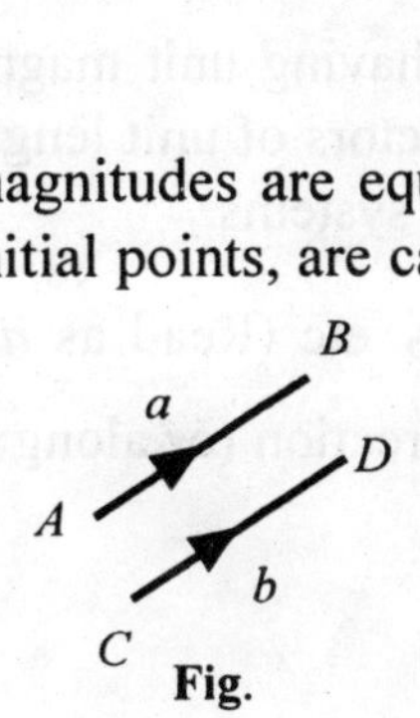

Fig.

It can be seen that for three vectors a, b and c,

$r = a + b + c = (a + b) + c = a + (b + c) + (a + c) + b$

This result can be extended to any number of vectors.

Free Vector

A vector which can be shifted along its line of action or parallel to itself and whose initial point can be anywhere, is called a free vector. Free vectors are distinguished only by their length and direction, and not by their position in space.

Unless stated, we always refer a vector as a free vector.

Localized Vector

It is also known as a sliding vector. It can be shifted only along its line of action.

Bounded Vector

It is called as a fixed vector. It must remain at the same point of action or application.

NEGATIVE OF A VECTOR

The negative of a vector u is the vector –u which has the same direction and inclination but is of the opposite sense (of direction).

The negative of the vector from A to B is the vector from B to A. It has the same length as $\overrightarrow{AB}$ but the opposite direction.

Thus the negative of the vector $\overrightarrow{AB}$ is $\overrightarrow{BA}$ and we also write $\overrightarrow{BA} = -\overrightarrow{AB}$.

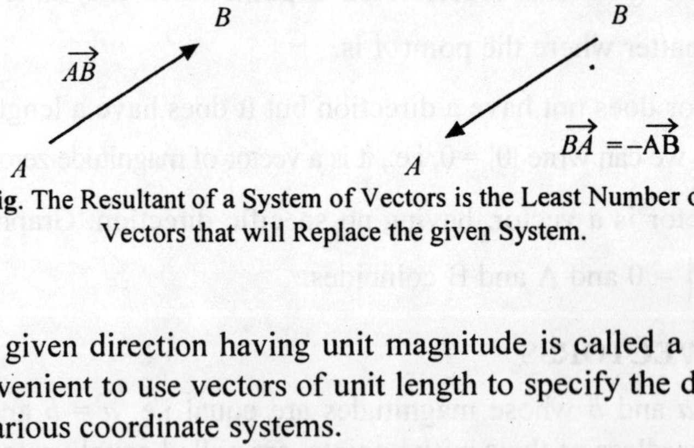

Fig. The Resultant of a System of Vectors is the Least Number of Vectors that will Replace the given System.

UNIT VECTOR

A vector in a given direction having unit magnitude is called a unit vector in that direction. It is convenient to use vectors of unit length to specify the directions of vector quantities in the various coordinate systems.

It is usually denoted by $\hat{n}, \hat{e}, \hat{a}, \hat{b}$ etc (Read as *n* Cap). Also, note that if $\vec{U}$ is given vector then the unit vector in the direction (or along or parallel) to $\vec{U}$ is given by

$$\hat{U} = \frac{\vec{U}}{|\vec{U}|}$$

in Cartesian coordinates it is typical to use $\hat{i}$, $\hat{j}$ and $\hat{k}$ to represent unit vectors along $\overrightarrow{OX}, \overrightarrow{OY}$ and $\overrightarrow{OZ}$ respectively.

Position vector $\vec{r} = x\hat{i} + y\hat{j} + z\hat{k}$

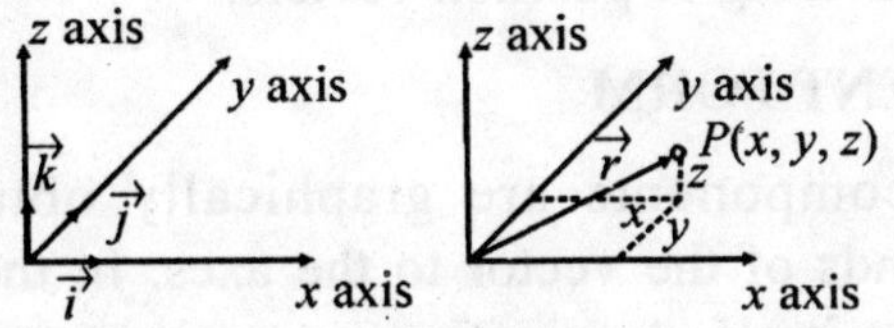

POSITION VECTOR OF A POINT

If A and B are any two points then $\overrightarrow{AB}$ is called the position vector of point B with reference to point A.

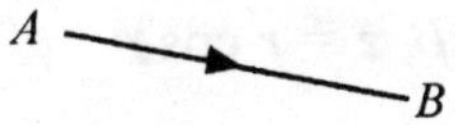

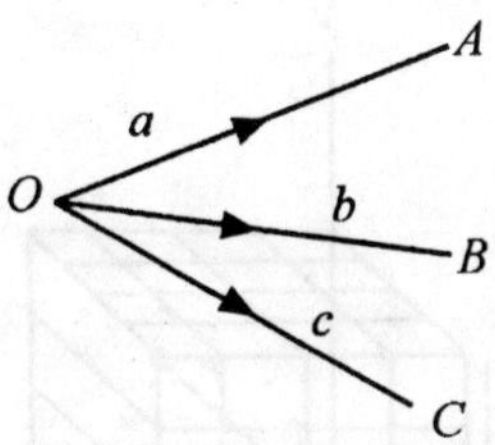

Usually we fix a point O and define the position vector of points with reference to O which is called the origin of reference.

DEFINITION: POSITION VECTOR

Position vector is a vector that extends from the reference point to the position of the particle.

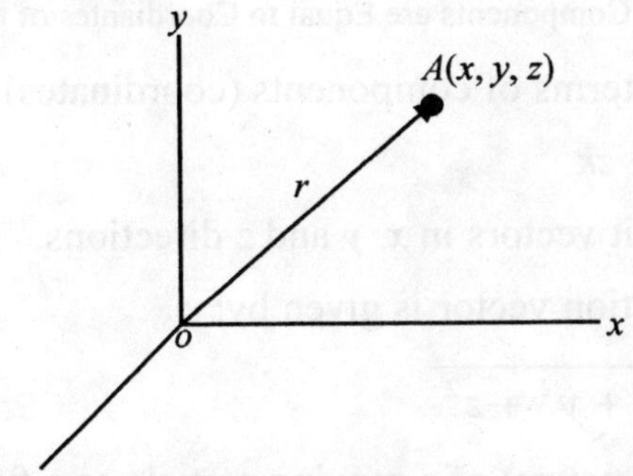

Fig. Position Vector is Represented by a Vector, Joining Origin to the Position of Point Object.

It is easy to realize that vector representation of position is appropriate, where directional properties of the motion are investigated.

The description of motion involving direction consists of position vector as a function of time. Consider the definitions: the "displacement" is equal to the change in position vector; the "velocity" is equal to the rate of change of position vector with respect to time; and "acceleration" is equal to the rate of change of velocity with respect to time, which, in turn, is the rate of change of position vector. Thus, all directional attributes of motion is based on the processing of position vectors.

VECTOR IN COMPONENT FORM

Recall that scalar components are graphically obtained by dropping two perpendiculars from the ends of the vector to the axes. In the case of position vector, one of the end is the origin itself. As position vector is rooted to the origin, the scalar components of position vectors in three mutually perpendicular directions of the coordinate system are equal to the coordinates themselves. Also, scalar components of position vector, *r*, by definition in the designated directions of the axes are:

$$x = r\cos\alpha,\ y = r\cos\beta,\ z = r\cos\gamma$$

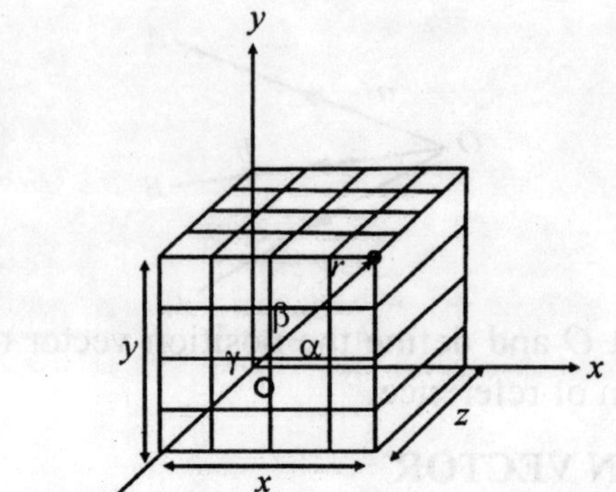

Fig. Scalar Components are Equal to Coordiantes of the Position

and position vector in terms of components (coordinates) is:

$$r = xi + yj + zk$$

where *i*, *j* and *k* are unit vectors in *x*, *y* and *z* directions.

The magnitude of position vector is given by:

$$r = |r| = \sqrt{x^2 + y^2 + z^2}$$

Example: Position (in meters) of a moving particle as a function of time (in seconds) is given by:

$$r = (3t^2 - 3)i + (4 - 7t)j + (-t^3)k$$

Find the coordinates of the positions of the particle at the start of the motion and at time $t = 2$ *s*. Also, determine the linear distances of the positions of the particle from the origin of the coordinate system at these time instants.

Solution. The coordinates of the position are projection of position vector on three mutually perpendicular axes. Whereas linear distance of the position of the particle from the origin of the coordinate system is equal to the magnitude of the position vector.

Now, When t = 0 (start of the motion)

$$r = (3 \times 0 - 3)\, i + (4 - 7 \times 0) j + (-0) k$$

The coordinates and distance are:

$$x = -1m \text{ and } y = 5m$$

$$r = |r| = \sqrt{(-3)^2 + 4^2} = \sqrt{25} = 5m$$

When t = 2*s*

$$r = (3 \times 2^2 - 3) i + (4 - 7 \times 2)\, j + (-2^3) k$$

$$= 9i - 10j - 8k$$

The coordinates and distance are:

$$x = 9\text{m},\ y = -10\text{m, and } z = -8\text{m}.$$

$$r = |r| = \sqrt{9^2 + (-10)^2 + (-8)^2} = \sqrt{245}$$

$$= 15.65m$$

MOTION TYPES

Position is a three dimensional concept, requiring three coordinate values to specify it.

In two dimensional motion, two of the three coordinates change with time. The remaining third coordinate is constant.

In one dimensional motion, only one of the three coordinates is changing with time. Other two coordinates are constant through out the motion. As such, it would be suffice to describe positions of the particle with the values of changing coordinate and neglecting the remaining coordinates.

$$x,\ y = 2,\ z = 2$$

A motion along x –axis or parallel to x – axis is, thus, described by *x* - component of the position vector i.e. *x* – coordinate of the position as shown in the figure.

The description of one dimensional motion is further simplified by shifting axis to the path of motion as shown below.

x, y = 0, z = 0

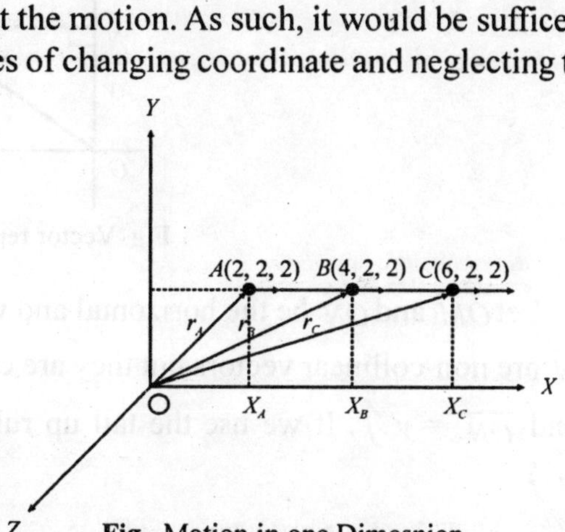

Fig. Motion in one Dimesnion

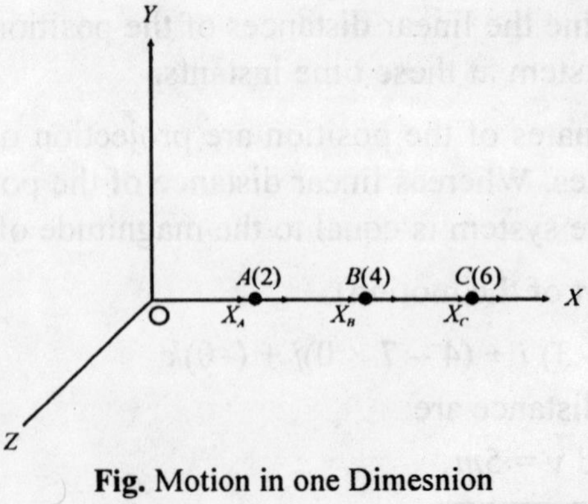

Fig. Motion in one Dimesnion

In this case, position vector itself is along x – axis and, therefore, its magnitude is equal to x – coordinate.

We shall use a, b, c etc. to denote position vectors of points A, B, C etc. respectively, relative to origin O.

REPRESENTATION OF A VECTOR IN A PLANE

Let $P(x, y)$ be any point on the coordinate plane. With $\overrightarrow{OX}$ and $\overrightarrow{OY}$ as coordinate axes and O as the origin. Then $\overrightarrow{OP}$ stands for the position vector of point P relative to origin. Let $P = \overrightarrow{OP}$.

Any such vector can be broken down into two components, a horizontal (along $\overrightarrow{OX}$) and a vertical (along $\overrightarrow{OY}$) component.

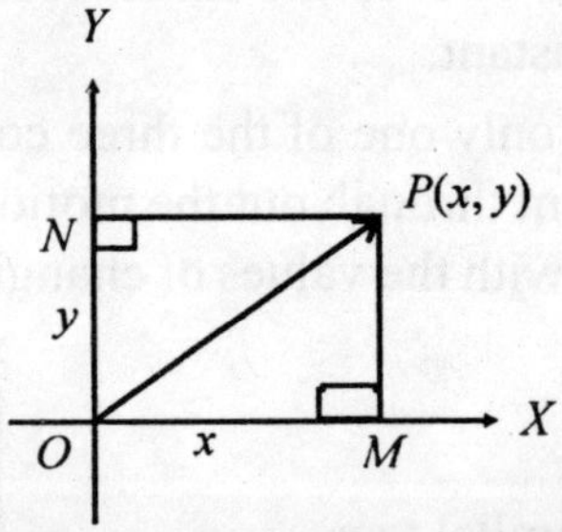

Fig. Vector represented in plane.

Let $\overrightarrow{OM}$ and $\overrightarrow{ON}$ be the horizontal and vertical component of $\overrightarrow{OP}$ i.e., P. Now $\overrightarrow{OM}$ and $\overrightarrow{ON}$ are non-collinear vectors but they are collinear with $\hat{i}$ and $\hat{j}$ respectively. So $\overrightarrow{OM} = x\hat{i}$ and $\overrightarrow{ON} = y\hat{j}$. If we use the tail up rule we have, $\overrightarrow{OP} = \overrightarrow{OM} + \overrightarrow{MP}$. But $\overrightarrow{MP} = \overrightarrow{ON} = y\hat{j}$.

$\therefore \quad \overrightarrow{OP} = x\hat{i} + y\hat{j}$ i.e., $p = x\hat{i} + y\hat{j}$

Here x and y are components of p along x-axis and y-axis respectively. This ordered pair (x, y) associated with p is known as the rectangular components of p.

Let $\angle POX = \theta$ then

$$\cos\theta = \frac{|x|}{|p|} \text{ and } \sin\theta = \frac{|y|}{|p|} \text{ so that } p = \sqrt{x^2 + y^2}$$

which is the magnitude of p.

This concept can be extended to higher dimensions such as three dimension.

Also, note that any vector $\overrightarrow{AB}$ can be shifted so that its initial point A is set at 'O', then $\overrightarrow{AB}$ is in 'standard position'. Now $\overrightarrow{AB} = \overrightarrow{OP}$ then $\overrightarrow{OP}$ replaces $\overrightarrow{AB}$.

Thus $\overrightarrow{OP}$ is the standard vector in the plane for all vectors in the plane of the same magnitude and direction as $\overrightarrow{OP}$.

$\therefore \qquad OP = OM + ON$

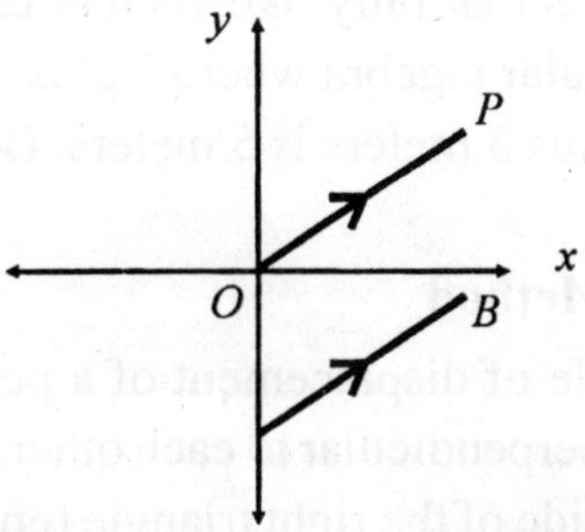

Fig. Vectors of same Magnitude

If coordinates of A are (xa, ya) and that of B are (xb, yb) then shifting A to O and making $B = P$ we have coordinates of P are $(xb - xa, yb - ya)$

$\therefore \overrightarrow{AB}$ or $\overrightarrow{OP} \equiv (xb - xa, yb - ya)$

Thus an algebraic vector is an ordered pair of real numbers that corresponds to a standard geometric vector $\overrightarrow{OP}$.

OPERATIONS ON VECTORS

ADDITION OF VECTORS

Vector addition represents the net effect of the directional quantities

Vectors operate with other scalar or vector quantities in a particular manner. Unlike scalar algebraic operation, vector operation draws on graphical representation to incorporate directional aspect.

Basically, vector addition is dealt in two conceptually equivalent ways:

1. Graphical and
2. Algebraic (vector)

The resulting vector after addition is termed as sum or resultant vector. The resultant vector corresponds to the "resultant" or "net" effect of a physical quantity having directional attribute. The effect of a force system on a body, is determined by the resultant force acting on it. The idea of resultant force, in this case, conveys that the resulting force (vector) has the same effect as that of the forces (vectors) being added.

$$R = F_1 + F_2 + F_3 + F_4$$

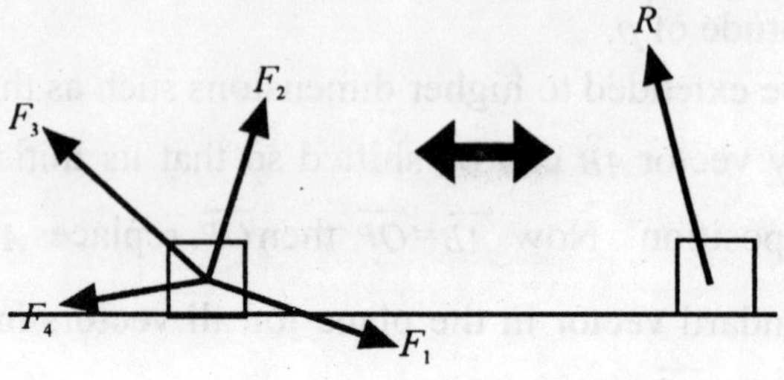

Fig. Resultant Force

Vector addition rule restricts that only like vectors can be added. It is expected also. The requirement is similar to scalar algebra where 2 plus 3 is always 5, but we need to add similar quantity like 2 meters plus 3 meters is 5 meters. But, we can not add, for example, distance and temperature.

Vector Addition: Graphical Method

Let us examine the example of displacement of a person in two different directions. The two displacement vectors, perpendicular to each other, are added to give the "resultant" vector. In this case, the closing side of the right triangle represents the sum (i.e. resultant) of individual displacements AB and BC.

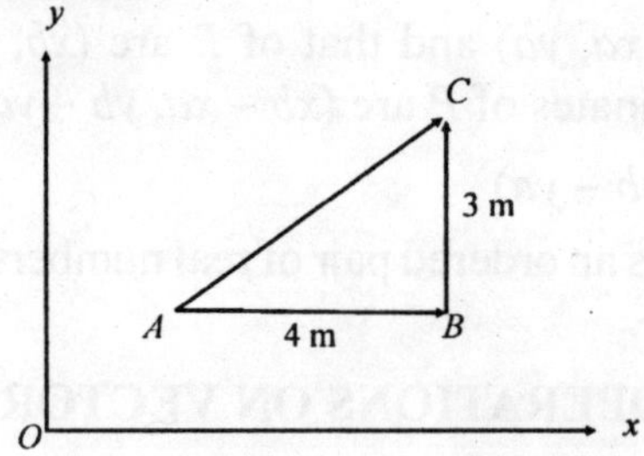

Fig. Displacement $AC = AB + BC$

The method used to determine the sum in this particular case (in which, the closing side of the triangle represents the sum of the vectors in both magnitude and direction) forms the basic consideration for various rules dedicated to implement vector addition.

Triangle Law

In most of the situations, we are involved with the addition of two vector quantities. Triangle law of vector addition is appropriate to deal with such situation.

Defenation: Parallelogram law, matter of fact, is an alternate statement of triangle law of vector addition. A graphics representation of the parallelogram law and its interpretation in terms of the triangle is shown in the figure:

Converting parallelogram sketch to that of triangle law requires shifting vector, *b*, from the position *OB* to position *AC* laterally as shown, while maintaining magnitude and direction.

Polygon Law

The polygon law is an extension of earlier two laws of vector addition.

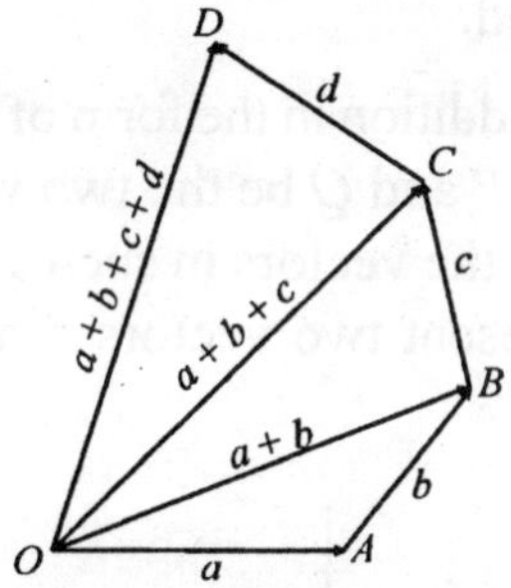

Fig. Successive Application of Triangle Law

Definition: Polygon law polygon law of vector addition: If (n –1) numbers of vectors are represented by (n-1) sides of a polygon in sequence, then n^{th} side, closing the polygon in the opposite direction, represents the sum of the vectors in both magnitude and direction.

In the figure 15 shown below, four vectors namely a, b, c and d are combined to give their sum. Starting with any vector, we add vectors in a manner that the subsequent vector begins at the arrow end of the preceding vector. The illustrations in figures 15 i, iii and iv begin with vectors a, d and c respectively.

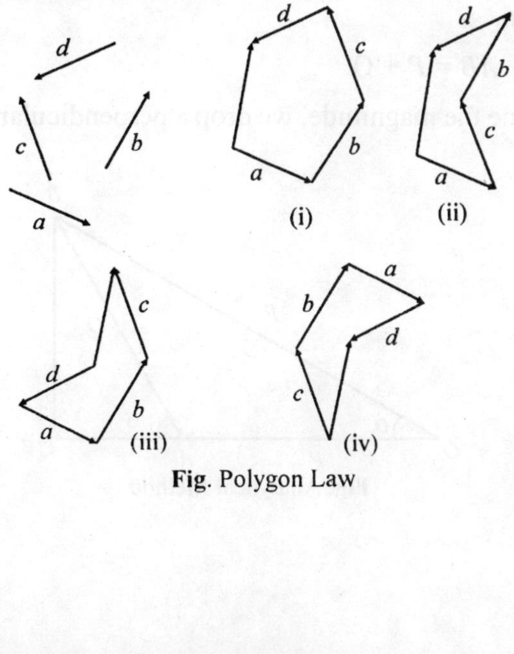

Fig. Polygon Law

Matter of fact, polygon formation has great deal of flexibility. It means that the order of grouping of vectors for addition has no consequence on the result. This characteristic of vector addition is known as "associative" property of vector addition and is expressed mathematically as:

$$(a + b) + c = a + (b + c)$$

ANALYTICAL METHOD

Analytical method, based on geometry, provides a solution in this regard. It allows us to accurately determine the sum or the resultant of the addition, provided velues of magnitudes and angles are supplied.

Here, we shall analyze vector addition in the form of triangle law to obtain the magnitude of the sum of the two vectors. Let *P* and *Q* be the two vectors to be added, which make an angle, with each other. We arrange the vectors in such a manner that two adjacent sides *OA* and *AB* of the triangle *OAB*, represent two vectors *P* and *Q* respectively as shown in the figure.

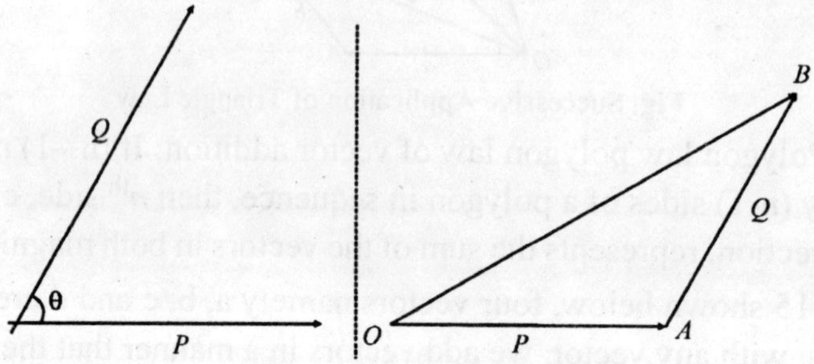

Fig. Analytical Method

According to triangle law, the closing side *OB* represent sum of the vectors in both magnitude and direction.

$$OB = OA + AB = P + Q$$

In order to determine the magnitude, we drop a perpendicular *BC* on the extended line *OC*.

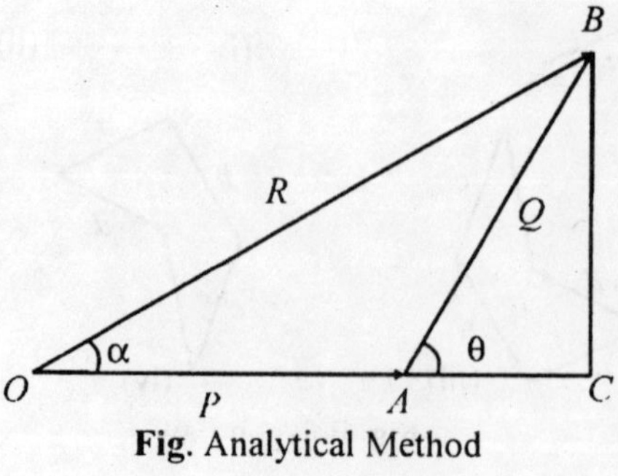

Fig. Analytical Method

In Δ ACB,

$$AC = AB \cos\theta = Q\cos\theta$$

$$BC = AB \sin\theta = Q\sin\theta$$

In right Δ *OCB*, we have:

$$OB = \sqrt{OC^2 + BC^2} = \sqrt{(OA + AC)_2 + BC^2}$$

Substituting for AC and BC,

$$|OB| = \sqrt{(P + Q\cos\theta)^2 + (Q\sin\theta)^2}$$

$$P = \sqrt{P^2 + Q^2\cos^2\theta + 2PQ\cos\theta + Q^2\sin^2\theta}$$

$$R = OB = \sqrt{P^2 + Q^2 + 2PQ\cos\theta}$$

Let "α" be the angle that line *OA* makes with *OC*, then

$$\tan a = \frac{BC}{OC} = \frac{Q\sin\theta}{P + Q\cos\theta}$$

The equations give the magnitude and direction of the sum of the vectors. The above equation reduces to a simpler form, when two vectors are perpendicular to each other. In that case, $\theta = 90°$; sinq = sin90° = 1; cos θ = cos90° = 0

and,

$$OB = \sqrt{P^2 + Q^2}$$

$$\tan\alpha = \frac{Q}{P}$$

These results for vectors at right angle are exactly same as determined, using Pythagoras theorem.

VECTOR ADDITION: ALGEBRAIC METHOD

Graphical method is meticulous and tedious as it involves drawing of vectors on a scale and measurement of angles. We can, however, extend algebraic techniques to vectors, *provided vectors are represented on a rectangular coordinate system*. The representation of a vector on coordinate system uses the concept of unit vector and component.

Components of a Vector

A scalar component, also known as projection, of a vector *AB* in the positive direction of a straight line *C'C* is defined as:

$$AC = |AB|\cos\theta = AB\cos\theta$$

where θ , is the angle that vector AB makes with the specified direction C'C as shown in the figure.

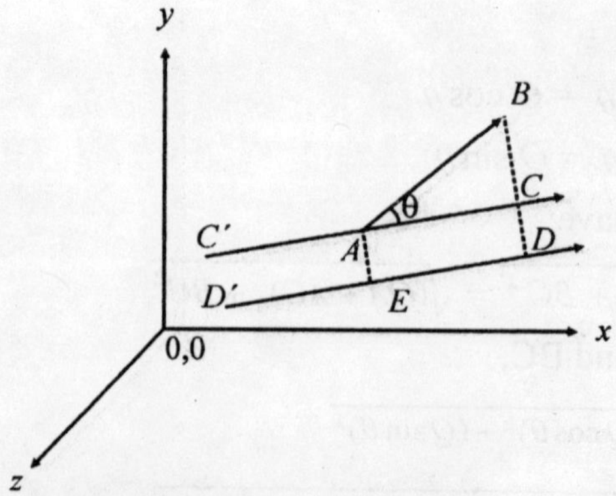

Fig. Vectors are Represented on a Rectangular Coordinate System.

It should be understood that the component so determined is not rooted to the common point "*A*" or any specific straight line like *C′C* in a given direction. Matter of fact, component of vector *AB* drawn on any straight lines like *D′D* parallel to *C′C* are same. Note that the component i.e. projection of the vector is graphically obtained by drawing two perpendicular lines from the ends of the vector on the straight line in the specified direction.

$$ED = AC = AB\cos\theta,$$

Following the above definition of component of vector (specifically known as scalar component), we can resolve a given vector in three components in three mutually perpendicular directions of rectangular coordinate system. Note that we measure angle with respect to parallel lines to the axes. By convention, we denote components by using the non-bold type face of the vector symbol with a suffix representing direction (*x* or *y* or *z*).

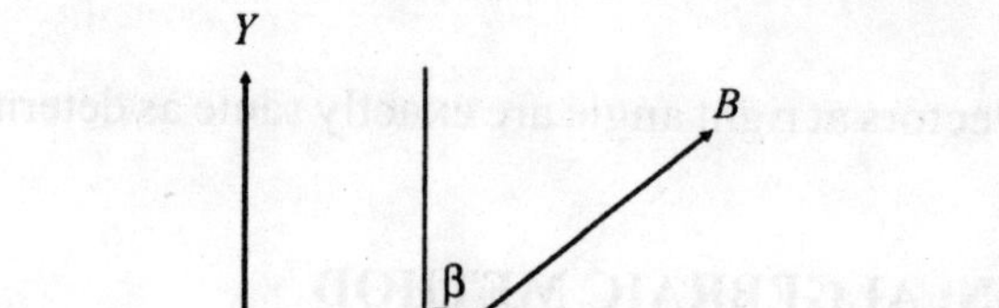

Fig. Components of a Vector

$$AB_x = \text{AB}\cos\alpha$$

$$AB_y = \text{AB}\cos\beta$$

$$AB_y = \text{AB}\cos\gamma$$

where α, β and γ are the angles that vector AB makes with the positive directions of x, y and z directions respectively.

As a vector can be resolved in a set of components, the reverse processing of components in a vector is also expected. The vector component is the vector counterpart of the scalar component, which is obtained by multiplying the scalar component with the unit vector in axial direction. The vector components of a vector a, are $a_x i$, $a_y j$ and $a_z k$.

A vector a, is equal to the vector sum of component vectors in three mutually perpendicular directions of rectangular coordinate system.

$a = normal\ a_x i + normal\ a_y j + normal\ a_z k$

where i, j and k are unit vectors in the respective directions.

Magnitude of the sum of two vectors (a + b) in x and y direction, which are perpendicular to each other, is:

$$|\vec{a}| = \sqrt{a_x^{\,2} + a_y^{\,2}} = a_x^{\,2} + a_y^{\,2}$$

and the magnitude of the sum of vectors (a + b) and third vector c, which are perpendicular to each other, is:

$$|\vec{a}| = \sqrt{a_x^{\,2} + a_y^{\,2}} = \left(a_x^{\,2} + a_y^{\,2}\right)^{1/2}$$

$$|\vec{c}| = \sqrt{a_x^{\,2} + a_y^{\,2} + a_z^{\,2}}$$

VECTOR ADDITION USING COMPONENT OF VECTOR

The frame work for vector addition draws on two important concepts. The first concept is that a vector can be equivalently expressed in terms of three component vectors.

$a = a_x i + a_y j + a_z k$ and $b = b_x i + b_y j + b_z k$

The component vector form has important significance. It ensures that component vectors to be added are restricted to three known directions only. If two vectors are along the same line, then resultant is given as:

When , $\theta = 0°$, $\cos\theta = \cos 0° = 1$ and,

$$|\vec{R}| = \sqrt{P^2 + 2PQ + Q^2} = \sqrt{(P+Q)^2} = P + Q$$

When θ , $= 180°$, $\cos\theta = \cos 180° = -1$

and,

$$|\vec{R}| = \sqrt{P^2 - 2PQ + Q^2} = \sqrt{(P-Q)^2} = P - Q$$

Thus, we see that the magnitude of the resultant is equal to algebraic sum of the magnitudes of the two vectors.

Using these two concepts, the addition of vectors is affected as outlined here:

1. Represent the vectors (*a* and *b*) to be added in terms of components:

$$a = a_x i + a_y j + a_z k \text{ and } b = b_x i + b_y j + b_z k$$

2. Group components in a given direction:

$$a + b = a_x i + a_y j + a_z k + b_x i + b_y j + b_z k$$
$$a+b = (a_x+b_x)\, i + (a_y+b_y)\, j + (a_z+b_z)\, k$$

3. Find the magnitude and direction of the sum, using analytical method

Properties of Vector Addition

1. Vector addition is commutative i.e. $a + b = b + a$ Extension to sums of two or more vectors is immediate.

2. Vector addition is associative. Observe the adjoining figure we have $\overrightarrow{OA} = a$, $\overrightarrow{AB} = b$, $\overrightarrow{BC} = c$ and $\overrightarrow{OC} = r$ using tail-tip rule, for triangle *OAB*, $\overrightarrow{OB} = a + b$, for triangle ABC, $\overrightarrow{AC} = b + c$ for triangle *OBC*, *c* i.e. $r = (a + b) + c$. Also, for triangle OAC, $\overrightarrow{OC} = (a + c) + c$ i.e $r = a + (b + c)$.

   Thus, $r = (a + b) + c = a + (b + c)$.

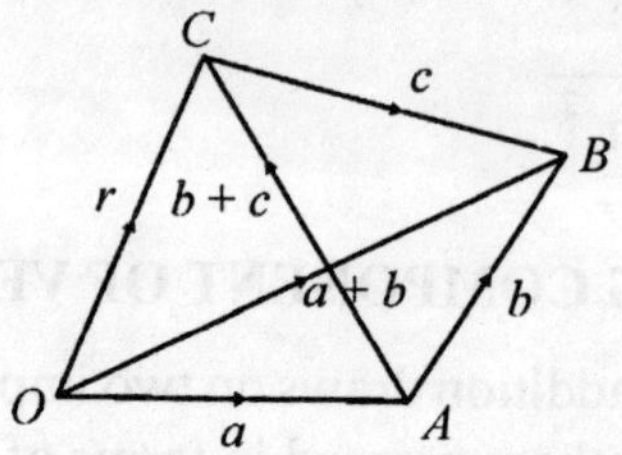

Fig. Vector Addition by Tail-tip Rule.

Subtraction of a Vector

It is accomplished by adding the negative of vectors i.e.

$a - b = a + (-b)$. Note also that $-(a + b) = -a - b$.

Subtraction is considered an addition process with one modification that the second vector after reversing its direction is added to the first vector.

Let us consider the problem of subtracting vector, b, from, a. Using graphical techniques, we first reverse the direction of vector, b, and obtain the sum applying triangle or parallelogram law.

Symbolically,

$$a - b = a + (-b)$$

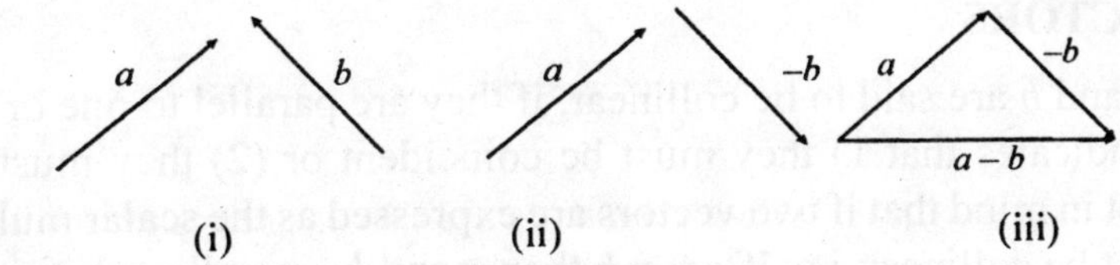

Fig. Subtraction of Vector

COMPOSITION OF VECTORS

It is the process of determining the resultant of a system of vectors. For this we have to draw a vector polygon, placing tail of each vector at the tip of the preceding vector. Then draw a vector from the tail of the first vector to tip of the last vector in the system. Later on we will show that not all vector systems reduce to a single vector. Also note that the order of vectors are drawn is immaterial.

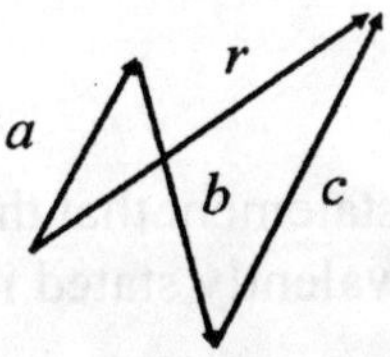

Fig. Vector Composition.

MULTIPLICATION OF VECTORS BY SCALARS

Let m be a scalar and a be a vector then the product ma whose magnitude is $|m|$ times of $| a |$ and which is similarly or oppositely directed to a depending upon whether m is positive or negative.

Then

1. For $m > 0$, m a is a vector of magnitude m $|a|$ and is in the direction of a.
2. For $m = 0$, ma is a null vector.
3. For $m < 0$, ma is the vector of magnitude $-m\ |a|$ and is exactly opposite to the vector a

From this, it follows that $(-1)\ a = -a$ and $0\ a = 0$

(B) Other operations with scalars m and n are

$(m + n)\ a = m\ \mathrm{a} + n\ \mathrm{a}$

$m\ (\mathrm{a} + \mathrm{b}) = m\ \mathrm{a} + m\ \mathrm{b}$

$mn\ (a) = n\ (m\mathrm{a}) = (m\ n\ \mathrm{a})$

LINEAR COMBINATION OF VECTORS

Let a and b be two vectors and x and y be any real numbers. Then the vector $r = x\ a + y\ b$ is called the linear combination of vectors a and b.

COLLINEAR VECTORS

Two vectors a and b are said to be collinear, if they are parallel to one or on the same line. This clearly indicates that 1) they must be coincident or (2) they must be parallel. Also, it must be kept in mind that if two vectors are expressed as the scalar multiple of each other then they must be collinear. i.e. if $a = x\,b$ then a and b are collinear if $x \neq 0$ scalar.

COPLANAR VECTORS

Geometric objects lying in a common plane are said to be coplanar. Three non-collinear points determine a plane and so are trivially coplanar. Four points are coplanar iff the volume of the tetrahedron defined by them is 0,

$$\begin{vmatrix} x_1 & y_1 & z_1 & 1 \\ x_2 & y_2 & z_2 & 1 \\ x_3 & y_3 & z_3 & 1 \\ x_4 & y_4 & z_4 & 1 \end{vmatrix} = 0$$

co-planarity is equivalent to the statement that the pair of lines determined by the four points are not skew, and can be equivalently stated in vector form as

$$(x_3 - x_1)\cdot[(x_2 - x_1)\times(x_4 - x_3)] = 0$$

An arbitrary number of n points $x_1, x_2, \ldots, x_n$ can be tested for co-planarity by finding the point-plane distances of the points $x_1, \ldots, x_n$ from the plane determined by (x_1, x_2, x_3) and checking if they are all zero. If so, the points are all coplanar.

A set of n vectors V is coplanar if the nullity of the linear mapping defined by V has dimension 1, the matrix rank of V (or equivalently, the number of its singular values) is $(n - 1)$.

If vectors are parallel to a given plane, then they are coplanar. Note that if a and b be two non-collinear vectors which can be uniquely expressed as $r = x\,a + y\,b$, $x, y \neq 0$ scalars, then they must be coplanar.

Parallel lines in three-dimensional space are coplanar, but skew lines are not.

MULTIPLICATION

Vector multiplication provides concise and accurate representation of natural laws, which involve vectors.

The vector multiplication, however, is not an unique mathematical construct like scalar multiplication. The multiplication depends on the nature of quantities (vector or scalar) and on the physical process, necessitating scalar or vector multiplication.

The rules of vector multiplication have been formulated to encapsulate physical processes in their completeness. In order to explore this aspect, let us find out the direction of acceleration in the case of parabolic motion of a particle. There may be two ways to deal with the requirement. We may observe the directions of velocities at two points along the path and find out the direction of the change of velocities. Since we know that the direction of change of velocity is the direction of acceleration, we draw the vector diagram and find out the direction of acceleration. We can see that the direction of acceleration turns out to act in vertically downward direction.

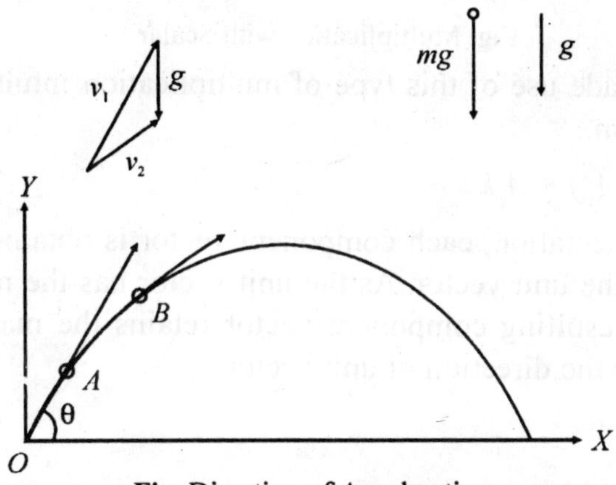

Fig. Direction of Acceleration

The conceptualization of physical laws in vector form, however, provides us with powerful means to arrive at the result in relatively simpler manner. If we look at the flight of particle in parabolic motion, then we observe that the motion of particle is under the force of gravity, which is acting vertically downward. There is no other force (neglecting air resistance). Now, from second law of motion, we know that:

$$F_{\text{Resultant}} = ma$$

This equation reveals that the direction of acceleration is same as that of the resultant force acting on the particle. Thus, acceleration of the particle in parabolic motion is acting vertically downward. We see that this second approach is more elegant of the two methods. Using the fundamental law of nature in vector form involving multiplication of the vector with a scalar in this case, we could arrive at the correct answer in a very concise manner, without getting into the details of the motion.

Multiplication of Vector with Scalar

Multiplication of a vector *A*, with another scalar quantity, a, results in another vector, *B*. The magnitude of the resulting vector is equal to the product of the magnitude of vector with the scalar quantity. The direction of the resulting vector, however, is same as that of the original vector.

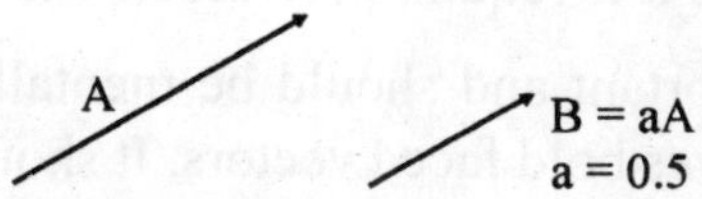

(i)

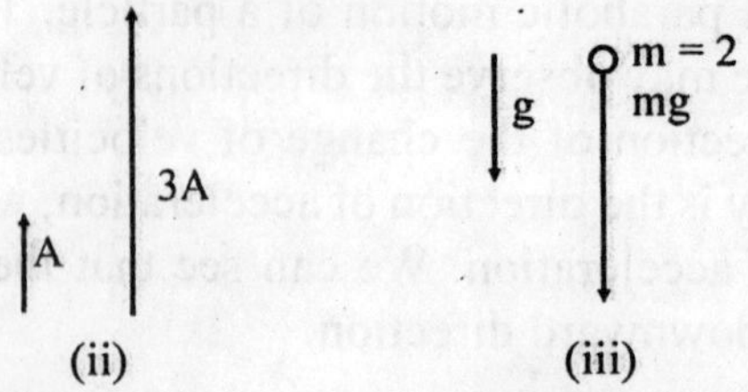

Fig. Multiplication with Scalar

We have already made use of this type of multiplication intuitively in expressing a vector in component form.

$$A = A_x i + A_y j + A_z k$$

In this vector representation, each component vector is obtained by multiplying the scalar component with the unit vector. As the unit vector has the magnitude of 1 with a specific direction, the resulting component vector retains the magnitude of the scalar component, but acquires the direction of unit vector.

$$A_x = A_x i$$

PRODUCTS OF VECTORS

Some physical quantities are itself a scalar quantity, but composed from the product of vector quantities. One such example is "work", which is a product of force (a vector) and displacement (a vector), while "work" itself is a scalar quantity. On the other hand, magnetic force is a vector, composed from the product of velocity of a moving charge (a vector) and magnetic field (vector).

Thus, products of vectors are defined in two distinct manner – one resulting in a scalar quantity and the other resulting in a vector quantity.

SCALAR PRODUCT

Scalar product of two vectors a and b is a scalar quantity defined as:

$$a.b = ab \cos\theta,$$

where "a" and "b" are the magnitudes of two vectors and "θ" is the angle between the direction of two vectors.

It is important to note that vectors have two angles θ and 2π – θ. We can use either of them as cosine of both "θ" and "2π – θ" are same. However, it is suggested to use the smaller of the enclosed angle as it is required for vector (i.e. cross) product.

The notation "a.b" is important and should be mentally noted to represent a scalar quantity – even though it involves bold faced vectors. It should be noted that all quantities on the right hand side are scalars.

MEANING OF SCALAR PRODUCT

We can read the definition of scalar product in either of the following manners:

$$a.b = a\ (b \cos\theta)$$

$$a.b = b\ (a \cos\theta)$$

Recall that "$b \cos \theta$" is the scalar component of vector b along the direction of vector a and "$a \cos \theta$" is the scalar component of vector a along the direction of vector *b*. Thus, we may consider the scalar product of vectors *a* and *b* as the product of the magnitude of one vector and the scalar component of other vector along the first vector.

The figure below shows drawing of scalar components. The scalar component of vector b in figure (i) is obtained by drawing perpendicular from the tip of the vector, b, on the direction of vector, a. Similarly, the scalar component of vector a in figure (i) is obtained by drawing perpendicular from the tip of the vector, a, on the direction of vector, *b*.

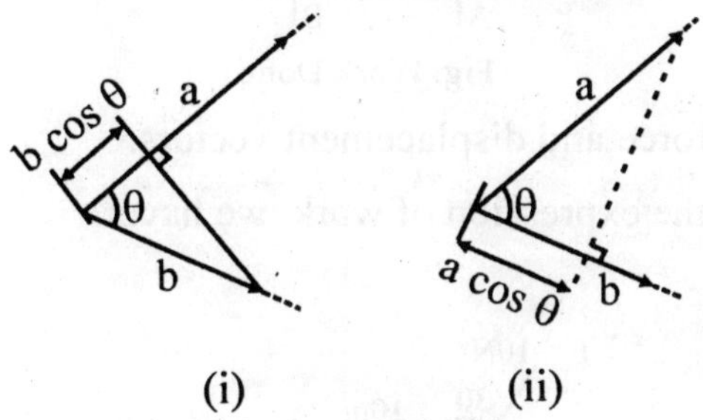

Fig. Scalar Product

The two alternate ways of evaluating dot product of two vectors indicate that the product is commutative i.e. independent of the order of two vectors:

$$a.\ b = b.\ a$$

Values of Scalar Product

The value of dot product is maximum for the maximum value of $\cos\theta$. Now, the maximum value of cosine is $\cos 0° = 1$. For this value, dot product simply evaluates to the product of the magnitudes of two vectors.

$$(a.b)_{max} = ab$$

For $\theta = 180°, \cos 180° = -1$ and

$$a.b = -ab$$

Thus, we see that dot product can evaluate to negative value as well. This is a significant result as many scalar quantities in physics are given negative value. The work done, for example, can be negative, when the work is done on the system; instead of work being done by the system (positive).

The scalar product evaluates to zero for , $\theta = 90^0$ and 270^0 as cosine of these angles are zero. These results have important implication for unit vectors. The dot products of same unit vector evaluates to 1.

$$i.i = j.j = k.k = 1$$

The dot products of combination of different unit vectors evaluate to zero.

$$i.j = j.k = k.i = 0$$

Example: A block of mass "m" moves from point A to B along a smooth plane surface under the action of force as shown in the figure. Find the work done if it is defined as:

$W = F.\Delta x$

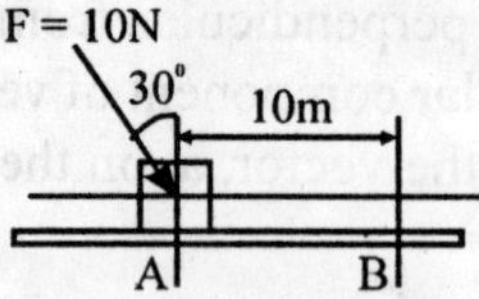

Fig. Work Done

where F and Δ x are force and displacement vectors.

Solution. Expanding the expression of work, we have:

Work done

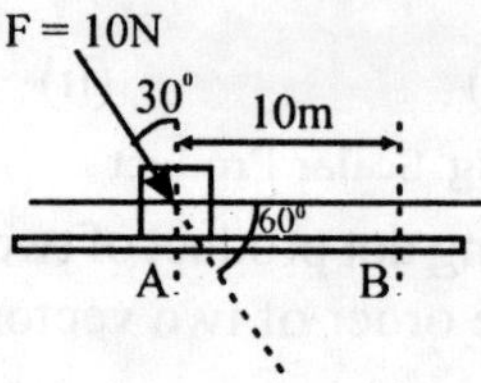

Fig. Work Done

$$W = F.\Delta x = F\Delta x \cos,$$

Here, $F = 10\ N$, $\Delta x = 3$ m and $\cos\theta = \cos 60° = 0.5$.

$$W = 10 \times 10 \times 0.5 = 50 \text{ J}$$

SCALAR PRODUCT IN COMPONENT FORM

Two vectors in component forms are written as:

$$a = a_x i + a_y j + a_z k$$

$$b = b_x i + b_y j + b_z k$$

In evaluating the product, we make use of the fact that multiplication of the same unit vectors is 1, while multiplication of different unit vectors is zero. Finally, the dot product evaluates to scalar terms:

$$\vec{a} \cdot \vec{b} = (a_x i + a_y j + a_z k).(b_x i + b_y j + b_z k)$$

$$\vec{a} \cdot \vec{b} = a_x i.b_x i + a_y j.b_y j + a_z k.b_z k$$

$$\vec{a} \cdot \vec{b} = a_x b_x + a_y b_y + a_z b_z$$

IMPORTANCE OF DOT PRODUCT

An important use of the dot product is to test whether or not two vectors are orthogonal. Two vectors are orthogonal if the angle between them is 90 degrees. Thus, using we see that the dot product of two orthogonal vectors is zero. Conversely, the only way the dot product can be zero is if the angle between the two vectors is 90 degrees (or trivially if one or both of the vectors is the zero vector). Thus, two non-zero vectors have dot product zero if and only if they are orthogonal.

PROJECTIONS OF A VECTOR

One important use of dot products is in projections. The scalar projection of *b* onto *a* is the *length* of the segment *AB* shown in the figure below. The vector projection of *b* onto *a* is the *vector* with this length that begins at the point *A* points in the same direction (or opposite direction if the scalar projection is negative) as *a*.

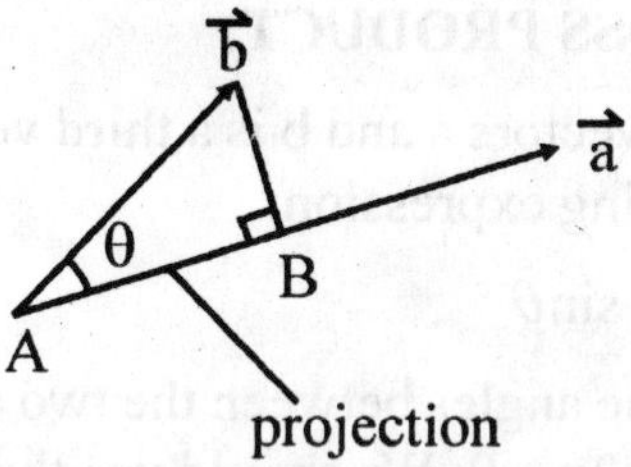

Fig. Projection of Vector

Thus, mathematically, the scalar projection of *b* onto a is $|b| \cos \theta$ (where θ is the angle between *a* and *b*) which from (*) is given by

$$\text{Projection of } b \text{ on } a = \frac{\vec{a} \cdot \vec{b}}{|\vec{a}|}$$

This quantity is also called the component of *b* in the *a* direction (hence the notation comp). And, the vector projection is merely the unit vector $a/|a|$ times the scalar projection of *b* onto *a*:

$$proj_a \vec{b} = \frac{\vec{a} \cdot \vec{b}}{|\vec{a}|} \frac{\vec{a}}{|\vec{a}|} = \frac{\vec{a} \cdot \vec{b}}{|\vec{a}|^2} \vec{a}$$

Thus, the scalar projection of b onto a is the magnitude of the vector projection of b onto a.

Example: Suppose you wish to find the work W done in moving a particle from one point to another. From physics we know $W = Fd$ where F is the magnitude of the force moving the particle and d is the distance between the two points. However, this relation is only valid when the force acts in the direction the particle moves.

Suppose this is not the case. Let the force vector be $F = <2,3,4>$ and the displacement vector be $d = <1,2,3>$. In this case, the work is the product of the distance moved (the magnitude of the displacement vector) and the magnitude of the component of the force that acts in the direction of displacement (the scalar projection of F onto d):

$$W = \left|\vec{d}\right|\left|\vec{F}\right| \cos\theta = \vec{F} \cdot \vec{d}$$

Thus, the work done by the force to displace the particle from say the origin to the point (1,2,3) is

$$W = \vec{F} \cdot \vec{d} = 1(2) + 2(3) + 3(4) = 20 \text{ Joules}$$

Note that this is the easiest way to compute the dot product since the angle between the vectors F and d is unknown.

VECTOR PRODUCT (CROSS PRODUCT)

The cross product of two vectors a and b is a third vector. The magnitude of the vector product is given by the following expression:

$$|c| = |a \times b| = ab \sin\theta$$

where , is the smaller of the angles between the two vectors. It is important to note that vectors have two angles θ and 2π – θ. We should use the smaller of the angles as sine of θ and 2π – θ are different.

DIRECTION OF VECTOR PRODUCT

The direction of the vector product is perpendicular to the plane containing vectors, a, and, b. We must understand that two vectors (a and b) define an unique plane. The vector product is perpendicular to this plane defined by the vectors as shown in the figure. The angle, θ, impacts the magnitude of the cross product an.d not its direction.

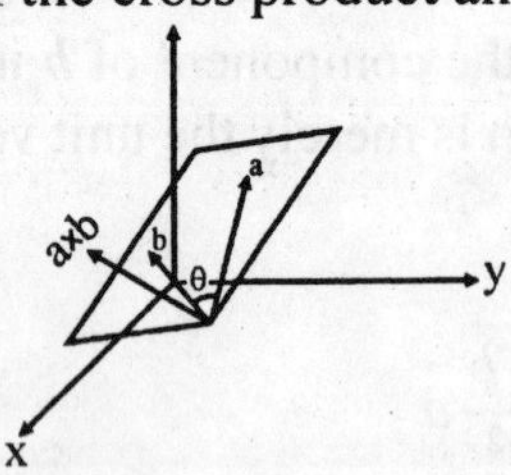

Fig. Direction of Vector Product

We know that rectangular coordinate system comprises of three planes, which are at right angles to each other. It is, therefore, suggested that we must orient our coordinate system in such a manner that vectors lie in one of the three planes defined by the rectangular coordinate system. This arrangement renders the cross product oriented in the direction of the remaining axis as shown in the figure.

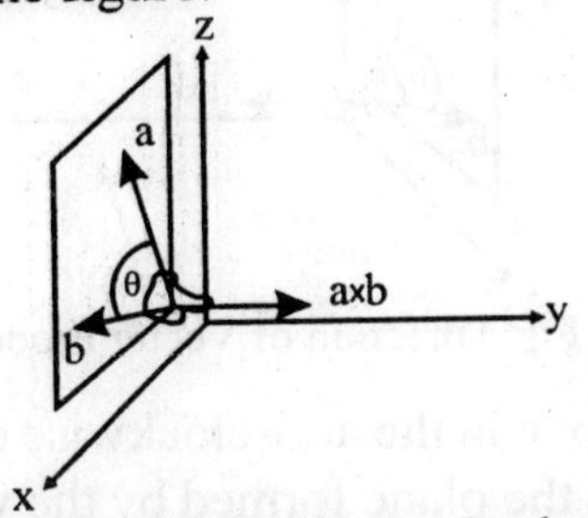

Fig. Direction of Vector Product

We can draw the vector product perpendicular to the plane on either of the two sides. For example, the product can be drawn either along the positive direction of y – axis or along the negative direction of y-axis direction.

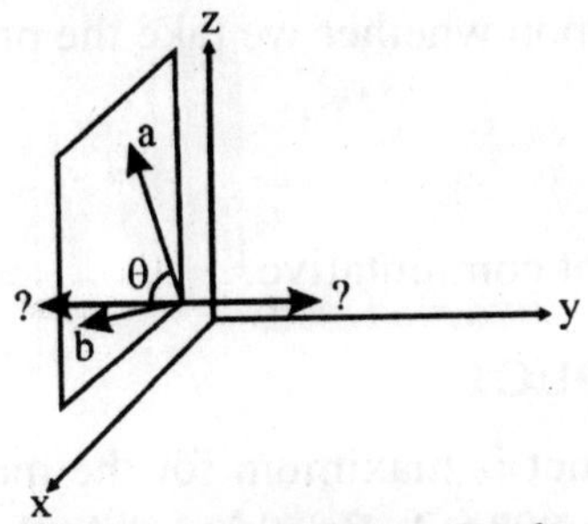

Fig. Direction of Vector Product

The direction of the vector product, including which side of the plane, is determined by right hand rule for vector products. According to this rule, we place right fist such that the curl of the fist follows as we proceed from the first vector, a, to the second vector, b. The stretched thumb, then gives the direction of vector product.

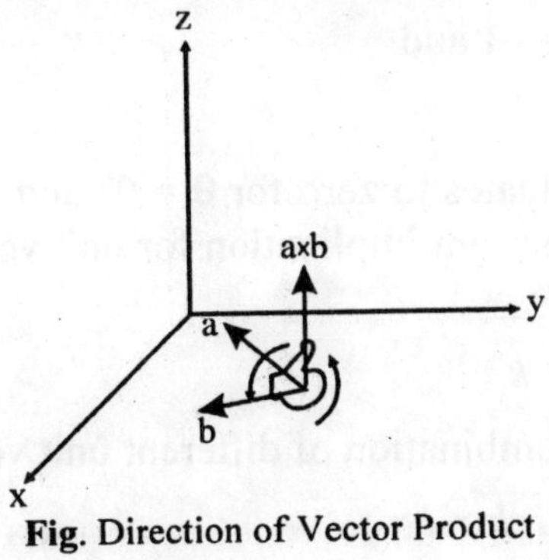

Fig. Direction of Vector Product

When we apply this rule to the case considered before, we find that the vector product is in the positive y – direction as shown below:

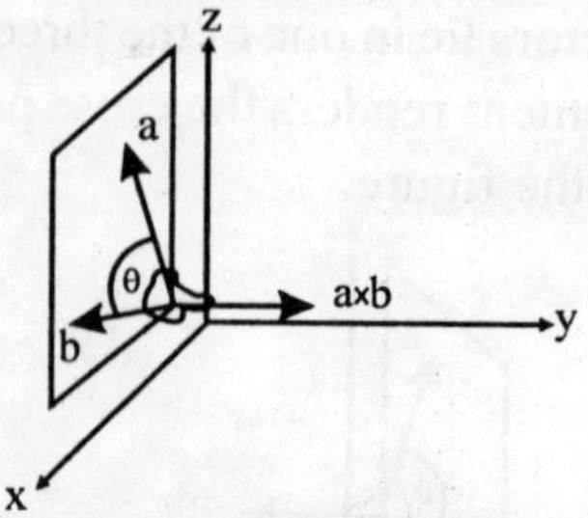

Fig. Direction of Vector Product

Here, we notice that we move in the anti-clockwise direction as we move from vector, a, to vector, b, while looking at the plane formed by the vectors. This fact can also be used to determine the direction of the vector product. If the orientation is anticlockwise, then the vector product is directed towards us; otherwise the vector product is directed away on the other side of the plane.

Further, it is important to note that the direction of cross product can be on a particular side of the plane, depending upon whether we take the product from a to b or from b to a. Implying:

$$a \times b = -b \times a$$

Thus, vector product is not commutative.

VALUES OF CROSS PRODUCT

The value of vector product is maximum for the maximum value of sinθ. Now, the maximum value of $\sin\theta$ is $\sin 90° = 1$. For this value, the vector product evaluates to the product of the magnitude of two vectors multiplied.

Thus, we see that cross product evaluates to negative value, which corresponds to the negative direction of the vector product i.e. other side of the planar formed by the vectors.

$$|a \times b|_{\max} = ab$$

For $\theta = 270°$, $\sin 270° = -1$ and

$$|a \times b|_{\max} = -ab$$

The vector product evaluates to zero for $\theta = 0°$ and $180°$ as sine of these angles are zero. These results have important implication for unit vectors. The dot products of same unit vector evaluates to 0.

$$i \times i = j \times j = k \times k$$

The cross products of combination of different unit vectors evaluate as:

$$i \times j = k; j \times k = i; k \times i = j$$

$j \times i = -k;\ k \times j = -i;\ i \times k = -j$

There is a simple rule to determine the sign of the cross product. We write the unit vectors in sequence *i,j,k*. Now, we form pair of vectors as we move from left to right like $i \times j, j \times k$ and right to left at the end like $k \times i$ in cyclic manner.

The cross products of these pairs result in the remaining unit vector with positive sign. Cross products of other pairs result in the remaining unit vector with negative sign.

CROSS PRODUCT IN COMPONENT FORM

Two vectors in component forms are written as:

$$a = a_x i + a_y j + a_z k$$

$$b = b_x i + b_y j + b_z k$$

In evaluating the product, we make use of the fact that multiplication of the same unit vectors gives the value of 0, while multiplication of two different unit vectors result in remaining vector with appropriate sign. Finally, the vector product evaluates to vector terms:

$$a \times b = (a_x i + a_y j + a_z k) \times (b_x i + b_y j + b_z k)$$

$$a \times b = a_x i \times b_y j + a_x i \times b_z k + a_y j \times b_x i + a_y j \times b_z k + a_z k \times b_x i + a_z k \times b_y j$$

$$a \times b = a_x b_x k - a_x b_z j \times a_y b_x k + a_y b_z i + a_z b_x j - a_z b_y i$$

$$a \times b = (a_y b_z - a_z b_y) i + (a_z b_x - a_x b_z) j + (a_x b_y - a_y b_x) k$$

Evidently, it is difficult to remember above expression. If we know expanding matrix, then we can write above expression in the matrix form, which is easy to remember.

$$\vec{a} \times \vec{b} = \begin{vmatrix} i & j & k \\ a_x & a_y & c_x \\ b_x & b_y & c_y \end{vmatrix}$$

Example: If $a = 2i + 3j$ and $b = -3i - 2j$, find $a \times b$.

Solution.

$$a \times b = (2i + 3j) \times (-3i - 2j)$$

Neglecting terms involving same unit vectors, we expand the multiplication algebraically as:

$$a \times b = (2i) \times (-2j) + (3j) \times (-3i)$$

$$a \times b = -4k + 9k = 9k$$

Example: Consider the force given as:

$F = q\,(v \times B)$

Given $q = 10^{-6}$ C, $v = (3i + 4j)$ m/s, $B = 1j$ Tesla. Find the force.

Solution.

$$F = q\,(v \times B) = 10^{-6}(3i + 4j) \times (-1i)$$

$$F = 4 \times 10^{-6}k$$

PROPERTIES OF THE CROSS PRODUCT

1. The length of the cross product of two vectors is $|a \times b| = ab \sin\theta$
2. The length of the cross product of two vectors is equal to the area of the parallelogram determined by the two vectors.
3. Anticommutativity: $\vec{a} \times \vec{b} = -\vec{b} \times \vec{a}$
4. Multiplication by scalars: $\left(c\vec{a}\right) \times \vec{b} = c\left(\vec{a} \times \vec{b}\right) = \vec{a} \times \left(c\vec{b}\right)$
5. Distributivity: $\vec{a} \times \left(\vec{b} \times \vec{c}\right) = \vec{a} \times \vec{b} + \vec{a} \times \vec{c}$

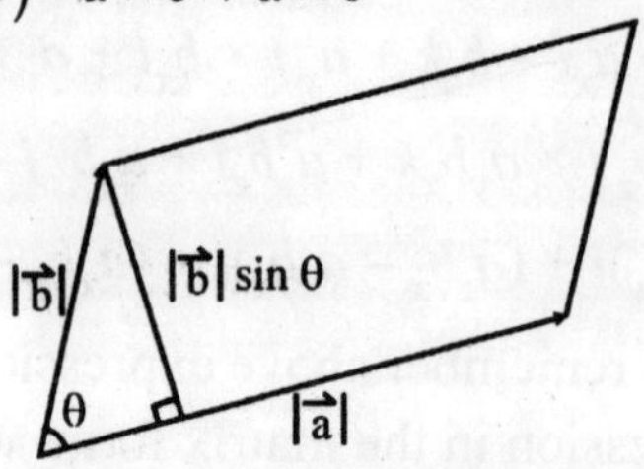

Fig. Cross Product of Vectors.

Note that the result for the length of the cross product leads directly to the fact that two vectors are parallel if and only if their cross product is the zero vector. This is true since two vectors are parallel if and only if the angle between them is 0 degrees (or 180 degrees).

Example: To find the area of the triangle with vertices (1,1,3), (4,-1,1), and (0,1,8), one could find the length of one of the altitudes of the triangle and proceed to find *A*=1/2(altitude)(base).

However, this may not be an easy task in three dimensions. So consider finding the one-half the area of the parallelogram determined by the vector *a* from (1,1,3) to (4,–1,1) and the vector *b* from (1,1,3) to (0,1,8). Then $a = <2-1, 2-(-1), 2-3> = <3,-2,-2>$ and $b = <-1,0,5>$. Using the result from the previous example and property number two above, we have the length of the cross product (and therefore the parallelogram determined by *a* and *b*) as

$$\left|\vec{a} \times \vec{b}\right| = \sqrt{(-10)^2 + (-13)^2 + (-2)^2} = \sqrt{273} = 16.52$$

So, the area of the triangle is one-half this quantity, or 8.26.

The cross product occurs in many formulas in physics, such as curl of a vector field, torque, and many integrals over surfaces.

PRODUCT OF THREE VECTORS

SCALAR TRIPLE PRODUCT

The scalar triple product of three vectors a, b, and c is $(a \times b) \cdot c$. It is a scalar product because it evaluates to a single number (unlike the cross product). The scalar triple product is important because its

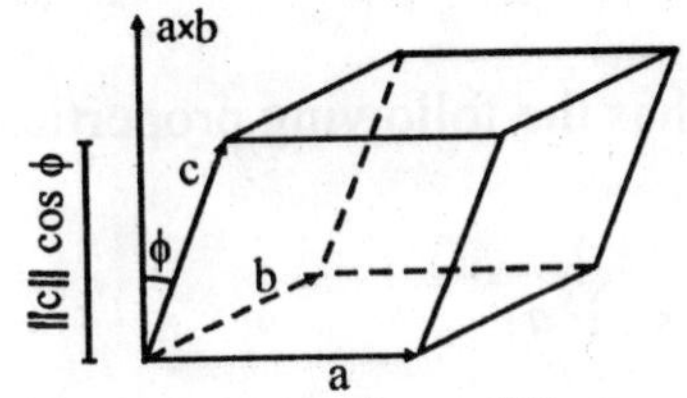

Fig. Triple Product of Vectors.

absolute value $|(a \times b) \cdot c|$ is the volume of the parallelepiped spanned by a, b, and c (i.e., the parallelepiped whose adjacent sides are the vectors a, b, and c).

This results from simple geometry. The volume is the area of the base times the height. We already know the area of the parallelogram base: $\|a \times b\|$. The height is the component of c in the direction normal to the base, i.e., in the direction of $a \times b$. Hence the height is $\left|\|c\| \cos\phi\right|$, where ϕ is the angle between c and $a \times b$. (Why do we need the absolute value? $\cos\phi$ would be negative if $\phi > \pi/2$.)

The volume of the parallelepiped is therefore

$$\text{Volume} = \left\|\vec{a} \times \vec{b}\right\| \left|\|\vec{c}\| \cos\phi\right| = \left|\left(\vec{a} \times \vec{b}\right) \cdot \vec{c}\right|$$

Using the formula for the cross product in components form, we can write the scalar triple product in component form as Write scalar triple product in component form as

$$\left(\vec{a} \times \vec{b}\right) \cdot \vec{c} = \begin{vmatrix} a_2 & a_3 \\ b_2 & b_3 \end{vmatrix} c_1 - \begin{vmatrix} a_1 & a_3 \\ b_1 & b_3 \end{vmatrix} c_2 + \begin{vmatrix} a_1 & a_2 \\ b_1 & b_2 \end{vmatrix} c_3 = \begin{vmatrix} c_1 & c_2 & c_3 \\ a_1 & a_2 & a_3 \\ b_1 & b_2 & b_3 \end{vmatrix}$$

The triple product is also invariant under any cyclic permutation of a, b, and c,

$$a \cdot b \times c = b \cdot c \times a = c \cdot a \times b$$

but any anti-cyclic permutation causes it to change sign,

$$a \cdot b \times c = -b \cdot a \times c$$

The scalar triple product is zero if any two of a, b, and c are parallel, or if a, b, and c are co-planar.

If a, b, and care non-coplanar, then any vector r can be written in terms of them:

$$r = \alpha a + \beta b + \gamma c$$

Forming the dot product of this equation with $b \times c$, we then obtain

$$r \cdot b \times c = \alpha a \cdot b \times c$$

$$\text{so } \alpha = \frac{r \cdot b \times c}{a \cdot b \times c}$$

Analogous expressions can be written for β and γ. The parameters α, and γ are uniquely determined provided $a \cdot b \times c \neq 0$: *i.e.*, provided that the three basis vectors are not co-planar

The scalar triple product has the following properties

$$a \cdot (b \times c) = (b \times c) \cdot a$$
$$a \cdot (b \times c) = c \cdot (a \times b) = (b \times c) \cdot a$$
$$a \cdot (b \times c) = -a \cdot (c \times b)$$

VECTOR TRIPLE PRODUCT

The vector triple product is defined as the cross product of one of the vectors with the cross product of the other two. This is known as the *vector triple product* because it results in a vector.

$$a \times (b \times c) = b(a \cdot c) - c(a \cdot b)$$

For three vectors a, b, and c, the vector triple product is defined $a \times (b \times c)$

- The brackets are important because $a \times (b \times c) \neq (a \times b) \times c$
- In fact, it can be demonstrated that

$$a \times (b \times c) = (a \cdot c)b - (a \cdot b)c$$

and

$$(a \times b) \times c = (a \cdot c) - (b \cdot c)a$$

Let us try to prove the first of the above theorems. The left-hand side and the right-hand side are both proper vectors, so if we can prove this result in one particular coordinate system then it must be true in general. Let us take convenient axes such that the x-axis lies along b, and c lies in the x-y plane. It follows that $b = (b_x, 0, 0)$, $c = (c_x, c_y, 0)$ and $a = (a_x, a_y, a_z)$.

The vector $b \times c$ is directed along the z-axis: $b \times c = (0, 0, b_x c_y)$.

It follows that $a \times (b \times c)$ lies in the x-y plane:

$$a \times (b \times c) = (a_y b_x c_y, - a_x b_x c_y, 0).$$

To evaluate the right-hand side, of the above equation we need

$$a \cdot c = a_x c_x + a_y c_y \text{ and } a \cdot b = a_x b_x .$$

It follows that the right-hand side is

$$\text{R.H.S} = ([a_x c_y + a_y c_y] b_x, 0) - (a_x b_x c_y, a_x b_x c_y, 0)$$

$$= (a_x c_y b_x, -a_x b_x c_y, 0) = \text{LH.S}$$

which proves the theorem.

The vector triple product has the properties

$$(a \times b) \times c = -c \times (a \times b) = c \times (b \times a)$$
$$a \times b(b \times c) = -a \cdot (c \times b)$$
$$a \times (b \times c) = (a \cdot c)b - (a \cdot b)c$$

Chapter 4

Vectors and Direction

A vector quantity is a quantity which is fully described by both magnitude and direction. On the other hand, a scalar quantity is a quantity which is fully described by its magnitude. The emphasis of this unit is to understand some fundamentals about vectors and to apply the fundamentals in order to understand motion and forces which occur in two dimensions.

Examples of vector quantities which have been previously discussed include displacement, velocity, acceleration, and force. Each of these quantities are unique in that a full description of the quantity demands that both a magnitude and a direction are listed. For example, suppose that your teacher tells you that "a bag of gold is located outside the classroom; to find it, displace yourself 20 meters." This statement may provide yourself enough information to pique your interest; yet, not enough information to find the bag of gold.

The displacement required to find the bag of gold has not been fully described. On the other hand, suppose your teacher tells you that "a bag of gold is located outside the classroom; to find it, displace yourself from the center of the classroom door 20 meters in a direction 30 degrees to the west of north." This statement provides a complete description of the displacement vector - it lists both magnitude (20 meters) and direction (30 degrees to the west of north) relative to a reference or starting position (the center of the classroom door). Vector quantities are not fully described unless both magnitude and direction are listed.

Vector quantities are often represented by scaled vector diagrams. Vector diagrams depict a vector by use of an arrow drawn to scale in a specific direction. Vector diagrams were introduced and used in earlier units to depict the forces acting upon an object; such diagrams are known as free-body diagrams. An example of a scaled vector diagram is shown in the diagram at the right to depict a displacement vector. Observe that there are several characteristics of this diagram which make it an appropriately drawn vector diagram.

- A scale is clearly listed
- An arrow (with arrowhead) is drawn in a specified direction; thus, the vector has a head and a tail.
- The magnitude and direction of the vector is clearly labeled; in this case, the diagram shows the magnitude is 20 m and the direction is (30 degrees West of North).

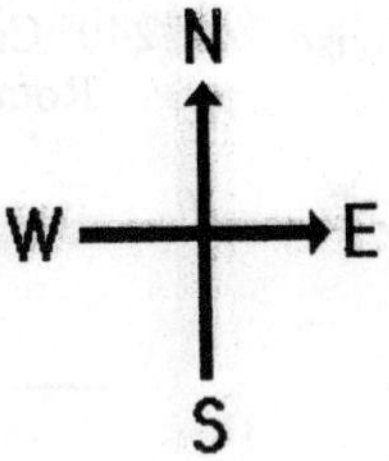

Vectors can be directed due East, due West, due South, and due North. But some vectors are directed northeast (at a 45 degree angle); and some vectors are even directed northeast, yet more north than east.

Thus, there is a clear need for some form of a convention for identifying the direction of a vector which is not due East, due West, due South, or due North. There are a variety of conventions for describing the direction of any vector. The two conventions which will be discussed and used in this unit are described below:

1. The direction of a vector is often expressed as an angle of rotation of the vector about its "tail" from either east, west, north, or south. For example, a vector can be said to have a direction of 40 degrees North of West (meaning a vector pointing West has been rotated 40 degrees towards the northerly direction) of 65 degrees East of South (meaning a vector pointing South has been rotated 65 degrees towards the easterly direction).
2. The direction of a vector is often expressed as an counterclockwise angle of rotation of the vector about its "tail" from due East. Using this convention, a vector with a direction of 30 degrees is a vector which has been rotated 30 degrees in a counterclockwise direction relative to due east. A vector with a direction of 160 degrees is a vector which has been rotated 160 degrees in a counterclockwise direction relative to due east. A vector with a direction of 270 degrees is a vector which has been rotated 270 degrees in a counterclockwise direction relative to due east. This is one of the most common conventions for the direction of a vector and will be utilized throughout this unit.

Two illustrations of the second convention (discussed above) for identifying the direction of a vector are shown below.

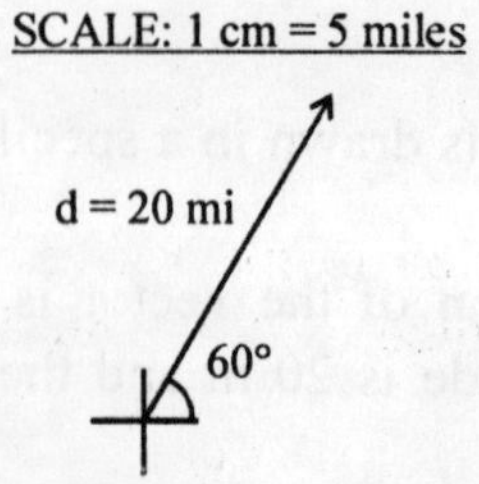

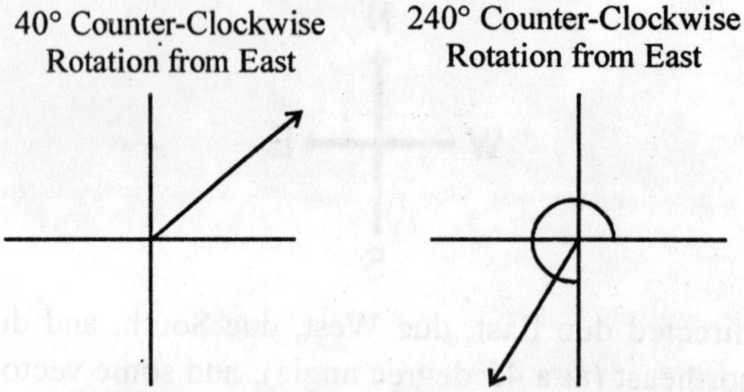

Observe in the first example that the vector is said to have a direction of 40 degrees. This means that the tail of the vector was pinned down, and the vector was rotated an angle of 40 degrees in the counterclockwise direction beginning from due east. Observe in the second example that the vector is said to have a direction of 240 degrees. This means that the tail of the vector was pinned down, and the vector was rotated an angle of 240 degrees in the counterclockwise direction beginning from due east; a rotation of 240 degrees is equivalent to rotating the vector through two quadrants (180 degrees) and then an additional 60 degrees into the third quadrant.

The magnitude of a vector in a scaled vector diagram is depicted by the length of the arrow. The arrow is drawn a precise length in accordance with a chosen scale. For example, the diagram at the right shows a vector with a magnitude of 20 miles. Since the scale used for constructing the diagram is 1 cm = 5 miles, the vector arrow is drawn with a length of 4 cm. That is, 4 cm x (5 miles/1 cm) = 20 miles. Using the same scale (1 cm = 5 miles), a displacement vector which is 15 miles will be represented by a vector arrow which is 3 cm in length. Similarly, a 25 mile displacement vector is represented by a 5 cm long vector arrow. And finally, an 18 mile displacement vector is represented by a 3.6 cm long arrow. See the examples shown below.

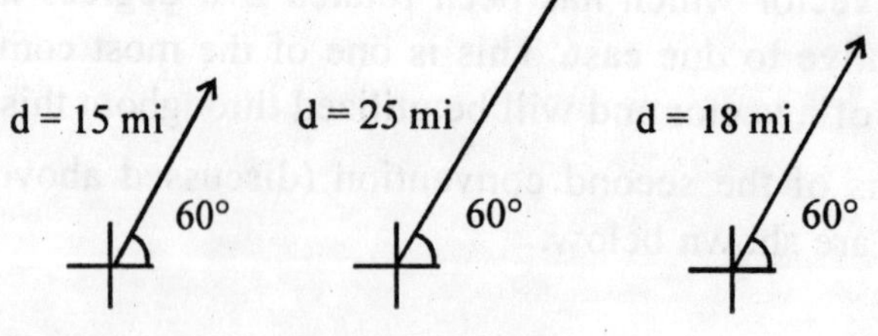

In conclusion, vectors can be represented by use of a scaled vector diagram. On such a diagram, an arrow is drawn to represent the vector; the arrow has an obvious tail and arrowhead. The magnitude of a vector is represented by the length of the arrow; a scale is indicated (such as, 1 cm = 5 miles) and the arrow is drawn the proper length according to the chosen scale. The arrow points in the precise direction. Directions are described by the use of some convention; the most common convention is that the direction of a vector is the counterclockwise angle of rotation which that vector makes with due East.

In the remainder of this lesson, in the entire unit, and even in future units, scaled vector diagrams and the above convention for the direction of a vector will be commonly used to describe motion and solve problems concerning motion. For this reason, it is critical that you have a comfortable understanding of the means of representing and describing vector quantities.

VECTOR ADDITION

A variety of mathematical operations can be performed with and upon vectors. One such operation is the addition of vectors. Two vectors can be added together to determine the result (or resultant). This process of adding two or more vectors has already been discussed in an earlier unit. Recall in our discussion of Newton's laws of motion, that the net force acting upon an object was determined by computing the vector sum of all the individual forces acting upon that object. That is the net force was the result (or resultant) of adding up all the force vectors. During that unit, the rules for summing vectors (such as force vectors) was kept relatively simple. Observe the following summations of two forces:

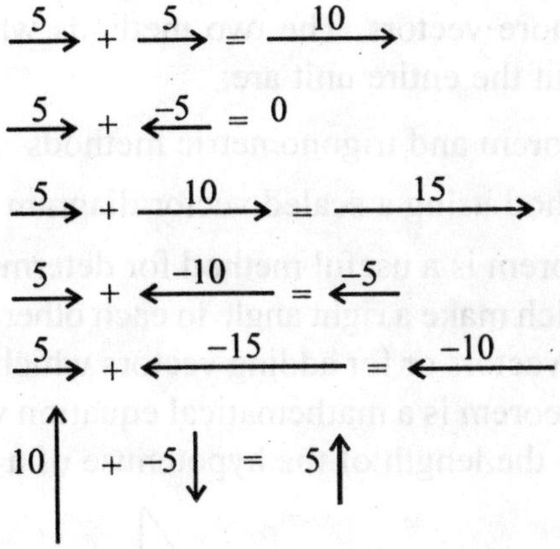

These rules for summing vectors were applied to free-body diagrams in order to determine the net force (i.e., the vector sum of all the individual forces). Sample applications are shown in the diagram below.

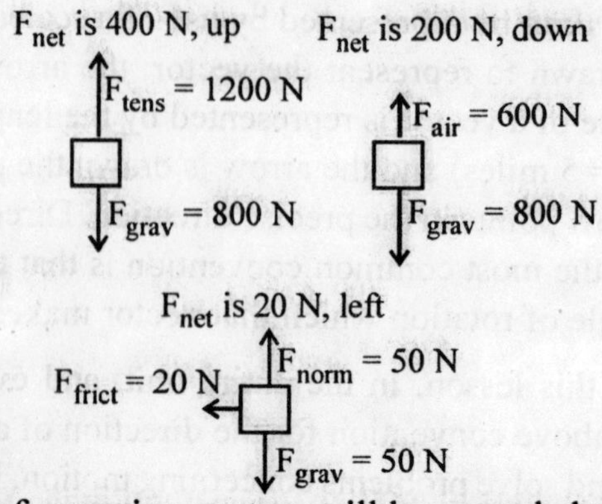

In this unit, the task of summing vectors will be extended to more complicated cases in which the vectors are directed in directions other vertical and horizontal directions. For example, a vector directed up and to the right will be added to a vector directed up and to the left. For cases such as those shown in the diagrams below, what is the resultant and how can it be determined?

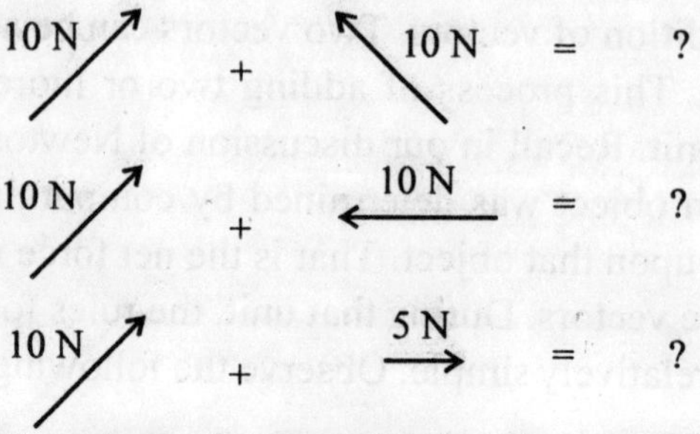

There are a variety of methods for determining the magnitude and direction of the result of adding two or more vectors. The two methods which will be discussed in this lesson and used throughout the entire unit are:

- The Pythagorean theorem and trigonometric methods
- The head-to-tail method using a scaled vector diagram

The Pythagorean theorem is a useful method for determining the result of adding two (and only two) vectors which make a right angle to each other. The method is not applicable for adding more than two vectors or for adding vectors which are not at 90-degrees to each other. The Pythagorean theorem is a mathematical equation which relates the length of the sides of a right triangle to the length of the hypotenuse of a right triangle.

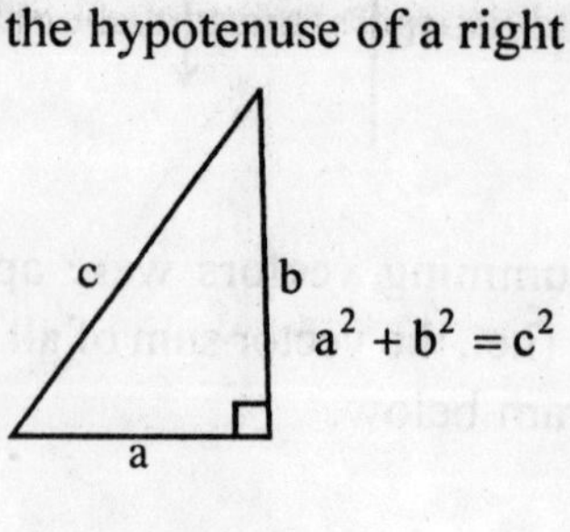

PYTHAGOREAN THEOREM

To see how the method works, consider the following problem:

A hiker leaves camp and hikes 11 km, north and then hikes 11 km east. Determine the resulting displacement of the hiker.

This problem asks to determine the result of adding two displacement vectors which are at right angles to each other. The result (or resultant) of walking 11 km north and 11 km east is a vector directed northeast as shown in the diagram to the right. Since the northward displacement and the eastward displacement are at right angles to each other, the Pythagorean theorem can be used to determine the resultant (i.e., the hypotenuse of the right triangle).

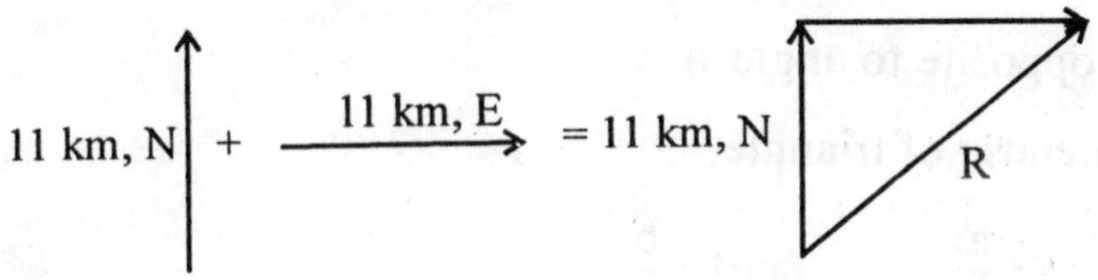

$$11^2 + 11^2 = R^2$$

$$242 = R^2$$

$$15.6 = R$$

The resultant of 11 km, north plus 11 km, east is a vector with a magnitude of 15.6 km. Later, the method of determining the direction of the vector will be discussed. For now, let's test your understanding with the following two practice problems. In each case, use the Pythagorean theorem to determine the magnitude of the resultant. When finished, depress the mouse on the "pop-up menu" to view the answer.

Practive A	Practice B
10 km, North	30 km, West
+	+
5 km, West	40 km, South

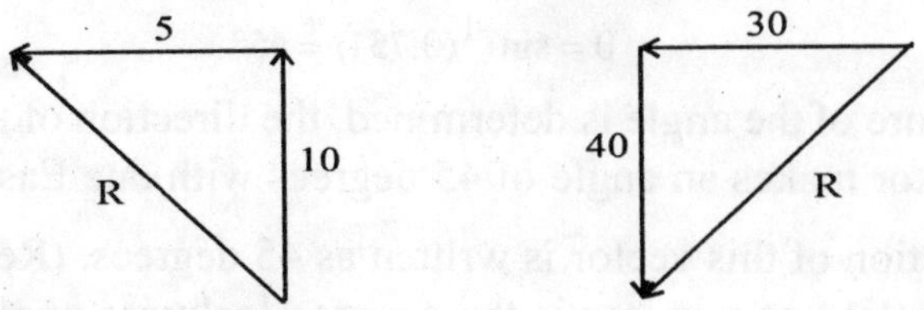

The direction of vector R in the diagram above can be determined by use of three common trigonometric functions - sine, cosine, and tangent functions. These three functions relate the angle of a right triangle to the ratio of the lengths of two of the sides of a right triangle. The sine function relates the sine of an angle to the ratio of the length of the side opposite the angle to the length of the hypotenuse.

The cosine function relates the cosine of an angle to the ratio of the length of the side adjacent the angle to the length of the hypotenuse. The tangent function relates the tangent of an angle to the ratio of the length of the side opposite the angle to the length of the side adjacent to the angle. The three equations below summarize these three functions in equation form.

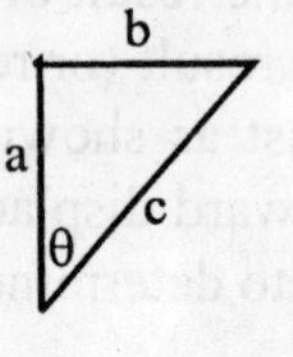

a = side adjacent to angle θ

b = side opposite to angle θ

c = hypotenuse of triangle

$$\sin\theta = \frac{b}{c} \qquad \cos\theta = \frac{a}{c} \qquad \tan\theta = \frac{b}{a}$$

These three trigonometric functions can be applied to the hiker problem in order to determine the direction of the hiker's displacement. The process begins by the selection of one of the two angles (other the right angle) of the triangle. Once the angle is selected, any of the three functions can be used to find the measure of the angle. Write the function and proceed with the proper algebraic steps to solve for the measure of the angle. The work is shown below.

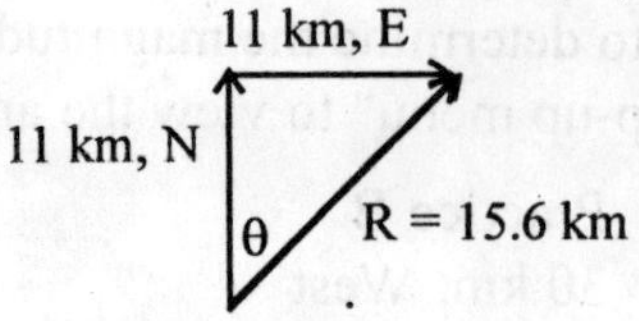

$$\sin\theta = \frac{11\text{km}}{15.6\text{km}} = 0.7051$$

$$\theta = \sin^{-1}(0.751) = 45°$$

Once the measure of the angle is determined, the direction of the vector can be found. In this case the vector makes an angle of 45 degrees with due East.

Thus, the direction of this vector is written as 45 degrees. (Recall from earlier in this lesson, that the direction of a vector is the counterclockwise angle of rotation which the vector makes with due East.)

The measure of an angle as determined is not always the direction of the vector. For example, study the following situation in which the angle of the triangle is determined to be 26.6 degrees, yet the direction of the vector is 206.6 degrees.

10 km, West + 5 km, South

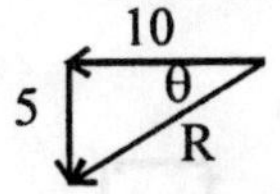

$$\tan\theta = \frac{5\,\text{km}}{10\,\text{km}} = 0.5$$

$$\theta = \tan^{-1}(0.5) = 26.6°$$

Direction of R is 180° + 26.6°

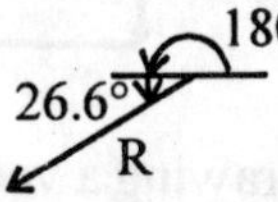

Test your understanding of the use of SOH CAH TOA to determine the vector direction by trying the following two practice problems. In each case, use SOH CAH TOA to determine the direction of the resultant. When finished, depress the mouse on the "pop-up menu" to view the answer.

In the above problems, the magnitude and direction of the sum of two vectors is determined using the Pythagorean theorem and trigonometric methods.

Practice A
10 km, North
+
5 km, West

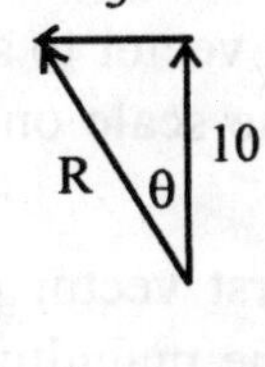

Practice B
30 km, West
+
40 km, South

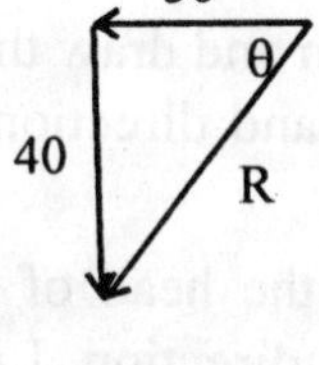

The procedure is restricted to the addition of two vectors which make right angles to each other. When the two vectors which are to be added do not make right angles to one another, or when there are more than two vectors to add together, we will employ a method known as the head-to-tail vector addition method. This method is described below.

The magnitude and direction of the sum of two or more vectors can also be determined by use of an accurately drawn scaled vector diagram. Using a scaled diagram, the head-to-tail method is employed to determine the resultant. You were given a set of 18 directions to follow, these 18 displacement vectors were added together in consecutive fashion to determine the result of adding the set of 18 directions. Each time, where one measurement ended, the next measurement began. In essence, you were using the head-to-tail method of vector addition.

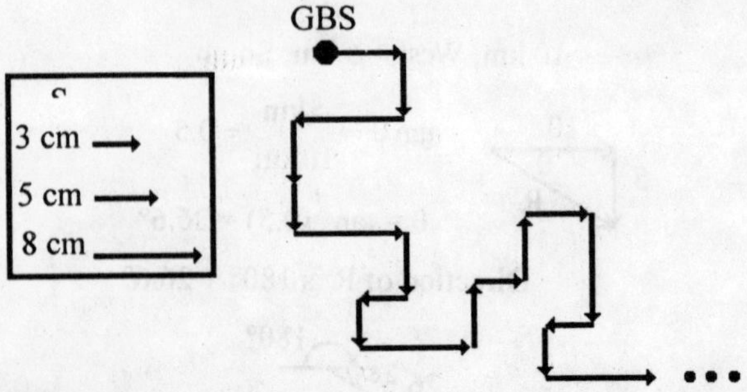

The head-to-tail method involves drawing a vector to scale on a sheet of paper beginning at a designated starting position; where the head of this vector ends the tail of the next vector begins (thus, head-to-tail method). The process is repeated for all vectors which are added. Once all vectors have been added head-to-tail, the resultant is drawn from the tail of the first vector to the head of the last vector; i.e., from start to finish. Once the resultant is drawn, its length can be measured and converted to real units using the given scale. The direction of the resultant can be determined by using a protractor and measuring its counterclockwise angle of rotation from due East.

The following step-by-step method for applying the head-to-tail method to determine the sum of two or more vectors is given below.

1. Choose a scale and indicate it on a sheet of paper (the best choice of scale is one which will result in a diagram which is as large as possible, yet fits on the sheet of paper).
2. Pick a starting location and draw the first vector to scale in the indicated direction. Label the magnitude and direction of the scale on the diagram (e.g., SCALE: 1 cm = 20 m).
3. Starting from where the head of the first vector ends, draw the first vector to scale in the indicated direction. Label the magnitude and direction of the vector on the diagram.
4. Repeat steps 2 and 3 for all vectors which are to be added
5. Draw the resultant from the tail of the first vector to the head of the last vector. Label this vector as "Resultant" or simply "R."
6. Using a ruler, measure the length of the resultant and determine its magnitude by converting to real units using the scale (4.4 cm x 20 m/1 cm = 88 m).
7. Measure the direction of the resultant using the counterclockwise convention discussed earlier in this lesson.

An example of the use of the head-to-tail method is illustrated below. The problem involves the addition of three vectors:

20 m, 45 deg. + 25 m, 300 deg. + 15 m, 210 deg.

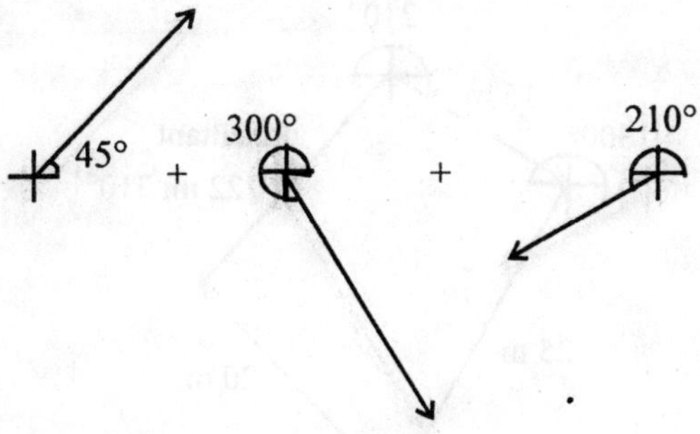

Scale: 1 cm = 5 m

The head-to-tail method is employed as described above and the resultant is determined (drawn in red). Its magnitude and direction is labeled on the diagram.

Scale: 1 cm = 5 m

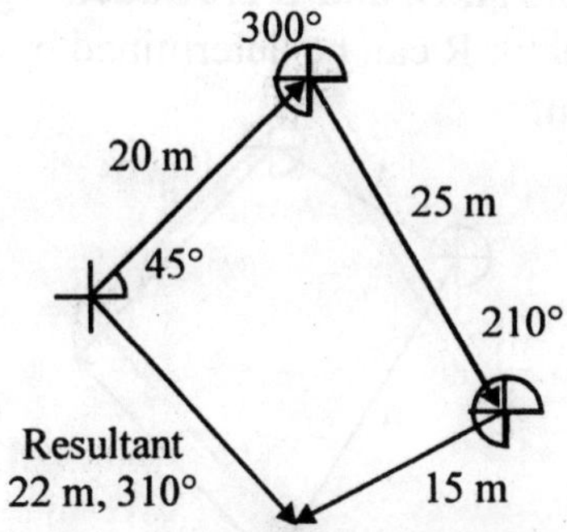

Interestingly enough, the order in which three vectors are added is insignificant; the resultant will still have the same magnitude and direction. For example, consider the addition of the same three vectors in a different order.

15 m, 210 deg. + 25 m, 300 deg. + 20 m, 45 deg.

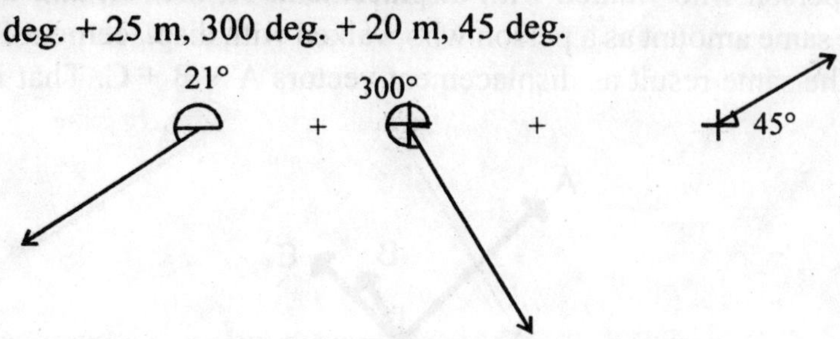

Scale: 1 cm = 5 m

When added together in this different order, these same three vectors still produce a resultant with the same magnitude and direction as before (22 m, 310 deg.). The order in which vectors are added using the head-to-tail method is insignificant.

Scale: 1 cm = 5 m

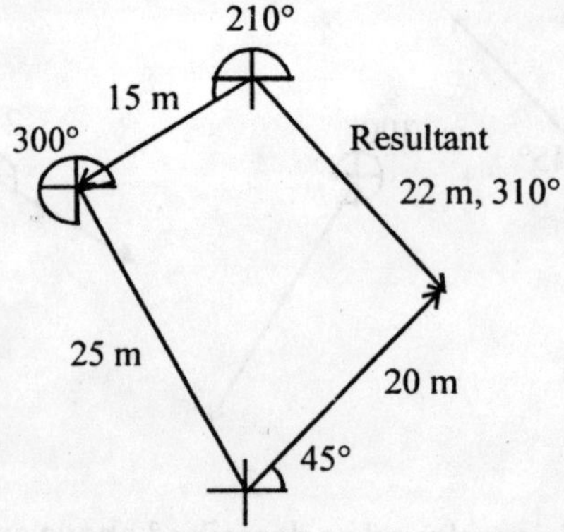

RESULTANTS OF VECTOR ADDITION

The resultant is the vector sum of two or more vectors. It is *the result* of adding vectors together. If displacement vectors A, B, and C are added together, the result will be vector R. As shown in the diagram, vector R can be determined by the use of an accurately drawn, scaled, vector addition diagram.

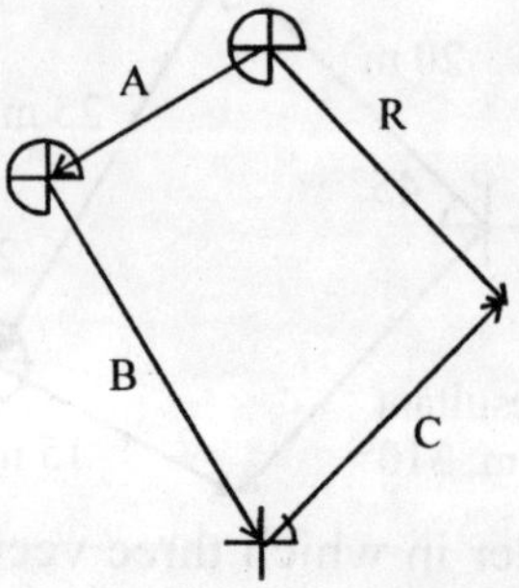

To say that vector R is the resultant displacement of displacement vectors A, B, and C is to say that a person who walked with displacements A, then B, and then C would be displaced by the same amount as a person who walked with displacement R. Displacement vector R gives the same result as displacement vectors A + B + C. That is why it can be said that

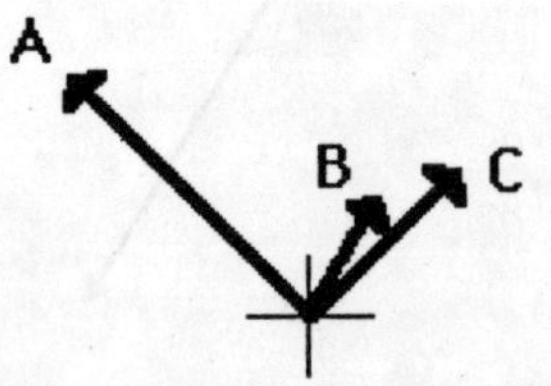

A + B + C = R

The above discussion pertains to the result of adding displacement vectors; when displacement vectors are added, the result is a resultant displacement. But any two vectors can be added as long as they are the same vector quantity.

Add two or more velocity vectors and the result is a resultant velocity; add two or more force vectors and the result is a resultant force; add two or more momentum vectors and the result is...

In all such cases, the resultant vector (whether a displacement vector, force vector, velocity vector, etc.) is the result of adding the individual vectors. It is the same thing as adding A + B + C +.... "To do A + B + C is the same as to do R." As an example, consider a football player who gets hit simultaneously by three players on the opposing team (players A. B. and C). The football player experiences three different applied forces.

Each applied force contributes to a total or resulting force. If the three forces are added together using methods of vector addition (discussed earlier), then the resultant vector R can be determined.

In this case, to experience the three forces A, B and C is the same as experiencing force R. To be hit by players A, B, and C would result in the same force as being hit by one player applying force R. "To do A + B + C is the same as to do R." Vector R is the same result as vector A + B + C!!

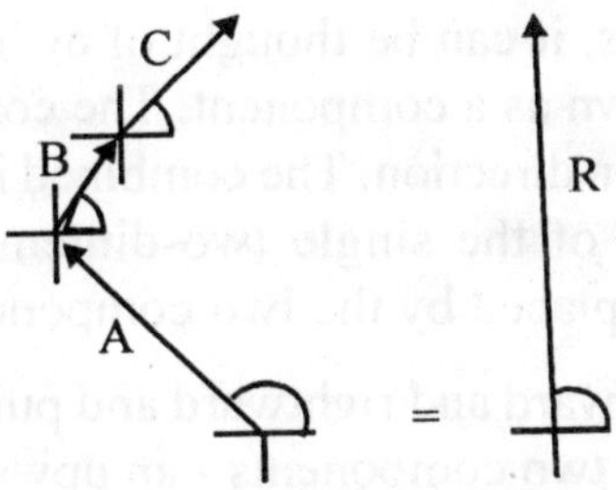

In summary, the resultant is the vector sum of all the individual vectors. The resultant is the result of combining the individual vectors together. The resultant can be determined by adding the individual forces together using vector addition methods.

VECTOR COMPONENTS

A vector is a quantity which has magnitude and direction. Displacement, velocity, acceleration, and force are the vector quantities which we have discussed thus far in our course. In the first couple of units of our course, all vectors which we discussed were simply directed up, down, left or right. When there was a free-body diagram depicting the forces acting upon an object, those forces were directed in one dimension - up, down, left or right. When an object had an acceleration and we described its direction, it was directed in one dimension - up, down, left or right. Now in this unit, we begin to see examples of vectors which are directed in two dimensions - upward and rightward, northward and westward, eastward and southward, etc.

In situations in which vectors are directed at angles to the customary coordinate axes, a useful mathematical trick will be employed to transform the vector into two parts, with each part being directed along the coordinate axes. For example, a vector which is directed Northwest can be thought of as having two parts - a Northward and a Westward part.

A vector which is directed upward and rightward can be thought of as having two parts - an upward and a rightward part.

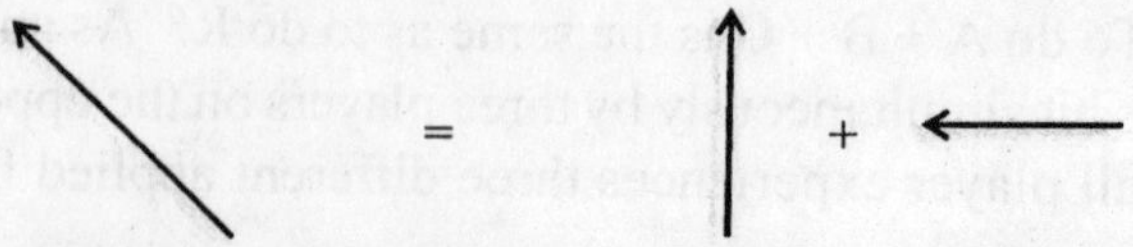

Northwest vectors have a northward and westward part.

An upword and rightward vector has an upward and rightward part.

Any vector directed in two dimensions can be thought of as having an influence in two different directions. That is, it can be thought of as having two parts. Each part of a two-dimensional vector is known as a component. The components of a vector depict the influence of that vector in a given direction. The combined influence of the two components is equivalent to the influence of the single two-dimensional vector. The single two-dimensional vector could be replaced by the two components.

If dog chain is stretched upward and rightward and pulled tight by his master, then the tensional force in the chain has two components - an upward and a rightward component. The influence of the chain on dog's body is equivalent to the influence of two chains on his body - one pulling upward and the other pulling rightward. If the single chain were replaced by two chains (each one having the magnitude and direction of the components), the dog would not know the difference. This is not because dog is dumb (a quick glance at his picture reveals that he is certainly not that), but rather because the combined influence of the two components is equivalent to the influence of the single two-dimensional vector.

Consider a picture which is hung to a wall by means of two wires which are stretched vertically and horizontally. Each wire exerts a tensional force upon the picture to support its weight. Since each wire is stretched in two dimensions (both vertically and horizontally), the tensional force of each wire has two components - a vertical and a horizontal component.

Focusing on the wire on the left, we could say that the wire has a leftward and an upward component. This is to say that the wire on the left could be replaced by two wires, one pulling leftward and the other pulling upward. If the single wire were replaced by two wires (each one having the magnitude and direction of the components), then there would be no effect upon the stability of the picture. The combined influence of the two components is equivalent to the influence of the single two-dimensional vector.

Consider an airplane which is flying from O'Hare International Airport to a destination in Canada. Suppose that the plane is flying in such a manner that its resulting displacement vector is northwest. If this is the case, then the displacement of the plane has two components - a component in the northward direction and a component in the westward direction. This is to say that the plane would have the same displacement if it were to take the trip into Canada in two segments - one directed due North and the other directed due West.

If the single displacement vector were replaced by these two individual displacement vectors, then the passengers in the plane would end up in the same final position. The combined influence of the two components is equivalent to the influence of the single two-dimensional displacement.

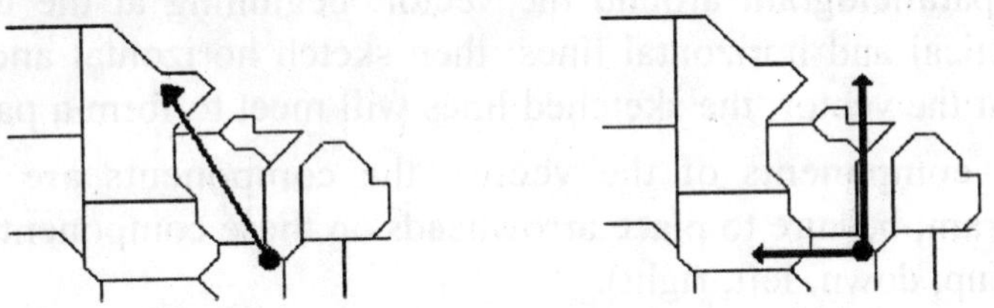

The palne's northwest displacement is equivalent to a northward plus a westward displacement.

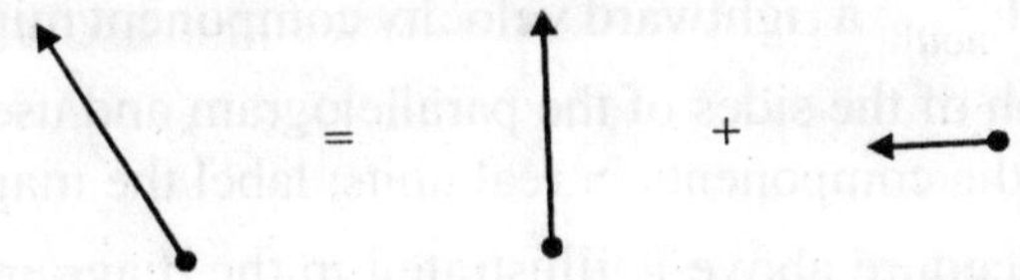

Any vector directed in two dimensions can be thought of as having two different components. The component of a single vector describes the influence of that vector in a given direction.

VECTOR RESOLUTION

As mentioned earlier in this lesson, any vector directed at an angle to the horizontal (or the vertical) can be thought of as having two parts (or components). That is, any vector directed in two dimensions can be thought of as having two components. For example, if a chain pulls upward at an angle on the collar of a dog, then there is a tensional force directed in two dimensions. This tensional force has two components: an upward and a rightward component. As another example, consider an airplane which is displaced northwest from O'Hare International Airport (in Chicago) to a destination in Canada. The displacement vector of the plane is in two dimensions (northwest). Thus, this displacement vector has two components: a northward and a westward component.

In this unit, we learn two basic methods for determining the magnitudes of the components of a vector directed in two dimensions. The process of determining the magnitude of a vector is known as vector resolution. The two methods of vector resolution which we will examine are

- The parallelogram method
- The trigonometric method

The parallelogram method of vector resolution involves using an accurately drawn, scaled vector diagram to determine the components of the vector. Briefly put, the method involves drawing the vector to scale in the indicated direction, sketching a parallelogram around the vector such that the vector is the diagonal of the parallelogram, and determining the magnitude of the components (the sides of the parallelogram) using the scale. A step-by-step procedure for using the parallelogram method of vector resolution is:

1. Select a scale and accurately draw the vector to scale in the indicated direction.
2. Sketch a parallelogram around the vector: beginning at the tail of the vector, sketch vertical and horizontal lines; then sketch horizontal and vertical lines at the head of the vector; the sketched lines will meet to form a parallelogram.
3. Draw the components of the vector; the components are the sides of the parallelogram; be sure to place arrowheads on these components to indicate their direction (up, down, left, right).
4. Meaningfully label the components of the vectors with symbols to indicate which component is being represented by which side; a northward force component would be labeled F_{north}; a rightward velocity component might be labeled v_x; etc.
5. Measure the length of the sides of the parallelogram and use the scale to determine the magnitude of the components in real units; label the magnitude on the diagram.

The step-by-step procedure above is illustrated in the diagram below to show how a velocity vector with a magnitude of 50 m/s and a direction of 60 degrees above the horizontal may be resolved into two components.

The diagram shows that the vector is first drawn to scale in the indicated direction; a parallelogram is sketched about the vector; the components are labeled on the diagram; and the result of measuring the length of the vector components and converting to m/s using the scale.

Note: Because different computer monitors have different resolutions, the actual length of the vector on your monitor may not be 5 cm.

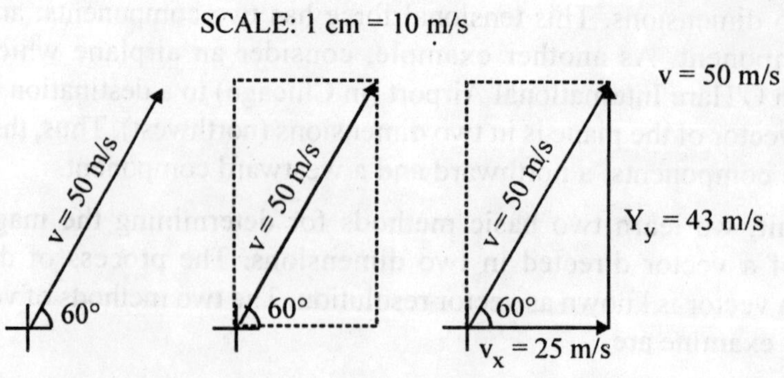

The trigonometric method of vector resolution involves using trigonometric functions to determine the components of the vector.

Earlier in lesson 1, the use of trigonometric functions to determine the direction of a vector was described. Now, in this part of lesson 1, trigonometric functions will be used to determine the components of a single vector.

Recall from the earlier discussion that trigonometric functions relate the length of the sides of a right triangle to the angle of a right triangle.

As such, trigonometric functions can be used to determine the length of the sides of a right triangle if one angle and the length of one side are known.

The method of employing trigonometric functions to determine the components of a vector are as follows:

1. Construct a sketch (no scale needed) of the vector in the indicated direction; label its magnitude and the angle which it makes with the horizontal.
2. Draw a rectangle about the vector such that the vector is the diagonal of the rectangle; beginning at the tail of the vector, sketch vertical and horizontal lines; then sketch horizontal and vertical lines at the head of the vector; the sketched lines will meet to form a parallelogram.
3. Draw the components of the vector; the components are the sides of the rectangle; be sure to place arrowheads on these components to indicate their direction (up, down, left, right).
4. Meaningfully label the components of the vectors with symbols to indicate which component is being represented by which side; a northward force component would be labeled F-north; a rightward force velocity component might be labeled v-x; etc.
5. To determine the length of the side opposite the indicated angle, use the sine function; substitute the magnitude of the vector for the length of the hypotenuse; use some algebra to solve the equation for the length of the side opposite the indicated angle.
6. Repeat the above step using the cosine function to determine the length of the side adjacent to the indicated angle.

RELATIVE VELOCITY AND RIVERBOAT PROBLEMS

On occasion objects move within a medium which is moving with respect to an observer. For example, an airplane usually encounters a wind - air which is moving with respect to an observer on the ground below. As another example, a motor boat in a river is moving amidst a river current - water which is moving with respect to an observer on dry land. In such instances as this, the magnitude of the velocity of the moving object (whether it be a plane or a motor boat) with respect to the observer on land will not be the same as the speedometer reading of the vehicle.

That is to say, the speedometer on the motor boat might read 20 mi/hr; yet the motor boat might be moving relative to the observer on shore at a speed of 25 mi/hr. Motion is relative to the observer. The observer on land, often named (or misnamed) the "stationary observer" would measure the speed to be different than that of the person on the boat.

To illustrate this principle, consider a plane flying amidst a tailwind. A tailwind is merely a wind which approaches the plane from behind, thus increasing its resulting velocity. If the plane is traveling at a velocity of 100 km/hr with respect to the air, and if the wind velocity is 25 km/hr, then what is the velocity of the plane relative to an observer on the ground below?

The resultant velocity of the plane (that is, the result of the wind velocity contributing to the velocity due to the plane's motor) is the vector sum of the velocity of the plane and the velocity of the wind. This resultant velocity is quite easily determined if the wind approaches the plane directly from behind. As shown in the diagram below, the plane travels with a resulting velocity of 125 km/hr relative to the ground.

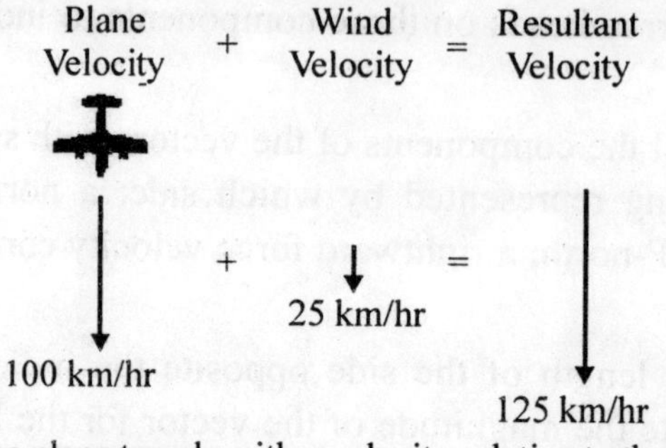

The plane travels with a velocity relative to the ground which is the vector sum of the plane's velocity (relative to the air) plus the wind velocity.

If the plane encounters a headwind, the resulting velocity will be less than 100 km/hr. Since a headwind is a wind which approaches the plane from the front, such a wind would decrease the plane's resulting velocity.

Suppose a plane traveling with a velocity of 100 km/hr with respect to the air meets a headwind with a velocity of 25 km/hr. In this case, the resultant velocity would be 75 km/hr; this is the velocity of the plane relative to an observer on the ground. This is depicted in the diagram below.

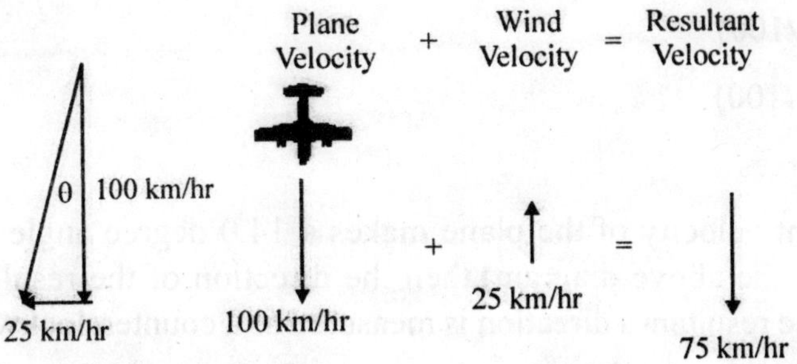

Now consider a plane traveling with a velocity of 100 km/hr, South which encounters a side wind of 25 km/hr, West. Now what would the resulting velocity of the plane be? This question can be answered in the same manner as the previous questions. The resulting velocity of the plane is the vector sum of the two individual velocities. To determine the resultant velocity, the plane velocity (relative to the air) must be added to the wind velocity.

This is the same procedure which was used above for the headwind and the tailwind situations; only now, the resultant is not as easily computed. Since the two vectors to be added - the southward plane velocity and the westward wind velocity - are at right angles to each other, the Pythagorean theorem can be used. This is illustrated in the diagram below.

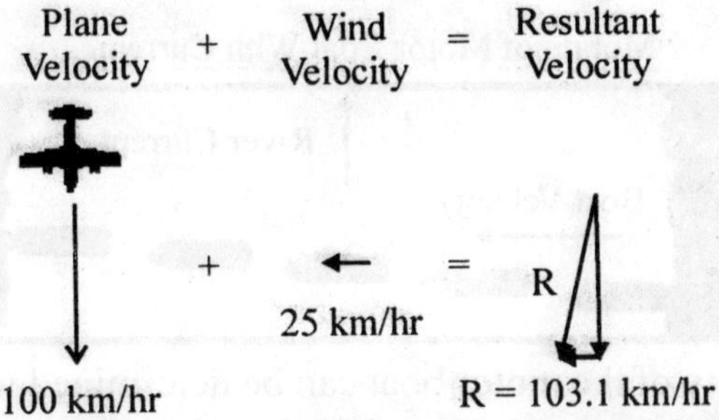

In this situation of a side wind, the southward vector can be added to the westward vector using the usual methods of vector addition. The magnitude of the resultant velocity is determined using Pythagorean theorem. The algebraic steps are as follows:

$(100 \text{ km/hr})^2 + (25 \text{ km/hr})^2 = R^2$

$10\,000 \text{ km}^2/\text{hr}^2 + 625 \text{ km}^2/\text{hr}^2 = R^2$

$10\,625 \text{ km}^2/\text{hr}^2 = R^2$

$\Rightarrow \; R = \sqrt{10625} \text{ km/hr} = 103.1 \text{ km/hr}$

The direction of the resulting velocity can be determined using a trigonometric function. Since the plane velocity and the wind velocity form a right triangle when added together in head-to-tail fashion, the angle between the resultant and the southward vector can be determined using either the sine, cosine, or tangent functions. The tangent function can be used; this is shown below:

$\tan(\theta) = (25/100)$

$\theta = \tan^{-1}(25/100)$

$\theta = 14.0$ degrees

If the resultant velocity of the plane makes a 14.0 degree angle with the southward direction (theta in the above diagram),then the direction of the resultant is 256 degrees. Like any vector, the resultant's direction is measured as a counterclockwise angle of rotation from due East.

The effect of the wind upon the plane is similar to the effect of the river current upon the motor boat. If a motor boat were to head straight across a river (that is, if the boat were to point its bow straight towards the other side), it would not reach the shore directly across from its starting point. The river current influences the motion of the boat and carries it downstream.

The motor boat may be moving with a velocity of 4 m/s directly across the river, yet the resultant velocity of the boat will be greater than 4 m/s and at an angle in the downstream direction. While the speedometer of the boat may read 4 m/s, its speed with respect to an observer on the shore will be greater than 4 m/s.

Motion of Motor Boat With Current

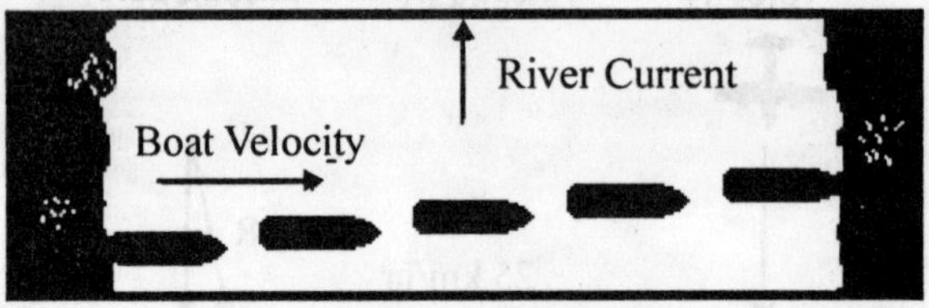

The resultant velocity of the motor boat can be determined in the same manner as was done for the plane. The resultant velocity of the boat is the vector sum of the boat velocity and the river velocity. Since the boat heads straight across the river and since the current is always directed straight downstream, the two vectors are at right angles to each other.

Thus, the Pythagorean theorem can be used to determine the resultant velocity. Suppose that the river was moving with a velocity of 3 m/s, North and the motor boat was moving with a velocity of 4 m/s, East. What would be the resultant velocity of the motor boat (i.e., the velocity relative to an observer on the shore)? The magnitude of the resultant can be found as follows:

$(4.0 \text{ m/s})^2 + (3.0 \text{ m/s})^2 = R^2$

$16 \text{ m}^2/\text{s}^2 + 9 \text{ m}^2/\text{s}^2 = R^2$

$\Rightarrow R = $ m/s

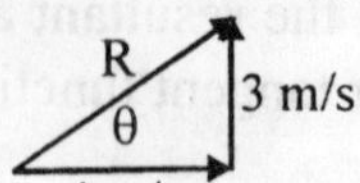

5.0 m/s = R

The direction of the resultant is the counterclockwise angle of rotation which the resultant vector makes with due East. This angle can be determined using a trigonometric function as shown below.

$\tan\theta = (3/4)$

$\theta = \tan^{-1}(3/4)$

$\theta = 36.9$ degrees

Given a boat velocity of 4 m/s, East and a river velocity of 3 m/s, North, the resultant velocity of the boat will be 5 m/s at 36.9 degrees.

Motor boat problems such as these are typically accompanied by three separate questions:

1. What is the resultant velocity (both magnitude and direction) of the boat?
2. If the width of the river is X meters wide, then how much time does it take the boat to travel shore to shore?
3. What distance downstream does the boat reach the opposite shore?

The first of these three questions was answered above; the resultant velocity of the boat can be determined using the Pythagorean theorem (magnitude) and a trigonometric function (direction). The second and third of these questions can be answered using the average speed equation (and a lot of logic).

Average speed = distance/time

Consider the following example.

Example:

A motor boat traveling 4 m/s, East encounters a current traveling 3.0 m/s, North.

1. What is the resultant velocity of the motor boat?
2. If the width of the river is 80 meters wide, then how much time does it take the boat to travel shore to shore?
3. What distance downstream does the boat reach the opposite shore?

The solution to the first question has already been shown in the above discussion. The resultant velocity of the boat is 5 m/s at 36.9 degrees. We will start in on the second question.

The river is 80-meters wide; that is, the distance from shore to shore as measured straight across the river is 80 meters. The time to cross this 80-meter wide river can be determined by rearranging and substituting into the average speed equation.

time = distance/(Average speed)

The distance is 80 m can be substituted into the numerator. But what about the denominator? What value should be used for average speed? Should 3 m/s (the current velocity), 4 m/s (the boat velocity), or 5 m/s (the resultant velocity) be used as the average speed value for covering the 80 meters? With what average speed is the boat traversing the 80 meter wide river? Most students want to use the resultant velocity in the equation since that is the actual velocity of the boat with respect to the shore. Yet the value of 5 m/s is the speed at which the boat covers the diagonal dimension of the river. And the diagonal distance across the river is not known in this case.

If one knew.the distance C in the diagram below, then the average speed C could be used to calculate the time to reach the opposite shore. Similarly, if one knew the distance B in the diagram below, then the average speed B could be used to calculate the time to reach the opposite shore. And finally, if one knew the distance A in the diagram below, then the average speed A could be used to calculate the time to reach the opposite shore.

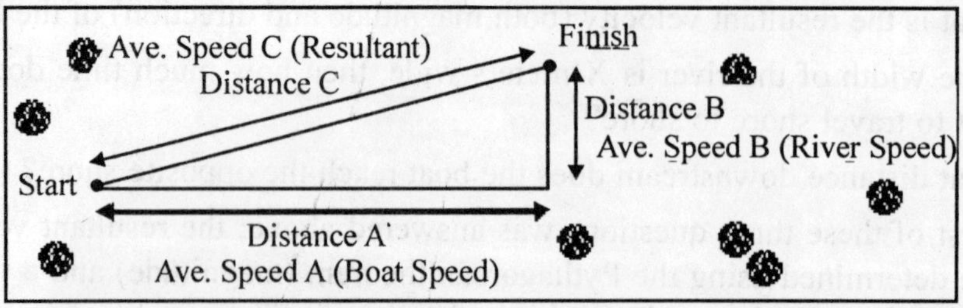

In our problem, the 80 m corresponds to the distance A, and so the average speed of 4 m/s (average speed in the direction straight across the river) should be substituted into the equation to determine the time.

time = (80 m)/(4 m/s) = 20 s

It requires 20 s for the boat to travel across the river. During this 20 s of crossing the river, the boat also drifts downstream. Part c of the problem asks "What distance downstream does the boat reach the opposite shore?" The same equation must be used to calculate this downstream distance. And once more, the question arises, which one of the three average speed values must be used in the equation to calculate the distance downstream? The distance downstream corresponds to Distance B on the above diagram.

The speed at which the boat covers this distance corresponds to Average Speed B on the diagram above (i.e., the speed at which the current moves - 3 m/s). And so the average speed of 3 m/s (average speed in the downstream direction) should be substituted into the equation to determine the time.

distance = ave. speed * time = (3 m/s) * (20 s)

distance = 60 m

The boat is carried 60 meters downstream during the 20 seconds it takes to cross the river.

The mathematics of the above problem is no more difficult than dividing or multiplying two numerical quantities by each other. The mathematics is easy! The difficulty of the problem is conceptual in nature; the difficulty lies in deciding which numbers to use in the equations. That decision emerges from one's conceptual understanding (or unfortunately, one's misunderstanding) of the complex motion which is occurring. The motion of the river boat can be divided into two simultaneous parts - a motion in the direction straight across the river and and a motion in the downstream direction.

These two parts (or components) of the motion occur simultaneously for the same time duration (which was 20 seconds in the above problem). The decision as to which velocity value or distance value to use in the equation must be consistent with the diagram above. The boat's motor is what carries the boat across the river the Distance A; and so any calculation involving the Distance A must involve the speed value labeled as Speed A (the boat speed relative to the water). Similarly, it is the current of the river which carries the boat downstream for the Distance B; and so any calculation involving the Distance B must involve the speed value labeled as Speed B (the river speed). Together, these two parts (or components) add up to give the resulting motion of the boat.

That is, the across-the-river component of displacement adds to the downstream displacement to equal the resulting displacement. And likewise, the boat velocity (across the river) adds to the river velocity (down the river) to equal the resulting velocity. And so any calculation of the Distance C or the Average Speed C ("Resultant Velocity") can be performed using the Pythagorean theorem.

Now to illustrate an important point, let's try a second example problem which is similar to the first example problem. Make an attempt to answer the three questions and then use the pop-up menu to check your answer.

Example:

A motor boat traveling 4 m/s, East encounters a current traveling 7.0 m/s, North.

1. What is the resultant velocity of the motor boat?
2. If the width of the river is 80 meters wide, then how much time does it take the boat to travel shore to shore?
3. What distance downstream does the boat reach the opposite shore?

An important concept emerges from the analysis of the two example problems above. In Example, the time to cross the 80-meter wide river (when moving 4 m/s) was 20 seconds. This was in the presence of a 3 m/s current velocity. In Example, the current velocity was much greater - 7 m/s - yet the time to cross the river remains unchanged. In fact, the current velocity itself has no effect upon the time required for a boat to cross the river. The river moves downstream parallel to the banks of the river; as such, there is no way that the current is capable of assisting a boat in crossing a river.

While the increased current may effect the resultant velocity - making the boat travel with a greater speed with respect to an observer on the ground - it does not increase the speed in the direction across the river. The increased resultant velocity is only a direction pointing down the river. It is often said that "perpendicular components of motion are independent of each other." As applied to river boat problems, this would mean that an across-the-river variable would be independent of (i.e., not be effected by) a downstream variable. The time to cross the river is dependent upon the velocity at which the boat crosses the river.

It is only the component of motion directed across the river (i.e., the boat velocity) which effects the time to travel the distance directly across the river (80 m in this case). The component of motion perpendicular to this direction - the current velocity - only effects the distance which the boat travels downstream.

Exercise

1. Show that the vectors $a = i + 2k$, $b = 2i - j + 3k$ and $c = i - 3j + k$ are position vectors of vertices of a right angled triangle.
2. $a = 3i + 7j + 6k$, $b = i + 2j + 9k$, $c = i + 95j$, find

   (*a*) 3a −2 b + 7c (*b*) $|2a + 3b + 4c|$

   (*c*) Unit vector petrpendicular to $3a + 2b + 4c$.
3. Using dot product, find the angle between the lines AB and AC where A, B, C are the points (6, 4, 1), (1, 1, 1) and (5, 1, 1) respectively.
5. Given two vectors $a = i + j + 5k$, $b = 3i + 5j + 7k$, find $a.\ b$
6. Find the area of the parallelogram whose two adjacent sides are represented by the vectors

   $a = 4i + 7j$, $b = 2i - 9j - 6k$.
7. If $a = 3i - 4j - 6k$, $b = -5i - 6j + 4k$, $i + j + 6k$, find $(a \times b) \cdot c$
8. Find the volume of parallelopiped whose coterminous edges are represented by the following vectors:

   (i) $a = i - j - k$, $b = -2 - 3j + 6k$, $c = 4i - 6j - 7k$

   (ii) $a = 4i + 5j + 6k$, $b = 12 + 9j + 11k$, $c = i + j - 9k$
9. Find μ so that the vectors

   $a = 7i + \mu j - k$, $b = i + j - k$, $c = \mu i - 8j + 12k$ are coplanar.
10. Show that $[a - b\ b - c\ c - a] = 0$
11. A plane can travel with a speed of 80 mi/hr with respect to the air. Determine the resultant velocity of the plane (magnitude only) if it encounters a

(*a*) 10 mi/hr headwind. (*b*) 10 mi/hr tailwind.

(*c*) 10 mi/hr crosswind. (*d*) 60 mi/hr crosswind.

12. A motor boat traveling 5 m/s, East encounters a current traveling 2.5 m/s, North.
    (*a*) What is the resultant velocity of the motor boat?
    (*b*) If the width of the river is 80 meters wide, then how much time does it take the boat to travel shore to shore?
    (*c*) What distance downstream does the boat reach the opposite shore?
13. A motor boat traveling 5 m/s, East encounters a current traveling 2.5 m/s, South.
    (*a*) What is the resultant velocity of the motor boat?
    (*b*) What distance downstream does the boat reach the opposite shore?
    (*c*) If the width of the river is 80 meters wide, then how much time does it take the boat to travel shore to shore?
14. A motor boat traveling 6 m/s, East encounters a current traveling 3.8 m/s, South.
    (*a*) What is the resultant velocity of the motor boat?
    (*b*) If the width of the river is 120 meters wide, then how much time does it take The boat to travel shore to shore?
    (*c*) What distance downstream does the boat reach the opposite shore?
15. If the current velocity in question #4 were increased to 5 m/s, then
    (*a*) How much time would be required to cross the same 120-m wide river?
    (*b*) What distance downstream would the boat travel during this time?

VECTOR SPACES

Definition: An R-module is a set M together with an addition operation on M and a function $\mu: R \times M \to M$, such that the set M under addition forms an abelian group, and moreover, for all $c, d \in R$ and $\alpha, \beta \in M$, we have:

1. $\mu(c, \mu(d, \alpha)) = \mu(cd, \alpha)$,
2. $\mu(c + d, \alpha) = \mu(c, \alpha) + \mu(d, \mu)$,
3. $\mu(c, \alpha + \beta) = \mu(c, \alpha) + \mu(c, \beta)$,
4. $\mu(1_R, \alpha) = \alpha$.

One may also call an R-module M a module over R, and elements of R are sometimes called scalars. The function μ in the definition is called a scalar multiplication map, and the value $\mu(c, \alpha)$ is called the scalar product of c and α.

Usually, we shall simply write $c\alpha$ (or $c.\alpha$) instead of $\mu(c, \alpha)$. When we do this, properties (i) – (iv) of the definition may be written as follows:

$$c(d\alpha) = (cd)\alpha,\ (c + d)\alpha = c\alpha + d\alpha,\ c(\alpha + \beta) = c\alpha + c\beta,\ 1_R\alpha = \alpha.$$

Note that there are two addition operations at play here: addition in R (such as $c + d$) and addition in M (such as $\alpha + \beta$). Likewise, there are two multiplication operations at play: multiplication in R (such as cd) and scalar multiplication (such as c). Note that by property (i), we may write cd without any ambiguity, as both possible interpretations, $c(d\alpha)$ and $(cd)\alpha$, yield the same value.

For fixed $c \in R$, the map that sends $\alpha \in M$ to $c\alpha \in M$ is a group homomorphism with respect to the additive group operation of M (by property (iii) of the definition); likewise, for fixed $\alpha \in M$, the map that sends $c \in R$ to $c\alpha \in M$ is a group homomorphism from the additive group of R into the additive group of M (by property (ii)). Combining these observations with basic facts about group homomorphisms, we may easily derive the following basic facts about R-modules:

Theorem: If M is a module over R, then for all $c \in R$, $\alpha \in M$, and $k \in \mathbb{Z}$, we have:

1. $0_R. \alpha = 0_M$,
2. $c. 0_M = 0_M$,
3. $(-c)\alpha = -(c\alpha) = c(-\alpha)$,
4. $(kc)\alpha = k(c\alpha) = c(k\alpha)$.

Proof. Exercise.

The definition of a module includes the trivial module, consisting of just the zero element 0_M. If R is the trivial ring, then any R-module is trivial, since for every $\alpha \in M$, we have $\alpha = 1_R\, \alpha = 0_R\, \alpha = 0_M$.

Example: If E is any ring containing R as a subring (i.e., E is an extension ring of R), then E is a module over R, with addition and scalar multiplication defined in terms of the addition and multiplication operations of E.

Example: The ring of polynomials $R[X]$ over R forms an R-module in the natural way, with addition and scalar multiplication defined in terms of the addition and multiplication operations of the polynomial ring.

Example: If $M_1,..., M_k$ are R-modules, then so is their direct product $M_1 \times ... \times M_k$, where addition and scalar product are defined component-wise. If $M = M_1 = ... = M_k$, we write this as $M^{\times k}$.

Example: Any abelian group G, written additively, can be viewed as a $\mathbb{Z}$-module, with scalar multiplication defined in terms of the usual integer multiplication map.

Example. If I is an arbitrary set, and M is an R-module, then Map(I, M), which is the set of all functions $f: I \rightarrow M$, may be naturally viewed as an R-module, with point-wise addition and scalar multiplication: for $f, g \in$ Map(I, M) and $c \in R$, we define

$$(f+g)(i) := f(i) + g(i) \text{ and } (cf)(i) := cf(i) \text{ for all } i \in I.$$

Example: Let G be any group, written additively, whose exponent divides n. Then we may define a scalar multiplication that maps $[k]_n \in \mathbb{Z}_n$ and $\alpha \in G$ to k. That this map is unambiguously defined follows from the fact that G has exponent dividing n, so that if $k \equiv k'\alpha \pmod{n}$, we have $k\alpha - k'\alpha = (k - k')\alpha = 0_G$, since $n \mid (k - k')$. It is easy to check that this scalar multiplication map indeed makes G into a $\mathbb{Z}_n$-module.

Example: The set $R^{\times n}$, which consists of all of n-tuples of elements of R, forms an R-module, with addition and scalar multiplication defined componentwise: for $\alpha = (a_1,..., a_n) \in R^{\times n}$ $\beta, = (b_1,..., bn) \in R^{\times n}$, and $c \in R$, we define

$$\alpha + \beta := (a_1 + b_1,..., a_n + b_n) \text{ and } c\alpha := (ca_1,..., ca_n).$$

Example: Of course, viewing a group as a module does not depend on whether or not we happen to use additive notation for the group operation. If we specialize the previous example to the group $G = \mathbb{Z}^*_p$, where p is prime, then we may view G as a $\mathbb{Z}_{p-1}$-module. However, since the group operation itself is written multiplicatively, the "scalar product" of $[k]_{p-1} \in \mathbb{Z}_{p-1}$ and $\alpha \in \mathbb{Z}^*_p$ is the power k.

Example: The ring R itself can be viewed as an R-module in the obvious way, with addition and scalar multiplication defined in terms of the addition and multiplication operations of R.

MODULE HOMOMORPHISMS AND ISOMORPHISMS

Again, throughout this section, R is a ring. The notion of a group homomorphism extends in the obvious way to R-modules.

Definition: Let M and M' be modules over R. An R-module homomorphism from M to M' is a function $\rho: M \to M'$, such that

1. ρ is a group homomorphism from M to M', and
2. For all $c \in R$ and $\alpha \in M$, we have $\rho(c\alpha) = c\rho(\alpha)$.

An R-module homomorphism is also called an R-linear map. We shall use this terminology from now on. Expanding the definition, we see that a map $\rho: M \to M'$ is an R-linear map if and only if $\alpha(\alpha + \beta) = \rho(\alpha) + \rho(\alpha)$ and $\rho(c\alpha) = c\rho(\alpha)$ for all $\alpha, \beta \in M$ and all $c \in R$.

Example: Let E be an extension ring of R. As we saw in Example, E may be viewed as an R-module in a natural way. Let $\alpha \in E$, and consider the -multiplication map on E, which sends $\beta \in E$ to $\alpha\beta \in E$. Then it is easy to see that this is an R-linear map.

Example: Let M be an R-module, and let k be an integer. Then the k-multiplication map on M is not only a group homomorphism, but it is also easily seen to be an R-linear map. Its image is the submodule kM, and its kernel the submodule$M\{k\}$.

Example: Let M be an R-module, and let c be an element of R. The map

$$\rho: M \to M$$

$$\alpha \mapsto c\alpha$$

is called c-multiplication map on M, and is easily seen to be an R-linear map whose image is the submodule cM, and whose kernel is the submodule $M\{c\}$. The set of all $c \in R$ for which $cM = \{0_M\}$ is called the R-exponent of M, and is easily seen to be an ideal of R.

Example: 1Let M be an R-module, and let be an element of$M$. The map

$$\rho: R \to M$$

$$c \mapsto c\alpha$$

is easily seen to be an R-linear map whose image is the submodule R (i.e., the submodule generated by α). The kernel of this map is called the R-order of α, and is easily seen to be an ideal of R.

Example: If N is a submodule of an R-module M, then the inclusion map $i: N \to M$ is obviously an R-linear map.

Example: Generalizing the previous example, let M be an R-module, and let $\alpha_1,..., \alpha_k$ be elements of M. The map

$$\rho: R^{\times k} \to M$$
$$(c_1,..., c_k) \mapsto c_1\alpha_1 +... + c_k\alpha_k$$

is easily seen to be an R-linear map whose image is the submodule $R\alpha_1 +... + R\alpha_k$ (i.e., the submodule generated by $\alpha_1,..., \alpha_k$).

Example: Suppose N is a submodule of an R-module$M$. Then the natural map

$$\rho: M \to M/N$$
$$\alpha \mapsto [\alpha]_N$$

is not just a group homomorphism, it is also easily seen to be an R-linear map.

Example: Suppose that $M_1,..., M_k$ are submodules of an R-module M.

Then the map

$$\rho: M_1 \times ... \times M_k \to M$$

$$(\alpha_1,..., \alpha_k) \mapsto \alpha_1 +... + \alpha_k$$

is easily seen to be an R-linear map whose image is the submodule $M_1 +... + M_k$.

Theorem: If ρ is an R-module isomorphism of M with M', then the inverse function ρ^{-1} is an R-module isomorphism of M' with M.

Theorem: Let $\rho: M \to M'$ be an R-linear map.

1. For every submodule N of M, $\rho(N)$ is a submodule of M'; in particular (setting $N:=M$), Im ρ is a submodule of M'.
2. For every submodule N' of M', $\rho^{-1}(N')$ is a submodule of M'; in particular (setting $N' := \{0_{M'}\}$), Ker ρ is a submodule of M.

Internal Direct Product

Theorem: Let M be an R-module with submodules N_1, N_2, where $N_1 \cap N_2 = \{0_M\}$. Then we have an R-module isomorphism

$$N_1 \times N_2 \cong N_1 + N_2$$

given by the map

$$\rho: N_1 \times N_2 \to N_1 + N_2$$
$$(\alpha_1, \alpha_2) \mapsto \alpha_1 + \alpha_2.$$

Theorem: Let $\rho_i: M \to M_i'$, for $i = 1, \ldots, k$, be R-linear maps. Then the map

$$\rho: M \to M_1' \times \ldots \times M_k'$$
$$\alpha \mapsto (\rho_1(\alpha), \ldots, \rho_k(\alpha))$$

is an R-linear map.

If an R-linear map $\rho: M \to M'$ is bijective, then it is called an R-module isomorphism of M with M'. If such an R-module isomorphism ρ exists, we say that M is isomorphic to M', and write $M \cong M'$. Moreover, if $M = M'$, then ρ is called an R-module automorphism on M.

Theorem: If $\rho: M \to M'$ and $\rho': M' \to M''$ are R-linear maps, then so is their composition $\rho' \circ \rho: M \to M''$.

First Isomorphism Theorem

Theorem: Let $\rho: M \to M'$ be an R-linear map with kernel K and image N'. Then we have an R-module isomorphism

$$M/K \cong N'.$$

Specifically, the map

$$\bar{\rho}: M/K \to M'$$
$$[\alpha]_K \mapsto \rho(\alpha)$$

is an injective R-linear map whose image is N'.

Theorem: Let M and M' be R-modules, and consider the R-module of functions Map(M, M'). Then

$$\text{Hom}_R(M, M') := \{\sigma \in \text{Map}(M, M') : \sigma \text{ is an } R\text{-linear mapg}$$

is a submodule of Map(M, M').

Theorem: Let $\rho: M \to M'$ be an R-linear map. Then for every submodule N of M with $N \subseteq \text{Ker } \rho$, we may define an R-linear map

$$\bar{\rho} : M/N \to M'$$

$$[\alpha]_N \mapsto \rho(\alpha).$$

Moreover, $\text{Im } \bar{\rho} = \text{Im } \rho$, and $\bar{\rho}$ is injective if and only if $N = \text{Ker } \rho$.

EXERCISE

1. Let $\rho: M \to M'$ be a surjective R-linear map. Let S be the set of all submodules of M that contain Ker ρ, and let S' be the set of all submodules of M'. Show that the sets S and S' are in one-to-one correspondence, via the map that sends $N \in$ S to $\rho(N) \in$ S'.
2. Verify that the "is isomorphic to" relation on R-modules is an equivalence relation; that is, for all R-modules M_1, M_2, M_3, we have:
   (*a*) $M_1 \cong M_1$;
   (*b*) $M_1 \cong M_2$ implies $M_2 \cong M_1$;
   (*c*) $M_1 \cong M_2$ and $M_2 \cong M_3$ implies $M_1 \cong M_3$.
3. Let $\rho: M \to M'$ be an R-linear map with kernel K. Let N be a submodule of M. Show that we have an R-module isomorphism $M/(N + K) \cong \rho(M)/\rho(N)$.
4. Let $\rho: M \to M'$ be an R-linear map, and let $c \in R$. Show that $\rho(cM) = c\rho(M)$.
5. Show that if $M = M_1 \times M_2$ for R-modules M_1 and M_2, and N_1 is a subgroup of M_1 and N_2 is a subgroup of M_2, then we have an R-module isomorphism$M/(N_1 \times N_2) \cong M_1/N_1 \times M_2/N_2$.
6. Let M be an R-module with submodules N_1 and N_2. Show that we have an R-module isomorphism $(N_1 + N_2)/N_2 \cong N_1/(N_1 \cap N_2)$.
7. Let $\rho: M \to M'$ be an R-linear map. Let N be a submodule of M, and let $\tau: N \to M'$ be the restriction of ρ to N. Show that τ is an R-linear map and that Ker τ = Ker $\rho \cap N$.
8. Suppose $M_1, ..., M_k$ are R-modules. Show that for each $i = 1, ..., k$, the projection map $\pi_i: M_1 ... M_k \to M_i$ that sends $(\alpha_1, ..., \alpha_k)$ to α_i is a surjective R-linear map.
9. Let $\rho_i: M_i \to M_i'$, for $i = 1, ..., k$, be R-linear maps. Show that the map

$$\rho: M_1 \times ... \times M_k \to M_1' \times ... \times M_k'$$

$$(\alpha_1,\ldots,\alpha_k) \mapsto (\rho_1(\alpha_1),\ldots,\rho_k(\rho_k))$$

is an R-linear map.

10. Let M be an R-module with submodules N_1, N_2, and A, where $N_2 \subseteq N_1$. Show that $(N_1 \cap A)/(N_2 \cap A)$ is isomorphic to a submodule of N_1/N_2.

SUBMODULES AND QUOTIENT MODULES

Again, throughout this section, R denotes a ring. The notions of subgroups and quotient groups extend in the obvious way to R-modules.

Definition: Let M be an R-module. A subset N of M is a submodule (over R) of M if

1. N is closed under scalar multiplication; that is, for all $c \in R$ and $\alpha \in N$, we have $c \in N$.
2. N is a subgroup of the additive group M.

It is easy to see that a submodule N of an R-module M is also an R-module in its own right, with addition and scalar multiplication operations inherited from M.

Expanding the above definition, we see that a non-empty subset N of M is a submodule if and only if for all $c \in R$ and all $\alpha, \beta \in N$, we have

$$\alpha + \beta \in N, -\alpha \in N, \text{ and } c\alpha \in N.$$

Observe that the condition $-\alpha \in N$ is redundant, as it is implied by the condition $c \in$

N

with $c = -1_R$.

Clearly, $\{0_M\}$ and M are submodules of M. For $k \in \mathbb{Z}$, it is easy to see that not only are kM and $M\{k\}$ subgroups of M, they are also submodules of M. Moreover, for $c \in R$,

$$cM := \{c\alpha : \alpha \in M\} \text{ and } M\{c\} := \{\alpha \in M : c\alpha = 0_M\}$$

are also submodules of M. Further, for $\alpha \in M$,

$$R\alpha := \{c\alpha : c \in R\}$$

is a submodule of M. Finally, if N_1 and N_2 are submodules of M, then $N_1 + N_2$ and $N_1 \cap N_2$ are not only subgroups of M, they are also submodules of M. We leave it to the reader to verify all these facts: they are quite straightforward.

Let $\alpha_1,\ldots,\alpha_k \in M$. The submodule

$$R\alpha_1 + \ldots + R\alpha_k$$

is called the submodule (over R) generated by $\alpha_1,\ldots,\alpha_k$. It consists of all R-linear combinations

$$c_1\alpha_1 + \ldots + c_k\alpha_k,$$

where the c_i's are elements of R, and is the smallest submodule of M that contains the elements $\alpha_1,..., \alpha_k$. We shall also write this submodule as $\langle \alpha_1,...,\alpha_k \rangle_R$. As a matter of definition, we allow $k = 0$, in which case this submodule is $\{0_M\}$. We say that M is finitely generated (over R) if $M = \langle \alpha_1,...,\alpha_k \rangle_R$ for some $\alpha_1,..., \alpha_k \in M$.

Example: Let G be an abelian group. We can view G as a $\mathbb{Z}$-module in a natural way. Subgroups of G are just the same thing as submodules of G, and for $a_1,..., a_k \in G$, the subgroup $\langle a_1,...,a_k \rangle$ is the same as the submodule $\langle a_1,...,a_k \rangle_{\mathbb{Z}}$.

Example: Suppose E is an extension ring of R, and I is an ideal of E. Viewing E as an R-module, I is a submodule of E, and hence the quotient ring E/I may naturally be viewed as an R-module, with scalar multiplication defined by $c.\ [\alpha]_I := [c]_I$ for $c \in R$ and $\alpha \in E$. Example is a special case of this, applied to the extension ring $R[X]$ and the ideal (f).

Example: For a given integer $\ell \geq 0$, define $R[X]_{<\ell}$ to be the set of polynomials of degree less than ℓ. The reader may verify that $R[X]_{<\ell}$ is a submodule of the R-module $R[X]$, and indeed, is the submodule generated by $1, X,..., X^{\ell-1}$. If $\ell = 0$, then this submodule is the trivial submodule $\{0_R\}$.

EXERCISE

1 Show that if R is non-trivial, then $R[X]$ is not finitely generated.

2. Let M_1 and M_2 be R-modules, and let N_1 be a submodule of M_1 and N_2 a submodule of M_2. Show that $N_1 \times N_2$ is a submodule of $M_1 \times M_2$.

VECTOR SPACES AND DIMENSION

Throughout this section, F denotes a field.

A module over a field is also called a vector space. In particular, an F-module is called an F-vector space, or a vector space over F.

For vector spaces over F, one typically uses the terms subspace and quotient space, instead of (respectively) submodule and quotient module; likewise, one usually uses the terms F-vector space homomorphism, isomorphism and automorphism, as appropriate.

We now develop the basic theory of dimension for finitely generated vector spaces. Recall that a vector space V over F is finitely generated if

$$V = \langle \alpha_1,...,\alpha_n \rangle_F$$

for some $\alpha_1,..., \alpha_n$ of V. The main results here are that

- Every finitely generated vector space has a basis,

- And all bases have the same number of elements.

Throughout the rest of this section, V denotes a vector space over F. We begin with a technical fact that will be used several times throughout this section:

Theorem: If V is spanned by some family of n elements of V, then every family of $n + 1$ elements of V is linearly dependent.

Proof. We prove this by induction on n. If $n = 0$, the theorem is clear, so assume that $n > 0$. Let $\{\alpha_i\}_{i=1}^{n}$ be a family that spans V, and let $\{\beta_i\}_{i=1}^{n+1}$ be an arbitrary family of elements of V. We wish to show that $\{\beta_i\}_{i=1}^{n+1}$ is linearly dependent.

We know that β_{n+1} is a linear combination of the β_i's, say,

$$\beta_{n+1} = c_1\beta_1 + ... + c_n\beta_n.$$

If all the c_i's were zero, then we would have $\beta_{n+1} = 0_V$, and so trivially, $\{\beta_i\}_{i=1}^{n+1}$ is linearly dependent. So assume that some c_i is non-zero, and for concreteness, say $c_n \neq 0$. Dividing equation through by c_n, it follows that α_n is an F-linear combination of $\alpha_1, ..., \alpha_{n-1}, \beta_{n+1}$. Therefore,

$$\langle \alpha_1, ..., \alpha_{n-1}, \beta_{n+1} \rangle_F \supseteq \langle \alpha_1, ..., \alpha_{n-1} \rangle_F + \langle \alpha_n \rangle_F = V.$$

Now consider the subspace $W := \langle \beta_{n+1} \rangle_F$ and the quotient space V/W. Since the family of elements $\alpha_1, ..., \alpha_{n-1}, \beta_{n+1}$ spans V, it is easy to see that $\{[\alpha_i]_W\}_{i=1}^{n-1}$ spans V/W; therefore, by induction, $\{[\beta_i]_W\}_{i=1}^{n}$ is linearly dependent. This means that there exist $d_1, ..., dn \in F$, not all zero, such that $d_1\beta_1 + ... + d_n\beta_n \equiv 0 \pmod{W}$, which means that for some $d_{n+1} \in F$, we have $d_1\beta_1 + ... + d_n\beta_n = d_{n+1}\,\beta_{n+1}$. That proves that $\{\beta_i\}_{i=1}^{n+1}$ is linearly dependent.

Theorem: Suppose $\{\alpha_i\}_{i=1}^{n}$ is a family of elements that spans V. Then for some subset $J \subseteq \{1, ..., n\}$, the subfamily $\{\alpha_j\}_{j \in J}$ is a basis for V.

Proof. We prove this by induction on n. If $n = 0$, the theorem is clear, so assume $n > 0$. Consider the subspace W of V spanned by $\{\alpha_i\}_{i=1}^{n-1}$. By the induction hypothesis, for some $K \subseteq \{1, ..., n - 1\}$, the subfamily $\{\alpha_k\}_{k \in K}$ is a basis for W. There are two cases to consider.

Case 1: $\alpha_n \in W$. In this case, $W = V$, and the theorem clearly holds with $J := K$.

Case 2: $\alpha_n \notin W$. We claim that setting $J := K \cup \{n\}$, the subfamily $\{\alpha_j\}_{j \in J}$ is a basis for V. Indeed, since $\{\alpha_k\}_{k \in K}$ is linearly independent, and $\alpha_n \notin W$, Theorem immediately implies that $\{\alpha_j\}_{j\in J}$ is linearly independent. Also, since $\{\alpha_k\}_{k\in K}$ spans W, it is clear that $\{\alpha_j\}_{j\in J}$ spans $W + \langle\alpha_n\rangle_F = V$.

Theorem: If V is finitely generated, then any two bases have the same size.

Theorem: Suppose that $\{\alpha_i\}_{i=1}^{n}$ is a linearly independent family of elements that spans a subspace $W \subseteq V$, and that $\alpha_{n+1} \in V \setminus W$. Then $\{\alpha_i\}_{i=1}^{n+1}$ is also linearly independent.

Proof. Suppose we have a linear relation

$$0_V = c_1\alpha_1 + ... + c_n\alpha_n + c_{n+1}\,\alpha_{n+1},$$

where the c_i's are in F. We want to show that all the c_i's are zero. If $c_{n+1} \neq 0$, then we have

$$\alpha_{n+1} = -c_{n+1}^{-1}\,(c_1\alpha_1 + ... + c_n\alpha_n) \in W,$$

contradicting the assumption that $\alpha_{n+1} \notin W$. Therefore, we must have $c_{n+1} = 0$, and the linear independence of $\{\alpha_i\}_{i=1}^{n+1}$ implies that $c_1 = ... = c_n = 0$.

The next theorem says that every finitely generated vector space has a basis, and in fact, any family that spans a vector space contains a subfamily that is a basis for the vector space.

Definition: If V is finitely generated, the common size of any basis is called the dimension of V, and is denoted $\dim_F(V)$.

Note that from the definitions, we have $\dim_F(V) = 0$ if and only if V is the trivial vector space (i.e., $V = \{0_V\}$). We also note that one often refers to a finitely generated vector space as a finite dimensional vector space. We shall give preference to this terminology from now on.

To summarize the main results in this section up to this point: if V is finite dimensional, it has a basis, and any two bases have the same size, which is called the dimension of V.

Theorem: Suppose that V is finite dimensional and W is a subspace of V. Then W is also finite dimensional, with $\dim_F(W) \leq \dim_F(V)$. Moreover, $\dim_F(W) = \dim_F(V)$ if and only if $W = V$.

Proof. Suppose $\dim_F(V) = n$. Consider the set S of all linearly independent families of the form $\{\alpha_i\}_{i=1}^{m}$, where $m \geq 0$ and each α_i is in W. The set S is certainly non-empty, as it contains the empty family. Moreover, by Theorem, every member of S must have at most n elements. Therefore, we may choose some particular element $\{\alpha_i\}_{i=1}^m$ of S, where m is as large as possible. We claim that this family $\{\alpha_i\}_{i=1}^m$ is a basis for W. By definition, $\{\alpha_i\}_{i=1}^m$ is linearly independent and spans a subspace of W. If this is a proper subspace $W' \subseteq W$, we can choose an element $\alpha_{m+1} \in W \setminus W'$, and by Theorem, the family $\{\alpha_i\}_{i=1}^{m+1}$ is linearly independent, and therefore, this family also belongs to S, contradicting the assumption that m is as large as possible.

That proves that W is finite dimensional with $\dim_F(W) \leq \dim_F(V)$. It remains to show that these dimensions are equal if and only if $W = V$. Now, if $W = V$, then clearly $\dim_F(W) = \dim_F(V)$. Conversely, if $\dim_F(W) = \dim_F(V)$, then by Theorem, any basis for W must already span V.

Theorem: If V is finite dimensional, then every linearly independent family of elements of V can be extended to form a basis for V.

Proof. One can prove this by generalizing the proof of Theorem. Alternatively, we can adapt the proof of the previous theorem. Let $\{\beta_i\}_{j=1}^m$ be a linearly independent family of elements that spans a subspace W of V. As in the proof of the previous theorem, if $\{[\alpha_i]_W\}_{i=1}^{\ell}$ is a basis for the quotient space V/W, then the elements

$$\alpha_1, \ldots, \alpha_\ell, \beta_1, \ldots, \beta_m$$

form a basis for V.

Example: Suppose that F is finite, say $|F| = q$, and that V is finite dimensional, say $\dim_F(V) = n$. Then clearly $|V| = q^n$. If W is a subspace with $\dim_F(W) = m$, then $|W| = q^m$, and by Theorem, $\dim_F(V/W) = n - m$, and hence $|V/W| = q^{n-m}$. Just viewing V and W as additive groups, we know that the index of W in V is $[V: W] = |V/W| = |V|/|W| = q^{n-m}$, which agrees with the above calculations.

We next consider the relation between the notion of dimension and linear maps. First, observe that by Theorem, if two finite dimensional vector spaces have the same dimension, then they are isomorphic.

Theorem: If V is finite dimensional, and W is a subspace of V, then the quotient space V/W is also finite dimensional, and

$$\dim_F(V/W) = \dim_F(V) - \dim_F(W).$$

Proof. Suppose that $\{\alpha_i\}_{i=1}^n$ spans V. Then it is clear that $\{[\alpha_i]_W\}_{i=1}^n$ spans V/W. By Theorem, we know that V/W has a basis of the form $\{[\alpha_i]_W\}_{i=1}^{\ell}$, where $\ell \leq n$

(renumbering the α_i's as necessary). By Theorem, we know that W has a basis, say $\{\beta_i\}_{j=1}^{m}$. The theorem will follow immediately from the following:

Claim. The elements

$$\alpha_1,..., \alpha_\ell, \beta_1,..., \beta_m$$

form a basis for V.

To see that this family spans V, consider any element γ of V. Then since $\{[\alpha_i]_W\}_{i=1}^{\ell}$ spans V/W, we have $\gamma \equiv \sum_i c_i\alpha_i$ (mod W) for some $c_1,..., c_\ell \in F$. If we set $\beta := \gamma - \sum_i c_i\alpha_i \in W$, then since $\{\beta_j\}_{j=1}^{m}$ spans W, we have $\beta = \sum_j d_j\beta_j$ for some $d_1,..., d_m \in F$, and hence $\gamma = \sum_i c_i\alpha_i + \sum_j d_j\beta_j$. That proves that the family of elements spans V. To prove this family is linearly independent, suppose we have a relation of the form $\sum_i c_i\alpha_i + \sum_j d_j\beta_j = 0_V$, where $c_1,..., c_\ell \in F$ and $d_1,..., d_m \in F$. If any of the c_i's were non-zero, this would contradict the assumption that $\{[\alpha_i]_W\}_{i=1}^{\ell}$ is linearly independent. So assume that all the c_i's are zero. If any of the d_j's were non-zero, this would contradict the assumption that $\{\beta_i\}_{j=1}^{m}$ is linearly independent. Thus, all the c_i's and d_j's must be zero, which proves that the family of elements is linearly independent. That proves the claim.

Theorem: Suppose that $\dim_F(V) = n$, and that $\{\alpha_i\}_{i=1}^{n}$ is a family of n elements of V. The following are equivalent:

1. $\{\alpha_i\}_{i=1}^{n}$ is linearly independent;
2. $\{\alpha_i\}_{i=1}^{n}$ spans V;
3. $\{\alpha_i\}_{i=1}^{n}$ is a basis for V.

Proof. Let W be the subspace of V spanned by $\{\alpha_i\}_{i=1}^{n}$.

First, let us show that (i) implies (ii). Suppose $\{\alpha_i\}_{i=1}^{n}$ is linearly independent. Also, by way of contradiction, suppose that $W \subseteq V$, and choose $\alpha_{n+1} \in V \setminus W$. Then Theorem implies that $\{\alpha_i\}_{i=1}^{n+1}$ is linearly independent. But then we have a linearly independent family of $n + 1$ elements of V, which is impossible by Theorem.

Second, let us prove that (ii) implies (i). Let us assume that $\{\alpha_i\}_{i=1}^{n}$ is linearly dependent, and prove that $W \subseteq V$. By Theorem, we can find a basis for W among the α_i's, and since

$\{\alpha_i\}_{i=1}^n$ is linearly dependent, this basis must contain strictly fewer than n elements. Hence, $\dim_F(W) < \dim_F(V)$, and therefore, $W \subseteq V$.

The theorem now follows from the above arguments, and the fact that, by definition, (iii) holds if and only if both (i) and (ii) hold.

We next examine the dimension of subspaces of finite dimensional vector spaces.

Theorem: If V is of finite dimension n, and V is isomorphic to V', then V' is also of finite dimension n.

Proof. If $\{\alpha_i\}_{i=1}^n$ is a basis for V, then by Theorem, $\{\rho(\alpha_i)\}_{i=1}^n$ is a basis for V'.

Thus, two finite dimensional vector spaces are isomorphic if and only if they have the same dimension.

We next illustrate one way in which the notion of dimension is particularly useful. In general, if we have a function $f: A \to B$, injectivity does not imply surjectivity, nor does surjectivity imply injectivity. If A and B are finite sets of equal size, then these implications do indeed hold. The following theorem gives us another important setting where these implications hold, with finite dimensionality playing the role corresponding to finite cardinality.

Theorem: If V is finite dimensional, and $\rho: V \to V'$ is an F-linear map, then Im ρ is a finite dimensional vector space, and

$$\dim_F(V) = \dim_F(\text{Im } \rho) + \dim_F(\text{Ker } \rho).$$

Intuitively, one way to think of Theorem is as a "law of conservation" for dimension: any "dimensionality" going into ρ that is not "lost" to the kernel of ρ must show up in the image of ρ.

Theorem: If ρ: V $\to$ V' is an F-linear map, and if V and V' are finite dimensional with $\dim_F(V) = \dim_F(V')$, then we have:

ρ is injective if and only if ρ is surjective.

Proof. Let $\{\alpha_i\}_{i=1}^n$ be a basis for V. Then

$$\rho \text{ is injective} \Leftrightarrow \{\rho(\alpha_i\}_{i=1}^n \text{ is linearly independent}$$

$$\Leftrightarrow \{\rho(\alpha_i\}_{i=1}^n \text{ spans } V'$$

$$\Leftrightarrow \rho \text{ is surjective}$$

This theorem may be generalized as follows:

EXERCISE

1. Show that if V is a finite dimensional vector space over F with subspaces W_1 and W_2, then
$\dim_F(W_1 + W_2) = \dim_F(W_1) + \dim_F(W_2) - \dim_F(W_1 \cap W_2)$.
2. Show that if $V_1, \ldots, V_n$ are finite dimensional vector spaces over F, then $V_1 \times \ldots \times V_n$ has dimension $\sum_{i=1}^{n} \dim_F(V_i)$.
3. From the previous exercise, one might be tempted to think that a more general "inclusion/exclusion principle" for dimension holds. Determine if the following statement is true or false: if V is a finite dimensional vector space over F with subspaces W_1, W_2, and W_3, then

$$\begin{aligned}\dim_F(W_1 + W_2 + W_3) &= \dim_F(W_1) + \dim_F(W_2) + \dim_F(W_3)\\ &- \dim_F(W_1 \cap W_2) - \dim_F(W_1 \cap W_3) - \dim_F(W_2 \cap W_3)\\ &+ \dim_F(W_1 \cap W_2 \cap W_3).\end{aligned}$$

4. Let V be a vector space over F with basis $\{\alpha_i\}_{i=1}^{n}$. Let S be a finite, non-empty subset of F, and define

$$B := \left\{\sum_{i=1}^{n} c_i\alpha_i : c_1, \ldots, c_n \in S\right\}.$$

Show that if W is a subspace of V, with $W \subseteq V$, then $|B \cap W| \leq |S|^{n-1}$.

5. Suppose that V and W are vector spaces over F, V is finite dimensional, and $\{\alpha_i\}_{i=1}^{k}$ is a linearly independent family of elements of V. In addition, let $\beta_1, \ldots, \beta_k$ be arbitrary elements of W. Show that there exists an F-linear map $\rho: V \to W$ such that $\rho(\alpha_i) = \beta_i$ for $i = 1, \ldots, k$.
5. The theory of dimension for finitely generated vector spaces is quite elegant and powerful. There is a theory of dimension (of sorts) for modules over an arbitrary, non-trivial ring R, but it is much more awkward and limited. This exercise develops a proof of one aspect of this theory: if an R-module M has a basis at all, then any two bases have the same size. To prove this, we need the fact that any non-trivial ring has a maximal ideal. Let n, m be positive integers, let $\alpha_1, \ldots, \alpha_m$ be elements of $R^{\times n}$, and let I be an ideal of R.

   (*a*) Show that if $\{\alpha_i\}_{i=1}^{m}$ spans $R^{\times n}$, then every element of $I^{\times n}$ can be expressed as $c_1\alpha_1 + \ldots + c_m\alpha_m$, where $c_1, \ldots, c_m$ belong to I.

(*b*) Show that if $m > n$ and I is a maximal ideal, then there exist $c_1, ..., c_m$ α R, not all in I, such that $c_1\alpha_1 + ... + c_m\alpha_m \in I^{\times n}$.

(c) From (a) and (b), deduce that if $m > n$, then $\{\alpha_i\}_{i=1}^m$ cannot be a basis for $R^{\times n}$.

(*d*) From (c), conclude that any two bases for a given R-module M must have the same size.

LINEAR INDEPENDENCE AND BASES

Throughout this section, R denotes a ring.

Definition: Let M be an R-module, and let $\{\alpha_i\}_{i=1}^n$ be a family of elements of M. We say that $\{\alpha_i\}_{i=1}^n$

1. Is linearly dependent (over R) if there exist $c_1, ..., c_n \in R$, not all zero, such that $c_1\alpha_1 + ... + c_n\alpha_n = 0_M$;
2. Is linearly independent (over R) if it is not linearly dependent;
3. Spans M (over R) if for every $\alpha \in M$, there exist $c_1, ..., c_n \in R$ such that $c_1\alpha_1 + ... + c_n\alpha_n = \alpha$;
4. Is a basis for M (over R) if it is linearly independent and spans M.

The family $\{\alpha_i\}_{i=1}^n$ always spans some submodule ofM, namely, the submodule N generated by $\alpha_1, ..., \alpha_n$. In this case, we may also call N the submodule (over R) spanned by $\{\alpha_i\}_{i=1}^n$.

The family $\{\alpha_i\}_{i=1}^n$ may contain duplicates, in which case it is linearly dependent (unless R is trivial). Indeed, if, say, $\alpha_1 = \alpha_2$, then setting $c_1 := 1$, $c_2 := -1$, and $c_3 := ... := c_n := 0$, we have the linear relation $\sum_{i=1}^n c_i\alpha_i = 0_M$.

If the family $\{\alpha_i\}_{i=1}^n$ contains 0_M, then it is also linear dependent (unless R is trivial). Indeed, if, say, $\alpha_1 = 0_M$, then setting $c_1 := 1$ and $c_2 := ... := c_n := 0$, we have the linear relation $\sum_{i=1}^n c_i\alpha_i = 0_M$.

The family $\{\alpha_i\}_{i=1}^n$ may also be empty (i.e., $n = 0$), in which case it is linearly independent, and spans the submodule $\{0_M\}$.

In the above definition, the ordering of the elements $\alpha_1, ..., \alpha_n$ makes no difference. As such, when convenient, we may apply the terminology in the definition to any family $\{\alpha_i\}_{i \in I}$, where I is an arbitrary, finite index set.

Example: Consider the R-module $R^{\times n}$. Define $\alpha_1,\ldots, \alpha_n \in R^{\times n}$ as follows:

$$\alpha_1 := (1, 0,\ldots, 0), \alpha_2 := (0, 1, 0,\ldots, 0),\ldots, \alpha_n := (0,\ldots, 0, 1);$$

that is, α_i has a 1 in position i and is zero everywhere else. It is easy to see that $\{\alpha_i\}_{i=1}^n$ is a basis for $R^{\times n}$. Indeed, for all $c_1,\ldots, cn \in R$, we have

$$c_1\alpha_1 + \ldots + c_n\alpha_n = (c_1,\ldots, c_n),$$

from which it is clear that $\{\alpha_i\}_{i=1}^n$ spans $R^{\times n}$ and is linearly independent. The family $\{\alpha_i\}_{i=1}^n$ is called the standard basis for $R^{\times n}$.

Example: Consider the $\mathbb{Z}$-module $\mathbb{Z}^{\times 3}$. In addition to the standard basis, which consists of the tuples

$$(1, 0, 0), (0, 1, 0), (0, 0, 1),$$

the tuples

$$\alpha_1 := (1, 1, 1), \alpha_2 := (0, 1, 0), \alpha_3 := (2, 0, 1)$$

also form a basis. To see this, first observe that for all $c_1, c_2, c_3, d_1, d_2, d_3 \in \mathbb{Z}$, we have

$$(d_1, d_2, d_3) = c_1\alpha_1 + c_2\alpha_2 + c_3\alpha_3$$

if and only if

$$d_1 = c_1 + 2c_3,\ d_2 = c_1 + c_2, \text{ and } d_3 = c_1 + c_3.$$

If holds with $d_1 = d_2 = d_3 = 0$, then subtracting the equation $c_1 + c_3 = 0$ from $c_1 + 2c_3 = 0$, we see that $c_3 = 0$, from which it easily follows that $c_1 = c_2 = 0$. This shows that the family $\{\alpha_i\}_{i=1}^3$ is linearly independent. To show that it spans $\mathbb{Z}^{\times 3}$, the reader may verify that for any given $d_1, d_2, d_3 \in \mathbb{Z}$, the values

$$c_1 := -d_1 + 2d_3,\ c_2 := d_1 + d_2 - 2d_3,\ c_3 := d_1 - d_3$$

The family of tuples

$$(1, 1, 1), (0, 1, 0), (1, 0, 1)$$

is not a basis, as it is linearly dependent: the third tuple is equal to the first minus the second.

The family of tuples (1, 0, 12), (0, 1, 30), (0, 0, 18) is linearly independent, but does not span $\mathbb{Z}^{\times 3}$: the last component of any $\mathbb{Z}$-linear combination of these tuples must be divisible by gcd(12, 30, 18) = 6. However, this family of tuples is a basis for the $\mathbb{Q}$-module $\mathbb{Q}^{\times 3}$.

Example: Consider again the submodule $R[X]_{<\ell}$ of $R[X]$, where $\ell \geq 0$, consisting of all polynomials of degree less than ℓ. Then $\{X^{i-1}\}_{i=1}^{\ell}$ is a basis for $R[X]_{<\ell}$ over R.

Example: Consider again the ring $E = R[X]/(f)$, where $f \in R[X]$ with $\deg(f) = \ell \geq 0$ and $\mathrm{lc}(f) \in R^*$. As in Example, we may naturally view E as a module over R. From the observations in Example, it is clear that $\{\xi^{i-1}\}_{i=1}^{\ell}$ is a basis for E over R, where $\xi := [X]_f \in E$.

The next theorem highlights a critical property of bases:

Theorem: If $\{\alpha_i\}_{i=1}^{n}$ is a basis for an R-module M, then the map

$$\varepsilon: R^{\times n} \to M$$
$$(c_1, \ldots, c_n) \mapsto c_1\alpha_1 + \ldots + c_n\alpha_n$$

is an R-module isomorphism. In particular, every element of M can be expressed in a unique way as $c_1\alpha_1 + \ldots + c_n\alpha_n$, for $c_1, \ldots, c_n \in R$.

Proof. We already saw that ε is an R-linear map in Example. Since $\{\alpha_i\}_{i=1}^{n}$ is linearly independent, it follows that the kernel of ε is trivial, so that ε is injective. That ε is surjective follows immediately from the fact that $\{\alpha_i\}_{i=1}^{n}$ spans M.

The following is an immediate corollary of this theorem:

Theorem: Any two R-modules that have bases of the same size are isomorphic.

The following theorem develops an important connection between bases and linear maps.

Theorem: Let $\{\alpha_i\}_{i=1}^{n}$ be a basis for an R-module M, and let $\rho: M \to M'$ be an R-linear map.

1. ρ is surjective if and only if $\{\rho(\alpha_i)\}_{i=1}^{n}$ spans M'.
2. ρ is injective if and only if $\{\rho(\alpha_i)\}_{i=1}^{n}$ is linearly independent.
3. ρ is an isomorphism if and only if $\{\rho(\alpha_i)\}_{i=1}^{n}$ is a basis for M'.

Proof.

1. By the previous theorem, we know that every element of M can be written uniquely as $\sum_i c_i\alpha_i$, where the c_i's are in R. Therefore, every element in Im ρ can be expressed as $\rho\left(\sum_i c_i\alpha_i\right) = \sum_i c_i\rho(\alpha_i)$. It follows that Im ρ is equal to the subspace of M' spanned by $\{\rho(\alpha_i)\}_{i=1}^{n}$. From this, (i) is clear.

2. Consider a non-zero element $\sum_i c_i\alpha_i$ of M, so that not all c_i's are zero. Now, $\sum_i c_i\alpha_i \in \text{Ker}\,\rho$ if and only if $\sum_i c_i\rho(\alpha_i) = 0_{M'}$, and thus, Ker ρ is non-trivial if and only if $\{\rho(\alpha_i)\}_{i=1}^n$ is linearly dependent. That proves (ii).
3. Follows from (i) and (ii). 2

EXERCISE

1. Let M be an R-module. Suppose $\{\alpha_i\}_{i=1}^n$ is a linearly independent family of elements of M. Show that for every $J \subseteq \{1,..., n\}$, the subfamily $\{\alpha_j\}_{j\in J}$ is also linearly independent.
2. Suppose $\rho: M \to M'$ is an R-linear map. Show that if $\{\alpha_i\}_{i=1}^n$ is a linearly dependent family of elements of M, then $\{\rho(\alpha_i)\}_{i=1}^n$ is also linearly dependent.
3. Suppose $\rho: M \to M'$ is an injective R-linear map and that $\{\alpha_i\}_{i=1}^n$ is a linearly independent family of elements of M. Show that $\{\rho(\alpha_i)\}_{i=1}^n$ is linearly independent.
4. Suppose that $\{\alpha_i\}_{i=1}^n$ spans an R-module M and that $\rho: M \to M'$ is an R-linear map. Show that:

   (*a*) ρ is surjective if and only if $\{\rho(\alpha_i)\}_{i=1}^n$ spans M';

   (*b*) If $\{\rho(\alpha_i)\}_{i=1}^n$ is linearly independent, then ρ is injective.

Chapter 5

Vector Space

MOTIVATION

Let us consider three examples.

First Example The general solution of a system of linear equations

$$\begin{aligned} x_1 + x_2 - 5x_4 &= 3 \\ x_3 + 2x_4 &= 4 \end{aligned}$$

is usually written as expressions of nonfree variables (say x_1, x_3) in terms of free variables (say x_2, x_4). The solution can also be written in the vector form

$$x = \begin{bmatrix} x_1 \\ x_2 \\ x_3 \\ x_4 \end{bmatrix} = \begin{bmatrix} 3 - x_2 + 5x_4 \\ x_2 \\ 4 - 2x_4 \\ x_4 \end{bmatrix} = \begin{bmatrix} 3 \\ 0 \\ 4 \\ 0 \end{bmatrix} + x_2 \begin{bmatrix} -1 \\ 1 \\ 0 \\ 0 \end{bmatrix} + x_4 = u + c_1 v_1 + c_2 v_2,$$

where u, v_1, v_2 are fixed vectors and c_1, c_2 are arbitrary constants.

Second Example Consider a differential equation

$f'' - 4f' + 3f = 3t - 1$, for differentiable functions $f = f(t)$.

The general solution is

$$f = (t + 1) + c_1 e^t + c_2 e^{3t} = u(t) + c_1 v_1(t) + c_2 v_2(t),$$

where $u(t)$, $v_1(t)$, $v_2(t)$ are fixed functions, and c_1, c_2 are arbitrary constants.

Third Example Consider a matrix equation

$$\begin{bmatrix} 1 & -2 & 3 \\ -2 & 3 & -4 \end{bmatrix} X = I, \text{ for 2 by 3 matrices } X.$$

The general solution is

$$X = \begin{bmatrix} 1 & -2 \\ 0 & 0 \\ 0 & 0 \end{bmatrix} + c_1 \begin{bmatrix} 1 & 0 \\ 2 & 0 \\ 1 & 0 \end{bmatrix} + c_2 \begin{bmatrix} 0 & 1 \\ 0 & 2 \\ 0 & 1 \end{bmatrix} = U + c_1 V_1 + c_2 V_2,$$

where U, V_1, V_2 are fixed vectors and c_1, c_2 are arbitrary constants.

Although the three problems are about different kinds of objects (euclidean vectors, differentiable functions, matrices), the solutions have similar structures. This suggests something common among these objects, which we will see is the linearity of the left side of the equations.

The purpose of this chapter is to explore the similarity illustrated by the three examples. The similarity enables us to extend the earlier discussion about euclidean vectors (such as systems of linear equations, linear transformations, existence, uniqueness, etc.) to much wider context.

HOW TO MAKE THE DEFINITION

Mathematical notions are often defined by detailed descriptions, which usually depends on the *shape* or *appearance*. For example, a polynomial is an expression of the form $p(t) = a_0 + a_1 t + a_2 t^2 + ... + a_n t^n$, a matrix is a rectangular array of numbers, the sine function is the ratio of two edges in certain triangle, etc.

We would like to define a concept leading to the phenomenon shared by the three examples. The problem is that the phenomenon occurs to objects that may have very different appearance. Therefore the concept cannot be defined by detailed description of the appearance.

To illustrate the difficulty, think about how you would define the concept of numbers. For example, you cannot define natural numbers simply as the collection N = {1, 2, 3,...} because that would exclude, by the appearance, the words "one", "two", "three" as numbers (even if you also include these, what about "un/une", "deux", "trois",.., or Chinese, Japanese, binary expression, etc). A more sensible (but not yet mathematical) definition of natural numbers would be any symbolic system used for everyday counting. Note that this is a definition in terms of the *property* (counting) instead of the appearance.

The natural numbers were found to be inadequate long time ago, so that integers Z, rational numbers Q, real numbers R, complex numbers C, etc., were invented. With the introduction of more sophisticated number systems, the question of the meaning of numbers arises again.

A naive definition of numbers may simply list of all the number systems invented so far. This approach has two problems. However, the list will be changing as new number systems are discovered, so that the definition is not *absolute* or *permanent*, as is required by mathematics. The right definition, then, still lies in the properties that characterize numbers.

What properties do we expect the numbers to have? Naturally, numbers can be added and multiplied. Moreover, these operations must be reasonable, in the sense that they must have some nice properties such as $a + b = b + a$. In summary, the definition of numbers may go like the following three steps.

Terminology We have a set (or collection) N, called *number system*. We also call the elements of N *numbers*.

Operation Given any two numbers a, b, we can produce numbers $a + b$, ab, called addition and multiplication of the two numbers.

Usually we denote the operations by

$a + b$: $N \times N \to N$, ab: $NXN \to N$,

where NXN means that the operations start with pairs of numbers, and $\to N$ means that the results of operations are single numbers.

Properties The two operations has the following properties.

- $a + b = b + a$
- $(a + b) + c = a + (b + c)$
- $(a + b)c = ac + bc$
- ...
- ...

The rigorous definition and the theory of numbers can be found here.

Definition: The definition of vector spaces is based on the addition and scalar multiplications, which were used in the expressions $u + c_1v_1 + c_2v_2$, $u(t) + c_1v_1(t) + c_2v_2(t)$, $U + c_1V_1 + c_2V_2$.

A vector space is a set V, its elements u called vectors, with two operations

addition: $u + v$: $VXV \to V$

scalar multiplication: cu: $RXV \to V$

satisfying the following properties

commutative

$$u + v = v + u$$

associative

$$(u + v) + w = u + (v + w)$$

zero

There is a special vector $0 \in V$ such that $u + 0 = u = 0 + u$

negative

For any $u \in V$ there is a vector $-u$ such that $u + (-u) = 0$

distributive

$$c(u + v) = cu + cv$$

distributive

$$(c + d)u = cu + du$$

associative

$$(cd)u = c(du)$$

one

$$1u = u$$

The eight properties comes from the properties satisfied by the euclidean vectors. Moreover, the notation $R \times V \to V$ means that the scalar multiplication takes a (number, vector)–pair and produces a vector.

Briefly speaking, a vector space is any set up in which addition and scalar multiplication can be done in a reasonable way (as measured by the eight properties).

Example: The euclidean space R^n, with the following operations

$$(x_1, x_2,..., x_n) + (y_1, y_2,..., y_n) = (x_1 + y_1, x_2 + y_2,..., x_n + y_n),$$

$$c(x_1, x_2,..., x_n) = (cx_1, cx_2,..., cx_n),$$

satisfies the eight properties and is the primary example of vector spaces.

Example: Let $F(X)$ be all the real valued functions on a set X. For example, $F(R)$ is all the real functions defined on the real line, and $F[0,1]$ is all the real functions defined on the interval $[0,1]$. With the usual operations

$$(f + g)(t) = f(t) + g(t),$$

$$(cf)(t) = c(f(t)),$$

$F(X)$ is a vector space. In fact, two functions can also be multiplied together. However, this operation is not needed for the vector space.

By imposing additional conditions on functions, we get

$C(R)$ = all continuous functions,

$C^1(R)$ = all functions $f(t)$ with continuous derivative $f'(t)$,

$C^k(R)$ = all functions with continuous k–th order derivative.

We also use $C^0(R)$ to denote $C(R)$ and use $C^\infty(R)$ to denote functions with derivatives of any order. With the same addition and scalar multiplication, these are also vector spaces. The important thing to verify here is that if $f(t)$ and $g(t)$ satisfy the additional conditions (say, continuous), then $f(t) + g(t)$, $cf(t)$ also satisfy the additional conditions (also continuous). The verification follows from the usual theorems in calculus.

Example: Another special collection of functions is polynomials. For any n, let P_n be the collection of polynomials

$$p(t) = a_0 + a_1t + a_2t^2 + ... + a_nt^n,$$

of degree d” n. The usual addition and scalar multiplication

$$(a_0 + a_1t + a_2t^2 + ... + a_nt^n) + (b_0 + b_1t + b_2t^2 + ... + b_nt^n) = (a_0 + b_0) + (a_1 + b_1)t + (a_2 + b_2)t^2 + ... + (a_n + b_n)t^n,$$

$$c(a_0 + a_1t + a_2t^2 + ... + a_nt^n) = ca_0 + ca_1t + ca_2t^2 + ... + ca_nt^n,$$

are consistent with the similar operations for general functions, and have the eight properties. Therefore P_n is a vector space.

Example: Given m and n, all the m by n matrices, together with the usual addition and scalar multiplication, form a vector space $M(m, n)$. The special collection $Sym(n)$ of symmetric matrices, with the same operations, is also a vector space. Similar to the case of special collections of functions, the key reason for $Sym(n)$ to be a vector space is A and B symmetric $\Rightarrow A + B$ and cA are symmetric.

Note that the additions and scalar multiplications in the examples have to be the routine one because otherwise some of the eight properties will fail.

Finally, in a vector space we may combine addition and scalar multiplication and form the linear combination $c_1u_1 + c_2u_2 + ... + c_ku_k$.

PROOFS IN VECTOR SPACE

Logically, the further development of the theory of vector spaces will depend entirely on the eight properties. In other words, all the proofs will trace their reasons back to the properties. Since writing all such details down is rather tedious and may not contribute much to the understanding of linear algebra, we will only present here a sample of how strict logic works. We will not be concerned too much with such details later on.

We will prove the following simple and "apparently" obvious fact.

$$cu = 0 \Leftrightarrow c = 0 \text{ or } u = 0.$$

The key problem here is that the statement has to be justified by using the eight properties (and the usual properties about numbers).

Proof. We break the statement into three parts.

1. $0u = \mathbf{0}$. (note that the plain 0 on the left means the *number* zero, and the bold 0 on the right means the *vector* zero)

$$\begin{aligned} 0u &= 0u + \mathbf{0} \text{ (property 3: } u + \mathbf{0} = u) \\ &= 0u + (u + (-u)) \\ &= (0u + u) + (-u) \\ &= (0u + 1u) + (-u) \\ &= (0 + 1)u + (-u) \\ &= 1u + (-u) \\ &= u + (-u) \\ &= \mathbf{0}. \end{aligned}$$

We emphasis that the simple property takes five of the eight properties to prove.

2. $c\mathbf{0} = \mathbf{0}$.

$$\begin{aligned} \mathbf{0} &= c\mathbf{0} + (-c\mathbf{0}) \\ &= c(\mathbf{0} + \mathbf{0}) + \\ &= (c\mathbf{0} + c\mathbf{0}) + (-c\mathbf{0}) \\ &= c\mathbf{0} + (c\mathbf{0} + (-c\mathbf{0})) \\ &= c\mathbf{0} + \mathbf{0} \\ &= c\mathbf{0}. \end{aligned}$$

3. $cu = \mathbf{0} \Rightarrow c = 0$ or $u = \mathbf{0}$

The statement is equivalent to $cu = \mathbf{0}$ and c "' $0 \Rightarrow u = \mathbf{0}$, which is what we will actually prove..

$$\begin{aligned} u &= 1u \\ &= (c^{-1}c)\, u \\ &= c^{-1}(cu) \\ &= c^{-1}\mathbf{0} \\ &= \mathbf{0}. \end{aligned}$$

Exercise: Prove that $(-1)u = -u$ in any vector space. Note that $(-1)u$ means the number -1 is multiplied to the vector u, and $-u$ means the negative vector in the fourth property of the definition of vector spaces.

Answer:

$$\begin{aligned} -u &= \mathbf{0} + (-u) \\ &= 0u + (-u) \end{aligned}$$

$$
\begin{aligned}
&= ((-1) + 1)u + (-u) \\
&= ((-1)u + 1u) + (-u) \\
&= (-1)u + (1u + (-u)) \\
&= (-1)u + (u + (-u)) \\
&= (-1)u + 0 \\
&= (-1)u.
\end{aligned}
$$

Remark: The exercise justifies the notations such as $u - 2v$, which could mean $u + (-2)v$, $u + (-1)(2v)$, or $u + (-2v)$. The exercise implies that the three expressions are equal, so that the meaning of $u - 2v$ is not ambiguous.

Exercise: Prove that $(a_1u_1 + a_2u_2) + (b_1u_1 + b_2u_2) = (a_1 + b_1)u_1 + (a_2 + b_2)u_2$ in any vector space.

Answer:

$$
\begin{aligned}
(a_1u_1 + a_2u_2) + (b_1u_1 + b_2u_2) &= ((a_1u_1 + a_2u_2) + b_1u_1) + b_2u_2 \\
&= (a_1u_1 + (a_2u_2 + b_1u_1)) + b_2u_2 \\
&= (a_1u_1 + (b_1u_1 + a_2u_2)) + b_2u_2 \\
&= ((a_1u_1 + b_1u_1) + a_2u_2) + b_2u_2 \\
&= ((a_1 + b_1)u_1 + a_2u_2) + b_2u_2 \\
&= (a_1 + b_1)u_1 + (a_2u_2 + b_2u_2) \\
&= (a_1 + b_1)u_1 + (a_2 + b_2)u_2.
\end{aligned}
$$

Exercise: Give a detailed reason why, in any vector space,

- $u + v = 0 \Rightarrow u = -v$.
- $3u + 2v - 4w = 0 \Rightarrow v = -3/2\, u + 2w$.

Answer: $u + v = 0$ implies

$$
\begin{aligned}
-v &= 0 + (-v) \text{ (property 3: } u + 0 = u) \\
&= (u + v) + (-v) \text{ (assumption: } u + v = 0) \\
&= u + (v + (-v)) \text{ (property 2: } (u + v) + w = u + (v + w)) \\
&= u + 0 \text{ (property 4: } u + (-u) = 0) \\
&= u. \text{ (property 3: } u + 0 = u)
\end{aligned}
$$

This completes the proof of the first part.

Note that in the second part, we used the notation $u + v + w$ to denote either $(u + v) + w$ or $u + (v + w)$, which are equal by the associativity. In fact, the notation for the linear combination $c_1u_1 + c_2u_2 + ... + c_ku_k$ is not ambiguous because of the associativity of the addition operation.

Assume $3u + 2v - 4w = 0$. Then the commutativity of the addition implies $2v + 3u - 4w = 0$. Multiplying 1/2 to the equality, the left side is

$$(1/2)(2v + 3u - 4w) = (1/2)(2v) + (1/2)(3u) + (1/2)(-4w)$$
$$= ((1/2)X2)v + ((1/2)X3)u + ((1/2)X(-4))w$$
$$= 1v + (3/2)u + (-2)w$$
$$= v + (3/2)u + (-2)w,$$

while the right side is $(1/2)0 = 0$. Therefore we get $v + (3/2)u + (-2)w = 0$. Now by the first part and an earlier exercise, we have

$$v = -((3/2)u + (-2)w)$$
$$= (-1)((3/2)u + (-2)w)$$
$$= (-1)((3/2)u) + (-1)((-2)w)$$
$$= ((-1)(3/2))u + ((-1)(-2))w$$
$$= -3/2\, u + 2w.$$

SUBSPACE

One number system may be included in other. For example, the integers Z is contained in the real numbers R. Moreover, the addition and multiplication are compatible: The equalities $1 + 2 = 3$, $2 \times 3 = 6$ hold either as integer operations or real number operations. Therefore we call Z a sub–number system of the number system R. The same remark applies to other inclusions such as $Z \subset Q$ or $R \subset C$. The similar situation may happen for one vector space contained in another in a compatible way.

A subset H of a vector space V is a subspace if it is closed under two operations:

- $u, v \in H \Rightarrow u + v \in H$.
- $u \in H, c \in R \Rightarrow cu \in H$.

By taking $c = 0$ in the second property, we see that a subspace must contain the zero vector 0. Moreover, by combining the two properties, a subspace is also closed under linear combinations.

Example: In any vector space V, the zero vector form the zero subspace $\{0\}$: $u, v \in \{0\}$ $u = v = 0 \Rightarrow u + v = 0 \in \{0\}$, $cu = 0 \in \{0\}$.

Example: $C(R)$ is a subspace of $F(R)$ because the addition and scalar multiplication of continuous functions are still continuous. More generally, $C^k(R)$ is also a subspace of $F(R)$. On the other hand, if $k > k'$, then $C^{k'}(R)$ is a subspace of $C^k(R)$.

The subset $H = \{f \in F(\mathrm{R}): f(2) = 0\}$ is also a subspace: If $f(2) = g(2) = 0$, then $(f + g)(2) = f(2) + g(2) = 0$ and $(cf)(2) = c(f(2)) = 0$.

The subset $K = \{f \in F(\mathrm{R}): f(t) = f(-t)\}$ of even functions is also a subspace: If $f(t) = f(-t)$ and $g(t) = g(-t)$, then $(f + g)(t) = f(t) + g(t) = f(-t) + g(-t) = (f + g)(-t)$ and $(cf)(t) = c(f(t)) = c(f(-t)) = (cf)(-t)$.

On the other hand, the subset $H' = \{f \in F(\mathrm{R}): f(2) = 1\}$ is not a subspace of $F(\mathrm{R})$ because it does not contain the zero function, which is the zero vector of the vector space $F(\mathrm{R})$.

Example: P_3 is a subspace of P_5. On the other hand, $H = \{\text{polynomials of degree} \geq 3 \text{ and} \leq 5\}$ is not a subspace of P_5. A counterexample is given by $p(t) = 1 + t^4$ and $q(t) = -t^4$. We have $p, q \in H$ but $(p + q)(t) = 1 \notin H$.

Example: The solutions $H_1 = \{(x, y, z): x + y + z = 0\}$ of a homogeneous equation is a subspace of R^3: $(x, y, z), (x', y', z') \in H_1 \Rightarrow x + y + z = x' + y' + z' = 0 \Rightarrow (x + x') + (y + y') + (z + z') = 0$ and $cx + cy + cz = 0$.

The solutions $H_2 = \{(x, y, z): x + y + z = 1\}$ of a non–homogeneous equation is not a subspace of R^3 because it does not contain the zero vector $(0, 0, 0)$.

The first quadrant $H_3 = \{(x, y): x \geq 0, y \geq 0\}$ of R^2 is not a subspace because $u = (1, 1) \in H_3$ but $(-1)u = -u = (-1, -1) \notin H_3$. The following shows more subsets of R^2 that are not subspaces.

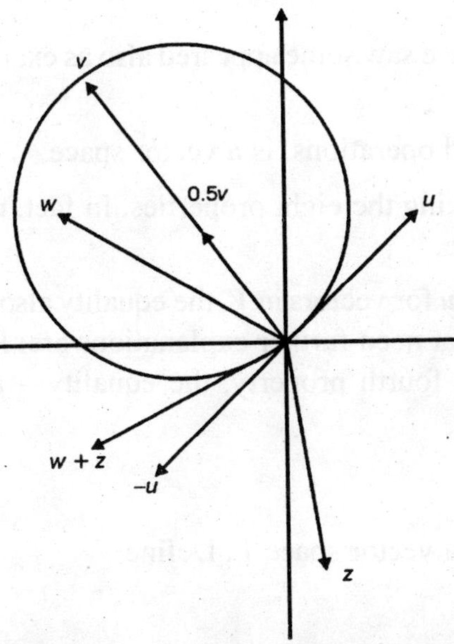

The following shows that a (infinite) straight line passing through the origin is a subspace, while the one not passing through the origin is not a subspace.

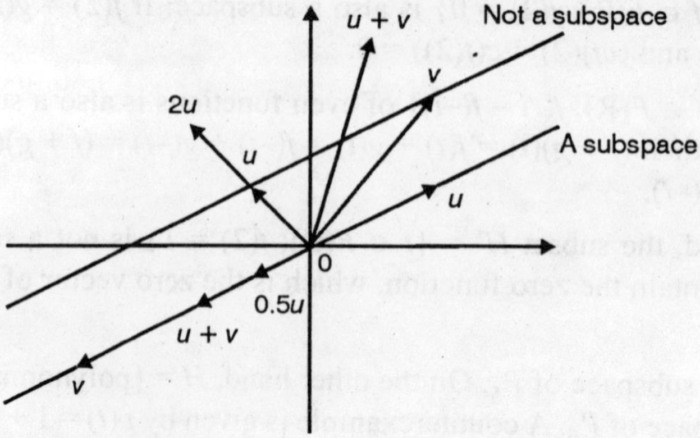

Example: R^2 is *not* a subspace of R^3. In fact, R^2 is defined as pairs of numbers, while R^3 is triples of numbers, so that R^2 is not contained in R^3.

On the other hand, R^2 can be *embedded* into R^3 as (but not is) subspaces by maps such as $(x, y) \Rightarrow (x, y, 0)$. This is perhaps where our intuition that R^2 is somehow contained in R^3 comes from.

However, the intuition is ambiguous because of the existence of other embeddings, such as $(x, y) \Rightarrow (0, x, y)$ or $(x, y) \Rightarrow (x, x, y)$. Therefore, to be on the safe side, it is better not to confuse R^2 with (2–dimensional) subspaces of R^3.

Example: The set *Sym*(n) of symmetric matrices is a subspace of the vector space $M(n, n)$ of n by n matrices.

In our examples of subspaces, we saw some appeared also as examples of vector spaces. Indeed we have

A subspace, with the inherited operations, is a vector space.

The proof is a matter of checking the eight properties. In fact, most of the properties are inherited from the bigger space.

For example, since $u + v = v + u$ for vectors in V, the equality also holds when restricted to vectors in H. The only things that need further explanations are: For the third property, we already know $0 \in H$. For the fourth property, the equality $-u = (-1)u$ shows that $u\, H \Rightarrow -u \in H$.

SUM AND INTERSECTION

Let H and K be subspaces of a vector space V. Define

sum

$H + K = \{u + v: u \in H, v \in K\}$,

intersection

$H \cap K = \{u: u \in H, u \in K\}$.

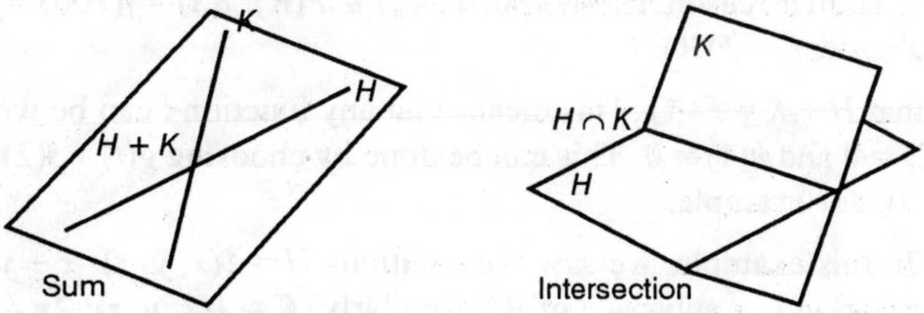

Sum Intersection

The sum and intersection of subspaces are still subspaces.

Proof For $H + K$, we have

$u, v \in H + K$

$\Rightarrow u = u_H + u_K, u_H \in H, u_K \in K; v = v_H + v_K, v_H \in H, v_K \in K$

$\Rightarrow u + v = u_H + u_K + v_H + v_K = (u_H + v_H) + (u_K + v_K) \in H + K.$

The verification of $u \in H + K \Rightarrow cu \in H + K$ is similar.

For $H \cap K$, we have

$u, v \in H \cap K$

$\Rightarrow u \in H, v \in H, u \in K, v \in K$

$\Rightarrow u + v \in H$ (since H is a subspace), $u + v \in K$ (since K is a subspace)

$\Rightarrow u + v \in H \cap K.$

The verification of $u \in H \cap K \Rightarrow cu \in H \cap K$ is similar.

We note that the sum and the intersection are the adaptations of the union and the intersection of sets to vector spaces. While the intersection remains the same, the sum is different form the union, because the union of two subspaces is not a subspace.

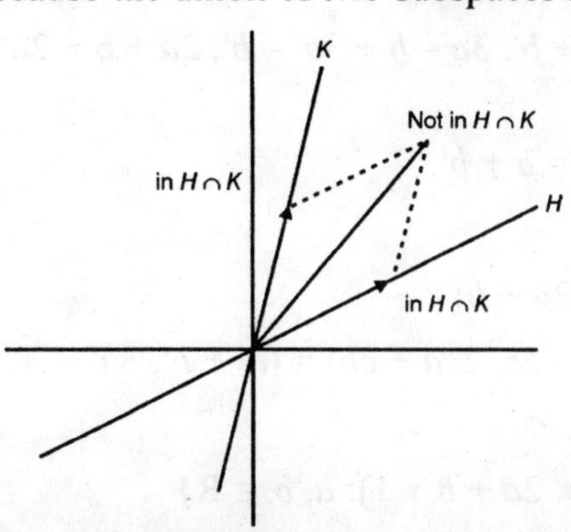

Example: In this example, we saw that $H = \{f \in F(R): f(2) = 0\}$ is a subspace of F(R). Similarly, $K = \{f \in F(R): f(1) = 0\}$ is also a subspace of $F(R)$. The intersection $H \cap K$

consists of all functions f satisfying $f(2) = f(1) = 0$. In particular, such functions form a subspace. By the similar reason, subsets such as $\{f \in F(R): f(3) = f(100) = f(-1) = f(-0.5) = 0\}$ are also subspaces of $F(R)$.

We claim that $H + K = F(R)$. This means that any function f can be written as $f = g + h$, such that $g(2) = 0$ and $h(1) = 0$. This can be done by choosing $g(t) = f(2)(t-1)$ and $h(t) = f(t) - f(2)(t-1)$, for example.

Example: In this example, we saw the solutions $H = \{(x, y, z): x + y + z = 0\}$ of a homogeneous equation is a subspace of R^3. Similarly, $K = \{(x, y, z): 2x - y + 3z = 0\}$ is also a subspace. Their intersection $H \cap K$ consists of all solutions of the 2 by 3 homogeneous system $x + y + z = 0$, $2x - y + 3z = 0$. By a simple computation, the solutions of the system are found to be $y(-4, 1, 3)$, with y arbitrary. Thus the subspace $H \cap K$ actually consists of all scalar mutiples of $(-4, 1, 3)$.

We claim the sum $H + K = R^3$. This means that any (x, y, z) can be written as $(x, y, z) = (x_1, y_1, z_1) + (x_2, y_2, z_2)$, satisfying $x_1 + y_1 + z_1 = 0$ and $2x_2 - y_2 + 3z_2 = 0$. We try to find such a combination by taking $z_2 = 0$. Then $y_2 = 2x_2$, the combination becomes $(x, y, z) = (x - x_2, y - 2x_2, z) + (x_2, 2x_2, 0)$, and the problem is reduced to $(x - x_2) + (y - 2x_2) + z = 0$. Thus we find $x_2 = (x + y + z)/3$ and finally get the combination

$$(x, y, z) = ((2x - y - z)/3, (-2x + y - 2z)/3, z) + ((x + y + z)/3, (2x + 2y + 2z)/3, 0)$$

we are looking for.

Exercise: Which of the following are subspaces of some euclidean spaces?

1. $H = \{(a + b, 3a - b, 2a + b): a, b \in R\}$

Answer: H is a subspace.

$u, v \in H$

$\Rightarrow u = (a + b, 3a - b, 2a + b)$, $v = (a' + b', 3a' - b', 2a' + b')$

$\Rightarrow u + v = (a + b + a' + b', 3a - b + 3a' - b', 2a + b + 2a' + b') = (a'' + b'', 3a'' - b'', 2a'' + b'') \in H$,

where $a'' = a + a'$, $b'' = b + b'$.

$u \in H, c \in R$

$\Rightarrow u = (a + b, 3a - b, 2a + b)$

$\Rightarrow cu = (ca + cb, 3ca - cb, 2ca + cb) = (a' + b', 3a' - b', 2a' + b') \in H$

where $a' = ca$, $b' = cb$.

2. $H = \{(a + b, 3a - b, 2a + b + 1): a, b \in R\}$

Answer: If H is a subspace, then it must contain the zero vector. In other words, we can find a and b, such that

$a + b = 0, 3a - b = 0, 2a + b + 1 = 0.$

However, the first two equations imply $a = b = 0$, which does not satisfy the third equation. We conclude that H is not a subspace.

3. $H = \{(a + b, 3a - b, 2a + b): a, b \in Z\}$

Answer: By taking $a = 1, b = 0$, we see that $(1, 2, 3) \in H$. If H is a subspace, then we should also have $1/2(1, 2, 3) = (1/2, 1, 3/2) H$. On the other hand, we cannot find two *integers* a and b such that $a + b = 1/2$. Therefore H is not a subspace.

4. $H = \{(x, y, z): x + y - z = 0, 2x - 3y + z = 0\}$

Answer: H is a subspace.

$u, v \in H$

$\Rightarrow u = (x, y, z), x + y - z = 0, 2x - 3y + z = 0; v = (x', y', z'), x' + y' - z' = 0, 2x' - 3y' + z' = 0$

$\Rightarrow u + v = (x + x', y + y', z + z'), (x + x') + (y + y') + (z + z') = x + y - z + x' + y' - z' = 0, 2(x + x') - 3(y + y') + (z + z') = 2x - 3y + z + 2x' - 3y' + z' = 0$

$\Rightarrow u + v \in H.$

The verification of $u \in H, c \in R \Rightarrow cu \in H$ is similar.

5. $H = \{(x, y, z): x + y - z = 0, 2x - 3y + z = 1\}$

Answer: Since $x = y = z = 0$ does not satisfy the condition $2x - 3y + z = 1$, H does not contain the zero vector. Therefore H is not a subspace.

6. $H = \{(x, y, z): x^2 + y^2 - z^2 = 0\}$

Answer: Since $(1, 0, 1), (0, 1, 1) \in H$ and $(1, 0, 1) + (0, 1, 1) = (1,1,2) \notin H$, we conclude that H is not a subspace.

7. $H = \{(x, y): xy \geq 0\}$

Answer: Since $(2, 0), (-1, -1) \in H$ and $(2, 0) + (-1, -1) = (1, -1) \notin H$, we conclude that H is not a subspace.

8. $H = \{(x, y, 0): x, y \in R\}$

Answer: H is a subspace.

$u, v \in H$

$\Rightarrow u = (x, y, 0), v = (x', y', 0)$

$\Rightarrow u + v = (x + x', y + y', 0 + 0) = (x + x', y + y', 0) \in H.$

The verification of $u \in H, c \in R \Rightarrow cu \in H$ is similar.

Exercise: Which of the following are subspaces of $F(R)$?

1. $H = \{f \in F(\mathrm{R}): f(1) + f(2) = 0\}$

Answer: H is a subspace.

$f, g \in H$

$\Rightarrow f(1) + f(2) = 0, g(1) + g(2) = 0$

$\Rightarrow (f+g)(1) + (f+g)(2) = f(1) + g(1) + f(2) + g(2) = 0$

$\Rightarrow f + g \in H.$

The verification of $f \in H$, c R $\Rightarrow cf \in H$ is similar.

2. $H = \{f \in F(\mathrm{R}): f(1) = f(2) = 0\}$

Answer: H is a subspace.

$f, g \in H$

$\Rightarrow f(1) = f(2) = 0, g(1) = g(2) = 0$

$\Rightarrow (f+g)(1) = f(1) + g(1) = 0, (f+g)(2) = f(2) + g(2) = 0$

$\Rightarrow f + g \in H.$

The verification of $f \in H$, c R $\Rightarrow cf \in H$ is similar.

3. $H = \{f \in F(\mathrm{R}): f(1) + f(2) = 1\}$

Answer: H is not a subspace because it does not contain the zero function.

4. $H = \{a \cos t + \sin t: a \in \mathrm{R}\}$

Answer: H is not a subspace because it does not contain the zero function.

5. $H = \{a \cos t + b \sin t: a, b \in \mathrm{R}\}$

Answer: H is a subspace.

$f, g \in H$

$\Rightarrow f = a \cos t + b \sin t, g = a' \cos t + b' \sin t$

$\Rightarrow f + g = a \cos t + b \sin t + a' \cos t + b' \sin t = a'' \cos t + b'' \sin t \in H.$

where $a'' = a + a'$, $b'' = b + b'$. The verification of $f \in H, c \in \mathrm{R} \Rightarrow cf \in H$ is similar.

6. $H = \{f \in C^2(\mathrm{R}): f'' - 4f' + 3f = 3t - 1\}$

Answer: H is not a subspace because it does not contain the zero function.

7. $H = \{f \in C^2(\mathrm{R}): f'' - 4f' + 3f = 0\}$

Answer: H is a subspace.

$f, g \in H$

$\Rightarrow f'' - 4f' + 3f = 0, g'' - 4g' + 3g = 0$

$\Rightarrow (f+g)'' - 4(f+g)' + 3(f+g) = f' - 4f + 3f + g'' - 4g' + 3g = 0$

$\Rightarrow f + g \in H.$

The verification of $f \in H$, c R $\Rightarrow cf \in H$ is similar.

8. $H = \{f \in C^2(\text{R}): f' - 4f + 3f = 0, f(0) = 0\}$

Answer: H is a subspace. The verification is similar to the previous problem. We need to additionally verify that $f(0) = 0$ and $g'(0) = 0$ imply $(f+g)'(0) = 0$ and $(cf)'(0) = 0$.

Exercise: Which of the following are subspaces of some matrix vector spaces?

1. $H = \{X \in M(n, n): X^{\text{T}} = -X\}$

Answer: H is a subspace.

$X, Y \in H$

$\Rightarrow X^{\text{T}} = -X, Y^{\text{T}} = -Y$

$\Rightarrow (X+Y)^{\text{T}} = X^{\text{T}} + Y^{\text{T}} = -X - Y = -(X+Y)$

$\Rightarrow X + Y \in H.$

The verification of $X \in H$, $c \in \text{R} \Rightarrow cX \in H$ is similar.

2. $H = \{X \in M(n, n): AX = XA\}$, where A is a fixed n by n matrix

Answer: H is a subspace.

$X, Y \in H$

$\Rightarrow AX = XA, AY = YA$

$\Rightarrow A(X+Y) = AX + AY = XA + YA = (X+Y)A$

$\Rightarrow X + Y \in H.$

The verification of $X \in H$, $c \in \text{R} \Rightarrow cX \in H$ is similar.

3. $H = \{X\ M(m, n): AX = I\}$, where A is a fixed n by m matrix

Answer: H is not a subspace because it does not contain the zero matrix.

4. $H = \{X \in M(m, n): AX = O\}$, where A is a fixed n by m matrix

Answer: H is a subspace.

$X, Y \in H$

$\Rightarrow AX = O, AY = O$

$\Rightarrow A(X+Y) = AX + AY = O + O = O$

$\Rightarrow X + Y \in H.$

The verification of $X \in H$, $c \in \text{R} \Rightarrow cX \in H$ is similar.

5. $H = \{X \in M(n, n): X^T X = I\}$

Answer: H is not a subspace because it does not contain the zero matrix.

6. $H = \{X \in M(n, n): X \text{ is invertible}\}$

Answer: H is not a subspace because it does not contain the zero matrix.

7. $H = \{X \in M(n, n): X \text{ is upper triangular}\}$

Answer: H is a subspace because adding two upper triangular matrices still gives an upper triangular matrix, and multiplying a number to an upper triangular matrix still gives an upper triangular matrix.

8. $H = \{X \in M(n, n): X \text{ is diagonal}\}$

Answer: H is a subspace because adding two diagonal matrices still gives a diagonal matrix, and multiplying a number to a diagonal matrix still gives a diagonal matrix.

Exercise: Describe the sum and intersection.

1. $H = \{(a + b, 3a - b, 2a + b): a, b \in \mathrm{R}\}$, $K = \{(3a, a, 2a): a \in \mathrm{R}\}$

Answer: $H + K$ consists of vectors of the form $(a + b, 3a - b, 2a + b) + (3c, c, 2c)$ for some $a, b, c \in \mathrm{R}$. In other words, $(x, y, z) \in H + K$ means the following system

$$a + b + 3c = x$$

$$3a - b + c = y$$

$$2a + b + 2c = z$$

has solutions. Row operations on the coefficient matrix give $\begin{bmatrix} 1 & 1 & 3 \\ 0 & -1 & -4 \\ 0 & 0 & 8 \end{bmatrix}$

Thus the solutions exist for any right side, and we have $H + K = \mathrm{R}^3$.

$H \cap K$ consists of vectors that can be written in the form $(a + b, 3a - b, 2a + b)$ and in the form $(3c, c, 2c)$ for some $a, b, c \in \mathrm{R}$. Thus they are obtained by solving the following system

$$a + b = 3c$$

$$3a - b = c$$

$$2a + b = 2c$$

This is equivalent to the homogeneous system

$$a + b - 3c = 0$$

$$3a - b - c = 0$$

$$2a + b - 2c = 0$$

The computation for $H + K$ tells us that all columns of the coefficient matrix are pivot. Therefore the only solution is the trivial one, and we conclude $H \cap K = \{0\}$.

2. $H = \{(x, y, z): x + y - z = 0, 2x - 3y + z = 0\}$, $K = \{(x, y, z): x + y - z = 0, x - 2y + 3z = 0\}$

Answer: By solving the systems defining H and K, we get

$H = \{(2a, 3a, 5a): a \in \mathrm{R}\}$, $K = \{(-a, 4a, 3a): a \in \mathrm{R}\}$.

Thus $H + K$ consists of vectors of the form $(2a, 3a, 5a) + (-b, 4b, 3b)$ for some $a, b \in \mathrm{R}$. In other words, $(x, y, z) \in H + K$ means the following system

$$2a - b = x$$

$$3a + 4b = y$$

$$5a + 3b = z$$

has solutions. Row operations on the augmented matrix give $\begin{bmatrix} 1 & 5 & y-x \\ 0 & -11 & 3x-2y \\ 0 & 0 & x+y-z \end{bmatrix}$.

Thus the condition for the solutions to exist is $x + y - z = 0$, and we have $H = \{(x, y, z): x + y - z = 0\}$.

$H \cap K$ consists of vectors satisfying both conditions defining H and K. In other words, they are the solutions of the system

$$x + y - z = 0$$

$$2x - 3y + z = 0$$

$$x + y - z = 0$$

$$x - 2y + 3z = 0$$

It is easy to see that all columns of the coefficient matrix are pivot. Therefore the only solution is the trivial one, and we conclude $H \cap K = \{0\}$.

3. $H = \{(a + b, 3a - b, 2a + b): a, b \in \mathrm{R}\}$, $K = \{(x, y, z): x + y - z = 0, x - 2y + 3z = 0\}$

Answer: From the previous problem, we have $K = \{(-a, 4a, 3a): a \in \mathrm{R}\}$. Thus $H + K$ consists of vectors of the form $(a + b, 3a - b, 2a + b) + (-c, 4c, 3c)$ for some $a, b, c \in \mathrm{R}$. In other words, $(x, y, z) \in H + K$ means the following system

$$a + b - c = x$$

$$3a - b + 4c = y$$

$$2a + b + 3c = z$$

has solutions. It is easy to see that all rows of the coefficient matrix are pivot. Thus the solutions exist for any right side, and we have $H + K = \mathrm{R}^3$.

$H \cap K$ consists of vectors of the form $(a + b, 3a - b, 2a + b)$ and satisfying the system that defines K. By substituting the form $(a + b, 3a - b, 2a + b)$ into the two equations defining K, we get

$$(a + b) + (3a - b) - (2a + b) = 2a - b = 0,$$

$$(a + b) - 2(3a - b) + 3(2a + b) = a + 6b = 0.$$

The only a, b satisfying both conditions is $a = b = 0$, with the corresponding vector $(a + b, 3a - b, 2a + b) = \{0\}$. Therefore we conclude $H \cap K = \{0\}$.

4. $H = \{f \in F(\mathrm{R}): f(1) + f(2) = 0\}$, $K = \{a \cos(\pi t/2) + b \sin(\pi t/2): a, b \in \mathrm{R}\}$

Answer: $H + K$ consists of functions of the form $f = g + a \cos(\pi t/2) + b \sin(\pi t/2)$, satisfying $g(1) + g(2) = 0$ and $a, b \in \mathrm{R}$. Thus

$$f(1) + f(2) = g(1) + g(2) + a \cos(\pi/2) + b \sin(\pi/2) + a \cos(\pi) + b \sin(\pi) = -a + b.$$

This suggests us to take $a = 0$, $b = f(1) + f(2)$. Then it is easy to verify that for any function f,

$$f = [\, f - (f(1) + f(2))\sin(\pi t/2)\,] + [\,(f(1) + f(2))\sin(\pi t/2)\,]$$

is a decomposition into the sum of a function in H and another function in K. Thus we conclude $H + K = F(\mathrm{R})$.

$H \cap K$ consists of functions $f = a \cos(\pi t/2) + b \sin(\pi t/2)$ satisfying

$$f(1) + f(2) = a \cos(\pi/2) + b \sin(\pi/2) + a \cos(\pi) + b \sin(\pi) = -a + b = 0.$$

In other words, $f = a \cos(\pi t/2) - a \sin(\pi t/2)$, and $H \cap K$ is all the scalar multiples of $\cos(\pi t/2) - \sin(\pi t/2)$.

5. $H = \{X \in M(n, n): X^{\mathrm{T}} = -X\}$, $K = \{X \in M(n, n): X^{\mathrm{T}} = X\}$

Answer: This earlier exercise shows that $H + K = M(n, n)$. On the other hand, $X \in H \cap K$ means $X^{\mathrm{T}} = -X$ and $X^{\mathrm{T}} = X$. Clearly, only the zero matrix satisfies both. Thus $H \cap K = \{O\}$.

6. $H = \{X \in M(n, n): X$ is upper triangular$\}$, $K = \{X \in M(n, n): X$ is lower triangular$\}$

Answer: Any n by n matrix is a sum of an upper trangular matrix and a lower triangular matrix. For example,

$$\begin{bmatrix} 1 & 2 & 3 & 4 \\ 5 & 6 & 7 & 8 \\ 9 & 10 & 11 & 12 \\ 13 & 14 & 15 & 16 \end{bmatrix} = \begin{bmatrix} 1 & 2 & 3 & 4 \\ 0 & 6 & 7 & 8 \\ 0 & 0 & 11 & 12 \\ 0 & 0 & 0 & 16 \end{bmatrix} + \begin{bmatrix} 0 & 0 & 0 & 0 \\ 5 & 0 & 0 & 0 \\ 9 & 10 & 0 & 0 \\ 13 & 14 & 15 & 0 \end{bmatrix}$$

Therefore $H + K = M(n, n)$. On the other hand, the only matrix that is both upper and lower triangular is the diagonal matrix. Therefore $H \cap K = \{X \in M(n, n): X \text{ is diagonal}\}$.

Exercise: Let H and K be subspaces. Prove that $H \cap K = H \Leftrightarrow H \subset K \Leftrightarrow H + K = K$.

Answer: The fact $H \cap K = H \Leftrightarrow H \subset K$ follows from set theory. Note that by writing $v \in K$ as $v = 0 + v$ and the fact $0 \in H$, we always have

1. $H + K \supset K$.

Similarly, we have

$$H + K \supset H.$$

Now if $H \subset K$, then

$x \in H + K$

$x = u + v$ for some $u \in H, v \in K$

$x = u + v$ with $u \in K, v \in K$

$x \in K$. (K is a subspace)

This completes the proof that

2. $H \subset K \Rightarrow H + K \subset K$.

Conversely, if $H + K \subset K$, then by $H + K \supset H$, we have $H \subset K$. Thus we also proved

3. $H + K \subset K \Rightarrow H \subset K$.

Combining 1, 2, 3, we get $H \subset K \Leftrightarrow H + K = K$.

LINEAR TRANSFORMATION

Having extended the euclidean spaces to general vector spaces, we may also extend linear transformations.

A transformation $T: V \to W$ between two vector spaces is a linear transformation if

$T(u + v) = T(u) + T(v)$, $T(cu) = cT(u)$.

Since linear transformations between euclidean spaces are equivalent to matrices, the extension can be thought of as a generalization of matrices. Moreover, by taking $c = 0$ in $T(cu) = cT(u)$, we have

$T(0) = 0$.

By combining the two properties, T also preserves linear combinations

$T(c_1u_1 + c_2u_2 + ... + c_ku_k) = c_1T(u_1) + c_2T(u_2) + ... + c_kT(u_k)$.

Example: For any vector space V, the identity transformation

$id(u) = u: V \rightarrow V$

is linear. More generally, any scaling map, such as $u \rightarrow 2u$ is also linear.

For any vector spaces V and W, the zero transformation

$O(u) = 0: V \rightarrow W$

sending any vector in V to the zero vector in W is linear.

Example: The transformation

$D(f) = f': C^1(R) \rightarrow C(R)$

of taking derivatives is linear. The transformation

$M(f) = (t^2+1)f: F(R) \rightarrow F(R)$

of multiplying the function t^2+1 is also linear.

The exponential transformation

$P(f) = e^f: F(R) \rightarrow F(R)$

is not linear because $P(0) = e^0 = 1 \neq 0$, where 0 is the constant zero function, and is the zero vector of the vector space $F(R)$.

Example: The evaluation

$E(f) = f(2): F(R) \rightarrow R$

at $t = 2$ is linear. The multiple evaluations

$E(f) = (f(0), f(1), f(2)): F(R) \rightarrow R^3$

at $t = 0, 1, 2$ is also linear. The transformation

$C(a, b) = a \cos t + b \sin t: R^2 \rightarrow F(R)$

of combining sine and cosine is linear. A more general statement about the linearity of C can be found here.

Example: The transpose transformation

$T(X) = X^T: M(m, n) \rightarrow M(n, m)$

is linear. The transformation

$M(X) = AX: M(k, n) \rightarrow M(m, n)$

of multiplying a fixed m by k matrix A on the left is also linear. More generally, for fixed A and B, the transformation $X \rightarrow AXB$ is also linear. The transformation

$S(X) = X^2: M(n, n) \rightarrow M(n, n)$

of taking squares is not linear because $S(2X) = 4X^2 \neq 2X^2 = 2S(X)$.

OPERATIONS OF LINEAR TRANSFORMATIONS AND HOM SPACE

The operations for matrices/linear transformations between euclidean spaces may be generalized.

The addition of two linear transformations $T, S: V \to W$ is given by

$(T + S)(u) = T(u) + S(u)$.

The scalar multiplication of a number c to a linear transformation $T: V \to W$ is given by

$(cT)(u) = cT(u)$.

The composition of linear transformations $T: V \to W$ and $S: U \to V$ is given by

$(TS)(u) = T(S(u))$.

The verification of the linearity of $T + S$, cT, TS can be found in this exercise. The generalization of the transpose will be the dual transformation.

Example The transformation $T(f) = f' - 4f + 3f: C^2(R) \to C(R)$ is linear because

$T = DD + 4D - 3id$

is a linear combination of transformations DD, D, id. The identity id and the derivative D are known to be linear. As the composition of two linear transformations, DD is also linear. Note that since T is linear, the differential equation $f' - 4f + 3f = 3t - 1$ is a linear equation.

Example: Fixing an n by n matrix A, matrices X commuting with A are the solutions of the equation $AX = XA$. The equation is linear because it can be written as $AX - XA = O$. The left side $AX - XA$ is a linear transformation of X because it is a linear combination of $X \to AX$ and $X \to XA$, which are known to be linear.

Denote by $Hom(V, W)$ the collection of all linear transformations from V to W. It is easy to verify that the two operations satisfy the eight properties in the definition of vector spaces. For example, for T, S $Hom(V, W)$ and u V, we have

$$\begin{aligned}(T + S)(u) &= T(u) + S(u)\\ &= S(u) + T(u)\\ &= (S + T)(u).\end{aligned}$$

This verifies the first property. The other properties can be similarly verified. Therefore $Hom(V, W)$ is a vector space.

In the special case $V = R^n$, $W = R^m$, $Hom(R^n, R^m)$ can be naturally identified with the collection $M(m, n)$ of m by n matrixes. Moreover, it is easy to see that the addition and scalar multiplication in $Hom(R^n, R^m)$ and $M(m, n)$ match in the identification.

Exercise: Which of the following are linear transformations?

1. $T(f) = f(1) + f(2)$: $F(R) \to R$

Answer: T is linear.

$T(f + g) = (f + g)(1) + (f + g)(2) = (f(1) + g(1)) + (f(2) + g(2)) = f(1) + f(2) + g(1) + g(2) = T(f) + T(g)$,

$T(cf) = (cf)(1) + (cf)(2) = c(f(1)) + c(f(2)) = c(f(1) + f(2)) = cT(f)$.

2. $T(f) = f(1) + f(2) - 1$: $F(R) \to R$

Answer: T is not linear because $T(0) = -1 \neq 0$.

3. $T(f) = f(1)f(2)$: $F(R) \to R$

Answer: T is not linear because $T(2f) = (2f)(1)(2f)(2) = 4f(1)f(2) \neq 2f(1)f(2) = 2T(f)$.

4. $T(a) = a \cos t + \sin t$: $R \to F(R)$

Answer: T is not linear because $T(0) = \sin t$ is not the zero function.

5. $T(a, b) = a \cos t + b \sin t$: $R^2 \to F(R)$

Answer: T is linear.

$T((a, b) + (a', b')) = (a + a') \cos t + (b + b') \sin t = a \cos t + a' \cos t + b \sin t + b' \sin t = T(a, b) + T(a', b')$.

The verification of $T(c(a, b)) = cT(a, b)$ is similar.

6. $T(f) = f' - f$: $C^2(R) \to C(R)$

Answer: We already know that the identity *id* is linear and the derivative D is linear. Now $T = DD - 3id$ is obtained by composition, addition, and scalar multiplication of linear transformations. Thus T is linear.

7. $T(f) = +\int_0^1 f(t)dt$: $C(R) \to R$

Answer: T is linear by the standard properties of integration from calculus.

$+\int(f(t) + g(t))dt = +\int f(t)dt + +\int g(t)dt$,

$+\int cf(t)dt = c+"f(t)dt$.

8. $T(f(t)) = f(-t)$: $F(R) \to F(R)$

Answer: T is linear by changing the variable in the usual definition of $f + g$ and cf from t to $-t$.

$(f + g)(t) = f(t) + g(t)\ (f + g)(-t) = f(-t) + g(-t),\ (cf)(-t) = c(f(-t))$.

Remark Any change of variable, e.g., $T(f(t)) = f(t^2 + \sin t)$: $F(R) \to F(R)$, is also linear.

9. $T(a, b, c) = a + (a - 2b)t + (3a + c)t^2$: $R^3 \to P_2$

Answer: T is linear.

$T((a, b, c) + (a', b', c')) = (a + a') + ((a + a') - 2(b + b'))t + (3(a + a') + (c + c'))t^2 = a + (a - 2b)t + (3a + c)t^2 + a' + (a' - 2b')t + (3a' + c')t^2 = T(a, b, c) + T(a', b', c')$.

The verification of $T(d(a, b, c)) = dT(a, b, c)$ is similar.

10. $T(p(t)) = (p(t), p'(t))$: $P_4 \to P_4$ E P_3

Answer: T is linear.

$T(p(t) + q(t)) = (p(t) + q(t), (p(t) + q(t)') = (p(t) + q(t), p'(t) + q'(t)) = T(p(t)) + T(q(t))$.

The verification of $T(cp(t)) = cT(p(t))$ is similar.

11. $T(p(t)) = (p(0), p'(0))$: $P_4 \to \mathrm{R}^2$

Answer: T is linear. The reason is similar to the previous problem.

12. $T(X) = AX - XA$: $M(n, n) \to M(n, n)$; where A is a fixed n by n matrix

Answer: T is linear. In this example, we have seen that $X \to AX$ is linear. By the similar reason, $X \to XA$ is also linear. As a linear combination of linear transformations, T is linear.

13. $T(X) = X - X^{\mathrm{T}}$: $M(n, n) \to M(n, n)$

Answer: T is linear, because it is a linear combination of the identity and the transpose, both being obviously linear.

14. $T(X) = AX - I$: $M(n, n) \to M(n, n)$, where A is a fixed n by n matrix

Answer T is not linear because $T(O) = -I \neq O$.

15. $T(X)$ = first column of X: $M(m, n) \to R^m$

Answer: T is linear because addition and scalar multiplication of matrices is compatible with addition and scalar multiplication of columns of matrices.

Exercise: Let u, v, w be three vectors in a vector space V. Show that the transformation $T(a, b, c) = au + bv + cw$: $\mathrm{R}^3 \to V$ is a linear transformation.

Answer: The following verifies T preserves addition.

$$\begin{aligned} T(\,(a, b, c) + (a', b', c')\,) &= T(a + a', b + b', c + c') \\ &= (a + a')u + (b + b')v + (c + c')w \\ &= (au + bv + cw) + (a'u + b'v + c'w) \\ &= T(a, b, c) + T(a', b', c'). \end{aligned}$$

The verification of $T(\,d(a, b, c)\,) = dT(a, b, c)$ is similar.

Remark: The conclusion has the obvious generalization: The transformation

$T(x_1, x_2, \ldots, x_n) = x_1v_1 + x_2v_2 + \ldots + x_nv_n$: $\mathrm{R}^n \to V$ is linear.

Exercise: Verify that the operations $T + S$, cT, TS on linear transformations indeed produce linear transformations.

Answer: The following shows that $T + S$ preserves addition.

$$\begin{aligned}(T+S)(u+v) &= T(u+v) + S(u+v) \\ &= T(u) + T(v) + S(u) + S(v) \\ &= T(u) + S(u) + T(v) + S(v) \\ &= (T+S)(u) + (T+S)(v).\end{aligned}$$

The following shows that cT preserves addition.

$$\begin{aligned}(cT)(u+v) &= c(T(u+v)) \\ &= c(T(u) + T(v)) \\ &= c(T(u)) + c(T(v)) \\ &= (cT)(u) + (cT)(v).\end{aligned}$$

The following shows that TS preserves addition.

$$\begin{aligned}(TS)(u+v) &= T(S(u+v)) \\ &= T(S(u) + S(v)) \\ &= T(S(u)) + T(S(v)) \\ &= (TS)(u) + (TS)(v).\end{aligned}$$

The verification for preserving scalar multiplication is similar.

Exercise: Prove that T is an invertible linear transformation □The inverse transformation T^{-1} is also linear.

Answer: Denote $S = T^{-1}$. Then $TS = id$, $ST = id$, and

$$\begin{aligned}T(S(u+v)) &= (TS)(u+v) \\ &= u + v \\ &= (TS)(u) + (TS)(v) \\ &= T(S(u)) + T(S(v)) \\ &= T(S(u) + S(v)).\end{aligned}$$

Applying S to both sides and using $ST = id$, we get $S(u + v) = S(u) + S(v)$. The equality $S(cu) = cS(u)$ can be proved in a similar way.

Exercise: Let P: $U \to V$ be a linear transformation. Prove that the transformation $P^*(T) = TP$: $Hom(V, W) \to Hom(U, W)$ is linear.

Answer: Let $T, S \in Hom(V, W)$, $u \in U$. Then we have

$$(P^*(T+S))(u) = ((T+S)P)(u)$$
$$= (T+S)(P(u))$$
$$= T(P(u)) + S(P(u))$$
$$= (TP)(u) + (SP)(u)$$
$$= (P^*(T))(u) + (P^*(S))(u)$$
$$= (P^*(T) + P^*(S))(u).$$

The verification of $P^*(cT) = cP^*(T)$ is similar.

Remark: In case U, V, W are euclidean spaces, the transformation P^* means multiplying a matrix on the right: $X \to XA$.

Exercise: Let $P: W \to U$ be a linear transformation. Prove that the transformation $P_*(T) = PT: Hom(V, W) \to Hom(V, U)$ is linear.

Answer: Let $T, S \in Hom(V, W)$, $v \in V$. Then we have

$$(P_*(T+S))(v) = (P(T+S))(v)$$
$$= P((T+S)(v))$$
$$= P(T(v) + S(v))$$
$$= P(T(v)) + P(S(v))$$
$$= (PT)(u) + (PS)(u)$$
$$= (P_*(T))(u) + (P_*(S))(u)$$
$$= (P_*(T) + P_*(S))(u).$$

The verification of $P_*(cT) = cP_*(T)$ is similar.

Remark: In case U, V, W are euclidean spaces, the transformation P_* means multiplying a matrix on the left: $X \to AX$.

RANGE AND COLUMN SPACE

RANGE

Recall the range of a transformation $T: X \to Y$ is all the images under the transformation

$rangeT = \{T(x): \text{any } x \in X\}$

$= \{b \in Y: \text{There is } x \in X, \text{ such that } T(x) = b\}$

$= \{b\ Y: T(x) = b \text{ has solutions}\}$.

For a *linear* transformation, we have

The range of a linear transformation $T: V \to W$ is a subspace of W.

Proof. $w, w' \in rangeT \Rightarrow w = T(v), w' = T(v')$ for some $v, v' \in V \Rightarrow w + w' = T(v) + T(v') = T(v + v') \in rangeT$, where the linearity of T is used in the red equality. The proof of $w \in rangeT \Rightarrow cw \in rangeT$ is similar.

Example: The subset

$\{(a + b, 3a - b, 2a + b): a, b \in R\} \subset R^3$

is a subspace because it is the range of the linear transformation

$T(a, b) = (a + b, 3a - b, 2a + b): R^2 \to R^3.$

The subset

$\{a \cos t + b \sin t: a, b \in R\} \subset F(R)$

is a subspace because it is the range of the linear transformation

$C(a, b) = a \cos t + b \sin t: R^2 \to F(R).$

We also recall that a transformation is onto (surjective) if and only if the range is the whole target space. In other words, any element of Y is the image of some element of X.

Example: Consider the derivative transformation

$D(f) = f': C^1(R) \to C(R).$

By the fundamental theorem of calculus, any continuous function $f \in C(R)$ has an antiderivative (for example, given by $F(t) = +\int_0^t f(t)dt$). This implies that D is onto. However, if we change the target space from $C(R)$ to $F(R)$, then D is no longer onto.

Example: Consider the linear transformation

$$M(f) = (t^2+1)f: F(R) \to F(R)$$

of multiplying the function t^2+1. For any function $f \in F(R)$, we may construct another function $F(t) = f(t)/(t^2+1)$. Then we have $M(F) = f$. Thus the transformation M is onto. Note that the key here is that $t^2 + 1$ never vanishes, so that it can be "inverted". If we change the function to $t^2 - 1$, then M is no longer onto.

For the similar reason, if X is an invertible n by n matrix, then the linear transformation

$$M(X) = AX: M(k, n) \to M(m, n)$$

is onto.

COLUMN SPACE

For a linear transformation $T: R^n \to R^m$ given by an m by n matrix $A = [a_1\ a_2 \dots a_n]$, we have

$$T(x) = \begin{bmatrix} a_{11}x_1 + a_{12}x_2 + \dots + a_{1n}x_n \\ a_{21}x_1 + a_{22}x_2 + \dots + a_{2n}x_n \\ a_{m1}x_1 + a_{m2}x_2 + \dots + a_{mn}x_n \end{bmatrix}$$

$$= \begin{bmatrix} a_{11} \\ a_{21} \\ \vdots \\ a_{m1} \end{bmatrix} + x_2 \begin{bmatrix} a_{12} \\ a_{22} \\ \vdots \\ a_{m2} \end{bmatrix} + \dots + x_n \begin{bmatrix} a_{1n} \\ a_{2n} \\ \vdots \\ a_{mn} \end{bmatrix}$$

$$= x_1 a_1 + x_2 a_2 + \dots + x_n a_n$$

Thus *rangeT* is given by all the linear combinations (i.e., for all choices of the coefficients $x_1, x_2, \dots, x_n$) of the columns of A.

The column space of a matrix A is

$$colA = \{\text{all the linear combinations of columns of } A\}$$
$$= \{Ax: \text{any } x \in \mathrm{R}^n\}$$
$$= \{b \in \mathrm{R}^n: Ax = b \text{ has solutions}\}.$$

We remark that the column space of an m by n matrix is a subspace of R^m.

Example: Because

$(a + b, 3a - b, 2a + b) = a(1, 3, 2) + b(1, -1, 1),$

the subspace

$\{(a + b, 3a - b, 2a + b): a, b \in \mathrm{R}\} \subset \mathrm{R}^3$

given in an earlier example is in fact the columns space of the matrix

$$A = \begin{bmatrix} 1 & 1 \\ 3 & -1 \\ 2 & 1 \end{bmatrix}$$

Notice that the linear transformation in the earlier example is also given by the matrix.

We also note that since the system $Ax = b$ has solutions for $b = (3, 1, 4)$ and has no solution for $b = (1, 0, 1)$, the vector $(3, 1, 4)$ is in *colA*, and the vector $(1, 0, 1)$ is not in *colA*.

Exercise: Find the ranges and determine whether the transformations are onto.

1. $T(f) = f(1) + f(2)$: $F(\mathrm{R}) \rightarrow \mathrm{R}$

Answer: rangeT = *R*. *T* is onto.

2. $T(a, b) = a \cos t + b \sin t$: $\mathrm{R}^2 \rightarrow F(\mathrm{R})$

Answer: The range of T is all the linear combinations of cos*t* and sin*t*. In particular the functions in the range are periodic with period 2π. Since not all functions are periodic, the range is not all the functions, and T is not onto.

3. $T(f) = f' - f$: $C^2(\mathrm{R}) \to C(\mathrm{R})$

Answer: The range of T is the whole space $C(\mathrm{R})$, because it is a standard fact from the theory of differential equations that any continuous function h can be written as $f' - f$. Specifically, f is constructed as follows:

- Solve $g' - g = h$: From $(e^{-t}g) = e^{-t}g' - e^{-t}g = e^{-t}h$, we have $g = e^t + \int e^{-t}hdt$.
- Solve $f' + f = g$: By multiplying e^t instead of e^{-t}, we get $f = e^{-t} + \int e^t g dt$.

You may verify by yourself that the function f thus obtained indeed satisfies $f' - f = h$. Thus T is onto.

4. $T(f) = + \int_0^1 f(t)dt$: $C(\mathrm{R}) \to \mathrm{R}$

Answer: Any number a is the integration from 0 to 1 of the constant function a. Thus the range of T is the whole R, and T is onto.

5. $T(f(t)) = f(-t)$: $F(\mathrm{R}) \to F(\mathrm{R})$

Answer: Given any function $g(t)$, we may construct a function $f(t) = g(-t)$ and get $g(t) = f(-t)$. Thus any function $g(t)$ is the image of the funciton $g(-t)$. Thus the range of T is the whole $F(\mathrm{R})$, and T is onto.

Remark: The range seriously depends on the change of variable. For example, any image $g(t) = f(t^2)$ of the linear transformation $T(f(t)) = f(t^2)$: $F(\mathrm{R}) \to F(\mathrm{R})$ satisfies $g(t) = g(-t)$, i.e., g is an even function. In particular, the function $f(t) = t$ is not in the range of T and T is not onto.

6. $T(a, b, c) = a + (a - 2b)t + (3a + c)t^2$: $\mathrm{R}^3 \to P_2$

Answer: For any d_0, d_1, d_2, the system

$a = d_0$, $a - 2b = d_1$, $3a + c = d_2$

always have solutions. This implies the range of T is whole P_2, and T is onto.

7. $T(p(t)) = (p(t), p'(t))$: $P_4 \to P_4 \mathrm{E}\, P_3$

Answer: The range of T consists of pairs $(p(t), q(t))$, where $p(t) \in P_4$ and $q(t) \in P_3$ satisfy $q(t) = p'(t)$. In particular, the pair (t, t) is not in the range, and T is not onto.

8. $T(p(t)) = (p(0), p'(0))$: $P_4 \to \mathrm{R}^2$

Answer: For any two numbers a, b, the polynomial $p(t) = a + bt$ satisfies $T(p(t)) = (a, b)$. Therefore the range of T is R^2 and T is onto.

9. $T(p(t)) = (p(0), p'(0), p'''(0), p \neq (0))$: $P_2 \to \mathrm{R}^4$

Answer: A polynomial of degree 2 is of the form $p(t) = a + bt + ct^2$, and we have $T(p(t)) = (a, b, 2c, 0)$. Thus *rangeT* consists of those vectors in R^4 with last coordinate zero. In particular, (1, 1, 1, 1) is not in the range, and T is not onto.

Remark: Comparing problems 7 and 8, what can you say about $T(p(t)) = (p(0), p'(0),..., p^{(n)}(0))$: P_m R^{n+1}?

10.$T(X) = AX - XA$: $M(n, n) \to M(n, n)$, where A is a fixed n by n matrix

Answer: The range of T can be as small as the zero space, by taking $X = O$ for example. T cannot be onto, because the sum of the diagonal entries (the trace) of $AX - XA$ must be zero.

Let us consider the 2 by 2 case in more detail. For

$$A = \begin{bmatrix} a & b \\ c & d \end{bmatrix}, X \begin{bmatrix} x & z \\ y & w \end{bmatrix}$$

we have

$$AX - XA = \begin{bmatrix} bz - cy & (a-d)y + b(w-x) \\ c(x-w) + (d-a) & zcy - bz \end{bmatrix} = \begin{bmatrix} h_{11} & h_{12} \\ h_{21} & h_{22} \end{bmatrix}$$

In addition to the trace constraint

$$h_{11} + h_{12} = (bz - cy) + (cy - bz) = 0,$$

we have another constraint

$$ch_{12} + bh_{21} = c(a-d)y + b(d-a)z = (d-a)(bz - cy) = (d-a)h_{11}.$$

These are all the constraints on the matrices in the range.

11.$T(X) = X - X^{\mathrm{T}}$: $M(n, n) \to M(n, n)$

Answer: Since $(X - X^{\mathrm{T}})^{\mathrm{T}} = X^{\mathrm{T}} - (X^{\mathrm{T}})^{\mathrm{T}} = -(X - X^{\mathrm{T}})$, the matrices in the range of T are skew–symmetric. Conversely, if $A = -A^{\mathrm{T}}$, then for $X = A/2$, we have $A = X - X^{\mathrm{T}}$. Thus we conclude that *rangeT* = skew–symmetric matrices. In particular, T is not onto.

12.$T(X)$ = first column of X: $M(m, n) \to \mathrm{R}^m$

Answer: The range of T is the whole R^m, and T is onto.

Exercise: Show the following are subspaces by identifying them as the ranges of linear transformations.

1. $H = \{(y - 2z, 2x - 4y + 7z): x, y, z \in \mathrm{R}\}$

Answer: H is the range of the linear transformation $(x, y, z) \to (y - 2z, 2x - 4y + 7z)$ given by the matrix $\begin{bmatrix} 01 & -2 \\ 2 & -47 \end{bmatrix}$

Note that H is also the column space of the matrix.

2. $H = \{r \cos t + s \sin t + (r + s)t: r, s \in R\}$

Answer: H is the range of the linear transformation $T(f) = r \cos t + s \sin t + (r + s)t = r(\cos t + t) + s(\sin t + t)$: $R^2 \to F(R)$. The verification of the linearity of T is similar to this exercise.

3. H = all polynomials of degree ≤ 5 with $t = 2$ and 4 as two of the five roots

Answer: H consists of all the polynomials of the form $(t-2)(t-4)q(t)$, with the degree of $q(t) \leq 3$. Therefore H is the range of the linear transformation $T(q(t)) = (t-2)(t-4)q(t)$: $P_3 \to P_5$.

4. H = all polynomials of degree ≤ 5 with $t = 2$ as a multiple root

Answer: H consists of all the polynomials of the form $(t-2)^2 q(t)$, with the degree of $q(t) \leq 3$. Therefore H is the range of the linear transformation $T(q(t)) = (t-2)^2 q(t)$: $P_3 \to P_5$.

Exercise: Find out whether the given vectors are in the range of the linear transformations.

1. $T(x, y, z) = (x - 2y + z, -x + 3y, y + z, x + 2y + 5z)$: $R^3 \to R^4$

$u = (1, 2, 3, 13)$, $v = (2, -2, 4, 5)$

Answer: $u \in rangeT$ means we can find x, y, z, such that $T(x, y, z) = u$. In other words, the system

$$\begin{aligned} x - 2y + z &= 1 \\ -x + 3y &= 2 \\ y + z &= 3 \\ x + 2y + 5z &= 13 \end{aligned}$$

has solutions. Similar interpretation can be made for $T(x, y, z) = v$. Thus we consider the combined augmented matrix $\begin{bmatrix} 1 & -2 & 1 & 1 & 2 \\ -1 & 3 & 0 & 2 & -2 \\ 0 & 1 & 1 & 3 & 4 \\ 1 & 2 & 5 & 13 & 5 \end{bmatrix}$.

In this exercise, we have computed the row echelon form $\begin{bmatrix} 1 & -2 & 1 & 1 & 2 \\ 0 & 1 & 1 & 3 & 0 \\ 0 & 0 & 0 & 0 & 4 \\ 0 & 0 & 0 & 0 & 0 \end{bmatrix}$.

From the first four columns, $T(x, y, z) = u$ has solutions. By deleting [col 4], we see that $T(x, y, z) = v$ has no solution. Therefore we conclude

$u \in rangeT, v \notin rangeT.$

By the way, from the first four columns of the reduced row echelon form

$$\begin{bmatrix} 1 & 0 & 3 & 7 & 0 \\ 0 & 1 & 1 & 3 & 0 \\ 0 & 0 & 0 & 0 & 1 \\ 0 & 0 & 0 & 0 & 0 \end{bmatrix},$$

we further have $T(7, 3, 0) = (1, 2, 3, 13)$.

Remark: The conclusion can also be interpreted as $u \in colA$ and v O $colA$, where

$$A = \begin{bmatrix} 1 & -2 & 1 \\ -1 & 3 & 0 \\ 0 & 1 & 1 \\ 1 & 2 & 5 \end{bmatrix}$$

is the matrix of the transformation T.

2. $T(a, b, c) = (a - 2b + c) + (-a + 3b)t + (b + c)t^2 + (a + 2b + 5c)t^3$: $R^3 \to P_3$

$u = 1 + 2t + 3t^2 + 13t^3$, $v = 2 - 2t + 4t^2 + 5t^3$

Answer: $u \in rangeT$ means we can find a, b, c, such that $T(a, b, c) = u$. By comparing the coefficients of $1, t, t^2, t^3$ on both sides, the problem is equivalent to the existence of solutions to the system

$$a - 2b + c = 1$$
$$-a + 3b = 2$$
$$b + c = 3$$
$$a + 2b + 5c = 13$$

In the first problem, we have seen the solutions exist. Therefore we have $u \in rangeT$. In fact, we also know from the first part that $T(7, 3, 0) = 1 + 2t + 3t^2 + 13t^3$.

Similar discussion can be carried out for the existence problem about $T(a, b, c) = v$. Again the situation is similar to the first problem. Thus we conclude v O $rangeT$.

3. $T(a + bt + ct^2) = \begin{bmatrix} a - 2b + c & -a + 3b \\ b + c & a + 2b + 5c \end{bmatrix}: P_2 \to M(2, 2)$

$$u = \begin{bmatrix} 1 & 2 \\ 3 & 13 \end{bmatrix}, v = \begin{bmatrix} 2 & -2 \\ 4 & 5 \end{bmatrix}$$

Answer: By considering the four entries of the 2 by 2 matrices, the equalities $T(a + bt + ct^2) = u$ and $T(a + bt + ct^2) = v$ are equivalent to the two systems discussed in the first problem. Therefore we have the same conclusion.

$u \in rangeT$, $v \notin rangeT$.

Remark: The equivalence between the three problems indicates the equivalence between general vector spaces and the euclidean space. Practically, many general linear algebra problems can be translated into euclidean space problems, which can be usually solved by row operations. We will see more such equivalencies and will set up a theory for such equivalencies.

4. $T(a, b) = a \cos t + b \sin t$: $R^2 \to F(R)$

   $u = 1 + 2t$, $v = \sin 2t$

Answer: If $u \in rangeT$, then there are numbers a and b such that $a \cos t + b \sin t = 1 + 2t$ for all t. However, for any fixed numbers a and b, $a \cos t + b \sin t$ is a bounded function, while $1 + 2t$ is not bounded. Therefore the equality cannot hold, and u O $rangeT$.

Similarly, $v \in rangeT$ means there are numbers a and b such that $a \cos t + b \sin t = \sin 2t$ for all t. Taking $t = 0$, we get

$a = a \cos 0 + b \sin 0 = \sin(2 \times 0) = 0.$

Thus we have $b \sin t = \sin 2t$ for all t. Now taking $t = \pi/2$, we get

$b = b \sin(\pi/2) = \sin(2(\pi/2)) = \sin\pi = 0.$

Thus we have $0 = \sin 2t$. Since this is wrong, we conclude that v $range\ T$.

5. $T(f) = f'' - f$: $C^2(R) \to C(R)$

   $u = 1 + 2t$, $v = \sin 2t$

Answer: $u \in rangeT$ means $f'' - f = 1 + 2t$ has a solution. From $(a + bt)'' = 0$, we see that $f(t) = -1 - 2t$ is a solution. Thus u is in the range.

Similarly, to find a solution of $f'' - f = \sin 2t$, we use $(\sin 2t)'' = -4\sin 2t$. Then from $(c \sin 2t)'' - (c \sin 2t) = -5c \sin 2t$, we see that $f(t) = (-1/5) \sin 2t$ is a solution. Thus v is in the range.

6. $T(p(t)) = (p(0), p(1), p(2), p(3))$: $P_2 \to R^4$

   $u = (1, 2, 3, 4)$, $v = (1, 2, 3, 0)$

Answer: A polynomial of degree 2 is of the form $p(t) = a + bt + ct^2$, and $T(p(t)) = (a + b + c, a + 2b + 4c, a + 3b + 9c, a + 4b + 16c)$. Thus $u \in rangeT$ means the system

$$a + b + c = 1$$

$$a + 2b + 4c = 2$$

$$a + 3b + 9c = 3$$
$$a + 4b + 16c = 4$$

has solutions. From the first 3 equations, we get a *unique* solution $a = 0, b = 1, c = 0$. This also satisfy the 4th equation. Thus $u \in rangeT$.

Similarly, $v \in rangeT$ means the system

$$a + b + c = 1$$
$$a + 2b + 4c = 2$$
$$a + 3b + 9c = 3$$
$$a + 4b + 16c = 0$$

has solutions. The first 3 equations imply $a = 0, b = 1, c = 0$, which does not satisfy the 4th equation. Thus $v \in rangeT$.

Exercise: Let A be an m by n matrix. Prove that $colA = R^m \Rightarrow n \geq m$.

Answer: By the third description of $colA$, $colA = R^m$ means $Ax = b$ has solutions for any $b \in R^m$. Then the inequality $n \geq m$ follows from this.

z For what fixed function $m(t)$ is the linear transformation $M(f) = mf$: $F(R) \rightarrow F(R)$ onto?

Answer: If $m(t) \neq 0$ for any $t \in R$, then $m(t)^{-1}$ is also a function on R, and for any $f \in F(R)$, we have $M(m^{-1}f) = f$. Thus M is onto.

If $m(t_0) = 0$ for some $t_0 \in R$, then we always have $M(f)(t_0) = m(t_0)f(t_0) = 0\, f(t_0) = 0$. Therefore functions not vanshing at t_0 are not in the range, and M is not onto.

We conclude that $M(f) = mf$: $F(R) \rightarrow F(R)$ is onto $\Leftrightarrow m(t) \neq 0$ for any $t \in R$.

Exercise: For what fixed matrix A is the linear transformation $M(X) = AX$ onto? What about the linear transformation $M(X) = XA$?

Answer: $M(X) = AX$ is onto means that $AX = B$ has solutions for any B. By looking at the columns of X and B, and making use of this formula, we see the condition is equivalent to $Ax = b$ has solutions for any b. In fact, if b is taken to be the i–th column of B, then the solution x is the i–th column of X. From computational perspective, $M(X) = AX$ is onto is equivalent to all rows of A are pivot.

Similarly, the transformation $M(X) = XA$ is onto means that $XA = B$ has solutions for any B. By taking transpose and using the property of matrix product under transpose, we see that this is equivalent to $A^T Y = C$ has solutions for any C (where $Y = X^T$ and $C = B^T$). Thus we conclude that

$M(X) = XA$ is onto $\Leftrightarrow A^T x = b$ has solutions for any b

We will see in future theory that this is actually equivalent to the uniqueness of the solution of $Ax = b$. Thus from computational perspective, $M(X) = XA$ is onto is equivalent to all columns of A are pivot.

Exercise: If A and B are matrices with same number of rows, how are the column spaces of A, B, $[A\ B]$ related?

Answer: By the second description of the columns space, $col[A\ B]$ consists of vectors of the form

$$[A\ B] = \begin{bmatrix} x \\ y \end{bmatrix} \quad Ax + By,$$

where x and y run through all vectors (in appropriate euclidean spaces). Since all Ax form $colA$ and all By form $colB$, we conclude that

$$col[A\ B] = colA + colB.$$

Exercise: Let $T, S: V \to W$ be linear transformations. How are the ranges of T, S, and $T + S$ related?

Answer: A vector in the range of $T + S$ is of the form $(T + S)(v) = T(v) + S(v)$ for some $v \in V$. In particular, it is the sum of a vector in *rangeT* and a vector in *rangeS*. Thus we conclude

$$range(T + S) \subset rangeT + rangeS.$$

Note that the two sides are not necessarily equal (take $S = -T$, for example).

Exercise: Let $T: V \to W$ and $S: U \to V$ be linear transformations. How are the ranges of T, S, and TS related?

Answer: An image of TS is of the form $T(S(u))$ for some $u \in U$. In particular, it is also the image of $S(u)$ under the transformation T. Thus we conclude

$$rangeTS \subset rangeT.$$

By taking $S = O$, the zero transformation, we see that the two sides may not be equal.

Chapter 6

Direct Sum of Vector Spaces

DIRECT SUM

The direct sum of two vector spaces V and W is the set V E W of pairs of vectors (v, w) in V and W, with the operations

$$(v, w) + (v', w') = (v + v', w + w'),$$

$$c(v, w) = (cv, cw).$$

The direct sum of more than two vector spaces can be similarly defined.

The direct sum of two vector spaces is a vector space.

The proof is a routine verification of the eight properties. The following verifies the first property

$$\begin{aligned}(v, w) + (v', w') &= (v + v', w + w') \text{ (definition)}\\ &= (v' + v, w' + w) \text{ (commutativity of additions in } V \text{ and } W)\\ &= (v', w') + (v, w). \text{ (definition)}\end{aligned}$$

The zero vector in V E W is given by $(0_V, 0_W)$. The nagative of (v, w) is $(-v, -w)$. The verification of the other properties are similar.

Example: By grouping the first three and the last two coordinates, the euclidean space R^5 is isomorphic to $R^3 \oplus R^2$.

Example: We often think of a 2 by 3 matrix as columns: $A = [u\ v\ w]$, where $u, v, w \in R^2$. This gives a natural isomorphism between $M(2, 3)$ and R^2 E R^2 E R^2. Moreover, this particular example of partitioning 4 by 5 matrices gives a natural isomorphism between $M(4, 5)$ and $M(3, 3) \oplus M(3, 2) \oplus M(1, 3)$ E $M(1, 2)$.

DIRECT SUM OF SUBSPACES

Let $H_1, H_2, \ldots, H_k$ be subspaces of V. Then

$$S(v_1, v_2,..., v_k) = v_1 + v_2 +... + v_k: H_1 \oplus H_2 \oplus ... \oplus H_k \to V$$

is a linear transformation with the subspace $H_1 + H_2 +... + H_k$ as the range.

The sum $H_1 + H_2 +... + H_k$ of subspaces of a vector space V is a direct sum if the natural linear transformation S is one-to-one. If we also have $V = H_1 + H_2 +... + H_k$, then we say V is a direct sum of the subspaces and denote $V = H_1 \mathrm{E}\, H_2 \oplus ... \oplus H_k$.

Thus a sum of subspaces is a direct sum simply means that $H_1 + H_2 +... + H_k$ is isomorphic to $H_1 \quad H_2 \quad ... \quad H_k$ in a natural way. By the definition, the condition for a sum to be direct is

$$v_1 + v_2 +... + v_k = 0, v_i \in H_i \Rightarrow v_1 = v_2 = ... = v_k = 0.$$

In the special case $H_i = span\{u_i\}$ are 1–dimensional lines (u_i “‘ 0), we have $v_i = c_i u_i$, and the condition above becomes the linear independence of $v_1, v_2,..., v_k$. Thus direct sum generalizes the concept of linear independence. Moreover, in case $k = 2$, we have *kernelS* $= \{(v, -v): v \Rightarrow H_1 \cap H_2\}$. Therefore

$$H_1 + H_2 \text{ is a direct sum} \Leftrightarrow H_1 \cap H_2 = \{0\}.$$

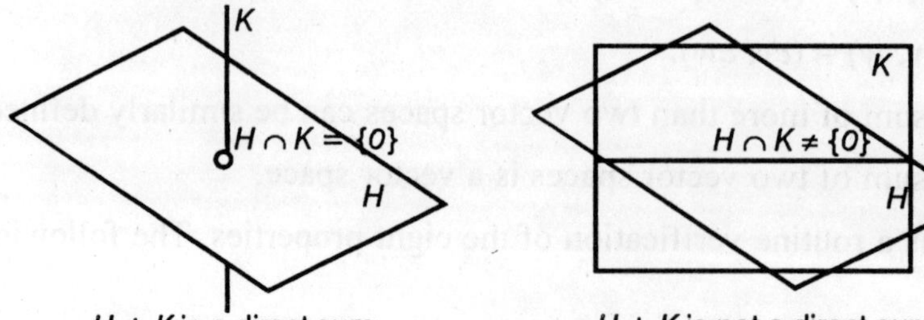

H + *K* is a direct sum   *H* + *K* is not a direct sum

Example: Consider subspaces

$$H_1 = span\{(1, -1, 0, 1)\}, H_2 = span\{(3, -1, 4, 3), (0, 1, 4, -2)\}, H_3 = span\{(1, 1, 3, 0)\}$$

of R^4. For any vectors

$$v_1 = c_1(1, -1, 0, 1), v_2 = c_2(3, -1, 4, 3) + c_3(0, 1, 4, -2), v_3 = c_4(1, 1, 3, 0)$$

in the three subspaces, we have

$$v_1 + v_2 + v_3 = c_1(1, -1, 0, 1) + c_2(3, -1, 4, 3) + c_3(0, 1, 4, -2) + c_4(1, 1, 3, 0).$$

By the row operations in this example (look at columns 1, 2, 4, 6), the four vectors are linearly independent. Therefore

$$v_1 + v_2 + v_3 = 0 \Rightarrow c_1 = c_2 = c_3 = c_4 = 0 \Rightarrow v_1 = v_2 = v_3 = 0,$$

and we conclude $H_1 + H_2 + H_3$ is a direct sum.

In general, if a basis of V is divided into a disjoint union of subsets, then the subspaces spanned by the subsets form a direct sum of V.

Example: Consider the matrix

$$A = \begin{bmatrix} 1 & 3 & 0 \\ -1 & -1 & 1 \\ 0 & 4 & 4 \\ 1 & 3 & -2 \end{bmatrix}$$

Using first two rows as the coefficient matrix, we have a kernel subspace

$$H = \{(x_1, x_2, x_3): x_1 + 3x_2 = 0, -x_1 - x_2 + x_3 = 0\}.$$

Using the last two rows, we also have

$$K = \{(x_1, x_2, x_3): 4x_2 + 4x_3 = 0, x_1 + 3x_2 - 2x_3 = 0\}.$$

The intersection $H \cap K$ consists of solutions of the system $Ax = 0$. From this example, we know the solution has to be trivial. Therefore $H \cap K = \{0\}$, and $H + K$ is a direct sum.

Example: Since the only symmetric and skew symmetric matrix is the zero matrix, the sum of the subspace of symmetric matrices and the subspace of skew–symmetric matrices is a direct one. In fact, by this exercise, we may say that the space of all *n* by *n* matrices is the *direct* sum of symmetric and skew symmetric matrices.

Example: By this example, the even functions form a subspace $H = \{f \in F(R): f(t) = f(-t)\}$ of $F(R)$. By the similar reason, the odd functions also form a subspace $K = \{f \in F(R): f(t) = -f(-t)\}$. Now for any function $f(t) \in F(R)$, we construct

$$f_E(t) = \left[\frac{f(t) + f(-t)}{2}\right], f_O(t) = \left[\frac{f(t) - f(-t)}{2}\right].$$

Then it is easy to verify that f_E is an even function, f_O is an odd function, and $f = f_E + f_O$. This means $H + K = F(R)$. Moreover, if f is both even and odd, then we will have $f(t) = f(-t) = -f(t)$, which implies $f(t) = 0$. Thus we conclude H E $K = F(R)$.

PROPERTIES OF DIRECT SUM

Recall that the sum of spans of single vectors being a direct sum is equivalent to the linear independence. The following is an extension of the observation, which complements this result about spanning vectors in a sum.

Let $H_1 + H_2 + ... + H_k$ be a direct sum of subspaces of V. Let $B_i = \{v_{i1}, v_{i2}, ...\}$ be linearly independent vectors in H_i. Then $B_1 \cup B_2 \cup ... \cup B_k = \{v_{ij}: \text{all } i, j\}$ are linearly independent.

Proof. For the case $k = 3$ and $B_1 = \{v_{11}, v_{12}\}$, $B_2 = \{v_{21}, v_{22}\}$, $B_3 = \{v_{31}, v_{32}, v_{33}\}$ are linearly independent vectors in H_1, H_2, H_3, we will prove that $B = \{v_{11}, v_{12}, v_{21}, v_{22}, v_{31}, v_{32}, v_{33}\}$ is linearly independent. The general case is similar.

Let a linear combination of vectors of B be trivial:

$$(c_{11}v_{11} + c_{12}v_{12}) + (c_{21}v_{21} + c_{22}v_{22}) + (c_{31}v_{31} + c_{32}v_{32} + c_{33}v_{33}) = v_1 + v_2 + v_3 = 0.$$

We need to conclude that all the seven coefficients are zero. Since

$$v_1 = c_{11}v_{11} + c_{12}v_{12} \in H_1, v_2 = c_{21}v_{21} + c_{22}v_{22} \in H_2, v_3 = c_{31}v_{31} + c_{32}v_{32} + c_{33}v_{33} \in H_3,$$

the fact that $H_1 + H_2 + H_3$ is a direct sum implies that $v_1 = v_2 = v_3 = 0$. Then by $c_{11}v_{11} + c_{12}v_{12} = v_1 = 0$ and $B_1 = \{v_{11}, v_{12}\}$ being linearly independent, we have $c_{11} = c_{12} = 0$. Similarly, we have $c_{21} = c_{22} = 0$ and $c_{31} = c_{32} = c_{33} = 0$.

In the special case B_i is a basis of H_i, we conclude the $\Rightarrow$ direction of the following.

$$H_1 + H_2 + ... + H_k \text{ is a direct sum} \Leftrightarrow dim(H_1 + H_2 + ... + H_k) = dimH_1 + dimH_2 + ... + dimH_k.$$

We remark that by this exercise, we have $dim(H_1 + H_2 + ... + H_k) \le dimH_1 + dimH_2 + ... + dimH_k$ in general.

Proof For the $\Leftarrow$ direction, we again take a basis B_i of H_i. Then $B = B_1 \cup B_2 \cup ... \Theta B_k$ contains $dimH_1 + dimH_2 + ... + dimH_k$ number of vectors and spans $H_1 + H_2 + ... + H_k$. In case the equality $dim(H_1 + H_2 + ... + H_k) = dimH_1 + dimH_2 + ... + dimH_k$ holds, the number of vectors in B is the same as the dimension of $H_1 + H_2 + ... + H_k$. By this result, B is a basis of $H_1 + H_2 + ... + H_k$.

Now for the special case $k = 3$ and $B_1 = \{v_{11}, v_{12}\}$, $B_2 = \{v_{21}, v_{22}\}$, $B_3 = \{v_{31}, v_{32}, v_{33}\}$, the linear independence of B means that

$$(c_{11}v_{11} + c_{12}v_{12}) + (c_{21}v_{21} + c_{22}v_{22}) + (c_{31}v_{31} + c_{32}v_{32} + c_{33}v_{33}) = 0$$

implies all seven coefficients are zero. Equivalently, for the arbitrary (becase B_1, B_2, B_3 are bases) vectors

$$v_1 = c_{11}v_{11} + c_{12}v_{12} \in H_1, v_2 = c_{21}v_{21} + c_{22}v_{22} \in H_2, v_3 = c_{31}v_{31} + c_{32}v_{32} + c_{33}v_{33} \in H_3,$$

$v_1 + v_2 + v_3 = 0$ implies $v_1 = v_2 = v_3 = 0$. This means exactly that $H_1 + H_2 + H_3$ is a direct sum.

DIRECT SUMMAND

Let V be a vector space. Let H be a subspace of V. A direct summand of H in V is a subspace K such that $V = H$ E K.

Example: From this example, the subspace of skew-symmetric matrices is a direct summand of the subspace of symmetric matrices.

From this example, the subspace of odd functions is a direct summand of the subspace of even functions.

Example: As explained in this example, the vectors (3, 1, 2), (1, –1, 2), (-1, 1, 1) form a basis of R^3. Let

$H = span\{(3, 1, 2), (1, -1, 2)\}, K = span\{(-1, 1, 1)\}$.

Then the fact that the three vectors span R^3 implies $H + K = R^3$. Moreover, an argument similar to this example shows that the sum $H + K$ is a direct one. Therefore K is a direct summand of H (and H is also a direct summand of K).

The argument in the example above can be extended. In general, if a basis $\{b_1, b_2,..., b_n\}$ of V is divided into two disjoint parts $\{b_1, b_2,..., b_k\}$ and $\{b_{k+1}, b_{k+2},..., b_n\}$, then $span\{b_{k+1}, b_{k+2},..., b_n\}$ is a direct summand of $span\{b_1, b_2,..., b_k\}$. The idea can be used to prove the existence of the direct summand.

Any subspace of a finite dimensional vector space has a direct summand.

Proof. Let H be a subspace of a finite dimensional vector space V. Let $B = \{b_1, b_2,..., b_k\}$ be a basis of H. If we can extend B to a basis $B' = \{b_1, b_2,..., b_n\}$ be a basis of V, then by the remark made before, the subspace $K = span\{b_{k+1}, b_{k+2},..., b_n\}$ is a direct summand of $H = span\{b_1, b_2,..., b_k\}$.

If $V = H = span\{b_1, b_2,..., b_k\}$, then B is already a basis of V. Otherwise, we can find b_{k+1} O H. By this result, $b_1, b_2,..., b_k, b_{k+1}$ are linearly independent.

If $V = H_1 = span\{b_1, b_2,..., b_k, b_{k+1}\}$, then we have extended B to a basis of V by adding one extra vector. Otherwise, we find another b_{k+2} O H_1. Again, $b_1, b_2,..., b_k, b_{k+1}, b_{k+2}$ are linearly independent.

Again we ask whether $V = H_2 = span\{b_1, b_2,..., b_k, b_{k+1}, b_{k+2}\}$. The process may continue, and we get larger and larger collection of linearly independent vectors. Since V is finite dimensional, the process will eventually stop at $V = H_{n-k} = span\{b_1, b_2,..., b_k, b_{k+1}, b_{k+2},..., b_n\}$. Then $B' = \{b_1, b_2,..., b_k, b_{k+1}, b_{k+2},..., b_n\}$ is a basis obtained by extending B.

The idea of the proof is by extending B one vector at a time (and hence extending H one dimension at a time) until V is "filled up".

LINEAR TRANSFORMATION BETWEEN DIRECT SUMS

Let $V = V_1 \oplus V_2$ and $W = W_1 \oplus W_2$ be direct sums. We express the vectors in V and W vertically

$$v = \begin{bmatrix} v_1 \\ v_2 \end{bmatrix} \in V, w = \begin{bmatrix} w_1 \\ w_2 \end{bmatrix} \in W, \text{ where } v_1 \in V_1, v_2 \in V_2, w_1 \in W_1, w_2 \in W_2.$$

Then a linear transformation $T: V \to W$ is of the form

$$T(v) = \begin{bmatrix} T_1(v) \\ T_2(v) \end{bmatrix}, \text{ where } T_1: V \to W_1 \text{ and } T_2: V \to W_2.$$

From

$$T(u+v)=\begin{bmatrix} T_1(v) \\ T_2(v) \end{bmatrix},\ T(u)+T(v)=\begin{bmatrix} T_1(u)+T_1(v) \\ T_2(u)+T_2(v) \end{bmatrix},$$

we get $T_1(u+v) = T_1(u) + T_1(v)$, $T_2(u+v) = T_2(u) + T_2(v)$. Similarly, we have $T_1(cu) = cT_1(u)$, $T_2(cu) = cT_2(u)$. Thus T_1 and T_2 are linear.

The linearity of T_1 implies $T_1(v) = T_{11}(v_1) + T_{12}(v_2)$, where

$$T_{11}(v_1) = T_1\begin{bmatrix} v_1 \\ 0 \end{bmatrix} : V_1 \to W_1,\ T_{12}(v_2) = T_1\begin{bmatrix} 0 \\ v_2 \end{bmatrix} : V_2 \to W_1$$

are also linear transformations. Similarly, we have $T_2(v) = T_{21}(v_1) + T_{22}(v_2)$ for linear transformations $T_{21}: V_1 \to W_2$ and $T_{22}: V_2 \to W_2$. In summary, we have

$$T(v)=\begin{bmatrix} T_1(v) \\ T_2(v) \end{bmatrix}=\begin{bmatrix} T_{11}(v_1)+T_{12}(v_2) \\ T_{21}(v_1)+T_{22}(v_2) \end{bmatrix}=\begin{bmatrix} T_{11} & T_{12} \\ T_{21} & T_{22} \end{bmatrix}\begin{bmatrix} v_1 \\ v_2 \end{bmatrix}.$$

In the last equality, we consider $T(v)$ as product of T and v.

The computation illustrates that linear transformations between direct sums are given by "partitioned transformations", comparable to partitioned matrix.

By writing vectors in direct sums vertically and thinking of $T(v)$ as product of T and v, a linear transformation

$T: V_1 \oplus V_2 \oplus ... \oplus V_k \to W_1 \oplus W_2 .. \oplus W_l$

is given by

$$T\begin{bmatrix} v_1 \\ v_2 \\ : \\ v_k \end{bmatrix}=\begin{bmatrix} T_{11} & T_{12} & .. & T_{1k} \\ T_{21} & T_{22} & .. & T_{2k} \\ : & : & & : \\ T_{l1} & T_{l2} & .. & v_k \end{bmatrix}\begin{bmatrix} v_1 \\ v_2 \\ : \\ v_k \end{bmatrix}$$

where $T_{ij}: V_j \to W_i$ are linear transformations.

Exercise: Determine whether the sum is a direct sum.

1. $span\{(1, -1, 0)\} + span\{(-2, 3, 1)\} + span\{(1, 0, 1)\}$

Answer: By this remark on the relation between the direct sum of 1-dimensional subspaces and the linear independence, we should consider whether the vectors $(1, -1, 0)$, $(-2, 3, 1)$, $(1, 0, 1)$ are linearly independent. The following row operation

$$\begin{bmatrix} 1 & -2 & 1 \\ -1 & 3 & 0 \\ 0 & 1 & 1 \end{bmatrix} \to \begin{bmatrix} 1 & -2 & 1 \\ 0 & 1 & 1 \\ 0 & 0 & 0 \end{bmatrix}$$

indicates that the three vectors are linearly dependent. Therefore the sum is not a direct one.

Remark: Geometrically, the three spans are three lines in R^3. The three lines lie in one plane. In order to have a direct sum, the three lines have to be in positions similar to coordinate axis.

2. $span\{(1, -1, 1)\} + span\{(-2, 3, 1)\} + span\{(1, 0, 1)\}$

Answer Similar to the previous problem, we should consider whether the vectors (1, –1, 1), (–2, 3, 1), (1, 0, 1) are linearly independent. The following row operation

$$\begin{bmatrix} 1 & -2 & 1 \\ -1 & 3 & 0 \\ 1 & 1 & 1 \end{bmatrix} \to \begin{bmatrix} 1 & -2 & 1 \\ 0 & 1 & 1 \\ 0 & 0 & -3 \end{bmatrix}$$

indicates that the three vectors are linearly independent. Therefore the sum is a direct one.

3. $Span\{(1, -1, 1), (-2, 3, 1)\} + span\{(1, 0, 1)\}$

Answer: For $u = c_1(1, -1, 1) + c_2(-2, 3, 1) \in span\{(1, -1, 1), (-2, 3, 1)\}$ and $v = c_3(1, 0, 1) \in span\{(1, 0, 1)\}$, we have

$$\begin{aligned} u + v = c_1(1, -1, 1) + c_2(-2, 3, 1) + c_3(1, 0, 1) = (0, 0, 0) \\ \Rightarrow c_1 = c_2 = c_3 = 0 \\ \Rightarrow u = v = 0, \end{aligned}$$

where the first $\Rightarrow$ is due to the fact that the three vectors are linearly independent, a conclusion of the previous problem. Thus the sum is a direct one.

Please note the similarity between the problem and this example.

4. $Span\{(1, -1, 0, 1)\} + span\{(-2, 3, 1, 2)\} + span\{(0, 1, 1, 4)\} + span\{(-1, 2, 1, 0)\} + span\{(0, 4, 1, 1)\}$

Answer: Whether the sum is a direct one is equivalent to whether the vectors (1, –1, 0, 1), (–2, 3, 1, 2), (0, 1, 1, 4), (–1, 2, 1, 0), (0, 4, 1, 1) are linearly independent. Since five vectors in R^4 are always linearly dependent, the sum cannot be a direct one.

5. $Span\{(1, -1, 0, 1), (-2, 3, 1, 2), (0, 1, 1, 4)\} + span\{(-1, 2, 1, 0), (0, 4, 1, 1)\}$

Answer: By this remark, we would like to find out whether the intersection of the two subspaces is zero. Thus we consider when a vector $v = a(-1, 2, 1, 0) + b(0, 4, 1, 1)$ in the second subspace also lies in the first subspace. We form a matrix with the five vectors as columns and carry out row operation

$$\begin{bmatrix} 1 & -2 & 0 & -1 & 0 \\ -1 & 3 & 1 & 2 & 4 \\ 0 & 1 & 1 & 1 & 1 \\ 1 & 2 & 4 & 0 & 1 \end{bmatrix} \to \begin{bmatrix} 1 & -2 & 0 & -1 & 0 \\ 0 & 1 & 1 & 1 & 1 \\ 0 & 0 & 0 & 1 & 1 \\ 0 & 0 & 0 & 0 & 1 \end{bmatrix}$$

This is equivalent to

$$\begin{bmatrix} 1 & -2 & 0 \\ -1 & 3 & 1 \\ 0 & 1 & 1 \\ 1 & 2 & 4 \end{bmatrix} \to \begin{bmatrix} 1 & -2 & 0 & -a \\ 0 & 1 & 1 & a+b \\ 0 & 0 & 0 & a+nb \\ 0 & 0 & 0 & b \end{bmatrix}$$

Thus the condition for v to lie in the first subspace is $a + b = b = 0$. In other words, v has to be the zero vector. We conclude the sum is a direct one.

6. $\{(a + b, 3a - b, 2a + b): a, b \in R\} + \{(3a, a, 2a): a \in R\}$

Answer: In an earlier exercise, we have argued that the intersection of the two subspaces is $\{0\}$. Therefore the sum is a direct one.

7. $\{(x, y, z): x + y - z = 0, 2x - 3y + z = 0\} + \{(x, y, z): x + y - z = 0, x - 2y + 3z = 0\}$

Answer: In an earlier exercise, we have argued that the intersection of the two subspaces is $\{0\}$. Therefore the sum is a direct one.

8. $\{(a + b, 3a - b, 2a + b): a, b \Xi R\} + \{(x, y, z): x + y - z = 0, x - 2y + 3z = 0\}$

Answer: In an earlier exercise, we have argued that the intersection of the two subspaces is $\{0\}$. Therefore the sum is a direct one.

9. $\{f \in F(R): f(1) + f(2) = 0\} + \{a \cos(\pi t/2) + b \sin(\pi t/2): a, b \in R\}$

Answer: In an earlier exercise, we have argued that the intersection of the two subspaces is is all the scalar multiples of $\cos(\pi t/2) - \sin(\pi t/2)$.. Therefore the sum is not a direct one.

10. $\{f \in C(R): f(1) = 0\} \dotplus \{f \in C(R): f(2) = 0\} + \{f \in C(R): \int_0^1 f(1) = 0\}$

Answer We can easily find a nonzero function $f \in C(R)$ satisfying $f(1) = 0$ and $f(2) = 0$. An example is $f(t) = (t - 1)(t - 2)$. Then we have $f \in \{f \in C(R): f(1) = 0\}$, $-f \in \{f \in C(R): f(2) = 0\}$, $0 \in \{f \in C(R):$ +"$_0{}^1 f(1) = 0\}$, satisfying $f + (-f) + 0 = 0$. The example and this remark shows us that the sum is not a direct one.

11. $span\{\cos^2 t\} + span\{\sin^2 t\} + \{f \in C^2(R): f' + 4f = 0\}$

Answer: By this remark, the problem is whether $a\cos^2 t + b\sin^2 t + f = 0$ and $f' + 4f = 0$ would imply the numbers $a = b = 0$ and the function $f = 0$. From the first equality, we get

$f = -a\cos^2 t - b\sin^2 t$. Substituting into the second equation, we get $(2a + 2b)\cos^2 t + (2a + 2b)\sin^2 t = 0$. Thus we find it is possible to take $a = 1$ and $b = -1$. Therefore the sum is not a direct one.

12.$Span\{\cos^2 t\} + span\{\sin^2 t\} + \{f \in C^2(\mathrm{R}): f'' + f = 0\}$

Answer: Similar to the provious problem, we ask whether $a\cos^2 t + b\sin^2 t + f = 0$ and $f'' + f = 0$ would imply the numbers $a = b = 0$ and the function $f = 0$. The computation gives us $(-a + 2b)\cos^2 t + (2a - b)\sin^2 t = 0$. From $-a + 2b = 0$ and $2a - b = 0$, we conclude that $a = b = 0$. Therefore the sum is a direct one.

Exercise: Prove that V E $\{0\} = \{(v, 0): v \in V\}$ and $\{0\} \in W = \{(0, w): w \in W\}$ are subspaces of the direct sum V E W. Moreover, prove that V E W is the direct sum of V E $\{0\}$ and $\{0\}$ E W.

Answer: The equalities

$$(v, 0) + (v', 0) = (v + v', 0), c(v, 0) = (cv, 0)$$

imply that $V \oplus \{0\}$ is a subspace of V E W. Similarly, $\{0\}$E W is a subspace of V E W. The equality

$$(v, w) = (v, 0) + (0, w)$$

shows that any vector of V E W is the sum of a vector in $V \oplus \{0\}$ and another vector in $\{0\} \oplus W$. Therefore $(V \oplus \{0\}) + (\{0\} \oplus W) = V \oplus W$. The quality also implies that if the sum of a vector $(v, 0) \in V$ E $\{0\}$ and a vector $(0, w) \in \{0\} \oplus W$ is zero, then$(v, 0) = (0, w) = (0, 0)$. Therefore the sum $(V \oplus \{0\}) + (\{0\} \oplus W)$ is a direct one.

Exercise: How are the dual spaces of V, W and V E W related?

Answer: Applying this result to linear transformations $f: V \oplus W \to \mathrm{R}$, we see that f is given by a pair of linear functions $g: V \to \mathrm{R}$, $h: W \to \mathrm{R}$ via the formula

$$f(v, w) = g(v) + h(w).$$

where $g(v) = f(v, 0)$ and $h(w) = f(v, w)$. Moreover, the correspondence $f \leftrightarrow (g, h)$ given by the formula above matches the addition and the scalar multiplication. Therefore we conclude $(V$ E $W)^* = V^* \oplus W^*$.

FIELD

NUMBER SYSTEM AND FIELD

In developing the theory of linear algebra, we almost always use real numbers R as the scalar (coefficients, coordinates, entries, etc). In fact, we may also use other number systems such as

rational numbers Q: numbers of the form m/n, where m, n are integers;

complex numbers C: numbers of the form $a + ib$, where a, b are real numbers and $i = \sqrt{-1}$.

Two things need to be explained in order to justify the use of number systems other than the real numbers in linear algebra.

- Why is it *necessary* to consider non–real numbers? In other words, why is real numbers good enough?
- Why is it *possible* to use non–real numbers? In other words, what are the key features of scalars that are used in linear algebra?

For the first question, a general philosophical answer is that linear algebra is the language used to model linear structures. The linear structures not only appear over real number, but also appear over other number systems. We would like to emphasis that the *complex* linear structures are especially important.

One key advantage of C over R is that C is algebraically closed, meaning that any polynomial has complex roots. As a result, complex numbers (but not real numbers) provide solutions to many problems. For example, it is necessary to include complex numbers in order to gain deeper understanding of eigenvalue and eigenvectors.

For the second question, we may revisit all the theories we have developed and find out that the four usual operations of numbers (summation, substraction, multiplicaiton, division) are all the ones we needed. Moreover, we also used the usual properties of these operations, such as $a + b = b + a$, $a(b + c) = ab + ac$. This leads to the following generalization of real numbers.

A field is a set F, with two operations

addition

$$a + b: F \times F \to F$$

multiplication

$$ab: F \times F \to F$$

satisfying the following properties

commutative

$$a + b = b + a,\ ab = ba$$

associative

$$(a + b) + c = a + (b + c),\ (ab)c = a(bc)$$

distributive

$$a(b + c) = ab + ac$$

zero

There is a special element $0 \in F$ such that $0 + a = a$

nagative

For any $a \in F$ there is an element $-a$ such that $a + (-a) = 0$

unit

There is a special element $1 \in F$ such that $1a = a$

inverse

For any $a \in F - \{0\}$ there is an element a^{-1} such that $aa^{-1} = 1$

The substraction is then $a - b = a + (-b)$, and the division is $a/b = ab^{-1}$. Note the similarity to the definition of vector spaces.

Example: With the usual addition and multiplicaiton, the real numbers form a field R. The rational numbers also form a field Q. The key reason here is that if a and b are rational, then $a + b$ and ab are also rational. This makes Q a subfield of R (compare with the definition of subspace and this result).

The integers Z is not a field because the inverse does not always exist. The positive real numbers R_+ is also not a field because zero and negative do not exist.

Example: A complex numbers is a number of the form $z = a + ib$, where $i = \sqrt{-1}$ is defined by the property $i^2 = -1$.

The addition and multiplication are defined by the usual method:

$$(a + ib) + (c + id) = (a + c) + i(b + d),$$

$$(a + ib)(c + id) = ac + iad + ibc + i^2bd = (ac - bd) + i(ad + bc).$$

With these operations, we may verify that the collection C of all complex numbers is a field. The inverse is, in particular, given by

$$(a + ib)^{-1} = \left[\frac{a - ib}{a^2 + b^2}\right]$$

Example: Polynomials do not form a field because the inverse of a polynomial is generally not a polynomial. To get a field out of the polynomials, therefore, we introduce rational functions, which are the quotient of two polynomials. The following are some examples of rational functions

$$\left[\frac{1}{1+t}\right], \left[\frac{1+t^2}{1-t^2}\right], \left[\frac{2-t+t^2}{1-t^2}\right], \left[\frac{1+t^2}{1+t-t^2-t^3}\right], \left[\frac{3-5t+5t^4}{1+2t+t^2+t^3}\right].$$

The addition and multiplication can be done in the usual way:

$$\begin{bmatrix}1\\1+t\end{bmatrix}+\begin{bmatrix}1+t^2\\1-t^2\end{bmatrix}=\begin{bmatrix}2-t+t^2\\1-t^2\end{bmatrix},\begin{bmatrix}1\\1+t\end{bmatrix}\times\begin{bmatrix}1+t^2\\1-t^2\end{bmatrix}=\begin{bmatrix}2-t+t^2\\1+t-t^2-t^3\end{bmatrix}$$

With the two operations, we get the field R(t) of (real) rational functions.

In general, for any field F, we have the field $F(t)$ of rational functions over F.

Example: Let $F = \{0, 1, 2, 3, 4\}$. For any integer n, define "n modulo 5" to mean dividing the number by 5 and keeping the remainder only. For example,

$$13 = 2X5 + 3 \rightarrow 13 \text{ modulo } 5 = 2,$$

$$-8 = (-2)X5 + 2 \rightarrow -8 \text{ modulo } 5 = 2.$$

Then we define the addition and multiplication in F to be the usual one modulo 5. For example

$$3 + 4 = 7 \text{ modulo } 5 = 2,\ 1 + 3 = 4 \text{ modulo } 5 = 4;$$

$$2 \times 2 = 4 \text{ modulo } 5 = 4,\ 3 \times 4 = 12 \text{ modulo } 5 = 2.$$

With these operations, F form a field, called finite field of order 5.

The only nontrivial properties to check is the existence of inverse, which is done below,

$$1 \times 1 = 1 \text{ modulo } 5 = 1 \rightarrow 1^{-1} = 1,$$

$$2 \times 3 = 6 \text{ modulo } 5 = 1 \rightarrow 2^{-1} = 3,$$

$$3 \times 2 = 6 \text{ modulo } 5 = 1 \rightarrow 3^{-1} = 2,$$

$$4 \times 4 = 16 \text{ modulo } 5 = 1 \rightarrow 4^{-1} = 4.$$

Although the formulae above are somewhat counter–intuitive, we obtain them, however, by strictly following the rules.

The key reason behind the existence of the inverse is the fact that 5 is a prime number. In general, for any prime number p, we may define the finite field of order p in similar way.

COMPLEX NUMBER

In this section, we will list important facts about the complex numbers.

Geometry

We may identify C with the real plane R^2 by $a + ib \leftrightarrow (a, b)$. We call $a = Re(z)$ the real part and $b = Im(z)$ the imaginary part. Correspondingly, C is a real plane with specific choices of real and imaginary axis.

The geometry of R^2 leads to two concepts.

Geometry of R^2		*Complex number*	*Formula*	
Concept	Notation	Concept	Notation	
length	r	absolute value/norm	$\|z\|$	$\sqrt{a^2 + b^2}$
angle	θ	argument	$arg(z)$	$tan^{-1}(b/a)$

Note that angle/argument is a number well–defined only up to $+2k\pi$

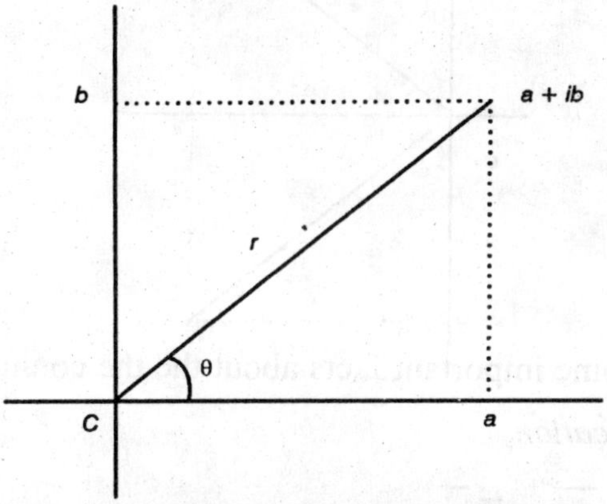

Clearly, the addition in C corresponds to the vector addition in R^2. To see the geometric meaning of multiplication, we note that the complex number is (the reason for the red equality is here)

$$z = a + ib = r\cos\theta + ir\sin\theta = r(\cos\theta + i\sin\theta) = re^{i\theta}.$$

Thus the product of two complex numbers is

$$z\,z' = re^{i\theta}\, r'e^{i\theta'} = rr'\, e^{i\theta}e^{i\theta'} = rr'\, e^{i(\theta+\theta')}.$$

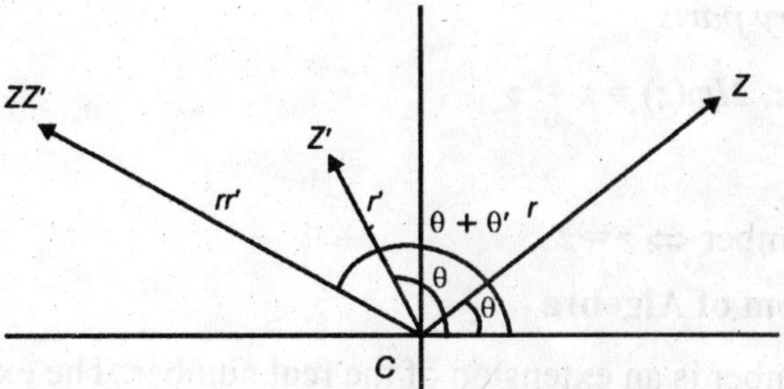

In other words, multiplying complex numbers means multiplying the lengths and adding the angles

Note: The exponential function $f(t) = e^{ct}$ for real t and c is characterized by the property that $f' = cf, f(0) = 1$.Now for $f(t) = \cos t + i\sin t$, we have $fA(t) = -\sin t + i\cos t = i(i\sin t + \cos t) = if(t)$ and $f(0) = \cos 0 + i\sin 0 = 1$. Thus $f(t) = \cos t + i\sin t$ fits into the similar characterization with c replaced by i. This suggests $\cos t + i\sin t = e^{it}$.

Conjugation Another important thing we need to know about complex numbers is the conjugation:

$a + ib = a - ib.$

Geometrically, this means the reflection in the real axis.

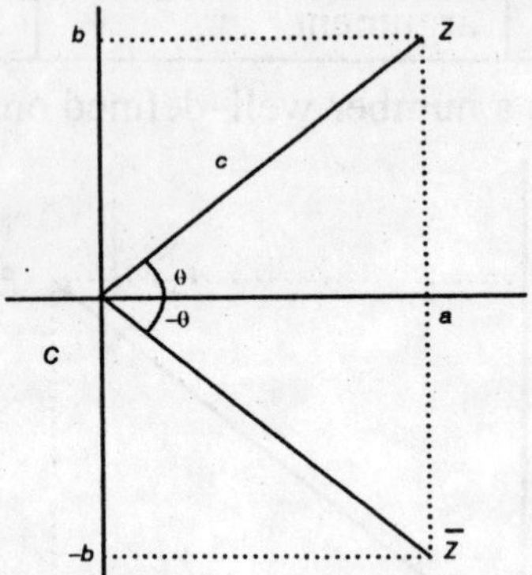

The following lists some important facts about the the conjugation.

addition and multiplication

$$\overline{w} = \overline{z} + \overline{z'} , \overline{zz'} = \overline{z}\,\overline{z'}$$

twice conjugation = identity

$$w = \overline{z} \Rightarrow z = \overline{w}$$

absolute value

$$z\overline{z} = |z|^2$$

real and imaginary parts

$$2Re(z) = z + z, 2Im(z) = z - \overline{z}$$

realness

$$z \text{ is a real number} \Leftrightarrow z = \overline{z}$$

Fundamental Theorem of Algebra

The complex number is an extension of the real number. The extension was invented in order to solve problems that require taking the square root of – 1. A typical such problem is to find roots (solutions of $p(t) = 0$) of quadratic polynomials $p(t) = a_0 + a_1 t + a_2 t^2$. In general, a (real) quadratic polynomial may not have real roots. On the other hand, (real or complex) quadratic polynomials always have complex roots.

In fact, the existence of complex roots is not restricted to quadratic polynomials. The following is the so called fundamental theorem of algebra.

· Any complex polynomial of positive degree has complex roots

For any field F, we may also ask whether any polynomial with coefficients in F always has solutions in F. If the answer is yes, then we say F is algebraically closed. Thus C is an algebraically closed field. It is often advantageous to develop theories (especially the ones involving polynomials) over an algebraically closed field.

LINEAR ALGEBRA OVER A FIELD

Linear algebra can be introduced over any field because we can define the concept of vector space over any field. All we need to do is to change real number R to a field F in the definition of vector spaces over R.

Let F be a field. A vector space over F is a set V, its elements u called vectors, with two operations

addition

$u + v: V \times V \rightarrow V$

scalar multiplication

$cu: F \times V \rightarrow V$

satisfying the following properties

commutative

$$u + v = v + u$$

associative

$$(u + v) + w = u + (v + w)$$

zero

There is a special vector $0 \in V$ such that $u + 0 = u = 0 + u$

negative

For any $u \in V$ there is a vector $-u$ such that $u + (-u) = 0$

distributive

$$c(u + v) = cu + cv$$

distributive

$$(c + d)u = cu + du$$

associative

$$(cd)u = c(du)$$

one

$$1u = u$$

The claim is that, with one exception, everything we did so far is still valid over field. The one exception has to do with fields of characteristic 2. We say an field has characteristic 2 if $2a = a + a = 0$ for any $a \in F$. There are certain arguments in which we need to divide the number 2 (= 1 + 1, where 1 is the unit element of the field). Such argument is only valid in a field of characteristic $\neq 2$.

We will do some examples to practice linear algebra over C.

Example: We try to solve the system

$$(2 + i)x_1 + (1 - i)x_2 = 2$$
$$(1 - 2i)x_1 + (-3 - 2i)x_2 = -5 - 2i$$

Noting that $1 - 2i = -i(2 + i)$, we do i[equation 1] + [equation 2] and get

$$(2 + i)x_1 + (1 - i)x_2 = 2$$
$$(-2 - i)x_2 = -5$$

Then

$$x_2 \frac{-5}{-2 - i} = \frac{-5(-2 + i)}{(-2 - i)(-2 + i)} = \frac{-5(-2 + i}{5} = 2 - i$$

and

$$x_1 = \frac{2 - (1 - i)(2 - i)}{2 + i} = \frac{(1 + 3i)(2 - i)}{(2 + i)(2 - i)} = \frac{5 + 5i}{5} = 1 + i$$

Alternatively, we may also use the inverse formula and get

$$\begin{bmatrix} 2+i & 1-i \\ 1-2i & -3-2i \end{bmatrix}^{-1} = \left[\frac{1}{(2+i)(-3-2i) - (1-i)(1-2i)}\right] \begin{bmatrix} -3-2i & -1+i \\ -1+2i & 2+i \end{bmatrix}$$

$$= \left[\frac{-3+5i}{(-3-4i)(-3+4i)}\right] \begin{bmatrix} -3-2i & -1+i \\ -1+2i & 2+i \end{bmatrix} = \frac{1}{25} \begin{bmatrix} 17-6i & -1-7i \\ -5-10i & -10+5i \end{bmatrix}$$

Then the solution is

$$x = \frac{1}{-3-4i} \begin{bmatrix} -3-2i & -1+i \\ -1+2i & 2+i \end{bmatrix} \begin{bmatrix} 2 \\ -5-2i \end{bmatrix} = \left[\frac{-3+4i}{(-3-4i)(-3+4i)} \begin{matrix} 1-7i \\ -10-5i \end{matrix}\right] = \begin{bmatrix} 1+i \\ 2-i \end{bmatrix}$$

Example: Consider a matrix

$$A = \begin{bmatrix} 1 & 0 & 1+i & 1 \\ 1 & -1+2i & -1 & i \\ 1 & 1+3i & -2+2i & -1 \\ 1 & -i & 2+i & -i \end{bmatrix}$$

By (–1)[row 1] + [row 2], (–1)[row 1] + [row 3], (–1)[row 1] + [row 3], we have

$$\begin{bmatrix} 1 & 0 & 1+i & 1 \\ 0 & -1+2i & -2-i & -1+i \\ 0 & 1+3i & -3+i & -2 \\ 0 & -i & 1 & -1-i \end{bmatrix}$$

By [row 2] ↔ [row 4], (3 – i)[row 2] + [row 3], (2 + i)[row 2] + [row 4], we have

$$\begin{bmatrix} 1 & 0 & 1+i & 1 \\ 0 & -i & 1 & -1-i \\ 0 & 0 & 0 & -6-2i \\ 0 & 0 & 0 & -2-2i \end{bmatrix}$$

By [row 3] ↔ [row 4], (–2 + i)[row 3] + [row 4], we have

$$\begin{bmatrix} 1 & 0 & 1+i & 1 \\ 0 & -i & 1 & -1-i \\ 0 & 0 & 0 & -2-2i \\ 0 & 0 & 0 & 0 \end{bmatrix}$$

Thus the four columns of A are linearly dependent and do not span C^4. Note that in the linear dependence relation

$(1 + i)[\text{col } 1] + i[\text{col } 2] + (-1)[\text{col } 3] + 0[\text{col } 4] = 0,$

it is often necessary to take complex coefficients. We also know $dim_C colA = 3$, $dim_C nulA = 1$, where dim_C is the complex dimension. The dimensions we used before were real dimensions.

The discussion in the complex euclidean space C^4 can also be translated to other complex vector spaces. For example, we know

$$p_1(t) = 1 + t + t^2 + t^3,$$
$$p_2(t) = (-1 + 2i)t + (1 + 3i)t^2 - it^3,$$
$$p_3(t) = (1 + i) - t + (-2 + 2i)t^2 + (2 + i)t^3,$$
$$p_4(t) = 1 + it - t^2 - it^3,$$

are linearly dependent and span a (complex) dimension 3 subspace of the vector space $P_3(C)$of complex polynomials of degree ≤ 3.

Chapter 7

Eigenvalue and Eigenvector

The linear transformation $T(x_1, x_2) = (5x_1, 15x_2)$ has a simple geometrical meaning: stretch x_1-direction 5 times and stretch x_2-direction 15 times. In this exercise, we saw other examples of linear transformations with clear geometrical meanings.

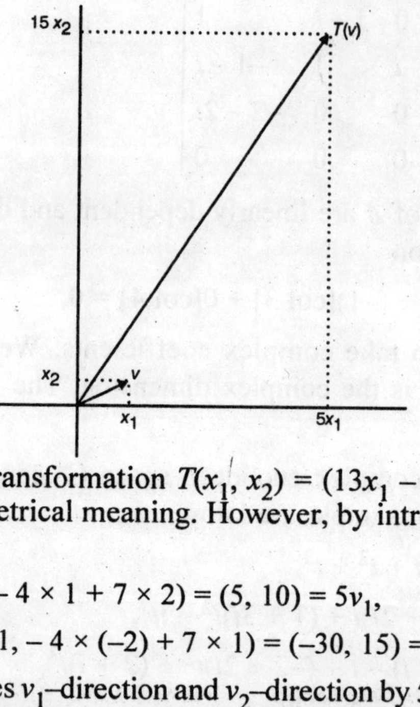

In contrast, the linear transformation $T(x_1, x_2) = (13x_1 - 4x_2, -4x_1 + 7x_2)$ does not appear to have simple geometrical meaning. However, by introducing a basis $v_1 = (1, 2)$, $v_2 = (-2, 1)$, we get

$T(v_1) = (13 \times 1 - 4 \times 2, -4 \times 1 + 7 \times 2) = (5, 10) = 5v_1,$

$T(v_2) = (13 \times (-2) - 4 \times 1, -4 \times (-2) + 7 \times 1) = (-30, 15) = 15v_2.$

In other words, T stretches v_1–direction and v_2–direction by 5 and 15 times, respectively. From the viewpoint of a new basis, T is geometrically as simple as the last one.

For another example, the linear transformation T: $R^3 \to R^3$ given by the matrix

$$A = \begin{bmatrix} 2/3 & -1/3 & -1/3 \\ -1/3 & 2/3 & -1/3 \\ -1/3 & -1/3 & 2/3 \end{bmatrix}$$

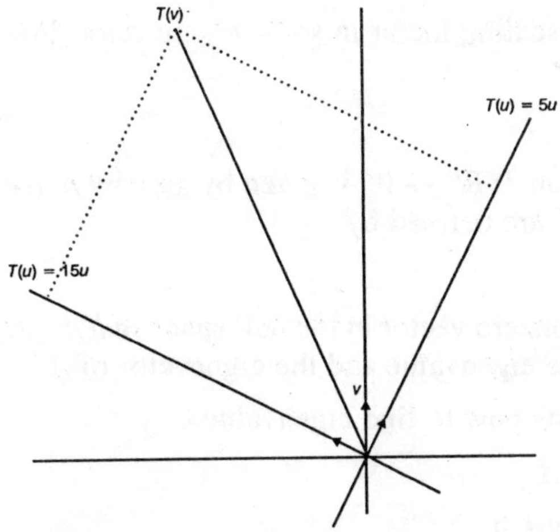

appears to be messy. From this exercise, however, we have $T(v_1) = v_1$, $T(v_2) = v_2$, $T(v_3) = 0$ for the basis $v_1 = (-1, 1, 0)$, $v_2 = (-1, 0, 1)$, $v_3 = (1, 1, 1)$. Thus T preserves (i.e., multiplying 1) in v_1 and v_2 directions and kills (i.e., multiplying 0) in v_3 direction.

The understanding above is very useful for answering many questions about the linear transformation. For example, the powers of $T(x_1, x_2) = (13x_1 - 4x_2, -4x_1 + 7x_2)$ is characterized by

$T^k(v_1) = 5^k v_1$, $T^k(v_2) = 15^k v_2$,

so that

$T^k(0, 5) = T^k(2v_1 + v_2) = 2T^k(v_1) + T^k(v_2) = 2\text{X}5^k(1, 2) + 15^k(-2, 1) = 5^k(2 - 3^k 2, 4 + 3^k)$.

A direct computation would be

$$\begin{bmatrix} 13 & -4 \\ -4 & 7 \end{bmatrix}^k \begin{bmatrix} 0 \\ 5 \end{bmatrix},$$

which is impractical to do.

Similarly, direct computation of the powers of the 3 by 3 matrix A above is almost impossible. However, the geometric meaning tells us that $A^k = A$ for any k.

The key for the good geometrical understanding of a linear transformation is the directions for which the transformation is simply a scalar multiple. This leads to the following definition.

Let $T: V \rightarrow V$ be a linear transformation. If λ is a number and v “‘ 0 is a nonzero vector such that

$T(v) = \lambda v$,

then λ is an eigenvalue and v is an eigenvector.

Note that $v \neq 0$ is required because the zero vector does not provide a direction. Moreover, an eigenvector is always associated with an eigenvalue. On the other hand, an engenvalue is always the scaling factor in some eigenvector direction. The two concepts are always bound together.

COMPUTATION

A linear transformation $T: R^n \to R^n$ is given by an n by n matrix A. The eigenvalue λ and the eigenvector v of T are defined by

$Av = \lambda v.$

Equivalently, v is a nonzero vector in the null space $nul(A - \lambda I)$. The number λ and the vector v are also called the eigenvalue and the eignevctor of A.

The following suggests how to find eigenvalues.

λ is an eigenvalue of A

$\Leftrightarrow Av = \lambda v$ for some $v \neq 0$

$\Leftrightarrow (A - \lambda I)x = 0$ has $x = v$ as a nontrivial solution

$\Leftrightarrow A - \lambda I$ is not invertible

$\Leftrightarrow det(A - \lambda I) = 0.$

The characteristic polynomial of a square matrix A is $det(A - \lambda I)$.

Thus the eigenvalues and the eigenvectors can be computed as follows.

Step 1: Solve the characteristic equation $det(A - \lambda I) = 0$ and get the eigenvalues $\lambda_1, \lambda_2, \ldots.$

Step 2: For each eigenvalue λ_i, solve the homogeneous system $(A - \lambda_i I)x = 0$ and get the eigenvectors with λ_i as the eigenvalue.

In the second step, the answer is actually presented as a basis of $nul(A - \lambda_i I)$, called the eigenspace of A with eigenvalue λ_i.

Example: Consider the matrix

$$A = \begin{bmatrix} 13 & -4 \\ -4 & 7 \end{bmatrix}$$

of the linear transformation $T(x_1, x_2) = (13x_1 - 4x_2, -4x_1 + 7x_2)$ on R^2. The characteristic polynomial is

$$det(A - \lambda I) = \begin{bmatrix} 13-\lambda & -4 \\ -4 & 7-\lambda \end{bmatrix} = (13 - \lambda)(7 - \lambda) - (-4)(-4) = 75 - 20\lambda + \lambda^2.$$

Since $75 - 20\lambda + \lambda^2 = (5 - \lambda)(15 - \lambda)$, we get two eigenvalues $\lambda_1 = 5, \lambda_2 = 15$.

For $\lambda_1 = 5$, we have

$$A - 5I = \begin{bmatrix} 13-5 & -4 \\ -4 & 7-5 \end{bmatrix} = \begin{bmatrix} 8 & -4 \\ -4 & 2 \end{bmatrix}$$

The solutions of $(A - 5I)x = 0$ are of the form $x = c(1, 2)$, c arbitrary. We get an eigenvector $v_1 = (1, 2)$, which is really a basis of the eigenspace $nul(A - 5I)$.

For $\lambda_2 = 15$, we have

$$A - 15I = \begin{bmatrix} 13-5 & -4 \\ -4 & 7-5 \end{bmatrix} = \begin{bmatrix} -2 & -4 \\ -4 & -8 \end{bmatrix}$$

The solutions of $(A - 15I)x = 0$ are of the form $x = c(-2, 1)$, c arbitrary. We get an eigenvector $v_2 = (-2, 1)$, which is really a basis of the eigenspace $nul(A - 15I)$.

Example: Consider the matrix

$$A = \begin{bmatrix} 1 & 3 & -3 \\ -3 & 7 & -3 \\ -6 & 6 & -2 \end{bmatrix}$$

The characteristic polynomial has been computed in an earlier example: $det(A - \lambda I) = (4 - \lambda)^2(-2 - \lambda)$. We have two eigenvalues $\lambda_1 = 4$, $\lambda_2 = -2$.

For $\lambda_1 = 4$, we have

$$A - 4I = \begin{bmatrix} 1-4 & 3 & -3 \\ -3 & 7-4 & -3 \\ -6 & 6 & -2-4 \end{bmatrix} = \begin{bmatrix} -3 & 3 & -3 \\ -3 & 3 & -3 \\ -6 & 6 & -6 \end{bmatrix}$$

The solutions of $(A - 4I)x = 0$ are of the form $x = x_2(1, 1, 0) + x_3(1, 0, -1)$, x_2 and x_3 arbitrary. We get two eigenvectors $v_1 = (1, 1, 0)$, $v_2 = (1, 0, -1)$, which form a basis of the eigenspace $nul(A - 4I)$.

For $\lambda_2 = -2$, we have

$$A + 2I = \begin{bmatrix} 1+2 & 3 & -3 \\ -3 & 7+2 & -3 \\ -6 & 6 & -2+3 \end{bmatrix} = \begin{bmatrix} 3 & 3 & -3 \\ -3 & 9 & -3 \\ -6 & 6 & 0 \end{bmatrix}$$

The solutions of $(A + 2I)x = 0$ are of the form $x = c(1, 1, 2)$, c arbitrary. We get an eigenvector $v_3 = (1, 1, 2)$, which is really a basis of the eigenspace $nul(A + 2I)$.

We may verify that v_1, v_2, v_3 actually form a basis of R^3. Geometrically, we have complete undestanding of the linear transformation given by A.

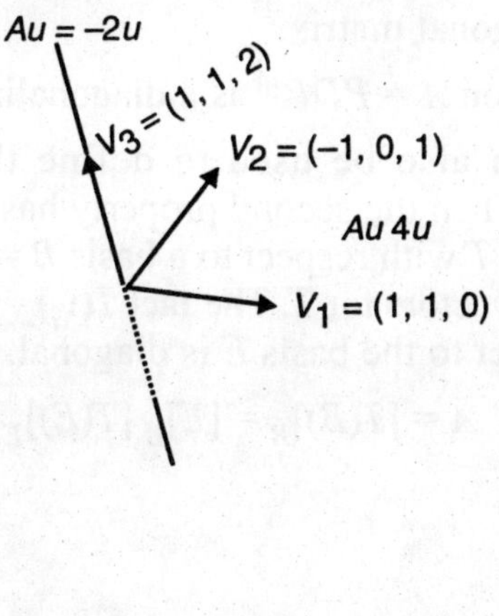

From the examples, we saw that for an n by n matrix A, $det(A - \lambda I)$ is a polynomial of degree n. The general reason can be found here.

If A is the matrix of a linear transformation $T: V \to V$ on a finite dimensional vector space with respect to one basis, then according to the discussion here, the matrix of T with respect to another basis is $B = PAP^{-1}$. By this property, we have

$$det(B - \lambda I) = det(PAP^{-1} - \lambda I) = det(P(A - \lambda I)P^{-1}) = det(P^{-1}P(A - \lambda I)) = det(A - \lambda I).$$

Thus the characteristic polynomial of a linear transformation does not depend on the choice of the basis. We also call $det(A - \lambda I)$ the characteristic polynomial of T.

DIAGONALIZATION

The eigenvalues and eigenvectors have been defined in terms of linear transformations and vectors. In this section, we give matrix interpretation.

Let A be an n by n matrix. Let $v_1, v_2,..., v_k$ be vectors and let $\lambda_1, \lambda_2,..., \lambda_k$ be numbers. Denote

$$P = [v_1\ v_2...\ v_k],\ D = \begin{bmatrix} \lambda_1 & 0 & .. & 0 \\ 0 & \lambda_2 & .. & 0 \\ : & : & & : \\ 0 & 0 & .. & \lambda_k \end{bmatrix}$$

Then the equalities

$$Av_1 = \lambda_1 v_1, Av_2 = \lambda_2 v_2,..., Av_k = \lambda_k v_k,$$

are equivalent to

$$AP = [Av_1\ Av_2...\ Av_k] = [\lambda_1 v_1\ \lambda_2 v_2...\ \lambda_k v_k] = [v_1\ v_2...\ v_k]D = PD.$$

Moreover, $v_1, v_2,..., v_k$ form a basis if and only if P is an invertible matrix, and the equality $AP = PD$ becomes $A = PDP^{-1}$.

We just proved that the two properties in the following definition are equivalent.

A matrix is diagonalizable if it has any of the following equivalent properties

- A has a basis of eigenvectors.
- A is similar to a diagonal matrix.

Moreover, any expression $A = PDP^{-1}$ is a diagonalization of A.

The first property can also be used to define the diagonalizability for linear transformations $T: V \to V$. Then the second property has the following interpretation. Let $A = [T(B)]_B$ be the matrix of T with respect to a basis $B = \{b_1, b_2,..., b_n\}$ of V. Let $E = \{v_1, v_2,..., v_n\}$ be a basis of eigenvectors for T. The fact $T(v_i) = \lambda_i v_i$ simply means that the matrix $D = [T(E)]_E$ of T with respect to the basis E is diagonal. By this discussion, we conclude

$$A = [T(B)]_B = [E]_B\ [T(E)]_E\ [E]_B^{-1} = PDP^{-1}.$$

Example: In an earlier example, we found the matrix

$$A = \begin{bmatrix} 13 & -4 \\ -4 & 7 \end{bmatrix}$$

has a basis $v_1 = (1, 2)$, $v_2 = (-2, 1)$ of eigenvectors, with eigenvalues $\lambda_1 = 5, \lambda_2 = 15$. Thus $A = PDP^{-1}$, with

$$P = [v_1\ v_2] = \begin{bmatrix} 1 & 2 \\ -2 & 1 \end{bmatrix}, D = \begin{bmatrix} \lambda_1 & 0 \\ 0 & \lambda_2 \end{bmatrix} = \begin{bmatrix} 5 & 0 \\ 0 & 15 \end{bmatrix}.$$

In another earlier example, we found the matrix

$$\begin{bmatrix} 1 & 3 & -3 \\ -3 & 7 & -3 \\ -6 & 6 & -2 \end{bmatrix}$$

has a basis $(1, 1, 0), (1, 0, -1), (1, 1, 2)$ of eigenvectors, with corresponding eigenvalues 4, 4, 2. This gives us

$$\begin{bmatrix} 1 & 3 & -3 \\ -3 & 7 & -3 \\ -6 & 6 & -2 \end{bmatrix} = \begin{bmatrix} 1 & 1 & 1 \\ 1 & 0 & 1 \\ 0 & -1 & 2 \end{bmatrix} \begin{bmatrix} 4 & 0 & 0 \\ 0 & 4 & 0 \\ 0 & 0 & 2 \end{bmatrix} \begin{bmatrix} 1 & 1 & 1 \\ 1 & 0 & 1 \\ 0 & -1 & 2 \end{bmatrix}^{-1}.$$

Note that by rearranging the order of the eigenvectors in the basis, with corresponding change of order for eigenvalues, we have other diagonalizations of the same matrix

$$\begin{bmatrix} 1 & 3 & -3 \\ -3 & 7 & -3 \\ -6 & 6 & -2 \end{bmatrix} = \begin{bmatrix} 1 & 1 & 1 \\ 1 & 0 & 1 \\ 0 & -1 & 2 \end{bmatrix} \begin{bmatrix} 2 & 0 & 0 \\ 0 & 4 & 0 \\ 0 & 0 & 4 \end{bmatrix} \begin{bmatrix} 1 & 1 & 1 \\ 1 & 0 & 1 \\ 2 & -1 & 0 \end{bmatrix}^{-1}$$

$$\begin{bmatrix} 1 & 3 & -3 \\ -3 & 7 & -3 \\ -6 & 6 & -2 \end{bmatrix} =\mathrm{m} \begin{bmatrix} 1 & 1 & 1 \\ 1 & 1 & 0 \\ 0 & 2 & -1 \end{bmatrix} \begin{bmatrix} 4 & 0 & 0 \\ 0 & 2 & 0 \\ 0 & 0 & 4 \end{bmatrix} \begin{bmatrix} 1 & 1 & 1 \\ 1 & 1 & 0 \\ 0 & 2 & -1 \end{bmatrix}^{-1}.$$

Example: Consider the matrix

$$A = \begin{bmatrix} 2 & 3 & -3 \\ 0 & 2 & -3 \\ 0 & 0 & 1 \end{bmatrix}$$

Being upper triangular, it is easy to see that the characteristic polynomial is $(2 - \lambda)^2(1 - \lambda)$ and there are two eigenvalues $\lambda_1 = 2, \lambda_2 = 1$.

For $\lambda_1 = 2$, we find $v_1 = (1, 0, 0)$ form a basis of $nul(A - 2I)$. For $\lambda_2 = 1$, we find $v_2 = (-6, 3, 1)$ form a basis of $nul(A - I)$. Since we only get two eigenvectors, A does not have a basis of eigenvectors and is not diagonalizable.

In terms of matrices, we have

$$AP = \begin{bmatrix} 2 & 3 & -3 \\ 0 & 2 & -3 \\ 0 & 0 & 1 \end{bmatrix} \begin{bmatrix} 1 & -6 \\ 0 & 3 \\ 0 & 1 \end{bmatrix} = \begin{bmatrix} 1 & -6 \\ 0 & 3 \\ 0 & 1 \end{bmatrix} \begin{bmatrix} 2 & 0 \\ 0 & 1 \end{bmatrix} = PD.$$

However, the matrix

$$P = \begin{bmatrix} 1 & -6 \\ 0 & 3 \\ 0 & 1 \end{bmatrix}$$

is not invertible, and we cannot multiply P^{-1} on the right of $AP = PD$ to get diagonalization.

Exercise: Find the eigenvalues and eigenvectors. Then find diagonalization if possible.

1. $\begin{bmatrix} 3 & 1 \\ 2 & 4 \end{bmatrix}$

Answer:

$$det\begin{bmatrix} 3-\lambda & 1 \\ 2 & 4-\lambda \end{bmatrix} = (3 - \lambda)(4 - \lambda) - 2 = (2 - \lambda)(5 - \lambda).$$

For $\lambda_1 = 2$, we have

$$A - 2I = \begin{bmatrix} 1 & 1 \\ 2 & 2 \end{bmatrix}$$

and $v_1 = (1, -1)$ is a basis of the eigenspace $nul(A - 2I)$. For $\lambda_2 = 5$, we have

$$A - 5I = \begin{bmatrix} -2 & 1 \\ 2 & -1 \end{bmatrix},$$

and $v_2 = (1, 2)$ is a basis of the eigenspace $nul(A - 5I)$. A diagonalization of A is

$$A = \begin{bmatrix} 1 & 1 \\ -1 & 2 \end{bmatrix} \begin{bmatrix} 2 & 0 \\ 0 & 5 \end{bmatrix} \begin{bmatrix} 1 & 1 \\ -1 & 2 \end{bmatrix}^{-1}.$$

2. $\begin{bmatrix} 2 & 1 \\ -1 & 4 \end{bmatrix}$

Answer:

$$det\begin{bmatrix} 2-\lambda & 1 \\ -1 & 4-\lambda \end{bmatrix} = (2-\lambda)(4-\lambda)+1 = (3-\lambda)^2.$$

The only eigenvalue is $\lambda = 3$. We have

$$A - 3I = \begin{bmatrix} -1 & 1 \\ -1 & 1 \end{bmatrix}$$

and $v = (1, 1)$ is a basis of the eigenspace $nul(A - 3I)$. A is not diagonalizable.

3. $\begin{bmatrix} 2 & 1 \\ 0 & 2 \end{bmatrix}$

Answer:

$$det\begin{bmatrix} 2-\lambda & 1 \\ 0 & 2-\lambda \end{bmatrix} = (2-\lambda)^2.$$

The only eigenvalue is $\lambda = 2$. We have

$$A - 2I = \begin{bmatrix} 0 & 1 \\ 0 & 0 \end{bmatrix}$$

and $v = (1, 0)$ is a basis of $nul(A - 2I)$. A is not diagonalizable.

4. $\begin{bmatrix} 0 & 0 \\ 1 & 0 \end{bmatrix}$

Answer:

$$det\begin{bmatrix} -\lambda & 0 \\ 1 & -\lambda \end{bmatrix} = \lambda^2.$$

The only eigenvalue is $\lambda = 0$, and $v = (0, 1)$ is a basis of $nul(A - 0I) = nulA$. A is not diagonalizable.

5. $\begin{bmatrix} 1 & 0 \\ 0 & 2 \end{bmatrix}$

Answer: One may follow the steps and carry out the computation. On the other hand, the matrix represents multiplying 1 to x_1-direction and multiplying 2 to x_2-direction. So the standard basis vectors are eigenvectors, with 1 and 2 as eigenvalues.

A is already diagonal, and is diagonalizable by tautology: $A = IAI^{-1}$.

Remark: Any diagonal matrix has the standard basis as eigenvectors. The process of diagonalization is trying to find a basis so that, *by setting the new basis to be the standard one*, the matrix becomes a diagonal one.

6. $\begin{bmatrix} 1 & 0 & 0 \\ 0 & 2 & 0 \\ 0 & 0 & 3 \end{bmatrix}$

Answer: The situation is the same as the previous one. A has the standard basis e_1, e_2, e_3 as eigenvectors, with eigenvalues 1, 2, 3. A is diagonalizable by tautology.

7. $\begin{bmatrix} 2 & 1 & 0 \\ 0 & 2 & 1 \\ 0 & 0 & 2 \end{bmatrix}$

Answer:

$$det\begin{bmatrix} 2-\lambda & 1 & 0 \\ 0 & 2-\lambda & 1 \\ 0 & 0 & 2-\lambda \end{bmatrix} = (2-\lambda)^3.$$

The only eigenvalue is $\lambda = 2$. We have

$$A - 2I = \begin{bmatrix} 0 & 1 & 0 \\ 0 & 0 & 1 \\ 0 & 0 & 0 \end{bmatrix}$$

and $v = (1, 0, 0)$ is a basis of $nul(A - 2I)$. A is not diagonalizable.

Remark: The matrices in problems 3 and 7 are typical non-diagonalizable matrices. The general form is the Jordan block

$$J = \begin{bmatrix} a & 1 & 0 & .. & 0 \\ 0 & a & 1 & .. & 0 \\ 0 & 0 & a & .. & 0 \\ \vdots & \vdots & \vdots & & \vdots \\ 0 & 0 & 0 & .. & a \end{bmatrix}$$

8. $\begin{bmatrix} 1 & 1 & 0 \\ 0 & 2 & 1 \\ 0 & 0 & 3 \end{bmatrix}$

Answer:

$$det\begin{bmatrix}1-\lambda & 1 & 0\\ 0 & 2-\lambda & 1\\ 0 & 0 & 3-\lambda\end{bmatrix} = (1-\lambda)(2-\lambda)(3-\lambda).$$

The eigenvalues are $\lambda_1 = 1, \lambda_2 = 2, \lambda_3 = 3$. We have

$$A-I=\begin{bmatrix}0 & 1 & 0\\ 0 & 1 & 1\\ 0 & 0 & 2\end{bmatrix}, A-2I=\begin{bmatrix}-1 & 1 & 0\\ 0 & 0 & 1\\ 0 & 0 & 1\end{bmatrix}, A-3I=\begin{bmatrix}-2 & 1 & 0\\ 0 & -1 & 1\\ 0 & 0 & 0\end{bmatrix}$$

and $v_1 = (1, 0, 0)$, $v_2 = (1, 1, 0)$, $v_3 = (1, 2, 2)$ span the respective eigenspaces. A diagonalization is

$$A=\begin{bmatrix}1 & 1 & 1\\ 0 & 1 & 2\\ 0 & 0 & 2\end{bmatrix}\begin{bmatrix}1 & 0 & 0\\ 0 & 2 & 0\\ 0 & 0 & 3\end{bmatrix}\begin{bmatrix}1 & 1 & 1\\ 0 & 1 & 2\\ 0 & 0 & 2\end{bmatrix}^{-1}$$

9. $\begin{bmatrix}1 & 10 & 5\\ 0 & -4 & 0\\ 5 & 10 & 1\end{bmatrix}$

Answer: By cofactor expansion along [row 2], we have

$$det\begin{bmatrix}1-\lambda & 10 & 5\\ 0 & -4-\lambda & 0\\ 5 & 10 & 1-\lambda\end{bmatrix} = (-4-\lambda)\ det\begin{bmatrix}1-\lambda & 5\\ 5 & 1-\lambda\end{bmatrix} = (-4-\lambda)^2(6-\lambda).$$

The eigenvalues are $\lambda_1 = -4, \lambda_2 = 6$, and

$$A+4I=\begin{bmatrix}5 & 10 & 5\\ 0 & 0 & 0\\ 5 & 10 & 5\end{bmatrix}, A-6I=\begin{bmatrix}-5 & 10 & 5\\ 0 & -10 & 0\\ 5 & 10 & -5\end{bmatrix}.$$

The vectors $v_1 = (1, 0, -1)$, $v_2 = (0, 1, -2)$ form a basis of $nul(A + 4I)$. The vector $v_3 = (1, 0, 1)$ forms a basis of $nul(A - 6I)$. A diagonalization is

$$A=\begin{bmatrix}1 & 0 & 1\\ 0 & 1 & 0\\ -1 & -2 & 1\end{bmatrix}\begin{bmatrix}-4 & 0 & 0\\ 0 & -4 & 0\\ 0 & 0 & 6\end{bmatrix}\begin{bmatrix}1 & 0 & 1\\ 0 & 1 & 0\\ -1 & -2 & 1\end{bmatrix}^{-1}$$

10. $\begin{bmatrix} -5 & 2 & -1 \\ 2 & -2 & -2 \\ -1 & -2 & -5 \end{bmatrix}$

Answer: By an earlier exercise, the characterisitc equation is $-\lambda(6+\lambda)^2 = 0$. The eigenvalues are $\lambda_1 = 0$, $\lambda_2 = -6$, and

$$A - 0I = \begin{bmatrix} -5 & 2 & -1 \\ 2 & -2 & -2 \\ -1 & -2 & -5 \end{bmatrix}, A + 6I = \begin{bmatrix} 1 & 2 & -1 \\ 2 & 4 & -2 \\ -1 & -2 & 1 \end{bmatrix}$$

The vector $v_1 = (-1, -2, 1)$ forms a basis of *nulA*. The vectors $v_2 = (1, 0, 1)$, $v_3 = (-1, 1, 1)$ form a basis of *nul*$(A + 6I)$. A diagonalization is

$$A = \begin{bmatrix} -1 & 1 & -1 \\ -2 & 0 & 1 \\ 1 & 1 & 1 \end{bmatrix} \begin{bmatrix} 0 & 0 & 0 \\ 0 & -6 & 0 \\ 0 & 0 & -6 \end{bmatrix} \begin{bmatrix} -1 & 1 & -1 \\ -2 & 0 & 1 \\ 1 & 1 & 1 \end{bmatrix}^{-1}$$

11. $\begin{bmatrix} 2 & 2 & -1 \\ 1 & 3 & -1 \\ 1 & 4 & -2 \end{bmatrix}$

Answer: By an earlier exercise, the charcaterisitc equation is $(1-\lambda)(3-\lambda)(-1-\lambda) = 0$. The eigenvalues are $\lambda_1 = 1$, $\lambda_2 = 3$, $\lambda_3 = -1$, and

$$A - I = \begin{bmatrix} 2 & 2 & -1 \\ 1 & 2 & -1 \\ 1 & 4 & -2 \end{bmatrix}, A - 3I = \begin{bmatrix} -1 & 2 & -1 \\ 1 & 0 & -1 \\ 1 & 4 & -5 \end{bmatrix}, A + I = \begin{bmatrix} 3 & 2 & -1 \\ 1 & 4 & -1 \\ 1 & 4 & -1 \end{bmatrix}$$

The vectors $v_1 = (-1, 1, 1)$, $v_2 = (1, 1, 1)$, $v_3 = (1, 1, 5)$ span the respective eigenspaces. A diagonalization is

$$A = \begin{bmatrix} -1 & 1 & 1 \\ 1 & 1 & 1 \\ 1 & 1 & 5 \end{bmatrix} \begin{bmatrix} 1 & 0 & 0 \\ 0 & 3 & 0 \\ 0 & 0 & -1 \end{bmatrix} \begin{bmatrix} -1 & 1 & 1 \\ 1 & 1 & 1 \\ 1 & 1 & 5 \end{bmatrix}^{-1}$$

12. $\begin{bmatrix} 4 & 1 & 1 \\ 1 & 4 & 1 \\ 1 & 1 & 4 \end{bmatrix}$

Answer: In the first equality, we use [row 2] + [row 1] and [row 3] + [row 1]. In the second equality, we use (–1)[col 1] + [col 2] and (–1)[col 1] + [col 3].

$$det\begin{bmatrix}4-\lambda & 1 & 1\\ 1 & 4-\lambda & 1\\ 1 & 1 & 4-\lambda\end{bmatrix} = det\begin{bmatrix}6-\lambda & 6-\lambda & 6-\lambda\\ 1 & 4-\lambda & 1\\ 1 & 1 & 4-\lambda\end{bmatrix} = det\begin{bmatrix}6-\lambda & 0 & 0\\ 1 & 3-\lambda & 0\\ 1 & 0 & 3-\lambda\end{bmatrix}$$

Thus the eigenvalues are $\lambda_1 = 6$, $\lambda_2 = 3$, and

$$A-6I=\begin{bmatrix}-2 & 1 & 1\\ 1 & -2 & 1\\ 1 & 1 & -2\end{bmatrix}, A-3I=\begin{bmatrix}1 & 1 & 1\\ 1 & 1 & 1\\ 1 & 1 & 1\end{bmatrix}$$

The vector $v_1 = (1, 1, 1)$ forms a basis of $nul(A - 6I)$. The vectors $v_2 = (1, 0, -1)$, $v_3 = (0, 1, -1)$ form a basis of $nul(A - 3I)$. A diagonalization is

$$A=\begin{bmatrix}1 & 1 & 0\\ 1 & 0 & 1\\ 1 & -1 & -1\end{bmatrix}\begin{bmatrix}6 & 0 & 0\\ 0 & 3 & 0\\ 0 & 0 & 3\end{bmatrix}\begin{bmatrix}1 & 1 & 0\\ 1 & 0 & 1\\ 1 & -1 & -1\end{bmatrix}^{-1}$$

13. $\begin{bmatrix}5 & -4 & -2 & 4\\ 3 & -2 & 0 & 2\\ 0 & 0 & 1 & 1\\ 0 & 0 & 1 & 1\end{bmatrix}$

Answer: By this property, we have

$$det\begin{bmatrix}5-\lambda & -4 & -2 & 4\\ 3 & -2-\lambda & 0 & 2\\ 0 & 0 & 1-\lambda & 1\\ 0 & 0 & 1 & 1-\lambda\end{bmatrix}$$

$$= det\begin{bmatrix}5-\lambda & -4\\ 3 & -2-\lambda\end{bmatrix} det\begin{bmatrix}1-\lambda & 1\\ 3 & 1-\lambda\end{bmatrix} = -\lambda(1-\lambda)(2-\lambda)^2.$$

The eigenvalues are 0, 1, 2, and

$$A-0I=\begin{bmatrix}5 & -4 & -2 & 4\\ 3 & -2 & 0 & 2\\ 0 & 0 & 1 & 1\\ 0 & 0 & 1 & 1\end{bmatrix}, A-1I=\begin{bmatrix}5 & -4 & -2 & 4\\ 3 & -3 & 0 & 2\\ 0 & 0 & 0 & 1\\ 0 & 0 & 1 & 0\end{bmatrix}, A-2I=\begin{bmatrix}3 & -4 & -2 & 4\\ 3 & -4 & 0 & 2\\ 0 & 0 & -1 & 1\\ 0 & 0 & 1 & -1\end{bmatrix}.$$

We then have the bases {(0, 0, 1, –1)}, {(1, 1, 0, 0)}, {(4, 3, 0, 0). (0, 0, 1, 1)} for the respective eigenspaces. A diagonalization is

$$A=\begin{bmatrix}0 & 1 & 4 & 0\\ 0 & 1 & 3 & 0\\ 1 & 0 & 0 & 1\\ -1 & 0 & 0 & 1\end{bmatrix}\begin{bmatrix}0 & 0 & 0 & 0\\ 0 & 1 & 0 & 0\\ 0 & 0 & 2 & 0\\ 0 & 0 & 0 & 2\end{bmatrix}\begin{bmatrix}0 & 1 & 4 & 0\\ 0 & 1 & 3 & 0\\ 1 & 0 & 0 & 1\\ -1 & 0 & 0 & 1\end{bmatrix}^{-1}$$

14. $\begin{bmatrix}0 & 0 & 1 & 1\\ 0 & 1 & 0 & 1\\ 1 & 0 & 1 & 0\\ 1 & 1 & 0 & 0\end{bmatrix}$

Answer: In the first equality, we use [row 2] + [row 1], [row 3] + [row 1], [row 4] + [row 1], (–1)[col 1] + [col 2], (–1)[col 1] + [col 3], (–1)[col 1] + [col 4]. In the second equality, we use the cofactor expansion along [row 1] and the direct computation of the 3 by 3 determinant.

$$det\begin{bmatrix}-\lambda & 0 & 1 & 1\\ 1 & 1-\lambda & 0 & 1\\ 1 & 0 & 1-\lambda & 0\\ 1 & 1 & 0 & -\lambda\end{bmatrix}=det\begin{bmatrix}2-\lambda & 0 & 0 & 0\\ 0 & 1-\lambda & 0 & 1\\ 1 & -1 & -\lambda & -1\\ 1 & 1 & -1 & -1-\lambda\end{bmatrix}$$

$$= (2-\lambda)[(1-\lambda)(-\lambda)(-1-\lambda)+1-(1-\lambda)] = -\lambda(2-\lambda)(2-\lambda^2).$$

The eigenvalues are 0, 2, $\sqrt{2}$, $-\sqrt{2}$, and the vectors (1, –1, –1, 1), (1, 1, 1, 1), ($\sqrt{2}-1$, –1, 1, $-\sqrt{2}+1$), ($-\sqrt{2}-1$, –1, 1, $\sqrt{2}+1$) span the respective eigenspaces. A diagonalization is

$$A=\begin{bmatrix}1 & 1 & \sqrt{2}-1 & -\sqrt{2}-1\\ -1 & 1 & -1 & -1\\ -1 & 1 & 1 & 1\\ 1 & 1 & -\sqrt{2}+1 & \sqrt{2}+1\end{bmatrix}\begin{bmatrix}0 & 0 & 0 & 0\\ 0 & 2 & 0 & 0\\ 0 & 0 & \sqrt{2} & 0\\ 0 & 0 & 0 & -\sqrt{2}\end{bmatrix}\begin{bmatrix}1 & 1 & \sqrt{2}-1 & -\sqrt{2}-1\\ -1 & 1 & -1 & -1\\ -1 & 1 & 1 & 1\\ 1 & 1 & -\sqrt{2}+1 & \sqrt{2}+1\end{bmatrix}$$

15. $\begin{bmatrix} 2 & 0 & 0 & 0 \\ 2 & 1 & 0 & 0 \\ 2 & 1 & 2 & 0 \\ 2 & 1 & 2 & 1 \end{bmatrix}$

Answer: The eigenvalues are 2 and 1, and

$$A-2I=\begin{bmatrix} 0 & 0 & 0 & 0 \\ 2 & -1 & 0 & 0 \\ 2 & 1 & 2 & 0 \\ 2 & 1 & 2 & -1 \end{bmatrix}, A-1I=\begin{bmatrix} 1 & 0 & 0 & 0 \\ 2 & 0 & 0 & 0 \\ 2 & 1 & 1 & 0 \\ 2 & 1 & 2 & 0 \end{bmatrix}.$$

We then have the bases {(0, 0, 1, 2)}, {(0, 0, 0, 1)} for the respective eigenspaces. A is not diagonalizable.

16. $\begin{bmatrix} 2 & 2 & 2 & 2 \\ 0 & 1 & 1 & 1 \\ 0 & 0 & 2 & 2 \\ 0 & 0 & 0 & 1 \end{bmatrix}$

Answer: The eigenvalues are 2 and 1, and

$$A-2I=\begin{bmatrix} 0 & 2 & 2 & 2 \\ 0 & -1 & 1 & 1 \\ 0 & 0 & 0 & 2 \\ 0 & 0 & 0 & -1 \end{bmatrix}, A-1I=\begin{bmatrix} 1 & 2 & 2 & 2 \\ 0 & 0 & 1 & 1 \\ 0 & 0 & 1 & 2 \\ 0 & 0 & 0 & 0 \end{bmatrix}.$$

We then have the bases {(1, 0, 0, 0)}, {(2, –1, 0, 0)} for the respective eigenspaces. A is not diagonalizable.

Exercise: Prove that the eigenvalues of upper or lower triangular matrices are the diagonal entries.

Answer: By this property of determinant, for upper trangular matrix

$$A=\begin{bmatrix} a_1 & \# & .. & \# \\ 0 & a_2 & .. & \# \\ : & : & & : \\ 0 & 0 & .. & a_n \end{bmatrix},$$

we have

$$det(A - \lambda I) = det\begin{bmatrix} a_1 - \lambda & \# & .. & \# \\ 0 & a_2 - \lambda & .. & \# \\ : & : & & : \\ 0 & 0 & .. & a_n - \lambda \end{bmatrix} = (a_1 - \lambda)(a_2 - \lambda)...(a_n - \lambda).$$

Clearly, the solutions of the characteristic equation $det(A - \lambda I) = 0$ are $a_1, a_2,..., a_n$.

Exercise: Use diagonalization to find the matrix for the orthogonal projection onto the plane $x + y + z = 0$ (compare with this exercise). Also find the matrix for the reflection with respect to the plane.

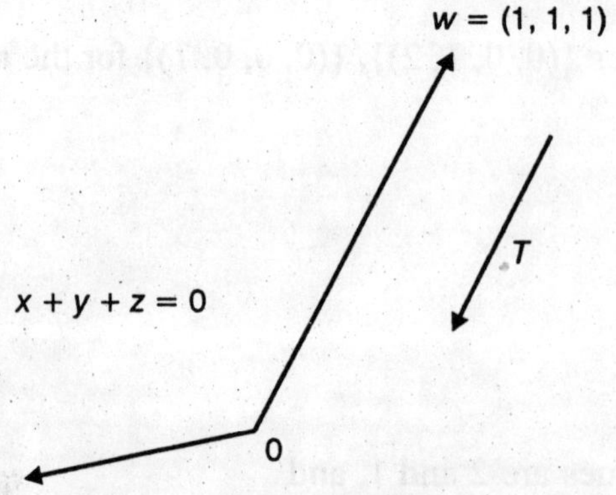

Answer: In an earlier exercise, we noticed that for the basis

$u = (-1, 1, 0)$ and $v = (-1, 0, 1)$, $w = (1, 1, 1)$

of R^3, the orthogonal projection satisfies $T(u) = u$, $T(v) = v$, $T(w) = 0$. In other words, u, v, w are eigenvectors with eigenvalues 1, 1, 0. Thus the matrix for T is

$$\begin{bmatrix} -1 & -1 & 1 \\ 1 & 0 & 1 \\ 0 & 1 & 1 \end{bmatrix}\begin{bmatrix} 1 & 0 & 0 \\ 0 & 1 & 0 \\ 0 & 0 & 0 \end{bmatrix}\begin{bmatrix} -1 & -1 & 1 \\ 1 & 0 & 1 \\ 0 & 1 & 1 \end{bmatrix} = \begin{bmatrix} 2/3 & -1/3 & -1/3 \\ -1/3 & 2/3 & -1/3 \\ -1/3 & -1/3 & 2/3 \end{bmatrix}$$

Similarly, the reflection with respect to the plane satisfies $R(u) = u$, $R(v) = v$, $T(w) = -w$. In other words, u, v, w are eigenvectors with eigenvalues 1, 1, −1. Thus the matrix for R is

$$\begin{bmatrix} -1 & -1 & 1 \\ 1 & 0 & 1 \\ 0 & 1 & 1 \end{bmatrix}\begin{bmatrix} 1 & 0 & 0 \\ 0 & 1 & 0 \\ 0 & 0 & -1 \end{bmatrix}\begin{bmatrix} -1 & -1 & 1 \\ 1 & 0 & 1 \\ 0 & 1 & 1 \end{bmatrix} = \begin{bmatrix} 1/3 & -2/3 & -2/3 \\ -2/3 & 1/3 & -2/3 \\ -2/3 & -2/3 & 1/3 \end{bmatrix}$$

Exercise: Find the matrix of the linear transformations given by the descriptions.

1. T: $R^2 \to R^2$, stretching (1, 2) twice and shrinking (3, 4) twice

Answer: (1, 2) and (3, 4) are eigenvectors with eigenvalues 2 and 1/2. The matrix for the linear transformation is

$$\begin{bmatrix} 1 & 3 \\ 2 & 4 \end{bmatrix}\begin{bmatrix} 2 & 0 \\ 0 & 1/2 \end{bmatrix}\begin{bmatrix} 1 & 3 \\ 2 & 4 \end{bmatrix}^{-1} = \begin{bmatrix} -5/2 & 9/4 \\ -6 & 5 \end{bmatrix}.$$

2. T: $R^2 \to R^2$, reflecting with respect to the line $3x - 5y = 0$

Answer: (3, –5) and (5, 3) are eigenvectors with eigenvalues –1 and 1. The matrix for the linear transformation is

$$\begin{bmatrix} 3 & 5 \\ -5 & 3 \end{bmatrix}\begin{bmatrix} -1 & 0 \\ 0 & 1 \end{bmatrix}\begin{bmatrix} 3 & 5 \\ -5 & 3 \end{bmatrix}^{-1} = \begin{bmatrix} 13/17 & 15/17 \\ 15/17 & 13/17 \end{bmatrix}.$$

3. T: $R^3 \to R^3$, stretching (1, 0, 0) twice, (1, 1, 0) three times, and killing (1, 2, 1)

Answer: (1, 0, 0), (1, 1, 0), (1, 2, 1) are eigenvectors with eigenvalues 2, 3, 0. The matrix for the linear transformation is

$$\begin{bmatrix} 1 & 1 & 1 \\ 0 & 1 & 2 \\ 0 & 0 & 1 \end{bmatrix}\begin{bmatrix} 2 & 0 & 0 \\ 0 & 3 & 0 \\ 0 & 0 & 0 \end{bmatrix}\begin{bmatrix} 1 & 1 & 1 \\ 0 & 1 & 2 \\ 0 & 0 & 1 \end{bmatrix}^{-1} = \begin{bmatrix} 2 & 1 & -4 \\ 0 & 3 & -6 \\ 0 & 0 & 0 \end{bmatrix}$$

4. T: $R^3 \to R^3$, stretching everything on the plane $x + 2y + 3z = 0$ twice, and reversing (0, –1, 1)

Answer: The vectors (2, –1, 0), (3, 0, –1) form a basis of the plane $x + 2y + 3z = 0$ and are eigenvectors with eigenvalue 2. The vector (0, 1, 1) is an eigenvector with eigenvalue –1. The matrix for the linear transformation is

$$\begin{bmatrix} 2 & 3 & 0 \\ -1 & 0 & 1 \\ 0 & -1 & 1 \end{bmatrix}\begin{bmatrix} 2 & 0 & 0 \\ 0 & 2 & 0 \\ 0 & 0 & -1 \end{bmatrix}\begin{bmatrix} 2 & 3 & 0 \\ -1 & 0 & 1 \\ 0 & -1 & 1 \end{bmatrix}^{-1} = \begin{bmatrix} 2 & 0 & 0 \\ 3 & 8 & 9 \\ -3 & -6 & -7 \end{bmatrix}.$$

Exercise: Prove that if A is diagonalizable and has only one eigenvalue, then A is a constant multiple of the identity matrix.

Answer: Let $B = \{v_1, v_2,..., v_n\}$ be a basis of eigenvectors with the same eigenvalue λ. Then for any vector u, we have

$$u = c_1v_1 + c_2v_2 + ... + c_nv_n$$

and

$$Au = c_1Av_1 + c_2Av_2 + ... + c_nAv_n = c_1\lambda v_1 + c_2\lambda v_2 + ... + c_n\lambda v_n = \lambda u.$$

Thus the linear transformation given by A is multiplying everything by λ. This means $A = \lambda I$.

Alternatively, we have $A = PDP^{-1}$. Since all the diagonal entries of D, being eigenvalues of A, are the same number λ, we have $D = \lambda I$. This implies $A = P\lambda IP^{-1} = \lambda PP^{-1} = \lambda I$.

Exercise: Find the eigenvalues and eigenvectors of the derivative transformation $D(f) = f': C^{\infty}(R) \to C^{\infty}(R)$.

Answer: By the definition, we are looking for a number λ and a nonzero function f such that $f' = \lambda f$. Given such an equality, we have

$$(e^{-\lambda t} f)' = -\lambda e^{-\lambda t} f + e^{-\lambda t} f' = -\lambda e^{-\lambda t} f + e^{-\lambda t}\lambda f = 0.$$

Thus $e^{-\lambda t} f = c$ is a constant, and $f = ce^{\lambda t}$ is a constant multiple of $e^{\lambda t}$.

We conclude that any number λ is an eigenvector. The corresponding eigenspace is the 1-dimensional subspace of $C^{\infty}(R)$ spanned by $e^{\lambda t}$.

Remark: We cannot use the usual method for finding the eigenvalues and eigenvectors because $C^{\infty}(R)$ is infinite dimensional.

Exercise: Find the eigenvalues and eigenvectors of the transpose transformation $T(X) = X^{T}: M(n, n) \to M(n, n)$.

Answer: By the definition, we are looking for a number λ and a nonzero matrix X such that $X^{T} = \lambda X$. Given such an equality, we have

$$X = (X^{T})^{T} = (\lambda X)^{T} = \lambda^2 X.$$

Since X is nonzero, we must have $\lambda^2 = 1$. Thus the eigenvalues of T must be either 1 or -1.

For the eigenvalue 1, the corresponding eigenspace consists of matrices satisfying $X^{T} = X$. In other words, the eigenspace consists of symmetric matrices, and any nonzero symmetric matrix is an eigenvector of T with eigenvalue 1.

For the eigenvalue -1, the corresponding eigenspace consists of matrices satisfying $X^{T} = -X$. In other words, the eigenspace consists of skew–symmetric matrices, and any nonzero skew–symmetric matrix is an eigenvector of T with eigenvalue -1.

Chapter 8

Kernel and Null Space

KERNEL

The definition of the kernel of a linear transformation between euclidean spaces can be adopted without any change to the more general case. For any linear transformation $T: V \to W$,

kernelT **$= \{u \in V: T(u) = 0\}$.**

We note that the concept of kernel is not defined for *nonlinear* transformations.

Similar to the range of a linear transformation, we have

The kernel of a linear transformation $T: V \to W$ is a subspace of V.

Proof. $u, v \in kernelT \Rightarrow T(u) = T(v) = 0 \Rightarrow T(u + v) = T(u) + T(v) = 0 + 0 = 0 \Rightarrow u + v \in kernelT$, where the linearity of T is used in the red equality. The proof for $u \in kernelT \Rightarrow cu \in kernelT$ is similar.

***Example*:** The subset

$$\{(x, y, z): x + y - z = 0, 2x - 3y + z = 0\} \subset R^3$$

is a subspace because it is the kernel of the linear transformation

$$T(x, y, z) = (x + y - z, 2x - 3y + z): R^3 \to R^2.$$

The subset

$$\{f \in F(R): f(1) = f(2) = 0\} \subset F(R)$$

is a subspace because it is the kernel of the linear transformation

$$E(f) = (f(1), f(2))\ F(R) \to R^2.$$

The subset

$$\{f \in C^2(R): f'' - 4f' + 3f = 0\} \supset C^2(R)$$

is a subspace because it is the kernel of the linear transformation

$T(f) = f'' - 4f' + 3f$: $C^2(\mathrm{R}) \rightarrow C(\mathrm{R})$.

We also recall the fact that a linear transformation between euclidean spaces is one–to–one (injective) if and only if the kernel contains only the zero vector. The proof of the fact can be applied to the general case without any change. Therefore we also have

A linear transformation $T: V \rightarrow W$ is one–to–one $\Leftrightarrow$ $kernelT = \{0\}$.

Example: Consider the derivative transformation

$D(f) = f'$: $C^1(\mathrm{R}) \rightarrow C(\mathrm{R})$

The kernel consists of those functions with zero derivative. From calculus, we know that such functions are exactly constant functions. Therefore $kernelD = \{c: c \in \mathrm{R}\}$. In particular, the kernel is not the zero subspace. Thus D is not one–to–one.

Example: Consider the linear transformation

$M(f) = (t^2+1)f$: $F(\mathrm{R}) \rightarrow F(\mathrm{R})$

of multiplying the function t^2+1. If $(t^2+1)f = 0$, then by thé fact that $t^2 + 1$ is never zero, we see that $f = 0$. Therefore $kernelM = \{0\}$ and M is one–to–one.

If we change $t^2 + 1$ to $t^2 - 1$, however, then M is no longer onto. In fact, the kernel of the linear transformation $f \rightarrow (t^2-1)f$ consists of functions of the form

$$f(t) = \begin{cases} 0 \text{ if } t \neq -1, 1 \\ c \text{ if } t = -1 \\ d \text{ if } t = 1 \end{cases}$$

where c and d are arbitrary numbers.

On the other hand, if we restrict the source and target by considering

$M(f) = (t^2-1)f$: $C(\mathrm{R}) \rightarrow C(\mathrm{R})$,

then the continuity implies that $c = d = 0$. Therefore the linear transformation is again one–to–one.

NULL SPACE

For a linear transformation T: $\mathrm{R}^n \rightarrow \mathrm{R}^m$ given by an m by n matrix A, we have a special name for the kernel.

The null space of a matrix A is

$nulA = \{$all solutions of the homogeneous system $Ax = 0\}$.

We remark that the null space of an m by n matrix is a subspace of R^n.

Example: The subspace

$\{(x, y, z): x + y - z = 0, 2x - 3y + z = 0\} \subset R^3$

given in an earlier example is in fact the null space of the matrix $\begin{bmatrix} 1 & 1 & -1 \\ 2 & -3 & 1 \end{bmatrix}$.

Moreover, by

$$\begin{bmatrix} 1 & 1 & -1 \\ 2 & -3 & 1 \end{bmatrix}\begin{bmatrix} 1 \\ 1 \\ 1 \end{bmatrix} = \begin{bmatrix} 1 \\ 0 \end{bmatrix} \neq 0,$$

we have (1, 1, 1)   *nulA*, and by

$$\begin{bmatrix} 1 & 1 & -1 \\ 2 & -3 & 1 \end{bmatrix}\begin{bmatrix} 2 \\ 3 \\ 5 \end{bmatrix} = \begin{bmatrix} 0 \\ 0 \end{bmatrix} = 0,$$

we have $(2, 3, 5) \in$ *nulA*.

STRUCTURE OF SOLUTIONS OF GENERAL LINEAR EQUATIONS

Let $T: V \to W$ be a linear transformation. Then the equation (where $b \in W$ is fixed and $x \in V$ is unknown)

$$T(x) = b$$

is a linear equation. The following tells us the structure of solutions of linear equations.

The solutions of a linear equation $T(x) = b$ are of the form $x = x_0 + x_h$, where

- x_0 is one particular solution of $T(x) = b$.
- x_h is any vector in *kernelT*.

Proof. Let x_0 be one particular solution of $T(x) = b$. Then for any x we have

$T(x - x_0) = T(x) - T(x_0) = T(x) - b,$

where the first equality is because T is linear, and the second equality is because x_0 is one solution. Consequently,

$T(x) = b \Leftrightarrow T(x - x_0) = 0 \Leftrightarrow x_h = x - x_0$ *kernelT*.

Geometrically, the result means that the general solution of $T(x) = b$ is obtained by shifting the subspace *kernelT* = {all x_h} by the amount x_0. More specific geometric pictures can be found in this exercise.

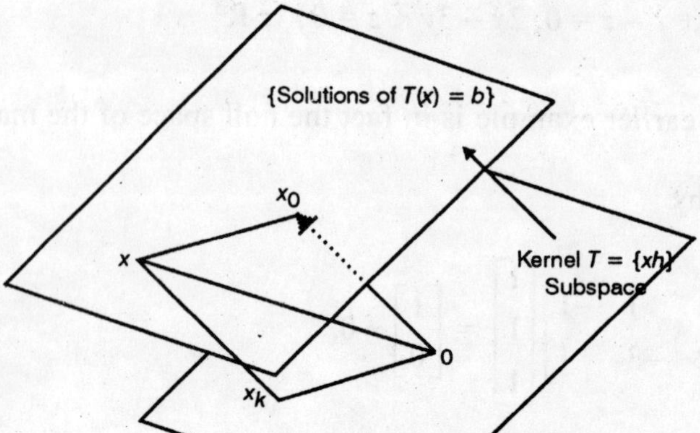

The result explains what we saw in the three examples at the beginning of this chapter.

Example: The system

$$x_1 + x_2 - 5x_4 = 3$$
$$x_3 - 2x_4 = 3$$

is linear because the left side is a linear transformation

$T(x_1, x_2, x_3, x_4) = (x_1 + x_2 - 5x_4, x_3 + 2x_4)$: $R^4 \to R^2$.

By taking $x_2 = x_4 = 0$, we find a special solution $x_0 = (3, 0, 4, 0)$ of the system. Moreover, the kernel of T, which is the general solution of the homogeneous system

$$x_1 + x_2 - 5x_4 = 0$$
$$x_3 + 2x_4 = 0$$

consists of vectors of the form

$x_h = (-x_2 + 5x_4, x_2, -2x_4, x_4) = x_2(-1, 1, 0, 0) + x_4(5, 0, -2, 1)$, x_2 and x_4 arbitrary.

Therefore the general solution is

$x = x_0 + x_h = (3, 0, 4, 0) + x_2(-1, 1, 0, 0) + x_4(5, 0, -2, 1)$.

We emphasize that the general solution is the sum of the fixed part $x_0 = (3, 0, 4, 0)$ and the free part $x_h = x_2(-1, 1, 0, 0) + x_4(5, 0, -2, 1)$, with the freedom in the free part given by x_2 and x_4. The free part is the solution of the homogeneous system.

Example: The differential equation

$f'' - 4f' + 3f = 3t - 1$

is linear because the left side is a linear transformation. The general solution is

$f = (t + 1) + c_1e^t + c_2e^{3t}$.

The fixed part $f_0 = t + 1$ is a particular solution of the equation. The free part $f_h = c_1e^t + c_2e^{3t}$ is all the solutions of the homogeneous equation

$f'' - 4f' + 3f = 0.$

We also note that, being all the linear combinations of e^t and e^{3t}, the free part form a subspace for the same reason as in this exercise.

Example: The matrix equation

$$\begin{bmatrix} 1 & -2 & 3 \\ -2 & 3 & -4 \end{bmatrix} X = I, \text{ for 2 by 3 matrix } X$$

is linear because the left side is a linear transformation. The general solution is

$$X = \begin{bmatrix} 1 & -2 \\ 0 & 0 \\ 0 & 0 \end{bmatrix} + c_1 \begin{bmatrix} 1 & 0 \\ 2 & 0 \\ 1 & 0 \end{bmatrix} + c_2 \begin{bmatrix} 0 & 1 \\ 0 & 2 \\ 0 & 1 \end{bmatrix}.$$

The fixed part

$$X_0 = \begin{bmatrix} 1 & -2 \\ 0 & 0 \\ 0 & 0 \end{bmatrix}$$

is a special solution, and the free part

$$X_h = c_1 \begin{bmatrix} 1 & 0 \\ 2 & 0 \\ 1 & 0 \end{bmatrix} + c_2 \begin{bmatrix} 0 & 1 \\ 0 & 2 \\ 0 & 1 \end{bmatrix}$$

is all the solutions of the homogeneous equation

$$\begin{bmatrix} 1 & -2 & 3 \\ -2 & 3 & -4 \end{bmatrix} X = O.$$

Exercise: Find the kernels and determine whether the transformations are one–to–one.

1. $T(f) = f(1) + f(2)$: $F(\mathrm{R}) \to \mathrm{R}$

Answer: The kernel of T consists of functions satisfying $f(1) + f(2) = 0$. For example, the nonzero function $f(t) = 2t - 3$ is in the kernel. Therefore T is not one–to–one.

2. $T(a, b) = a \cos t + b \sin t$: $\mathrm{R}^2 \to F(\mathrm{R})$

Answer: The kernel of T consists of a and b satisfying a cos$t$ + b sin$t$ = 0 for all t. By taking t = 0, we have a = 0. Thus b sin$t$ = 0 for any t, which implies b = 0. Therefore the kernel contains only the zero vector, and T is one-to-one.

3. $T(f) = f' - f: C^2(R) \to C(R)$

Answer The kernel of T consists of all the solutions of the equation $f' - f = 0$. It is a standard fact from the theory of differential equations that the solutions are of the form a $e^t + b$ e^{-t}, with a, b arbitrary constants. Therefore

kernelT = {a e^t + b *et*: any a, b}

and T is not one–to–one.

4. $T(f) = +\int_0^1 f(t)dt: C(R) \to R$

Answer: The kernel of T consists of those functions such that the areas of the "positive" and the "negative" parts of the graph are the same. In particular, T is not one–to–one.

5. $T(f(t)) = f(-t): F(R) \to F(R)$

Answer: The kernel of T consists of those functions f such that $f(-t) = 0$ for all t. The condition is the same as $f(t) = 0$ for all t. Therefore the kernel of T consists of the zero function only and T is one–to–one.

Remark: The kernel seriously depends on the change of variable. For example, the kernel of the linear transformation $T(f(t)) = f(t^2): F(R) \to F(R)$ consists of those functions f such that $f(t^2) = 0$ for all t. This is the same as $f(t) = 0$ for all t e" 0. In particular, the kernel contains nonzero functions such as

$$f(t) = \begin{cases} 0 \text{ if } t \geq 0 \\ t \text{ if } t < 0 \end{cases}$$

6. $T(a, b, c) = a + (a - 2b)t + (3a + c)t^2: R^3 \to P_2$

Answer: The kernel of T is obtained by solving the homogeneous system

$$a = 0,\ a - 2b = 0,\ 3a + c = 0$$

Since the only solution is the trivial one, the kernel is {(0, 0, 0)}, and T is one–to–one.

7. $T(p(t)) = (p(t), p'(t)): P_4 \to P_4 \quad P_3$

Answer The kernel of T consists of those polynomials $p(t)$ such that $p(t) = p'(t) = 0$. Thus the kernel contains only the zero polynomial, and T is one–to–one.

8. $T(p(t)) = (p(0), p'(0)): P_4 \to R^2$

Answer: The kernel consists of polynomials satisfying $p(0) = p'(0) = 0$. These are exactly the polynomials with 0 as a multiple root. Such polynomials are of the form $p(t) = t^2q(t)$. Since the degree of p is ≤ 4, we see that $p(t) = t^2(a + bt + ct^2)$, where a, b, c are arbitrary constants. Thus

$kernelT = \{at^2 + bt^3 + ct^4\text{: any } a, b, c\}$

and T is not one–to–one.

9. $T(p(t)) = (p(0), p'(0), p''(0), p'''(0))\colon P_2 \to R^4$

Answer: A polynomial of degree 2 is of the form $p(t) = a + bt + ct^2$, and we have $T(p(t)) = (a, b, 2c, 0)$. Thus kernelT consists of those vectors $p(t)$ satisfying $a = b = 2c = 0$. Therefore the kernel contains only the zero polynomial, and T is one–to–one.

Remark: Comparing problems 7 and 8, what can you say about $T(p(t)) = (p(0), p'(0),..., p^{(n)}(0))\colon P_m \to R^{n+1}$?

10. $T(X) = AX - XA\colon M(n, n) \to M(n, n)$, where A is a fixed n by n matrix

Answer: The kernel consists of all the matrices X commuting with A: $AX = XA$. In particular, the identity matrix is in the kernel. Therefore T is not one–to–one.

11. $T(X) = X - X^T\colon M(n, n) \to M(n, n)$

Answer: The kernel consists of matrices X satisfying $X = XT$. Thus kernelT = symmetric matrices. In particular, T is not one–to–one

12. $T(X)$ = first column of X: $M(m, n) \to R^m$

Answer: The kernel consists of those matrices with zero first column. Unless $n = 1$, T will not be one–to–one.

Exercise: Show the following are subspaces by identifying them as the kernels of linear transformations.

1. $H = \{(a, b, c, d)\colon a + b = c + d\}$

Answer: The condition $a + b = c + d$ is the same as $a + b - c - d = 0$. Thus H is the kernel of the linear transformation $(a, b, c, d) \to a + b - c - d\colon R^4 \to R$.

2. $H = \{f \in F(R)\colon f(1) + f(2) = 0\}$

Answer: H is the kernel of the linear transformation $T(f) = f(1) + f(2)\colon F(R) \to R$. The linearity of T has been verified in this exercise.

3. $H = \{f \in C[0,1]\colon +\int_0^1 f(t)dt = 0\}$

Answer: H is the kernel of the linear transformation $T(f) = +\int_0^1 f(t)dt\colon F(R) \to R$. The linearity of T has been discussed in this exercise.

4. $H = \{f \in C^2(R)\colon f'' - 4f' + 3f = 0, f'(0) = 0\}$

Answer: H is the kernel of the linear transformation $T(f) = (f'' - 4f' + 3f, f'(0))\colon C^2(R) \to C(R)$ E R. The linearity of T follows from the fact that each component of T is linear.

5. $H = \{f \in F(R)\colon f(t) = f(-t)\}$

Answer: H is the kernel of the linear transformation $T(f(t)) = f(t) - f(-t)\colon F(R) \to F(R)$, which is a linear combination of the identity and the linear transformation $T(f(t)) = f(-t)$: $F(\mathrm{R}) \to F(\mathrm{R})$.

6. H = all polynomials of degree ≤ 5 with $t = 2$ as a multiple root

Answer: A polynomial $p(t)$ has 2 as a multiple root $\Leftrightarrow p(2) = p'(2) = 0$. Thus H is the kernel of the linear transformation $T(p(t)) = (p(2), p'(2))$: $P \to \mathrm{R}^2$.

7. H = all matrices commuting with A

Answer: H is the kernel of the linear transformation $T(X) = AX - XA$: $M(n, n) \to M(n, n)$. The linearity of T has been discussed in this exercise.

8. H = all n by n symmetric matrices

Answer: H is the kernel of the linear transformation $T(X) = X - X^T$: $M(n, n) \to M(n, n)$. The linearity of T has been discussed in this exercise.

Exercise: Find out whether the given vectors are in the kernels of the linear transformations.

1. $T(x, y, z) = (x - 2y + z, -x + 3y, y + z, x + 2y + 5z)$: $\mathrm{R}^3 \to \mathrm{R}^4$

$u = (1, 2, -1)$, $v = (-3, -1, 1)$

Answer: By

$T(u) = (-4, 5, 1, 5)$ ‘™ 0, $T(v) = (0, 0, 0, 0) = 0$,

we have $u \in$ *kernelT*, $v \notin$ *kernelT*.

Remark: The conclusion can also be interpreted as u O *nulA* and $v \in$ *nulA*, where

$$A = \begin{bmatrix} 1 & -2 & 1 \\ -1 & 3 & 0 \\ 0 & 1 & 1 \\ 1 & 2 & 5 \end{bmatrix}$$

is the matrix of the transformation T.

2. $T(a, b, c) = (a - 2b + c) + (-a + 3b)t + (b + c)t^2 + (a + 2b + 5c)t^3$: $\mathrm{R}^3 \to P_3$

$u = (1, 2, -1)$, $v = (-3, -1, 1)$

Answer: By

$T(u) = -4 + 5t + t^2 + 5t^3$ ‘™ 0, $T(v) = 0 + 0t + 0t^2 + 0t^3 = 0$,

we have $u \notin$ *kernelT*, $v \in$ *kernelT*.

3. $T(a + bt + ct^2) = \begin{bmatrix} a - 2b + c & -a + 3b \\ b + c & a + 2b + 5c \end{bmatrix}$: $P_2 \to M(2, 2)$

$u = 1 + 2t - t^2$, $v = -3 - t + t^2$

Answer: By

$$T(u) = \begin{bmatrix} -4 & 5 \\ 1 & 5 \end{bmatrix} \neq O, T(v) = \begin{bmatrix} 0 & 0 \\ 0 & 0 \end{bmatrix} = O$$

we have $u \notin kernel T$, $v \in kernel T$.

Remark: The computation in the three problems are comparable, which indicates the equivalence between general vector spaces and the euclidean space.

4. $T(a, b, c) = a\cos^2 t + b\sin^2 t + c$: $R^3 \to F(R)$

   $u = (1, 1, -1)$, $v = (1, 2, 0)$

Answer: By

$T(u) = \cos^2 t + \sin^2 t - 1 = 0$, $T(v) = \cos^2 t + 2\sin^2 t + 0 \neq 0$,

we have $u \in kernel T$, v O $kernel T$.

5. $T(f) = f'' - f$: $C^2(R) \to C(R)$

   $u = 1 + 2t$, $v = e^t$

Answer: By

$T(u) = -1 - 2t \neq 0$, $T(v) = e^t - e^t = 0$,

we have u O $kernel T$, $v \in kernel T$.

6. $T(p(t)) = (p(0), p'(0))$: $P_3 \to R^2$

   $u = 1 + 2t$, $v = 3t^2 + 4t^3$

Answer: By

$T(u) = (1, 2) \leq 0$, $T(v) = (0, 0) = 0$,

we have $u \notin kernel T$, $v \in kernel T$.

Exercise: Draw pictures showing how the general solutions are related to the solutions of the homogeneous equation.

1. $x + 2y = 2$

Answer: The red line is the solutions of the homogeneous equation (which is orthogonal to the coefficient vector (1, 2)).

The blue line is the general solution, obtained by shifting the red line by the special solution (2,0).

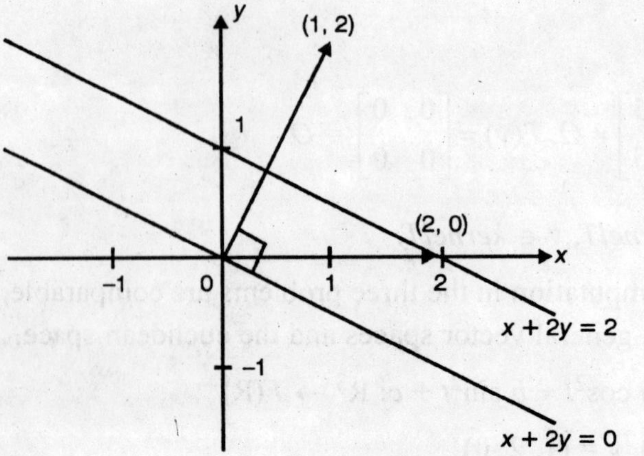

2. $x + 2y + 3z = 1$

Answer:

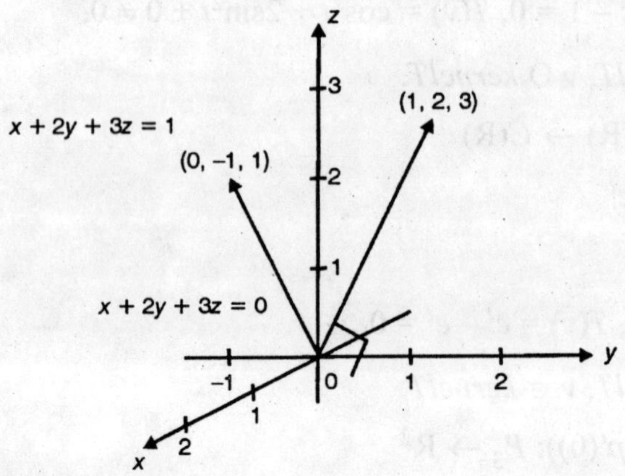

3. $x + 2y + 3z = 1, x + y = -1$

Answer:

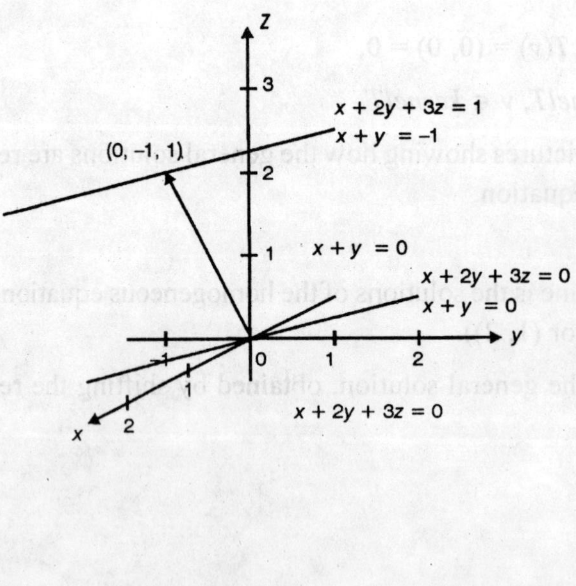

Exercise: **With the help of the kernel for the left side of the linear equations, find the general solutions.**

1. $f' = \sin t$, $kernel(f \to f') = \{\text{constant functions}\}$

Answer: A special solution of $f' = \sin t$ is $f_0 = -\cos t$. The general solutions is

$f = f_0 + \text{constant} = -\cos t + c$.

2. $f'' - 4f' + 3f = \sin t$, $kernel\,(f \to f'' - 4f' + 3f) = \{c_1e^t + c_2e^{3t}: \text{any } c_1, c_2\}$

Answer: We expect the equation $f'' - 4f' + 3f = \sin t$ to have a special solution of the form $f_0 = a\cos t + b\sin t$. Substituting into the equation, we get

$(-a\cos t - b\sin t) - 4(-a\sin t + b\cos t) + 3(a\cos t + b\sin t) = \sin t$.

Comparing the coefficients of $\cos t$, $\sin t$ and solving the resulting equation, we find $f_0 = (1/5)\cos t + (1/10)\sin t$. Thus the general solution is

$f = f_0 + \text{functions in the kernel} = (1/5)\cos t + (1/10)\sin t + c_1e^t + c_2e^{3t}$,

where c_1, c_2 are arbitrary constants.

3. $X - X^{\mathrm{T}} = \begin{bmatrix} 0 & 1 \\ -1 & 0 \end{bmatrix}$, $kernel(X \to X - X^{\mathrm{T}}) = \{\text{symmetric matrices}\}$

Answer: The matrix

$$X = \begin{bmatrix} 0 & 1 \\ 0 & 0 \end{bmatrix}$$

is clearly a solution of the equation. Thus the general solution is

$$X = \begin{bmatrix} 0 & 1 \\ 0 & 0 \end{bmatrix} + \text{symmetric matrices} = \begin{bmatrix} a & 1+b \\ b & c \end{bmatrix}$$

where a, b, c are arbitrary constants.

4. $\begin{bmatrix} 1 & -2 & 3 \\ -2 & 3 & -4 \end{bmatrix} X = \begin{bmatrix} 1 & -2 \\ -2 & 3 \end{bmatrix}$, for 2 by 3 matrix, X kernel given in this example.

Answer: Note that the right side is the same as the first two columns of the coefficient matrix on the left. Therefore

$$X_0 = \begin{bmatrix} I_2 \\ 0 \quad 0 \end{bmatrix} \begin{bmatrix} 1 & 0 \\ 0 & 1 \\ 0 & 0 \end{bmatrix}$$

is a solution of the equation. Then by the kernel given in the example, we have the general solution

$$X = \begin{bmatrix} 1 & 0 \\ 0 & 1 \\ 0 & 0 \end{bmatrix} + c_1 \begin{bmatrix} 1 & 0 \\ 2 & 0 \\ 1 & 0 \end{bmatrix} + c_2 \begin{bmatrix} 0 & 1 \\ 0 & 2 \\ 0 & 1 \end{bmatrix}$$

where c_1, c_2 are arbitrary constants.

Exercise: Let A be an m by n matrix. Show that $nulA = \{0\} \Rightarrow n \leq m$.

Answer: By the definition of *nulA*, $nulA = \{0\}$ means the homogeneous system $Ax = 0$ has only the trivial solution, which is equivalent to the uniqueness for the solution of $Ax = b$. Then the inequality *n* d" *m* follows from this.

Exercise: For what fixed function $m(t)$ is the linear transformation $M(f) = mf$: $F(R) \rightarrow F(R)$ one–to–one?

Answer: If $m(t) \leq 0$ for any $t \in R$, then $M(f(t)) = m(t)f(t) = 0$ for any t implies $f(t) = 0$ for any t. Thus $kernelM = \{0\}$, and M is one–to–one.

If $m(t_0) = 0$ for some $t_0 \in R$, then the nonzero function

$$f(t) = \begin{cases} 0, & \text{if } t \neq t_0 \\ 1, & \text{if } t = t_0 \end{cases}$$

is in the kernel of M. Thus M is not one–to–one.

We conclude that $M(f) = mf$: $F(R) \rightarrow F(R)$ is one–to–one $\Leftrightarrow m(t) \leq 0$ for any $t \in R$.

Exercise: For what fixed matrix A is the linear transformation $M(X) = AX$ one–to–one? What about the linear transformation $M(X) = XA$?

Answer: $M(X) = AX$ is one–to–one means the solution of $AX = B$ (X is the variable) is unique. By looking at the columns of X and B, and making use of this formula, we see the condition is equivalent to the solution of $Ax = b$ is unique for columns b of B. From computational perspective, $M(X) = AX$ is one–to–one is equivalent to all columns of A are pivot.

By using the transpose, we have ($Y = X^T$ and $C = B^T$)

$M(X) = XA$ is one–to–one $\Leftrightarrow M'(Y) = CY$ is one–to–one $\Leftrightarrow$ Solution of $A^Tx = b$ is unique

We will see in future theory that this is actually equivalent to the existence of the solution of $Ax = b$ for any b. Thus from computational perspective, $M(X) = XA$ is one–to–one is equivalent to all rows of A are pivot.

Exercise: If A and B are matrices with same number of columns, how are the null spaces of A, B, and $\begin{bmatrix} A \\ B \end{bmatrix}$ related?

Answer: Let n be the number of columns of A and B. Then

$$nul\begin{bmatrix} A \\ B \end{bmatrix} = \{x \in R^n: \begin{bmatrix} A \\ B \end{bmatrix} x = \begin{bmatrix} Ax \\ Bx \end{bmatrix} = 0\}$$

$$= \{x \in R^n: Ax = 0, Bx = 0\}$$

$$= nulA \cap nulB.$$

Remark: By using the second part of this exercise, the exercise has interpretation interms of linear transformations.

Exercise: Let $T, S: V \to W$ be linear transformations. How are the kernels of T, S, and $T + S$ related?

Answer: If v is in the kernel of both T and S, then we have $(T + S)(v) = T(v) + S(v) = 0 + 0 = 0$, i.e., v is in the kernel of $T + S$. Thus we conclude

kernel(T + S) ⊃ kernelT)™ kernelS.

Note that the two sides are not necessarily equal (take $S = -T$, for example).

Exercise: Let $T: V \to W$ and $S: U \to V$ be linear transformations. How are the kernels of T, S, and TS related?

Answer: If $u \in kernelS$, then $S(u) = 0$, which implies $T(S(u)) = 0$. In other words, $u \in kernelTS$. Thus we conclude

$$kernelTS \supset kernelS.$$

SPAN

DEFINITION AND RELATION TO EXISTENCE

The columns space *colA* of a matrix A is all the linear combinations of the columns of A. More generally, we have

The span of vectors $v_1, v_2, \ldots, v_k$ is

$span\{v_1, v_2, \ldots, v_k\} = \{c_1v_1 + c_2v_2 + \ldots + c_kv_k$: all $c_1, c_2, \ldots, c_k \in R\}$.

Example: Any vector in R^3 can be expressed as follows.

$x = (x_1, x_2, x_3) = x_1(1, 0, 0) + x_2(0, 1, 0) + x_3(0, 0, 1) = x_1e_1 + x_2e_2 + x_3e_3.$

Thus any vector is a linear combination of the standard basis e_1, e_2, e_3. In other words, the standard basis vectors span the euclidean space R^3. By the same reason, the standard basis $e_1, e_2, \ldots, e_n$ span R^n.

Example: Since any polynomial $a_0 + a_1t + a_2t^2 + ... + a_nt^n$ of degree d™ n is a linear combination of $1, t, t^2,..., t^n$, P_n is spanned by $1, t, t^2,..., t^n$.

Example: By the equality $1 = \cos^2 t + \sin^2$, the constant value 1 function is in the span of $\cos^2 t$ and $\sin^2 t$. Similarly, the function $\cos 2t = \cos^2 t - \sin^2 t \in span\{\cos^2 t, \sin^2 t\}$.

The fact that *colA* is a subspace can also be extended.

A span is a subspace.

Proof. We only show that $span\{v_1, v_2, v_3\}$ is a subspace. The general case is similar.

For $u, v \in span\{v_1, v_2, v_3\}$, we have $u = c_1v_1 + c_2v_2 + c_3v_3$, $v = d_1v_1 + d_2v_2 + d_3v_3$, so that

$u + v = (c_1 + d_1)v_1 + (c_2 + d_2)v_2 + (c_3 + d_3)v_3 \in span\{v_1, v_2, v_3\}$.

For $u \in span\{v_1, v_2, v_3\}$ and $d \in$ R, we have $u = c_1v_1 + c_2v_2 + c_3v_3$, so that

$du = dc_1v_1 + dc_2v_2 + dc_3v_3 \in span\{v_1, v_2, v_3\}$.

Example: By $(a + b, 3a - b, 2a + b) = a(1, 3, 2) + b(1, -1, 1)$, the subset

$\{(a + b, 3a - b, 2a + b): a, b \in \mathrm{R}\} \subset \mathrm{R}^3$

is the span of (1, 3, 2) and (1, –1, 1) and is a subspace.

The subset

$\{a \cos t + b \sin t: a, b \in \mathrm{R}\} \subset F(\mathrm{R})$

is a subspace because it is the span of the vectors cos*t* and sin*t*.

The question of whether a vector $b \in colA$ is the same as whether the system $Ax = b$ has solutions. Correspondingly, the question of whether a vector $b \in span\{v_1, v_2,..., v_k\}$ is the same as whether the vector equation

$c_1v_1 + c_2v_2 + ... + c_kv_k = b$,

where $c_1, c_2,..., c_k$ are the unknown variables, has solutions.

Example: If $\cos t \in span\{\cos^2 t, \sin^2 t\}$, then

$a \cos^2 t + b \sin^2 t = \cos t$

for some numbers a and b. By taking $t = 0$, we get $a = 1$. By taking $t = \pi$, we get $a = -1$. The contradiction shows that $\cos t \notin span\{\cos^2 t, \sin^2 t\}$.

Example: Consider polynomials $p_1 = 1 - t + t^3$, $p_2 = 3 - t + 4t^2 + 3t^3$, $p_3 = 2 - t + 2t^2 + 2t^3$, $p_4 = t + 4t^2 - 2t^3$, $p_5 = 1 + 3t^2$ in P_3. We would like to determine whether $q = t + 3t^2 \in span\ \{p_1, p_2, p_3, p_4, p_5\}$. The question is the same as whether we can find c_1, c_2, c_3, c_4, c_5 such that

$c_1(1 - t + t^3) + c_2(3 - t + 4t^2 + 3t^3) + c_3\ 2 - t + 2t^2 + 2t^3) + c_4(t + 4t^2 - 2t^3) + c_5(1 + 3t^2) = t + 3t^2$.

By comparing the coefficients of 1, t, t^2, and t^3, we get a system of four equations.

$$x_1 + 3x_2 + 2x_3 + x_5 = 0$$
$$-x_1 - x_2 - x_3 + x_4 = 1$$
$$4x_2 + 2x_3 + 4x_4 + 3x_5 = 3$$
$$x_1 + 3x_2 + 2x_3 - 2x_4 = 0$$

By this example, the system has no solution. Therefore q is not in the span.

Observe that the augmented matrix for the system is

$$\begin{bmatrix} 1 & 3 & 2 & 0 & 1 & 0 \\ -1 & -1 & -1 & 1 & 0 & 1 \\ 0 & 4 & 2 & 4 & 3 & 3 \\ 1 & 3 & 2 & -2 & 0 & 0 \end{bmatrix} = [[p_1]\ [p_2]\ [p_3]\ [p_4]\ [p_5]\ [q]\ ],$$

where we used the translation

$p = a + bt + ct^2 + dt^3 \in P_3 \leftrightarrow [p] = (a, b, c, d) \in \mathrm{R}^4$.

Thus problems about spans in general vector spaces can be translated into existence problems about usual systems of linear equations.

If we modify q to become $r = t + 2t^2$, then the augmented matrix of the corresponding existence problem is changed to

$$[[p_1]\ [p_2]\ [p_3]\ [p_4]\ [p_5]\ [r]\ ] = \begin{bmatrix} 1 & 3 & 2 & 0 & 1 & 0 \\ -1 & -1 & -1 & 1 & 0 & 1 \\ 0 & 4 & 2 & 4 & 3 & 2 \\ 1 & 3 & 2 & -2 & 0 & 0 \end{bmatrix}$$

By an earlier example, the system has solutions, so that $t + 2t^2 \in span\{p_1, p_2, p_3, p_4, p_5\}$.

In the earlier example, we also saw that the system with the augmented matrix

$$[\ [p_1]\ [p_2]\ [p_4]\ [r]\ ] = \begin{bmatrix} 1 & 3 & 0 & 0 \\ -1 & -1 & 1 & 1 \\ 0 & 4 & 4 & 2 \\ 1 & 3 & -2 & 0 \end{bmatrix}$$

has solutions. Correspondingly, we have $t + 2t^2 \in span\{p_1, p_2, p_4\}$.

GEOMETRIC INTUITION OF SPAN

The span of one vector is all the scalar multiples of the vector.

$$span\{u\} = \{cu: c \in \mathrm{R}\} = \begin{cases} \text{a line,} & \text{if } u \neq 0 \\ \text{a point} & \text{if } u = 0 \end{cases}$$

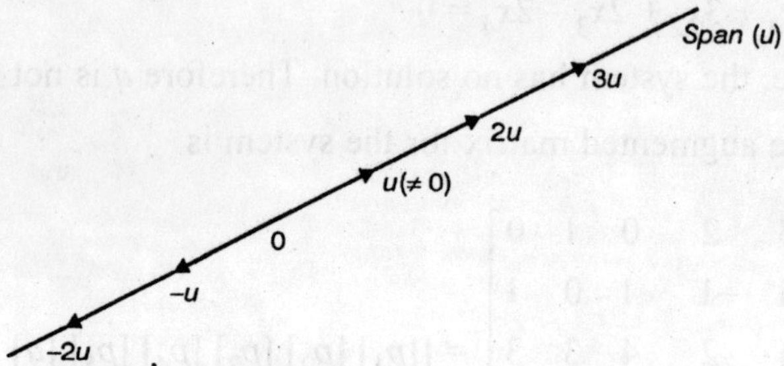

The span of two vectors is

$$span\{u, v\} = \{cu + dv: c, d \in \mathrm{R}\} =$$

$$\begin{cases} \text{a plane,} & \text{if } u, v \text{ are not parallel} \\ \text{a line,} & \text{if } u, v \text{ are not parallel and at least one nonzero} \\ \text{a point,} & \text{if } u = v = 0 \end{cases}$$

where parallel means that one vector is a scalar multiple of another. The following picture illustrates the span of two non–parallel vectors.

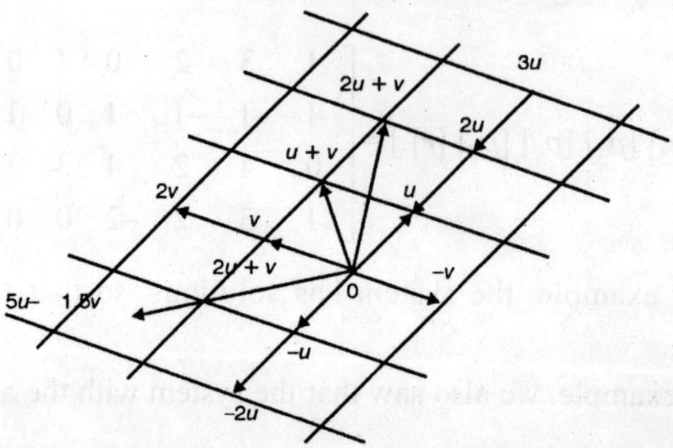

If $v = au$, then $cu + dv = (c + da)u$, so that $span\{u, v\} = span\{u\}$, which we know is a point or a line, depending on whether u is zero or not.

The intuition continues with the span of three vectors, which can be 0, 1, 2, or 3 dimensional. When no one vector is the linear combination of the other two (or independent, a concept we will study), the span is 3 dimensional. The following picture illustrates the situation.

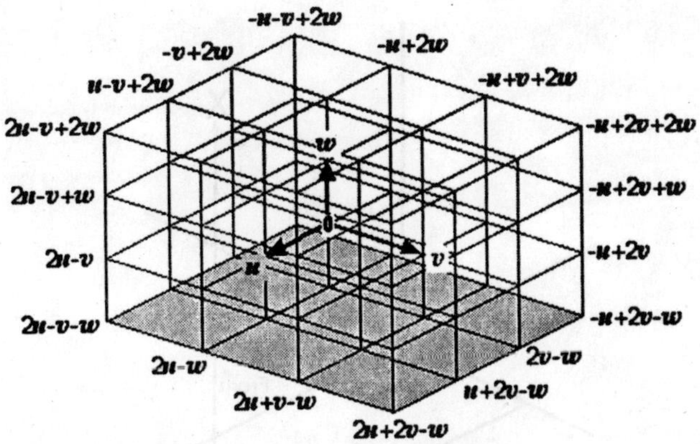

If one vector is a linear combination of the other two, say $w = au + bv$, then $cu + bv + ew = (c + ea)u + (d + eb)v$, so that $span\{u, v, w\} = span\{u, v\}$, which we know can be a plane, a line, or a point.

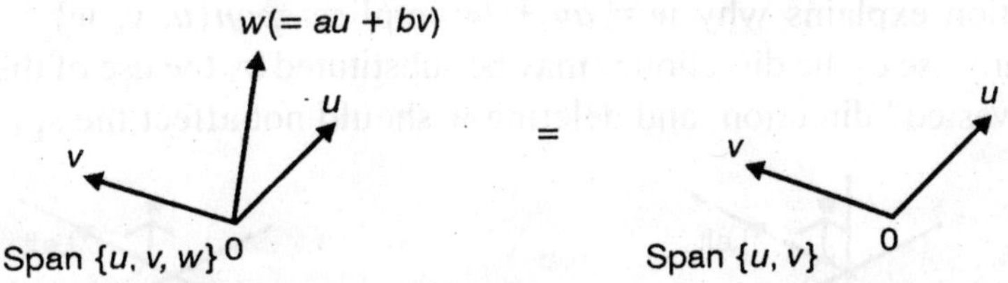

Intuitively, span is all the places one can go to by combining the given directions. For example, we can go from the origin to $2u - 3v$ by first moving in u–direction by 2 units and then moving in the opposite of v–direction by 3 units (or moving in v–direction by -3 units). The following pictures illustrate such intuition.

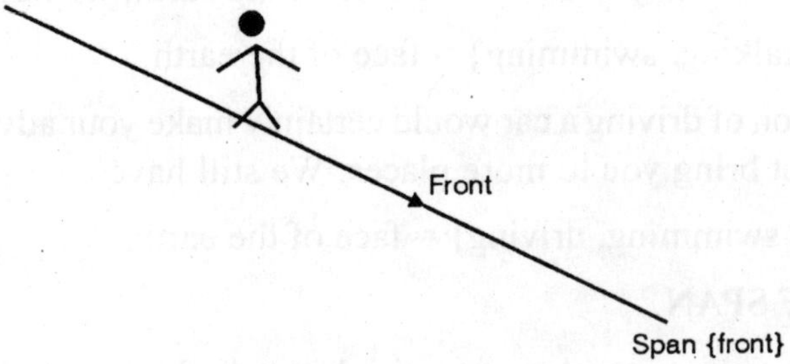

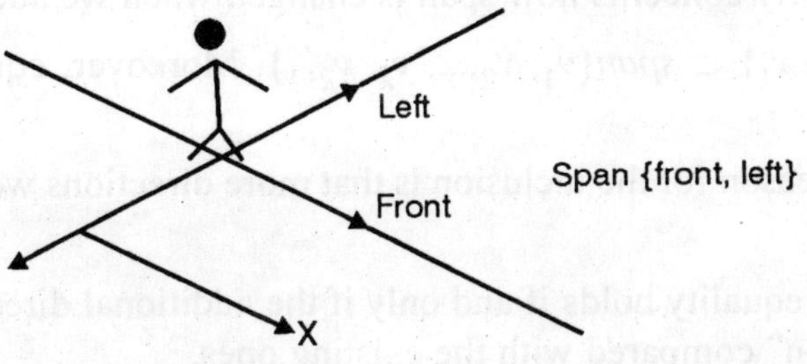

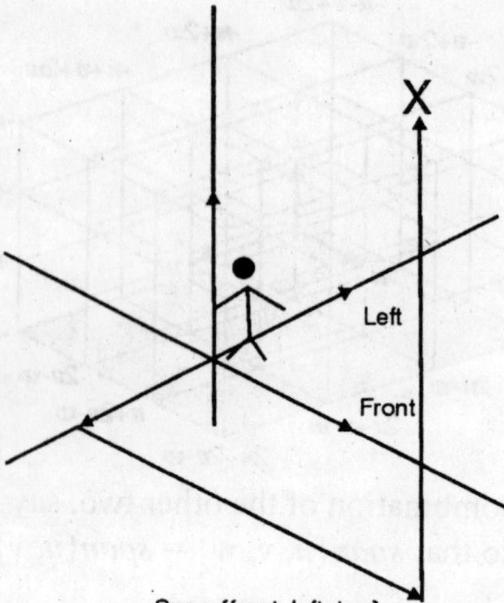

Span {front, left, top}

The intuition explains why $w = au + bv$ implies $span\{u, v, w\} = span\{u, v\}$. The reason is that any use of the direction w may be substituted by the use of the other directions. Thus w is a "wasted" direction, and deleting w should not affect the span.

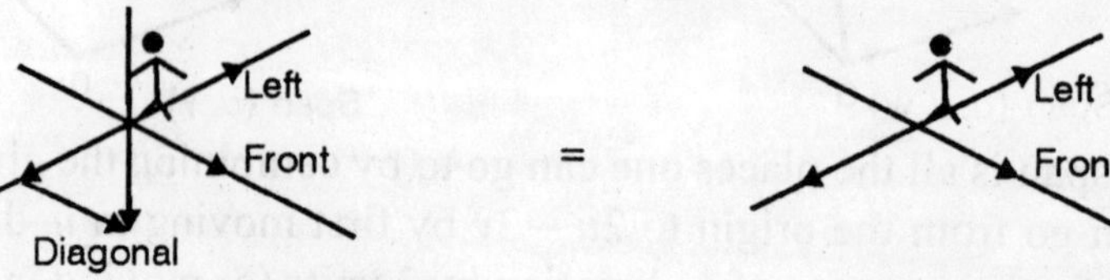

Span {front, left, diagonal} Span {Front, left}

To draw an analogue of the situation, since by walking and swimming, we may (at least theoretically) reach anywhere on the face of the earth, we have

$$span\{\text{walking, swimming}\} = \text{face of the earth.}$$

Now the addition of driving a car would certainly make your adventures more efficient. However, it will not bring you to more places. We still have

$$span\{\text{walking, swimming, driving}\} = \text{face of the earth.}$$

PROPERTIES OF SPAN

The first property concerns how span is changed when we add/delete vectors.

$span\{v_1, v_2,..., v_k\} \subset span\{v_1, v_2,..., v_k, v_{k+1}\}$. Moreover, equality holds $\Leftrightarrow v_{k+1} \in span\{v_1, v_2,..., v_k\}$.

The intuitive reason for the inclusion is that more directions we have, more places we can go.

Moreover, the equality holds if and only if the additional direction is "wasted", or is not a "new direction" compared with the existing ones.

The second property is related to the sum of subspaces.

$span\{v_1, v_2, ..., v_m, w_1, w_2, ..., w_n\} = span\{v_1, v_2, ..., v_m\} + span\{w_1, w_2, ..., w_n\}$.

The next property is useful for computational purpose. Recall that we used row operations to simplify a matrix and then *Answer:* many questions related to the matrix. We may simplify the span by introducing similar operations for a collection of vectors.

r[vector i] + [vector j]

add a multiple of one vector to another.

[vector i] $\leftrightarrow$ [vector j]

exchange two vectors.

d[vector i], $d \neq 0$

multiply a nonzero number to a vector.

The following result allows us to use three operations to simply the vectors that generate the span.

The three operations do not change the span.

The proof of the two properties can be found here.

Example: By using the three operations, we have

$$\begin{aligned} span\{\cos^2 t, \sin^2 t\} &= span\{\cos^2 t, 1\} \\ &= span\{-2\cos^2 t, 1\} \\ &= span\{\cos 2t, 1\} \\ &= span\{\cos 2t, 2\sin^2 t\} \\ &= span\{\cos 2t, \sin^2 t\} \end{aligned}$$

Example: We would like to find out whether the vectors $v_1 = (1, -1, 0, 1)$, $v_2 = (3, -1, 4, 3)$, $v_3 = (2, -1, 2, 2)$, $v_4 = (0, 1, 4, -2)$, $v_5 = (1, 0, 3, 0)$ span R^4 (i.e., whether the span of the five vectors is equal to R^4). Note that the span is the column space of the matrix

$$A = \begin{bmatrix} 1 & 3 & 2 & 0 & 1 \\ -1 & -1 & -1 & 1 & 0 \\ 0 & 4 & 2 & 4 & 3 \\ 1 & 3 & 2 & -2 & 0 \end{bmatrix}$$

Therefore the question is the same as whether $Ax = b$ has solutions for any b. From the computation in an earlier example, the row echelon form (obtained by *row operations* on A), contains a zero row. Then by the criterion for always existence, $Ax = b$ does not always have solutions. Therefore we conclude that $span\{v_1, v_2, v_3, v_4, v_5\} \neq R^4$.

Alternatively, we may use the properties of span to find the answer. The given vectors are too complicated for us to see $span\{v_1, v_2, v_3, v_4, v_5\}$ clearly. Thus we apply three operations to the five vectors, which are essentially *column operations* on A.

By applying (–3)[col 1] + [col 2], (–2)[col 1] + [col 3], (–1)[col 1] + [col 5] to A, we get

$$\begin{bmatrix} 1 & 0 & 0 & 0 & 0 \\ -1 & 2 & 1 & 1 & 1 \\ 0 & 4 & 2 & 4 & 3 \\ 1 & 0 & 0 & -2 & -1 \end{bmatrix}$$

By (–2)[col 3] + [col 2], (–1)[col 3] + [col 4], (–1)[col 3] + [col 5], and then [col 2] ↔ [col 3], we get

$$\begin{bmatrix} 1 & 0 & 0 & 0 & 0 \\ -1 & 1 & 0 & 0 & 0 \\ 0 & 2 & 0 & 2 & 1 \\ 1 & 0 & 0 & -2 & -1 \end{bmatrix}$$

By (–2)[col 5] + [col 4] and [col 3] ↔ [col 5], we get

$$\begin{bmatrix} 1 & 0 & 0 & 0 & 0 \\ -1 & 1 & 0 & 0 & 0 \\ 0 & 2 & 1 & 0 & 0 \\ 1 & 0 & -1 & 0 & 0 \end{bmatrix}$$

This is a column echelon form of A, the simplest we can get from the column operations. The computation tells us

$span\{v_1, v_2, v_3, v_4, v_5\}$

$= span\{(1, -1, 0, 1), (0, 1, 2, 0), (0, 0, 1, -1), (0, 0, 0, 0), (0, 0, 0, 0)\}$

$= span\{(1, -1, 0, 1), (0, 1, 2, 0), (0, 0, 1, -1)\}$

$\neq R^4$

The last step is due to the fact that three vectors cannot span R^4.

The example above leads to some interesting observations. First for euclidean vectors, the preservation of the span under the three operations can be rephrased in terms of column operations.

If $A \to B$ by column operations, then $colA = colB$.

Second is the numerical consequence of spanning the whole space.

$Span\{v_1, v_2,..., v_k\} = R^n \Rightarrow k \geq n.$

The equivalent statement in terms of the column space appeared in the first part of this exercise.

PROPERTIES OF SPAN

Theorem: $span\{v_1, v_2,..., v_k\} \subset span\{v_1, v_2,..., v_k, v_{k+1}\}$. Moreover, equality holds $\Leftrightarrow$ $v_{k+1} \in span\{v_1, v_2,..., v_k\}$.

Proof. In case $k = 2$, for $u \in span\{v_1, v_2\}$ we have

$u = c_1v_1 + c_2v_2 = c_1v_1 + c_2v_2 + 0v_3 \in span\{v_1, v_2, v_3\}.$

Thus we conclude $span\{v_1, v_2\} \subset span\{v_1, v_2, v_3\}$.

If $span\{v_1, v_2\} = span\{v_1, v_2, v_3\}$, then

$v_3 = 0v_1 + 0v_2 + 1v_3 \in span\{v_1, v_2, v_3\} = span\{v_1, v_2\}.$

Conversely, if $v_3 \in span\{v_1, v_2\}$, then $v_3 = av_1 + bv_2$, and for $u \in span\{v_1, v_2, v_3\}$ we have

$u = c_1v_1 + c_2v_2 + c_3v_3 = c_1v_1 + c_2v_2 + c_3(av_1 + bv_2) = (c_1 + ac_3)v_1 + (c_2 + bc_3)v_2 \in span\{v_1, v_2\}.$

Thus we conclude $span\{v_1, v_2, v_3\} \subset span\{v_1, v_2\}$. Combined with the first part, we have the equality.

This completes the proof for $k = 2$. The general case is similar.

Theorem: $span\{v_1, v_2,..., v_m, w_1, w_2,..., w_n\} = span\{v_1, v_2,..., v_m\} + span\{w_1, w_2,..., w_n\}$.

Proof. Any vector in $span\{v_1, v_2,..., v_m\} + span\{w_1, w_2,..., w_n\}$ is of the form $v + w$, where

$v = c_1v_1 + c_2v_2 +... + c_mv_m \in H, w = c'_1w_1 + c'_2w_2 +... + c'_nw_n \in K.$

Thus the vectors in the sum are

$v + w = c_1v_1 + c_2v_2 +... + c_mv_m + c'_1w_1 + c'_2w_2 +... + c'_nw_n,$

which are exactly vectors in $span\{v_1, v_2,..., v_m, w_1, w_2,..., w_n\}$.

Theorem: The three operations do not change the span.

Proof. We only consider special cases of the operations on three vectors. The general case is similar.

For the first operation, we will prove $span\{v_1, v_2, v_3\} = span\{v_1, v_2 + rv_1, v_3\}$. For $u \in span\{v_1, v_2, v_3\}$, we have

$u = c_1v_1 + c_2v_2 + c_3v_3 = (c_1 - rc_2)v_1 + c_2(v_2 + rv_1) + c_3v_3 \in span\{v_1, v_2 + rv_1, v_3\}.$

This proves $span\{v_1, v_2, v_3\} \subset span\{v_1, v_2 + rv_1, v_3\}$. Conversely, for $u \in span\{v_1, v_2 + rv_1, v_3\}$, we have

$u = c_1v_1 + c_2(v_2 + rv_1) + c_3v_3 = (c_1 + rc_2)v_1 + c_2v_2 + c_3v_3 \in span\{v_1, v_2, v_3\}$.

This proves $span\{v_1, v_2, v_3\} \supset span\{v_1, v_2 + rv_1, v_3\}$.

For the second operation, we will prove $span\{v_1, v_2, v_3\} = span\{v_2, v_1, v_3\}$. For $u \in span\{v_1, v_2, v_3\}$ we have

$u = c_1v_1 + c_2v_2 + c_3v_3 = c_2v_2 + c_1v_1 + c_3v_3 \in span\{v_2, v_1, v_3\}$.

This proves $span\{v_1, v_2, v_3\} \subset span\{v_2, v_1, v_3\}$. The inclusion in the other direction is similar.

For the third operation, we will prove $span\{v_1, v_2, v_3\} = span\{dv_1, v_2, v_3\}$ for d "" 0. For $u \in span\{v_1, v_2, v_3\}$ we have

$u = c_1v_1 + c_2v_2 + c_3v_3 = (c_1/d)dv_1 + c_2v_2 + c_3v_3 \in span\{dv_1, v_2, v_3\}$.

This completes the proof that $span\{v_1, v_2, v_3\} \subset span\{dv_1, v_2, v_3\}$. The inclusion in the other direction can be proved similarly

Exercise: Find the condition on the parameter so that the span is the whole space.

1. $Span\{(1, -1, 1), (1, 0, a), (-1, -a, 0)\}$ R^3

Answer: The problem is the same as whether the system

$$x_1 + x_2 - x_3 = b_1$$
$$x_1 - ax_3 = b_2$$
$$x_1 + ax_2 = b_3$$

has solutions for any right side. According to an earlier exercise, the condition for the span to be the whole space is $a \neq 0$ and 2.

2. $Span\{1 + t, 1 + t^2, 1 + at^2, 1 + at\}$ M P_2

Answer: By this earlier remark, we may use

$p = a + bt + ct^2 \in P_2 \leftrightarrow [p] = (a, b, c) \in R^3$

to translate the polynomial problem into whether the euclidean span

$span\{[1 + t], [1 + t^2], [1 + at^2], [1 + at]\} = span\{(1, 1, 0), (1, 0, 1), (1, 0, a), (1, a, 0)\} \subset R^3$

is the whole space. The euclidean problem is the same as whether the system

$$x_1 + x_2 + x_3 + x_4 = b_1$$
$$x_1 + ax_4 = b_2$$
$$x_2 + ax_3 = b_3$$

has solutions for any right side. According to an earlier exercise, the condition for the span to be the whole space is $a \neq 1$.

3. $Span\{1 + at + at^2 + at^3, 1 + t + at^2 + at^3, 1 + t + t^2 + at^3, 1 + t + t^2 + t^3\} \subset P_3$

Answer: Similar to the previous problem, this can be translated into whether the euclidean span

$Span\{(1, a, a, a), (1, 1, a, a), (1, 1, 1, a), (1, 1, 1, 1)\} \subset R^4$

is the whole space. The euclidean problem is the same as whether the system with the coefficient matrix (the columns are the four vectors)

$$\begin{bmatrix} 1 & 1 & 1 & 1 \\ a & 1 & 1 & 1 \\ a & a & 1 & 1 \\ a & a & a & 1 \end{bmatrix}$$

has solutions for any right side. By row operations, we get

$$\begin{bmatrix} 1 & 1 & 1 & 1 \\ 0 & 1-a & 1-a & 1-a \\ 0 & 0 & 1-a & 1-a \\ 0 & 0 & 0 & 1-a \end{bmatrix}$$

The condition for the matrix to have all rows pivot is a ‘‘‘ 1. This is the condition for the span to be the whole space.

4. $Span\left\{\begin{bmatrix} a & 1 \\ 1 & 0 \end{bmatrix}, \begin{bmatrix} 1 & a \\ 0 & 1 \end{bmatrix}, \begin{bmatrix} 1 & 0 \\ a & 1 \end{bmatrix}, \begin{bmatrix} 0 & 1 \\ 1 & a \end{bmatrix}\right\} \subset M(2, 2)$

Answer: As in this exercise, we may use

$$A = \begin{bmatrix} a & b \\ c & d \end{bmatrix} \leftrightarrow [A] = (a, b, c, d)$$

to translate the problem into whether the euclidean span

$span\{(a, 1, 1, 0), (1, a, 0, 1), (1, 0, a, 1), (0, 1, 1, a)\} \subset R^4$

is the whole space. Computationally, this is the same as whether all rows of the following matrix (with the four vectors as columns) are pivot.

$$\begin{bmatrix} a & 1 & 1 & 0 \\ 1 & a & 0 & 1 \\ 1 & 0 & a & 1 \\ 0 & 1 & 1 & a \end{bmatrix}$$

By row operations, we get

$$\begin{bmatrix} 1 & 0 & a & 1 \\ 0 & 1 & 1 & a \\ 0 & 0 & 2a & a^2 \\ 0 & 0. & 0 & 4a - a^3 \end{bmatrix}$$

The condition for the matrix to have all rows pivot is $a \neq 0, 2, -2$. This is the condition for the span to be the whole space.

Exercise: We already know that the column space is preserved by column operations. Is the column space preserved by row operations?

Answer: Applying [row 1] + [row 2] to the matrix

$$A = \begin{bmatrix} 1 \\ 0 \end{bmatrix}$$

gives us

$$B = \begin{bmatrix} 1 \\ 1 \end{bmatrix}.$$

The column space of A is the x_1–axis in R^2. The column space of B is the diagonal line. Therefore the column operation may change the column space.

Exercise: Let u, v, w be three vectors in a vector space V. Then we have a linear transformation $T(a, b, c) = au + bv + cw$: $R^3 \to V$. Show that $rangeT = span\{u, v, w\}$.

Answer: By definition of range, we have

$rangeT = \{T(a, b, c)$: all $a, b, c\} = \{au + bv + cw$: all $a, b, c\}$.

This is exactly all the linear combinations of u, v, w, i.e., the span of u, v, w.

Exercise: How are $span\{u, v, w\}$, $span\{x, y\}$, $span\{u, v, w, x, y\}$ related? How are $span\{u, v, w\}$, $span\{v, w, x\}$, $span\{u, v, w, x\}$ related?

Answer: $Span\{u, v, w, x, y\}$ consists of vectors of the form

$au + bv + cw + dx + ey$, where a, b, c, d, e are arbitrary.

The vectors $au + bv + cw$, for arbitrary a, b, c, form $span\{u, v, w\}$. Moreover, the vectors $dx + ey$, for arbitrary d, e, form $span\{x, y\}$. Therefore

$span\{u, v, w, x, y\} = span\{u, v, w\} + span\{x, y\}$.

By this property, we have

$span\{v, w\} \subset span\{u, v, w\}$,

$span\{v, w\} \subset span\{v, w, x\}$.

Therefore

$span\{v, w\} \subset span\{u, v, w\} \cap span\{v, w, x\}$.

The two sides may not be equal, however. A counterexample is given by the special case $x = u \notin span\{v, w\}$.

Exercise: Let $v_1, v_2, ..., v_k$ be vectors in V. Let $T: V \to W$ be a linear transformation.

1. Prove that T is onto and $v_1, v_2, ..., v_k$ span $V \Rightarrow T(v_1), T(v_2), ..., T(v_k)$ span W.

Answer: Since T is onto, for any $w \in W$ we have $w = T(v)$ for some $v \in V$. Since $v_1, v_2, ..., v_k$ span V, we can write v as a linear combination $v = c_1v_1 + c_2v_2 + ... + c_kv_k$. Then by the linearity of T, we have

$w = T(v) = T(c_1v_1 + c_2v_2 + ... + c_kv_k) = c_1T(v_1) + c_2T(v_2) + ... + c_kT(v_k)$.

Thus we have shown that any vector of W is a linear combination of $T(v_1), T(v_2), ..., T(v_k)$.

2. Is it true that $T(v_1), T(v_2), ..., T(v_k)$ span $W \Rightarrow v_1, v_2, ..., v_k$ span V?

Answer: For an extreme example, let $W = \{0\}$ be the zero space. Then for any $v_1, v_2, ..., v_k$ and T, the vectors $T(v_1), T(v_2), ..., T(v_k)$ always span W. However, $v_1, v_2, ..., v_k$ may not span V.

3. What if T is not assumed to be onto in the first part?

Answer: If T is not onto, then the span of $T(v_1), T(v_2), ..., T(v_k)$ is always contained in the range of T, which is always strictly smaller than W. Therefore the statement is no longer true.

Chapter 9

Eigenspace and Multiplicity

DIRECT SUM OF EIGENSPACES

The diagonalization process is actually the process of computing the eigenspaces $E_\lambda = nul(A - \lambda I)$. The following result says that the different eigenspaces are independent. See here for the definition of direct sum and the relation to linear independence.

Let $\lambda_1, \lambda_2,..., \lambda_k$ be distinct eigenvalues of A. Then the sum $E_{\lambda 1} + E_{\lambda 2} +... + E_{\lambda k}$ of the corresponding eigenspaces is a direct sum.

Proof. We prove by induction. Let us assume the statement is true for $k - 1$ eigenspaces. We prove the statement is still true for k eigenspaces. Specifically, for $v_1 \in E_{\lambda 1}$, $v_2 \in E_{\lambda 2},..., v_k \in E_{\lambda k}$ satisfying

$v_1 + v_2 +... + v_k = 0,$

we need to prove $v_1 = v_2 =... = v_k = 0$.

Multiplying λ_1 and applying A to the equality, respectively, we get

$\lambda_1 v_1 + \lambda_1 v_2 +... + \lambda_1 v_k = 0, \lambda_1 v_1 + \lambda_2 v_2 +... + \lambda_k v_k = 0.$

Substracting the two equalities, we get

$(\lambda_1 - \lambda_2)v_2 +... + (\lambda_1 - \lambda_k)v_k = 0.$

Since the eigenspaces are subspaces, we have $v'_2 = (\lambda_1 - \lambda_2)v_2 \in E_{\lambda 2},..., v'_k = (\lambda_1 - \lambda_k)v_k \in E_{\lambda k}$. By the inductive assumption, $E_{\lambda 2} +... + E_{\lambda k}$ is already a direct sum. Thus from

$v'_2 +... + v'_k = (\lambda_1 - \lambda_2)v_2 + (\lambda_1 - \lambda_k)v_k = 0,$

we have $v'_2 =... = v'_k = 0$. Since $\lambda_1, \lambda_2,..., \lambda_k$ are distinct, this further implies $v_2 =... = v_k = 0$. Finally, from $v_1 + v_2 +... + v_k = 0$, we have $v_1 = 0$.

By this property of direct sum, the result above has the following consequence (B_i consisting of one vector).

Eigenvectors with distinct eigenvalues must be linearly independent.

Recall that for an n by n matrix A, $det(A - \lambda I)$ is a polynomial of degree n. If the polynomial has no multiple root, then the n distinct roots are n distinct eigenvalues $\lambda_1, \lambda_2, \ldots, \lambda_n$. Then each eigenvalue λ_i must yield at least one eigenvector v_i. Thus we have n eigenvectors $v_1, v_2, \ldots, v_n \in R^n$ with distinct eigenvalues, which must be linearly independent and form a basis of R^n. The discussion leads to the following conclusion.

The characteristic polynomial has no multiple roots $\Rightarrow$ The matrix is diagonalizable.

In this example, once we know the eigenvalues of the 2 by 2 matrix are 5 and 15, we may conclude, without further computation, that the matrix has a basis of eigenvectors.

Example: In an earlier example, we know the eigenvalues of A must be from –5, –3, –1, 1, 3. Moreover, we also know that the eigenvalues must include –1, at least one from 1, –3, and at least one from 3, –5. Thus if A is 3 by 3, then A must have three distinct eigenvalues and must be diagonalizable.

Example: The following lower triangular matrix

$$\begin{bmatrix} 2.3 & 0 & 0 & 0 \\ \sqrt{2} & \pi & 0 & 0 \\ \sqrt{1.7} & 100 & \pi^2 & 0 \\ \pi^2 & 5 & 0.01 & 11 \end{bmatrix}$$

has 4 distinct eigenvalues 2.3, π, π^2, 11 and has a basis of eigenvectors.

ALGEBRAIC AND GEOMETRIC MULTIPLICITIES

In this section, we analyse the diagonalization process in detail.

To diagonalize an n by n matrix A, we first find the eigenvalues. As a root of the characteristic polynomial $p(\lambda) = det(A - \lambda I)$, any eigenvalue λ_i has a multiplicity m_i^A defined by

$$p(\lambda_i) = (\lambda_i - \lambda)^{m_i^A}\, s(\lambda) \text{ for integer } m_i^A > 0 \text{ and another polynomial } s \text{ satisfying } s(\lambda_i) \neq 0.$$

Basically, m_i^A is the number of times λ_i appears as a root of the characteristic polynomial. This multiplicity m_i^A is called the algebraic multiplicity of the eigenvalue λ_i. Moreover, this fact about polynomial and the fact that the top degree term of $det(A - \lambda I)$ is $(-1)^n \lambda^n$ tell us the following.

Let $\lambda_1, \lambda_2, \ldots, \lambda_k$ be all the distinct eigenvalues of A, with algebraic multiplicities $m_1^A, m_2^A, \ldots, m_k^A$. Then

$$det(A - \lambda I) = (\lambda_1 - \lambda)^{m_1^A} (\lambda_2 - \lambda)^{m_2^A} \ldots (\lambda_k - \lambda)^{m_k^A}.$$

In particular, $m_1^A + m_2^A + \ldots + m_k^A = n$.

In the second step of the diagonalization process, we compute a basis $B_i = \{v_{i1}, v_{i2}, ..., v_{imi}{}^G\}$ for the eigenspace $E_{\lambda i} = nul(A - \lambda_i I)$ of each eigenvalue λ_i. The number of vectors in B_i is $m_i{}^G = dimE_{\lambda i}$, called the geometrical multiplicity of the eigenvalue λ_i.

Example: Based on the computations, we have the following algebraic and geometric multiplicities.

Matrix	*Characteristic Polynomial*	*Eigenvalue*	*Algebraic Multiplicity*	*Eigenvector*	*Geometric Multiplicity*
$\begin{bmatrix} 13 & -4 \\ -4 & 7 \end{bmatrix}$	$(5-\lambda)(15-\lambda)$	$\lambda_1 = 5$	$m_1{}^A = 1$	(1,2)	$m_1{}^G = 1$
		$\lambda_2 = 15$	$m_2{}^A = 1$	(–2, 1)	$m_2{}^G = 1$
$\begin{bmatrix} 1 & 3 & -3 \\ -3 & 7 & -3 \\ -6 & 6 & -2 \end{bmatrix}$	$(4-\lambda)^2(-2-\lambda)$	$\lambda_1 = 4$	$m_1{}^A = 2$	(1, 1, 0), (1, 0, –1)	$m_1{}^G = 2$
		$\lambda_2 = -2$	$m_2{}^A = 1$	(1, 1, 2)	$m_2{}^G = 1$
$\begin{bmatrix} 2 & 3 & -3 \\ 0 & 2 & -3 \\ 0 & 0 & 1 \end{bmatrix}$	$(2-\lambda)^2(1-\lambda)$	$\lambda_1 = 2$	$m_1{}^A = 2$	(1, 0, 0)	$m_1{}^G = 1$
		$\lambda_2 = 1$	$m_2{}^A = 1$	(–6, 3, 1)	$m_2{}^G = 1$

We observe that the algebraic multiplicity is always no smaller than the geometric multiplicity: $m_i{}^A \geq m_i{}^G$. Moreover, the first two matrices are diagonalizable, for which we have $m_i{}^A = m_i{}^G$. However, the third matrix is not diagonalizable, for which we see the equality does not always hold. The observation will be generalized in the next part.

MULTIPLICITY AND DIAGONALIZATION

For an n by n matrix A, we first find eigenvalues $\lambda_1, \lambda_2, ..., \lambda_k$ and then a basis $B_i = \{v_{i1}, v_{i2}, ..., v_{imi}{}^G\}$ for each eigenspace $E_\lambda = nul(A - \lambda I)$. Since $E_{\lambda 1} + E_{\lambda 2} + ... + E_{\lambda k}$ is a direct sum, by this result, the collection $B = B_1 \cup B_2 \cup ... \cup B_k = \{v_{ij}$: all $i, j\}$ of all the eigenvectors we computed is linearly independent. The number of vectors in B is the sum $m_1{}^G + m_2{}^G + ... + m_k{}^G$ of geometric multiplicities.

The diagonalizability of A is equivalent to whether B is a basis of R^n. Since B is already linearly independent, by this result, we have the following condition for diagonalizability.

An n by n matrix is diagonalizable $\Leftrightarrow$ The sum of geometric multiplicities is n.

We already know the sum of algebraic multiplicities must be n. In view of the statement above, it is natural to ask how the algebraic and the geometric multiplicities are related.

The algebraic multiplicity is no less than the geometric multiplicity: $m_i^A \geq m_i^G$.

Proof Let us consider the first eigenvalue λ_1 of A and abbreviate its multiplicities: $m = m_1^G$, $m' = m_1^A$. Then we have m linearly independent eigenvectors $B_1 = \{v_1, v_2, \dots, v_m\}$ with eigenvalue λ_1. Extend the vectors to a basis $\{v_1, v_2, \dots, v_m, w_{m+1}, \dots, w_n\}$ of R^n. Then

$$P = [v_1\ v_2 \dots v_m\ w_{m+1} \dots w_n] = [P_1\ P_2]$$

(P_1 is the first m columns, P_2 is the last $n - m$ columns) is an invertible matrix. We partition $P^{-1}AP$ accordingly (m and $n - m$ columns, also m and $n - m$ rows):

$$P^{-1}AP = \begin{bmatrix} B_{11} & B_{12} \\ B_{21} & B_{22} \end{bmatrix}.$$

Comparing

$$AP = [\lambda_1 v_1\ \lambda_1 v_2 \dots \lambda_1 v_m\ Aw_{m+1} \dots Aw_n] = [\lambda_1 P_1\ AP_2]$$

with

$$P\begin{bmatrix} B_{11} & B_{12} \\ B_{21} & B_{22} \end{bmatrix} = [P_1B_{11} + P_2B_{21}\ P_1B_{12} + P_2B_{22}],$$

we get $B_{11} = \lambda_1 I$, $B_{21} = O$. Thus

$$P^{-1}AP = \begin{bmatrix} \lambda_1 I & B_{12} \\ O & B_{22} \end{bmatrix}.$$

and (note that $\lambda_1 I$ is an m by m matrix)

$$det(A - \lambda I) = det(P^{-1}AP - \lambda I) = det\begin{bmatrix} (\lambda_1 - \lambda)I & B_{12} \\ O & B_{22} - \lambda I \end{bmatrix} = (\lambda_1 - \lambda)^m\, det(B_{22} - \lambda I).$$

In particular, λ_1 appears as a root of the characteristic polynomial for at least m times. Since the algebraic multiplicity m' is the *total* number of times λ_1 appears as a root, we conclude that $m' \geq m$.

Based on $m_1^A + m_2^A + \dots + m_k^A = n$ and $m_i^A \geq m_i^G$, we conclude that $m_1^G + m_2^G + \dots + m_k^G = n \Leftrightarrow m_i^A = m_i^G$. Thus the earlier result on the diagonalizability has the following reformulation.

A square matrix is diagonalizable $\Leftrightarrow$ Algebraic and geometric multiplicities are equal.

Exercise: What are the algebraic and geometric multiplicities in case the characteristic polynomial has no multiple root?

Answer: If all the roots of the characteristic polynomial are single, then all the algebraic multiplicities are 1. Since the geometric multiplicy is between 1 and the corresponding algebraic multiplicity, the geometric multiplicities are also 1.

Exercise: Prove that the sum of eigenvectors with different eigenvalues is not an eigenvalue.

Answer: Let $Au = \lambda u$, $Av = \mu v$, with $u, v \neq 0$, and $\lambda \neq \mu$. If $u + v$ were an eigenvector with eigenvalue v, then

$$A(u + v) = vu + vv.$$

On the other hand,

$$A(u + v) = \lambda u + \mu v.$$

Comparing the two equalities, we have

$$(\lambda - v)u + (\mu - v)v = 0.$$

Since $\lambda \neq \mu$, the coefficients cannot be all zero. Thus we conclude u and v are linearly dependent. Since this contradicts with this result, we see that $u + v$ cannot be an eigenvector.

The proof clearly extends to the sum of more eigenvectors.

Exercise: Suppose A is a 3 by 3 matrix with a basis {(1, 2, 3), (3, 2, 1), (1, 0, 1)} of eigenvectors, with eigevalues 3, 3, –1. Which of the following are eigenvectors of A?

(2, 2, 2), (1, 3, 2), (3, 1, 2), (–1, –2, –3), (1, 0, –1), (3, 0, 3), (0, 0, 0).

Answer: By this exercise,

(2, 2, 2) = (1/2)(1, 2, 3) + (1/2)(3, 2, 1) is an eigenvector with eigenvalue 3

(–1, –2, –3) = –(1, 2, 3) is an eigenvector with eigenvalue 3

(1, 0, –1) = (–1/2)(1, 2, 3) + (1/2)(3, 2, 1) is an eigenvector with eigenvalue 3

(3, 0, 3) = 3(1, 0, 1) is an eigenvector with eigenvalue –1

On the other hand, (3, 2, 1) is neither a linear combination of (1, 2, 3), (3, 2, 1), nor a scalar multiple of (1, 0, 1). Thus when we expand it with respect to the basis

$$(3, 2, 1) = c_1(1, 2, 3) + c_2(3, 2, 1) + c_3(1, 0, 1) = u + v,$$

we must have $u = c_1(1, 2, 3) + c_2(3, 2, 1)$ and $v = c_3(1, 0, 1)$ nonzero. However, since u and v are eigenvectors with distinct eigenvalues 3 and –1, by this exercise, their sum (3, 2, 1) cannot be an eigenvector.

By the same reason, (3, 1, 2) is not an eigenvector.

Finally, (0, 0, 0) is not an eigenvector because eigenvectors must be nonzero.

Exercise: Determine the parameters so that the matrix is diagonalizable.

1. $\begin{bmatrix} 1 & -a \\ a & 3 \end{bmatrix}$

Answer: The characteristic polynomial is $(3 + a^2) - 4\lambda + \lambda^2$. If a^2 "" 1, then the polynomial has no multiple roots. By this result (or this exercise) the matrix is diagonalizable.

It remains to consider the following cases

$$a = 1: \begin{bmatrix} 1 & -1 \\ 1 & 3 \end{bmatrix}, a = -1: \begin{bmatrix} 1 & 1 \\ -1 & 3 \end{bmatrix}.$$

In both cases, there is only one eigenvalue, and the matrix is not a scalar multiple of the identity. By this exercise, the matrix is not diagonalizable in either case.

2. $\begin{bmatrix} -1 & a & b \\ 0 & 1 & 2 \\ 0 & 2 & 1 \end{bmatrix}$

Answer: The characteristic polynomial is $(-1 - \lambda)((1 - \lambda)^2 - 2^2) = (3 - \lambda)(-1 - \lambda)^2$. For the eigenvalue 3, we find one eigenvector $((a + b)/2, 1, 1)$. For the eigenvalue -1, we need to solve the system $ax_2 + bx_3 = 0$, $2x_2 + 2x_3 = 0$ of *three* variables x_1, x_2, x_3. The diagonalizability requires the system to have two free variables (i.e., the eigenspace for -1 has dimension 2). Since x_1 is already a free variable, the requirement is the same as $ax_2 + bx_3 = 0$ and $2x_2 + 2x_3 = 0$ are equivalent equations. The condiiton is exactly $a = b$.

The condition for the matrix to be diagonalizable is $a = b$.

3. $\begin{bmatrix} 0 & -3a & 0 \\ 1 & 2a & 0 \\ 0 & -3 & 1 \end{bmatrix}$

Answer: The characteristic polynomial is $(1 - \lambda)(3h - 2h\lambda + \lambda^2)$. By this result, if the characteristic polynomial has no multiple roots, the matrix is diagonalizable.

There are two ways the polynomial may have multiple roots. First, $3h - 2h\lambda + \lambda^2$ may have multiple root. This means $3h = h^2$, or $h = 3$ or 0. Second, $\lambda = 1$ may be a root of $3h - 2h\lambda + \lambda^2$. This means $h = -1$.

For $h = 3$, the eigenvalues are 3 and 1. We find one eigenvector $(3, -1, 3)$ for the eigenvalue 3. We find one eigenvector $(0, 0, 1)$ for the eigenvalue 1. The matrix is not diagonalizable.

For $h = 0$, the eigenvalues are 0 and 1. We find one eigenvector $(0, 1, 3)$ for the eigenvalue 0. We find one eigenvector $(0, 0, 1)$ for the eigenvalue 1. The matrix is not diagonalizable.

For $h = -1$, the eigenvalues are 1 and -3. We find two eigenvector (0, 0, 1) and (3, 1, 0) for the eigenvalue 1. We find one eigenvector (1, –1, –1) for the eigenvalue –3. The matrix is diagonalizable.

We onclude that the matrix is diagonalizable for *h* ‵‵‵ 3, 0.

Exercise: Given the following information, find possible eigenvalues of A. Moreover, if possible, determine the invertibility and the diagonalizability of *A*.

1. The eigenvalues of $3I - 4A + A^2$ are –1, 0, 3

Answer: By this result, any eigenvalue of *A* must be the solution of one of the following equations.

$3 - 4\lambda + \lambda^2 = -1$ (solution: $\lambda = 2$)

$3 - 4\lambda + \lambda^2 = 0$ (solution: $\lambda = 1, 3$)

$3 - 4\lambda + \lambda^2 = 3$ (solution: $\lambda = 0, 4$)

Since 0 is included in the five numbers, it may (and may not) be an eigenvalue of *A*. Thus by this result, we cannot say whether *A* is invertible or not. Moreover, the size of *A* is not given, and we do not know whether *A* is diagonalizable or not. In fact, both of the following matrices satisfy the description.

$$\begin{bmatrix} 2 & 0 & 0 & 0 \\ 0 & 2 & 0 & 0 \\ 0 & 0 & 1 & 0 \\ 0 & 0 & 0 & 0 \end{bmatrix}, \begin{bmatrix} 2 & 1 & 0 & 0 \\ 0 & 2 & 0 & 0 \\ 0 & 0 & 1 & 0 \\ 0 & 0 & 0 & 4 \end{bmatrix}$$

The first is diagonalizable and is not invertible. The second is not diagonalizable and is invertible.

2. The characteristic polynomial of $3I - 4A + A^2$ is $-\lambda(-1 - \lambda)(3 - \lambda) = 3\lambda + 2\lambda^2 - \lambda^3$

Answer: The information is the same as the first one, with the addition that the matrix is 3 by 3. By this result, the eigenvalues of *A* must come from the five candidates, and must include 2, at least one from {1, 3}, and at least one from {0, 4}. Thus *A* must have three distinct eigenvalues. In particular, *A* is diagonalizable. Moreover, *A* may or may not be invertible, by the similar reason as the previous problem.

3. The characteristic polynomial of $3I - 4A + A^2$ is $\lambda^2(-1 - \lambda)(3 - \lambda) = -3\lambda^2 - 2\lambda^3 + \lambda^4$

Answer: The information is the same as the first one, with the addition that the matrix is 4 by 4. By this result, the eigenvalues of *A* must come from the five candidates, and must include 2, at least one from {1, 3}, and at least one from {0, 4}. As we have seen in the first problem (especially the examples), A may or may not be diagonalizable, may or may not be invertible.

POLYNOMIALS OF MATRIX

CAYLEY–HAMILTON THEOREM

Let A be a diagonalizable matrix, with a basis $v_1, v_2, \ldots, v_n$ of eigenvectors. Let $p(\lambda) = det(A - \lambda I)$ be the characteristic polynomial of A. Then for the eigenvalue λ_k of v_k, we have

$$p(A)v_k = p(\lambda_k)v_k \text{ (this property)}$$
$$= 0v_k = 0. \text{ (eigenvalues are the roots of } det(A - \lambda I))$$

Since applying $p(A)$ to any vector in the basis $v_1, v_2, \ldots, v_n$ gives us the zero vector, we conclude $p(A)$ must be the zero matrix.

We just proved the following result for diagonalizable matrices. The result is called Cayley–Hamilton theorem, and is true for any square matrix.

Let $p(\lambda) = det(A - \lambda I)$ be the characteristic polynomial of a matrix A. Then $p(A) = O$.

Proof. Let

$$p(\lambda) = p_0 + p_1\lambda + \ldots + p_{n-1}\lambda^{n-1} + p_n\lambda^n.$$

Let $Q(\lambda)$ be the adjugate matrix of the square matrix $A - \lambda I$, which may be considered as a polynomial in λ and with matrix coefficients:

$Q(\lambda) = Q_0 + \lambda Q_1 + \ldots + \lambda^{q-1}Q_{q-1} + \lambda^q Q_q$, where Q_q are constant matrices.

By the formula $(adjA)A = (detA)I$, we have

$$Q(\lambda)(A - \lambda I) = p(\lambda)I = p_0 I + p_1\lambda I + \ldots + p_{n-1}\lambda^{n-1}I + p_n\lambda^n I.$$

On the other hand, we have

$$Q(\lambda)(A - \lambda I) = Q_0 A + \lambda(Q_1 A - Q_0) + \ldots + \lambda^q(Q_q A - Q_{q-1}) - \lambda^{q+1}Q_q.$$

Thus we get $q = n - 1$ and

$$Q_0 A = p_0 I,$$
$$Q_1 A - Q_0 = p_1 I,$$
$$\vdots$$
$$Q_{n-1}A - Q_{n-2} = p_{n-1}I,$$
$$-Q_{n-1} = p_n I.$$

Multiplying powers of A on the right sides, we get

$$Q_0 A = p_0 I,$$

$$Q_1A^2 - Q_0A = p_1A,$$

$$\vdots$$

$$Q_{n-1}A^n - Q_{n-2}A^{n-1} = p_{n-1}A^{n-1},$$

$$-Q_{n-1}A^n = p_nA^n.$$

Adding all the equalities together, we get

$$p(A) = p_0I + p_1A + ... + p_{n-1}A^{n-1} + p_nA^n = O.$$

By this discussion, the characteristic polynmomial of a 2 by 2 matrix A is

$$det(A - \lambda I) = detA - (trA)\lambda + \lambda^2.$$

Therefore the Cayley–Hamilton theorem tells us

$$(detA)I - (trA)A + A^2 = O.$$

GENERALIZED EIGENSPACE

Let $\lambda_1, \lambda_2,..., \lambda_k$ be all the distinct eigenvalues of an n by n matrix A, with algebraic multiplicities $m_1, m_2,..., m_k$. Then by this result, the characteristic polynomial is

$$p(t) = det(A - \lambda I) = (\lambda_1 - \lambda)^m{}_1 (\lambda_2 - \lambda)^m{}_2 ... (\lambda_k - \lambda)^m{}_k.$$

By Cayley–Hamilton theorem, we have

$$p(A) = (\lambda_1 I - A)^m{}_1 (\lambda_2 I - A)^m{}_2 ... (\lambda_k I - A)^m{}_k = O.$$

In particular, we have

In an earlier example, we considered the following diagonalizable matrices

$$A = \begin{bmatrix} 13 & -4 \\ -4 & 7 \end{bmatrix}, \text{ with eigenvalues 5, 15;}$$

$$B = \begin{bmatrix} 1 & 3 & -3 \\ -3 & 7 & -3 \\ -6 & 6 & -2 \end{bmatrix}, \text{ with eigenvalues 4, –2.}$$

For the polynomial $p(t) = (5 - t)(15 - t) = 75 - 20t + t^2$, applying the matrix $p(A) = 75I - 20A + A^2$ to the basis of eigenvectors of A is the same as multiplying $p(\lambda) = p(5 \text{ or } 15) = 0$ to the eigenvectors. In other words, applying $p(A)$ to the basis vectors always give us

Exercise: Suppose $AB = BA$, and v is an eigenvector of A with eigenvalue λ. Prove that if $Bv \neq 0$, then Bv is an eigenvector of A with eigenvalue λ.

Answer: $A(Bv) = (AB)v = (BA)v = B(Av) = B(\lambda v) = \lambda Bv.$

Exercise: Use the $A = PDP^{-1}$ definition of diagonalizability to prove that A is diagonalizable $\Leftrightarrow$ The polynomials and the inverse (if invertible) of A are diagonalizable.

Answer: If $A = PDP^{-1}$, then

$A^2 = PDP^{-1}PDP^{-1} = PD^2P^{-1}$,

$A^3 = PDP^{-1}PDP^{-1}PDP^{-1} = PD^3P^{-1}$,

...

By taking the linear combinations of above, for any polynomial $p(t)$ we get

$p(A) = P\,p(D)\,P^{-1}$.

It is also easy to verify

$$D = \begin{bmatrix} \lambda_1 & 0 & .. & 0 \\ 0 & \lambda_2 & .. & 0 \\ \vdots & \vdots & & \vdots \\ 0 & 0 & .. & \lambda_n \end{bmatrix} \Rightarrow p(D) = \begin{bmatrix} p(\lambda_1) & 0 & .. & 0 \\ 0 & p(\lambda_2) & .. & 0 \\ \vdots & \vdots & & \vdots \\ 0 & 0 & .. & p(\lambda_n) \end{bmatrix}$$

so that $p(D)$ is a diagonal matrix. Thus $p(A)$ is diagonalizable.

As for the inverse, we have

$A = PDP^{-1} \Rightarrow A^{-1} = PD^{-1}P^{-1}$,

where

$$D^{-1} = \begin{bmatrix} \lambda_1^{-1} & 0 & .. & 0 \\ 0 & \lambda_2^{-1} & .. & 0 \\ \vdots & \vdots & & \vdots \\ 0 & 0 & .. & \lambda_n^{-1} \end{bmatrix}$$

is still diagonal.

Exercise: (complex number required) Prove that for a 2 by 2 matrix A, the following are all the posibilities regarding diagonalizability

- A has two distinct eigenvalues and is diagonalizable,
- A has only one eigenvalue and is not diagonalizable,
- $A = cI$ for a constant c.

Answer: The characteristic polynomial of A has degree 2. Thus A has either one eigenvalue or two distinct eigenvalues.

Suppose A has two distinct eigenvalues λ_1 and λ_2. Then for each eigenvalue, we can find at least one eigenvector. Let $v_1, v_2 \in R^2$ be eigenvectors with λ_1, λ_2 as eigenvalues. Then we claim v_1, v_2 are linearly independent. Assume $c_1v_1 + c_2v_2 = 0$. Then applying A and multiplying λ_2 give us

$$c_1Av_1 + c_2Av_2 = c_1\lambda_1v_1 + c_2\lambda_2v_2 = 0,\ c_1\lambda_2v_1 + c_2\lambda_2v_2 = 0.$$

The difference between the two equalities is

$$c_1(\lambda_1 - \lambda_2)v_1 = 0.$$

Since $\lambda_1 \neq \lambda_2$ and $v_1 \neq 0$, we conclude that $c_1 = 0$. Similarly, we also have $c_2 = 0$. Thus v_1, v_2 are two linearly independent vectors in R^2. Then by this result, the two eigenvectors form a basis. In other words, A is diagonalizable.

If A has one eigenvalue λ and is diagonalizable, then $A = PDP^{-1}$. Because the two diagonal entries of D are the same eigenvalue λ, we have $D = \lambda I$ and $A = PDP^{-1} = \lambda I$.

The remaining case is that A has one eigenvalue and is not diagonalizable. This completes the discussion of the three cases as long as the exercise is concerned.

We continue with a more detailed study of the case that A has one eigenvalue λ and is not diagonalizable. Let v be the one eigenvector we can find for λ. Choose another vector w, such that $\{w, v\}$ form a basis of R^2. Then

$$Aw = aw + bv$$

for some numbers a, b. Since A has no basis of eigenvectors, w is not an eigenvector. This means $b \neq 0$. Let $u = b^{-1}w$. Then $\{u, v\}$ is still a basis, and by dividing the equality above by b, we get

$$Au = au + v$$

Combined with

$$Av = \lambda v,$$

we see the matrix for the linear transformation $x \to Ax$ with respect to the basis $\{u, v\}$ is

$$\begin{bmatrix} a & 1 \\ 0 & \lambda \end{bmatrix}.$$

In particular, a is an eigenvalue of A. Since λ is assumed to be the only eigenvalue, we have $a = \lambda$ and

$$Au = \lambda u + v,\ Av = \lambda v.$$

Thus the matrix for the linear transformation with respect to the basis $\{u, v\}$ is

$$\begin{bmatrix} \lambda & 1 \\ 0 & \lambda \end{bmatrix}.$$

This is equivalent to

$$A = P\begin{bmatrix} \lambda & 1 \\ 0 & \lambda \end{bmatrix}P^{-1}.$$

Remark: Any 2 by 2 matrix is similar to one of the following two matrices

$$\begin{bmatrix} \lambda_1 & 1 \\ 0 & \lambda_2 \end{bmatrix}, \begin{bmatrix} \lambda & 1 \\ 0 & \lambda \end{bmatrix}$$

These are the possible Jordan canonical forms of A. The vector u we found in the discussion is a generalized eigenvector.

Exercise: Given the following information, find possible eigenvalues of A. Moreover, if possible, determine the invertibility and the diagonalizability of A.

1. The eigenvalues of $3I - 4A + A^2$ are $-1, 0, 3$

Answer: By this result, any eigenvalue of A must be the solution of one of the following equations.

$3 - 4\lambda + \lambda^2 = -1$ (solution: $\lambda = 2$)

$3 - 4\lambda + \lambda^2 = 0$ (solution: $\lambda = 1, 3$)

$3 - 4\lambda + \lambda^2 = 3$ (solution: $\lambda = 0, 4$)

Since 0 is included in the five numbers, it may (and may not) be an eigenvalue of A. Thus by this result, we cannot say whether A is invertible or not.

Moreover, the size of A is not given, and we do not know whether A is diagonalizable or not. In fact, both of the following matrices satisfy the description.

$$\begin{bmatrix} 2 & 0 & 0 & 0 \\ 0 & 2 & 0 & 0 \\ 0 & 0 & 1 & 0 \\ 0 & 0 & 0 & 0 \end{bmatrix}, \begin{bmatrix} 2 & 1 & 0 & 0 \\ 0 & 2 & 0 & 0 \\ 0 & 0 & 1 & 0 \\ 0 & 0 & 0 & 4 \end{bmatrix}$$

The first is diagonalizable and is not invertible. The second is not diagonalizable and is invertible.

2. The characteristic polynomial of $3I - 4A + A^2$ is $-\lambda(-1 - \lambda)(3 - \lambda) = 3\lambda + 2\lambda^2 - \lambda^3$

Answer: The information is the same as the first one, with the addition that the matrix is 3 by 3. By this result, the eigenvalues of A must come from the five candidates, and must include 2, at least one from {1, 3}, and at least one from {0, 4}. Thus A must have three distinct eigenvalues. In particular, A is diagonalizable.

Moreover, A may or may not be invertible, by the similar reason as the previous problem.

3. The characteristic polynomial of

$$3I - 4A + A^2 \text{ is } \lambda^2(-1 - \lambda)(3 - \lambda) = -3\lambda^2 - 2\lambda^3 + \lambda^4$$

Answer: The information is the same as the first one, with the addition that the matrix is 4 by 4. By this result, the eigenvalues of A must come from the five candidates, and must include 2, at least one from {1, 3}, and at least one from {0, 4}.

As we have seen in the first problem (especially the examples), A may or may not be diagonalizable, may or may not be invertible.

Chapter 10

Dual Space

DUAL SPACE

A linear function on an euclidean space

$$f(x_1, x_2,..., x_n) = a_1x_1 + a_2x_2 +... + a_nx_n$$

is simply the left side of the usual linear equation. Note that adding and scalar multiplying such linear functions still produce linear functions. Therefore such linear functions form a vector space. More generally, we have

A linear function on a vector space V is $f: V \to R$ satisfying

$$f(u + v) = f(u) + f(v), f(cu) = c f(u).$$

The dual space V^* is the collection of all linear functions.

Using the concept of Hom space, we have

$$V^* = \text{Hom}(V, R).$$

In particular, V^* is a vector space with the following addition and scalar multiplication

$$(f + g)(v) = f(v) + g(v), (cf)(v) = c(f(v)).$$

The definition above must be justified by showing that a linear function $f: R^n \to R$ is given by the familier formula. Write a vector of R^n as

$$(x_1, x_2,..., x_n) = x_1e_1 + x_2e_2 +... + x_ne_n,$$

where $\{e_1, e_2,..., e_n\}$ is the standard basis of R^n. Then the linearity of f implies

$$\begin{aligned} f(x_1, x_2,..., x_n) &= f(x_1e_1 + x_2e_2 +... + x_ne_n) \\ &= x_1f(e_1) + x_2f(e_2) +... + x_nf(e_n) \\ &= a_1x_1 + a_2x_2 +... + a_nx_n, \end{aligned}$$

where $a_i = f(e_i)$. Thus we conclude that

$(R^n)^* = \{f(x_1, x_2,..., x_n) = a_1x_1 + a_2x_2 +... + a_nx_n: a_1, a_2,..., a_n$ are real numbers$\}$.

Note that the coefficients of the linear function gives an identification

ι_n: $(R^n)^* \leftrightarrow R^n$, $a_1x_1 + a_2x_2 + ... + a_nx_n \leftrightarrow (a_1, a_2,..., a_n)$.

The argument above is a special case of the proof that linear transformations between euclidean spaces are given by matrices. In fact, as a linear transformation from R^n to R^1, the function f is given by the 1 by n matrix $[f(e_1) f(e_2)... f(e_n)]$.

Example: For a linear function f: $P_3 \to R$, we have

$f(c_0 + c_1t + c_2t^2 + c_3t^3) = c_0f(1) + c_1f(t) + c_2f(t^2) + c_3f(t)^3 = c_0a_0 + c_1a_1 + c_2a_2 + c_3a_3$.

Therefore

$(P_3)^* = \{f(c_0 + c_1t + c_2t^2 + c_3t^3) = c_0a_0 + c_1a_1 + c_2a_2 + c_3a_3$: a_0, a_1, a_2, a_3 are real numbers}.

We actually see a pattern here. Basically, linear functions on (finite dimensional) vector spaces are determined by the values on basis vectors. For example, we also have

$$(M(2, 2))^* = \{f\begin{bmatrix} x & y \\ z & w \end{bmatrix} = ax + by + cz + dw: a, b, c, d \text{ are real numbers}\}.$$

Example: In this example, we saw the evaluation

$f \to f(2)$: $F(R) \to R$

at $t = 2$ is a linear function. Clearly the evaluations at the other places and the linear combinations of evaluations such as

$f \to 2f(2) - 3f(1)$: $F(R) \to R$

is also linear.

Another linear function on function spaces is given by the integration

$f \to \int_0^1 f(t)\, dt$: $C(R) \to R$.

Integration after multiplying a function such as

$f \to \int_0^1 f(t)\cos\pi t\, dt$: $C(R) \to R$

is also a linear function.

DUAL BASIS

A basis of a vector space can be used to construct an isomorphism, via the coordinates, between the vector space and the euclidean space. The basis can also be used to identify the dual space with the euclidean space.

Let $\{b_1, b_2,..., b_n\}$ be a basis of V. Then

$f \to (f(b_1), f(b_2),..., f(b_n))$, $V^* \to R^n$

is an isomorphism. In particular, we have $dimV = dimV^*$.

Proof. Denote the transformation by $E(f) = (f(b_1), f(b_2),..., f(b_n))$. The following verifies $E(f + g) = E(f) + E(g)$.

$$\begin{aligned} &E(f+g) \\ &= ((f+g)(b_1), (f+g)(b_2),..., (f+g))(b_n)) \\ &= (f(b_1) + g(b_1), f(b_2) + g(b_2),..., f(b_n) + g(b_n)) \text{ (definition of the addition } f+g) \\ &= (f(b_1), f(b_2),..., f(b_n)) + (g(b_1), g(b_2),..., g(b_n)) \\ &= E(f) + E(g). \end{aligned}$$

We may verify $E(cf) = cE(f)$ in the similar way. Therefore E is linear.

Next, we construct an inverse for E. For any $a = (a_1, a_2,..., a_n) \in R^n$, we construct a linear function on V via the formula

$$f_a(c_1b_1 + c_2b_2 +... + c_nb_n) = a_1c_1 + a_2c_2 +... + a_nc_n.$$

Strictly speaking, we need to explain that f_a is well–defined and linear. This can be proved directly, or can be seen from the fact that f_a is a composition

$$v = c_1b_1 + c_2b_2 +... + c_nb_n \to (c_1, c_2,..., c_n) \to a_1c_1 + a_2c_2 +... + a_nc_n,$$

where the first $\to$ is the coordinates with respect to the basis $\{b_1, b_2,..., b_n\}$, and the second $\to$ is a well-known linear function on the euclidean space.

By the way f_a is constructed, it is easy to verify that the transformation $a \to f_a$, $R^n \to V^*$ is the inverse of E.

In the special case $V = R^n$ and $\{b_1, b_2,..., b_n\}$ is the standard basis, the isomorphism $f \to (f(b_1), f(b_2),..., f(b_n))$ is simply the isomorphism ι_n: $a_1x_1 + a_2x_2 +... + a_nx_n \to (a_1, a_2,..., a_n)$ constructed early on.

Under the isomorphism $f \to (f(b_1), f(b_2),..., f(b_n))$, $V^* \to R^n$, the standard basis on the right side corresponds to a basis $\{\beta_1, \beta_2,..., \beta_n\}$ of V^*. In other words, we should have $(\beta_i(b_1), \beta_i(b_2),..., \beta_i(b_n)) = e_i$. This leads to the following concept.

The dual basis $\{\beta_1, \beta_2,..., \beta_n\}$ of a basis $\{b_1, b_2,..., b_n\}$ of V consists of linear functions satisfying

$$\beta_i(b_j) = \begin{cases} 1, & \text{if } i = j \\ 0, & \text{if } i \neq j \end{cases}$$

Example: The linear functions

$$\begin{aligned} &\varepsilon_1(x_1, x_2, x_3) = x_1, \\ &\varepsilon_2(x_1, x_2, x_3) = x_2, \\ &\varepsilon_3(x_1, x_2, x_3) = x_3, \end{aligned}$$

form a basis of $(R^3)^*$, which is in fact the dual basis of the standard basis of R^3.

The linear functions

$$\alpha(a + bt + ct^2 + dt^3) = a,$$

$$\beta(a + bt + ct^2 + dt^3) = b,$$

$$\gamma(a + bt + ct^2 + dt^3) = c,$$

$$\delta(a + bt + ct^2 + dt^3) = d,$$

form the dual basis of the basis $\{1, t, t^2, t^3\}$ of P_3.

Example In an earlier example, we had a basis $B = \{(3, 1, 2), (1, -1, 2), (-1, 1, 1)\}$ of R^3. Let $B^* = \{a_{11}x_1 + a_{12}x_2 + a_{13}x_3, a_{21}x_1 + a_{22}x_2 + a_{23}x_3, a_{31}x_1 + a_{32}x_2 + a_{33}x_3\}$ be its dual basis. Then the definition of the dual basis can be reformulated exactly as

$$\begin{bmatrix} a_{11} & a_{12} & a_{13} \\ a_{21} & a_{22} & a_{23} \\ a_{31} & a_{32} & a_{33} \end{bmatrix} \begin{bmatrix} 3 & 1 & -1 \\ 1 & -1 & 1 \\ 2 & 2 & 1 \end{bmatrix} = \begin{bmatrix} 1 & 0 & 0 \\ 0 & 1 & 0 \\ 0 & 0 & 1 \end{bmatrix}$$

Then by this example, we get

$$\begin{bmatrix} a_{11} & a_{12} & a_{13} \\ a_{21} & a_{22} & a_{23} \\ a_{31} & a_{32} & a_{33} \end{bmatrix} \begin{bmatrix} 3 & 1 & -1 \\ 1 & -1 & 1 \\ 2 & 2 & 1 \end{bmatrix} = \begin{bmatrix} 1/4 & 1/4 & 0 \\ -1/12 & -5/12 & 1/3 \\ -1/3 & 1/3 & 1/3 \end{bmatrix}$$

We conclude that the dual basis is $B^* = \{(1/4)x_1 + (1/4)x_2, -(1/12)x_1 - (5/12)x_2 + (1/3)x_3, -(1/3)x_1 + (1/3)x_2 + (1/3)x_3\}$.

DUAL TRANSFORMATION

A linear transformation $T: V \to W$ induces a transformation

$$T^*(f) = fT: W^* \to V^*$$

between the dual spaces, called the dual transformation. Specifically, the formula means that $T^*(f)$ is a linear function on V, and the value of the function at $v \in V$ is $T^*(f)(v) = f(T(v))$. The generalization of T^* to Hom spaces is given in this exercise.

The following verifies that $T^*(f + g) = T^*(f) + T^*(g)$.

$$\begin{aligned} T^*(f + g)(v) &= (f + g)(T(v)) \\ &= f(T(v)) + g(T(v)) \\ &= T^*(f)(v) + T^*(g)(v) \\ &= (T^*(f) + T^*(g))(v). \end{aligned}$$

The equality $T^*(cf) = cT^*(f)$ can be similarly verified. Thus the dual transformation is linear.

Next we compute the dual transformation of a linear transformation $T(x) = Ax$: $\mathrm{R}^n \to \mathrm{R}^m$ given by an m by n matrix A. By making use of the identification ι: $(\mathrm{R}^n)^* \leftrightarrow \mathrm{R}^n$, the dual transformation T^*: $(\mathrm{R}^m)^* \to (\mathrm{R}^n)^*$ may be identified with a linear transformation $T' = \iota T^* \iota^{-1}$: $\mathrm{R}^m \to \mathrm{R}^n$. The matrix of this transformation (the red question in the picture) may be regarded as the matrix of the dual transformation.

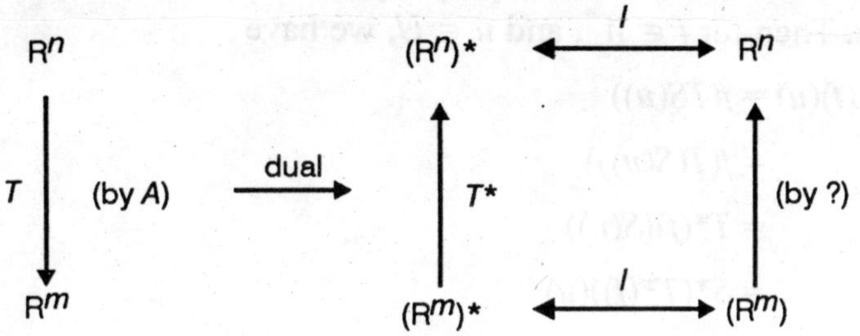

To find the matrix for T', we try the case that

$$A = \begin{bmatrix} a_{11} & a_{12} \\ a_{21} & a_{22} \\ a_{31} & a_{32} \end{bmatrix}, \text{ or } T(x_1, x_2) = (a_{11}x_1 + a_{12}x_2, a_{21}x_1 + a_{22}x_2, a_{31}x_1 + a_{32}x_2): \mathrm{R}^2 \to \mathrm{R}^3.$$

For the inear function $f(y_1, y_2, y_3) = c_1y_1 + c_2y_2 + c_3y_3 \in (\mathrm{R}^3)^*$, we have

$$\begin{aligned} T^*(f)(x_1, x_2) &= f(T(x_1, x_2)) \\ &= f(a_{11}x_1 + a_{12}x_2, a_{21}x_1 + a_{22}x_2, a_{31}x_1 + a_{32}x_2) \\ &= c_1(a_{11}x_1 + a_{12}x_2) + c_2(a_{21}x_1 + a_{22}x_2) + c_3(a_{31}x_1 + a_{32}x_2) \\ &= (a_{11}c_1 + a_{21}c_2 + a_{31}c_3)x_1 + (a_{12}c_1 + a_{22}c_2 + a_{32}c_3)x_2 \in (\mathrm{R}^2)^*. \end{aligned}$$

The transformation T' takes $\iota(f) = (c_1, c_2, c_3)$ to $\iota(T^*(f)) = (a_{11}c_1 + a_{21}c_2 + a_{31}c_3, a_{12}c_1 + a_{22}c_2 + a_{32}c_3)$:

$$T'\begin{bmatrix} c_1 \\ c_2 \\ c_3 \end{bmatrix} = \begin{bmatrix} a_{11}c_1 + a_{21}c_2 + a_{31}c_3 \\ a_{11}c_1 + a_{21}c_2 + a_{31}c_3 \end{bmatrix} = \begin{bmatrix} a_{11} & a_{21} & a_{31} \\ a_{12} & a_{22} & a_{32} \end{bmatrix}\begin{bmatrix} c_1 \\ c_2 \\ c_3 \end{bmatrix} = A^T\begin{bmatrix} c_1 \\ c_2 \\ c_3 \end{bmatrix}.$$

Thus the matrix for T' is A^{T}. In general, we have the following.

Under the identification ι: $(\mathrm{R}^n)^* \leftrightarrow \mathrm{R}^n$, if a linear transformation is given by a matrix A, then the dual transformation is given by the matrix A^{T}.

The fact tells us that the dual transformation is, up to the identification ι, a generalization of the transpose. The linearity and product properties of the transpose then follow from the following properties of the dual. The explanation for the involution property is much more complicated and appears here.

linearity

$$(T + S)^* = T^* + S^*, (cT)^* = cT^*$$

product

$$(TS)^* = S^*T^*$$

Similar properties can be proved for the generalization to the Hom space.

Proof We prove the the product property. Let $S: U \to V$ and $T: V \to W$ be linear transformations. Then for $f \in W^*$, and $u \in U$, we have

$$\begin{aligned}(TS)^*(f)(u) &= f(TS(u)) \\ &= f(T(S(u))) \\ &= T^*(f)(S(u)) \\ &= S^*(T^*(f))(u) \\ &= (S^*T^*)(f)(u).\end{aligned}$$

Thus $(TS)^*(f)(u) = (S^*T^*)(f)(u)$ for any f and u. This means $(TS)^* = S^*T^*$.

The proof of linearity is similar.

DOUBLE DUAL

Let V be a vector space. Its dual V^* is also a vector space. The double dual of V is the dual V^{**} of the dual space V^*.

There is a natural relation between V and V^{**}. In contrast, there is no natural relation between V and V^*. For example, the relation such as $\iota: (\mathrm{R}^n)^* \leftrightarrow \mathrm{R}^n$ *depends on the choice* of the standard basis.

A vector $v \in V$ induces a linear function $v^{**}(f) = f(v)$ on V^*. The vector $v^{**} \in V^{**}$ is the double dual of v. The transformation $v \to v^{**}: V \to V^{**}$ is the double dual transformation.

Of course we need to justify the definition by showing that v^{**} is indeed a linear function on V^* and $v \to v^{**}$ is linear. The following verifies $v^{**}(f + g) = v^{**}(f) + v^{**}(g)$.

$$\begin{aligned}v^{**}(f + g) &= (f + g)(v) \\ &= f(v) + g(v) \\ &= v^{**}(f) + v^{**}(g).\end{aligned}$$

The equality $v^{**}(cf) = c\, v^{**}(f)$ can be similarly verified. As for the linearity of $v \to v^{**}$, the following verifies $(u + v)^{**} = u^{**} + v^{**}$.

$$\begin{aligned}(u + v)^{**}(f) &= f(u + v) \text{ (definition of } (u + v)^{**}) \\ &= f(u) + f(v) \text{ } (f \text{ is linear})\end{aligned}$$

$= u^{**}(f) + v^{**}(f)$ (definition of u^{**} and v^{**})

$= (u^{**} + v^{**})(f)$. (definition of the addition $u^{**} + v^{**}$ of linear functions)

The equality $(cv)^{**} = c(v^{**})$ can be similarly verified.

The double dual transformation takes a basis to a basis.

Let $\{b_1, b_2,..., b_n\}$ be a basis of V. Let $\{\beta_1, \beta_2,..., \beta_n\}$ be its dual basis. Then $\{b_1^{**}, b_2^{**},..., b_n^{**}\}$ is the dual basis of $\{\beta_1, \beta_2,..., \beta_n\}$.

Proof We simply need to verify the defining property of the dual basis for $\{b_1^{**}, b_2^{**},..., b_n^{**}\}$:

$$b_i^{**}(\beta_j) = \beta_j(b_i) = \begin{cases} 1, & \text{if } i = j \\ 0, & \text{if } i \neq j \end{cases}$$

where the first = is due to the definition of v^{**}, and the second = is due to the assumption that $\{\beta_1, \beta_2,..., \beta_n\}$ is the dual of the basis $\{b_1, b_2,..., b_n\}$.

Given the basis and its double dual, the linearity of $v \to v^{**}$ then implies that the double dual transformation is the same as

$$v = c_1b_1 + c_2b_2 + ... + c_nb_n \to v^{**} = c_1b_1^{**} + c_2b_2^{**} + ... + c_nb_n^{**}.$$

This is the composition of two linear transformations

$$v = c_1b_1 + c_2b_2 + ... + c_nb_n \to (c_1, c_2,..., c_n) \to v^{**} = c_1b_1^{**} + c_2b_2^{**} + ... + c_nb_n^{**}.$$

where the first $\to$ is the coordinates with respect to $\{b_1, b_2,..., b_n\}$ and the second $\to$ is the inverse to the coordinates with respect to $\{b_1^{**}, b_2^{**},..., b_n^{**}\}$. Since both are isomorphisms, the compisition is an isomorphism.

For finite dimensional vector spaces, the double dual transformation $v \to v^{**}: V \to V^{**}$ is an isomorphism.

Finally, for a linear transformation $T: V \to W$, we may the dual twice and get the double dual $T^{**}: V^{**} \to W^{**}$ of T. Prove that $T^{**}(v^{**}) = T(v)^{**}$.

$T^{**}(v^{**}) = T(v)^{**}$.

Proof. We only need to show that test $T^{**}(v^{**})(f) = T(v)^{**}(f)$ for any linear function $f \in V^*$.

$$\begin{aligned} T^{**}(v^{**})(f) &= v^{**}(T^*(f)) \\ &= T^*(f)(v) \\ &= f(T(v)) \\ &= T(v)^{**}(f). \end{aligned}$$

The equality $T^{**}(v^{**}) = T(v)^{**}$ means that, after V^{**} and W^{**} are naturally identified with V and W via the double dual transformation, $T^{**}: V^{**} \to W^{**}$ is identified with the original linear transformation $T: V \to W$. The situation is illustrated by the following diagram.

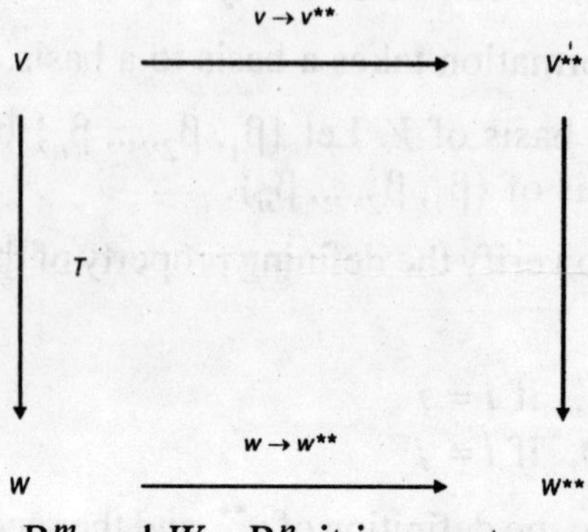

In the special case $V = \mathrm{R}^m$ and $W = \mathrm{R}^n$, it is easy to verify that $v \to v^{**}: \mathrm{R}^m \to (\mathrm{R}^m)^{**}$ is the same as combining the identification $\imath$ and its dual together: $\imath^* \imath^{-1}: \mathrm{R}^m \to (\mathrm{R}^m)^* \to (\mathrm{R}^m)^{**}$. Then the natural identification between T^{**} and T explains the involutive property of the transpose of matrices.

Exercise: Determine whether the linear functions span the dual space and whether they are linearly independent.

1. $x_1 + 2x_2 + 3x_3, 4x_1 + 5x_2 + 6x_3, 7x_1 + 8x_2 + 9x_3 \; \Xi \; (\mathrm{R}^3)^*$

Answer: By using the identification $\imath$, the problem is the same as whether the three vectors

$$\imath(x_1 + 2x_2 + 3x_3) = (1, 2, 3)$$
$$\imath(4x_1 + 5x_2 + 6x_3) = (4, 5, 6)$$
$$\imath(7x_1 + 8x_2 + 9x_3) = (7, 8, 9)$$

span R^3 and are linear independent. It is easy to see that (1, 2, 3), (4, 5, 6), (7, 8, 9) do not span R^3 and are linearly dependent. Therefore $x_1 + 2x_2 + 3x_3, 4x_1 + 5x_2 + 6x_3, 7x_1 + 8x_2 + 9x_3$ do not span $(\mathrm{R}^3)^*$ and are linearly dependent.

2. $x_1 + x_2 + 3x_3 - x_4, 2x_2 + 2x_3 - 2x_4, x_1 + 3x_2 + 4x_3 - 4x_4, -x_1 + 2x_3 - 2x_4 \in (\mathrm{R}^4)^*$

Answer: Similar to the first problem, the problem is the same as whether the vectors

$$\imath(x_1 + x_2 + 3x_3 - x_4) = (1, 1, 3, -1)$$
$$\imath(2x_2 + 2x_3 - 2x_4) = (0, 2, 2, -2)$$
$$\imath(x_1 + 3x_2 + 4x_3 - 4x_4) = (1, 3, 4, -4)$$
$$\imath(-x_1 + 2x_3 - 2x_4) = (-1, 0, 2, -2)$$

span R^4 and are linear independent. By the computation in this exercise, the answer is yes.

3. $p(0), p(1), p(-1), p(2) \in P_2^*$

Answer: By this property, we have $dimP_2^* = dimP_2 = 3$. Therefore the four linear functions must be linearly dependent. To see whether they span P_2^*, we pick a basis $\{1, t, t^2\}$ of P_2 and then use this property to convert the span problem into whether the vectors

values of $\{1, t, t^2\}$ at $0 = (1, 0, 0)$

values of $\{1, t, t^2\}$ at $1 = (1, 1, 1)$

values of $\{1, t, t^2\}$ at $-1 = (1, -1, 1)$

values of $\{1, t, t^2\}$ at $2 = (1, 2, 4)$

span R^3. By the row operation

$$\begin{bmatrix} 1 & 1 & 1 & 1 \\ 0 & 1 & -1 & 2 \\ 0 & 1 & 1 & 4 \end{bmatrix} \to \begin{bmatrix} 1 & 1 & 1 & 1 \\ 0 & 1 & -1 & 2 \\ 0 & 0 & 2 & 2 \end{bmatrix}$$

we conclude that the four vectors indeed span R^4. Therefore the four linear functions (of evaluating at 0, 1, –1, 2) span the dual space.

4. $p(0), p(1), p(-1), p(2) \in P_3^*$

Answer: Similar to the previous problem, we pick a basis $\{1, t, t^2, t^3\}$ of P_3 and then use this property to convert the problem into whether the vectors

values of $\{1, t, t^2, t^3\}$ at $0 = (1, 0, 0, 0)$

values of $\{1, t, t^2, t^3\}$ at $1 = (1, 1, 1, 1)$

values of $\{1, t, t^2, t^3\}$ at $-1 = (1, -1, 1, -1)$

values of $\{1, t, t^2, t^3\}$ at $2 = (1, 2, 4, 8)$

span R^4 and are linearly independent. By the row operation

$$\begin{bmatrix} 1 & 1 & 1 & 1 \\ 0 & 1 & -1 & 2 \\ 0 & 1 & 1 & 4 \\ 0 & 1 & -1 & 8 \end{bmatrix} \to \begin{bmatrix} 1 & 1 & 1 & 1 \\ 0 & 1 & -1 & 2 \\ 0 & 0 & 2 & 2 \\ 0 & 0 & 0 & 6 \end{bmatrix}$$

we conclude that the four linear functions span the dual space and are linearly independent.

5. $f(0), f(1), f(-1), f(2) \in F(R)^*$

Answer: Since the dimension of $F(R)$ is infinite, the dimension of the dual space $F(R)^*$

is also infinite. In particular, the four linear functions cannot span the dual space. To check the linear independence, we note the following: Suppose we have a linear relation

$$c_1f(0) + c_2f(1) + c_3f(-1) + c_4f(2) = 0.$$

Then the relation holds for all functions $f \in F(\mathrm{R})$. By making some special choices of f, we get

$$f = 1: c_1 + c_2 + c_3 + c_4 = 0$$
$$f = t: c_2 - c_3 + 2c_4 = 0$$
$$f = t^2: c_2 + c_3 + 4c_4 = 0$$
$$f = t^3: c_2 - c_3 + 8c_4 = 0$$

The four equations is equivalent to a linear relation

$$c_1\begin{bmatrix}1\\0\\0\\0\end{bmatrix} + c_2\begin{bmatrix}1\\1\\1\\1\end{bmatrix} + c_3\begin{bmatrix}1\\-1\\1\\-1\end{bmatrix} + c_4\begin{bmatrix}1\\2\\4\\8\end{bmatrix} = \begin{bmatrix}0\\0\\0\\0\end{bmatrix}.$$

From the row operation in the previous problem, we know that the relation implies $c_1 = c_2 = c_3 = c_4 = 0$. Thus we conclude that

$$c_1f(0) + c_2f(1) + c_3f(-1) + c_4f(2) = 0 \text{ for all } f \to c_1 = c_2 = c_3 = c_4 = 0.$$

In other words, the four linear functions (of evaluating at 0, 1, –1, 2) are linearly independent.

Remark: The choice of the function 1, t, t^2, t^3 in the proof of linear independence is just one possibility. We may also choose other functions such as sin$t$, e^t, etc. The only requirement is that, in order to successfully prove the independence, we need to choose at least four functions.

Remark: The theoretical explanation for the similarity between this problem and the previous one can be found here.

6. $\int_0^1 f(t)\,dt, \int_0^2 f(t)\,dt, \int_0^3 f(t)\,dt \in C(\mathrm{R})^*$

Answer Since the dimension of $C(\mathrm{R})$ is infinite, the dimension of the dual space $C(\mathrm{R})^*$ is also infinite. Therefore the three linear functions cannot span the dual space. As for the linear independence, we use the same idea as in the previous problem. Suppose we have a linear relation

$$c_1\int_0^1 f(t)\,dt + c_2\int_0^2 f(t)\,dt + c_3\int_0^3 f(t)\,dt = 0.$$

Then by choosing f to be 1, $\pi\sin\pi t$, $2t$, we get three equations for c_1, c_2, c_3, which can be combined to become the vector equation

$$c_1\begin{bmatrix}1\\2\\1\end{bmatrix}+c_2\begin{bmatrix}2\\0\\4\end{bmatrix}+c_3\begin{bmatrix}3\\2\\9\end{bmatrix}=\begin{bmatrix}0\\0\\0\end{bmatrix}.$$

It is then easy to see that this implies $c_1 = c_2 = c_3 = 0$. This shows that the three linear functions (of integrating on [0, 1], [0, 2], [0, 3]) are linearly independent.

Remark: We could have chosen the integration on 1, t, t^2 in the proof of linear independence. The different choice here is simply to show the other possibilities.

7. $\int_0^1 f(t)\,dt, \int_0^2 f(t)\,dt, \int_1^2 f(t)\,dt \in C(\mathrm{R})^*$

Answer: Similar to the previous two problems, because of the infinite dimension, the three linear functions cannot span the dual space. On the other hand, we always have

$$\int_0^1 f(t)\,dt - \int_0^2 f(t)\,dt + \int_1^2 f(t)\,dt = 0$$

for any $f \in C(\mathrm{R})$. Thus there is a nontrivial linear relation between the three linear functions, so that they are linearly dependent.

8. $\int_0^1 f(t)\,dt, \int_0^2 f(t)\,dt, 2f(2) - 3f(1), f(2) + 2f(1) \in C(\mathrm{R})^*$

Answer: Again because of infinite dimension, the four linear functions cannot span the dual space. On the other hand, by applying each linear function to the vectors in $\{1, t, t^2, t^3\}$, we get four vectors

$\int_0^1 f(t)\,dt$ applied to $\{1, t, t^2, t^3\} \Rightarrow (1, 1/2, 1/3, 1/4)$

$\int_0^2 f(t)\,dt$ applied to $\{1, t, t^2, t^3\} \Rightarrow (2, 2, 8/3, 4)$

$2f(2) - 3f(1)$ applied to $\{1, t, t^2, t^3\} \Rightarrow (-1, 1, 5, 13)$

$f(2) + 2f(1)$ applied to $\{1, t, t^2, t^3\} \Rightarrow (3, 4, 6, 10)$

By the similar argument as in the previous problems, since the four euclidean vectors are linearly independent, the four linear functions are linearly independent.

9. $trX, tr(\begin{bmatrix}1 & 2\\1 & 0\end{bmatrix}X), tr(X\begin{bmatrix}0 & 1\\1 & 2\end{bmatrix}), tr\begin{bmatrix}1 & 2\\1 & 0\end{bmatrix}(X\begin{bmatrix}0 & 1\\1 & 2\end{bmatrix}) \in M(2, 2)^*$

Answer: Let

$$X=\begin{bmatrix}x & y\\z & w\end{bmatrix}.$$

Then the four linear functions are $x + w$, $x + y + 2z$, $y + z + 2w$, $x + 3y + w$. By considering these as linear functions of $(x, y, z, w) \in \mathrm{R}^4$, as in the first problem, the span and independence problem is the same as the span and the independence problem for (1, 0, 0, 0), (1, 1, 2, 0), (0, 1, 1, 2), (1, 3, 0, 1) $\in \mathrm{R}^4$ (these are the coefficients of the four linear functions. Since the four euclidean vectors indeed span R^4 and are linearly independent, the original four linear functions (defined by trace) span $M(2, 2)^*$ and are linearly independent.

Exercise: The space of linear transformations $f: V \to R$ is the dual space. What about the space of linear transformations $f: R \to V$?

Answer For a linear transformation $f: R \to V$, denote $v = f(1)$. Any number $x \in R$ can be considered as the vector $1 \in R$ scalar multiplied by x. Thus by the linearity of f, we have $f(x) = f(x1) = xf(1) = xv$. Conversely, given any $v \in V$, it is easy to verify that $f(x) = xv$ is a linear transformation from R to V. Thus we have a one–to–one correspondence

linear transformation $f: R \to V \Leftrightarrow$ vector $v \in V$

given by $f \leftrightarrow f(1)$.

Moreover, the addition $f + g$ corresponds to the addition $(f + g)(1) = f(1) + g(1)$, and the scalar multiplication cf also corresponds to the scalar multiplication $(cf)(1) = c\, f(1)$. Therefore the space of all linear transformations $f: R \to V$ is naturally identified with V.

Remark: The identification

$\iota: (R^n)^* \leftrightarrow R^n,\ a_1x_1 + a_2x_2 + \ldots + a_nx_n \leftrightarrow (a_1, a_2, \ldots, a_n)$

makes use of the standard basis $e_1, e_2, \ldots, e_n$. Since an abstract vector space V is generally not equipped with a standard basis, we do not have a natrual identification between V^* and V.

In contrast, the identification $f \leftrightarrow f(1)$ does not involve any choice about V. Therefore the identification is natrual with respect to V.

Remark A linear transformation $f: R^3 \to V$ corresponds to a triple of vectors $u = f(1, 0, 0)$, $v = f(0, 1, 0)$, $w = f(0, 0, 1)$ via

$f(x_1, x_2, x_3) = x_1u + x_2v + x_3w.$

Thus the space of linear transformations $f: R^3 \to V$ can be identified with V E V E V (E is the direct sum).

Exercise: For a linear transformation $T: V \to W$ between finite dimensional vector spaces. How are the onto/one–to–one for T and the onto/one–to–one for the dual $T^*: W^* \to V^*$ are related?

Answer Without loss of generality, we may consider the case $V = R^n$ and $W = R^m$. Then T is given by an m by n matrix A. Moreover, by this result, $T^*: W^* \to V^*$ is given by the transpose A^T, after suitable identification. Then

T is onto

$\Leftrightarrow$ All rows of A are pivot

$\Leftrightarrow$ All columns of A^T are pivot

$\Leftrightarrow T^*$ is one–to–one.

By similar reason, we have

T is one–to–one $\Leftrightarrow T^*$ is onto.

Remark: The relation is another form of the complementarity principal.

Exercise: Let V be a finite dimensional vector space. Prove that a vector $v = 0 \Leftrightarrow f(v) = 0$ for all $f \in V^*$.

Answer: If $v = 0$, then by thinking of a linear function f as a linear transformation, we always have $f(v) = 0$. Conversely, assume $f(v) = 0$ for all $f \in V^*$. Since V is finite dimensional, we can find an isomorphism $T: V \to \mathrm{R}^n$. Then

T is linear

$\Rightarrow T(u) = (f_1(u), f_2(u),..., f_n(u))$ with $f_1, f_2,..., f_n \in V^*$

$\Rightarrow T(v) = (f_1(v), f_2(v),..., f_n(v)) = (0, 0,..., 0)$

$\Rightarrow v = 0$.

Exercise: Let H be a subspace of V. Let $f_1, f_2,..., f_n$ be linear functions on V. Then the restrictions on the subspace gives us the linear functions $f'_1, f'_2,..., f'_n$ on H.

1. Prove that $f'_1, f'_2,..., f'_n$ are linearly independent $\Rightarrow f_1, f_2,..., f_n$ are linearly independent.

Answer: A linear function $f \in V^*$ is the zero vector in the dual space if and only if $f(v) = 0$ for all $v \in V$. Thus if $f' \in H^*$ is the restriction of f on H, then $f = 0 \Rightarrow f' = 0$ (where 0 denotes the zero vector in the respective dual spaces). Since $c_1 f'_1 + c_2 f'_2 +... + c_n f'_n$ is clearly the restriction of $c_1 f_1 + c_2 f_2 +... + c_n f_n$ to the subspace H, we have

$c_1 f_1 + c_2 f_2 +... + c_n f_n = 0 \Rightarrow c_1 f'_1 + c_2 f'_2 +... + c_n f'_n = 0.$

Consequently, if $f'_1, f'_2,..., f'_n$ are linearly independent, then

$c_1 f_1 + c_2 f_2 +... + c_n f_n = 0 \Rightarrow c_1 f'_1 + c_2 f'_2 +... + c_n f'_n = 0 \Rightarrow c_1 = c_2 =... = c_n = 0,$

which means $f_1, f_2,..., f_n$ are linearly independent. Therefore we conclude

$f'_1, f'_2,..., f'_n$ are linearly independent $\Rightarrow f_1, f_2,..., f_n$ are linearly independent.

2. Prove that $f_1, f_2,..., f_n$ span $V^* \Rightarrow f'_1, f'_2,..., f'_n$ span H^*.

Answer: On the other hand, suppose $f_1, f_2,..., f_n$ span V^*. By this exercise, V^* is finite dimensional. Since $\dim V = \dim V^*$, V is also finite dimensional. By this exercise, any linear function f' on H can be extended to a linear function f on V. Since $f_1, f_2,..., f_n$ span V^*, f is a linear combination of these linear functions:

$$f = c_1 f_1 + c_2 f_2 +... + c_n f_n.$$

Restricting to the subspace H, we see that f' is a linear combination of $f'_1, f'_2, ..., f'_n$:

$$f' = c_1 f'_1 + c_2 f'_2 + ... + c_n f'_n.$$

Therefore we conclude

$$f_1, f_2, ..., f_n \text{ span } V^* \Rightarrow f'_1, f'_2, ..., f'_n \text{ span } H^*.$$

Remark: An alternative way of proving the relation on the span is to compare the linear transformations

$$T(v) = (f_1(v), f_2(v), ..., f_n(v)): V \to \mathrm{R}^n,\ T'(v) = (f'_1(v), f'_2(v), ..., f'_n(v)): H \to \mathrm{R}^n.$$

Then T' is really the restriction of T to H. Thus we have

$f_1, f_2, ..., f_n$ span V^*

$\Leftrightarrow T$ is one-to-one

$\Leftrightarrow T(v) = 0,\ v\ V$ implies $v = 0$

$\Leftrightarrow T(v) = 0,\ v \in H$ implies $v = 0$

$\Leftrightarrow T'$ is one–to–one

$\Leftrightarrow f'_1, f'_2, ..., f'_n$ span H^*.

The proof essentially makes use of the complementarity principle.

Remark: Let $T: V \to W$ be a linear transformation. Let $f_1, f_2, ..., f_n$ be linear functions on W. Then we have linear functions $f'_1 = f_1 T = T^*(f_1), f'_2 = T^*(f_2), ..., f'_n = T^*(f_n)$ on V. By using this exercise, this exercise, and this exercise, we may derive some relations between span and linear independence for $f_1, f_2, ..., f_n$ and $f'_1, f'_2, ..., f'_n$.

Note that in the special case $T(v) = v: H \to V$ is the inclusion transformation from the subspace H to the vector space V, T is one–to–one and we are in the situation discussed in the exercise.

Chapter 11

Basis

Consider the vector equation

$$c_1b_1 + c_2b_2 + \dots + c_nb_n = v$$

in a vector space V, where $c_1, c_2, \dots, c_n$ are the unknown variables. The existence of solutions for all $v \in V$ is equivalent to the vectors $b_1, b_2, \dots, b_n$ spanning V. Moreover, the concept of linear independence among $b_1, b_2, \dots, b_n$ was introduced to describe the uniqueness of the solution. Thus the best case for a collection of vectors is described in the following definition.

Vectors $b_1, b_2, \dots, b_n \in V$ form a basis of V if

- $b_1, b_2, \dots, b_n$ span V.
- $b_1, b_2, \dots, b_n$ are linearly independent.

The discussion at the beginning immediately gives us the following meaning of basis.

$B = \{b_1, b_2, \dots, b_n\}$ is a basis $\Leftrightarrow c_1b_1 + c_2b_2 + \dots + c_nb_n = v$ has a unique solution for any v.

Example: The standard basis vectors $e_1 = (1, 0, 0)$, $e_2 = (0, 1, 0)$, $e_3 = (0, 0, 1)$ span R^3 and are linearly independent. Therefore the standard basis $\{e_1, e_2, e_3\}$ is indeed a basis of R^3. The fact clearly extends to the standard basis of general R^n.

The monomials $1, t, t^2, \dots, t^n$ span P_n and are linearly independent. Therefore the monomials form a basis of P_n.

Similarly, the equality

$$A = \begin{bmatrix} a & b \\ c & d \end{bmatrix} = a\,s\begin{bmatrix} 1 & 0 \\ 0 & 0 \end{bmatrix} + b\begin{bmatrix} 0 & 1 \\ 0 & 0 \end{bmatrix} + c\begin{bmatrix} 0 & 0 \\ 1 & 0 \end{bmatrix} + d\begin{bmatrix} 0 & 0 \\ 0 & 1 \end{bmatrix}$$

$$= aE_{11} + bE_{12} + cE_{21} + dE_{22}$$

has the following two consequences

- Any 2 by 2 matrix A is a linear combination of $E_{11}, E_{12}, E_{21}, E_{22}$.

- If a linear combination A of $E_{11}, E_{12}, E_{21}, E_{22}$ (with a, b, c, d as coefficients) is the zero matrix, then all the entries of A (which are a, b, c, d) must be zero.

The first consequence means exactly that $M(2, 2)$ is spanned by $E_{11}, E_{12}, E_{21}, E_{22}$. The second consequence means exactly that $E_{11}, E_{12}, E_{21}, E_{22}$ are linearly independent. Thus $\{E_{11}, E_{12}, E_{21}, E_{22}\}$ is a basis of $M(2, 2)$.

In general, let E_{ij} be the m by n matrix with 1 as the ij–entry and 0 as all the other entries. Then $\{E_{ij}: 1 \le i \le m, 1 \le j \le n\}$ is a basis of $M(m, n)$.

Example: In an earlier example, the system

$$\begin{aligned} 3x_1 + x_2 - x_3 &= 2 \\ x_1 - x_2 + x_3 &= 2 \\ 2x_1 + 2x_2 + x_3 &= 6 \end{aligned}$$

was shown to have a unique solution. Since the number of rows is the same as the number of columns, by the basic principal of linear algebra, the general system

$$\begin{aligned} 3x_1 + x_2 - x_3 &= b_1 \\ x_1 - x_2 + x_3 &= b_2 \\ 2x_1 + 2x_2 + x_3 &= b_3 \end{aligned}$$

has a unique solution for any right side. Since the general system is the same as the vector equation

$$x_1(3, 1, 2) + x_2(1, -1, 2) + x_3(-1, 1, 1) = (b_1, b_2, b_3),$$

the existence for the general system means that (3, 1, 2), (1, –1, 2), (–1, 1, 1) span R^3, and the uniqueness means the three vectors are linearly independent. Thus we conclude that the three vectors form a basis of R^3.

The argument in the second example applies to general collections of euclidean vectors. After all, being a basis of R^n means always existence (span R^n) plus uniqueness (independence). According to the study on the basic principal of linear algebra, the exact condition is described as follows.

If the columns of a matrix A is a basis of R^n, then A has to be an n by n matrix. Moreover, for an n by n matrix A, the following are equivalent.

- Columns of A form a basis of R^n.
- $Ax = b$ has a unique solution for any b.
- $Ax = b$ has solutions for any b.
- The solution of $Ax = b$ is unique.
- A is invertible.

COORDINATE

Let $B = \{b_1, b_2,..., b_n\}$ be a basis of a vector space V. Then for any v " V, we can find unique numbers $c_1, c_2,..., c_n$, such that

$c_1b_1 + c_2b_2 +... + c_nb_n = v.$

The uniqueness of the coefficients make the following definition unambiguous.

The coordinates of v with respect to the basis B is the euclidean vector

$[v]_B = (c_1, c_2,..., c_n).$

Example: The formula

$(x_1, x_2, x_3) = x_1(1, 0, 0) + x_2(0, 1, 0) + x_3(0, 0, 1) = x_1e_1 + x_2e_2 + x_3e_3$

indicates that

$[(x_1, x_2, x_3)]_{\{e1, e2, e3\}} = (x_1, x_2, x_3).$

In other words, when we say x_1 is the first coordinate of the vector (x_1, x_2, x_3), we really mean it is the first coordinate with regard to the standard basis $\{e_1, e_2, e_3\}$. To illustrate the subtlety here, the formula

$(x_1, x_2, x_3) = x_2e_2 + x_1e_1 + x_3e_3$

tells us

$[(x_1, x_2, x_3)]_{\{e2, e1, e3\}} = (x_2, x_1, x_3),$

so that x_1 is the *second* coordinate of (x_1, x_2, x_3) with regard to the basis $\{e_2, e_1, e_3\}$.

By the definition of the coordinates, we also have

$[a_0 + a_1t + a_2t^2 +... + a_nt^n]_{\{1, t, t^2,..., t^n\}} = (a_0, a_1, a_2,..., a_n).$

In other words, the coefficients of a polynomial are the coordinates with regard to the standard monomial basis.

Finally, the equality

$$\begin{bmatrix} a & b \\ c & d \end{bmatrix} = a\begin{bmatrix} 1 & 0 \\ 0 & 0 \end{bmatrix} + b\begin{bmatrix} 0 & 1 \\ 0 & 0 \end{bmatrix} + c\begin{bmatrix} 0 & 0 \\ 1 & 0 \end{bmatrix} + d\begin{bmatrix} 0 & 0 \\ 0 & 1 \end{bmatrix} = aE_{11} + bE_{12} + cE_{21} + dE_{22}$$

tells us

$$\begin{bmatrix} a & b \\ c & d \end{bmatrix}_{\{E_{11}, E_{12}, E_{21}, E_{22}\}} = (a, b, c, d).$$

Example: In an earlier example, the system

$$3x_1 + x_2 - x_3 = 2$$

$$x_1 - x_2 + x_3 = 2$$
$$2x_1 + 2x_2 + x_3 = 6$$

was shown to have a unique solution $x_1 = 1, x_2 = 1, x_3 = 2$. This means

$1(3, 1, 2) + 1(1, -1, 2) + 2(-1, 1, 1) = (2, 2, 6),$

Recall that in this case, the uniqueness of the solution actually implies that {(3, 1, 2), (1, –1, 2), (–1, 1, 1)} is a basis of R^3. Moreover, the equality above means that the coordinates of (2, 2, 6) with regard to this basis is

$$[(2, 2, 6)]_{\{(3, 1, 2), (1, -1, 2), (-1, 1, 1)\}} = (1, 1, 2).$$

In general, the coordinates of a vector (b_1, b_2, b_3) with regard to the basis {(3, 1, 2), (1, –1, 2), (–1, 1, 1)} is obtained by solving the following system.

$$3x_1 + x_2 - x_3 = b_1$$
$$x_1 - x_2 + x_3 = b_2$$
$$2x_1 + 2x_2 + x_3 = b_3$$

The solution is

$$\begin{bmatrix} b_1 \\ b_2 \\ b_3 \end{bmatrix}_{\{(3, 1, 2), (1, -1, 2), (-1, 1, 1)\}} = \begin{bmatrix} 3 & 1 & -1 \\ 1 & -1 & 1 \\ 2 & 2 & 1 \end{bmatrix} \begin{bmatrix} b_1 \\ b_2 \\ b_3 \end{bmatrix}$$

$$= \begin{bmatrix} 1/4 & 1/4 & 0 \\ -1/12 & -5/12 & 1/3 \\ -1/3 & 1/3 & 1/3 \end{bmatrix} \begin{bmatrix} b_1 \\ b_2 \\ b_3 \end{bmatrix}$$

$$= \begin{bmatrix} (1/4)b_1 + (1/4)b_2 \\ -(1/12)b_1 - (5/12)b_2 + (1/3)b_3 \\ -(1/3)b_1 + (1/3)b_2 + (1/3)b_3 \end{bmatrix}$$

In the example above, the coordinates of a vector v with respect to a basis of R^n is the unique solution of $Bx = v$, where the columns of B is the basis. Recall that columns being a basis is equivalent to the invertibility of B, and the solution of $Bx = v$ is $B^{-1}v$ (for the invertible B). We then have the following general formula for coordinates with regard to any basis of R^n.

Let the columns of $B \doteq [b_1\ b_2 ...\ b_n]$ be a basis of R^n, then

$$[v]_{\{b1, b2, ..., bn\}} = B^{-1}v.$$

ISOMORPHISM

The coordinates with regard to a basis $B = \{b_1, b_2, ..., b_n\}$ of a vector space V may be considered as a transformation $[?]_B: V \to R^n$.

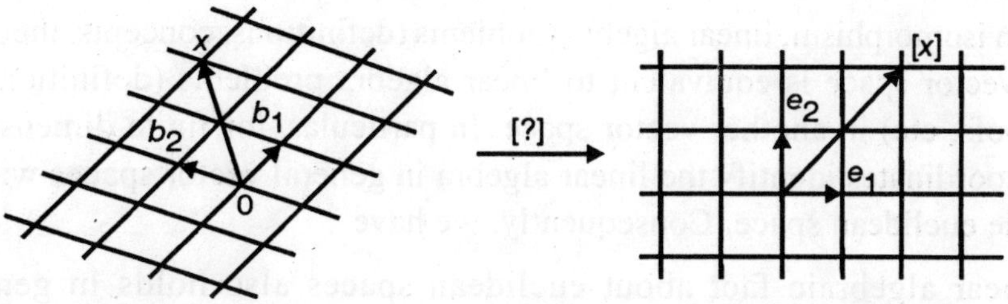

It is easy to verify that $[?]_B$ is an invertible linear transformation with

$T(x_1, x_2, ..., x_n) = x_1 b_1 + x_2 b_2 + ... + x_n b_n: R^n \to V$

as the inverse.

Two vector spaces are isomorphic if there is an invertible linear transformation between them. Any such invertible linear transformation is an isomorphism.

In particular, if a vector space has a finite basis, then it is isomorphic to the euclidean space. We also note that if a vector space is spanned by finitely many vectors, then it has a finite basis. Such vector spaces are called finite dimensional.

Example: The coordinates with regard to the basis $\{1, t, t^2, ..., t^n\}$ is the isomorphism

$a_0 + a_1 t + a_2 t^2 + ... + a_n t^n \leftrightarrow (a_0, a_1, a_2, ..., a_n)$.

This has been used in converting some problems about polynomials to euclidean vectors.

The coordinates with regard to the basis

$$\left\{\begin{bmatrix}1 & 0\\0 & 0\end{bmatrix}, \begin{bmatrix}0 & 1\\0 & 0\end{bmatrix}, \begin{bmatrix}0 & 0\\1 & 0\end{bmatrix}, \begin{bmatrix}0 & 0\\0 & 1\end{bmatrix}\right\}$$

is the isomorphism

$$\begin{bmatrix}a & b\\c & d\end{bmatrix} \leftrightarrow (a, b, c, d).$$

This has been used in converting some problems about matrices to euclidean vectors.

Example: By an earlier example, the coordinates with regard to the basis $\{(3, 1, 2), (1, -1, 2), (-1, 1, 1)\}$ gives us the isomorphism

$(b_1, b_2, b_3) \leftrightarrow ((1/4)b_1 + (1/4)b_2, -(1/12)b_1 - (5/12)b_2 + (1/3)b_3, -(1/3)b_1 + (1/3)b_2 + (1/3)b_3)$.

This is the inverse of the isomorphism

$(x_1, x_2, x_3) \leftrightarrow (3x_1 + x_2 - x_3, x_1 - x_2 + x_3, 2x_1 + 2x_2 + x_3)$,

whose matrix has the given basis vectors as columns.

Given an isomorphism, linear algebra problems (definitions, concepts, theories, proofs, etc) in one vector space is equivalent to linear algebra problems (definitions, concepts, theories, proofs, etc) in another vector space. In particular, for finite dimensional vector spaces, the coordinates identify the linear algebra in general vector spaces with the linear algebra in the euclidean space. Consequently, we have

Any linear algebraic fact about euclidean spaces also holds in general (finite dimensional) vector spaces.

For example, the following is the translation of this fact about euclidean spaces into general vector spaces.

For a linear transformation $T: V \to W$, the following are equivalent.

- T is invertible.
- T is onto and one–to–one.
- $rangeT = W$, $kernelT = \{0\}$.

The concept of isomorphism can be extended to any equivalence between two systems. For example, different languages are largely isomorphic, with the isomorphisms given by dictionaries. Suppose a French asks a Spanish to do some work, the Spanish may translate the job description into Spanish, finish the job in a Spanish environment, and then report back in French after finishing the job.

For another example, the numbers people use may vary in notation, the base (binary, decimal), etc. However, they are all equivalent number systems. The same also applies to the measurement systems such as length (meter, inch), weight (gram, pound), money (US dollar, Hong Kong dollar, Yen, Yuan, Euro).

DIMENSION

We used coordinates to construct isomorphisms between vector spaces V and euclidean spaces R^n. This produces a dictionary that translates the general linear algebra in V to the more concrete linear algebra in R^n.

The number n in R^n is the dimension of the euclidean space. It is also the number of vectors in the basis of V used for constructing the coordinates. The dimension of a vector space is the number of vectors in a basis. The dimension is infinite if there is no finite basis. From this exercise, we see that a vector space is finite dimensional if and only if it can be spanned by finitely many vectors.

Recall that the columns of any n by n invertible matrix is a basis of R^n. Thus the basis is not unique, and the definition of dimension must be justified by showing that different choices of bases will always have the same number of vectors. Since this is a linear algebra problem, we may translate the problem (by fixing any one basis and considering the corresponding coordinates) into the euclidean space. Therefore it is sufficient to show that any basis of R^n must contain n vectors. This special case has been proved by the characterization of bases of R^n.

Example: The standard basis of R^n contains n vectors. Therefore $dim R^n = n$.

The basis $\{1, t, t^2, ..., t^n\}$ of P_n consists of $n + 1$ monomials. Therefore $dim P_n = n + 1$.

The basis $\{E_{ij}$: 1 d" i d" m, 1 d" j d" $n\}$ of $M(m, n)$ contains mn matrices. Therefore $dim M(m, n) = mn$.

Example: All the 2 by 2 symmetric matrices form a vector space $Sym(2)$ (because it is a subspace of $M(2, 2)$). By

$$\begin{bmatrix} a & b \\ c & d \end{bmatrix} = a\begin{bmatrix} 1 & 0 \\ 0 & 0 \end{bmatrix} + b\begin{bmatrix} 0 & 1 \\ 1 & 0 \end{bmatrix} + c\begin{bmatrix} 0 & 0 \\ 0 & 1 \end{bmatrix} = aE_{11} + b(E_{12} + cE_{21}) + cE_{22},$$

and an argument similar to this example, we conclude that $\{E_{11}, E_{12} + E_{21}, E_{22}\}$ is a basis of $Sym(2)$, and $dim Sym(2) = 3$.

In general, the vector space $Sym(n)$ of symmetric n by n matrices has a basis consisting of the following vectors

- E_{ii}: $1 \leq i \leq n$
- $E_{ij} + E_{ji}$: $1 \leq i < j \leq n$

From this it is easy to conclude that $dim Sym(n) = n(n + 1)/2$.

PROPERTIES OF DIMENSION

Since vector spaces of dimension n may be treated just like R^n, all properties of euclidean vectors concerning the dimension n can be extended to general vector spaces. For example, the property on the size of spanning sets and the property on the size of linearly independent sets may be extended.

Let $dim V = n$. Then

- $v_1, v_2, ..., v_k$ span $V \Rightarrow k \leq n$.
- $v_1, v_2, ..., v_k \in V$ are linearly independent $\Rightarrow k \leq n$.

Moreover, the property on the size of onto linear transformations and the property on the size of one–to–one linear transformations may also be extended.

Let T: $V \rightarrow W$ be a linear transformation. Then

- T is onto $\Rightarrow dimV \geq dimW$.
- T is one-to-one $\Rightarrow dimV \leq dimW$.

In particular, isomorphic vector spaces have the same dimension.

The following extends the basic principal of linear algebra.

Let $v_1, v_2,..., v_n$ be vectors in a space V of dimension n. Then the following are equivalent.

- $v_1, v_2,..., v_n$ form a basis of V.
- $v_1, v_2,..., v_n$ span V.
- $v_1, v_2,..., v_n$ are linearly independent.

For linear transformations, the basic principal takes the following form.

Let $T: V \to W$ be a linear transformation between finite dimensional vector spaces. If $dimV = dimW$, then the following are equivalent.

- T is invertible.
- $TS = id$ for some transformation S.
- $ST = id$ for some transformation S.
- T is onto.
- T is on–to–one.

Finally, as a consequence of the properties above, we have

Let H be a subspace of V. Then

- $dimH \leq dimV$.
- If $dimH = dimV < \infty$ then $H = V$.

The property is easy to understand if H and V are replaced by finite sets and dimension is replaced by number of elements.

Proof. Let $\{v_1, v_2,..., v_k\}$ be a basis of H. Then $k = dimH$ and $v_1, v_2,..., v_k$ are linearly independent as vectors of H. Clearly, $v_1, v_2,..., v_k$ are also linearly independent as vectors of V. Thus by this property, we have $k \leq dimV$. Since $k = dimH$, the first part is proved.

For the second part, note that $v_1, v_2,..., v_k$ being linearly independent and $k = dimH = dimV$ implies that $v_1, v_2,..., v_k$ form a basis of V. In particular, the vectors span V. On the other hand, they also form a basis of H and hence span H. Thus we conclude

$H = span\{v_1, v_2,..., v_k\} = V$.

Exercise: Determine whether the vectors form a basis of the vector space.

1. $(-2) \in R$

Answer: The 1–dimensional vector (–2) is a basis of R.

2. $(1, 2) \in R^2$

Answer: A basis of R^2 must contain two vectors. Thus one vector (1, 2) is not a basis of R^2.

3. $(1, 2), (2, 3) \in R^2$

Answer: For two vectors to be a basis of $\mathbf{R}^2$, it is sufficient for them to be linearly independent . The two vectors are independent because they are clearly not parallel. Thus they form a basis.

Alternatively, we may consider the matrix

$$\begin{bmatrix} 1 & 2 \\ 2 & 3 \end{bmatrix}$$

with the two vectors as columns. By this criterion, the two vectors form a basis of $R^2 \Leftrightarrow$ the matrix is invertible. Since $1 \times 3 - 2 \times 2 \neq 0$, the matrix is indeed invertible. Thus we conclude that the two vectors form a basis of R^2.

4. $(1, 2), (2, 3), (3, 4) \in R^2$

Answer: A basis of R^2 must contain two vectors. Thus three vectors cannot form a basis of R^2.

5. $(1, 2, 3), (4, 5, 6), (7, 8, 9) \in R^3$

Answer: By this criterion, the three vectors form a basis of $R^3 \Leftrightarrow$ the matrix

$$\begin{bmatrix} 1 & 4 & 7 \\ 2 & 5 & 8 \\ 3 & 6 & 9 \end{bmatrix}$$

with given vectors as columns is invertible. Row operations on the matrix gives us

Therefore the matrix is not invertible, and the three vectors do not form a basis of R^3.

Alternatively, we note that $(1, 2, 3) - 2(4, 5, 6) + (7, 8, 9) = (0, 0, 0)$. Thus the three vectors are linearly dependent and do not form a basis.

6. $(1, 2, 3), (4, 5, 6), (7, 8, 9), (10, 11, 12) \in R^3$

Answer: A basis of R^3 must contain three vectors. Thus the four vectors cannot form a basis of R^3.

Alternatively, from the previous problem we already know (1, 2, 3), (4, 5, 6), (7, 8, 9) are linearly dependent. Thus the four vectors are also linearly dependent and do not form a basis of R^3.

7. $(-1, 1, 0), (-1, 0, 1), (1, 1, 1) \in R^3$

Answer: Similar to the fifth problem, by this criterion, the three vectors form a basis of $R^3 \Leftrightarrow$ the matrix

$$\begin{bmatrix} -1 & -1 & 1 \\ 1 & 0 & 1 \\ 0 & 1 & 1 \end{bmatrix}$$

with given vectors as columns is invertible. Row operations on the matrix gives us

$$\begin{bmatrix} -1 & -1 & 1 \\ 0 & -1 & 2 \\ 0 & 0 & 3 \end{bmatrix}$$

Therefore the matrix is invertible, and the three vectors form a basis of R^3.

8. $(-1, 1, 0), (-1, 0, 1), (0, -1, 1) \in R^3$

Answer: Similar to earlier problems, by this criterion, the three vectors form a basis of $R^3 \Leftrightarrow$ the matrix

$$\begin{bmatrix} -1 & -1 & 0 \\ 1 & 0 & -1 \\ 0 & 1 & 1 \end{bmatrix}$$

is invertible. Row operations on the matrix gives us

$$\begin{bmatrix} -1 & -1 & 0 \\ 0 & -1 & -1 \\ 0 & 0 & 0 \end{bmatrix}$$

Thus the three vectors do not form a basis of R^3. In fact, from $(-1, 1, 0) - (-1, 0, 1) + (0, -1, 1) = (0, 0, 0)$, the three vectors are linearly dependent.

9. $-1 + t, -1 + t^2, -t + t^2 \in P_2$

Answer: The basis $\{1, t, t^2\}$ of P_2 gives us a translation

$p = a + bt + ct^2 \in P_2$ "! $[p] = (a, b, c) \in R^3$

that converts the problem to whether the corresponding euclidean vectors

$[-1 + t] = (-1, 1, 0), [-1 + t^2] = (-1, 0, 1), [-t + t^2] = (0, -1, 1)$

form a basis of R^3. By the previous problem, the *Answer:* is no. Correspondingly, the three polynomials do not form a basis of P_2.

10. $t - t^2, 1 - t^2, 1 - t \in P_2$

Answer: Using the translation in the previous problem, the three polynomials form a basis ⇔ the euclidean vectors

$[t - t^2] = (0, 1, -1), [1 - t^2] = (1, 0, -1), [1 - t] = (1, -1, 0)$

form a basis of R^3. We may draw the conclusion by forming a matrix with the three vectors and then doing row operations. Alternatively, we may use a different basis $\{t^2, t, 1\}$ to introduce a different translation

$p = a + bt + ct^2 \in P_2 \leftrightarrow [p]' = (c, b, a) \in R^3$

Then our problem is equivalent to

$[t - t^2]' = (-1, 1, 0), [1 - t^2]' = (-1, 0, 1), [1 - t]' = (0, -1, 1)$

being a basis of R^3. It was shown in the eighth problem that the three euclidean vectors do not form a basis of R^3. Correspondingly, the three polynomials do not form a basis of P_2.

11. $(1, 1, 3, -1), (0, 2, 2, -2), (1, 3, 4, -4), (-1, 0, 2, -2) \in R^4$

Answer: Using the given vectors as columns, we form a matrix.

$$\begin{bmatrix} 1 & 0 & 1 & -1 \\ 1 & 2 & 3 & 0 \\ 3 & 2 & 4 & 2 \\ -1 & -2 & -4 & -2 \end{bmatrix}$$

By row operations, we get

$$\begin{bmatrix} 1 & 0 & 1 & -1 \\ 0 & 2 & 2 & 1 \\ 0 & 0 & -1 & 4 \\ 0 & 0 & 0 & -3 \end{bmatrix}$$

Since all columns (rows) are pivot, the matrix is invertible. Thus the four vectors form a basis of R^4.

12. $(1, 1, 3, -1), (0, 2, 2, -2), (1, 3, 7, -4), (-1, 0, 2, -2) \in R^4$

Answer: Using the given vectors as columns, we form a matrix.

$$\begin{bmatrix} 1 & 0 & 1 & -1 \\ 1 & 2 & 3 & 0 \\ 3 & 2 & 7 & 2 \\ -1 & -2 & -4 & -2 \end{bmatrix}$$

By row operations, we get

$$\begin{bmatrix} 1 & 0 & 1 & -1 \\ 0 & 2 & 2 & 1 \\ 0 & 0 & 2 & 4 \\ 0 & 0 & 0 & 0 \end{bmatrix}$$

Thus the matrix is not invertible, and the four vectors do not form a basis of R^4.

13.(2, 5, 3, 0), (4, 2, 5, 9), (–2, 3, 7, –2), (–6, 5, 2, 3), (7, 2, –5, 1) $\in R^4$

Answer: A basis of R^4 must contain four vectors. The five vectors cannot form a basis of R^4.

14.(1.22, 3.45, –3.56, 5.73), (2.11, 3.44, 5.23, –7.89) $\in R^4$

Answer: A basis of R^4 must contain four vectors. The two vectors cannot form a basis of R^4.

15. $\begin{bmatrix} 1 & 1 \\ 3 & -1 \end{bmatrix}, \begin{bmatrix} 0 & 2 \\ 2 & -2 \end{bmatrix}, \begin{bmatrix} 1 & 3 \\ 4 & -4 \end{bmatrix}, \begin{bmatrix} -1 & 0 \\ 2 & -2 \end{bmatrix} \in M(2, 2)$

Answer: The basis

$$\left\{ \begin{bmatrix} 1 & 0 \\ 0 & 0 \end{bmatrix}, \begin{bmatrix} 0 & 1 \\ 0 & 0 \end{bmatrix}, \begin{bmatrix} 0 & 0 \\ 1 & 0 \end{bmatrix}, \begin{bmatrix} 0 & 0 \\ 0 & 1 \end{bmatrix} \right\}$$

of $M(2, 2)$ gives us a translation

$$\begin{bmatrix} a & b \\ c & d \end{bmatrix} \in M(2, 2) \leftrightarrow (a, b, c, d) \in R^4$$

that converts the problem into whether (1, 1, 3,–1), (0, 2, 2, –2), (1, 3, 7, –4), (–1, 0, 2, –2) form a basis R^4. By the eleventh problem, the four euclidean vectors indeed form a basis. Correspondingly, the four matrices form a basis of $M(2, 2)$.

16. $\begin{bmatrix} 1 & -1 \\ 3 & 1 \end{bmatrix}, \begin{bmatrix} 0 & -2 \\ 2 & 2 \end{bmatrix}, \begin{bmatrix} 1 & -4 \\ 7 & 3 \end{bmatrix}, \begin{bmatrix} -1 & -2 \\ 2 & 0 \end{bmatrix} \in M(2, 2)$

Answer: If we use the translation of the previous problem, then the problem is converted to whether (1, –1, 3, 1), (0, –2, 2, 2), (1, –4, 7, 3), (–1, –2, 2, 0) form a basis of R^4. Alternatively, we may use the basis

$$\left\{ \begin{bmatrix} 1 & 0 \\ 0 & 0 \end{bmatrix}, \begin{bmatrix} 0 & 0 \\ 0 & 1 \end{bmatrix}, \begin{bmatrix} 0 & 0 \\ 1 & 0 \end{bmatrix}, \begin{bmatrix} 0 & 1 \\ 0 & 0 \end{bmatrix} \right\}$$

to get a translation

$$\begin{bmatrix} a & b \\ c & d \end{bmatrix} \in M(2, 2) \leftrightarrow (a, b, c, d) \in R^4$$

so that the problem is converted into whether (1, 1, 3,–1), (0, 2, 2, –2), (1, 3, 7, –4), (–1, 0, 2, –2) span R^4. By the twelfth problem, the euclidean vectors do not span R^4. Thus the four matrices do not span $M(2, 2)$.

17. $\begin{bmatrix} -1 & 4 \\ 4 & 3 \end{bmatrix}, \begin{bmatrix} -5 & 3 \\ 3 & 5 \end{bmatrix}, \begin{bmatrix} 1 & -1 \\ -1 & -1 \end{bmatrix} \in M(2, 2)$

Answer: Any basis of $M(2, 2)$ must contain four matrices. Thus the three matrices do not form a basis.

18. $\begin{bmatrix} -1 & 4 \\ 4 & 3 \end{bmatrix}, \begin{bmatrix} -5 & 3 \\ 3 & 5 \end{bmatrix}, \begin{bmatrix} 1 & -1 \\ -1 & -1 \end{bmatrix} \in Sym(2)$

Answer: By making use of the basis in this example, we have a translation

$$\begin{bmatrix} a & b \\ b & c \end{bmatrix} \in Sym(2) \leftrightarrow (a, b, c) \in R^3$$

and the problem is converted into whether (–1, 4, 3), (–5, 3, 5), (1, –1, 1) form a basis of R^3. By row operations on the corresponding 3 by 3 matrix, it is easy to see that the three euclidean vectors do form a basis. Therefore the three symmetric matrices form a basis of the vector spaces of all 2 b 2 symmetric matrices.

Exercise: Find the coordinates of vectors with respect to the basis.

1. $B = \{(1, 2), (2, 3)\}$, $u = (3, 4)$, $v = (0, 1)$

Answer: Using the basis vectors as the columns, we have a matrix.

$$A = \begin{bmatrix} 1 & 2 \\ 2 & 3 \end{bmatrix}$$

The coordinates of a vector b is the solution of $Ax = b$, which can be found by row operations on $[A\ u\ v]$. We may also compute A^{-1} and then calculate $A^{-1}u$, $A^{-1}v$. By either method, we have

$[u]_B = (-1, 2)$, $[v]_B = (2, -1)$.

Note that the answer can be verified by checking

(3, 4) = (–1) (1, 2) + 2 (2, 3), (0, 1) = 2 (1, 2) + (–1) (2, 3).

2. $B = \{(-1, 1, 0), (-1, 0, 1), (1, 1, 1)\}$, $u = (1, 0, 0)$, $v = (0, 1, 0)$, $w = (0, 0, 1)$

Answer: The coordinates are obtained by solving the system $Ax = b$, where A has vectors in B as columns, and b is one of u, v, w. Thus we form the expanded augmented matrix.

$$[A\ u\ v\ w] = \begin{bmatrix} -1 & -1 & 1 & 1 & 0 & 0 \\ 1 & 0 & 1 & 0 & 1 & 0 \\ 0 & 1 & 1 & 0 & 0 & 1 \end{bmatrix}$$

Row operations give us

$$R = \begin{bmatrix} 1 & 0 & 0 & -1/3 & 2/3 & -1/3 \\ 0 & 1 & 0 & -1/3 & -1/3 & 2/3 \\ 0 & 0 & 1 & 1/3 & 1/3 & 1/3 \end{bmatrix}$$

By looking at the first four columns, we see that [col 4] of R is the solution of $Ax = u$. Thus

$$[u]_B = (-1/3, -1/3, 1/3).$$

Similarly,

$$[v]_B = (2/3, -1/3, 1/3),\ [w]_B = (-1/3, 2/3, 1/3).$$

Remark: **In an euclidean space, finding the coordinates of the standard basis vectors with regard to any basis B is the same as computing the inverse of the matrix formed by the vectors in B.**

3. $B = \{(1, 1, 3, -1), (0, 2, 2, -2), (1, 3, 4, -4), (-1, 0, 2, -2)\}$, $u = (1, 6, 11, -9)$, $v = (1, 1, 3, -1)$

Answer: The coordinates of u with regard to B is the solutions of

$$\begin{aligned} x_1 + x_3 - x_4 &= 1 \\ x_1 + 2x_2 + 3x_3 &= 6 \\ 3x_1 + 2x_2 + 4x_3 + 2x_4 &= 11 \\ -x_1 - 2x_2 - 4x_3 - 2x_4 &= -9 \end{aligned}$$

where the left side has vectors in B as the column vectors of the coefficient matrix, and the right side is u. The solution is $x_1 = x_2 = x_3 = x_4 = 1$. Therefore we have

$[u]_B = (1, 1, 1, 1)$.

One may do the same for $v = (1, 1, 3, -1)$. Alternatively, noting that v is the first vector in the basis, we have

$v = 1(1, 1, 3, -1) + 0(0, 2, 2, -2) + 0(1, 3, 4, -4) + 0(-1, 0, 2, -2)$,

which means

$[v]_B = (1, 0, 0, 0)$.

4. $B = \{1 + t + 3t^2 - t^3, 2t + 2t^2 - 2t^3, 1 + 3t + 4t^2 - 4t^3, -1 + 2t^2 - 2t^3\}$, $p(t) = 1 + 6t + 11t^2 - 9t^3$, $q(t) = 1 + t + 3t^2 - t^3$

Answer: We may use

$p = a + bt + ct^2 + dt^3 \in P_3 \leftrightarrow [p] = (a, b, c, d) \in \mathbb{R}^4$

to translate the problem into the coordinates of

$[p(t)] = (1, 6, 11, -9)$, $[q(t)] = (1, 1, 3, -1)$

with regard to the basis

$[B] = \{(1, 1, 3, -1), (0, 2, 2, -2), (1, 3, 4, -4), (-1, 0, 2, -2)\}$.

This was solved in the previous problem. Corresponding to $[\,[p(t)]\,]_{[B]} = (1, 1, 1, 1)$, we have $[p(t)]_B = (1, 1, 1, 1)$. Corresponding to $[\,[q(t)]\,]_{[B]} = (1, 0, 0, 0)$, we have $[q(t)]_B = (1, 0, 0, 0)$.

5. $B = \left\{\begin{bmatrix} -1 & 4 \\ 4 & 3 \end{bmatrix}, \begin{bmatrix} -5 & 3 \\ 3 & 5 \end{bmatrix}, \begin{bmatrix} 1 & -1 \\ -1 & -1 \end{bmatrix}\right\}$, $A = \begin{bmatrix} 6 & -1 \\ -1 & 0 \end{bmatrix}$, $B = \begin{bmatrix} -1 & 0 \\ 0 & 4 \end{bmatrix}$

Answer: The translation

$$\begin{bmatrix} a & b \\ b & c \end{bmatrix} \in Sym(2) \leftrightarrow (a, b, c) \in \mathbb{R}^3$$

converts the problem into the computation of the coordinates of

$[A] = (6, -1, 0)$, $[B] = (-1, 0, 4)$

with regard to the basis

$[B] = \{(-1, 4, 3), (-5, 3, 5), (1, -1, 1)\}$.

The solution of the system

$$\begin{aligned} -x_1 - 5x_3 + x_3 &= 6 \\ 4x_1 + 3x_2 - x_3 &= -1 \\ 3x_1 + 5x_2 + x_3 &= 0 \end{aligned}$$

is $x_1 = 1$, $x_2 = -1$, $x_3 = 2$, which gives us the coordinates

$[A]_B = (1, -1, 2)$.

We note that this means

$$\begin{bmatrix} 6 & -1 \\ -1 & 0 \end{bmatrix} = 1\begin{bmatrix} -1 & 4 \\ 4 & 3 \end{bmatrix} - 1\begin{bmatrix} -5 & 3 \\ 3 & 5 \end{bmatrix} + \begin{bmatrix} 1 & -1 \\ -1 & -1 \end{bmatrix}$$

Similarly, the solution of the system

$$\begin{aligned} -x_1 - 5x_3 + x_3 &= -1 \\ 4x_1 + 3x_2 - x_3 &= 0 \\ 3x_1 + 5x_2 + x_3 &= 4 \end{aligned}$$

gives us the coordinates

$[B]_B = (0, 1/2, 3/2)$.

Exercise: Find the condition on the parameter so that the vectors form a basis.

1. $(1, -1, 1), (1, 0, a), (-1, -a, 0) \in \mathbb{R}^3$

Answer: We use the vectors as columns and form a matrix.

$$\begin{bmatrix} 1 & 1 & -1 \\ -1 & 0 & -a \\ 1 & a & 0 \end{bmatrix}$$

By the computation in an earlier exercise, row operations give us

$$\begin{bmatrix} 1 & 1 & 1 \\ 0 & -1 & 1-a \\ 0 & 0 & 2a-a^2 \end{bmatrix}$$

Being a basis means all rows (and columns) are pivot. The condition is a ‘“ 0 and 2.

2. $\begin{bmatrix} a & 1 \\ 1 & 0 \end{bmatrix}, \begin{bmatrix} 1 & a \\ 0 & 1 \end{bmatrix}, \begin{bmatrix} 1 & 0 \\ a & 1 \end{bmatrix}, \begin{bmatrix} 0 & 1 \\ 1 & a \end{bmatrix} \in M(2, 2)$

Answer: By the usual translation, the problem is the same as whether the euclidean vectors

$(a, 1, 1, 0), (1, a, 0, 1), (1, 0, a, 1), (0, 1, 1, a)$

form a basis of $\mathbb{R}^4$. The criterion is all columns (and rows) of the following matrix (with the four vectors as columns) are pivot.

$$\begin{bmatrix} a & 1 & 1 & 0 \\ 1 & a & 0 & 1 \\ 1 & 0 & a & 1 \\ 0 & 1 & 1 & a \end{bmatrix}$$

By row operations, we get

$$\begin{bmatrix} 1 & 0 & a & 1 \\ 0 & 1 & 1 & a \\ 0 & 0 & 2a & a^2 \\ 0 & 0 & 0 & 4a - a^3 \end{bmatrix}$$

The condition for the matrix to have all columns (and rows) pivot is $a \neq 0, 2, -2$. This is the condition for the four matrices to form a basis of $M(2, 2)$.

3. $t + t^2, 1 + t^2 + at^3, 1 + at - t^3, a + t + t^3 \in P_3$

Answer: By the usual translation, the problem is the same as whether the euclidean vectors

$[t + t^2] = (0, 1, 1, 0), [1 + t^2 + at^3] = (1, 0, 1, a), [1 + at - t^3] = (1, a, 0, -1), [a + t + t^3] = (a, 1, 0, 1)$

form a basis of R^4. The criterion is all columns (and rows) of the following matrix (with the four vectors as columns) are pivot.

$$\begin{bmatrix} 0 & 1 & 1 & a \\ 1 & 0 & a & 1 \\ 1 & 1 & 0 & 0 \\ 0 & a & -1 & 1 \end{bmatrix}$$

By the computation in an earlier exercise, row operations give us

$$\begin{bmatrix} 1 & 0 & a & 1 \\ 0 & 1 & 1 & a \\ 0 & 0 & -a-1 & -a-1 \\ 0 & 0 & 0 & -a^2 + a + 2 \end{bmatrix}$$

The condition for the matrix to have all columns (and rows) pivot is a “‘“ –1, 2. This is the condition for the four polynomials to form a basis of P_3.

4. $1 + t + at^2, a + t^2 - at^3, -1 + at - 2at^2 + t^3, a + at + t^3, 1 + at \in P_3$

Answer: Any basis of P_3 must contain four polynomials. Thus the five polynomials can never be a basis of P_3.

Exercise: The following vectors span the vector space. Find a basis by deleting some vectors.

1. $(1, -1, 0, 1), (-2, 3, 1, 2), (1, 0, 1, 5), (2, -2, 4, 5), (0, 0, 0, 1) \in R^4$

Answer: Using the given vectors as columns, we form a matrix.

$$\begin{bmatrix} 1 & -2 & 1 & 2 & 0 \\ -1 & 3 & 0 & -2 & 0 \\ 0 & 1 & 1 & 4 & 0 \\ 1 & 2 & 5 & 5 & 1 \end{bmatrix}$$

By the same row operations as in this exercise, we get a row echelon form.

$$\begin{bmatrix} 1 & -2 & 1 & 2 & 0 \\ 0 & 1 & 1 & 0 & 0 \\ 0 & 0 & 0 & 4 & 0 \\ 0 & 0 & 0 & 0 & 1 \end{bmatrix}$$

If we delete [col 3], which is the only nonpivot column, then the row operations on the remaining columns

$$\begin{bmatrix} 1 & -2 & 2 & 0 \\ -1 & 3 & -2 & 0 \\ 0 & 1 & 4 & 0 \\ 1 & 2 & 5 & 1 \end{bmatrix}$$

gives us a row echelon form

$$\begin{bmatrix} 1 & -2 & 2 & 0 \\ 0 & 1 & 0 & 0 \\ 0 & 0 & 4 & 0 \\ 0 & 0 & 0 & 1 \end{bmatrix}$$

in which all columns (and rows) are pivot. Correspondingly, by deleting the 3rd vector, we get a basis (1, –1, 0, 1), (–2, 3, 1, 2), (2, –2, 4, 5), (0, 0, 0, 1) of R^4.

Remark: To find a basis from any given vectors $v_1, v_2, \ldots, v_k$ that span R^n, we may form a matrix $A = [v_1\ v_2 \ldots v_k]$. Then the spanning assumption means that the row echelon form of A has all rows pivot. Thus by deleting the vectors corresponding to the nonpivot columns (in the problem above, this is the 3rd vector), we get a basis of R^n.

For the spanning vectors of more general vector spaces, we may translate the problem into an euclidean one and solve it.

2. $\begin{bmatrix} 1 & 0 \\ -1 & 1 \end{bmatrix}, \begin{bmatrix} -2 & 1 \\ 3 & 2 \end{bmatrix}, \begin{bmatrix} 1 & 1 \\ 0 & 5 \end{bmatrix}, \begin{bmatrix} 2 & 4 \\ -2 & 5 \end{bmatrix}, \begin{bmatrix} 0 & 0 \\ 0 & 1 \end{bmatrix} \in M(2, 2)$

Answer: By the translation

$$\begin{bmatrix} a & c \\ b & d \end{bmatrix} \leftrightarrow (a, b, c, d),$$

the problem becomes picking a basis of R^4 from (1, –1, 0, 1), (–2, 3, 1, 2), (1, 0, 1, 5), (2, –2, 4, 5), (0, 0, 0, 1). This is exactly the previous problem, whose conclusion tells us that by deleting the 3rd matrix,

form a basis of $M(2, 2)$.

3. $1 + 2t^2, 2 + 4t^2, 1 + 2t + 3t^2, 2t + t^2, 1 + 4t + 4t^2, 1 + t + t^2 \in P_2$

Answer: By the usual translation $[a + bt + ct^2] = (a, b, c)$, the problem becomes picking a basis of R^3 from (1, 0, 2), (2, 0, 4), (1, 2, 3), (0, 2, 1), (1, 4, 4), (1, 1, 1). Thus we use the vectors as columns and form a matrix.

$$\begin{bmatrix} 1 & 2 & 1 & 0 & 1 & 1 \\ 0 & 0 & 2 & 2 & 4 & 1 \\ 2 & 4 & 3 & 1 & 4 & 1 \end{bmatrix}$$

After row operations, we get

$$\begin{bmatrix} 1 & 2 & 1 & 0 & 1 & 1 \\ 0 & 0 & 1 & 1 & 2 & -1 \\ 0 & 0 & 0 & 0 & 0 & 3 \end{bmatrix}$$

Corresponding to the pivot columns, we see that the 1st, 3rd, 6th vectors form a basis of R^3. Correspondingly, $1 + 2t^2, 1 + 2t + 3t^2, 1 + t + t^2$ form a basis of P_2.

Exercise: Find a basis by adding some vectors to the linearly independent vectors.

1. $(1, -1, 0), (-2, 3, 1) \in R^3$

Answer: Using the two vectors and the standard basis $e_1 = (1, 0, 0)$, $e_2 = (0, 1, 0)$, $e_3 = (0, 0, 1)$ as columns, we form a matrix.

$$\begin{bmatrix} 1 & -2 & 1 & 0 & 0 \\ -1 & 3 & 0 & 1 & 0 \\ 0 & 1 & 0 & 0 & 1 \end{bmatrix}$$

By row operations, we get

$$\begin{bmatrix} 1 & -2 & 1 & 0 & 0 \\ 0 & 1 & 1 & 1 & 0 \\ 0 & 0 & -1 & -1 & 1 \end{bmatrix}$$

Considering the first three columns only, we see that adding (1, 0, 0) to the two given vectors gives us a basis {(1, –1, 0), (–2, 3, 1), (1, 0, 0)} of R^3.

Remark: If we add any basis $b_1, b_2,..., b_n$ to any vectors $v_1, v_2,..., v_k \in V$, then by $v_k \in V = span\{b_1, b_2,..., b_n\}$ and this property, we have

$span\{v_1, v_2,..., v_k, b_1, b_2,..., b_n\} = span\{b_1, b_2,..., b_n\} = V.$

If $V = R^n$, this means that the matrix $A = [v_1, v_2,..., v_k, b_1, b_2,..., b_n]$ has all rows pivot. Thus the vectors corresponding to the pivot columns (in the problem above, these are the first three vectors) form a basis of R^n.

If $v_1, v_2,..., v_k$ are already linearly independent, then the first k columns will be pivot and our pick will include $v_1, v_2,..., v_k$. In other words, the method produces a basis by expanding the given vectors.

For more general vector spaces, we may translate the problem into an euclidean one and solve it.

2. $1 + 2t + t^2 + 2t^3 + t^4, 1 + t + 2t^3 + t^4 \in P_4$

Answer: By the usual translation $[a + bt + ct^2 + dt^3 + et^4] = (a, b, c, d, e)$, the problem becomes expanding (1, 2, 1, 2, 1), (1, 1, 0, 2, 1) to a basis of R^5. Similar to the previous problem, we add the standard basis (in fact, any basis of R^5 will do) and consider the matrix

$$\begin{bmatrix} 1 & 1 & 1 & 0 & 0 & 0 & 0 \\ 2 & 1 & 0 & 1 & 0 & 0 & 0 \\ 1 & 0 & 0 & 0 & 1 & 0 & 0 \\ 2 & 2 & 0 & 0 & 0 & 1 & 0 \\ 1 & 1 & 0 & 0 & 0 & 0 & 1 \end{bmatrix}$$

After row operations, we get

$$\begin{bmatrix} 1 & 1 & 1 & 0 & 0 & 0 & 0 \\ 0 & -1 & -2 & 1 & 0 & 0 & 0 \\ 0 & 0 & 1 & -1 & 1 & 0 & 0 \\ 0 & 0 & 0 & -1 & 1 & 0 & 1 \\ 0 & 0 & 0 & 0 & 0 & 1 & -2 \end{bmatrix}$$

Corresponding to the pivot columns, the vectors (1, 2, 1, 2, 1), (1, 1, 0, 2, 1), (1, 0, 0, 0, 0), (0, 1, 0, 0, 0), (0, 0, 0, 1, 0) form a basis of $\mathbb{R}^5$. Translating back, adding $1, t, t^2$ to $1 + 2t + t^2 + 2t^3 + t^4$, $1 + t + 2t^3 + t^4$ gives us a basis of P_4.

3. $\begin{bmatrix} 1 & 0 & 0 \\ 1 & 2 & 3 \end{bmatrix}, \begin{bmatrix} 1 & 1 & 0 \\ 1 & 2 & 3 \end{bmatrix}, \begin{bmatrix} 1 & 1 & 1 \\ 1 & 2 & 3 \end{bmatrix} \in M(2, 3)$

Answer: By the usual translation

$$\begin{bmatrix} a & b & c \\ d & e & f \end{bmatrix} \leftrightarrow (a, b, c, d, e, f)$$

the problem becomes expanding (1, 0, 0, 1, 2, 3), (1, 1, 0, 1, 2, 3), (1, 1, 1, 1, 2, 3) into a basis of $\mathbb{R}^6$. Thus we form a matrix.

$$\begin{bmatrix} 1 & 1 & 1 & 1 & 0 & 0 & 0 & 0 & 0 \\ 0 & 1 & 1 & 0 & 1 & 0 & 0 & 0 & 0 \\ 0 & 0 & 1 & 0 & 0 & 1 & 0 & 0 & 0 \\ 1 & 1 & 1 & 0 & 0 & 0 & 0 & 1 & 0 \\ 2 & 2 & 2 & 0 & 0 & 0 & 0 & 1 & 0 \\ 3 & 3 & 3 & 0 & 0 & 0 & 0 & 0 & 1 \end{bmatrix}$$

After row operations, we get

$$\begin{bmatrix} 1 & 1 & 1 & 1 & 0 & 0 & 0 & 0 & 0 \\ 0 & 1 & 1 & 0 & 1 & 0 & 0 & 0 & 0 \\ 0 & 0 & 1 & 0 & 0 & 1 & 0 & 0 & 0 \\ 0 & 0 & 0 & -1 & 0 & 0 & 1 & 0 & 0 \\ 0 & 0 & 0 & 0 & 0 & 0 & -2 & 1 & 0 \\ 0 & 0 & 0 & 0 & 0 & 0 & 0 & -3/2 & 1 \end{bmatrix}$$

Thus (1, 0, 0, 1, 2, 3), (1, 1, 0, 1, 2, 3), (1, 1, 1, 1, 2, 3), (1, 0, 0, 0, 0, 0), (0, 0, 0, 1, 0, 0), (0, 0, 0, 0, 1, 0) form a basis of $\mathbb{R}^6$. Translating back, we conclude that adding

$$\begin{bmatrix} 1 & 0 & 0 \\ 0 & 0 & 0 \end{bmatrix}, \begin{bmatrix} 0 & 0 & 0 \\ 1 & 0 & 0 \end{bmatrix}, \begin{bmatrix} 0 & 0 & 0 \\ 0 & 1 & 0 \end{bmatrix}$$

to

$$\begin{bmatrix} 1 & 0 & 0 \\ 1 & 2 & 3 \end{bmatrix}, \begin{bmatrix} 1 & 1 & 0 \\ 1 & 2 & 3 \end{bmatrix}, \begin{bmatrix} 1 & 1 & 1 \\ 1 & 2 & 3 \end{bmatrix}$$

gives us a basis of $M(2, 3)$.

Exercise: Prove that the dimension of $F(R)$ is infinite.

Answer: The implication

$$a_0 + a_1t + a_2t^2 + ... + a_nt^n = 0 \Rightarrow a_0 = a_1 = a_2 = ... = a_n = 0$$

means that $1, t, t^2, ..., t^n$ are linearly independent for any n. Thus in $F(R)$, we can find arbitrary number of linearly independent vectors. Then it follows from this property that the dimension of F(R) cannot be finite.

Exercise: Let $T: R^3 \to R^3$ be the orthogonal projection onto the plane $x + y + z = 0$. The linear transformation is characterized by

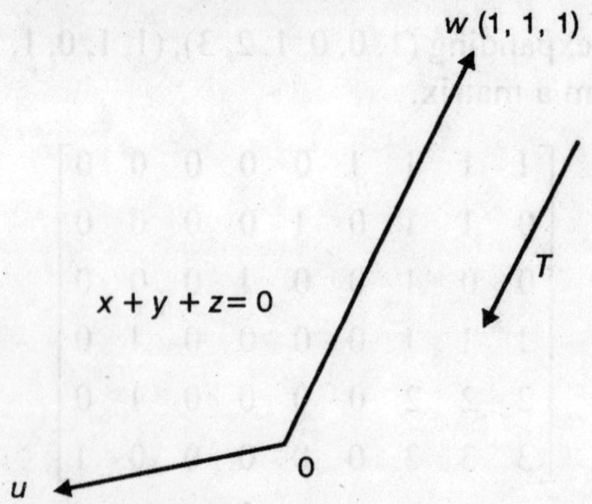

- $T(u) = u$ for any u on the plane $x + y + z = 0$.
- $T(w) = 0$ for the "coefficient vector" $w = (1, 1, 1)$ of the plane equation.

Find the matrix for T.

Answer: Since $u = (-1, 1, 0)$ and $v = (-1, 0, 1)$ satisfy $x + y + z = 0$, we have $T(u) = u$ and $T(v) = v$. Together with $T(w) = 0$, we know the value of T on a basis $B = \{u, v, w\}$.

The matrix of T is $[T(e_1)\ T(e_2)\ T(e_3)]$, where e_1, e_2, e_3 are the standard basis vectors. To find the value of T on the standard basis, we may try to express e_1, e_2, e_3 as linear combinations of u, v, w and then apply T.

Specifically, writing e_1, e_2, e_3 as linear combinations of u, v, w is the same as finding the coordinates of e_1, e_2, e_3 with regard to the basis B. In an earlier exercise, this has been done.

$$[e_1]_B = (-1/3, -1/3, 1/3),\ [e_2]_B = (2/3, -1/3, 1/3),\ [e_3]_B = (-1/3, 2/3, 1/3).$$

Then

$$T(e_1) = T(-(1/3)u + (1/3)v - (1/3)w) \text{ (meaning of coordinates of } e_1)$$

$$= -(1/3)T(u) + (1/3)T(v) - (1/3)T(w) \text{ (} T \text{ is linear)}$$

$$= -(1/3)u + (1/3)v - (1/3)0 \text{ (value of } T \text{ on } B)$$

$= -(1/3)(-1, 1, 0) + (1/3)(-1, 0, 1)$

$= (2/3, -1/3, -1/3).$

Similarly,

$T(e_2) = (2/3)(-1, 1, 0) - (1/3)(-1, 0, 1) = (-1/3, 2/3, -1/3),$

$T(e_3) = -(1/3)(-1, 1, 0) + (2/3)(-1, 0, 1) = (2/3, -1/3, -1/3).$

We conclude the following matrix for T.

$$\begin{bmatrix} 2/3 & -1/3 & -1/3 \\ -1/3 & 2/3 & -1/3 \\ -1/3 & -1/3 & 2/3 \end{bmatrix}$$

Exercise: Let $p_1 = 1 + 2t + t^2$, $p_2 = 1 + 3t$, $p_3 = 1 + 2t^2$. Let T: $P_2 \to \mathrm{R}^3$ be a linear transformation such that

$T(p_1) = (1, 0, 1)$, $T(p_2) = (0, 1, 1)$, $T(p_3) = (1, -1, 0)$.

Find the general formula for $T(a + bt + ct^2)$. Is T onto? Is T one–to–one?

Answer: Since $T(a + bt + ct^2) = aT(1) + bT(t) + cT(t^2)$. The problem becomes the computation of $T(1)$, $T(t)$, $T(t^2)$. Similar to this exercise, this can be done by finding the coordinates of $1, t, t^2$ with regard to the basis $B = \{p_1, p_2, p_3\}$. By the usual translation $[a + bt + ct^2] = (a, b, c)$, we form a matrix.

$$[\,[p_1]\,[p_2]\,[p_3]\,[1]\,[t]\,[t^2]\,] = \begin{bmatrix} 1 & 1 & 1 & 1 & 0 & 0 \\ 2 & 3 & 0 & 0 & 1 & 0 \\ 1 & 0 & 2 & 0 & 0 & 1 \end{bmatrix}$$

Row operations give us

$$\begin{bmatrix} 1 & 0 & 0 & -6 & 2 & 3 \\ 0 & 1 & 0 & 4 & -1 & -2 \\ 0 & 0 & 1 & 3 & -1 & -1 \end{bmatrix}$$

From the first 4 columns, we get $1 = -6p_1 + 4p_2 + 3p_3$. Therefore

$T(1) = -6T(p_1) + 4T(p_2) + 3T(p_3) = -6(1, 0, 1) + 4(0, 1, 1) + 3(1, -1, 0) = (-3, 1, -2).$

Similarly, we have $t = 2p_1 - p_2 - p_3$, $t^2 = 3p_1 - 2p_2 - p_3$ and

$T(t) = 2T(p_1) - T(p_2) - T(p_3) = 2(1, 0, 1) - (0, 1, 1) - (1, -1, 0) = (1, 0, 1),$

$T(t^2) = 3T(p_1) - 2T(p_2) - T(p_3) = 3(1, 0, 1) - 2(0, 1, 1) - (1, -1, 0) = (2, -1, 1).$

Finally, we have

$T(a + bt + ct^2) = at(1) + bT(t) + cT(t^2) = a(-3, 1, 2) + b(1, 0, 1) + c(2, -1, 1) = (-3a + b + 2c, a - c, -2a + b + c)$.

For onto and one–to–one, we note that the translation $[a + bt + ct^2] = (a, b, c)$ also converts the linear transformation T: $P_2 \to R^3$ given by the above formula to the linear translation $[T]$: $R^3 \to$ given by the matrix

$$\begin{bmatrix} -3 & 1 & 2 \\ 1 & 0 & -1 \\ -2 & 1 & 1 \end{bmatrix}$$

Then whether T is onto or one–to–one is the same as whether $[T]$ is onto or one–to–one. Row operations on the matrix give us

$$\begin{bmatrix} 1 & 0 & -1 \\ 0 & 1 & -1 \\ 0 & 0 & 0 \end{bmatrix}$$

Therefore T is neither onto nor one–to–one.

Alternatively, the equality $(1, 0, 1) - (0, 1, 1) - (1, -1, 0) = (0, 0, 0)$ means that $T(p_1 - p_2 - p_3) = (0, 0, 0)$. Since p_1, p_2, p_3 form a basis, the vector $p_1 - p_2 - p_3$ is nonzero. Thus the kernel of T is nontrivial, and T is not one-to-one. Then by $dimP_2 = 3 = dim R^3$, we also know that T is not onto.

Remark: A more direct method for checking onto and one–to–one for T is the following. The basis $B = \{p_1, p_2, p_3\}$ induces a translation

$[ap_1 + bp_2 + cp_3]' = (a, b, c)$

that converts the linear transformation T to

$$\begin{aligned} [T]'(a, b, c) &= T(ap_1 + bp_2 + cp_3) \\ &= aT(p_1) + bT(p_2) + cT(p_3) \\ &= a(1, 0, 1) + b(0, 1, 1) + c(1, -1, 0): R^3 \to R^3. \end{aligned}$$

Then the onto and one–to–one problems for T is equivalent to the same problems for $[T]'$. The later problems are further equivalent to whether the vectors

$$T(p_1) = (1, 0, 1),\ T(p_2) = (0, 1, 1),\ T(p_3) = (1, -1, 0)$$

span R^3 or are linearly independent.

In general, if $B = \{b_1, b_2, ..., b_n\}$ is a basis of V and $T : V \to W$ is a linear transformation, then by using B to translate T to $[T]'$: $R^n \to W$, we find

T is onto $\Leftrightarrow T(b_1), T(b_2),..., T(b_n)$ span W,

T is one–to–one $\Leftrightarrow T(b_1), T(b_2),..., T(b_n)$ are linearly independent.

Exercise: Let $B = \{v_1, v_2, v_3\}$ be a basis of V. Let $w_1, w_2, w_3 \in W$.

1. Prove that there is a unique linear transformation $T: V \to W$ such that $T(v_1) = w_1, T(v_2) = w_2, T(v_3) = w_3$.
2. Describe the range and the kernel of T.
3. When is T onto? When is T one–to–one?
4. What if B is not a basis?

Answer: Since B is a basis, any vector $v \in V$ is a unique linear combination $v = c_1v_1 + c_2v_2 + c_3v_3$ and we must have

$$T(v) = c_1T(v_1) + c_2T(v_2) + c_3T(v_3) = c_1w_1 + c_2w_2 + c_3w_3.$$

This formula is the unique way of defining T.

When v runs though all vectors in V, the coordinates c_1, c_2, c_3 take all numbers. Therefore

$$rangeT = span\{w_1, w_2, w_3\}.$$

This implies that T is onto $\Leftrightarrow w_1, w_2, w_3$ span W.

The formula for T also tells us

$$kernelT = \{c_1v_1 + c_2v_2 + c_3v_3\text{: any } c_1, c_2, c_3 \text{ satisfying } c_1w_1 + c_2w_2 + c_3w_3 = 0\}.$$

Thus one–to–one for T means $kernelT = \{0\}$, which also means

$$c_1w_1 + c_2w_2 + c_3w_3 = 0 \Rightarrow c_1v_1 + c_2v_2 + c_3v_3 = 0.$$

Since v_1, v_2, v_3 are linearly independent, the right side is the same as $c_1 = c_2 = c_3 = 0$. Thus we have

$$c_1w_1 + c_2w_2 + c_3w_3 = 0 \Rightarrow c_1 = c_2 = c_3 = 0,$$

which means w_1, w_2, w_3 are linearly independent. We conclude T is one–to–one $\Leftrightarrow$ w_1, w_2, w_3 are linearly independent.

Now consider the case B is not a basis. B is either linearly dependent or does not span V.

If B is linearly dependent, say $v_3 = av_1 + bv_2$, then by applying T we get $w_3 = aw_1 + bw_2$. Consequently, for the linear transformation T to exist, w_1, w_2, w_3 must satisfy whatever relations satisfied by v_1, v_2, v_3.

If B does not span V, then the formula $T(c_1v_1 + c_2v_2 + c_3v_3) = c_1w_1 + c_2w_2 + c_3w_3$ only specifies T for vectors in $span\{v_1, v_2, v_3\}$. For the vectors not in $span\{v_1, v_2, v_3\}$, the value of T can have many choices.

Thus the lack of linear independence for B may affect the existence of T, while the lack of spanning V may affect the uniqueness of T.

Exercise: Let V and W be vector spaces. Let H be a subspace of V. Then any linear transformation $T: V \to W$ can be restricted to become a linear transformation $T' : H \to W$. Conversely, prove that if V is finite dimensional, then any linear transformation $T': H \to W$. can be extended to a linear transformation $T: V \to W$.

Answer: Since V is finite dimensional, by this remark, we may construct a basis $\{v_1, v_2, ..., v_k\}$ of H and then extend to a basis $\{v_1, v_2, ..., v_k, v_{k+1}, ..., v_n\}$ of V. To illustrate the argument more clearly, we might as well assum $k = 3$ and $n = 5$.

By the first part of this exercise, linear transformations $T' : H \to W$ are determined by the vectors $w_1 = T'(v_1)$, $w_2 = T'(v_2)$, $w_3 = T'(v_3) \in W$ via the formula

$$T'(c_1v_1 + c_2v_2 + c_3v_3) = c_1w_1 + c_2w_2 + c_3w_3.$$

and linear transformations $T' : H \to W$ are determined by the vectors $w_1 = T(v_1)$, $w_2 = T(v_2)$, $w_3 = T(v_3)$, $w_4 = T(v_4)$, $w_5 = T(v_5) \in W$ via the formula

$$T(c_1v_1 + c_2v_2 + c_3v_3 + c_4v_4 + c_5v_5) = c_1w_1 + c_2w_2 + c_3w_3 + c_4w_4 + c_5w_5.$$

Thus being given T' is the same as being given three vectors $w_1, w_2, w_3 \in W$. Then by choosing two more vectors, for example $w_4 = w_5 = 0$, we get the extension

$$T(c_1v_1 + c_2v_2 + c_3v_3 + c_4v_4 + c_5v_5) = c_1w_1 + c_2w_2 + c_3w_3$$

of T'.

Exercise: What is the dimension of $Hom(V, W)$?

Answer: Since any vector space is isomorphic to a euclidean space, we only need to consider the case $V = \mathrm{R}^n$ and $W = \mathrm{R}^m$. Then $Hom(V, W)$ is the same as the space of m by n matrices. Then by this example, the dimension of $Hom(V, W)$ is mn. In general, we have

$$dim(Hom(V, W)) = dimV\ dimW.$$

Exercise: Let H and K be finite dimensional subspaces of a vector space. Prove that $dim(H + K) + dim(H \cap K) = dimH + dimK$.

Answer: Let $B_{H\cap K} = \{b_1, b_2, ..., b_n\}$ be a basis of $H \cap K$. Then $B_{H\cap K}$ is also a collection of linearly independent vectors in H and in K. By this remark, the linearly independent collection can be extended to a basis $B_H = \{b_1, b_2, ..., b_n, b'_1, b'_2, ..., b'_p\}$ of H and a basis $B_K = \{b_1, b_2, ..., b_n, b''_1, b''_2, ..., b''_q\}$ of K.

Now we claim that $B_H \cup B_K = \{b_1, b_2, ..., b_n, b'_1, b'_2, ..., b'_p, b''_1, b''_2, ..., b''_q\}$ is a basis of $H + K$. Then by counting the number of vectors in bases, we have

$$dim(H \cap K) = n,$$

$$dimH = n + p,$$

$dimK = n + q,$

$dim(H + K) = n + p + q,$

which implies

$dim(H + K) + dim(H \cap K) = 2n + p + q = dimH + dimK.$

By this result, we already know that $B_H \cup B_K$ spans $H + K$. It remains to prove that $B_H \cup B_K$ is linearly independent. To simplify the proof, we will assume $n = p = q = 2$.

Suppose

(*) $c_1b_1 + c_2b_2 + c'_1b'_1 + c'_2b'_2 + c''_1b''_1 + c''_2b''_2 = 0.$

Then

$c''_1b''_1 + c''_2b''_2 = -c_1b_1 - c_2b_2 - c'_1b'_1 - c'_2b'_2 \in H.$

Since b''_1 and b''_2 span K, we have $c''_1b''_1 + c''_2b''_2 \in H \cap K = span\ \{b_1, b_2\}$. Since b_1, b_2, b''_1, b''_2 are linearly independent (they form a basis of H), the fact $c''_1b''_1 + c''_2b''_2 \in span\{b_1, b_2\}$ implies

$c''_1 = c''_2 = 0.$

By similar argument, we have

$c'_1 = c'_2 = 0.$

Substituting into the equality (*), we have $c_1b_1 + c_2b_2 = 0$. Then by $\{b_1, b_2\}$ being a basis of $H \cap K$, we conclude that

$c_1 = c_2 = 0.$

This completes the proof that B is linearly independent.

Exercise: Prove that a vector space is finite dimensional $\Leftrightarrow$ The space can be spanned by finitely many vectors.

Answer: An earlier exercise suggests that if a finitely set B of vectors spans V, then by deleting some vectors from B, we may find a basis of V. In particular, the dimension of V is the number of vectors in a subset of B, so that V is finite dimensional.

So assume $B = \{b_1, b_2, ..., b_k\}$ is a finite set of vectors that spans V. Then we ask whether B is linearly independent. If B is not linearly independent, then one vector, say b_k, is a linear combination of the others. By this property, the subset $\{b_1, b_2, ..., b_{k-1}\}$ still spans V. Then we ask again whether $\{b_1, b_2, ..., b_{k-1}\}$ is linearly independent. If not, we can delete a vector, say b_{k-1}, so that the rest $\{b_1, b_2, ..., b_{k-2}\}$ still spans V. We continue asking the same question until the answer is yes. Then we get a linearly independent subset of B, say $\{b_1, b_2, ..., b_n\}$, that still spans V. The subset $\{b_1, b_2, ..., b_n\}$ is then a basis of V.

Remark: A basis of V must span V and must be linearly independent. In case the spanning condition if satisfied, we may deleting "wasted" vectors one by one, until there is no more waste. Since the deleted vectors are "wasted", the span remains to be V. Since there is no more waste in the end, we get a linearly independence after deletion. This is the key idea in the argument above. This exercise gives a concrete example on the process.

Remark: Since a basis can be obtained by deleting vectors from a spanning set, we have this numerical relation for a spanning set as a consequence. Similarly, a basis can be obtained by adding vectors to a linearly independent set. as a result, we have this numerical relation for a linearly independent set.

Exercise: For a linear transformation $T: V \to W$, we know the following are equivalent

- T is invertible.
- T is onto and one–to–one.
- $rangeT = W$, $kernelT = \{0\}$.

by translating this fact for euclidean spaces. Try to give a direct explanation to the equivalence.

Answer: The equivalence between the first two properties is a fact about general transformations. Moreover, onto for T is defined as $rangeT = W$. Finally, the equivalence between one–to–one and $kernelT = \{0\}$ can be found here.

Remark: There is a subtle point here. Notice that when we say a linear transformation T is invertible, we implicitly require the inverse to be also a linear transformation (such requirement is used when we defined the inverse matrix). This is important when this fact about general transformations is used, because the fact only gives us inverse transformation and says nothing about the linearity of the inverse.

Fortunately, the inverses of linear transformations are automatically linear. Thus we can ignore the subtlely in linear algebra. The subtlety may become an issue when other types of transformaitons are studied.

COMPUTATION OF BASIS

BASIS OF NULL SPACE

In the next example, we find a basis of the null space of a matrix.

Example: The null space of the matrix

$$A = \begin{bmatrix} 1 & 3 & 2 & 0 & 1 \\ -1 & -1 & -1 & 1 & 0 \\ 0 & 4 & 2 & 4 & 3 \\ 1 & 3 & 2 & -2 & 0 \end{bmatrix}$$

consists of all the solutions of the homogeneous system $Ax = 0$. By the row operations in this example and this example, we have the reduced row echelon form.

$$\begin{bmatrix} 1 & 0 & 1/2 & 0 & -1/4 \\ 0 & 1 & 1/2 & 0 & 1/4 \\ 0 & 0 & 0 & 1 & 1/2 \\ 0 & 0 & 0 & 0 & 0 \end{bmatrix}$$

Then we derive the solutions of the homogeneous system.

$x_1 = -(1/2)x_3 - (1/4)x_5$

$x_2 = -(1/2)x_3 - (1/4)x_5$

x_3 is arbitrary

$x_4 = -(1/2)x_5$

x_5 is arbitrary

Thus *nulA* consists of vectors of the form

$$x = \begin{bmatrix} x_1 \\ x_2 \\ x_3 \\ x_4 \\ x_5 \end{bmatrix} = \begin{bmatrix} -(1/2)x_3 - (1/4)x_5 \\ -(1/2)x_3 - (1/4)x_5 \\ x_3 \\ -(1/2)x_5 \\ x_5 \end{bmatrix} = x_3 \begin{bmatrix} -1/2 \\ -1/2 \\ 1 \\ 0 \\ 0 \end{bmatrix} + x_5 \begin{bmatrix} -1/4 \\ -1/4 \\ 0 \\ -1/2 \\ 1 \end{bmatrix} = x_3 u + x_5 v.$$

Since x_3 and x_5 are arbitrary, we have $nulA = span\{u, v\}$. Moreover, the following argument shows that u and v are linearly independent.

$x_3 u + x_5 v = 0$

$\Rightarrow$ The solution $x = 0$

$\Rightarrow$ All coordinates of the solution $x = 0$

$\Rightarrow$ The 3rd and 5th coordinates $x_3 = x_5 = 0$.

Consequently, u and v form a basis of *nulA*.

The discussion in the example applies to the general case. Thus we have the following method for finding bases of the null spaces.

If the general solution of $Ax = 0$ is

$x = c_1 v_1 + c_2 v_2 + \ldots + c_k v_k$,

where $c_1, c_2, \ldots, c_k$ are the free variables. Then $v_1, v_2, \ldots, v_k$ form a basis of *nulA*.

The bases of kernels of linear transformations can be found in a similar way.

Example: Consider the linear transformation

$T(X) = AX - XA$: $M(2, 2) \rightarrow M(2, 2)$, where $A = \begin{bmatrix} 1 & 2 \\ 3 & 4 \end{bmatrix}$

Denote

$$X = \begin{bmatrix} x & y \\ z & w \end{bmatrix}.$$

Then by looking at the four entries, $T(X) = O$ is the same as

$$\begin{aligned} 3y - 2z &= 0 \\ 2x + 3y - 2w &= 0 \\ -3x - 3z + 3w &= 0 \\ -3y + 2z &= 0 \end{aligned}$$

The general solution is $x = -z + w$, $y = (2/3)z$, with z and w arbitrary. In the matrix form, the solution is

$$X = \begin{bmatrix} -z + w & (2/3)z \\ z & w \end{bmatrix} = Z \quad = \quad \begin{bmatrix} -1 & 2/3 \\ 1 & 0 \end{bmatrix} + w \begin{bmatrix} 1 & 0 \\ 0 & 1 \end{bmatrix}$$

Thus a basis of *kernelT* is given by the matrices

$$\begin{bmatrix} -1 & 2/3 \\ 1 & 0 \end{bmatrix} \begin{bmatrix} 1 & 0 \\ 0 & 1 \end{bmatrix}$$

BASIS OF COLUMN SPACE, FIRST METHOD

In the next example, we find a basis of the column space of a matrix.

Example: The column space of the matrix

$$A = \begin{bmatrix} 1 & 3 & 2 & 0 & 1 \\ -1 & -1 & -1 & 1 & 0 \\ 0 & 4 & 2 & 4 & 3 \\ 1 & 3 & 2 & -2 & 0 \end{bmatrix}$$

is the span of the five column vectors. By the row operations in this example, we get a row echelon form.

$$\begin{bmatrix} 1 & 3 & 2 & 0 & 1 \\ 0 & 2 & 1 & -1 & 0 \\ 0 & 0 & 0 & 2 & 1 \\ 0 & 0 & 0 & 0 & 0 \end{bmatrix}$$

Since [col 3] and [col 5] are not pivot, [col 5] is a linear combination of [col 1] through [col 4], and [col 3] is a linear combination of [col 1] and [col 2]. Then by this property, we have

$$\begin{aligned} colA &= span\{[\text{col } 1], [\text{col } 2], [\text{col } 3], [\text{col } 4], [\text{col } 5]\} \\ &= span\{[\text{col } 1], [\text{col } 2], [\text{col } 3], [\text{col } 4]\} \\ &= span\{[\text{col } 1], [\text{col } 2], [\text{col } 4]\}. \end{aligned}$$

On the other hand, if we pick [col 1], [col 2], [col 4] only, then

$$\begin{bmatrix} 1 & 3 & 0 \\ -1 & -1 & 1 \\ 0 & 4 & 4 \\ 1 & 3 & -2 \end{bmatrix}$$

has the following row echelon form

$$\begin{bmatrix} 1 & 3 & 0 \\ 0 & 2 & -1 \\ 0 & 0 & 2 \\ 0 & 0 & 0 \end{bmatrix}$$

in which all columns are pivot. Thus the three columns are linearly independent. Since they also span *colA*, we conclude that

[col 1] of $A = (1, -1, 0, 1)$

[col 2] of $A = (3, -1, 4, 3)$

[col 4] of $A = (0, 1, 4, -2)$

form a basis of *colA*.

The argument in the example can be applied to any matrix A. We have

- The nonpivot columns of A are linear combinations of the preceeding columns. Therefore they can be deleted without affecting the columns space.
- Tautologically, the matrix formed by the pivot columns of A has all columns pivot. Therefore the pivot columns are linearly independent.

This leads to the following method for finding a basis of the column space.

For a matrix A, the pivot columns of A form a basis of $colA$.

We emphasis that it is the columns of A, not the columns of the row echelon form, that form the basis.

Example: We would like to find a basis of

$$H = span\left\{\begin{bmatrix}1 & -1\\ 1 & 1\end{bmatrix}, \begin{bmatrix}2 & 1\\ -1 & 3\end{bmatrix}, \begin{bmatrix}4 & 2\\ -2 & 1\end{bmatrix}, \begin{bmatrix}3 & -3\\ 3 & 2\end{bmatrix}\right\}.$$

By the translation

$$A = \begin{bmatrix}a & b\\ c & d\end{bmatrix} \leftrightarrow [A] = (a, b, c, d),$$

the problem becomes the basis of the column space of

$$\begin{bmatrix}1 & 2 & 4 & 3\\ -1 & 1 & 2 & -3\\ 1 & -1 & -2 & 3\\ 1 & 3 & 1 & 2\end{bmatrix}$$

Row operations give us the row echelon form.

$$\begin{bmatrix}1 & 2 & 4 & 3\\ 0 & 1 & 2 & 0\\ 0 & 0 & 5 & 1\\ 0 & 0 & 0 & 0\end{bmatrix}$$

Thus the first three columns of the matrix form a basis of the column space. Correspondingly, the first three 2 by 2 matrices

$$\begin{bmatrix}1 & -1\\ 1 & 1\end{bmatrix}, \begin{bmatrix}2 & 1\\ -1 & 3\end{bmatrix}, \begin{bmatrix}4 & 2\\ -2 & 1\end{bmatrix}$$

form a basis of the span H.

Example: We would like to find a basis for the range of the linear transformation

$T(p(t)) = p(0) + p(1)t + (p'(0) + p'(1))t^2 : P_2 \to P_2$.

The range is spanned by

$T(1) = 1 + t,\ T(t) = t + 2t^2,\ T(t^2) = t + 2t^2$.

By the translation $[a + bt + ct^2] = (a, b, c)$, the problem becomes the basis of the column space of

$$[\ [T(1)]\ [T(t)]\ [T(t^2)]\ ] = \begin{bmatrix} 1 & 0 & 0 \\ 1 & 1 & 1 \\ 0 & 2 & 2 \end{bmatrix}$$

By row operations, we get

$$\begin{bmatrix} 1 & 0 & 0 \\ 0 & 1 & 1 \\ 0 & 0 & 0 \end{bmatrix}$$

Thus the first two columns form a basis of the column space. Correspondingly, $T(1) = 1 + t$ and $T(t) = t + 2t^2$ form a basis of *rangeT*.

BASIS OF COLUMN SPACE, SECOND METHOD

We illustrate another method for finding a basis of a column space through the same example.

Example: The column space of

$$A = \begin{bmatrix} 1 & 3 & 2 & 0 & 1 \\ -1 & -1 & -1 & 1 & 0 \\ 0 & 4 & 2 & 4 & 3 \\ 1 & 3 & 2 & -2 & 0 \end{bmatrix}$$

is the span of the five column vectors. The five spanning vectors can be simplified by column operations. The computation in this example gives us a column echelon form.

$$B = \begin{bmatrix} 1 & 0 & 0 & 0 & 0 \\ -1 & 1 & 0 & 0 & 0 \\ 0 & 2 & 1 & 0 & 0 \\ 1 & 0 & -1 & 0 & 0 \end{bmatrix}$$

Since zero columns do not contribute to the span, we have

$colA = colB = span\{\text{first three columns of } B\}$.

On the other hand, the first three columns of B are linearly independent.

$c_1[\text{col } 1] + c_2[\text{col } 2] + c_3[\text{col } 3] = 0$

$\Rightarrow c_1 = 0$

$\Rightarrow c_2 = 0$

$\Rightarrow c_3 = 0.$

Thus the nonzero columns of B

[col 1] of $B = (1, -1, 0, 1)$

[col 2] of $B = (0, 1, 2, 0)$

[col 3] of $B = (0, 0, 1, -1)$

form a basis of $colB = colA$.

The argument in the example can be applied to any matrix A. We have

- If B is the column echelon form of A, then $colB = colA$.
- The nonzero columns in B are linearly independent and span $colB$.

This leads to the following method for finding a basis of the column space.

For a matrix A, the nonzero columns of a column echelon form is a basis of $colA$.

We emphasis that it is the columns of the column echelon form, not the columns of A, that form the basis.

Example: To find a basis of

$$H = span\left\{\begin{bmatrix}1 & -1\\1 & 1\end{bmatrix}, \begin{bmatrix}2 & 1\\-1 & 3\end{bmatrix}, \begin{bmatrix}4 & 2\\-2 & 1\end{bmatrix}, \begin{bmatrix}3 & -3\\3 & 2\end{bmatrix}\right\}$$

we use the translation

$$A = \begin{bmatrix}a & b\\c & d\end{bmatrix} \leftrightarrow [A] = (a, b, c, d)$$

to convert the problem to the basis of the column space of

$$\begin{bmatrix}1 & 2 & 4 & 3\\-1 & 1 & 2 & -3\\1 & -1 & -2 & 3\\1 & 3 & 1 & 2\end{bmatrix}$$

Column operations give us a column echelon form.

$$\begin{bmatrix}1 & 0 & 0 & 0\\-1 & 3 & 0 & 0\\1 & -3 & 0 & 0\\1 & 1 & 1 & 0\end{bmatrix}$$

The nonzero columns (1, – 1, 1, 1), (0, 3, – 3, 1), (0, 0, 0, 1) form a basis of the column space. The corresponding 2 by 2 matrices

$$\begin{bmatrix} 1 & -1 \\ 1 & 1 \end{bmatrix}, \begin{bmatrix} 0 & 3 \\ -3 & 1 \end{bmatrix}, \begin{bmatrix} 0 & 0 \\ 0 & 1 \end{bmatrix}$$

form a basis of the span H.

Example: To find a basis for the range of the linear transformation

$T(p(t)) = p(0) + p(1)t + (p'(0) + p'(1))t^2 : P_2 \to P_2,$

we use the translation $[a + bt + ct^2] = (a, b, c)$ to convert the problem to the basis of the column space of

$$[\,[T(1)]\ [T(t)]\ [T(t^2)]\,] = \begin{bmatrix} 1 & 0 & 0 \\ 1 & 1 & 1 \\ 0 & 2 & 2 \end{bmatrix}.$$

By column operations, we get a column echelon form

$$\begin{bmatrix} 1 & 0 & 0 \\ 1 & 1 & 0 \\ 0 & 2 & 0 \end{bmatrix}$$

and the nonzero columns (1, 1, 0), (0, 1, 2) of the column echelon form is a basis of the column space. Correspondingly, $1 + t$ and $t + 2t^2$ form a basis of *rangeT*.

Exercise: For each matrix A, find bases for *colA*, *nulA*, $colA^T$, $nulA^T$.

1. $\begin{bmatrix} 1 & 2 & 3 \\ 2 & 3 & 4 \end{bmatrix}$

Answer: From

$$\text{row operation: } A = \begin{bmatrix} 1 & 2 & 3 \\ 2 & 3 & 4 \end{bmatrix} \to \begin{bmatrix} 1 & 0 & -1 \\ 0 & 1 & 2 \end{bmatrix};$$

we find the solution of $Ax = 0$ to be $x_1 = x_3, x_2 = -2x_3, x_3$ arbitrary. Therefore $x = (x_3, -2x_3, x_3) = x_3(1, -2, 1)$, and by the method for computing basis of null space, {(1, –2, 1)} is a basis of *nulA*.

The same row operation also tells us that the first two columns are pivot, so that by the first method for computing basis of column space, {(1, 2), (2, 3)} is a basis of *colA*.

By taking transpose, the row operation becomes

$$\text{column operation: } A^{\mathrm{T}} = \begin{bmatrix} 1 & 2 \\ 2 & 3 \\ 3 & 4 \end{bmatrix} \rightarrow \begin{bmatrix} 1 & 0 \\ 0 & 1 \\ -1 & 2 \end{bmatrix}.$$

By the second method for computing basis of column space, {(1, 0, –1), (0, 1, 2)} is a basis of $colA^{\mathrm{T}}$.

To find a basis for $nulA^{\mathrm{T}}$, we use

$$\text{row operation: } A^{\mathrm{T}} = \begin{bmatrix} 1 & 2 \\ 2 & 3 \\ 3 & 4 \end{bmatrix} \rightarrow \begin{bmatrix} 1 & 0 \\ 0 & 1 \\ 0 & 0 \end{bmatrix}$$

to find $A^{\mathrm{T}}x = 0$ has only the trivial solution. Thus $nulA^{\mathrm{T}} = \{0\}$ and the basis is the empty set.

Remark: One row operation on A can give us bases for three of the four induced subspaces. Similarly, the row operation on A^{T} also gives us a basis {(1, 2, 3), (2, 3, 4)} of $colA^{\mathrm{T}}$ and a basis Z of $nulA^{\mathrm{T}}$.

2. $\begin{bmatrix} 1 & 2 & 3 \\ 4 & 5 & 6 \\ 7 & 8 & 9 \end{bmatrix}$

Answer: From

$$\text{row operation: } A = \begin{bmatrix} 1 & 2 & 3 \\ 4 & 5 & 6 \\ 7 & 8 & 9 \end{bmatrix} \rightarrow \begin{bmatrix} 1 & 0 & -1 \\ 0 & 1 & 2 \\ 0 & 0 & 0 \end{bmatrix}$$

the solution of $Ax = 0$ is $x = (x_3, -2x_3, x_3) = x_3(1, -2, 1)$, so that by the method for computing basis of null space, {(1, –2, 1)} is a basis of $nulA$.

The same row operation also tells us, by the first method for computing basis of column space, the first two columns of A form a basis {(1, 4, 7), (2, 5, 8)} of $colA$.

By taking the transpose of the same row operation and applying second method for computing basis of column space, we get a basis {(1, 0, –1), (0, 1, 2)} of $colA^{\mathrm{T}}$.

Finally, from

$$\text{row operation: } A^{\mathrm{T}} = \begin{bmatrix} 1 & 4 & 7 \\ 2 & 5 & 8 \\ 3 & 6 & 9 \end{bmatrix} \rightarrow \begin{bmatrix} 1 & 0 & -1 \\ 0 & 1 & 2 \\ 0 & 0 & 0 \end{bmatrix}$$

the solution of $A^Tx = 0$ is $x = (x_3, -2x_3, x_3) = x_3(1, -2, 1)$, so that by the method for computing basis of null space, $\{(1, -2, 1)\}$ is a basis of $nulA^T$.

Remark: By the later theory on rank (= dimension of column space) and nullity (= dimension of null space), we know $dim(nulA^T) = 3 - dim(colA^T) = 1$. If we also know $(1, -2, 1)$ is a solution of $A^Tx = 0$, then we may also conclude $\{(1, -2, 1))$ is a basis of $nulA^T$.

3. $\begin{bmatrix} 1 & 0 & 1 & 2 \\ 1 & 4 & -3 & -2 \\ 1 & -1 & 2 & 3 \end{bmatrix}$

Answer: From

$$\text{row operation: } A = \begin{bmatrix} 1 & 0 & 1 & 2 \\ 1 & 4 & -3 & -2 \\ 1 & -1 & 2 & 3 \end{bmatrix} \to \begin{bmatrix} 1 & 0 & 1 & 2 \\ 0 & 1 & -1 & -1 \\ 0 & 0 & 0 & 0 \end{bmatrix},$$

the solution of $Ax = 0$ is $x = (-x_3 - 2x_4, x_3 + x_4, x_3, x_4) = x_3(-1, 1, 1, 0) + x_4(-2, 1, 0, 1)$, and $\{(-1, 1, 1, 0), (-2, 1, 0, 1)\}$ is a basis of $nulA$. Moreover, the first two columns of A form a basis $\{(1, 1, 1), (0, 4, -1)\}$ of $colA$, and the nonzero rows of the row echelon form (which are the nonzero columns of the column form of A^T) form a basis $\{(1, 0, 1, 2), (0, 1, -1, -1)\}$ of $colA^T$.

From

$$\text{row operation: } A^T = \begin{bmatrix} 1 & 1 & 1 \\ 0 & 4 & -1 \\ 1 & -3 & 2 \\ 2 & -2 & 3 \end{bmatrix} \to \begin{bmatrix} 1 & 1 & 1 \\ 0 & 4 & -1 \\ 0 & 0 & 0 \\ 0 & 0 & 0 \end{bmatrix}$$

the solution of $A^Tx = 0$ is $x = (x_3/4)(-5, -1, 4)$, so that $\{(-5, -1, 4)\}$ is a basis of $nulA^T$.

4. $\begin{bmatrix} 1 & -2 & 1 & 1 & 2 \\ -1 & 3 & 0 & 2 & -2 \\ 0 & 1 & 1 & 3 & 4 \\ 1 & 2 & 5 & 13 & 5 \end{bmatrix}$

Answer: From an earlier exercise, we have

$$\text{row operation: } A = \begin{bmatrix} 1 & -2 & 1 & 1 & 2 \\ -1 & 3 & 0 & 2 & -2 \\ 0 & 1 & 1 & 3 & 4 \\ 1 & 2 & 5 & 13 & 5 \end{bmatrix} \to \begin{bmatrix} 1 & 0 & 3 & 7 & 0 \\ 0 & 1 & 1 & 3 & 0 \\ 0 & 0 & 0 & 0 & 1 \\ 0 & 0 & 0 & 0 & 0 \end{bmatrix}$$

Then the solution of $Ax = 0$ is $x = (-3x_3 - 7x_4, -x_3 - 3x_4, x_3, x_4, 0) = x_3(-3, -1, 1, 0, 0) + x_4(-7, -3, 0, 1, 0)$, and $\{(-3, -1, 1, 0, 0), (-7, -3, 0, 1, 0)\}$ is a basis of *nulA*. Moreover, [col 1], [col 2], [col 5] of A form a basis $\{(1, -1, 0, 1), (-2, 3, 1, 2), (2, -2, 4, 5)\}$ of *colA*, and the nonzero rows of the row echelon form form a basis $\{(1, 0, 3, 7, 0), (0, 1, 1, 3, 0)\}$ of *col*A^T.

From another earlier exercise, we have

$$\text{row operation: } A^T = \begin{bmatrix} 1 & -1 & 0 & 1 \\ -2 & 3 & 1 & 2 \\ 1 & 0 & 1 & 5 \\ 1 & 2 & 3 & 13 \\ 2 & -2 & 4 & 5 \end{bmatrix} \to \begin{bmatrix} 1 & -1 & 0 & 1 \\ 0 & 1 & 1 & 4 \\ 0 & 0 & 4 & 3 \\ 0 & 0 & 0 & 0 \\ 0 & 0 & 0 & 0 \end{bmatrix}$$

Then we find the solution of $A^T x = 0$ to be $x = (x_4/4)(-17, -13, -3, 4)$, so that $\{(-17, -13, -3, 4)\}$ is a basis of *nul*A^T.

Exercise: Find a basis for the span. Then extend the basis to a basis of the whole space.

1. *Span*$\{1 + t + t^2 + t^3, t + t^2 + t^3, 1 + 2t + 2t^2 + 2t^3, 1 + t^2 + 2t^3\}$ M P_3

Answer: We use the translation $[a + bt + ct^2 + dt^3] = (a, b, c, d)$ to convert the problem to the column space of the matrix

$$A = \begin{bmatrix} 1 & 0 & 1 & 1 \\ 1 & 1 & 2 & 0 \\ 1 & 1 & 2 & 1 \\ 1 & 1 & 2 & 2 \end{bmatrix}.$$

Row operations on A gives us a row echelon form.

$$R = \begin{bmatrix} 1 & 0 & 1 & 1 \\ 0 & 1 & 1 & -1 \\ 0 & 0 & 0 & 1 \\ 0 & 0 & 0 & 0 \end{bmatrix}$$

The pivot columns [col 1], [col 2], [col 4] of A form a basis of *colA*. Correspondingly, the 1st, 2nd, 4th polynomials form a basis $\{1 + t + t^2 + t^3, t + t^2 + t^3, 1 + t^2 + 2t^3\}$ of the span.

To extend the basis to a basis of the whole space, we use the idea in this remark and append the standard basis of R^4 to the basis of the *colA* and consider

$$[\,[\text{col } 1]\ [\text{col } 2]\ [\text{col } 4]\ I\,] = \begin{bmatrix} 1 & 0 & 1 & 1 & 0 & 0 & 0 \\ 1 & 1 & 0 & 0 & 1 & 0 & 0 \\ 1 & 1 & 1 & 0 & 0 & 1 & 0 \\ 1 & 1 & 2 & 0 & 0 & 0 & 1 \end{bmatrix}$$

Row operations give us

$$\begin{bmatrix} 1 & 0 & 1 & 1 & 0 & 0 & 0 \\ 0 & 1 & -1 & -1 & 1 & 0 & 0 \\ 0 & 0 & 1 & 0 & -1 & 1 & 0 \\ 0 & 0 & 0 & 0 & 1 & -2 & 1 \end{bmatrix}.$$

Noting the first three columns and the 5th column are the pivot ones, [col 1], [col 2], [col 4] of A and $e_2 = (0, 1, 0, 0)$ form a basis of R^4. Correspondingly, the extension $\{1 + t + t^2 + t^3, t + t^2 + t^3, 1 + t^2 + 2t^3, t\}$ is a basis of P_3.

Remark: We may also use the column operation on A to get

$$C = \begin{bmatrix} 1 & 0 & 0 & 0 \\ 1 & 1 & 0 & 0 \\ 1 & 1 & 1 & 0 \\ 1 & 1 & 2 & 0 \end{bmatrix}$$

Corresponding to the nonzero columns of C, we get a basis $\{1 + t + t^2 + t^3, t + t^2 + t^3, t^2 + 2t^3\}$ of the span. The basis can be extended to a basis of P_3 by adding t^3.

2. $Span \left\{ \begin{bmatrix} -1 & 1 \\ -2 & 1 \end{bmatrix}, \begin{bmatrix} 2 & -2 \\ 4 & -2 \end{bmatrix}, \begin{bmatrix} 2 & -1 \\ 3 & 1 \end{bmatrix}, \begin{bmatrix} 2 & -3 \\ 5 & -4 \end{bmatrix}, \begin{bmatrix} -5 & 4 \\ -9 & -1 \end{bmatrix} \right\} \subset M(2, 2)$

Answer: We use the translation

$$\begin{bmatrix} a & b \\ c & d \end{bmatrix} \leftrightarrow (a, b, c, d)$$

to convert the problem to the column space of the matrix

$$A = \begin{bmatrix} -1 & 2 & 2 & 2 & -5 \\ 1 & -2 & -1 & -3 & 4 \\ -2 & 4 & 3 & 5 & -9 \\ 1 & -2 & 1 & -4 & -1 \end{bmatrix}$$

Instead of row operations, we compute a column echelon form.

$$C = \begin{bmatrix} -1 & 0 & 0 & 0 & 0 \\ 1 & 1 & 0 & 0 & 0 \\ -2 & -1 & 0 & 0 & 0 \\ 1 & 3 & 1 & 0 & 0 \end{bmatrix}$$

The nonzero columns of C form a basis of *colA*. Correspondingly, the matrices

$$\begin{bmatrix} -1 & 1 \\ -2 & 1 \end{bmatrix}, \begin{bmatrix} 0 & 1 \\ -1 & 3 \end{bmatrix}, \begin{bmatrix} 0 & 0 \\ 0 & 1 \end{bmatrix}$$

form a basis of the span.

Noting that adding a "missing pivot" to the nonzero columns of C will give us an invertible matrix

$$\begin{bmatrix} -1 & 0 & 0 & 0 \\ 1 & 1 & 0 & 0 \\ -2 & -1 & * & 0 \\ 1 & 3 & \# & 1 \end{bmatrix}$$

Thus we may extend (by taking * = 1, # = 0 for example) the basis of the span to the following basis

$$\begin{bmatrix} -1 & 1 \\ -2 & 1 \end{bmatrix}, \begin{bmatrix} 0 & 1 \\ -1 & 3 \end{bmatrix}, \begin{bmatrix} 0 & 0 \\ 1 & 0 \end{bmatrix}, \begin{bmatrix} 0 & 0 \\ 0 & 1 \end{bmatrix}$$

of $M(2, 2)$.

3. *Span* $\begin{bmatrix} 0 & 1 & 2 \\ 1 & 0 & 1 \\ 2 & 1 & 0 \end{bmatrix}, \begin{bmatrix} 1 & 1 & 0 \\ 1 & 1 & 1 \\ 0 & 1 & 1 \end{bmatrix}, \begin{bmatrix} -1 & 0 & 2 \\ 0 & -1 & 0 \\ 2 & 0 & -1 \end{bmatrix}, \begin{bmatrix} 2 & 1 & -2 \\ 1 & 2 & 1 \\ -2 & 1 & 2 \end{bmatrix}, \begin{bmatrix} 2 & 2 & 0 \\ 2 & 1 & 2 \\ 0 & 2 & 1 \end{bmatrix} \subset Sym(3)$

Answer: We use the translation

$$\begin{bmatrix} a & b & c \\ b & d & e \\ c & e & f \end{bmatrix} \to (a, b, c, d, e, f)$$

to convert the problem to the column space of the matrix

$$A = \begin{bmatrix} 0 & 1 & -1 & 2 & 2 \\ 1 & 1 & 0 & 1 & 2 \\ 2 & 0 & 2 & -2 & 0 \\ 0 & 1 & -1 & 2 & 1 \\ 1 & 1 & 0 & 1 & 2 \\ 0 & 1 & -1 & 2 & 1 \end{bmatrix}$$

Column operations give us

$$C = \begin{bmatrix} 1 & 0 & 0 & 0 & 0 \\ 1 & 1 & 0 & 0 & 0 \\ 0 & 2 & 0 & 0 & 0 \\ 1 & 0 & -1 & 0 & 0 \\ 1 & 1 & 0 & 0 & 0 \\ 1 & 0 & -1 & 0 & 0 \end{bmatrix}$$

Corresponding to nonzero columns of C, we get a basis

$$\begin{bmatrix} 1 & 1 & 0 \\ 1 & 1 & 1 \\ 0 & 1 & 1 \end{bmatrix}, \begin{bmatrix} 0 & 1 & 2 \\ 1 & 0 & 1 \\ 2 & 1 & 0 \end{bmatrix}, \begin{bmatrix} 0 & 0 & 0 \\ 0 & -1 & 0 \\ 0 & 0 & -1 \end{bmatrix}$$

of the span. Moreover, by expanding the nonzero columns of C in the following way

$$C = \begin{bmatrix} 1 & 0 & 0 & 0 & 0 & 0 \\ 1 & 1 & 0 & 0 & 0 & 0 \\ 0 & 2 & 1 & 0 & 0 & 0 \\ 1 & 0 & 0 & -1 & 0 & 0 \\ 1 & 1 & 0 & 0 & 1 & 0 \\ 1 & 0 & 0 & -1 & 0 & 1 \end{bmatrix}$$

we get an invertible matrix. Thus the following basis of $Sym(3)$

$$\begin{bmatrix}1&1&0\\1&1&1\\0&1&1\end{bmatrix},\begin{bmatrix}0&1&2\\1&0&1\\2&1&0\end{bmatrix},\begin{bmatrix}0&0&0\\0&-1&0\\0&0&-1\end{bmatrix},\begin{bmatrix}0&0&1\\0&0&0\\1&0&0\end{bmatrix},\begin{bmatrix}0&0&0\\0&0&1\\0&1&0\end{bmatrix},\begin{bmatrix}0&0&0\\0&0&0\\0&0&1\end{bmatrix}$$

is an extension of the basis of the span.

Exercise: Find bases for the range and kernel.

1. $T(p(t)) = (t + 1)p'(t) - p(t): P_3 \to P_3$

Answer: By

$T(a + bt + ct^2 + dt^3) = (-a + b) + 2ct + (c + 3d)t^2 + 2dt^3$,

and the translation $[a + bt + ct^2 + dt^3] = (a, b, c, d)$, the problem is converted to the column space and the null space of the matrix

$$A = \begin{bmatrix}-1&1&0&0\\0&0&2&0\\0&0&1&3\\0&0&0&2\end{bmatrix}.$$

The pivot columns are [col 1], [col 3], [col 4]. Correspondingly, $\{-1, 2t + t^2, 3t^2 + 2t^3\}$ is a basis of the range. On the other hand, it is easy to see the solution of $Ax = 0$ is $x = b(1, 1, 0, 0)$. Correspondingly, $\{1 + t\}$ is a basis of the kernel.

2. $T(p(t), q(t)) = (p'(t) + tq'(t) - 2q(t), q'(t) + p(t) - q(t)): P_2 \oplus P_2 \to P_2 \oplus P_2$

Answer: Let $p(t) = a_1 + b_1t + c_1t^2$, $q(t) = a_2 + b_2t + c_2t^2$. Then

$T(p(t), q(t)) = ((b_1 - 2a_2) + (2c_1 - b_2)t, (a_1 - a_2 + b_2) + (b_1 - b_2 + 2c_2)t + (c_1 - c_2)t^2))$.

Thus by the usual translation $[(p(t), q(t))] = (a_1, b_1, c_1, a_2, b_2, c_2)$, the problem is converted to the column space and the null space of the matrix

$$\begin{bmatrix}0&1&0&-2&0&0\\0&0&2&0&-1&0\\0&0&0&0&0&0\\1&0&0&-1&1&0\\0&1&0&0&-1&2\\0&0&1&0&0&-1\end{bmatrix}$$

Row operations change the matrix into

$$\begin{bmatrix} 1 & 0 & 0 & -1 & 1 & 0 \\ 0 & 1 & 0 & 0 & -1 & 2 \\ 0 & 0 & 1 & 0 & 0 & -1 \\ 0 & 0 & 0 & 2 & -1 & 2 \\ 0 & 0 & 0 & 0 & 1 & -3 \\ 0 & 0 & 0 & 0 & 0 & 0 \end{bmatrix}$$

All except [col 6] are pivot columns. Corresponding to the pivot columns of A, $\{(0, 1), (1, t), (2t, t^2), (-2, -1), (-1, 1-t), (0, 2t-t^2)\}$ is a basis of the range. On the other hand, it is easy to see the solution of $Ax = 0$ is $x = c(5, -1, -2, -1, -6, 2)$. Correspondingly, $\{(5 - t - 2t^2, -1 - 6t + 2t^2)\}$ is a basis of the kernel.

3. $T(X) = \begin{bmatrix} 1 & 2 \\ -1 & -2 \end{bmatrix} X \begin{bmatrix} 2 & 4 \\ 3 & 6 \end{bmatrix} : M(2,2) \to M(2,2)$

Answer: By

$$T\begin{bmatrix} x & y \\ z & w \end{bmatrix} = (2x + 3y + 4z + 6w)\begin{bmatrix} 1 & 2 \\ -1 & -2 \end{bmatrix},$$

we see the range of T is spanned by

$$\begin{bmatrix} 1 & 2 \\ -1 & -2 \end{bmatrix},$$

and has the matrix as a basis. The kernel of T consists of matrices of the form

$$\begin{bmatrix} -(3/2)y - 2z - 3w & y \\ z & w \end{bmatrix} = y\begin{bmatrix} -3/2 & 1 \\ 0 & 0 \end{bmatrix} + z\begin{bmatrix} -2 & 0 \\ 1 & 0 \end{bmatrix} + w\begin{bmatrix} -3 & 0 \\ 0 & 1 \end{bmatrix}$$

and the three matrices in the linear combination form a basis.

RANK AND NULLITY

RANK AND NULLITY FOR MATRIX

The system

$$x + y + z = 1$$
$$2x + 2y + 2z = 2$$
$$-x - y - z = -1$$
$$x + y + z = 1$$

consists of four equations and appears to be "large". However, we know the system *essentially* consists of just one equation: $x + y + z = 1$. The true size of the system is reflected in the fact that the augmented matrix

$$\begin{bmatrix} 1 & 1 & 1 & 1 \\ 2 & 2 & 2 & 2 \\ -1 & -1 & -1 & -1 \\ 1 & 1 & 1 & 1 \end{bmatrix}$$

has only one pivot in the row echelon form.

$$\begin{bmatrix} 1 & 1 & 1 & 1 \\ 0 & 0 & 0 & 0 \\ 0 & 0 & 0 & 0 \\ 0 & 0 & 0 & 0 \end{bmatrix}$$

Moreover, by this method of computing the basis of the column space, the number of pivots is the dimension of the column space of the matrix. The observation leads to the definition of the rank of a matrix.

$rankA$ = number of pivots in the row echelon form of $A = dim(colA)$.

Example: Consider the matrix

$$\begin{bmatrix} 1 & 3 & 2 & 0 & 1 & 0 \\ -1 & -1 & -1 & 1 & 0 & 1 \\ 0 & 4 & 2 & 4 & 3 & 3 \\ 1 & 3 & 2 & -2 & 0 & 0 \end{bmatrix}$$

In this example, we computed a row echelon form.

$$\begin{bmatrix} 1 & 3 & 2 & 0 & 1 & 0 \\ 0 & 2 & 1 & -1 & 0 & 1 \\ 0 & 0 & 0 & 2 & 1 & 0 \\ 0 & 0 & 0 & 0 & 1 & 1 \end{bmatrix}$$

Thus

$$rank \begin{bmatrix} 1 & 3 & 2 & 0 & 1 & 0 \\ -1 & -1 & -1 & 1 & 0 & 1 \\ 0 & 4 & 2 & 4 & 3 & 3 \\ 1 & 3 & 2 & -2 & 0 & 0 \end{bmatrix} = 4.$$

Moreover, if we take only the first four or five columns, then

$$rank\begin{bmatrix} 1 & 3 & 2 & 0 & 1 \\ -1 & -1 & -1 & 1 & 0 \\ 0 & 4 & 2 & 4 & 3 \\ 1 & 3 & 2 & -2 & 0 \end{bmatrix} = rank\begin{bmatrix} 1 & 3 & 2 & 0 \\ -1 & -1 & -1 & 1 \\ 0 & 4 & 2 & 4 \\ 1 & 3 & 2 & -2 \end{bmatrix}$$

If we take only the first two or three columns, then

$$rank\begin{bmatrix} 1 & 3 \\ -1 & -1 \\ 0 & 4 \\ 1 & 3 \end{bmatrix} = rank\begin{bmatrix} 1 & 3 & 2 \\ -1 & -1 & -1 \\ 0 & 4 & 2 \\ 1 & 3 & 2 \end{bmatrix} = 2.$$

A matrix also has the null space. Its dimension is the nullity.

$nullityA = dim(nulA)$.

By this method of computing the basis of the null space, we have

$nullityA$

= number of free variables in the solution of $Ax = 0$

= number of nonpivot columns in (the row echelon form of) A,

where the second equality follows from this correspondence. Compared with the definition of *rankA,* we get the following equality.

$rankA + nullityA$ = number of columns of A.

Example: By the computation in the earlier example and the relation between the rank and nullity, we have

$$nullity\begin{bmatrix} 1 & 3 & 2 & 0 & 1 & 0 \\ -1 & -1 & -1 & 1 & 0 & 1 \\ 0 & 4 & 2 & 4 & 3 & 3 \\ 1 & 3 & 2 & -2 & 0 & 0 \end{bmatrix} = 6 - 4 = 2.$$

$$nullity\begin{bmatrix} 1 & 3 & 2 & 0 & 1 \\ -1 & -1 & -1 & 1 & 0 \\ 0 & 4 & 2 & 4 & 3 \\ 1 & 3 & 2 & -2 & 0 \end{bmatrix} = 5 - 3 = 2.$$

$$nullity\begin{bmatrix} 1 & 3 & 2 & 0 \\ -1 & -1 & -1 & 1 \\ 0 & 4 & 2 & 4 \\ 1 & 3 & 2 & -2 \end{bmatrix} = 4 - 3 = 1.$$

PROPERTIES OF RANK AND NULLITY

The first property is that the rank is not changed by the transpose. The proof can be found here.

$rankA = rankA^T$.

Example: From the computation

$$rank\begin{bmatrix} 1 & 3 & 2 & 0 & 1 & 0 \\ -1 & -1 & -1 & 1 & 0 & 1 \\ 0 & 4 & 2 & 4 & 3 & 3 \\ 1 & 3 & 2 & -2 & 0 & 0 \end{bmatrix}$$

in an earlier example, we have

$$rank\begin{bmatrix} 1 & -1 & 0 & 1 \\ 3 & -1 & 4 & 3 \\ 2 & -1 & 2 & 2 \\ 0 & 1 & 4 & -2 \\ 1 & 0 & 3 & 0 \\ 0 & 1 & 3 & 0 \end{bmatrix} = 4,\ nullity\begin{bmatrix} 1 & -1 & 0 & 1 \\ 3 & -1 & 4 & 3 \\ 2 & -1 & 2 & 2 \\ 0 & 1 & 4 & -2 \\ 1 & 0 & 3 & 0 \\ 0 & 1 & 3 & 0 \end{bmatrix} = 4 - 4 = 0.$$

Note that since the nullity of the 6 by 4 matrix is zero, the null space consists of only the zero vector. This means that the four column vectors (of R^6) are linearly independent.

Since the rank means the "essential size", it is always no bigger than the "apparent size".

For an m by n matrix A, $rankA \le m$ and n. Moreover,

- $rankA = m \Leftrightarrow$ Columns of A span $R^m \Leftrightarrow Ax = b$ has solutions for all b.
- $rankA = n \Leftrightarrow$ Columns of A are linearly independent $\Leftrightarrow$ Solution of $Ax = b$ is unique.

The proof can be found here. As a consequence of the property, we may characterize the invertibility of a square matrix by the rank and nullity.

For an n by n matrix A, the following are equivalent.

- A is invertible.
- $colA = \mathrm{R}^m$.
- $rankA = n$.
- $nulA = \{0\}$.
- $nullityA = 0$.

PROPERTIES OF RANK AND NULLITY

Theroem: $rankA = rankA^{\mathrm{T}}$.

Proof. The equality can be derived by comparing the two methods for computing a basis of the column space.

$$\begin{aligned} rankA &= \text{number of pivots in the column echelon form of } A \\ &= \text{number of pivots in the row echelon form of } A^{\mathrm{T}} \\ &= rankA^{\mathrm{T}}. \end{aligned}$$

Theorem: For an m by n matrix A, $rankA \leq m$ and n. Moreover

- $rankA = m \Leftrightarrow$ Columns of A span $\mathrm{R}^m \Leftrightarrow Ax = b$ has solutions for all b.
- $rankA = n \Leftrightarrow$ Columns of A are linearly independent $\Leftrightarrow$ Solution of $Ax = b$ is unique.

Proof. First,

$rankA$ = number of pivot columns of $A \leq$ number of all columns of A.

Then,

$rankA = rankA^{\mathrm{T}} \leq$ number of all columns of A^{T} = number of all rows of A.

Moreover,

$rankA = m$

$\Leftrightarrow dim(colA) = dim\mathrm{R}^m$

$\Leftrightarrow colA = \mathrm{R}^m$

$\Leftrightarrow$ The column vectors span R^m.

$rankA = n$

$\Leftrightarrow dim(nulA) = nullityA = n - rankA = 0$

$\Leftrightarrow nulA = \{0\}$

$\Leftrightarrow Ax = 0$ has only the trivial solution

$\Leftrightarrow$ The column vectors are linearly independent.

RANK AND NULLITY FOR LINEAR TRANSFORMATION

The discussion about the rank and nullity can be extended to linear transformations.

Let $T: V \rightarrow W$ be a linear transformation between vector spaces of finite dimensions. Then

$rankT = dim(rangeT)$, $nullityT = dim(kernelT)$.

By identifying abstract vector spaces with the euclidean ones, the properties about the rank and the nullity of a matrix can be translated to linear transformations.

Let $T: V \rightarrow W$ be a linear transformation between finite dimensional vector spaces. Then

- $rankT = rankT^*$.
- $rankT + nullity(T) = dimV$.
- $rankT \leq dimV$ and $dimW$.
- $rankT = dimW \Leftrightarrow T$ is onto.
- $rankT = dimV \Leftrightarrow T$ is one-to-one.

The following is the characterization of invertibility in terms of the rank and nullity.

Let $T: V \rightarrow W$ be a linear transformation and $\dim V = \dim W = n$. Then the following are equivalent.

- T is invertible.
- $rangeT = W$.
- $rankT = n$.
- $kernelT = \{0\}$.
- $nullityT = 0$.

Example: In an earlier example, we studied the linear transformation

$$T(X) = AX - XA: M(2, 2) \rightarrow M(2, 2), \text{ where } A = \begin{bmatrix} 1 & 2 \\ 3 & 4 \end{bmatrix}$$

and found $kernelT$ consists of two matrices. Therefore

$nullityT = 2$, $rankT = dimM(2, 2) - 2 = 2$.

In particular, T is not invertible.

Example: The kernel of the linear transformation $D(p(t)) = p'(t): P_n \rightarrow P_{n-1}$ consists of the constant polynomials. Therefore $nullity\ D = 1$ and $rankD = dim\ P_n - nullityD = n - 1 = dimP_{n-1}$. The computation on the rank tells us that D is onto.

Exercise: Find the rank and nullity of A and A^T.

1. [0]

Answer: $rankA = rankA^T = 0$, $nullityA = nullityA^T = 1$.

2. [0 1]

Answer: $rankA = rankA^T = 1$, $nullityA = 2 - 1 = 1$, $nullityA^T = 1 - 1 = 0$.

3. $\begin{bmatrix} 1 & 2 & 3 \\ 2 & 3 & 4 \end{bmatrix}$

Answer: The two rows are linearly independent. Therefore the two rows form a basis of $colA^T$ and $rankA = rankA^T = 2$, $nullityA = 3 - 2 = 1$, $nullityA^T = 2 - 2 = 0$.

4. $\begin{bmatrix} 1 & 2 & 3 \\ 4 & 5 & 6 \\ 7 & 8 & 9 \end{bmatrix}$

Answer: Row operations give us

$$\begin{bmatrix} 1 & 2 & 3 \\ 0 & -3 & -6 \\ 0 & 0 & 0 \end{bmatrix}$$

Therefore $rankA = rankA^T = 2$, $nullityA = 3 - 2 = 1$, $nullityA^T = 3 - 2 = 1$.

5. $\begin{bmatrix} 1 & 0 & 1 & 0 \\ 1 & 0 & 0 & 1 \\ 0 & 1 & 1 & 0 \end{bmatrix}$

Answer: Row operations give us

$$\begin{bmatrix} 1 & 0 & 1 & 0 \\ 0 & 1 & 1 & 0 \\ 0 & 0 & -1 & -1 \end{bmatrix}$$

Therefore $rankA = rankA^T = 3$, $nullityA = 4 - 3 = 1$, $nullityA^T = 3 - 3 = 0$.

6. $\begin{bmatrix} 1 & 2 & 3 & 0 & 1 & 1 \\ 0 & 3 & 2 & 1 & -1 & 0 \\ -1 & 1 & 2 & -2 & 1 & 0 \\ 1 & -2 & -2 & 1 & 0 & 0 \\ 4 & 1 & 3 & 2 & 2 & 1 \end{bmatrix}$

Answer: In an earlier exercise, we computed a row echelon form.

$$\begin{bmatrix} 1 & 0 & 0 & 1 & 0 & 0 \\ 0 & 1 & 0 & 1 & -1 & 0 \\ 0 & 0 & 1 & -1 & 1 & 0 \\ 0 & 0 & 0 & 0 & 0 & 1 \\ 0 & 0 & 0 & 0 & 0 & 0 \end{bmatrix}$$

Therefore $rankA = rankA^T = 4$, $nullityA = 6 - 4 = 2$, $nullityA^T = 5 - 4 = 1$.

Exercise: Study the rank of a general 2 by 2 matrix.

Answer: Let

$$A = \begin{bmatrix} a & b \\ c & d \end{bmatrix}$$

By this property, we have $rankA \leq 2$. Moreover,

$rankA = 0 \Leftrightarrow A = O$ (this is true for a matrix of any size).

$rankA = 2 \Leftrightarrow A$ is invertible $\Leftrightarrow ad \neq bc$ (by this criterion).

The remaining case is

$rankA = 1 \Leftrightarrow ad = bc$ and at least one of a, b, c, d is nonzero.

Exercise: Find the rank and the nullity. Then find bases for the range and the kernel. (You may try $n = 2$ or 3 first if you have difficulty with general case)

1. $T(p) = (p(0), p(1), p(2)): P_n \to R^3$

Answer The range of T is spanned by

$T(1) = (1, 1, 1)$

$T(t) = (0, 1, 2)$

$T(t^2) = (0, 1, 4)$

...

$T(t^n) = (0, 1, 2^n)$

Note that for $n \geq 2$, we have

$span\{T(1), T(t), T(t^2)\} \subset rangeT \subset R^3$.

On the other hand, it is easy to see that

$span\{T(1), T(t), T(t^2)\} = span\{(1, 1, 1), (0, 1, 2), (0, 1, 4)\} = R^3$.

Thus for $n \geq 2$, we have $rangeT = R^3$ and
$rankT = dim(rangeT) = 3$, $nullityT = dimP_n - rankT = n - 2$.

Since $rangeT = R^3$, any invertible 3 by 3 matrix gives a basis of the range. For example, $\{(1, 0, 0), (0, 1, 0), (0, 0, 1)\}$ is a basis.

The kernel consists of polynomials satisfying $p(0) = p(1) = p(2) = 0$. These are (degree d" n) polynomials of the form

$$p(t) = t(t-1)(t-2)(b_0 + b_1 t + b_2 t^2 + ... + b_{n-3} t^{n-3})$$
$$= b_0 t(t-1)(t-2) + b_1 t^2(t-1)(t-2) + b_2 t^3(t-1)(t-2) + ... + b_{n-3} t^{n-2}(t-1)(t-2).$$

Thus $\{t(t-1)(t-2), t^2(t-1)(t-2), t^3(t-1)(t-2), ..., t^{n-2}(t-1)(t-2)\}$ is a basis of the kernel.

It remains to consider $n = 0$ or 1. For $n = 0$, we have

$rankT = dim(span\{(1, 1, 1)\}) = 1$, $nullityT = dimP_0 - rankT = 1 - 1 = 0$.

The range has $\{(1, 1, 1)\}$ as a basis. The kernel consists of the zero vector only.

For $n = 1$, we have

$rankT = dim(span\{(1, 1, 1), (0, 1, 2)\}) = 2$, $nullityT = dimP_1 - rankT = 2 - 2 = 0$.

The range has $\{(1, 1, 1), (0, 1, 2)\}$ as a basis. The kernel consists of the zero vector only.

2. $T(p) = \begin{bmatrix} p(0) & p(0)+p(1) \\ p(1) & p(0)-p(1) \end{bmatrix} : P_n \to M(2, 2)$

Answer: From

$$T(p) = p(0)\begin{bmatrix} 1 & 1 \\ 0 & 1 \end{bmatrix} + p(1)\begin{bmatrix} 0 & 1 \\ 1 & -1 \end{bmatrix} \in span\left\{\begin{bmatrix} 1 & 1 \\ 0 & 1 \end{bmatrix}, \begin{bmatrix} 0 & 1 \\ 1 & -1 \end{bmatrix}\right\}$$

the range of T is contained in the span of the two matrices, and

$$rankT = dim(rangeT) \le dim\ span\left\{\begin{bmatrix} 1 & 1 \\ 0 & 1 \end{bmatrix}, \begin{bmatrix} 0 & 1 \\ 1 & -1 \end{bmatrix}\right\} = 2.$$

On other other hand, for $n \ge 1$, we have $rangeT = span\{T(1), T(t), T(t^2), ..., T(t^n)\}$ I $span\{T(1), T(t)\}$, and

$$rankT = dim(rangeT) \le dim\ span\{T(1), T(t)\} = dim\ span\left\{\begin{bmatrix} 1 & 2 \\ 1 & 2 \end{bmatrix}, \begin{bmatrix} 0 & 1 \\ 1 & -1 \end{bmatrix}\right\} = 2.$$

Thus we conclude

$rankT = 2$, $nullityT = dimP_n - rankT = n - 1$, for $n \ge 1$.

The kernel consists of polynomials satisfying $p(0) = p(1) = 0$. Similar to the discussion in the previous problem, $\{t(t-1), t^2(t-1), t^3(t-1),..., t^{n-1}(t-1)\}$ is a basis of the kernel.

In the remaining case $n = 0$, we have

$$rankT = dim(span\{T(1)\}) = dim(span\ \{\begin{bmatrix} 1 & 2 \\ 1 & 2 \end{bmatrix}\}) = 1,\ nullityT = dimP_0 - rankT = 0.$$

The one matrix forms a basis of the range. The kernel consists of the zero vector only.

3. $T(p) = (t+1)p' - p: P_n \to P_n$

Answer The range of T is spanned by

$$T(1) = -1$$

$$T(t) = 1$$

$$T(t^2) = 2t + t^2$$

$$T(t^3) = 3t^2 + 2t^3$$

...

$$T(t^n) = (t+1)nt^{n-1} - t^n = nt^{n-1} + (n-1)t^n$$

By the translation$[a_0 + a_1 t + a_2 t^2 + ... + a_n t^n] = (a_0, a_1, a_2,..., a_n)$, the range becomes the column space of the matrix

$$A = \begin{bmatrix} -1 & 1 & 0 & 0 & 0 & .. & 0 & 0 \\ 0 & 0 & 2 & 0 & 0 & .. & 0 & 0 \\ 0 & 0 & 1 & 3 & 0 & .. & 0 & 0 \\ 0 & 0 & 0 & 2 & 4 & .. & 0 & 0 \\ \vdots & \vdots & \vdots & \vdots & \vdots & & \vdots & \vdots \\ 0 & 0 & 0 & 0 & 0 & .. & n-2 & n \\ 0 & 0 & 0 & 0 & 0 & .. & 0 & n-1 \end{bmatrix}$$

which is an upper triangular matrix with $-1, 0, 1, 2,..., n-2; n-1$ as diagonal entries. All columns except the second are pivot. Thus by deleting [col 2], $\{T(1), T(t^2), T(t^3),..., T(t^n)\} = \{1, 2t + t^2, 3t^2 + 2t^3,..., nt^{n-1} + (n-1)t^n\}$ is a basis of the range, and

$$rankT = rankT = n,\ nullityT = nullityA = 1.$$

The kernel, being 1-dimensional, has $\{1 + t\}$ as a basis.

Strictly speaking, however, the argument works only if the second column exists, which means the above holds only for $n \geq 1$. For $n = 0$, we have $nullityT = 0$ and $rankT = 1$.

4. $T(p) = (t^2 - 1)p'' + (t+1)p' - 4p: P_n \to P_n$

Answer: The range of T is spanned by

$$T(1) = -4$$
$$T(t) = 1 - 3t$$
$$T(t^2) = -2 + 2t$$
$$T(t^3) = -6t - 3t^2 + 5t^3$$
$$T(t^4) = -12t^2 - 4t^3 + 12t^4$$
$$\ldots$$
$$T(t^n) = (t^2 - 1)n(n-1)t^{n-2} + (t+1)nt^{n-1} - 4t^n = -n(n-1)t^{n-2} - nt^{n-1} + (n^2 - 4)t^n$$

Similar to the previous problem, the range is translated into the span of the column space of an upper triangular matrix

$$A = \begin{bmatrix} -4 & 1 & -2 & 0 & 0 & .. & 0 & 0 \\ 0 & -3 & 2 & 6 & 0 & .. & 0 & 0 \\ 0 & 0 & 0 & -3 & -12 & .. & 0 & 0 \\ 0 & 0 & 0 & 5 & -4 & .. & 0 & 0 \\ 0 & 0 & 0 & 0 & 12 & .. & 0 & 0 \\ \vdots & \vdots & \vdots & \vdots & \vdots & & \vdots & \vdots \\ 0 & 0 & 0 & 0 & 0 & .. & (n-1)^2 - 4 & n \\ 0 & 0 & 0 & 0 & 0 & .. & 0 & n^2 - 4 \end{bmatrix}$$

with $-4, -3, 0, 5, 12, \ldots, (n-1)^2 - 4, n^2 - 4$ as diagonal entries. Then it is not difficult to see that for n e" 2, $\{-4, 1 - 3t, -6t - 3t^2 + 5t^3, -12t^2 - 4t^3 + 12t^4, \ldots, -n(n-1)t^{n-2} - nt^{n-1} + (n^2 - 4)t^n\}$ is a basis of the range, and

$$rankT = rankA = n, \; nullityT = nullityA = 1.$$

The nonzero polynomial $-1 + 2t + 3t^2$ is in the kernel and (because the kernel has dimension 1) form a basis of the kernel.

For $n = 0$, we have $rankT = 1$ and $nullityT = 0$. For $n = 1$, we have $rankT = 2$ and $nullityT = 0$.

Exercise: Prove the complementarity principal: $Ax = b$ has solutions for any $b \Leftrightarrow$ Solution of $A^{T}y = c$ is unique.

Answer: Let A be an m by n matrix. Then A^T is an n by m matrix, and

$Ax = b$ has solutions for any b

$\Leftrightarrow rankA = m$

$\Leftrightarrow rankA^{T} = m$

$\Leftrightarrow$ *nullity*$A^{\mathrm{T}} = m - m = 0$

$\Leftrightarrow$ Solution of $A^{\mathrm{T}}y = c$ is unique.

Exercise: How are the ranks of *T*, *S*, and *T* + *S* are related?

Answer: By an earlier exercise, we have

range$(T + S)$ M *rangeT* + *rangeS*.

Thus

rank$(T + S)$

= *dim*(*range*$(T + S)$) (definition of rank)

$\leq$ *dim*(*rangeT* + *rangeS*) (this exercise, and subspace has smaller dimension)

$\leq$ *dim*(*rangeT*) + *dim*(*rangeS*) (this exercise)

= *rankT* + *rankS*. (definition of rank)

Remark: The second $\leq$ becomes equality if and only if *rangeT* $\cap$ *rangeS* = $\{0\}$. Thus *rank*$(T + S)$ = *rankT* + *rankS* $\Rightarrow$ *rangeT* $\cap$ *rangeS* = $\{0\}$.

Exercise: How are the ranks of *T*, *S*, and *TS* are related?

Answer: By an earlier exercise, we have

rangeTS M *rangeT*.

Thus *rankTS* d" *rankT*. In fact, by taking *T* and *S* to be matrices *A* and *B*, we also have

$$\begin{aligned} rankAB &= rank(AB)^{\mathrm{T}} \\ &= rank(B^{\mathrm{T}}A^{\mathrm{T}}) \\ &\leq rankB^{\mathrm{T}} \\ &= rankB. \end{aligned}$$

Thus we also have *rankTS* $\leq$ *rankS*. In summary,

rankTS $\leq$ min{*rankT*, *rankS*}.

Chapter 12

Change of Basis

MATRIX FOR CHANGE OF COORDINATES

Let $B = \{b_1, b_2, b_3\}$ and $C = \{c_1, c_2, c_3\}$ be two bases of a vector space V. Then the vectors in B may be written as linear combinations of vectors in C.

$$b_1 = a_{11}c_1 + a_{12}c_2 + a_{13}c_3$$

$$b_2 = a_{21}c_1 + a_{22}c_2 + a_{23}c_3$$

$$b_3 = a_{31}c_1 + a_{32}c_2 + a_{33}c_3$$

The matrix

$$[B]_C = [\,[b_1]_C\,[b_2]_C\,[b_3]_C\,] = \begin{bmatrix} a_{11} & a_{21} & a_{31} \\ a_{21} & a_{22} & a_{32} \\ a_{13} & a_{23} & a_{33} \end{bmatrix}$$

with a very suggestive notation, is the "coordinates of B with respect to C". The next result suggests that $[B]_C$ is the matrix for changing the coordinates in B to the coordinates in C.

The coordinates of $v \in V$ in the bases B and C are related by

$$[v]_C = [B]_C[v]_B.$$

Proof. Without loss of generality, we may assume $V = \mathrm{R}^3$. We use the bases vectors as columns and construct invertible matrices $[B] = [b_1\, b_2\, b_3]$, $[C] = [c_1\, c_2\, c_3]$. Then

$$[B] = [a_{11}c_1 + a_{12}c_2 + a_{13}c_3\;\; a_{21}c_1 + a_{22}c_2 + a_{23}c_3\;\; a_{31}c_1 + a_{32}c_2 + a_{33}c_3]$$

$$= [c_1\, c_2\, c_3] \begin{bmatrix} a_{11} & a_{21} & a_{31} \\ a_{12} & a_{22} & a_{32} \\ a_{13} & a_{23} & a_{33} \end{bmatrix} = [C][B]_C.$$

Let the coordinates of $v \in V$ to be $[v]_B = (x_1, x_2, x_3)$ and $[v]_C = (y_1, y_2, y_3)$. Then $v = x_1b_1 + x_2b_2 + x_3b_3 = [B][v]_B = [C][B]_C[v]_B$.

Compared with

$v = y_1c_1 + y_2c_2 + y_3c_3 = [C][v]_C,$

we have $[B]_C[v]_B = [v]_C$.

We list some properties of the matrix for changing the coordinates.

transitivity

$$[C]_D\,[B]_C = [B]_D$$

inverse

$$[B]_C^{-1} = [C]_B$$

identity

$$[B]_B = I$$

The transitivity property means that changing the B-coordinates to the C-coordinates and then further changing to the D-coordinates is the same as changing the B-coordinates to the D-coordinates. The identity property means that the change from the B-coordinates to the B-coordinates is no change at all. The inverse follows from the transitivity property (take $D = B$) and the identity property.

COMPUTING THE MATRIX FOR CHANGE OF COORDINATES

Let $B = \{b_1, b_2, b_3\}$ and $C = \{c_1, c_2, c_3\}$ be two bases of the euclidean space R^3 (or a subspace of R^3). Then they induce matrices $[B] = [b_1\ b_2\ b_3]$, $[C] = [c_1\ c_2\ c_3]$. As we have seen in the proof of the formula $[v]_C = [B]_C[v]_B$, the matrix $[B]_C$ for change of coordinates fits into

$$[B] = [C][B]_C.$$

In particular, if we take B to be the standard basis of R^3, so that $[B] = I$ and $[v]_B = v$, then the formula $[v]_C = [B]_C[v]_B$ becomes $[v]_C = [C]^{-1}v$. This is the earlier formula for the coordinates with respect to any basis of R^n.

The equality $[B] = [C][B]_C$ shows that the columns of the matrix $[B]_C$ are the solutions of the system $[C]x = b_i$, where b_i are the columns of $[B]$. The combined augmented matrix of these systems is the matrix $[\ [C]\ [B]\ ]$. This suggests that the computation of $[B]_C$ is reduced to row operations on $[\ [C]\ [B]\ ]$.

Example: Let

$$A = \begin{bmatrix} 1 & 3 & 2 & 0 & 1 \\ -1 & -1 & -1 & 1 & 0 \\ 0 & 4 & 2 & 4 & 3 \\ 1 & 3 & 2 & -2 & 0 \end{bmatrix}$$

In this example, we found a basis $B = \{(1, -1, 0, 1), (3, -1, 4, 3), (0, 1, 4, -2)\}$ of *colA*. In this example, we found another basis $C = \{(1, -1, 0, 1), (0, 1, 2, 0), (0, 0, 1, -1)\}$ of *colA*. To find the matrix $[B]_C$ for changing the *B*–coordinates to *C*–coordiantes, we use the fact that the *i*–th column of $[B]_C$ is the solution of $[C]x$ = *i*–th column of $[B]$. Thus we consider the combined augmented matrix.

$$[\,[C]\,[B]\,] = \begin{bmatrix} 1 & 0 & 0 & 1 & 3 & 0 \\ -1 & 1 & 0 & -1 & -1 & 1 \\ 0 & 2 & 1 & 0 & 4 & 4 \\ 1 & 0 & -1 & 1 & 3 & -2 \end{bmatrix}$$

Row operations change this into

$$\begin{bmatrix} 1 & 0 & 0 & 1 & 3 & 0 \\ 0 & 1 & 0 & 0 & 2 & 1 \\ 0 & 0 & 1 & 0 & 0 & 2 \\ 0 & 0 & 0 & 0 & 0 & 0 \end{bmatrix}$$

Then it is easy to see that the columns in heavily shaded submatrix are the solutions of $[C]x = b_i$, and

$$[B]_C = \begin{bmatrix} 1 & 3 & 0 \\ 0 & 2 & 1 \\ 0 & 0 & 2 \end{bmatrix}$$

Now by an earlier computation, we have

$[(0, 1, 2, 0)]_B = (-3/2, 1/2, 0)$.

Then

$$[(0, 1, 2, 0)]_C = \begin{bmatrix} 1 & 3 & 0 \\ 0 & 2 & 1 \\ 0 & 0 & 2 \end{bmatrix} \begin{bmatrix} -3/2 \\ 1/2 \\ 0 \end{bmatrix} = \begin{bmatrix} 0 \\ 1 \\ 0 \end{bmatrix}.$$

This is consistent with the fact that the second vector in *C* is (0, 1, 2, 0).

Example: The following are two bases for the vector space P_3.

Chebyshev $= \{1, t, 2t^2, 4t^3 - 3t\}$, *Hermite* $= \{1, t, t^2 - 1, t^3 - 3t\}$

To find the matrix for changing the Chebyshev coordinates to Hermite coordinates, we use the translation $[a + bt + ct^2 + dt^3] = (a, b, c, d)$ and the method of the previous example. Thus we form the matrix

$$[\,[Hermite]\ [Chebychev]\,] = \begin{bmatrix} 1 & 0 & -1 & 0 & 1 & 0 & -1 & 0 \\ 0 & 1 & 0 & -3 & 0 & 1 & 0 & -3 \\ 0 & 0 & 1 & 0 & 0 & 0 & 2 & 0 \\ 0 & 0 & 0 & 1 & 0 & 0 & 0 & 4 \end{bmatrix}$$

Row operations give us

$$\begin{bmatrix} 1 & 0 & 0 & 0 & 1 & 0 & 2 & 0 \\ 0 & 1 & 0 & 0 & 0 & 1 & 0 & 9 \\ 0 & 0 & 1 & 0 & 0 & 0 & 2 & 0 \\ 0 & 0 & 0 & 1 & 0 & 0 & 0 & 4 \end{bmatrix}$$

Thus we conclude that the Hermite coordinates may be obtained by multiplying

$$[Chebychev]_{Hermite} = \begin{bmatrix} 1 & 0 & 2 & 0 \\ 0 & 1 & 0 & 9 \\ 0 & 0 & 2 & 0 \\ 0 & 0 & 0 & 4 \end{bmatrix}$$

to (the left of) the Chebychev coordinates.

MATRIX FOR LINEAR TRANSFORMATION

Let $T: V \to W$ be a linear transformation. Let $B = \{b_1, b_2, \ldots, b_n\}$ and $C = \{c_1, c_2, \ldots, c_m\}$ be bases of V and W. Then by isomorphisms $[?]_B: V \to R^n$, $[?]_C: W \to R^m$, T is translated to a linear transformation $[?]_C\, T\, [?]_B^{-1}: R^n \to R^m$. The linear transformation $[?]_C\, T\, [?]_B^{-1}$ is given by a matrix A, which we call matrix of the linear transformation with respect to the bases B and C.

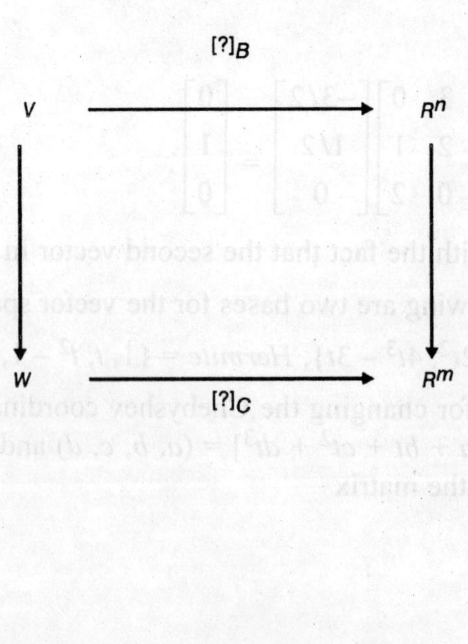

Since

i-th column of A

$= [?]_C T [?]_B^{-1}(e_i)$
$= [?]_C T(b_i)$
$= [T(b_i)]_C,$

we conclude the following formula for A.

The matrix of a linear transformation $T: V \to W$ with respect to bases $B = \{b_1, b_2, ..., b_n\}$ and $C = \{c_1, c_2,..., c_m\}$ of V and W is

$$[T(B)]_C = [\ [T(b_1)]_C\ [T(b_2)]_C ...\ [T(b_n)]_C\ ].$$

The notation $[T(B)]_C$ in introduced to suggest that the matrix is obtained by the following steps:

- Apply T to vectors in B.
- Take the coordinates with respect to C.

For an explicit example, in case $m = 3$, $n = 2$ and

$$T(b_1) = a_{11}c_1 + a_{12}c_2 + a_{13}c_3,$$
$$T(b_2) = a_{21}c_1 + a_{22}c_2 + a_{23}c_3,$$

$$\begin{bmatrix} a_{11} & a_{21} \\ a_{12} & a_{22} \\ a_{13} & a_{23} \end{bmatrix}$$

the matrix of T is

More concrete examples are given here, in which $V = W = P_2$ and $B = C = \{1, t, t^2\}$, and given in this exercise and this exercise, in which $V = W = P_n$ and $B = C = \{1, t, t^2,..., t^n\}$.

Example: Consider the linear transformation

$$T(p(t)) = \begin{bmatrix} p(0) & p(1) \\ p'(0) & p'(1) \end{bmatrix} : P_3 \to M(2, 2).$$

We use the bases

$$B = Hermite = \{1, t, t^2 - 1, t^3 - 3t\}, C = \left\{ \begin{bmatrix} 1 & 0 \\ 0 & 0 \end{bmatrix}, \begin{bmatrix} 0 & 1 \\ 0 & 0 \end{bmatrix}, \begin{bmatrix} 0 & 0 \\ 1 & 0 \end{bmatrix}, \begin{bmatrix} 0 & 0 \\ 0 & 1 \end{bmatrix} \right\}$$

for P_3 and $M(2, 2)$. Then

$$T(1) = \begin{bmatrix} 1 & 1 \\ 0 & 0 \end{bmatrix}, T(t) = \begin{bmatrix} 0 & 1 \\ 1 & 1 \end{bmatrix}, T(t^2 - 1) = \begin{bmatrix} -1 & 0 \\ 0 & 0 \end{bmatrix}, T(t^3 - 3t) = \begin{bmatrix} 0 & -2 \\ -3 & -3 \end{bmatrix}$$

tell us the matrix for T with respect to the bases is

$$[T(Hermite)]_C = [\ [T(1)]_C\ [T(t)]_C\ [T(2t^2)]_C\ [T(4t^3 - 3t)]_C\ ] = \begin{bmatrix} 1 & 0 & -1 & 0 \\ 1 & 1 & 0 & -2 \\ 0 & 1 & 0 & -3 \\ 0 & 1 & 0 & -3 \end{bmatrix}$$

We may conclude, for example, that $rankT = 3$ from the matrix.

The following lists some properties of the matrix for linear transformation.

linearity

$$[(T + S)(B)]_C = [T(B)]_C + [S(B)]_C,\ [cT(B)]_C = c[T(B)]_C$$

transitivity

$$[T(C)]_D\ [T(B)]_C = [T(B)]_D$$

inverse

$$[T(B)]_C^{-1} = [T^{-1}(C)]_B$$

identity

$$[id(B)]_B = I$$

CHANGE OF THE MATRIX FOR LINEAR TRANSFORMATION

The matrix $[T(B)]_C$ of a linear transformation $T: V \to W$ depends on the choices of bases B and C. If B' and C' are other choices, then everything fits into a diagram.

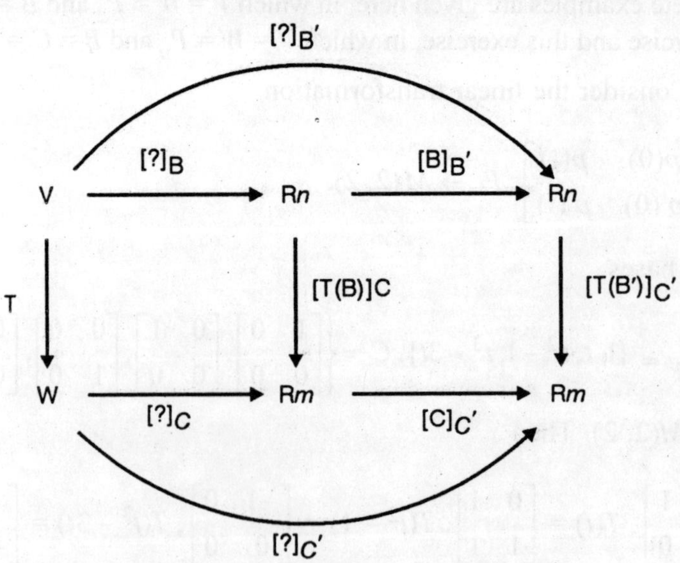

From the red arrows in the diagram, we conclude the following (the second equality follows from this formula).

For a linear transformation T, the matrices with regard to different bases are related by

$$[T(B')]_{C'} = [C]_{C'}\,[T(B)]_C\,[B]_{B'}^{-1} = [C]_{C'}\,[T(B)]_C\,[B']_B.$$

Example: For the linear transformation and bases in this example, we found the matrix for the transformation.

$$[T(Hermite)]_C = \begin{bmatrix} 1 & 0 & -1 & 0 \\ 1 & 1 & 0 & -2 \\ 0 & 1 & 0 & -3 \\ 0 & 1 & 0 & -3 \end{bmatrix}$$

Now we change the basis B to $Chebyshev = \{1, t, 2t^2, 4t^3 - 3t\}$ and keep the basis C as before. To find the new matrix $[T(Chebyshev)]_C$, we recall the matrix for the change of coordinates from another example.

$$[Chebychev]_{Hermite} = \begin{bmatrix} 1 & 0 & 2 & 0 \\ 0 & 1 & 0 & 9 \\ 0 & 0 & 2 & 0 \\ 0 & 0 & 0 & 4 \end{bmatrix}$$

Thus we haves

$$[T(Chebychev)]|_C = [T(Hermite)]_C\,[Chebychev]_{Hermite} = \begin{bmatrix} 1 & 0 & 0 & 0 \\ 1 & 1 & 2 & 1 \\ 0 & 1 & 0 & -3 \\ 0 & 1 & 0 & -3 \end{bmatrix}$$

We may compare the result with the direct application of T to *Chebychev*.

$$T(1) = \begin{bmatrix} 1 & 1 \\ 0 & 0 \end{bmatrix}, T(t) = \begin{bmatrix} 0 & 1 \\ 1 & 1 \end{bmatrix}, T(t^2) = \begin{bmatrix} 0 & 2 \\ 0 & 0 \end{bmatrix}, T(4t^3 - 3t) = \begin{bmatrix} 0 & 1 \\ -3 & -3 \end{bmatrix}$$

SIMILAR MATRIX

To build the matrix for a linear transformation $T: V \to V$ from a vector space to itself, we naturally want to choose one basis $B = \{b_1, b_2, \ldots, b_n\}$ for both copies of V. Then by this formula, the matrix for T with respect to B is $[T(B)]_B$.

Let B' be another basis of V. Then by this formula, the matrices of T with respect to B and B' are related by the equality

$[T(B')]_{B'} = [B]_{B'} [T(B)]_B [B]_{B'}^{-1}$.

The relation leads to the following definition.

Two square matrices A and B are said to be similar if there is an invertible P, such that $B = PAP^{-1}$.

Thus the matrix of a *self* linear transformation is changed by similarity when the basis is changed.

Example: Consider the linear transformation

$T(x, y) = (13x - 4y, -4x + 7y): R^2 \to R^2$

from R^2 to itself. Since x and y are the coordinates with respect to the standard basis $B = \{e_1, e_2\} = \{(1, 0), (0, 1)\}$, the matrix

$$A = [T(B)]_B = \begin{bmatrix} 13 & -4 \\ -4 & 7 \end{bmatrix}$$

that we usually associate to T is really the matrix of T with respect to the standard basis. If we choose a different basis $B' = \{u_1, u_2\} = \{(1, 2), (2, -1)\}$. Then

$$T(u_1) = (13 \times 1 - 4 \times 2, -4 \times 1 + 7 \times 2) = (5, 10) = 5u_1$$

$$T(u_2) = (13 \times 2 - 4 \times (-1), -4 \times 2 + 7 \times (-1)) = (30, -15) = 15u_2$$

tell us the matrix for T with respect to B' is

$$B = [T(B')]_{B'} = \begin{bmatrix} 5 & 0 \\ 0 & 15 \end{bmatrix}.$$

Denote

$$P = [B']_B = [u_1\ u_2] = \begin{bmatrix} 1 & 2 \\ 2 & -1 \end{bmatrix}.$$

It is easy to directly verify that $A = PBP^{-1}$.

We note that the original formula for T (with respect to $\{e_1, e_2\}$) is too complicated for us to have a good understanding of what T is doing on the plane R^2. However, if we look through the basis $\{u_1, u_2\}$ instead, then we have a very good understanding of the transformation. Basically T rescale the u_1 direction by 5 times and the u_2 direction by 15 times.

The example shows the importance of similarity. To gain a good undertanding of a self linear transformation, we find a basis (a suitable invertible matrix P) so that the the transformation behaves particularly nice with respect to the basis (the similarity change via P gives a diagonal matrix).

Example Consider the linear transformation

$$T(p) = (-1/3)(1 - t)^2p''' + (-1 + t)p'' - 2(1 + t)p' + 6p: P_3 \to P_3.$$

From

$$T(1) = 6$$
$$T(t) = -2 + 4t$$
$$T(t^2) = -2 - 2t + 2t^2$$
$$T(t^3) = -2 - 2t - 2t^2$$

the linear transformation has the following matrix with respect to the usual choice $\{1, t, t^2, t^3\}$ of basis.

$$A = \begin{bmatrix} 6 & -2 & -2 & -2 \\ 0 & 4 & -2 & -2 \\ 0 & 0 & 2 & -2 \\ 0 & 0 & 0 & 0 \end{bmatrix}$$

To see more clearly what the transformation does to polynomials, we consider a new basis $\{1, 1 + t, 1 + t + t^2, 1 + t + t^2 + t^3\}$. From

$$T(1) = 6 = 6(1)$$
$$T(1 + t) = 4(1 + t)$$
$$T(1 + t + t^2) = 2(1 + t + t^2)$$
$$T(t^3) = 0 = 0(1 + t + t^2 + t^3)$$

the matrix with respect to the new basis is

$$B = \begin{bmatrix} 6 & 0 & 0 & 0 \\ 0 & 4 & 0 & 0 \\ 0 & 0 & 2 & 0 \\ 0 & 0 & 0 & 0 \end{bmatrix}$$

Geometrically, this means that T is rescaling by factors 6, 4, 2, 0 in the new basis directions.

The matrix for changing the coordinates in the new basis to the coordinates in the usual basis is

$$Q = [\{1, 1 + t, 1 + t + t^2, 1 + t + t^2 + t^3\}]_{\{1, t, t^2, t^3\}} = \begin{bmatrix} 1 & 1 & 1 & 1 \\ 0 & 1 & 1 & 1 \\ 0 & 0 & 1 & 1 \\ 0 & 0 & 0 & 1 \end{bmatrix}$$

Direct verification shows $B = Q^{-1}AQ$ (by $[B]_{B'} = [B']_B^{-1}$, this Q is the inverse of P above).

Exercise: Find the matrix for the change of coordinates.

1. $B = \{(1, 2), (3, 4)\}$, $C = \{(3, 4), (1, 2)\}$

Answer: The equalities

$$(1, 2) = 0(3, 4) + 1(1, 2)$$

$$(3, 4) = 1(3, 4) + 0(1, 2)$$

tell us that the matrix for changing B–coordinates to C–coordinates is

$$[B]_C = \begin{bmatrix} 0 & 1 \\ 1 & 0 \end{bmatrix}$$

The matrix for changing C–coordinates to B–coordinates is

$$[C]_B = [B]_C^{-1} = \begin{bmatrix} 0 & 1 \\ 1 & 0 \end{bmatrix}$$

2. $B = \{(1, 2), (3, 4)\}$, $C = \{(1, 3), (2, 4)\}$, $D = \{(1, 4), (2, 3)\}$

Answer: Row operations on

$$[[C]\ [B]] = \begin{bmatrix} 1 & 2 & 1 & 3 \\ 3 & 4 & 2 & 4 \end{bmatrix}$$

give us

$$\begin{bmatrix} 1 & 0 & 0 & -2 \\ 0 & 1 & 1/2 & 5/2 \end{bmatrix}$$

This means

$$(1, 2) = 0(1, 3) + 1/2(2, 4)$$

$$(3, 4) = -2(1, 3) + 5/2(2, 4)$$

and

$$[B]_C = \begin{bmatrix} 0 & -2 \\ 1/2 & 5/2 \end{bmatrix}$$

Similarly, row operations on

$$[\ [D]\ [C]\ ] = \begin{bmatrix} 1 & 2 & 1 & 2 \\ 4 & 3 & 3 & 4 \end{bmatrix}$$

give us

$$\begin{bmatrix} 1 & 0 & 3/5 & 2/5 \\ 0 & 1 & 1/5 & 4/5 \end{bmatrix}$$

and

$$[C]_D = \begin{bmatrix} 3/5 & 2/5 \\ 1/5 & 4/5 \end{bmatrix}$$

We may also use row operations on [$[D]$ $[B]$] to compute $[B]_D$. Alternatively, we have

$$[B]_D = [C]_D[B]_C = \begin{bmatrix} 1/5 & -1/5 \\ 2/5 & 8/5 \end{bmatrix}.$$

The other matrices $[C]_B$, $[D]_C$, $[D]_B$ may be obtained by taking the inverses of $[B]_C$, $[C]_D$, $[B]_D$.

3. $B = \{(1, 0, 1), (1, -1, 0), (0, -1, 1)\}$, $C = \{(1, -1, 0), (0, -1, 1), (1, 0, 1)\}$

Answer: Note that C is obtained by rearranging the vectors in B. This easily leads to the following.

$$(1, 0, 1) = 0(1, -1, 0) + 0(0, -1, 1) + 1(1, 0, 1)$$
$$(1, -1\ 0) = 1(1, -1, 0) + 0(0, -1, 1) + 0(1, 0, 1)$$
$$(0, -1, 1) = 0(1, -1, 0) + 1(0, -1, 1) + 0(1, 0, 1)$$

Thus

$$[B]_C = \begin{bmatrix} 0 & 1 & 0 \\ 0 & 0 & 1 \\ 1 & 0 & 0 \end{bmatrix}, [C]_B = [B]_C^{-1} = \begin{bmatrix} 0 & 1 & 0 \\ 0 & 0 & 1 \\ 1 & 0 & 0 \end{bmatrix}.$$

4. $B = \{(1, 0, 0), (0, 1, 0), (0, 0, 1)\}$, $C = \{(2, 1, 0), (0, 1, 1), (1, 0, 1)\}$

Answer: Row operations on

$$[\ [C]\ [B]\ ] = \begin{bmatrix} 2 & 0 & 1 & 1 & 0 & 0 \\ 1 & 1 & 0 & 0 & 1 & 0 \\ 0 & 1 & 1 & 0 & 0 & 1 \end{bmatrix}$$

give us

$$\begin{bmatrix} 1 & 0 & 0 & 1/3 & 1/3 & -1/3 \\ 0 & 1 & 0 & -1/3 & 2/3 & 1/3 \\ 0 & 0 & 1 & 1/3 & -2/3 & 2/3 \end{bmatrix}$$

Thus

$$[B]_C = \begin{bmatrix} 1/3 & 1/3 & -1/3 \\ -1/3 & 2/3 & 1/3 \\ 1/3 & -2/3 & 2/3 \end{bmatrix}$$

On the other hand, since B is the standard basis, we have

$$[C]_B = [C] = \begin{bmatrix} 2 & 0 & 1 \\ 1 & 1 & 0 \\ 0 & 1 & 1 \end{bmatrix}$$

5. $B = \{1, t, t^2\}, C = \{2 + t, t + t^2, 1 + t^2\}, D = \{1, 1 + t, 1 + t + t^2\}$

Answer: Since B is the standard basis of P_2, we clearly have

$$[C]_B = \begin{bmatrix} 2 & 0 & 1 \\ 1 & 1 & 0 \\ 0 & 1 & 1 \end{bmatrix}, [D]_B = \begin{bmatrix} 1 & 1 & 1 \\ 0 & 1 & 1 \\ 0 & 0 & 1 \end{bmatrix}.$$

Then based on the computation in the previous problem, we have

$$[B]_C = [C]_B^{-1} = \begin{bmatrix} 1/3 & 1/3 & -1/3 \\ -1/3 & 2/3 & 1/3 \\ 1/3 & -2/3 & 2/3 \end{bmatrix}.$$

On the other hand, the equalities

$$1 = 1(1) + 0(1 + t) + 0(1 + t + t^2)$$
$$t = -1(1) + 1(1 + t) + 0(1 + t + t^2)$$
$$t^2 = 0(1) - 1(1 + t) + 1(1 + t + t^2)$$

tell us (of course we may also use $[B]_D = [D]_B^{-1}$)

$$[B]_D = \begin{bmatrix} 1 & -1 & 0 \\ 0 & 1 & -1 \\ 0 & 0 & 1 \end{bmatrix}.$$

Finally, we have

$$[C]_D = [B]_D[C]_B = \begin{bmatrix} 1 & -1 & 1 \\ 0 & 1 & -1 \\ 0 & 0 & 1 \end{bmatrix}, [D]_C = [C]_D^{-1} = [B]_C[D]_B = \begin{bmatrix} 1/3 & 2/3 & 1/3 \\ -1/3 & 1/3 & 2/3 \\ 1/3 & -1/3 & 1/3 \end{bmatrix}$$

6. $B = \left\{ \begin{bmatrix} 1 & 0 \\ 0 & 0 \end{bmatrix}, \begin{bmatrix} 1 & 2 \\ 0 & 0 \end{bmatrix}, \begin{bmatrix} 1 & 2 \\ 3 & 0 \end{bmatrix}, \begin{bmatrix} 1 & 2 \\ 3 & 4 \end{bmatrix} \right\}, C = \left\{ \begin{bmatrix} 0 & 0 \\ 0 & 1 \end{bmatrix}, \begin{bmatrix} 0 & 1 \\ 0 & 1 \end{bmatrix}, \begin{bmatrix} 0 & 1 \\ 1 & 1 \end{bmatrix}, \begin{bmatrix} 1 & 1 \\ 1 & 1 \end{bmatrix} \right\}$

Answer: We use

$$\begin{bmatrix} a & b \\ c & d \end{bmatrix} \to (a, b, c, d)$$

to translate the problem to the euclidean space. Row operations on

$$[\,[C]\,[B]\,] = \begin{bmatrix} 0 & 0 & 0 & 1 & 1 & 1 & 1 & 1 \\ 0 & 1 & 1 & 1 & 0 & 2 & 2 & 2 \\ 0 & 0 & 1 & 1 & 0 & 0 & 3 & 3 \\ 1 & 1 & 1 & 1 & 0 & 0 & 0 & 4 \end{bmatrix}$$

give us

$$\begin{bmatrix} 1 & 0 & 0 & 0 & 0 & -2 & -2 & 2 \\ 0 & 1 & 0 & 0 & 0 & 2 & -1 & -1 \\ 0 & 0 & 1 & 0 & -1 & -1 & 2 & 2 \\ 0 & 0 & 0 & 1 & 1 & 1 & 1 & 1 \end{bmatrix}$$

Therefore

$$[B]_C = \begin{bmatrix} 0 & -2 & -2 & 2 \\ 0 & 2 & -1 & -1 \\ -1 & -1 & 2 & 2 \\ 1 & 1 & 1 & 1 \end{bmatrix}.$$

Similarly, row operations on

$$[\,[B]\,[C]\,] = \begin{bmatrix} 1 & 1 & 1 & 1 & 0 & 0 & 0 & 1 \\ 0 & 2 & 2 & 2 & 0 & 1 & 1 & 1 \\ 0 & 0 & 3 & 3 & 0 & 0 & 1 & 1 \\ 0 & 0 & 0 & 4 & 1 & 1 & 1 & 1 \end{bmatrix}$$

give us

$$\begin{bmatrix} 1 & 0 & 0 & 0 & 0 & -1/2 & -1/2 & 1/2 \\ 0 & 1 & 0 & 0 & 0 & 1/2 & 1/6 & 1/6 \\ 0 & 0 & 1 & 0 & -1/4 & -1/4 & 1/12 & 1/12 \\ 0 & 0 & 0 & 1 & 1/4 & 1/4 & 1/4 & 1/4 \end{bmatrix}$$

and

$$[C]_B = \begin{bmatrix} 0 & -1/2 & -1/2 & 1/2 \\ 0 & 1/2 & 1/6 & 1/6 \\ -1/4 & -1/4 & 1/12 & 1/12 \\ 1/4 & 1/4 & 1/4 & 1/4 \end{bmatrix}$$

Exercise: **Find different bases for the following subspaces and the matrices for changing the coordinates between them.**

1. H = null space of $\begin{bmatrix} 1 & 2 & 3 \\ 2 & 3 & 4 \end{bmatrix}$

Answer: In an earlier exercise, we use

$$\text{row operation: } \begin{bmatrix} 1 & 2 & 3 \\ 2 & 3 & 4 \end{bmatrix} \to \begin{bmatrix} 1 & 0 & -1 \\ 0 & 1 & 2 \end{bmatrix}$$

to express the solution of $Ax = 0$ as $x = x_3(1, -2, 1)$ and get a basis $B = \{(1, -2, 1)\}$ of the null space H. If we use x_2 as the free variable, we may also express the solution of $Ax = 0$ as $x = x_2(-1/2, 1, -1/2)$ and get another basis $C = \{(-1/2, 1, -1/2)\}$ of the null space H. Then

$$(1, -2, 1) = -2(-1/2, 1, -1/2) \Rightarrow [B]_C = [-2] \text{ (1 by 1 matrix)}$$
$$(-1/2, 1, -1/2) = -1/2(1, -2, 1) \Rightarrow [C]_B = [-1/2]$$

2. H = column space of $\begin{bmatrix} 1 & 2 & 3 \\ 2 & 3 & 4 \end{bmatrix}$

Answer: In the earlier exercise, the same row operation as in the previous problem also gives us a basis $B = \{(1, 2), (2, 3)\}$ of the column space H. The row operations also tells us that the 1st and the 3rd vectors form another basis $C = \{(1, 2), (3, 4)\}$ of H. Then

$$\begin{matrix} (1, 2) = 1(1, 2) + 0(3, 4) \\ (2, 3) = 1/2(1, 2) + 1/2(3, 4) \end{matrix} \Rightarrow [B]_c = \begin{bmatrix} 1 & 1/2 \\ 0 & 1/2 \end{bmatrix}$$

and

$$\begin{matrix} (1, 2) = 1(1, 2) + 0(2, 3) \\ (3, 4) = -1(1, 2) + 2(2,3) \end{matrix} \Rightarrow [C]_B = \begin{bmatrix} 1 & -1 \\ 0 & 2 \end{bmatrix}$$

On the other hand, we may also use

$$\text{column operation: } \begin{bmatrix} 1 & 2 & 3 \\ 2 & 3 & 4 \end{bmatrix} \to \begin{bmatrix} 1 & 0 & 0 \\ 2 & 1 & 0 \end{bmatrix}$$

and the second method to find a basis $D = \{(1, 2), (0, 1)\}$ of H. Then

$$\begin{matrix}(1,2) = 1(1,2) + 0(0,1) \\ (2,3) = 2(1,2) - 1(0,1)\end{matrix} \Rightarrow [B]_D = \begin{bmatrix} 1 & 2 \\ 0 & -1 \end{bmatrix},$$

and

$$\begin{matrix}(1,2) = 1(1,2) + 0(2,3) \\ (0,1) = 2(1,2) - 1(2,3)\end{matrix} \Rightarrow [D]_B = \begin{bmatrix} 1 & 2 \\ 0 & -1 \end{bmatrix}.$$

The change between C and D can be similarly computed.

3. H = null space of $\begin{bmatrix} 1 & -2 & 1 & 1 & 2 \\ -1 & 3 & 0 & 2 & -2 \\ 0 & 1 & 1 & 3 & 4 \\ 1 & 2 & 5 & 13 & 5 \end{bmatrix}$

Answer: In an earlier exercise, we use

$$\text{row operation: } \begin{bmatrix} 1 & -2 & 1 & 1 & 2 \\ -1 & 3 & 0 & 2 & -2 \\ 0 & 1 & 1 & 3 & 4 \\ 1 & 2 & 5 & 13 & 5 \end{bmatrix} \to \begin{bmatrix} 1 & 0 & 3 & 7 & 0 \\ 0 & 1 & 1 & 3 & 0 \\ 0 & 0 & 0 & 0 & 1 \\ 0 & 0 & 0 & 0 & 0 \end{bmatrix}$$

to express the solution of $Ax = 0$ as $x = x_3(-3, -1, 1, 0, 0) + x_4(-7, -3, 0, 1, 0)$ and get a basis $B = \{(-3, -1, 1, 0, 0), (-7, -3, 0, 1, 0)\}$ of the null space H. If we use x_2 and x_4 as the free variables, we may also express the solution as $x = x_2(3, 1, -1, 0, 0) + x_4(2, 0, -3, 1, 0)$ and get another basis $C = \{(3, 1, -1, 0, 0), (2, 0, -3, 1, 0)\}$ of H. Then

$$\text{row operation: } [\,[C]\ [B]\,] = \begin{bmatrix} 3 & 2 & -3 & -7 \\ 1 & 0 & -1 & -3 \\ -1 & -3 & 1 & 0 \\ 0 & 1 & 0 & 1 \\ 0 & 0 & 0 & 0 \end{bmatrix} \to \begin{bmatrix} 1 & 0 & -1 & -3 \\ 0 & 1 & 0 & 1 \\ 0 & 0 & 0 & 0 \\ 0 & 0 & 0 & 0 \\ 0 & 0 & 0 & 0 \end{bmatrix}$$

tells us

$$[B]_C = \begin{bmatrix} -1 & -3 \\ 0 & 1 \end{bmatrix}$$

The change $[C]_B$ from C–coordinates to B–coordinates can be computed similarly.

4. H = column space of $\begin{bmatrix} 1 & -2 & 1 & 1 & 2 \\ -1 & 3 & 0 & 2 & -2 \\ 0 & 1 & 1 & 3 & 4 \\ 1 & 2 & 5 & 13 & 5 \end{bmatrix}$

Answer: In the earlier exercise, the same row operation as in the previous problem also gives us a basis $B = \{(1, -1, 0, 1), (-2, 3, 1, 2), (2, -2, 4, 5)\}$ of the column space H. The answer to the earlier exercise also contains (presented as row operation on A^T)

$$\text{column operation: } \begin{bmatrix} 1 & -2 & 1 & 1 & 2 \\ -1 & 3 & 0 & 2 & -2 \\ 0 & 1 & 1 & 3 & 4 \\ 1 & 2 & 5 & 13 & 5 \end{bmatrix} \to \begin{bmatrix} 1 & 0 & 0 & 0 & 0 \\ -1 & 1 & 0 & 0 & 0 \\ 0 & 1 & 4 & 0 & 0 \\ 1 & 4 & 3 & 0 & 0 \end{bmatrix}$$

which by the second method gives us a basis $C = \{(1, -1, 0, 1), (0, 1, 1, 4), (0, 0, 4, 3)\}$ of H. Then

$$\text{row operation: } [[C]\ [B]] = \left[\begin{array}{ccc|ccc} 1 & 0 & 0 & 1 & -2 & 2 \\ -1 & 1 & 0 & -1 & 3 & -2 \\ 0 & 1 & 4 & 0 & 1 & 4 \\ 1 & 4 & 3 & 1 & 2 & 5 \end{array}\right] \to \left[\begin{array}{ccc|ccc} 1 & 0 & 0 & 1 & -2 & 2 \\ 0 & 1 & 0 & 0 & 1 & 0 \\ 0 & 0 & 1 & 0 & 0 & 1 \\ 0 & 0 & 0 & 0 & 0 & 0 \end{array}\right]$$

tells us

$$[B]_C = \begin{bmatrix} 1 & -2 & 2 \\ 0 & 1 & 0 \\ 0 & 0 & 1 \end{bmatrix}$$

5. H = span of $\begin{bmatrix} -1 & 1 \\ -2 & 1 \end{bmatrix}, \begin{bmatrix} 2 & -2 \\ 4 & -2 \end{bmatrix}, \begin{bmatrix} 2 & -1 \\ 3 & 1 \end{bmatrix}, \begin{bmatrix} 2 & -3 \\ 5 & -4 \end{bmatrix}, \begin{bmatrix} -5 & 4 \\ -9 & -1 \end{bmatrix}$

Answer: In an earlier exercise, we converted the problem to the column space of the matrix

$$\begin{bmatrix} -1 & 2 & 2 & 2 & -5 \\ 1 & -2 & -1 & -3 & 4 \\ -2 & 4 & 3 & 5 & -9 \\ 1 & -2 & 1 & -4 & -1 \end{bmatrix}$$

and used a column echelon form to find a basis

$$B = \left\{ \begin{bmatrix} -1 & 1 \\ -2 & 1 \end{bmatrix}, \begin{bmatrix} 0 & 1 \\ -1 & 3 \end{bmatrix}, \begin{bmatrix} 0 & 0 \\ 0 & 1 \end{bmatrix} \right\}$$

of H. We may also use

$$\text{row operation: } \begin{bmatrix} -1 & 2 & 2 & 2 & -5 \\ 1 & -2 & -1 & -3 & 4 \\ -2 & 4 & 3 & 5 & -9 \\ 1 & -2 & 1 & -4 & -1 \end{bmatrix} \to \begin{bmatrix} -1 & 2 & 2 & 2 & -5 \\ 0 & 0 & 1 & -1 & -1 \\ 0 & 0 & 0 & 1 & -3 \\ 0 & 0 & 0 & 0 & 0 \end{bmatrix}$$

to conclude that the 1st, 3rd, 4th matrices form a basis

$$C = \left\{ \begin{bmatrix} -1 & 1 \\ -2 & 1 \end{bmatrix}, \begin{bmatrix} 2 & -1 \\ 3 & 1 \end{bmatrix}, \begin{bmatrix} 2 & -3 \\ 5 & -4 \end{bmatrix} \right\}$$

of H. To find the change of B–coordinates to C–coordinates, we use

$$\text{row operation: } [\,[C]\,[B]\,] = \begin{bmatrix} -1 & 2 & 2 & -1 & 0 & 0 \\ 1 & -1 & -3 & 1 & 1 & 0 \\ -2 & 3 & 5 & -2 & -1 & 0 \\ 1 & 1 & -4 & 1 & 3 & 1 \end{bmatrix} \to \begin{bmatrix} 1 & 0 & 0 & 1 & 2 & 4 \\ 0 & 1 & 0 & 0 & 1 & 1 \\ 0 & 0 & 1 & 0 & 0 & 1 \\ 0 & 0 & 0 & 0 & 0 & 0 \end{bmatrix}$$

to get

$$[B]_C = \begin{bmatrix} 1 & 2 & 4 \\ 0 & 1 & 1 \\ 0 & 0 & 1 \end{bmatrix}$$

6. $H = \text{kernel of } T(X) = \begin{bmatrix} 1 & 2 \\ -1 & -2 \end{bmatrix} X \begin{bmatrix} 2 & 4 \\ 3 & 6 \end{bmatrix} : M(2, 2) \to M(2, 2)$

Answer: In an earlier exercise, based on

$$T = \begin{bmatrix} x & y \\ z & w \end{bmatrix} (2x + 3y + 4z + 6w) \begin{bmatrix} 1 & 2 \\ -1 & -2 \end{bmatrix}$$

we found H consists of matrices of the form

$$\begin{bmatrix} -(3/2)y - 2z - 3w & y \\ z & w \end{bmatrix} = y \begin{bmatrix} -3/2 & 1 \\ 0 & 0 \end{bmatrix} + z \begin{bmatrix} -2 & 0 \\ 1 & 0 \end{bmatrix} + w \begin{bmatrix} -3 & 0 \\ 0 & 1 \end{bmatrix}$$

and the three 2 by 2 matrices form a basis

$$B = \left\{ \begin{bmatrix} -3/2 & 1 \\ 0 & 0 \end{bmatrix}, \begin{bmatrix} -2 & 0 \\ 1 & 0 \end{bmatrix}, \begin{bmatrix} -3 & 0 \\ 0 & 1 \end{bmatrix} \right\}$$

However, if we choose x, y, z as the free variables, then H consists of matrices of the form

$$\begin{bmatrix} x & y \\ z & -(1/3)x - (1/2)y - (2/3)z \end{bmatrix} = x \begin{bmatrix} 1 & 0 \\ 0 & -1/3 \end{bmatrix} + y \begin{bmatrix} 0 & 1 \\ 0 & -1/2 \end{bmatrix} + \begin{bmatrix} 0 & 0 \\ 1 & -2/3 \end{bmatrix}$$

and we get another basis

$$C = \left\{ \begin{bmatrix} 1 & 0 \\ 0 & -1/3 \end{bmatrix}, \begin{bmatrix} 0 & 1 \\ 0 & -1/2 \end{bmatrix}, \begin{bmatrix} 0 & 0 \\ 1 & -2/3 \end{bmatrix} \right\}$$

Moreover, we may use

row operation:

$$[[C]\,[B]\,] = \begin{bmatrix} 1 & 0 & 0 & -3/2 & -2 & -3 \\ 0 & 1 & 0 & 1 & 0 & 0 \\ 0 & 0 & 1 & 0 & 1 & 0 \\ -1/3 & -1/2 & -2/3 & 0 & 0 & 1 \end{bmatrix} \rightarrow \begin{bmatrix} 1 & 0 & 0 & -3/2 & -2 & -3 \\ 0 & 1 & 0 & 1 & 0 & 0 \\ 0 & 0 & 1 & 0 & 1 & 0 \\ 0 & 0 & 0 & 0 & 0 & 0 \end{bmatrix}$$

to get

$$[B]_C = \begin{bmatrix} -3/2 & -2 & -3 \\ 1 & 0 & 0 \\ 0 & 1 & 0 \end{bmatrix}$$

Exercise: Study the change of coordinates when the basis is changed by the three basic operations.

Answer: Let use work with a basis $B = \{v_1, v_2, v_3\}$ of three vectors.

The basis $B' = \{v_1, v_2 + rv_1, v_3\}$ is obtained by adding r multiple of the first vector to the second. The equality

$$u = c_1v_1 + c_2v_2 + c_3v_3 = (c_1 - rc_2)v_1 + c_2(v_2 + rv_1) + c_3v_3.$$

tells us

$$[u]_B = (c_1, c_2, c_3) \Rightarrow [u]_{B'} = (c_1 - rc_2, c_2, c_3).$$

In other words, we have

$$[u]_{B'} = \begin{bmatrix} 1 & -r & 0 \\ 0 & 1 & 0 \\ 0 & 0 & 1 \end{bmatrix} [u]_B.$$

The basis $B'' = \{v_2, v_1, v_3\}$ is obtained by exchanging the first and the second vectors. The equality

$$u = c_1v_1 + c_2v_2 + c_3v_3 = c_2v_2 + c_1v_1 + c_3v_3$$

tells us

$$[u]_B = (c_1, c_2, c_3) \Rightarrow [u]_{B''} = (c_2, c_1, c_3).$$

In other words, we have

$$[u]_{B''} = \begin{bmatrix} 0 & 1 & 0 \\ 1 & 0 & 0 \\ 0 & 0 & 1 \end{bmatrix} [u]_B.$$

The basis $B''' = \{dv_1, v_2, v_3\}$ is obtained by multiplying a number d "" 0 to the first vectors. The equality

$$u = c_1v_1 + c_2v_2 + c_3v_3 = (c_1/d)dv_1 + c_2v_2 + c_3v_3$$

tells us

$$[u]_B = (c_1, c_2, c_3) \Rightarrow [u]_{B'''} = (c_1/d, c_2, c_3).$$

In other words, we have

$$[u]_{B'''} = \begin{bmatrix} 1/d & 0 & 0 \\ 0 & 1 & 0 \\ 0 & 0 & 1 \end{bmatrix} [u]_B.$$

In general, we see that when the three operations are applied to a basis, the coordinates are modified by multiplying a matrix to the left of the coordinates. Specifically the matrices are the following.

Operation	Matrix differs from I by
r[vector i] + [vector j]	ij-entry = 1
[vector i] → [vector j]	ii-entry = jj–entries = 0, ij-entry = ji-entries = 1
d[vector i], $d \neq 0$	ii-entry = $1/d$

Exercise: Find the matrix for the following linear transformations with respect to the given bases (B denotes the bases for the source space, C denotes the bases for the target space).

1. $T(x_1, x_2, x_3) = (x_1 + 2x_2 + 3x_3, 2x_1 + 3x_2 + 4x_3)$: $R^3 \to R^2$
   $B_1 = \{(1, 0, 0), (0, 1, 0), (0, 0, 1)\}$, $C_1 = \{(1, 0), (0, 1)\}$, $B_2 = \{(1, 2, -1), (2, -1, 2), (3, 1, -1)$, $C_2 = \{(1, 2), (2, 3)\}$

Answer: By

$$[T(1, 0, 0)]_{C1} = [(1, 2)]_{\{(1, 0), (0, 1)\}} = (1, 2)$$
$$[T(0, 1, 0)]_{C1} = [(2, 3)]_{\{(1, 0), (0, 1)\}} = (2, 3)$$
$$[T(0, 0, 1)]_{C1} = [(3, 4)]_{\{(1, 0), (0, 1)\}} = (3, 4)$$

the matrix of T with respect to B_1 and C_1 is

$$A = [T(B_1)]_{C1} = \begin{bmatrix} 1 & 2 & 3 \\ 2 & 3 & 4 \end{bmatrix}.$$

By

$$[T(1, 2, -1)]_{C2} = [(2, 4)]_{\{(1, 2), (2, 3)\}} = (2, 0)$$
$$[T(2, -1, 2)]_{C2} = [(6, 9)]_{\{(1, 2), (2, 3)\}} = (0, 3)$$
$$[T(3, 1, -1)]_{C2} = [(2, 5)]_{\{(1, 2), (2, 3)\}} = (4, -1)$$

The matrix of T with respect to B_2 and C_2 is

$$[T(B_2)]_{C2} = \begin{bmatrix} 2 & 0 & 4 \\ 0 & 3 & -1 \end{bmatrix}$$

Alternatively, by

$$Q = [B_2]_{B1} = \begin{bmatrix} 1 & 2 & 3 \\ 2 & -1 & 1 \\ -1 & 2 & -1 \end{bmatrix}$$

and

$$[C_2]_{C1} = \begin{bmatrix} 1 & 2 \\ 2 & 3 \end{bmatrix}, P = [C_1]_{C2} = [C_2]_{C1}^{-1} = \begin{bmatrix} -3 & 2 \\ 2 & -1 \end{bmatrix}$$

we have

$$A' = [T(B_2)]_{C2} = PAQ = [C_1]_{C2} [T(B_1)]_{C1} [B_2]_{B1} = \begin{bmatrix} 2 & 0 & 4 \\ 0 & 3 & -1 \end{bmatrix}.$$

Remark: **For the matrix A of a linear transformation T: $V \to W$, changing basis of V means multiplying a matrix Q to the right, and changing basis of W means multiplying a matrix P to the left.**

2. $T(a + bt + ct^2) = (-3a + b + 2c, a - c, -2a + b + c): P_2 \to R^3$

$B_1 = \{1, t, t^2\}, B_2 = \{1 + 2t + t^2, 1 + 3t, 1 + 2t^2\}, C = \{(1, 0, 0), (0, 1, 0), (0, 0, 1)\}$

Answer: By

$$[T(1)]_C = [(-3, 1, -2)]_{\{(1, 0, 0), (0, 1, 0), (0, 0, 1)\}} = (-3, 1, -2)$$

$$[T(t)]_C = [(1, 0, 1)]_{\{(1, 0, 0), (0, 1, 0), (0, 0, 1)\}} = (1, 0, 1)$$

$$[T(t^2)]_C = [(2, -1, 1)]_{\{(1, 0, 0), (0, 1, 0), (0, 0, 1)\}} = (2, -1, 1)$$

the matrix of T with respect to B_1 and C is

$$[T(B_1)]_C = \begin{bmatrix} -3 & 1 & 2 \\ 1 & 0 & -1 \\ -2 & 1 & 1 \end{bmatrix}.$$

By

$$[T(1 + 2t + t^2)]_C = [(-3 + 2 + 2, 1 - 1, -2 + 2 + 1)]_{\{(1, 0, 0), (0, 1, 0), (0, 0, 1)\}} = (1, 0, 1)$$

$$[T(1 + 3t)]_C = [((-3 + 3, 1, -2 + 3)]_{\{(1, 0, 0), (0, 1, 0), (0, 0, 1)\}} = (0, 1, 1)$$

$$[T(1 + 2t^2)]_C = [(-3 + 4, 1 - 2, -2 + 2)]_{\{(1, 0, 0), (0, 1, 0), (0, 0, 1)\}} = (1, -1, 0)$$

The matrix of T with respect to B_2 and C is

$$[T(B_2)]_C = \begin{bmatrix} 1 & 0 & 1 \\ 0 & 1 & -1 \\ 1 & 1 & 0 \end{bmatrix}$$

Alternatively,

$$[T(B_2)]_C = [T(B_1)]_C\,[B_2]_{B1} = \begin{bmatrix} -3 & 1 & 2 \\ 1 & 0 & -1 \\ -2 & 1 & 1 \end{bmatrix}\begin{bmatrix} 1 & 1 & 1 \\ 2 & 3 & 0 \\ 1 & 0 & 2 \end{bmatrix} = \begin{bmatrix} 1 & 0 & 1 \\ 0 & 1 & -1 \\ -2 & 1 & 1 \end{bmatrix}$$

Remark: Let $B = \{b_1, b_2,..., b_n\}$ be a basis of V, and let $C = \{e_1, e_2,..., e_m\}$ be a basis of R^m. Then the matrix of a linear transformation $T: V \to R^m$ with respect to B and C is

$$[T(B)]_C = [T(b_1)\; T(b_2) ...\; T(b_n)].$$

3. T: $R^3 \to R^3$, the orthogonal projection onto the plane $x + y + z = 0$

$B_1 = C_1 = \{(-1, 1, 0), (-1, 0, 1), (1, 1, 1)\}$, $B_2 = C_2 = \{(1, 0, 0), (0, 1, 0), (0, 0, 1)\}$

Answer: Since $(-1, 1, 0)$ and $(-1, 0, 1)$ are on the plane, and $(1, 1, 1)$ is orthogonal to the plane, we have,

$$[T(-1, 1, 0)]_{C1} = [(-1, 1, 0)]_{\{(-1, 1, 0), (-1, 0, 1), (1, 1, 1)\}} = (1, 0, 0)$$
$$[T(-1, 0, 1)]_{C1} = [(-1, 0, 1)]_{\{(-1, 1, 0), (-1, 0, 1), (1, 1, 1)\}} = (0, 1, 0)$$
$$[T(1, 1, 1)]_{C1} = [(0, 0, 0)]_{\{(-1, 1, 0), (-1, 0, 1), (1, 1, 1)\}} = (0, 0, 0)$$

and the matrix of T with respect to B_1 and C_1 is

$$A = [T(B_1)]_{C1} = \begin{bmatrix} 1 & 0 & 0 \\ 0 & 1 & 0 \\ 0 & 0 & 0 \end{bmatrix}$$

Using the matrix

$$P = [B_1]_{B2} = [C_1]_{C2} = \begin{bmatrix} -1 & -1 & 1 \\ 1 & 0 & 1 \\ 0 & 1 & 1 \end{bmatrix}$$

for changing coordinates from $B_1 = C_1$ to $B_2 = C_2$, we have

$$[T(B_2)]_{C2} = PAP^{-1} = \begin{bmatrix} -1 & -1 & 1 \\ 1 & 0 & 1 \\ 0 & 1 & 1 \end{bmatrix}\begin{bmatrix} 1 & 0 & 0 \\ 0 & 1 & 0 \\ 0 & 0 & 0 \end{bmatrix}\begin{bmatrix} -1/3 & 2/3 & -1/3 \\ -1/3 & -1/3 & 2/3 \\ 1/3 & 1/3 & 1/3 \end{bmatrix}$$

$$= \begin{bmatrix} 2/3 & -1/3 & -1/3 \\ -1/3 & 2/3 & -1/3 \\ -1/3 & -1/3 & 2/3 \end{bmatrix}.$$

You may compare this with another approach.

4. $T(p(t)) = (1 + t)p'(t) - p(t): P_3 \to P_3$

$B_1 = C_1 = \{1, 1 + t, (1 + t)^2, (1 + t)^3\}, B_2 = C_2 = \{1, t, t^2, t^3\}$

Answer: By

$$[T(1)]_{C1} = [-1]_{\{(1,\, 1+t,\, (1+t)^2,\, (1+t)^3\}} = (-1, 0, 0, 0)$$
$$[T(1 + t)]_{C1} = [0]_{\{1,\, 1+t,\, (1+t)^2,\, (1+t)^3\}} = (0, 0, 0, 0)$$
$$[T((1 + t)^2)]_{C1} = [(1 + t)^2]_{\{1,\, 1+t,\, (1+t)^2,\, (1+t)^3\}} = (0, 0, 1, 0)$$
$$[T((1 + t)^3)]_{C1} = [2(1 + t)^3]_{\{1,\, 1+t,\, (1+t)^2,\, (1+t)^3\}} = (0, 0, 0, 2)$$

the matrix of T with respect to B_1 and C_1 is

$$A = [T(B_1)]_{C1} = \begin{bmatrix} -1 & 0 & 0 & 0 \\ 0 & 0 & 0 & 0 \\ 0 & 0 & 1 & 0 \\ 0 & 0 & 0 & 2 \end{bmatrix}$$

By

$$P = [B_1]_{B2} = [C_1]_{C2} = [1, 1 + t, 1 + 2t + t^2, 1 + 3t + 3t^2 + t^3]_{\{1, t, t^2, t^3\}}$$
$$= \begin{bmatrix} 1 & 1 & 1 & 1 \\ 0 & 1 & 2 & 3 \\ 0 & 0 & 1 & 3 \\ 0 & 0 & 0 & 1 \end{bmatrix}$$

we have

$$[T(B_2)]_{C2} = PAP^{-1} = \begin{bmatrix} 1 & 0 & 0 & 0 \\ 0 & 0 & 0 & 0 \\ 0 & 0 & 1 & 0 \\ 0 & 0 & 0 & 2 \end{bmatrix}$$

Alternatively, we have direct computation.

$$[T(B_2)]_{C2} = [-1, 1, 2t + t^2, 3t^2 + 2t^3]_{\{1, t, t^2, t^3\}} = \begin{bmatrix} -1 & 1 & 0 & 0 \\ 0 & 0 & 2 & 0 \\ 0 & 0 & 1 & 3 \\ 0 & 0 & 0 & 2 \end{bmatrix}.$$

5. $S(p(t)) = T^2(p(t)) + 2T(p(t)) = (1 + t)^2p''(t) + (1 + t)p'(t) - p(t)$: $P_3 \to P_3$, where T is from previous problem

   $B_1 = C_1 = \{1, 1 + t, (1 + t)^2, (1 + t)^3\}$, $B_2 = C_2 = \{1, t, t^2, t^3\}$

Answer: Using the results of the previous problem, we have

$$[S(B_1)]_{C1} = [T(B_1)]_{C1}{}^2 + 2[T(B_1)]_{C1} = \begin{bmatrix} -1 & 0 & 0 & 0 \\ 0 & 0 & 0 & 0 \\ 0 & 0 & 3 & 0 \\ 0 & 0 & 0 & 8 \end{bmatrix}$$

$$[S(B_2)]_{C2} = [T(B_2)]_{C2}{}^2 + 2[T(B_2)]_{C2} = \begin{bmatrix} -1 & 1 & 2 & 0 \\ 0 & 0 & 6 & 6 \\ 0 & 0 & 3 & 15 \\ 0 & 0 & 0 & 8 \end{bmatrix}$$

One may also compute the matrices directly, without using the results for T.

Remark: We have $[S(B_1)]_{C1} = [T(B_1)]_{C1}{}^2 + 2[T(B_1)]_{C1}$ because $B_1 = C_1$. If the two bases are not equal, then the equality no longer holds.

131-C

Encyclopaedia of

LINEAR ALGEBRA

Encyclopaedia of LINEAR ALGEBRA

Volume 1

R.K. Pandey

ANMOL PUBLICATIONS PVT. LTD.

NEW DELHI - 110 002 (INDIA)

ANMOL PUBLICATIONS PVT. LTD.

H.O.: 4374/4B, Ansari Road, Daryaganj,
New Delhi-110 002 (India)
Ph.: 23278000, 23261597

B.O.: No. 1015, Ist Main Road, BSK IIIrd Stage
IIIrd Phase, IIIrd Block,
Bangalore - 560 085 (India)
Visit us at: www.anmolpublications.com

Encyclopaedia of Linear Algebra

First Edition, 2009

ISBN 978-81-261-3914-9 (Set)

PRINTED IN INDIA

Printed at Mehra Offset Press, Delhi.

Contents

Preface

Linear Algebra is one of the foundations of modern Natural Science: wherever we are interested in calculations, from Engineering to String Theory, we use Linear Equations, Quadratic Forms, Matrices, Linear maps and their Cohomologies. There is a widespread feeling that the non-linear world is very different, and it is usually studied as a sophisticated phenomenon of Interpolation between different approximately-Linear Regimes.

Our aim in this book is basically a very simple one: to show how the familiar methods and ideas used in the analysis of the system of "Numbers" have been and are being extended and generalized to enable more diverse mathematical systems to be analyzed. In particular the concept of Factorization or Decomposition into "simpler" pieces underlies essentially all of our efforts. However, before we can proceed with this programme of study we must try to establish a common starting a point and path of approach. Even before doing this we must agree upon some basic assumptions, terms and notations, and these are set down in this book.

The book is written for a first-year graduate student. Very few undergraduates would have the necessary background and mathematical sophistication to appreciate this book. Not only can the book be used as a textbook but engineers and scientists will frequently consult it as a reference.

Author

List of Symbols

± ⊄ ⊂ ⊆ ∈ ∉ ℜ ∂ ≠ ≈ ∠ ∇

Σ Π Ω ϕ α β χ ε γ η λ μ

ν ⊥ π θ ρ σ τ ℜ ω ξ ψ ζ

ϒ ∃ ∫ ∬ ∭ × √ ∵ ≥ ∮ ⇒

Chapter 1

System of Linear Equations

LINEAR ALGEBRA

Linear algebra is the branch of mathematics concerned with the study of vectors, vector spaces (also called linear spaces), linear maps (also called linear transformations), and systems of linear equations. Vector spaces are a central theme in modern mathematics; thus, linear algebra is widely used in both abstract algebra and functional analysis. Linear algebra also has a concrete representation in analytic geometry and it is generalized in operator theory. It has extensive applications in the natural sciences and the social sciences, since nonlinear models can often be approximated by linear ones.

ELEMENTARY INTRODUCTION

Linear algebra had its beginnings in the study of vectors in Cartesian 2-space and 3-space. A vector, here, is a directed line segment, characterized by both its magnitude, represented by length, and its direction. Vectors can be used to represent physical entities such as forces, and they can be added to each other and multiplied with scalars, thus forming the first example of a real vector space.

Modern linear algebra has been extended to consider spaces of arbitrary or infinite dimension. A vector space of dimension n is called an n-space. Most of the useful results from 2- and 3-space can be extended to these higher dimensional spaces. Although people cannot easily visualize vectors in n-space, such vectors or n-tuples are useful in representing data. Since vectors, as n-tuples, are ordered lists of n components, it is possible to summarize and manipulate data efficiently in this framework.

For example, in economics, one can create and use, say, 8-dimensional vectors or 8-tuples to represent the Gross National Product of 8 countries. One can decide to display the GNP of 8 countries for a particular year, where the countries' order is specified, for example, (United States, United Kingdom, France, Germany, Spain, India, Japan, Australia), by using a vector (*v*1, *v*2, *v*3, *v*4, *v*5, *v*6, *v*7, *v*8) where each country's GNP is in its respective position.

A vector space (or linear space), as a purely abstract concept about which theorems are proved, is part of abstract algebra, and is well integrated into this discipline. Some striking examples of this are the group of invertible linear maps or matrices, and the ring of linear maps of a vector space. Linear algebra also plays an important part in analysis, notably, in the description of higher order derivatives in vector analysis and the study of tensor products and alternating maps.

In this abstract setting, the scalars with which an element of a vector space can be multiplied need not be numbers. The only requirement is that the scalars form a mathematical structure, called a field. In applications, this field is usually the field of real numbers or the field of complex numbers. Linear maps take elements from a linear space to another (or to itself), in a manner that is compatible with the addition and scalar multiplication given on the vector space(s).

The set of all such transformations is itself a vector space. If a basis for a vector space is fixed, every linear transform can be represented by a table of numbers called a matrix. The detailed study of the properties of and algorithms acting on matrices, including determinants and eigenvectors, is considered to be part of linear algebra.

One can say quite simply that the linear problems of mathematics - those that exhibit linearity in their behavior - are those most likely to be solved. For example differential calculus does a great deal with linear approximation to functions. The difference from nonlinear problems is very important in practice.

The general method of finding a linear way to look at a problem, expressing this in terms of linear algebra, and solving it, if need be by matrix calculations, is one of the most generally applicable in mathematics.

Some Useful Theorems

- Every vector space has a basis.
- Any two bases of the same vector space have the same cardinality; equivalently, the dimension of a vector space is well-defined.
- A matrix is invertible if and only if its determinant is nonzero.
- A matrix is invertible if and only if the linear map represented by the matrix is an isomorphism.
- If a square matrix has a left inverse or a right inverse then it is invertible.
- A matrix is positive semidefinite if and only if each of its eigenvalues is greater than or equal to zero.
- A matrix is positive definite if and only if each of its eigenvalues is greater than zero.
- The spectral theorem (regarding diagonalizable matrices).

SYSTEM OF LINEAR EQUATIONS

A linear equation in variables $x_1, x_2, \ldots, x_n$ is of the form

$$a_1x_1 + a_2x_2 + \ldots + a_nx_n = b,$$

where the coefficients $a_1, a_2, \ldots, a_n$ and the right side b are numbers. A system of linear equations is a collection of linear equations involving the same variables.

$$a_{11}x_1 + a_{12}x_2 + \ldots + a_{1n}x_n = b_1$$
$$a_{21}x_1 + a_{22}x_2 + \ldots + a_{2n}x_n = b_2$$
$$\ldots\ldots\ldots$$
$$a_{m1}x_1 + a_{m2}x_2 + \ldots + a_{mn}x_n = b_m$$

Example: The following 3 equations are linear.

$$2x_1 + 3x_2 - x_3 = 5,$$
$$x_1 - 2x_2 + 5x_3 + 10x_4 - 4x_5 = 12,$$
$$2x + 3y - z = 1.$$

The equations

$$x_1 + x_2 = x_3 + x_4,$$
$$2(x_1 - 3) = 2x_3 + 3(x_1 + 1) - 10x_2,$$
$$u + v = 2(u - 1) - 3v + 5,$$

are also linear because they can be rewritten as

$$x_1 + x_2 - x_3 - x_4 = 0,$$
$$-x_1 + 10x_2 - 2x_3 = 9,$$
$$-u + 4v = 3.$$

The equations

$$x_1x_2 = x_1 + x_2,$$
$$2\sin x_1 + 3\cos x_2 - \sin x_3 = 4,$$
$$x^3 + x^2y - xy^2 - y^3 = 1,$$

are not linear.

Example: The following is a system of 3 linear equations in 3 variables.

$$x_1 - x_2 + x_3 = 2$$
$$3x_1 + x_2 - x_3 = 2$$
$$2x_1 + 2x_2 + x_3 = 6$$

The following is a system of 4 linear equations in 5 variables.

$$x_1 + 3x_2 + 2x_3 + x_5 = 0$$
$$-x_1 - x_2 - x_3 + x_4 = 1$$
$$4x_2 + 2x_3 + 4x_4 + 3x_5 = 3$$
$$x_1 + 3x_2 + 2x_3 - 2x_4 = 0$$

Here is a system of 3 linear equations in 1 variable.

$$u = 2$$
$$3u = 2$$
$$2u = 6$$

Here is a system of 3 linear equations in 3 variables.

$$u = 2$$
$$3v = 2$$
$$2w = 6$$

Finally, $3x = 2$ is a system of 1 equation in 1 variable.

The following system is not linear.

$$x_1 + x_2 + x_3 = 1$$
$$x_1^2 + x_2^2 + x_3^2 = 1$$

The following is also not linear.

$$xy + yz + zx = 0$$
$$x^{1/2} - y^{1/2} + z^{1/2} = 2$$

Exercise: Which of the following are (systems of) linear equations?

1. $x_1^2 + x_2^2 = 1$

Answer: No

2. $x - y = y - z = z - x$

Answer: Yes

3. $x_1 + x_2 = x_3 + x_4 + 1, x_1 + x_3 = x_2 + x_4 + 2, x_1 + x_4 = x_2 + x_3 + 3$

Answer: Yes

4. $xy = 1$

Answer: No

5. $u + v + w = 2^{1/2}, u^2 + v^2 + w^2 = 3^{2/3}$
Answer: No

AUGMENTED MATRIX

In systems of linear equations, the notations for the variables are not essential.

Example: The essential information of the following system

$$3x_1 + x_2 - x_3 = 2$$
$$x_1 - x_2 + x_3 = 2$$
$$2x_1 + 2x_2 + x_3 = 6$$

is contained in the matrix

$$\begin{bmatrix} 3 & 1 & -1 & 2 \\ 1 & -1 & 1 & 2 \\ 2 & 2 & 1 & 6 \end{bmatrix}$$

made up of the coefficients on the left side and the numbers on the right side. Conversely, the system can be recovered from the matrix. For example, the following matrix

$$\begin{bmatrix} 1 & -2 & 0 & 3 \\ 0 & 1 & -1 & -2 \\ -2 & 0 & -5 & 1 \end{bmatrix}$$

comes from the following system (provided x_1, x_2, x_3 are chosen as the variables).

$$x_1 - 2x_2 = 3$$
$$x_2 - x_3 = -2$$
$$-2x_1 - 5x_3 = 1$$

In general, the essential information of a system of linear equations

$$a_{11}x_1 + a_{12}x_2 + \dots + a_{1n}x_n = b_1$$
$$a_{21}x_1 + a_{22}x_2 + \dots + a_{2n}x_n = b_2$$
$$\dots\dots\dots$$
$$a_{m1}x_1 + a_{m2}x_2 + \dots + a_{mn}x_n = b_m$$

is contained in the augmented matrix

$$[A\ b] = \begin{bmatrix} a_{11} & a_{12} & \dots & a_{1n} & b_1 \\ a_{21} & a_{22} & \dots & a_{2n} & b_2 \\ \vdots & \vdots & & \vdots & \vdots \\ a_{m1} & a_{m2} & \dots & a_{mn} & b_m \end{bmatrix}$$

which is made up of the coefficient matrix

$$A = \begin{bmatrix} a_{11} & a_{12} & \cdots & a_{1n} \\ a_{21} & a_{22} & \cdots & a_{2n} \\ \vdots & \vdots & \vdots & \vdots \\ a_{m1} & a_{m2} & \cdots & a_{mn} \end{bmatrix}$$

and the right side vector.

$$b = \begin{bmatrix} b_1 \\ b_2 \\ \vdots \\ b_m \end{bmatrix}$$

We also denote the left side of the system by

$$Ax = \begin{bmatrix} a_{11}x_1 + a_{12}x_2 + \ldots + a_{1n}x_n \\ a_{21}x_1 + a_{22}x_2 + \ldots + a_{2n}x_n \\ \vdots \\ a_{m1}x_1 + a_{m2}x_2 + \ldots + a_{nm}x_n \end{bmatrix}$$

so that the system of linear equations is $Ax = b$.

Example: Consider the system

$$x + 2y = 3$$
$$4x + 5y = 6$$
$$7x + 8y = 9$$
$$10x + 11y = 12$$

The coefficient matrix is

$$A = \begin{bmatrix} 1 & 2 \\ 4 & 5 \\ 7 & 8 \\ 10 & 11 \end{bmatrix}$$

the right side vector is

$$b = \begin{bmatrix} 3 \\ 6 \\ 9 \\ 12 \end{bmatrix}$$

And the augmented matrix is

$$[A\ b] = \begin{bmatrix} 1 & 2 & 3 \\ 4 & 5 & 6 \\ 7 & 8 & 9 \\ 10 & 11 & 12 \end{bmatrix}$$

Finally, we note the following correspondences between the augmented matrix and the system.

columns of $A \Leftrightarrow$ variables of the system $Ax = b$

rows of $[A\ b] \Leftrightarrow$ equations in the system $Ax = b$

Exercise: Find the augmented matrices.

1. $2x_1 - 3x_2 - x_3 = 3$

   $x_1 + x_2 + 2x_3 = -1$

Answer: $\begin{bmatrix} 2 & -3 & -1 & 3 \\ 1 & 1 & 2 & -1 \end{bmatrix}$

2. $x_1 + 2x_2 + 3x_3 + 4x_4 = 5$

Answer: $[1\ \ 2\ \ 3\ \ 4\ \ 5]$

3. $x_1 + 2x_2 = 3$

   $4x_1 + 5x_2 = 6$

   $7x_1 + 8x_2 = 9$

Answer: $\begin{bmatrix} 1 & 2 & 3 \\ 4 & 5 & 6 \\ 7 & 8 & 9 \end{bmatrix}$

4. $x_1 + 3x_2 + 2x_3 + x_5 = 0$

   $-x_1 - x_2 - x_3 + x_4 = 1$

   $4x_2 + 2x_3 + 4x_4 + 3x_5 = 3$

   $x_1 + 3x_2 + 2x_3 - 2x_4 = 0$

Answer: $\begin{bmatrix} 1 & 3 & 2 & 0 & 1 & 0 \\ -1 & -1 & -1 & 1 & 0 & 1 \\ 0 & 4 & 2 & 4 & 3 & 3 \\ 1 & 3 & 2 & -2 & 0 & 0 \end{bmatrix}$

Exercise: Recover systems of linear equations from the augmented matrices.

1. $\begin{bmatrix} 1 & 3 & 5 \\ 2 & 4 & 6 \end{bmatrix}$

Answer: $x_1 + 3x_2 = 5$

$2x_1 + 4x_2 = 6$

2. $[2 \quad 5]$

Answer: $2x_1 = 5$

3. $\begin{bmatrix} 1 & 1 \\ 2 & 2 \\ 3 & 3 \\ 4 & 4 \end{bmatrix}$

Answer: $x_1 = 1$

$2x_1 = 2$

$3x_1 = 3$

$4x_1 = 4$

4. $\begin{bmatrix} a & 1 & 1 \\ 1 & a & 1 \\ 1 & 1 & a \end{bmatrix}$

Answer: $ax_1 + x_2 = 1$

$x_1 + ax_2 = 1$

$x_1 + x_2 = a$

5. $\begin{bmatrix} 1 & 0 & 0 & 0 & 4 \\ 0 & 1 & 0 & 0 & 5 \\ 0 & 0 & 1 & 0 & 6 \\ 0 & 0 & 0 & 1 & 7 \end{bmatrix}$

Answer: $x_1 = 4$

$x_2 = 5$

$x_3 = 6$

$x_4 = 7$

6. $\begin{bmatrix} 0 & 0 & 0 & 0 & 0 \\ 0 & 2 & 0 & -3 & 4 \\ 1 & -1 & 5 & 0 & 8 \end{bmatrix}$

Answer: $0 = 0$

$$2x_2 - 3x_4 = 4$$

$$x_1 - x_2 + 5x_3 = 8$$

ROW OPERATION

GAUSSIAN ELIMINATION AND BACK SUBSTITUTION

To solve a system of linear equations, we first simplify the system and then do backward substitution.

Example: In the system

$$3x_1 + x_2 - x_3 = 2$$

$$x_1 - x_2 + x_3 = 2$$

$$2x_1 + 2x_2 + x_3 = 6$$

the variable x_1 appears in all 3 equations. We try to eliminate x_1 so that it appears in only 1 equation. Specifically, by multiplying – 3 to the 2nd equation and then add to the 1st equation, x_1 is eliminated from the 1st equation.

$$4x_2 - 4x_3 = -4$$

$$x_1 - x_2 + x_3 = 2$$

$$2x_1 + 2x_2 + x_3 = 6$$

Next the operation (– 2)[equation 2] + [equation 3] (multiply – 2 to the 2nd equation and then add to the 3rd equation) further eliminates x_1 from the 3rd equation.

$$4x_2 - 4x_3 = -4$$

$$x_1 - x_2 + x_3 = 2$$

The system is as simple as we can get as far as x1 is concerned.

Next we try to simplify the system regarding x2. The operation (– 1)[equation 1] + [equation 3] eliminates x2 from the 3rd equation.

$$4x_2 - 4x_3 = -4$$

$$x_1 - x_2 + x_3 = 2$$

$$3x_3 = 6$$

Although the system is simple enough to be solved. We may still make some cosmetic improvements. By exchanging the 1st and the 2nd equations (denoted [equation 1]? [equation 2]), we rearrange the equations from the most complicated down to the simplest.

$$x_1 - x_2 + x_3 = 2$$

$$4x_2 - 4x_3 = -4$$

$$3x_3 = 6$$

Finally, the coefficients can be further simplified by multiplying 1/4 and 1/3 to the 2nd and the 3rd equations (denoted (1/4)[equation 2] and (1/3)[equation 3]).

$$x_1 - x_2 + x3 = 2$$

$$x_2 - x_3 = -1$$

$$x_3 = 2$$

This completes the simplification process.

Now we start solving the system by backward substitution. From the 3rd equation (the simplest equation), we have

$$x_3 = 2.$$

Substituting this into the 2nd equation (the next simplest), we get

$$x_2 = -1 + x_3 = -1 + 2 = 1.$$

Further substituting into the 1st equation (the most complicated), we get

$$x_1 = 2 + x_2 - x_3 = 2 + 1 - 2 = 1.$$

We conclude that the system has a unique solution $x_1 = 1, x_2 = 1, x_3 = 2$.

In the example above, we have used the following operations to simplify a system.

r[equation i] + [equation j]

add a multiple of one equation to another.

[equation i] $\leftrightarrow$ [equation j]

exchange two equations.

d[equation i], $d \neq 0$

multiply a nonzero number to an equation.

In general, any system of linear equations can be simplified, by the three operations, to a simple system in which equations are arranged from the most complicated to the simplest. This process is called the gaussian elimination.

After completing the gaussian elimination, we may solve the simplest equation. Then we successively substitute the results into more and more complicated equations to get the whole solution. This process is called the backward substitution.

Example This example shows different choices may be taken in the gaussian elimination. However, the solution will not be affected by the choices.

For the same system just solved,

$$3x_1 + x_2 - x_3 = 2$$
$$x_1 - x_2 + x_3 = 2$$
$$2x_1 + 2x_2 + x_3 = 6$$

we try a slightly different gaussian elimination process. By (–3/2)[equation 3] + [equation 1] and (–1/2)[equation 3] + [equation 2], we have

$$-2x2 - 5/2x_3 = -7$$
$$-2x_2 + 1/2x_3 = -1$$
$$2x_1 + 2x_2 + x_3 = 6$$

Then (–1) [equation 2] + [equation 1] gives us

$$-3x_3 = -6$$
$$-2x_2 + 1/2x_3 = -1$$
$$2x_1 + 2x_2 + x_3 = 6$$

We further exchange [equation 1] ↔ [equation 3] to rearrange from the most complicated down to the simplest.

$$2x_1 + 2x_2 + x_3 = 6$$
$$-2x_2 + 1/2x_3 = -1$$
$$-3x_3 = -6$$

Now from the 3rd equation, we have

$$x_3 = (-6)/(-3) = 2.$$

Substituting into the 2nd equation, we have

$$x_2 = -(-1 - 1/2x_3)/2 = -(-1 - (1/2)\times 2)/2 = 1.$$

Substituting further into the 1st equation, we get

$$x_1 = (6 - 2x_2 - x_3)/2 = (6 - 2 - 2)/2 = 1.$$

THE THREE OPERATIONS DO NOT CHANGE SOLUTIONS

Theorem The following three operations do not change the solutions of systems of linear equations.

- r[equation i] + [equation j]: add a multiple of one equation to another
- [equation i] ↔ [equation j]: exchange two equations
- d[equation i], $d \neq 0$: multiply a nonzero number to an equation

Proof. For the *first* operation, let $\boldsymbol{x} = (x_1, x_2, \ldots, x_n)$ be a solution of the following system (denoted $Ax = b$).

$$a_{11}x_1 + a_{12}x_2 + \ldots + a_{1n}x_n = b_1$$
$$a_{21}x_1 + a_{22}x_2 + \ldots + a_{2n}x_n = b_2$$
$$\ldots \qquad \ldots \qquad \ldots$$
$$a_{m1}x_1 + a_{m2}x_2 + \ldots + a_{mn}x_n = b_m$$

The new system (denoted $A'x = b'$) obtained by the first operation is the same as the old one except that the j–th equation becomes

$$(ra_{i1}+a_{j1})x_1 + (ra_{i2}+a_{j2})x_2 + \ldots + (ra_{in}+a_{jn})x_n = rb_i + b_j.$$

To show $\boldsymbol{x}$ satisfies this, we note that $\boldsymbol{x}$ satisfies the i-th and the j-th equations in the old system.

$$a_{i1}x_1 + a_{i2}x_2 + \ldots + a_{in}x_n = b_i,$$
$$a_{j1}x_1 + a_{j2}x_2 + \ldots + a_{jn}x_n = b_j.$$

By multiplying r to the 1st equality and then add to the 2nd equality, we see that $\boldsymbol{x}$ satisfies the j-th equation in the new system.

We have shown that any solution of $Ax = b$ is a solution of $A'x = b'$. Conversely, the system $Ax = b$ can be recovered from the system $A'x = b'$ by the operation $(-r)$[equation i] + [equation j]. Therefore any solution of $A'x = b'$ is also a solution of $Ax = b$. This completes the proof that the solutions of the two systems $Ax = b$ and $A'x = b'$ are the same.

Since the order of the equations is irrelevant in whether or not $\boldsymbol{x}$ is a solution, the solution is not changed by the *second* operation.

For the *third* operations, let $\boldsymbol{x}$ satisfy the i-th equation

$$a_{i1}x_1 + a_{i2}x_2 + \ldots + a_{in}x_n = b_i.$$

Then it also satisfies the i–th equation after the third operation

$$da_{i1}x_1 + da_{i2}x_2 + \ldots + da_{in}x_n = db_i.$$

Conversely, if x satisfies the new i–th equation, then by dividing d, we see that x satisfies the old i-th equation.

ROW OPERATION

By the equivalence between systems of linear equations and augmented matrices, the three operations in the gaussian elimination are equivalent to three row operations.

$$r[\text{row } i] + [\text{row } j]$$

add a multiple of one row to another.

$$\begin{bmatrix} [\text{row } i] \\ \vdots \\ [\text{row } j] \end{bmatrix} \Rightarrow \begin{bmatrix} [\text{row } i] \\ \vdots \\ [\text{row } i] + [\text{row } j] \end{bmatrix}$$

$[\text{row } i] \leftrightarrow [\text{row } j]$

exchange rows.

$$\begin{bmatrix} [\text{row } i] \\ \vdots \\ [\text{row } j] \end{bmatrix} \Rightarrow \begin{bmatrix} [\text{row } j] \\ \vdots \\ [\text{row } i] \end{bmatrix}$$

$d[\text{row } i]$, $d \neq 0$

multiply a nonzero number to a row.

$$\begin{bmatrix} \vdots \\ [\text{row } i] \\ \vdots \end{bmatrix} \Rightarrow \begin{bmatrix} \vdots \\ d[\text{row } i] \\ \vdots \end{bmatrix}$$

***Example*: The operation (–3)[equation 2] + [equation 1] on the system**

$$\begin{matrix} 3x_1 + x_2 - x_3 = 2 & & 4x_2 - 4x_3 = -4 \\ x_1 - x_2 + x_3 = 2 & \Rightarrow & x_1 - x_2 + x_3 = 2 \\ 2x_1 + 2x_2 + x_3 = 6 & & 2x_1 + 2x_2 + x_3 = 6 \end{matrix}$$

is equivalent to the operation (–3)[row 2] + [row 1] on the augmented matrix.

$$\begin{bmatrix} 1 & -1 & 1 & 2 \\ 0 & 4 & -4 & -4 \\ 0 & 0 & 3 & 6 \end{bmatrix} \Rightarrow \begin{bmatrix} 1 & -1 & 1 & 2 \\ 0 & 1 & -1 & -1 \\ 0 & 0 & 1 & 2 \end{bmatrix}$$

The operation [equation 2] ”! [equation 3] on the system

$$\begin{matrix} x_1 - x_2 + x_3 = & & x_1 - x_2 + x_3 = 2 \\ 4x_2 - 4x_3 = -4 & \Rightarrow & x_2 - x_3 = -1 \\ 3x_3 = 6 & & x_3 = 6 \end{matrix}$$

is equivalent to the operation [row 2] ↔ [row 3] on the augmented matrix.

$$\begin{bmatrix} 0 & -4 & -4 & -4 \\ 1 & -1 & 1 & 2 \\ 0 & 0 & 3 & 6 \end{bmatrix} \Rightarrow \begin{bmatrix} 1 & -1 & 1 & 2 \\ 0 & 4 & -4 & -4 \\ 0 & 0 & 3 & 6 \end{bmatrix}$$

The operations (1/4)[equation 2] and (1/3)[equation 3] on the system

$$\begin{aligned} x_1 - x_2 + x_3 = 2 &\Rightarrow & x_1 - x_2 + x_3 = 2 \\ 4x_2 - 4x_3 = -4 & & x_2 - x_3 = -1 \\ 3x_3 = 6 & & x_3 = 2 \end{aligned}$$

is equivalent to the operation (1/4)[row 2] and (1/3)[row 3] on the augmented matrix.

$$\begin{bmatrix} 1 & -1 & 1 & 2 \\ 0 & 4 & -4 & -4 \\ 0 & 0 & 3 & 6 \end{bmatrix} \Rightarrow \begin{bmatrix} 1 & -1 & 1 & 2 \\ 0 & 1 & -1 & -1 \\ 0 & 0 & 1 & 2 \end{bmatrix}$$

The goal of gaussian elimination is to eliminate as many variables as possible. Correspondingly, the goal of row operation is to create as many zeros as possible.

Example: Consider the system (compare with this example).

$$\begin{aligned} 3x_1 + x_2 - x_3 &= 2 \\ x_1 - x_2 + x_3 &= 2 \\ 2x_1 - x_2 + x_3 &= 6 \end{aligned}$$

We simplify the system by doing row operations on the augmented matrix.

$$\begin{bmatrix} 3 & 1 & -1 & 2 \\ 1 & -1 & 1 & 2 \\ 2 & -1 & 1 & 6 \end{bmatrix}$$

Simplifying x_1 means creating two zeros in the [col 1] (1st column). By (–3)[row 2] + [row 1] and (–2)[row 2] + [row 3], we get

$$\begin{bmatrix} 0 & 4 & -4 & -4 \\ 1 & -1 & 1 & 2 \\ 0 & 1 & -1 & 2 \end{bmatrix}$$

By [row 1] ↔ [row 2] and [row 2] ↔ [row 3], we get

$$\begin{bmatrix} 1 & -1 & 1 & 2 \\ 0 & 1 & -1 & 2 \\ 0 & 4 & -4 & -4 \end{bmatrix}$$

Then (–4)[row 2] + [row 3] gives us

$$\begin{bmatrix} 1 & -1 & 1 & 2 \\ 0 & 1 & -1 & 2 \\ 0 & 0 & 0 & -12 \end{bmatrix}$$

This is the simplest shape we can get by row operations. The corresponding simple system is

$$\begin{aligned} x_1 - x_2 + x_3 &= 2 \\ x_2 - x_3 &= 2 \\ 0 &= -12 \end{aligned}$$

Since the 3rd equation $0 = -12$ is a contradiction, the original system has no solution.

Remark: Please do the problems by yourself and then compare with the given solution. Pay attention to the order the simplification is done: one column at a time, from left to right.

Please also note the three possible outcomes for the solutions: (1) no solution, (2) unique solution, (3) non–unique solution (with some arbitrary choice).

Exercise: Solve systems of linear equations.

1. $x_1 - x_2 + 3x_3 = 3$

$3x_1 + x_2 + x_3 = 5$

$x_1 + x_2 - x_3 = 1$

Answer: The augmented matrix is

$$\begin{bmatrix} 1 & -1 & 3 & 3 \\ 3 & 1 & 1 & 5 \\ 1 & 1 & -1 & 1 \end{bmatrix}$$

We start the simplification with the [col 1]. By (–1)[row 3] + [row 1] and (–3)[row 3] + [row 2], [col 1] is as simple as we can get.

$$\begin{bmatrix} 0 & -2 & 4 & 2 \\ 0 & -2 & 4 & 2 \\ 1 & 1 & -1 & 1 \end{bmatrix}$$

Then the exchange [row 1] ↔ [row 3] puts the most complicated row at the top.

$$\begin{bmatrix} 1 & 1 & -1 & 1 \\ 0 & -2 & 4 & 2 \\ 0 & -2 & 4 & 2 \end{bmatrix}$$

Now we turn our attention to [col 2]. By (–1)[row 2] + [row 3], we create one more zero in [col 2].

$$\begin{bmatrix} 1 & 1 & -1 & 1 \\ 0 & -2 & 4 & 2 \\ 0 & 0 & 0 & 0 \end{bmatrix}$$

Incidentally, [col 3] and [col 4] are also simplified. In fact, the shape is already the simplest we can get. Converted to a system of linear equations, we have

$$x_1 + x_2 - x_3 = 1$$
$$-2x_2 + 4x_3 = 2$$
$$0 = 0$$

Note that the last equation is trivial. The simplest nontrivial equation is the 2nd one, from which we get

$x_2 = -1 + 2x_3$, where x_3 is arbitrary.

Substituting this into the 1st equation, we get

$x_1 = 1 - x_2 + x_3 = 3 - 1 + 2x_3 - 3x_3 = 2 - x_3$.

Thus the general solution is given by the formulae that express x_1 and x_2 in terms of x_3, with x_3 arbitrary.

2. $2x_1 - 3x_2 - x_3 = 3$

$x_1 + x_2 + 2x_3 = -1$

Answer: The augmented matrix is

$$\begin{bmatrix} 2 & -3 & -1 & 3 \\ 1 & 1 & 2 & -1 \end{bmatrix}$$

We apply (–2)[row 2] + [row 1] to get

$$\begin{bmatrix} 0 & -5 & -5 & 5 \\ 1 & 1 & 2 & -1 \end{bmatrix}$$

Then by [row 1] ↔ [row 2] and (–1/5)[row 2], we have

$$\begin{bmatrix} 1 & 1 & 2 & -1 \\ 0 & 1 & 1 & -1 \end{bmatrix}$$

[row 2] corresponds to $x_2 + x_3 = -1$, which gives us $x_2 = -1 - x_3$, where x_3 is arbitrary. [row 1] corresponds to $x_1 + x_2 + 2x_3 = -1$. The general solution is

$x_1 = -1 - x_2 - 2x_3 = -1 - (-1 - x_3) - 2x_3 = -x_3$

$x_2 = -1 - x_3$

x_3 is arbitrary

3. $x_1 + 2x_2 = 3$

$4x_1 + 5x_2 = 6$

$7x_1 + 8x_2 = 9$

Answer: The augmented matrix is

$$\begin{bmatrix} 1 & 2 & 3 \\ 4 & 5 & 6 \\ 7 & 8 & 9 \end{bmatrix}$$

By (–4)[row 1] + [row 2] and (–7)[row 1] + [row 3], we get

$$\begin{bmatrix} 1 & 2 & 3 \\ 0 & -3 & -6 \\ 0 & -6 & -12 \end{bmatrix}$$

Then we apply (–2)[row 2] + [row 3] and (–1/3)[row 2] to get

$$\begin{bmatrix} 1 & 2 & 3 \\ 0 & 1 & 2 \\ 0 & 0 & 0 \end{bmatrix}$$

The last row corresponds to the trivial equality 0 = 0. The simplest nontrivial row is the second one, corresponding to the equation $x_2 = 2$. Substituting to the equation $x_1 + 2x_2 = 3$, we have $x_1 = 3 - 4 = 1$. Thus the system has a unique solution $x_1 = 1$, $x_2 = 2$.

$$x_1 + 3x_2 + 2x_3 + x_5 = 0$$
$$-x_1 - x_2 - x_3 + x_4 = 1$$
$$4x_2 + 2x_3 + 4x_4 + 3x_5 = 3$$
$$x_1 + 3x_2 + 2x_3 - 2x_4 = 0$$
$$x_1 + 3x_2 + 2x_3 + x_5 = 0$$
$$-x_1 - x_2 - x_3 + x_4 = 1$$
$$4x_2 + 2x_3 + 4x_4 + 3x_5 = 2$$
$$x_1 + 3x_2 + 2x_3 - 2x_4 = 0$$

Answer: The two systems have the same coefficient matrix A, with different right sides b and c. Therefore instead of doing row operations on the augmented matrices $[A\ b]$ and $[A\ c]$ separately, we may work on the combined augmented matrix.

$$[A\ b\ c] = \begin{bmatrix} 1 & 3 & 2 & 0 & 1 & 0 & 0 \\ -1 & -1 & -1 & 1 & 0 & 1 & 1 \\ 0 & 4 & 2 & 4 & 3 & 3 & 2 \\ 1 & 3 & 2 & -2 & 0 & 0 & 0 \end{bmatrix}$$

By first applying [row 2] + [row 4] and then applying [row 1] + [row 2], we have

$$\begin{bmatrix} 1 & 3 & 2 & 0 & 1 & 0 & 0 \\ 0 & 2 & 1 & 1 & 1 & 1 & 1 \\ 0 & 4 & 2 & 4 & 3 & 3 & 2 \\ 0 & 2 & 1 & -1 & 0 & 1 & 1 \end{bmatrix}$$

Next we apply (–2)[row 2] + [row 3] and (–1)[row 2] + [row 4] to get

$$\begin{bmatrix} 1 & 3 & 2 & 0 & 1 & 0 & 0 \\ 0 & 2 & 1 & 1 & 1 & 1 & 1 \\ 0 & 0 & 0 & 2 & 1 & 1 & 0 \\ 0 & 0 & 0 & -2 & -1 & 0 & 0 \end{bmatrix}$$

Finally we apply [row 3] + [row 4] to get

$$\begin{bmatrix} 1 & 3 & 2 & 0 & 1 & 0 & 0 \\ 0 & 2 & 1 & 1 & 1 & 1 & 1 \\ 0 & 0 & 0 & 2 & 1 & 1 & 0 \\ 0 & 0 & 0 & 0 & 0 & 1 & 0 \end{bmatrix}$$

If we take away the last column, then we see that the row operation changes the augmented matrix of the 4th system

$$\begin{aligned} x_1 + 3x_2 + 2x_3 \quad + x_5 &= 0 \\ -x_1 - x_2 - x_3 + x_4 &= 1 \\ 4x_2 + 2x_3 + 4x_4 + 3x_5 &= 3 \\ x_1 + 3x_2 + 2x_3 - 2x_4 &= 0 \end{aligned}$$

into

$$\begin{bmatrix} 1 & 3 & 2 & 0 & 1 & 0 \\ 0 & 2 & 1 & 1 & 1 & 1 \\ 0 & 0 & 0 & 2 & 1 & 1 \\ 0 & 0 & 0 & 0 & 0 & 1 \end{bmatrix}$$

The last row corresponds to the equation $0 = 1$. The contradiction shows that the fourth system has no solution.

Similarly, by taking away the 6th column, we see that the row operation changes the augmented matrix of the 5th system

$$\begin{aligned} x_1 + 3x_2 + 2x_3 + x_5 &= 0 \\ -x_1 - x_2 - x_3 + x_4 &= 1 \\ 4x_2 + 2x_3 + 4x_4 + 3x_5 &= 2 \\ x_1 + 3x_2 + 2x_3 - 2x_4 &= 0 \end{aligned}$$

into

$$\begin{bmatrix} 1 & 3 & 2 & 0 & 1 & 0 \\ 0 & 2 & 1 & 1 & 1 & 1 \\ 0 & 0 & 0 & 2 & 1 & 1 \\ 0 & 0 & 0 & 0 & 0 & 0 \end{bmatrix}$$

The last equation is trivial. The 3rd equation gives us

$x_4 = 1/2 - 1/2x_5$, x_5 arbitrary.

The 2nd equation gives us

$x_2 = 1/2 - 1/2x_3 - 1/2x_4 - 1/2x_5 = 1/4 - 1/2x_3 - 1/4x_5$, x_3 arbitrary.

The 1st equation then gives us

$x_1 = -3x_2 - 2x_3 - x_5 = -3/4 - 1/4x_3 - 1/4x_5$.

The general solution is given by the expressions of x_1, x_2, x_4 in terms of arbitrary x_3, x_5.

Exercise: Prove that if a matrix A can be changed to a matrix B by row operations, then the matrix B can also be changed to a matrix A by row operations.

Answer: If applying r[row i] + [row j] to A gives us B, then applying $(-r)$[row i] + [row j] to B gives us A back.

If applying [row i] $\leftrightarrow$ [row j] to A gives us B, then applying the same operations again to B gives us A back.

If applying d[row i] to A gives us B, then applying d^{-1}[row i] to B gives us A back.

ROW ECHELON FORM

SHAPE OF ROW ECHELON FORM

To understand the definition of row echelon form in more detail, we introduce some terminologies.

The leading entry of a row is the first nonzero entry counted from left. The leading entries are denoted by * in the example above. Moreover, the leading entries in the matrix

$$\begin{bmatrix} 1 & 0 & -1 & 0 \\ 0 & 0 & 3 & -3 \\ 0 & 2 & 0 & 0 \end{bmatrix}$$

are 1, 3, 2, and the matrix has the shape.

$$\begin{bmatrix} * & \# & \# & \# \\ 0 & 0 & * & \# \\ 0 & * & \# & \# \end{bmatrix}$$

The length of a nonzero row is the number of entries on the right of (and including) the leading entry. For example, the lengths of the three rows in the matrix above are 4, 2, and 3. The zero row (in which all entries are zero) has no leading entry and had length 0.

Example: Suppose a matrix has the following shape,

$$\begin{bmatrix} * & \# & \# & \# \\ 0 & * & \# & \# \\ 0 & * & \# & \# \end{bmatrix}$$

in which [row 2] and [row 3] have the same length. Denote the leading entries of the two rows by a and b (which are nonzero, by the definition of leading entries).

$$\begin{bmatrix} * & \# & \# & \# \\ 0 & a & \# & \# \\ 0 & b & \# & \# \end{bmatrix}$$

Then the operation $(-b/a)$[row 2] + [row 3] gives us

$$\begin{bmatrix} * & \# & \# & \# \\ 0 & a & \# & \# \\ 0 & 0 & \# & \# \end{bmatrix}$$

which has simpler shape because [row 3] has shorter length.

The example shows that if a matrix has the simplest shape, then any two *nonzero* rows must have different lengths. Thus the two properties in the definition of row echelon forms mean that, going from the top to the bottom, the lengths of the rows are *strictly* decreasing and then followed by the zero rows. Such a shape must be an *upside down staircase*.

Example: The following are some shapes of row echelon forms.

$$\begin{bmatrix} * & \# & \# & \# & \# \\ 0 & 0 & * & \# & \# \\ 0 & 0 & 0 & * & \# \end{bmatrix}$$

$$\begin{bmatrix} * & \# & \# & \# \\ 0 & 0 & * & \# \\ 0 & 0 & 0 & 0 \end{bmatrix}$$

$$\begin{bmatrix} * & \# & \# & \# \\ 0 & 0 & * & \# \\ 0 & 0 & 0 & 0 \\ 0 & 0 & 0 & 0 \end{bmatrix}$$

$$\begin{bmatrix} 0 & * & \# & \# & \# & \# & \# & \# \\ 0 & 0 & * & \# & \# & \# & \# & \# \\ 0 & 0 & 0 & 0 & * & \# & \# & \# \\ 0 & 0 & 0 & 0 & 0 & 0 & * & \# \end{bmatrix}$$

The length sequences are (5, 3, 2), (4, 2, 0), (4, 2, 0, 0), (7, 6, 4, 2). All of these are strictly decreasing sequences followed by zeros. The following matrices are more concrete examples of row echelon forms.

$$\begin{bmatrix} 9 \\ 32 \end{bmatrix}$$

$$\begin{bmatrix} 0 & 0 & 0 \\ 0 & 0 & 0 \end{bmatrix}$$

$$\begin{bmatrix} 0 & -1 \\ 0 & 0 \\ 0 & 0 \end{bmatrix}$$

$$\begin{bmatrix} 1 & 0 & 0 \\ 0 & 1 & 0 \\ 0 & 0 & 1 \end{bmatrix}$$

$$\begin{bmatrix} 1 & 3 & 2 & 0 & 1 & 0 \\ 0 & 2 & 1 & -1 & 0 & 1 \\ 0 & 0 & 0 & 2 & 1 & 0 \\ 0 & 0 & 0 & 0 & 0 & 1 \end{bmatrix}$$

Note that the one row matrices and the zero matrices are row echelon forms.

ROW ECHELON FORM

We use row operations to simplify matrices. A natural question is: How simple can a matrix become by row operations?

Example: The following matrix

$$\begin{bmatrix} 1 & 4 & 0 & 5 \\ 0 & 0 & 2 & 6 \\ 0 & 0 & 0 & 3 \end{bmatrix}$$

has the upside down staircase shape,

$$\begin{bmatrix} * & \# & \# & \# \\ 0 & 0 & * & \# \\ 0 & 0 & 0 & * \end{bmatrix}$$

in which * denotes nonzero numbers and # denotes any numbers. We claim that the shape cannot be improved by row operations.

First, the row operations r[row 1] + [row 2], r[row 1] + [row 3], r[row 2] + [row 3], with $r \neq 0$, produce the following shapes.

$$\begin{bmatrix} * & \# & \# & \# \\ * & \# & \# & \# \\ 0 & 0 & 0 & * \end{bmatrix}$$

$$\begin{bmatrix} * & \# & \# & \# \\ 0 & 0 & * & \# \\ * & \# & \# & \# \end{bmatrix}$$

$$\begin{bmatrix} * & \# & \# & \# \\ 0 & 0 & * & \# \\ 0 & 0 & * & \# \end{bmatrix}$$

All are more complicated than the original one. Moreover, adding multiples of lower rows to upper rows, such as *r*[row 3] + [row 1], will give us the same shape.

Next, the operation [row 2] ↔ [row 3] gives us

$$\begin{bmatrix} * & \# & \# & \# \\ 0 & 0 & 0 & * \\ 0 & 0 & * & \# \end{bmatrix}$$

Instead of simplifying the shape, this makes the shape uglier. You may check out the effects of the other exchange operations.

Finally, multiplying a nonzero number to a row does not change the shape, so that the shape is still not improved.

The following summarizes the simplest *shape* one can get by row operations.

A row echelon form is characterized by two properties

- The shape cannot be further improved by row operations
- The rows are arranged from the "longest" to the "shortest"

SHAPE OF ROW ECHELON FORM

To understand the definition of row echelon form in more detail, we introduce some terminologies.

The leading entry of a row is the first nonzero entry counted from left. The leading entries are denoted by * in the example above. Moreover, the leading entries in the matrix

$$\begin{bmatrix} 1 & 0 & -1 & 0 \\ 0 & 0 & 3 & -3 \\ 0 & 2 & 0 & 0 \end{bmatrix}$$

are 1, 3, 2, and the matrix has the shape.

$$\begin{bmatrix} * & \# & \# & \# \\ 0 & 0 & * & \# \\ 0 & * & \# & \# \end{bmatrix}$$

The length of a nonzero row is the number of entries on the right of (and including) the leading entry. For example, the lengths of the three rows in the matrix above are 4, 2, and 3. The zero row (in which all entries are zero) has no leading entry and had length 0.

Example: Suppose a matrix has the following shape,

$$\begin{bmatrix} * & \# & \# & \# \\ 0 & * & \# & \# \\ 0 & * & \# & \# \end{bmatrix}$$

in which [row 2] and [row 3] have the same length. Denote the leading entries of the two rows by a and b (which are nonzero, by the definition of leading entries).

$$\begin{bmatrix} * & \# & \# & \# \\ 0 & a & \# & \# \\ 0 & b & \# & \# \end{bmatrix}$$

Then the operation $(-b/a)$[row 2] + [row 3] gives us

$$\begin{bmatrix} * & \# & \# & \# \\ 0 & * & \# & \# \\ 0 & 0 & \# & \# \end{bmatrix}$$

which has simpler shape because [row 3] has shorter length.

The example shows that if a matrix has the simplest shape, then any two *nonzero* rows must have different lengths. Thus the two properties in the definition of row echelon forms mean that, going from the top to the bottom, the lengths of the rows are *strictly* decreasing and then followed by the zero rows. Such a shape must be an *upside down staircase*.

Example: The following are some shapes of row echelon forms.

$$\begin{bmatrix} * & \# & \# & \# & \# \\ 0 & 0 & * & \# & \# \\ 0 & 0 & 0 & * & \# \end{bmatrix}$$

$$\begin{bmatrix} * & \# & \# & \# \\ 0 & 0 & * & \# \\ 0 & 0 & 0 & 0 \end{bmatrix}$$

$$\begin{bmatrix} * & \# & \# & \# \\ 0 & 0 & * & \# \\ 0 & 0 & 0 & 0 \\ 0 & 0 & 0 & 0 \end{bmatrix}$$

$$\begin{bmatrix} 0 & * & \# & \# & \# & \# & \# & \# \\ 0 & 0 & * & \# & \# & \# & \# & \# \\ 0 & 0 & 0 & 0 & * & \# & \# & \# \\ 0 & 0 & 0 & 0 & 0 & 0 & * & \# \end{bmatrix}$$

The length sequences are (5, 3, 2), (4, 2, 0), (4, 2, 0, 0), (7, 6, 4, 2). All of these are strictly decreasing sequences followed by zeros. The following matrices are more concrete examples of row echelon forms.

$$[9]$$

$$[32]$$

$$\begin{bmatrix} 0 & 0 & 0 \\ 0 & 0 & 0 \end{bmatrix}$$

$$\begin{bmatrix} 0 & -1 \\ 0 & 0 \end{bmatrix}$$

$$\begin{bmatrix} 0 & 0 & 1 \\ 1 & 0 & 0 \\ 0 & 1 & 0 \\ 0 & 0 & 1 \end{bmatrix}$$

$$\begin{bmatrix} 1 & 3 & 2 & 0 & 1 & 0 \\ 0 & 2 & 1 & -1 & 0 & 1 \\ 0 & 0 & 0 & 2 & 1 & 0 \\ 0 & 0 & 0 & 0 & 0 & 1 \end{bmatrix}$$

Note that the one row matrices and the zero matrices are row echelon forms.

COMPUTING ROW ECHELON FORM

The leading entries in a row echelon form are called the pivots. The rows and the columns containing the pivots are the pivot rows and the pivot columns. For example, in the row echelon form

$$\begin{bmatrix} 1 & 3 & 4 & 5 \\ 0 & 0 & 2 & 6 \\ 0 & 0 & 0 & 0 \end{bmatrix}$$

the pivots are the entries 1, 2, the pivot rows are [row 1] and [row 2], the pivot columns are [col 1] and [col 3]. Moreover, since we may use row operations to change the matrix

$$A = \begin{bmatrix} -1 & -3 & 0 & 7 \\ 1 & 3 & 2 & -1 \\ 1 & 3 & 4 & 5 \end{bmatrix}$$

to the row echelon form above, we also say the pivot rows of A are [row 1] and [row 2], and the pivot columns of A are [col 1] and [col 3].

Example: The following is the augmented matrix of a system studied before.

$$\begin{bmatrix} 3 & 1 & -1 & 2 \\ 1 & -1 & 1 & 2 \\ 2 & 2 & 1 & 6 \end{bmatrix}$$

To find its row echelon form, we simplify one column at a time. We may create two zeros in [col 1] by (–3)[row 2] + [row 1] and (–2)[row 2] + [row 3].

$$\begin{bmatrix} 0 & 4 & -4 & -4 \\ 1 & -1 & 1 & 2 \\ 0 & 4 & -1 & 2 \end{bmatrix}$$

Then the exchange [row 1] $\leftrightarrow$ [row 2] moves the longest row to the top.

$$\begin{bmatrix} 1 & -1 & 1 & 2 \\ 0 & 4 & -4 & -4 \\ 0 & 4 & -1 & 2 \end{bmatrix}$$

Now [col 1] cannot be further improved, and we turn to [col 2]. By (–1)[row 2] + [row 3], we have

$$\begin{bmatrix} 1 & -1 & 1 & 2 \\ 0 & 4 & -4 & -4 \\ 0 & 0 & 3 & 6 \end{bmatrix}$$

This is a row echelon form.

Example: Consider a more complicated example.

$$\begin{bmatrix} 1 & 3 & 2 & 0 & 1 & 0 \\ -1 & -1 & -1 & 1 & 0 & 1 \\ 0 & 4 & 2 & 4 & 3 & 3 \\ 1 & 3 & 2 & -2 & 0 & 0 \end{bmatrix}$$

To simplify [col 1], we first apply [row 2] + [row 4] and then apply [row 1] + [row 2] to get

$$\begin{bmatrix} 1 & 3 & 2 & 0 & 1 & 0 \\ 0 & 2 & 1 & 1 & 1 & 1 \\ 0 & 4 & 2 & 4 & 3 & 3 \\ 1 & 2 & 1 & -1 & 0 & 1 \end{bmatrix}$$

Now [col 1] is simplest, and we try to simplify [col 2]. More precisely, we simplify the more heavily shaded 3 by 5 submatrix. By (–1)[row 4] + [row 2] and (–2)[row 4] + [row 3], we get

$$\begin{bmatrix} 1 & 3 & 2 & 0 & 1 & 0 \\ 0 & 2 & 1 & -1 & 0 & 1 \\ 0 & 0 & 0 & 2 & 1 & 0 \\ 0 & 0 & 0 & 6 & 3 & 1 \end{bmatrix}$$

By exchanging [row 2], [row 3], and [row 4], we get

$$\begin{bmatrix} 1 & 3 & 2 & 0 & 1 & 0 \\ 0 & 2 & 1 & -1 & 0 & 1 \\ 0 & 0 & 0 & 2 & 1 & 0 \\ 0 & 0 & 0 & 6 & 3 & 1 \end{bmatrix}$$

Note that [col 2] is now simplest. It turns out that [col 3] is also simplest. Therefore we should try to simplify [col 4], or the more heavily shaded 2 by 3 submatrix. By (–3)[row 3] + [row 4], we get

$$\begin{bmatrix} 1 & 3 & 2 & 0 & 1 & 0 \\ 0 & 2 & 1 & -1 & 0 & 1 \\ 0 & 0 & 0 & 2 & 1 & 0 \\ 0 & 0 & 0 & 0 & 0 & 1 \end{bmatrix}$$

This is a row echelon form.

We remark that different choices of row operations may produce different matrices at the end. However, the shape of the row echelon form is independent of the choices

REDUCED ROW ECHELON FORM

Although the row echelon form has the simplest shape, the individual entries can still be further simplified. The following summarizes the simplest *matrix* (as opposed to just *shape*) one can get by row operations.

A reduced row echelon form is characterized by two properties

- The matrix is a row echelon form
- The pivots are occupied by 1, and the entries above the pivots are occupied by 0

The following example justifies the definition.

***Example*:** Consider a row echelon form ($a \neq 0$)

$$\begin{bmatrix} * & \# & \# & c & \# \\ 0 & * & \# & 6 & \# \\ 0 & 0 & 0 & a & \# \end{bmatrix}$$

The operations $(-b/a)$[row 3] + [row 2], $(-c/a)$[row 3] + [row 1], and $(-1/a)$[row 3] give us

$$\begin{bmatrix} * & \# & \# & 0 & \# \\ 0 & * & \# & 0 & \# \\ 0 & 0 & 0 & 1 & \# \end{bmatrix}$$

Similar operations can further simplify [col 2] to become

$$\begin{bmatrix} 1 & 0 & \# & 0 & \# \\ 0 & 1 & \# & 0 & \# \\ 0 & 0 & 0 & 1 & \# \end{bmatrix}$$

This matrix not only has the simplest shape, but also has the simplest entries.

Given a matrix, we may first compute a row echelon form. Next we use the nonzero numbers at the pivots to cancel the numbers above the pivots. Then we may divide the numbers at the pivots to make them equal to 1. The result is the reduced row echelon form.

We remark that different choices of row operations will produce the same reduced row echelon form at the end.

Example: We simplify some row echelon forms given in a previous <u>example</u>. By (1/9)[row 1], the row echelon form

$$[9]$$

becomes a reduced row echelon form.

$$[1]$$

By (1/3)[row 1], the row echelon form

$$[32]$$

becomes a reduced row echelon form.

$$[12/3]$$

The row echelon form

$$\begin{bmatrix} 0 & 0 & 0 \\ 0 & 0 & 0 \end{bmatrix}$$

is already reduced. By (–1)[row 1], the row echelon form

$$\begin{bmatrix} 0 & -1 \\ 0 & 0 \\ 0 & 0 \end{bmatrix}$$

becomes a reduced row echelon form.

$$\begin{bmatrix} 0 & 1 \\ 0 & 0 \\ 0 & 0 \end{bmatrix}$$

The row echelon form

$$\begin{bmatrix} 1 & 0 & 0 \\ 0 & 1 & 0 \\ 0 & 0 & 1 \end{bmatrix}$$

is already reduced. To reduce the row echelon form

$$\begin{bmatrix} 1 & 3 & 2 & 0 & 1 & 0 \\ 0 & 2 & 1 & -1 & 0 & 1 \\ 0 & 0 & 0 & 2 & 1 & 0 \\ 0 & 0 & 0 & 0 & 0 & 1 \end{bmatrix}$$

we start from the rightmost pivot column. By (–1)[row 4] + [row 2], we get

$$\begin{bmatrix} 1 & 3 & 2 & 0 & 1 & 0 \\ 0 & 2 & 1 & -1 & 0 & 0 \\ 0 & 0 & 0 & 2 & 1 & 0 \\ 0 & 0 & 0 & 0 & 0 & 1 \end{bmatrix}$$

Then we use (1/2)[row 3] + [row 2] and (1/2)[row 3] to get

$$\begin{bmatrix} 1 & 3 & 2 & 0 & 1 & 0 \\ 0 & 2 & 1 & 0 & 1/2 & 0 \\ 0 & 0 & 0 & 1 & 1/2 & 0 \\ 0 & 0 & 0 & 0 & 0 & 1 \end{bmatrix}$$

Finally, we use (–3/2)[row 2] + [row 1] and (1/2)[Row 1] to get

$$\begin{bmatrix} 1 & 0 & 1/2 & 0 & -1/4 & 0 \\ 0 & 1 & 1/2 & 0 & 1/4 & 0 \\ 0 & 0 & 0 & 1 & 1/2 & 0 \\ 0 & 0 & 0 & 0 & 0 & 1 \end{bmatrix}$$

This is a reduced row echelon form.

Example: Let us compute the reduced row echelon form of the matrix

$$\begin{bmatrix} 1 & -1 & 1 & 0 & 2 \\ 2 & -2 & 0 & 2 & 2 \\ -1 & 1 & 2 & -3 & 1 \\ -2 & 2 & 1 & -3 & -1 \end{bmatrix}$$

By (–2)[row 1] + [row 2], [row 1] + [row 3], and 2[row 1] + [row 2], we get

$$\begin{bmatrix} 1 & -1 & 1 & 0 & 2 \\ 2 & 0 & -2 & 2 & -2 \\ 0 & 0 & 3 & -3 & 3 \\ 0 & 0 & 3 & -3 & 3 \end{bmatrix}$$

By (–1/2)[row 2], we get

$$\begin{bmatrix} 1 & -1 & 1 & 0 & 2 \\ 0 & 0 & 1 & -1 & 1 \\ 0 & 0 & 3 & -3 & 3 \\ 0 & 0 & 3 & -3 & 3 \end{bmatrix}$$

By (–3)[row 2] + [row 3] and (–3)[row 2] + [row 4], we get

$$\begin{bmatrix} 1 & -1 & 1 & 0 & 2 \\ 0 & 0 & 1 & -1 & 1 \\ 0 & 0 & 0 & 0 & 0 \\ 0 & 0 & 0 & 0 & 0 \end{bmatrix}$$

which is a row echelon form. By –[row 2] + [row 1], we get

$$\begin{bmatrix} 1 & -1 & 0 & 1 & 1 \\ 0 & 0 & 1 & -1 & 1 \\ 0 & 0 & 0 & 0 & 0 \\ 0 & 0 & 0 & 0 & 0 \end{bmatrix}$$

which is a reduced row echelon form.

It appears that reduced row echelon forms are more desirable than mere row echelon forms. However, the solution of many linear algebra problems only requires the row echelon form. For these problems, one may save much work by computing the row echelon form only.

REDUCED ROW ECHELON FORM VS. GENERAL SOLUTION

Suppose the augmented matrix of a system $Ax = b$ is simplified, by the row operations, to the following *reduced* row echelon form.

$$\begin{bmatrix} 1 & 2 & 0 & 0 & 3 \\ 0 & 0 & 1 & 0 & 4 \\ 0 & 0 & 0 & 1 & 5 \end{bmatrix}$$

This means that the simplified system is

$$x_1 + 2x_2 = 3$$
$$x_3 = 4$$
$$x_4 = 5$$

Then it is easy to get the general solution

$$x_1 = 3 - 2x_2$$
$$x_3 = 4$$
$$x_4 = 5$$

where x_2 is arbitrary. For a more general case, if the following is the reduced row echelon form of $[A\ b]$,

$$\begin{bmatrix} 0 & 1 & 0 & a_1 & 0 & b_1 & 0 & c_1 \\ 0 & 0 & 1 & a_2 & 0 & b_2 & 0 & c_2 \\ 0 & 0 & 0 & 0 & 1 & b_3 & 0 & c_3 \\ 0 & 0 & 0 & 0 & 0 & 0 & 1 & c_4 \end{bmatrix}$$

then the solution of the system $Ax = b$ is

$$x_2 = c_1 - a_1x_4 - b_1x_6$$
$$x_3 = c_2 - a_2x_4 - b_2x_6$$
$$x_5 = c_3 - b_3x_6$$
$$x_7 = c_4$$

where x_1, x_4, and x_6 are arbitrary.

We note that the coefficients in the general solution are explicitly given in the reduced row echelon form. Therefore the general solution of $Ax = b$ corresponds explicitly to the reduced row echelon form of $[A\ b]$. This has the following consequences:

- Since the general solution of a system of linear equations are independent of the elimination process, the reduced row echelon form of a matrix is independent of the row operations used in deriving it.
- Since the shape of the row echelon form is the same as the shape of the reduced row echelon form, the shape is also independent of the choice of the row operations.

Chapter 2

Existence and Uniqueness of Solution

EXISTENCE OF SOLUTIONS FOR FIXED RIGHT SIDE

The following discussion suggests the relation between the existence of solutions and the shape of the row echelon form of the augmented matrix.

Example: Suppose the following is the row echelon form of the augmented matrix $[A\ b]$ of a system $Ax = b$.

$$\begin{bmatrix} * & \# & \# & \# & \# \\ 0 & 0 & * & \# & \# \\ 0 & 0 & 0 & 0 & * \\ 0 & 0 & 0 & 0 & 0 \end{bmatrix}$$

The 3rd row is of the form $[0\ 0\ 0\ 0 \neq 0]$, which means that the 3rd equation is $0 = *$, where * is nonzero. The contradiction shows that the system $Ax = b$ has no solution.

On the other hand, if the row echelon form is

$$\begin{bmatrix} * & \# & \# & \# & \# \\ 0 & 0 & * & \# & \# \\ 0 & 0 & 0 & * & \# \\ 0 & 0 & 0 & 0 & 0 \end{bmatrix}$$

which has no rows like [0 0 0 0 “‘0], then the contradiction as above does not appear. In fact, further row operations will give us the reduced row echelon form.

$$\begin{bmatrix} 1 & a & 0 & 0 & b_1 \\ 0 & 0 & 1 & 0 & b_2 \\ 0 & 0 & 0 & 1 & b_3 \\ 0 & 0 & 0 & 0 & 0 \end{bmatrix}$$

The corresponding system has an obvious solution $x_1 = b_1, x_2 = 0, x_3 = b_2, x_4 = b_3$. The general solution can also be found by the discussion here. Thus the system has solutions.

You may also check out this example (no solution, row echelon form of the augmented matrix contains a row of the form [0 0... 0 ≠ 0]) and this example (has solutions, row echelon form of the augmented matrix contains no row of the form [0 0... 0 ≠ 0]). In general, we have the criterion for the existence of solutions.

A system of linear equations has solutions ⇔ There is no row like [0 0... 0 ≠ 0] in the row echelon form of the augmented matrix

When a system has solutions, we also say the system is consistent. Otherwise we say the system is inconsistent.

Example: Consider the following system of linear equations.

$$\begin{aligned} x_1 + 3x_2 + 2x_3 + x_5 &= 0 \\ -x_1 - x_2 - x_3 + x_4 &= 1 \\ 4x_2 + 2x_3 + 4x_4 + 3x_5 &= 3 \\ x_1 + 3x_2 + 2x_3 - 2x_4 &= 0 \end{aligned}$$

The augmented matrix is

$$\left[\begin{array}{ccccc|c} 1 & 3 & 2 & 0 & 1 & 0 \\ -1 & -1 & -1 & 1 & 0 & 1 \\ 0 & 4 & 2 & 4 & 3 & 3 \\ 1 & 3 & 2 & -2 & 0 & 0 \end{array}\right]$$

In this example, we computed its row echelon form.

$$\left[\begin{array}{ccccc|c} 1 & 3 & 2 & 0 & 1 & 0 \\ 0 & 2 & 1 & -1 & 0 & 1 \\ 0 & 0 & 0 & 2 & 1 & 0 \\ 0 & 0 & 0 & 0 & 0 & 1 \end{array}\right]$$

Since the last row is of the form [0 0 0 0 0 ≠ 0], the system has no solution. On the other hand, if we change the system a little bit,

$$\begin{aligned} x_1 + 3x_2 + 2x_3 + x_5 &= 0 \\ -x_1 - x_2 - x_3 + x_4 &= 1 \\ 4x_2 + 2x_3 + 4x_4 + 3x_5 &= 2 \\ x_1 + 3x_2 + 2x_3 - 2x_4 &= 0 \end{aligned}$$

then the same row operations give us

$$\begin{bmatrix} 1 & 3 & 2 & 0 & 1 & 0 \\ 0 & 2 & 1 & -1 & 0 & 1 \\ 0 & 0 & 0 & 2 & 1 & 0 \\ 0 & 0 & 0 & 0 & 0 & 0 \end{bmatrix}$$

Since there is no row of the form [0 0 0 0 0 $\neq$ 0], the system has solutions.

Next we modify the system again by taking away x_3 and x_5.

$$\begin{aligned} x_1 + 3x_2 &= 0 \\ -x_1 - x_2 + x_4 &= 1 \\ 4x_2 + 4x_4 &= 2 \\ x_1 + 3x_2 - 2x_4 &= 0 \end{aligned}$$

The row echelon form by the same row operations is obtained by dropping [col 3] and [col 5].

$$\begin{bmatrix} 1 & 3 & 0 & 0 \\ 0 & 2 & -1 & 1 \\ 0 & 0 & 2 & 0 \\ 0 & 0 & 0 & 0 \end{bmatrix}$$

Again there is no row of the form [0 0 0 ‘“0], and the system has solutions.

Example: We change the right side of the system in the previous example to arbitrary numbers,

$$\begin{aligned} x_1 + 3x_2 + 2x_3 + x_5 &= a \\ -x_1 - x_2 - x_3 + x_4 &= b \\ 4x_2 + 2x_3 + 4x_4 + 3x_5 &= c \\ x_1 + 3x_2 + 2x_3 - 2x_4 &= d \end{aligned}$$

and try to find the exact condition on *a*, *b*, *c*, *d* that makes the system consistent. By the same row operations as before, we get

$$\begin{bmatrix} 1 & 3 & 2 & 0 & 1 & a \\ 0 & 2 & 1 & -1 & 0 & b+d \\ 0 & 0 & 0 & 2 & 1 & a-d \\ 0 & 0 & 0 & 0 & 0 & -3a-2b+c+d \end{bmatrix}$$

For the system to be consistent, we need the last row to be [0 0 0 0 0 0]. Therefore the condition for the the system to have solutions is $-3a - 2b + c + d = 0$.

EXISTENCE OF SOLUTIONS FOR ANY RIGHT SIDE

For a fixed matrix A, the consistency of the system $Ax = b$ usually depends on the choice of b. In this part, we answer the following question: When the system $Ax = b$ has solutions for *any* choice of the right side b?

Example We have studied the following system before

$$3x_1 + x_2 - x_3 = b_1$$
$$x_1 - x_2 + x_3 = b_2$$
$$2x_1 + 2x_2 + x_3 = b_3$$

with a special choice on the right side. By applying the same row operations as before, the augmented matrix

$$\begin{bmatrix} 3 & 1 & -1 & b_1 \\ 1 & -1 & 1 & b_2 \\ 2 & 2 & 1 & b_3 \end{bmatrix}$$

becomes

$$\begin{bmatrix} 1 & -1 & 1 & b_2 \\ 0 & 4 & -4 & b_1 - 3b_2 \\ 0 & 0 & 3 & -b_1 + b_2 + b_3 \end{bmatrix}$$

Due to the fact that there is no zero row [0 0 0] in the darker part (which is a row echelon form of the coefficient matrix), there is no row of the form [0 0 0 ≠ 0] for all choice of b_1, b_2, b_3. Therefore the system is consistent for any choice of the right side.

Example: In an earlier example, we already know that the system

$$3x_1 + x_2 - x_3 = b_1$$
$$x_1 - x_2 + x_3 = b_2$$
$$2x_1 - x_2 + x_3 = b_3$$

has no solution for the special choice of $b_1 = 2$, $b_2 = 2$, $b_3 = 6$. Thus the system is *not consistent for some right side* (i.e., it is not true that the system is consistent for any right side).

We study the problem again by looking at the coefficient matrix only. The same row operations as before change the coefficient matrix

$$\begin{bmatrix} 3 & 1 & -1 \\ 1 & -1 & 1 \\ 2 & -1 & 1 \end{bmatrix}$$

to a row echelon form.

$$\begin{bmatrix} 1 & -1 & 1 \\ 0 & 1 & -1 \\ 0 & 0 & 0 \end{bmatrix}$$

The same row operations must also change the augmented matrix

$$\begin{bmatrix} 3 & 1 & -1 & b_1 \\ 1 & -1 & 1 & b_2 \\ 2 & -1 & 1 & b_3 \end{bmatrix}$$

to

$$\begin{bmatrix} 1 & -1 & 1 & b_1' \\ 0 & 1 & -1 & b_2' \\ 0 & 0 & 0 & b_3' \end{bmatrix}$$

where b'_1, b'_2, b'_3 are some combinations of b_1, b_2, b_3. Now we can choose some b_1, b_2, b_3 so that the last combination $b'_3 \neq 0$. Then the system has no solution for this choice. Thus we conclude that the system is *not always consistent*. Again we emphasis the *key reason* for this conclusion is that the row echelon form of the coefficient matrix contains a zero row [0 0 0].

You may also check out the always consistent problem for this example. From all these examples, it is easy to conclude the criterion for the existence of solutions for any right side.

The system $Ax = b$ has solutions for any right side b

⇔There is no zero row in the row echelon form of A

⇔ All rows of A are pivot

Exercise: Which of the following systems have solutions?

1. $2x_1 + 2x_3 + x_4 = 2$
   $x_1 + x_2 - x_3 = -1$
   $x_2 - 2x_3 + x_4 = -2$

Answer: The augmented matrix is

$$\begin{bmatrix} 2 & 0 & 2 & 1 & 1 \\ 1 & 1 & -1 & 0 & -1 \\ 0 & 1 & -2 & 1 & -2 \end{bmatrix}$$

By –2[row 2] + [row 1] and [row 1] ↔ [row 2], we get

$$\begin{bmatrix} 1 & 1 & -1 & 0 & -1 \\ 0 & -2 & 4 & 1 & 3 \\ 0 & 1 & -2 & 1 & -2 \end{bmatrix}$$

Then by 2[row 3] + [row 2] and [row 2] ↔ [row 3], we get

$$\begin{bmatrix} 1 & 1 & -1 & 0 & -1 \\ 0 & 1 & -2 & 1 & -2 \\ 0 & 0 & 0 & 3 & -1 \end{bmatrix}$$

Since there is no row of the form [0 0 0 0 ≠ 0], the system is consistent.

2. $x_1 + 2x_2 + x_3 = 1$

   $2x_1 + x_2 + x_3 = 0$

   $3x_2 + x_3 = 0$

   $3x_1 + x_3 = 0$

Answer: The augmented matrix

$$\begin{bmatrix} 1 & 2 & 1 & 1 \\ 2 & 1 & 1 & 0 \\ 0 & 3 & 1 & 0 \\ 3 & 0 & 1 & 0 \end{bmatrix}$$

appeared in this exercise, in which we have computed its row echelon form.

$$\begin{bmatrix} 1 & 2 & 1 & 1 \\ 0 & 3 & 1 & 0 \\ 0 & 0 & 0 & 1 \\ 0 & 0 & 0 & 0 \end{bmatrix}$$

Since [row 3] is of the form [0 0 0 ≠ 0], the system has no solution.

3&4

$$x_1 - x_2 + x_3 + 2x_5 = 1$$
$$2x_1 - 2x_2 + 2x_4 + 2x_5 = 0$$
$$-x_1 + x_2 + 2x_3 - 3x_4 + x_5 = 0$$
$$-2x_1 + 2x_2 + x_3 - 3x_4 - x_5 = 0$$
$$x_1 - x_2 + x_3 + 2x_5 = 3$$
$$2x_1 - 2x_2 + 2x_4 + 2x_5 = 2$$
$$-x_1 + x_2 + 2x_3 - 3x_4 + x_5 = 3$$
$$-2x_1 + 2x_2 + x_3 - 3x_4 - x_5 = 0$$

Answer: The two systems have the same coefficient matrices. We compute both systems at the same time. Specifically, for $Ax = b$ and $Ax = c$, we form the double augmented matrix.

$$[A\ b\ c] = \begin{bmatrix} 1 & -1 & 1 & 0 & 2 & 1 & 3 \\ 2 & -2 & 0 & 2 & 2 & 0 & 2 \\ -1 & 1 & 2 & -3 & 1 & 0 & 3 \\ -2 & 2 & 1 & -3 & -1 & 0 & 0 \end{bmatrix}$$

Now we carry out row operations, keeping in mind that the last two columns are the right sides. By –2[row 1] + [row 2], [row 1] + [row 3], 2[row 1] + [row 4], and (1/2)[row 2], we get

$$\begin{bmatrix} 1 & -1 & 1 & 0 & 2 & 1 & 3 \\ 0 & 0 & -1 & 1 & -1 & -1 & 0 \\ 0 & 0 & 3 & -3 & 3 & 1 & 6 \\ 0 & 0 & 3 & -3 & 3 & 2 & 6 \end{bmatrix}$$

Then by 3[row 2] + [row 3] and 3[row 2] + [row 4], we get

$$\begin{bmatrix} 1 & -1 & 1 & 0 & 2 & 1 & 3 \\ 0 & 0 & -1 & 1 & -1 & -1 & -2 \\ 0 & 0 & 0 & 0 & 0 & -2 & 0 \\ 0 & 0 & 0 & 0 & 0 & -1 & 0 \end{bmatrix}$$

Note that the coefficient matrix part has already become a row echelon form. If we delete the last column, we see that the augmented matrix $[A\ b]$ has been simplified to

$$\begin{bmatrix} 1 & -1 & 1 & 0 & 2 & 1 \\ 0 & 0 & -1 & 1 & -1 & -1 \\ 0 & 0 & 0 & 0 & 0 & -2 \\ 0 & 0 & 0 & 0 & 0 & -1 \end{bmatrix}$$

Because [row 3] is of the form $[0\ 0\ 0\ 0\ 0 \neq 0]$, the 3rd system has no solution. Similarly, the augmented matrix $[A\ c]$ has been simplified to

$$\begin{bmatrix} 1 & -1 & 1 & 0 & 2 & 3 \\ 0 & 0 & -1 & 1 & -1 & -2 \\ 0 & 0 & 0 & 0 & 0 & 0 \\ 0 & 0 & 0 & 0 & 0 & 0 \end{bmatrix}$$

From this we conclude that the 4th system has solutions.

Exercise: For what choices of the parameters are the following systems consistent?

1. $x + 2y = a$

   $3x + 4y = b$

Answer: The augmented matrix is

$$\begin{bmatrix} 1 & 2 & a \\ 3 & 4 & b \end{bmatrix}$$

By –3[row 1] + [row 2], we get

$$\begin{bmatrix} 1 & 2 & a \\ 0 & -2 & b-3a \end{bmatrix}$$

For any a and b, this is a row echelon form with no rows of the form $[0\ 0 \neq 0]$. Therefore the system is consistent for all choice of a and b.

2. $x + y = a$

   $2x + 2y = b$

   $3x + 3y = c$

   $4x + 4y = d$

Answer: The augmented matrix is

$$\begin{bmatrix} 1 & 1 & a \\ 2 & 2 & b \\ 3 & 3 & c \\ 4 & 4 & d \end{bmatrix}$$

By row operations, we get

$$\begin{bmatrix} 1 & 1 & a \\ 0 & 0 & b-2a \\ 0 & 0 & c-3a \\ 0 & 0 & d-4a \end{bmatrix}$$

The condition for no rows of the form [0 0 0] is $b = 2a$, $c = 3a$, $d = 4a$. This is exactly the condition for the system to be consistent.

3. $x_1 + x_2 + 2x_3 + x_4 = 1$

   $x_1 + 2x_3 = 0$

   $2x_1 + 2x_2 + 3x_3 = h$

   $x_2 + x_3 + 3x_4 = h$

Answer: The augmented matrix is

$$\begin{bmatrix} 1 & 1 & 2 & 1 & 1 \\ 1 & 0 & 2 & 0 & 0 \\ 2 & 2 & 3 & 0 & h \\ 0 & 1 & 1 & 3 & h \end{bmatrix}$$

By –[row 1] + [row 2] and 2–[row 1] + 3[row 2], we get

$$\begin{bmatrix} 1 & 1 & 2 & 1 & 1 \\ 0 & -1 & 0 & -1 & -1 \\ 0 & 0 & -1 & -2 & h-2 \\ 0 & 1 & 1 & 3 & h \end{bmatrix}.$$

By [row 2] + [row 4], we get

$$\begin{bmatrix} 1 & 1 & 2 & 1 & 1 \\ 0 & -1 & 0 & 0 & -1 \\ 0 & 0 & -1 & -2 & h-2 \\ 0 & 0 & 1 & 2 & h-1 \end{bmatrix}$$

By [row 3] + [row 4], we get

$$\begin{bmatrix} 1 & 1 & 2 & 1 & 1 \\ 0 & -1 & 0 & 0 & -1 \\ 0 & 0 & -1 & -2 & h-2 \\ 0 & 0 & 0 & 0 & 2h-3 \end{bmatrix}$$

The condition for the system to be consistent is $2h - 3 = 0$. In other words, $h = 3/2$.

4. $u + 2v = 1$

   $-2x + 2y - u + 4v = h$

   $x - y + u - v = 1$

Answer: The augmented matrix is

$$\begin{bmatrix} 0 & 0 & 1 & 2 & 1 \\ -2 & 2 & -1 & 4 & h \\ 1 & -1 & 1 & -1 & 1 \end{bmatrix}$$

By row operations, we get

$$\begin{bmatrix} 1 & -1 & 0 & -3 & 0 \\ 0 & 0 & 1 & 2 & 1 \\ 0 & 0 & 0 & 0 & h+1 \end{bmatrix}$$

The condition for the system to be consistent is $h = -1$.

5. $x_1 + x_2 - x_3 = -2$

   $x_1 - ax_3 = -1$

   $x_1 + ax_2 = -1$

Answer: The augmented matrix is

$$\begin{bmatrix} 1 & 1 & -1 & -2 \\ 1 & 0 & -a & -1 \\ 1 & 0 & -a & -1 \end{bmatrix}$$

By –[row 1] + [row 2] and –[row 1] + [row 3], we get

$$\begin{bmatrix} 1 & 1 & -1 & -2 \\ 0 & -1 & 1-a & 1 \\ 0 & a-1 & 1 & 1 \end{bmatrix}$$

Then by (a+1)[row 2] + [row 3], we have

$$\begin{bmatrix} 1 & 1 & -1 & -2 \\ 0 & -1 & 1-a & 1 \\ 0 & 0 & 2a-a^2 & a \end{bmatrix}$$

If $2a - a^2 \neq 0$, i.e., $a \neq 0$ or 2, then the system is clearly consistent.

If $a = 0$, then the last row consists of zeros. Thus the system is also consistent.

If $a = 2$, then the last row is [0 0 0 2]. Thus the system is inconsistent.

6. $x_1 + x_2 + x_3 + x_4 = 1$

   $x_1 + ax_4 = 1$

   $x_2 + ax_3 = 1$

Answer: The augmented matrix is

$$\begin{bmatrix} 1 & 1 & 1 & 1 & 1 \\ 1 & 0 & 0 & a & 1 \\ 0 & 1 & a & 0 & 1 \end{bmatrix}$$

By row operations, we get

$$\begin{bmatrix} 1 & 1 & 1 & 1 & 1 \\ 0 & 1 & 1 & 1-a & 0 \\ 0 & 0 & a-1 & a-1 & 1 \end{bmatrix}$$

The condition for the system to be consistent is $a \neq 1$.

7. $v + w = a$

   $u + ax = 1$

   $u + v = 0$

   $av - w = 1$

Answer: The augmented matrix is

$$\begin{bmatrix} 0 & 1 & 1 & a \\ 1 & 0 & a & 1 \\ 1 & 1 & 0 & 0 \\ 0 & a & -1 & 1 \end{bmatrix}$$

By row operations, we get

$$\begin{bmatrix} 1 & 0 & a & 1 \\ 0 & 1 & 1 & a \\ 0 & 0 & -a-1 & -a-1 \\ 0 & 0 & 0 & -a^2+a+3 \end{bmatrix}$$

The condition for the system to be consistent is $-a^2 + a + 2 = 0$. In other words, the system has solutions only for $a = -1$ or 2.

Exercise: Which of the following systems are consistent for any right side?

$$\begin{aligned} 2x_1 + 2x_3 + x_4 &= b_1 \\ x_1 + x_2 - x_3 &= b_2 \\ x_2 - 2x_3 + x_4 &= b_3 \end{aligned}$$

$$\begin{aligned} x_1 + 2x_2 + x_3 &= b_1 \\ 2x_1 + x_2 + x_3 &= b_2 \\ 3x_2 + x_3 &= b_3 \\ 3x_1 + x_3 &= b_4 \end{aligned}$$

$$\begin{aligned} x_1 - x_2 + x_3 + 2x_5 &= b_1 \\ 2x_1 - 2x_2 + 2x_4 + 2x_5 &= b_2 \\ -x_1 + x_2 + 2x_3 - 3x_4 + x_5 &= b_3 \\ -2x_1 + 2x_2 + x_3 - 3x_4 - x_5 &= b_4 \end{aligned}$$

Answer: By the row operations in an earlier exercise, we have the row echelon forms of the coefficient matrices (obtained by dropping the last columns).

$$\begin{bmatrix} 1 & 1 & -1 & 0 \\ 0 & 1 & -2 & 1 \\ 0 & 0 & 0 & 3 \end{bmatrix}$$

$$\begin{bmatrix} 1 & 2 & 1 \\ 0 & 3 & 1 \\ 0 & 0 & 0 \\ 0 & 0 & 0 \end{bmatrix}$$

$$\begin{bmatrix} 1 & -1 & 1 & 0 & 2 \\ 0 & 0 & -1 & 1 & -1 \\ 0 & 0 & 0 & 0 & 0 \\ 0 & 0 & 0 & 0 & 0 \end{bmatrix}$$

Since there is no zero row in the first row echelon form, the first system is consistent for any right side. Since there are zero rows in the second and the third row echelon forms, the second and the third systems are not always consistent.

Exercise: For what choices of the parameters are the following systems consistent for any right side?

$$\begin{aligned} x_1 + x_2 - x_3 &= b_1 \\ x_1 - ax_3 &= b_2 \\ x_1 + ax_2 &= b_3 \end{aligned}$$

$$\begin{aligned} x_1 + x_2 + x_3 + x_4 &= b_1 \\ x_1 + ax_4 &= b_2 \\ x_2 + ax_3 &= b_3 \end{aligned}$$

$$\begin{aligned} v + w &= b_1 \\ u + aw &= b_2 \\ u + v &= b_3 \\ av - w &= b_4 \end{aligned}$$

Answer: The coefficients of th4e systems have appeared in an earlier exercise. By the row operations there, we have the row echelon forms of the coefficient matrices (obtained by dropping the last columns).

$$\begin{bmatrix} 1 & 1 & -1 \\ 0 & -1 & 1-a \\ 0 & 0 & 2a-a^2 \end{bmatrix}$$

$$\begin{bmatrix} 1 & 1 & 1 & 1 \\ 0 & 1 & 1 & 1-a \\ 0 & 0 & a-1 & a-1 \end{bmatrix}$$

$$\begin{bmatrix} 1 & 0 & a \\ 0 & 1 & 1 \\ 0 & 0 & -a-1 \\ 0 & 0 & 0 \end{bmatrix}$$

The condition for the first system to be always consistent is $2a - a^2 \neq 0$, which is equivalent to $a \neq 0$ and 2. The condition for the second system to be always consistent is $a \neq 1$. Since [row 4] in the third row echelon form is a zero row, the third system is not always consistent for any choice of a.

Exercise: List all shapes of the row echelon forms the the augmented matrices of *consistent* systems of 2 equations and 2 variables. What about systems of other sizes?

Answer: For 2 equations in 2 variables, we look for all 2 by 3 matrices with no rows of the form $[0\ 0 \neq 0]$. The following are all such shapes.

$$\begin{bmatrix} * & \# & \# \\ 0 & * & \# \end{bmatrix}$$

$$\begin{bmatrix} * & \# & \# \\ 0 & 0 & 0 \end{bmatrix}$$

$$\begin{bmatrix} 0 & * & \# \\ 0 & 0 & 0 \end{bmatrix}$$

$$\begin{bmatrix} 0 & 0 & 0 \\ 0 & 0 & 0 \end{bmatrix}$$

Exercise: List all shapes of the row echelon forms of 2 by 2 matrices A so that $Ax = b$ has solutions for any b. What about matrices of other sizes?

Answer: We look for row echelon forms with no nonzero row (i.e., all rows pivot). For the 2 by 2 coefficient matrices, there is only one possibility.

$$\begin{bmatrix} * & \# \\ 0 & * \end{bmatrix}$$

For the 2 by 3 coefficient matrices, we have the following possibilities.

$$\begin{bmatrix} * & \# & \# \\ 0 & * & \# \end{bmatrix}$$

$$\begin{bmatrix} * & \# & \# \\ 0 & 0 & * \end{bmatrix}$$

$$\begin{bmatrix} 0 & * & \# \\ 0 & 0 & * \end{bmatrix}$$

We cannot find 3 by 2 matrices with all rows pivot. (This means that any system of 3 equations in 2 variables is inconsistent for some right side.)

In general, for a system of equations to be always consistent, the number of equations cannot be more than the number of variables.

Remark: The purpose of this exercise is to emphasis that one should not forget the original meaning of consistency.

Exercise: Without any computation, determine the consistency of the systems.

1. $2x_1 + 2x_3 + x_4 = 2$

   $x_1 + x_2 - x_3 = -1$

   $x_2 - 2x_3 + x_4 = -2$

Answer: The system has $x_1 = x_2 = x_4 = 0$, $x_3 = 1$ as an obvious solution. Therefore the system is consistent.

Note that in this exercise, we have used row operations to determine the system is consistent.

2. $x_1 + 2x_2 = 1$

   $2x_1 + 4x_2 = 2$

   $x_1 + 2x_2 = a$

   $-x_1 - 2x_2 = b$

Answer: As far as the coefficients on the left side are concerned, the 2nd, 3rd, and the 4th equations are respectively twice of, equal to, and negative of the 1st equation. Therefore the same should hold for the right side in order for the system to be consistent. The condition is $a = 1$, $b = -1$.

3. $x_1 + x_2 + x_3 = 1$

   $x_1 + 2x_2 + 3x_3 = 2$

   $2x_1 + 3x_2 + 4x_3 = 0$

Answer: As far as the coefficients on the left side are concerned, the sum of the 1st and the 2nd equations is the 3rd one. The same should hold for the right side in order for the system to be consistent. Since $1 + 2 \neq 0$, the system is not consistent.

Exercise: Show that the system $Ax = 0$, with the zero vector on the right side, is always consistent.

Answer: The system $Ax = 0$ always has the trivial solution $x = 0$ (all variables equal to zero). Since the solution exists, the system is consistent.

UNIQUENESS

STRUCTURE OF SOLUTIONS

Given a system, we first check the consistency. If the system is not consistent, then we cannot say anymore about the solutions. If the system is consistent, then we may further ask specific questions about the solutions.

In the subsequent discussion, we always assume the systems are consistent.

Example: Suppose the augmented matrix of a system $Ax = b$ has the following row echelon form.

$$\begin{bmatrix} 5 & 2 & 1 & -9 & 0 & -3 \\ 0 & 0 & -7 & 1 & 4 & 1 \\ 0 & 0 & 0 & 3 & 5 & -2 \end{bmatrix}$$

The last equation is $3x_4 + 5x_5 = -2$. Because the pivot entry 3 (the coefficient of x_4) is nonzero, we can move terms other than x_4 to the right side and then divide 3. The result is an expression of x_4 in terms of x_5: $x_4 = -2/3 - 5/3\ x_5$.

The middle equation is $-7x_3 + x_4 + 4x_5 = 1$. Because the pivot entry -7 (the coefficient of x_3) is nonzero, we can move terms other than x_3 to the right side and then divide -7. The result is an expression of x_3 in terms of x_4 and x_5. By further substituting the earlier expression $x_4 = -2/3 - 5/3\ x_5$, we see that x_3 is expressed in terms of x_5.

Similarly, in the top equation, because 5 (the coefficient of x_1) is a pivot, we can move terms other than x_1 to the right side and then divide the coefficient 5. The result is an expression of x_1 in terms of x_2, x_3, x_4, x_5. By further substituting the earlier expressions of x_3 and x_4 in terms of x_5, we see that x_1 is expressed in terms of x_2 and x_5.

The key observation here is that the pivot coefficient can be divided because it is *nonzero*. As a result, the corresponding "pivot variable" can be expressed in terms of "later variables". The back substitution process then further replace the "later variables" with "nonpivot variables".

Example: In this example, we studied a system with the augmented matrix having the following row echelon form.

$$\begin{bmatrix} * & \# & \# & \# & \# \\ 0 & 0 & * & \# & \# \\ 0 & 0 & 0 & * & \# \\ 0 & 0 & 0 & 0 & 0 \end{bmatrix}$$

By back substitution and successively dividing the coefficients of x_4, x_3, x_1, we may write the general solution as expressions of x_1, x_3, x_4 in terms of x_2, with x_2 arbitrary. Alternatively, by further row operations, we have the reduced row echelon form.

$$\begin{bmatrix} 1 & a & 0 & 0 & b_1 \\ 0 & 0 & 1 & 0 & b_2 \\ 0 & 0 & 0 & 1 & b_3 \\ 0 & 0 & 0 & 0 & 0 \end{bmatrix}$$

Then the general solution of the system is $x_1 = b_1 - ax_2$, $x_3 = b_2$, $x_4 = b_3$, with x_2 arbitrary.

Further examples can be found here. It is easy to conclude from the examples that, for a consistent system, the general solution can be presented as expressions of some nonfree variables (which cannot be arbitrarily chosen) in terms of some free variables (which can be arbitrarily chosen). Moreover, under the correspondence

variables of system $\Leftrightarrow$ columns of coefficient matrix,

we have

free variable $\Leftrightarrow$ nonpivot column

nonfree variable $\Leftrightarrow$ pivot column

Example: In an earlier example, we considered a system

$$\begin{aligned} x_1 + 3x_2 + 2x_3 + x_5 &= 0 \\ -x_1 - x_2 - x_3 + x_4 &= 1 \\ 4x_2 + 2x_3 + 4x_4 + 3x_5 &= 2 \\ x_1 + 3x_2 + 2x_3 - 2x_4 &= 0 \end{aligned}$$

The row echelon form of the augmented matrix is

$$\begin{bmatrix} 1 & 3 & 2 & 0 & 1 & 0 \\ 0 & 2 & 1 & -1 & 0 & 1 \\ 0 & 0 & 0 & 2 & 1 & 0 \\ 0 & 0 & 0 & 0 & 0 & 0 \end{bmatrix}$$

which implies the system is consistent. Moreover, the solution is not unique because we may express nonfree variables x_1, x_2, x_4 in terms of free variables x_3, x_5. Later in the same example, we took away the freedom x_3, x_5 (i.e., assuming both are zero) and considered the system

$$x_1 + 3x_2 = 0$$
$$-x_1 - x_2 + x_4 = 1$$
$$4x_2 + 4x_4 = 2$$
$$x_1 + 3x_2 - 2x_4 = 0$$

The row echelon form of the augmented matrix is

$$\begin{bmatrix} 1 & 3 & 0 & 0 \\ 0 & 2 & -1 & 1 \\ 0 & 0 & 2 & 0 \\ 0 & 0 & 0 & 0 \end{bmatrix}$$

The system is still consistent. Since all the columns of the coefficient matrix (i.e., the first three columns) are pivot, all the variables are nonfree, so that the solution is unique.

The uniqueness is to be expected because all the freedoms have been taken away.

UNIQUENESS OF SOLUTIONS

A consequence of our discussion on the structure of solutions is the criterion for the uniqueness of the solution.

Combining the following facts

- Uniqueness ⇔ no freedom (i.e., no free variables)
- Free variable ⇔ column
- Nonfree variable ⇔ pivot column

we have

A consistent system of linear equations has a unique solution

⇔ All variables are nonfree

⇔ All columns of the coefficient matrix are pivot

Example: In a previous example, we saw that the system

$$x_1 + 3x_2 + 2x_3 + x_5 = 0$$
$$-x_1 - x_2 - x_3 + x_4 = 1$$
$$4x_2 + 2x_3 + 4x_4 + 3x_5 = 2$$
$$x_1 + 3x_2 + 2x_3 - 2x_4 = 0$$

is consistent. Moreover, in the row echelon form of the coefficient matrix,

$$\begin{bmatrix} 1 & 3 & 2 & 0 & 1 \\ 0 & 2 & 1 & -1 & 0 \\ 0 & 0 & 0 & 2 & 1 \\ 0 & 0 & 0 & 0 & 0 \end{bmatrix}$$

[col 3] and [col 5] are not pivot. Therefore the solution is not unique.

The following system is obtained by taking away x_3 and x_5.

$$x_1 + 3x_2 = 0$$
$$-x_1 - x_2 + x_4 = 1$$
$$4x_2 + 4x_4 = 2$$
$$x_1 + 3x_2 - 2x_4 = 0$$

The system was also studied before and was known to be consistent. In the row echelon form of the coefficient matrix,

$$\begin{bmatrix} 1 & 3 & 0 \\ 0 & 2 & -1 \\ 0 & 0 & 2 \\ 0 & 0 & 0 \end{bmatrix}$$

all columns are pivot. Therefore the solution is unique.

We remark that the criterion for the uniqueness does not involve the right side. In other words, if both $Ax = b$ and $Ax = c$ are consistent, then the solution of $Ax = b$ is unique if and only if the solution of $Ax = c$ is unique. We also note that the homogeneous system $Ax = 0$ always has $x = 0$ as the trivial solution. The uniqueness for $Ax = 0$ simply means that the trivial solution is the only solution. In summary, we have the following important facts about uniqueness. .

The uniqueness is independent of the right side. Moreover,

A consistent system $Ax = b$ has a unique solution

$\Leftrightarrow$ The homogeneous system $Ax = \mathbf{0}$ has only the trivial solution

Example: The following system

$$x_1 + 3x_2 + 2x_3 + x_5 = 0$$
$$-x_1 - x_2 - x_3 + x_4 = 1$$
$$4x_2 + 2x_3 + 4x_4 + 3x_5 = 4$$
$$x_1 + 3x_2 + 2x_3 - 2x_4 = -2$$

has the obvious solution $x_1 = x_2 = x_3 = x_5 = 0, x_4 = 1$. In the previous example, we saw that another system with the same coefficient matrix has non–unique solutions. Therefore this system also has non–unique solutions. In other worlds, the system must have solutions other than the obvious one. Similarly, the homogeneous system

$$\begin{aligned} x_1 + 3x_2 + 2x_3 + x_5 &= 0 \\ -x_1 - x_2 - x_3 + x_4 &= 0 \\ 4x_2 + 2x_3 + 4x_4 + 3x_5 &= 0 \\ x_1 + 3x_2 + 2x_3 - 2x_4 &= 0 \end{aligned}$$

must have nontrivial (i.e., nonzero) solutions. On the other hand, the system

$$\begin{aligned} x_1 + 3x_2 + 2x_3 + x_5 &= 23 \\ -x_1 - x_2 - x_3 + x_4 &= -45 \\ 4x_2 + 2x_3 + 4x_4 + 3x_5 &= 78 \\ x_1 + 3x_2 + 2x_3 - 2x_4 &= 12 \end{aligned}$$

does not have an obvious solution. Since it has the same coefficient matrix as before, we conclude there are two possibilities.

- The system has no solution
- The system has more than one solutions

Of course, to find out which case exactly happens, one has to do more computations (in order to determine the consistency).

Exercise: Which of the following consistent systems have a unique solution (compare with this exercise)?

$$\begin{aligned} 2x_1 + 2x_3 + x_4 &= 2 \\ x_1 + x_2 - x_3 &= -1 \\ x_2 - 2x_3 + x_4 &= -2 \end{aligned}$$

$$\begin{aligned} x_1 + 2x_2 + x_3 &= 1 \\ 2x_1 + x_2 + x_3 &= 1 \\ 3x_2 + x_3 &= 1 \\ 3x_1 + x_3 &= 1 \end{aligned}$$

$$\begin{aligned} x_1 - x_2 + x_3 + 2x_5 &= 1 \\ 2x_1 - 2x_2 + 2x_4 + 2x_5 &= 0 \\ -x_1 + x_2 + 2x_3 - 3x_4 + x_5 &= 2 \end{aligned}$$

$$-x_1 + x_2 + 2x_3 - 3x_4 + x_5 = 3$$
$$-2x_1 + 2x_2 + x_3 - 3x_4 - x_5 = 0$$

$$x_1 - x_2 = 1$$
$$-2x_1 + 3x_2 + x_3 = 2$$
$$x_1 + x_3 = 5$$

$$x_1 + 2x_2 + 3x_3 = 13$$
$$2x_1 - 2x_2 + 4x_3 = 5$$

Answer: In this exercise, we have shown that the fourth system is consistent. Moreover, the other three systems have obvious solutions and are therefore also consistent.

By the computation in the earlier exercise, we have the following row echelon forms for the coefficient matrix (obtained by taking away the columns corresponding to the right sides)

$$\begin{bmatrix} 1 & 1 & -1 & 0 \\ 0 & 1 & -2 & 1 \\ 0 & 0 & 0 & 3 \end{bmatrix}$$

$$\begin{bmatrix} 1 & 2 & 1 \\ 0 & 3 & 1 \\ 0 & 0 & 0 \\ 0 & 0 & 0 \end{bmatrix}$$

$$\begin{bmatrix} 1 & -1 & 1 & 0 & 2 \\ 0 & 0 & -1 & 1 & -1 \\ 0 & 0 & 0 & 0 & 0 \\ 0 & 0 & 0 & 0 & 0 \end{bmatrix}$$

Note that the third and the fourth systems have the same coefficient matrix, and the row echelon form is the third matrix above. Since all row echelon forms have nonpivot columns, the solutions of all four systems are not unique. For the last system, we note that the augmented matrix appeared in an earlier exercise. The row echelon form was computed.

$$\begin{bmatrix} 1 & -1 & 0 & 1 \\ 0 & 1 & 1 & 4 \\ 0 & 0 & 4 & 3 \\ 0 & 0 & 0 & 0 \\ 0 & 0 & 0 & 0 \end{bmatrix}$$

Since the first three columns are pivot, and the last column is not pivot, the system has a unique solution.

Exercise: For what choices of the parameters do the following systems have a unique solution (compare with this exercise)?

1. $x + 2y = a$

   $3x + 4y = b$

Answer: The row operations change the augmented matrix into

$$\begin{bmatrix} 1 & 2 & a \\ 0 & -2 & b-3a \end{bmatrix}$$

The system is consistent for all a and b. Moreover, since the first two columns are pivot, the solution is unique.

2. $x_1 + x_2 - x_3 = -2$

   $x_1 - ax_3 = -1$

   $x_1 + ax_2 = -1$

Answer: The row operations change the augmented matrix into

$$\begin{bmatrix} 1 & 1 & -1 & -2 \\ 0 & -1 & 1-a & 1 \\ 0 & 0 & 2a-a^2 & a \end{bmatrix}$$

The system is consistent if and only if $a \neq 2$. However, for the solution to be unique, we further need $2a - a^2 \neq 0$. Therefore the condition for the system to have unique solution is $a \neq 0$ or 2.

3. $x_1 + x_2 + x_3 + x_4 = 1$

   $x_1 + ax_4 = 1$

   $x_2 + ax_3 = 1$s

Answer: The row operations change the augmented matrix into

$$\begin{bmatrix} 1 & 1 & 1 & 1 & 1 \\ 0 & 1 & 1 & 1-a & 0 \\ 0 & 0 & a-1 & a-1 & 1 \end{bmatrix}$$

The condition for the system is consistent is a "" 1. The solution is never unique.

4. $v + w = a$

   $u + aw = 1$

$u + v = 0$

$av - w = 1$

Answer: The row operations change the augmented matrix into

$$\begin{bmatrix} 1 & 0 & a & 1 \\ 0 & 1 & 1 & a \\ 0 & 0 & -a-1 & -a-1 \\ 0 & 0 & 0 & -a^2+a+2 \end{bmatrix}$$

The condition for the system to be consistent is $a = -1$ or 2. However, for the solution to be unique, we need $-a-1$ to be nonzero. Therefore only for $a = 2$ does the system have a unique solution.

Exercise: List all shapes of the row echelon forms of the augmented matrices of consistent systems of 2 equations in 2 variables with a unique solution. What about systems of other sizes?

Answer: In this exercise, we have listed all the consistent cases for 2 equations and 2 variables. Among these, only the following one has a unique solution.

$$\begin{bmatrix} * & \# & \# \\ 0 & * & \# \end{bmatrix}$$

For 3 equations in 2 variables, we are looking for 3 by 3 matrices with the first two columns pivot and the last column not pivot. The following is the only possibility.

$$\begin{bmatrix} * & \# & \# \\ 0 & * & \# \\ 0 & 0 & 0 \end{bmatrix}$$

For 2 equations in 3 variables, we need a 2 by 4 matrix with first three columns pivot and the last column not pivot. Since there are only two rows, this is impossible. Consequently, no system of 2 equations in 3 variables can have a unique solution.

In general, for a system of equations to have unique solution, the number of equations cannot be less than the number of variables.

Exercise: Let A be a 3 by 3 matrix. Prove that the following are equivalent

- $Ax = b$ is consistent for any b
- $Ax = 0$ has only the trivial solution
- A can be changed to the identity matrix by the row operations

What about the general case that A is an n by n matrix?

Answer: By this criterion, the first property means that all rows of A are pivot. In other words, the row echelon form has the following shape.

$$\begin{bmatrix} * & \# & \# \\ 0 & * & \# \\ 0 & 0 & * \end{bmatrix}$$

The corresponding reduced row echelon form is

$$\begin{bmatrix} 1 & 0 & 0 \\ 0 & 1 & 0 \\ 0 & 0 & 1 \end{bmatrix}$$

the identity matrix. Since the reduced row echelon form is obtained by row operations, we conclude that the first and the third properties are equivalent.

By combining this criterion and this result, the second property means that all columns of the row echelon form of A are pivot. In other words, the row echelon form has the following shape.

$$\begin{bmatrix} * & \# & \# \\ 0 & * & \# \\ 0 & 0 & * \end{bmatrix}$$

The corresponding reduced row echelon form is

$$\begin{bmatrix} 1 & 0 & 0 \\ 0 & 1 & 0 \\ 0 & 0 & 1 \end{bmatrix}$$

the identity matrix. By the similar reason as before, we conclude that the second and the third properties are equivalent.

The discussion above for the 3 by 3 matrix A can be applied to the general n by n case. Thus the equivalence of the three properties also holds in general.

A BASIC LINEAR ALGEBRA PRINCIPLE

IMPLICATION OF EXISTENCE/UNIQUENESS ON SIZE

In a row echelon form, different pivot entries are in different rows and different columns. In other words, any row or column can have at most one pivot entry. As a result, we have the following equality.

number of pivots = number of pivot rows = number of pivot columns

With the help of the equality, we have

$Ax = b$ has solutions for any right side b

$\Leftrightarrow$ All rows of A are pivot (by this criterion)

$\Leftrightarrow$ number of rows = number of pivots

$\Leftrightarrow$ number of rows = number of pivot columns (by the equality above)

Since number of pivot columns $\leq$ number of columns, we conclude the numerical consequence of always existence.

$Ax = b$ has solutions for any $b \Rightarrow$ number of rows of $A \leq$ number of columns of A

For a special case for the statement. The numerical consequence can also be rephrased.

A system has solutions for any right side $\Rightarrow$ number of equations $\leq$ number of variables

Example: Without any computation, we know the following system

$$\begin{aligned} 3.547x_1 + 5.782x_2 + 0.003x_3 &= b_1 \\ -2.358x_1 - 2.683x_2 - 3.907x_3 &= b_2 \\ -11.291x_1 + 4.902x_2 + 2.013x_3 &= b_3 \\ 7.912x_1 + 3.091x_2 + 7.892x_3 &= b_4 \end{aligned}$$

does not always have solutions, simply because 4 (number of rows) > 3 (number of columns).

Similarly, in an earlier exercise, we used row operations to conclude that the second and the third systems in the exercise do not always have solutions. In fact, the second system has more equations (number = 4) than variables (number = 3). Therefore without any computation, we may immediately conclude that the second system does not always have solutions.

Note that for the first and the third systems in the earlier exercise, we do have number of equations $\leq$ number of variables. In this case, we have to do row operations to determine whether solutions always exist.

Now we carry out a similar argument for the uniqueness case.

$Ax = b$ has a unique solution

$\Leftrightarrow$ All columns of A are pivot (by this criterion)

$\Leftrightarrow$ number of columns = number of pivots

$\Leftrightarrow$ number of columns = number of pivot rows (by the equality above)

Since number of pivot rows $\leq$ number of rows, we conclude the numerical consequence of uniqueness.

The solution of $Ax = b$ is unique $\Rightarrow$ number of rows of $A \geq$ number of columns of A

A special case can be found in this exercise. The numerical consequence can also be rephrased.

A system has a unique solution $\Rightarrow$ number of equations $\geq$ number of variables

Example: Because $3 < 4$, the following complicated system

$$\begin{aligned} 0.75x_1 + 3.55x_2 + 2.73x_3 + 0.39x_4 &= 1.22 \\ -3.24x_1 - 0.01x_2 - 6.01x_3 + 0.03x_4 &= 3.71 \\ 5.32x_1 + 4.22x_2 + 0.36x_3 - 7.56x_4 &= -0.19 \end{aligned}$$

cannot have a unique solution. In other words, it either has no solution, or has infinitely many solutions. Of course to find out which case happens, we need to do some complicated row operations.

Without any computation, we know that the solution to the following consistent system

$$\begin{aligned} x_1 + 3x_2 + 2x_3 + x_5 &= 0 \\ -x_1 - x_2 - x_3 + x_4 &= 1 \\ 4x_2 + 2x_3 + 4x_4 + 3x_5 &= 2 \\ x_1 + 3x_2 + 2x_3 - 2x_4 &= 0 \end{aligned}$$

is not unique, simply because 4 (number of rows) < 5 (number of columns).

IMPLICATION OF SIZE ON EXISTENCE/UNIQUENESS

By combining the numerical implications of the existence and the uniqueness on the size, we have

$Ax = b$ has a unique solution for any b

$\Rightarrow$ number of rows of A = number of columns of A (A is a square matrix)

$\Leftrightarrow$ number of equations = number of variables

Conversely, *assume A is an n by n matrix*. Then we have (if you find the general argument too difficult, try 3 by 3 matrix first)

$Ax = b$ has solutions for any b

$\Rightarrow$ All rows are pivot (by criterion for always existence)

$\Rightarrow$ Number of pivot rows is n (because there are n rows)

$\Rightarrow$ Number of pivot columns is n (by this equality)

$\Rightarrow$ All columns are pivot (because there are n columns)

$\Rightarrow$ The solution of a consistent system $Ax = b$ is unique (by criterion for uniqueness)

In other words, if the number of equations is equal to the number of variables, then *always existence implies uniqueness*. By similar argument, we can also prove that *uniqueness implies always existence*.

In summary, we have the following basic principle of linear algebra.

For a square matrix A, the following are equivalent

Always Existence + Uniqueness

$Ax = b$ has a unique solution for any b

Always Existence

$Ax = b$ has solutions for any b

Uniqueness

The solution of a consistent system $Ax = b$ is unique

Our discussion also tells us when the above happens from computational viewpoint.

For a square matrix A, the following are equivalent

- $Ax = b$ has a unique solution for any b
- All rows of A are pivot
- All columns of A are pivot
- A can be row operated to become I

For the claim that A can be row operated to become I, please check out more details in this exercise.

Example: The system

$$
\begin{aligned}
x_1 - x_2 + 2x_3 &= 1\\
3x_1 + x_2 - 2x_3 &= 3\\
2x_1 - x_2 + 2x_3 &= 2
\end{aligned}
$$

has $x_1 = 1, x_2 = x_3 = 0$ as an obvious solution. It is also easy to see that $x_1 = 1, x_2 = 2$, $x_3 = 1$ is another solution. Therefore the system has many solutions. Since the system

$$
\begin{aligned}
x_1 - x_2 + 2x_3 &= b_1\\
3x_1 + x_2 - 2x_3 &= b_2\\
2x_1 - x_2 + 2x_3 &= b_3
\end{aligned}
$$

has the same coefficient matrix, by the basic principle, it does not always have solutions.

The significance of the basic principle is the following: We may consider (always) existence and the uniqueness as two complementary aspects of systems of linear equations. In general, there is no relation between the two aspects. However, *in case the size is right* (square coefficient matrix, or number of variables = number of equations), the two aspects are equivalent.

Exercise: Without any computation, say as much as you can about the existence and the uniqueness for the solutions of the following systems?

1. $2x_1 + 2x_3 + x_4 = 2$

   $x_1 + x_2 - x_3 = -1$

   $x_2 - 2x_3 + x_4 = -2$

Answer: The system is consistent because of the obvious solution $x_1 = x_2 = x_4 = 0$, $x_3 = 1$. The solution is not unique (so there are infinitely many) because 3 (number of rows/equations) < 4 (number of columns/variables).

2. $2x_1 + 2x_3 = 2$

   $x_1 + x_2 - x_3 = -1$

   $x_2 - 2x_3 = -2$

Answer: The system is consistent for the the obvious solution $x_1 = x_2 = 0$, $x_3 = 1$. However, since the number of rows/equations (=3) is equal to the number of columns/variables (=3), we cannot say much about uniqueness of the solution without computation.

3 $x + y = 4$

$2x + 2y = 3$

$3x + 3y = 2$

$4x + 4y = 1$

Answer: The system is not consistent.

4. $x_1 - x_2 + x_3 + 2x_5 = a$

   $-2x_2 + 5x_3 + 2x_4 + 2x_5 = b$

   $-3x_4 + 7x_5 = c$

   $-x_5 = d$

Answer: Note that the coefficient matrix is already in row echelon form, from which we see that the solutions always exist for all *a, b, c, d*. Moreover, the solution cannot be unique because 4 < 5.

5. $1.23x_1 - 3.75x_2 + 0.02x_3 = a$

   $-4.57x_2 + 3.34x_3 = b$

   $-8.32x_1 - 3.05x_2 - 1.01x_3 = c$

   $+7.98x_3 = d$

   $4.71x_1 + 5.64x_2 + 11.26x_3 = e$

Answer: Since $5 > 3$, the system does not have solutions for any right side (i.e., we can find some specific values of a, b, c, d, e so that the system has no solution).

Moreover, if the system has solutions (for some choices of right side, such as $a = b = c = d = e = 0$), then the solutions are also the solutions of the following subsystem

$$\begin{aligned} 1.23x_1 - 3.75x_2 + 0.02x_3 &= a \\ -4.57x_2 + 3.34x_3 &= b \\ +7.98x_3 &= d \end{aligned}$$

Since the solution of the subsystem is clearly unique, we conclude the solution for the original system is also unique.

Exercise: Let A be a square matrix. Suppose for one vector b, the system $Ax = b$ has no solution. What can happen to the system $Ax = c$, where c is another vector?

Answer: If $Ax = c$ has a unique solution, then by the basic principle (the third item implies the first item), $Ax = b$ has a unique solution for any b.

Therefore under the assumption, this cannot happen. We are then left with two possibilities

- $Ax = c$ has no solution
- $Ax = c$ has infinitely many solutions

Exercise: Suppose $Ax = b$ has a unique solution for any b. What is the reduced row echelon form of a square matrix A?

Answer: The matrix must be a square matrix and the row echelon form must have all rows and all columns pivot. For the 3 by 3 case, this means

$$\begin{bmatrix} * & \# & \# \\ 0 & * & \# \\ 0 & 0 & * \end{bmatrix}$$

The corresponding reduced row echelon form is the identity matrix.

$$\begin{bmatrix} 1 & 0 & 0 \\ 0 & 1 & 0 \\ 0 & 0 & 1 \end{bmatrix}$$

In general, if $Ax = b$ has a unique solution for any b, then the reduced row echelon form of A is the identity matrix (i.e., A can be changed to the identity matrix by row operations).

Chapter 3

Linear Transformation and Matrix

VECTOR AND MATRIX

EUCLIDEAN VECTOR AND EUCLIDEAN SPACE

The n–dimensional (real) euclidean space R^n is the collection of n–tuples of numbers, called vectors. Vectors can be either presented horizontally,

$$x = (x_1, x_2, ..., x_n)$$

or vertically.

$$x = \begin{bmatrix} x_1 \\ x_2 \\ \vdots \\ x_n \end{bmatrix}$$

The only thing you need to keep in mind is that, when vectors are mixed with matrices, it is often more convenient to use vertical vectors.

The number n is the dimension of the vector x. The numbers $x_1, x_2, ..., x_n$ are called the coordinates of x. Strictly speaking, these are the coordinates with respect to the standard basis vectors

$$e_1 = (1, 0, ..., 0),$$
$$e_2 = (0, 1, ..., 0),$$
$$\vdots$$
$$e_n = (0, 0, ..., 1).$$

We also mention the special zero vector

$$0 = (0, 0, ..., 0).$$

Example: The 1–dimensional euclidean space

$R^1 = \{(x): x \text{ is a real number}\}$

can be identified with all real numbers R and be visualized as a straight line.

The 2–dimensional euclidean space

$R^2 = \{(x_1, x_2): x_1, x_2 \text{ are real numbers}\}$

can be visualized as the coordinate plane.

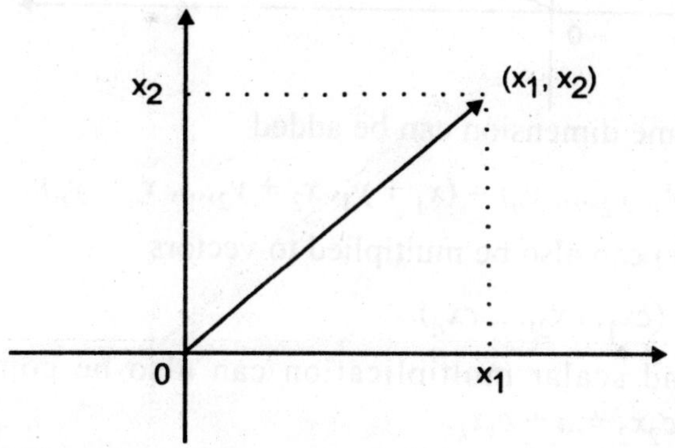

The 3–dimensional euclidean space

$R^3 = \{(x_1, x_2, x_3): x_1, x_2, x_3 \text{ are real numbers}\}$

can be visualized as the world we are living in.

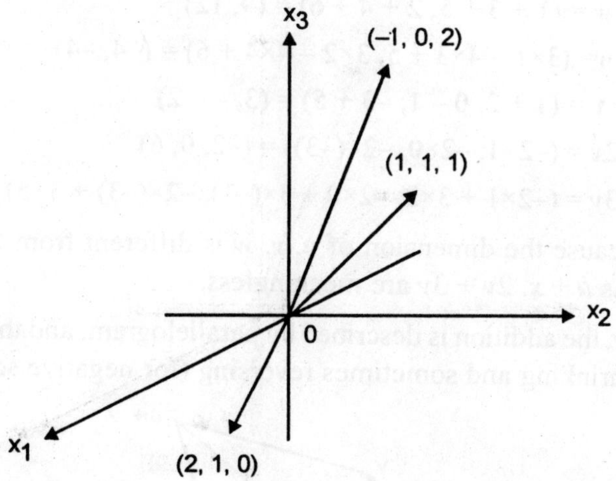

The 0–dimensional euclidean space R^0 consists of a single point – the origin.

The example illustrates that the euclidean vectors can be geometrically presented (in a coordinate system) as an arrow starting from the origin. If an arrow does not start from the origin, we need to parallel shift it so that it begins at the origin.

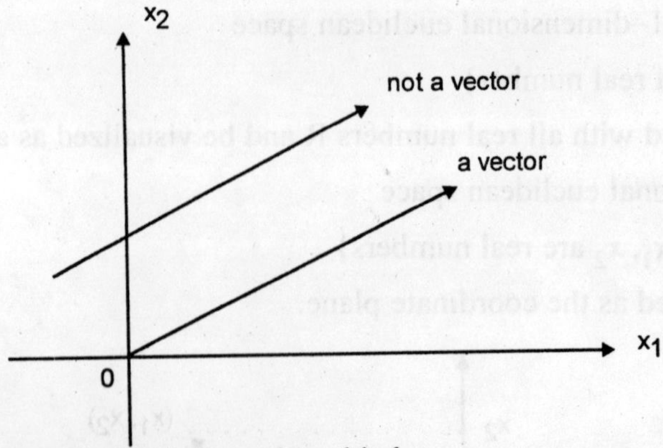

Vectors of the same dimension can be added

$(x_1, x_2,..., x_n) + (y_1, y_2,..., y_n) = (x_1 + y_1, x_2 + y_2,..., x_n + y_n)$.

A number (scalar) can also be multiplied to vectors

$c\,(x_1, x_2,..., x_n) = (cx_1, cx_2,..., cx_n)$.

The addition and scalar multiplication can also be combined to form linear combinations $c_1x_1 + c_2x_2 + ... + c_kx_k$.

Example: Let $u = (1, 2)$, $v = (3, 4)$, $w = (5, 6)$, $x = (1, 0, -3)$, $y = (2, -1, 5)$. Then

$$u + v = (1 + 3, 2 + 4) = (4, 6)$$

$$3u = (3\times1, 3\times2) = (3, 6)$$

$$u + v + w = (1 + 3 + 5, 2 + 4 + 6) = (9, 12)$$

$$3u - 4v + w = (3\times1 - 4\times3 + 5, 3\times2 - 4\times4 + 6) = (-4, -4)$$

$$x + y = (1 + 2, 0 - 1, -3 + 5) = (3, -1, 2)$$

$$-2x = (-2\times1, -2\times0, -2\times(-3)) = (-2, 0, 6)$$

$$-2x + 3y = (-2\times1 + 3\times2, -2\times0 + 3\times(-1), -2\times(-3) + 3\times5) = (4, -3, 21)$$

Moreover, because the dimension of u, v, w is different from the dimension of x, y, expressions such as $u + x$, $2v + 3y$ are meaningless.

Geometrically, the addition is described by parallelogram, and the scalar multiplication is the stretching/shrinking and sometimes reversing (for negative scalars) of vectors.

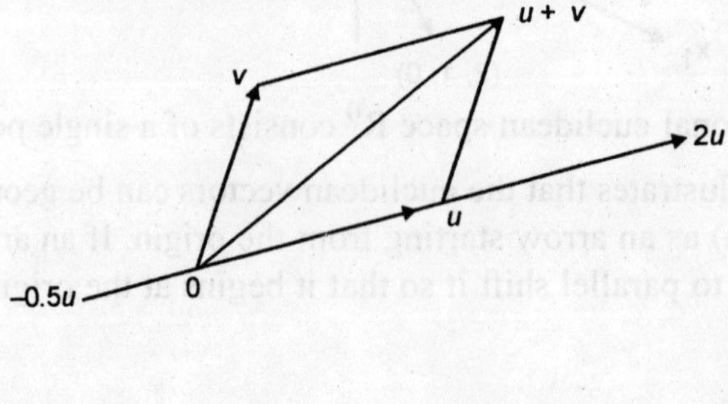

This picture shows the linear combinations of two (nonparallel) vectors. It suggests that all the linear combinations of the two vectors form a plane passing through the origin.

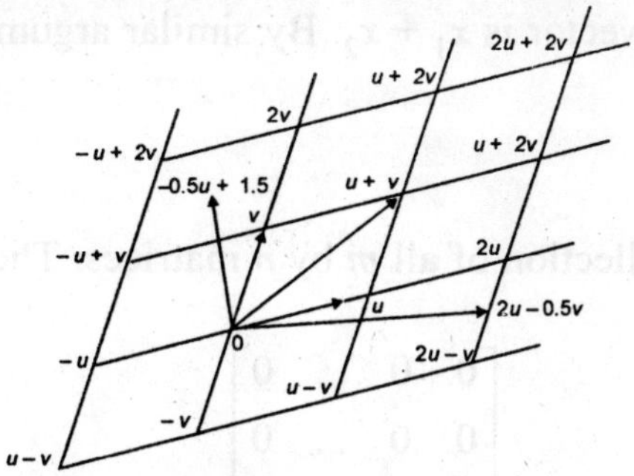

Finally, the operations have the following obvious properties.

commutative

$$u + v = v + u$$

associative

$$(u + v) + w = u + (v + w)$$

zero

$$u + 0 = u = 0 + u$$

distributive

$$r(u + v) = ru + rv$$

distributive

$$(r + s)u = ru + su$$

associative

$$(rs)u = r(su)$$

VECTOR ADDITION IS GIVEN BY PARALLELOGRAM

Let us look at the addition of two vectors (red and blue in the picture) in R^2. The green vector is constructed from the parallelogram.

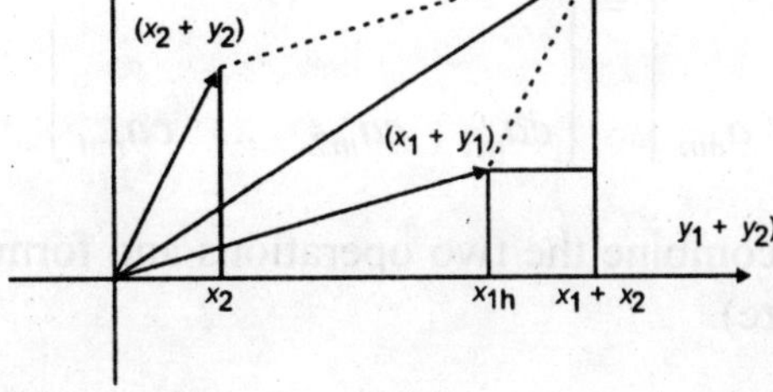

The parallelogram tells us that the two right triangles (in gray shade) are identical. In particular, the two pink segments of the x–axis have the same length. This implies that the x–coordinate of the green vector is $x_1 + x_2$. By similar argument, the y–coordinate of the green vector is $y_1 + y_2$.

MATRIX

Let $M(m, n)$ be the collection of all m by n matrices. The zero matrix is

$$O = \begin{bmatrix} 0 & 0 & \dots & 0 \\ 0 & 0 & \dots & 0 \\ \vdots & \vdots & & \vdots \\ 0 & 0 & \dots & 0 \end{bmatrix}$$

Like vectors, matrices of the same size may be added.

$$\begin{bmatrix} a_{11} & a_{12} & \dots & a_{1n} \\ a_{21} & a_{22} & \dots & a_{2n} \\ \vdots & \vdots & & \vdots \\ a_{m1} & a_{m2} & \dots & a_{mn} \end{bmatrix} + \begin{bmatrix} b_{11} & b_{12} & \dots & b_{1n} \\ b_{21} & b_{22} & \dots & b_{2n} \\ \vdots & \vdots & & \vdots \\ b_{m1} & b_{m2} & \dots & b_{mn} \end{bmatrix}$$

$$= \begin{bmatrix} a_{11}+b_{11} & a_{12}+b_{12} & \dots & a_{1n}+b_{1n} \\ a_{21}+b_{21} & a_{22}+b_{22} & \dots & a_{2n}+b_{2n} \\ \vdots & \vdots & & \vdots \\ a_{m1}+b_{m2} & a_{m2}+b_{m2} & \dots & a_{mn}+b_{mn} \end{bmatrix}$$

A number (scalar) can also be multiplied to matrices

$$c\begin{bmatrix} a_{11} & a_{12} & \dots & a_{1n} \\ a_{21} & a_{22} & \dots & a_{2n} \\ \vdots & \vdots & & \vdots \\ a_{m1} & a_{m2} & \dots & a_{mn} \end{bmatrix} = \begin{bmatrix} ca_{11} & ca_{12} & \dots & ca_{1n} \\ ca_{21} & ca_{22} & \dots & ca_{2n} \\ \vdots & \vdots & & \vdots \\ ca_{m1} & ca_{m2} & \dots & ca_{mn} \end{bmatrix}$$

Moreover, we may combine the two operations and form the linear combinations of matrices (of the same size).

Example: Let

$$A=\begin{bmatrix}1 & 1 & -2\\ 3 & 0 & 1\end{bmatrix}, B=\begin{bmatrix}-1 & 0 & 1\\ -2 & 1 & 1\end{bmatrix}, C=\begin{bmatrix}0 & 1 & 0\\ 1 & 0 & 0\end{bmatrix}, D=\begin{bmatrix}0 & 1\\ 1 & 2\end{bmatrix}$$

Then

$$A+B=\begin{bmatrix}1+(-1) & 1+0 & (-2)+1\\ 3+(-2) & 0+1 & 1+1\end{bmatrix}=\begin{bmatrix}0 & 1 & -1\\ 1 & 1 & 2\end{bmatrix}$$

$$A+B+C=\begin{bmatrix}1+(-1)+0 & 1+0+1 & (-2)+1+0\\ 3+(-2)+1 & 0+1+0 & 1+1+0\end{bmatrix}=\begin{bmatrix}0 & 2 & -1\\ 2 & 1 & 2\end{bmatrix}$$

$$2A=\begin{bmatrix}2\times 1 & 2\times 1 & 2\times(-2)\\ 2\times 3 & 2\times 0 & 2\times 1\end{bmatrix}=\begin{bmatrix}2 & 2 & -4\\ 6 & 0 & 2\end{bmatrix}$$

$$-3D=\begin{bmatrix}(-3)\times 0 & (-3)\times 1\\ (-3)\times 1 & (-3)\times 2\end{bmatrix}=\begin{bmatrix}0 & -3\\ -3 & -6\end{bmatrix}$$

$$3A-4B+2C=\begin{bmatrix}3+4+0 & 3+0+2 & -6-4+0\\ 9+8+2 & 0-4+0 & 3-4+0\end{bmatrix}=\begin{bmatrix}7 & 5 & -10\\ 19 & -4 & -1\end{bmatrix}$$

On the other hand, expressions such as $A+D, B+D, A+D$ are meaningless.

Euclidean vectors can be considered as matrices of special sizes. Specifically, a vertical n–dimensional euclidean vector is an n by 1 matrix, and the euclidean space R^n is identified with $M(n, 1)$. On the other hand, if we present vectors horizontally, then R^n is identified with $M(1, n)$.

Although euclidean vectors may be presented either vertically or horizontally, you are strongly advised to use vertical vectors when they are mixed with matrices. For example, in expressing a system of linear equation as $Ax = b$, because of the presence of the matrix A, we should consider x and b as vertical vectors.

Conversely, matrices can be considered as an (ordered) collection of vectors. Specifically, an m by n matrix is equivalent to n euclidean column vectors of dimension m.

$$A=[a_1\, a_2 \ldots\, a_n] \Leftrightarrow a_1, a_2, \ldots, a_n \in R^n.$$

Similarly, we may also consider row vectors, so that an m by n matrix is equivalent to m euclidean row vectors of dimension n.

Example: The matrix

$$A = \begin{bmatrix} 1 & 2 & 3 \\ 4 & 5 & 6 \end{bmatrix}$$

can be considered as three 2–dimensional column vectors

$$A = [a_1\ a_2\ a_3],\ a_1 = \begin{bmatrix} 1 \\ 4 \end{bmatrix},\ a_2 = \begin{bmatrix} 2 \\ 5 \end{bmatrix},\ a_3 = \begin{bmatrix} 3 \\ 6 \end{bmatrix}$$

It can also be considered as two 3–dimensional row vectors.

$$A = \begin{bmatrix} b_1 \\ b_2 \end{bmatrix},\ b_1 = (1, 2, 3),\ b_2 = (4, 5, 6)$$

The relations between the vectors and the matrices are compatible with addition and scalar multiplication. For example, for three column matrices $A = [a_1\ a_2\ a_3]$, $B = [b_1\ b_2\ b_3]$, we have

$$A + B = [a_1 + b_1\ a_2 + b_2\ a_3 + b_3],\ cA = [ca_1\ ca_2\ ca_3].$$

Because of this compatibility, the properties for the addition and scalar multiplication of vectors are also hold for matrices.

SOME MATRIX NOTATIONS AND TERMINOLOGIES

In theoretical discussions about matrices, we sometimes use the notation $A = (a_{ij})$ to denote a matrix, where a_{ij} is the entry at the i–th row and j–th column (the ij–entry). Sometimes we even indicate the size by writing $A = (a_{ij})_{mXn}$. In terms of this compact notation, we have

$$(a_{ij}) + (b_{ij}) = (a_{ij} + b_{ij}),\ c(a_{ij}) = (ca_{ij}).$$

Example:

$$(i+j)_{3\times 3} = \begin{bmatrix} 1+1 & 1+2 & 1+3 \\ 2+1 & 2+2 & 2+3 \\ 3+1 & 3+2 & 3+3 \end{bmatrix} = \begin{bmatrix} 2 & 3 & 4 \\ 3 & 4 & 5 \\ 4 & 5 & 6 \end{bmatrix}$$

$$(ij)_{4\times 5} = \begin{bmatrix} 1\times 1 & 1\times 2 & 1\times 3 & 1\times 4 & 1\times 5 \\ 2\times 1 & 2\times 2 & 2\times 3 & 2\times 4 & 2\times 5 \\ 3\times 1 & 3\times 2 & 3\times 3 & 3\times 4 & 3\times 5 \\ 4\times 1 & 4\times 2 & 4\times 3 & 4\times 4 & 4\times 5 \end{bmatrix} = \begin{bmatrix} 1 & 2 & 3 & 4 & 5 \\ 2 & 4 & 6 & 8 & 10 \\ 3 & 6 & 9 & 12 & 15 \\ 4 & 8 & 12 & 16 & 20 \end{bmatrix}$$

$$(j^2)_{4\times3} = \begin{bmatrix} 1^2 & 2^2 & 3^2 \\ 1^2 & 2^2 & 3^2 \\ 1^2 & 2^2 & 3^2 \\ 1^2 & 2^2 & 3^2 \end{bmatrix} = \begin{bmatrix} 1 & 4 & 9 \\ 1 & 4 & 9 \\ 1 & 4 & 9 \\ 1 & 4 & 9 \end{bmatrix}$$

For an n by n square matrix $A = (a_{ij})$, the terms $a_{11}, a_{22}, \ldots, a_{nn}$ are the diagonal entries of the matrix. A is a diagonal matrix if all the off–diagonal entries are zero. In other words, $a_{ij} = 0$ for $i \neq j$. The following are all the 2 by 2 and 3 by 3 diagonal matrices.

$$\begin{bmatrix} a & 0 \\ 0 & b \end{bmatrix}$$

$$\begin{bmatrix} a & 0 & 0 \\ 0 & b & 0 \\ 0 & 0 & c \end{bmatrix}$$

A is an upper triangular matrix if all the entries below the diagonal are zero. In other words, $a_{ij} = 0$ for $i > j$. The following are all the 2 by 2 and 3 by 3 upper triangular matrices.

$$\begin{bmatrix} a & b \\ 0 & c \end{bmatrix}$$

$$\begin{bmatrix} a & b & c \\ 0 & d & e \\ 0 & 0 & f \end{bmatrix}$$

Similarly, A is a lower triangular matrix if all the entries above the diagonal are zero. In other words, $a_{ij} = 0$ for $i < j$. The following are all the 2 by 2 and 3 by 3 upper triangular matrices.

$$\begin{bmatrix} a & 0 \\ b & c \end{bmatrix}$$

$$\begin{bmatrix} a & 0 & 0 \\ b & c & 0 \\ d & e & f \end{bmatrix}$$

Clearly, A is upper and lower triangular at the same time if and only if it is diagonal.

Exercise: Draw the vectors (1, 3), (0, 1), (0, 0), (–1, –3), (2, –2).

Answer:

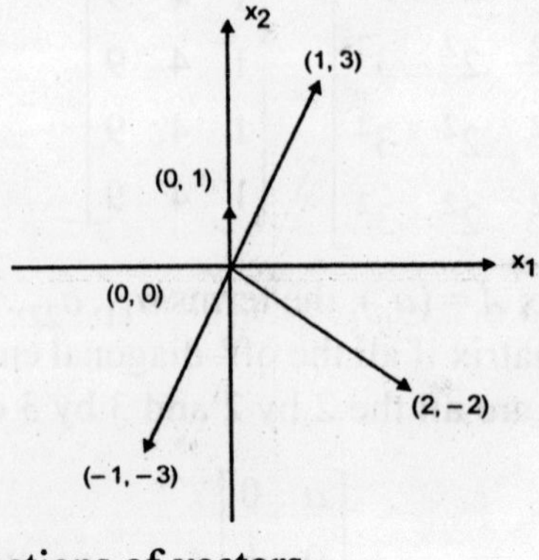

Exercise: Draw the collections of vectors.

$$X = \{(0, y): y \text{ is any number}\}$$
$$Y = \{(x, 0): x \text{ is any number}\}$$
$$S = \{(x, y): x^2 + y^2 = 1\}$$
$$D = \{(x, x): x \text{ is any number}\}$$
$$E = \{(x, y): x + 3y = 0\}$$

Answer:

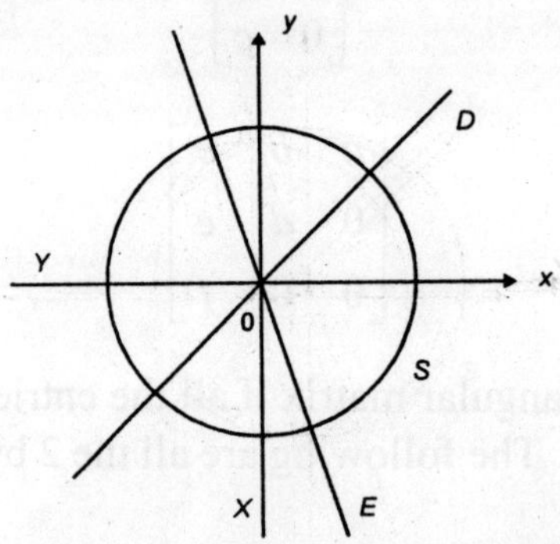

Exercise: Let $u = (1,-5)$, $v = (2, 4,1)$, $w = (3, 0, 1, -1, 4)$, $x = (2, -4, 3)$, $y = (-7, 2, 1)$, $z = (1, 2, 0, -3, 4)$. Compute.

$$u + v,\ v + y,\ y + v,\ w + z,\ z + w,\ x + y + v,$$
$$2u, -3v, -w, 0x,$$
$$2u + 5v, -2y + 3v,\ w - 3z,\ 4x + 2y + v.$$

Answer: $u + v$ is meaningless

$$v + y = (2 - 7, 4 + 2, 1 + 1) = (-5, 6, 2)$$
$$y + v = (-7 + 2, 2 + 4, 1 + 1) = (-5, 6, 2)$$

$$w + z = (3 + 1, 0 + 2, 1 + 0, -1 - 3, 4 + 4) = (4, 2, 1, -4, 8)$$

$$z + w = (1 + 3, 2 + 0, 0 + 1, -3 - 1, 4 + 4) = (4, 2, 1, -4, 8)$$

$$x + y + v = (2 - 7 + 2, -4 + 2 + 4, 3 + 1 + 1) = (-3, 2, 5)$$

$$2u = (2\times 1, 2\times(-5)) = (2, -10)$$

$$-3v = (-3\times 2, -3\times 4, -3\times 1) = (-6, -12, -3)$$

$$-w = -(3, 0, 1, -1, 4) = (-3, 0, -1, 1, -4)$$

$$0x = (0, 0, 0)$$

$2u + 5v$ is meaningless

$$-2y + 3v = -2(-7, 2, 1) + 3(2, 4, 1) = (-2\times(-7) + 3\times 2, -2\times 2 + 3\times 4, -2\times 1 + 3\times 1) = (20, 8, 1)$$

$$w - 3z = (3, 0, 1, -1, 4) - 3(1, 2, 0, -3, 4) = (3 - 3\times 1, 0 - 3\times 2, 1 - 3\times 0, -1 - 3\times(-3), 4 - 3\times 4) = (0, -6, 1, 8, -8)$$

$$4x + 2y + v = 4(2, -4, 3) + 2(-7, 2, 1) + (2, 4, 1) = (4\times 2 + 2\times(-7) + 2, 4\times(-4) + 2\times 2 + 4, 4\times 3 + 2\times 1 + 1) = (-4, 8, 15)$$

Exercise: Compute.

Answer:

$$\begin{bmatrix}1\\3\end{bmatrix} - \begin{bmatrix}5\\6\end{bmatrix} = \begin{bmatrix}-4\\-3\end{bmatrix}, \begin{bmatrix}2\\4\end{bmatrix} - \begin{bmatrix}7\\8\end{bmatrix} = \begin{bmatrix}-5\\-4\end{bmatrix}, 3\begin{bmatrix}1\\3\end{bmatrix} - 5\begin{bmatrix}5\\6\end{bmatrix}$$

$$= \begin{bmatrix}-22\\-21\end{bmatrix}, 3\begin{bmatrix}2\\4\end{bmatrix} - 5\begin{bmatrix}7\\8\end{bmatrix} = \begin{bmatrix}-29\\-28\end{bmatrix}$$

$$\begin{bmatrix}1 & 2\\3 & 4\end{bmatrix} - \begin{bmatrix}5 & 7\\6 & 8\end{bmatrix} = \begin{bmatrix}-4 & -5\\-3 & -4\end{bmatrix}, 3\begin{bmatrix}1 & 2\\3 & 4\end{bmatrix} - 5\begin{bmatrix}5 & 7\\6 & 8\end{bmatrix}$$

$$\begin{bmatrix}0 & -2 & 3\\1 & 0 & -3\\2 & -2 & 2\end{bmatrix} + \begin{bmatrix}2 & 0 & -1\\0 & 4 & -3\\3 & -3 & 0\end{bmatrix} = \begin{bmatrix}2 & -2 & 2\\1 & 4 & -6\\5 & -5 & 2\end{bmatrix}, -2\begin{bmatrix}0 & -2 & 3\\1 & 0 & -3\\2 & -2 & 2\end{bmatrix}$$

$$+3\begin{bmatrix}2 & 0 & -1\\0 & 4 & -3\\3 & -3 & 0\end{bmatrix} = \begin{bmatrix}4 & 4 & -9\\-2 & 12 & -15\\5 & -5 & -4\end{bmatrix}$$

$$a\begin{bmatrix}1 & 0 & 0\\0 & 0 & 0\\0 & 0 & 0\end{bmatrix} + b\begin{bmatrix}0 & 0 & 0\\0 & 1 & 0\\0 & 0 & 0\end{bmatrix} + c\begin{bmatrix}0 & 0 & 0\\0 & 0 & 0\\0 & 0 & 1\end{bmatrix} = \begin{bmatrix}a & 0 & 0\\0 & b & 0\\0 & 0 & c\end{bmatrix}$$

$$a_{11}\begin{bmatrix}1 & 0\\0 & 0\end{bmatrix}+a_{12}\begin{bmatrix}0 & 1\\0 & 0\end{bmatrix}+a_{21}\begin{bmatrix}0 & 0\\1 & 0\end{bmatrix}+a_{22}\begin{bmatrix}0 & 0\\0 & 1\end{bmatrix}=\begin{bmatrix}a_{11} & a_{12}\\a_{21} & a_{22}\end{bmatrix}$$

$$\begin{bmatrix}1 & -3 & 5 & 10\\0 & 5 & 1 & -7\\1 & 0 & 9 & -3\\2 & 0 & 1 & 0\\-6 & 3 & 0 & 0\end{bmatrix}+\begin{bmatrix}4 & -9 & 8\\4 & 0 & 3\\1 & 4 & 0\\0 & 0 & 4\\0 & 5 & 3\end{bmatrix}=?,12\begin{bmatrix}0 & 0 & 0 & 0 & 0\\0 & 0 & 0 & 0 & 0\\0 & 0 & 0 & 0 & 0\end{bmatrix}-13\begin{bmatrix}0 & 0 & 0 & 0 & 0\\0 & 0 & 0 & 0 & 0\\0 & 0 & 0 & 0 & 0\\0 & 0 & 0 & 0 & 0\\0 & 0 & 0 & 0 & 0\end{bmatrix}$$

The last two are meaningless because the sizes do not match.

Exercise: Take the standard basis vectors $e_1 = (1, 0, 0)$, $e_2 = (0, 1, 0)$, $e_3 = (0, 0, 1)$ of R^3. Compute the linear combination $x_1e_1 + x_2e_2 + x_3e_3$. Then generalize your result to other dimensions.

Answer: $x_1e_1 + x_2e_2 + x_3e_3 = x_1(1, 0, 0) + x_2(0, 1, 0) + x_3(0, 0, 1) = (x_1, 0, 0) + (0, x_2, 0) + (0, 0, x_3) = (x_1, x_2, x_3)$. In general, for the standard basis $e_1 = (1, 0,..., 0)$, $e_2 = (0, 1,..., 0)$,..., $e_n = (0, 0,..., 1)$ of R^n, we have $(x_1, x_2,..., x_n) = x_1e_1 + x_2e_2 +... + x_ne_n$.

Exercise: For $u = (1, 2)$, $v = (3, 4)$, $w = (5, 6)$, compute the linear combination $x_1u + x_2v + x_3w$. Do the same for $a_1 = (1, 2, 3)$, $a_2 = (4, 5, 6)$, $a_3 = (7, 8, 9)$. Express your computation in vertical vectors and draw a general observation.

Answer:

$$x_1\begin{bmatrix}1\\2\end{bmatrix}+x_2\begin{bmatrix}3\\4\end{bmatrix}+x_3\begin{bmatrix}5\\6\end{bmatrix}=\begin{bmatrix}x_1\\2x_1\end{bmatrix}+\begin{bmatrix}3x_2\\4x_2\end{bmatrix}+\begin{bmatrix}5x_3\\6x_3\end{bmatrix}=\begin{bmatrix}x_1+3x_2+5x_3\\2x_1+4x_2+6x_3\end{bmatrix}$$

$$x_1\begin{bmatrix}1\\2\\3\end{bmatrix}+x_2\begin{bmatrix}4\\5\\6\end{bmatrix}+x_3\begin{bmatrix}7\\8\\9\end{bmatrix}=\begin{bmatrix}x_1\\2x_1\\3x_1\end{bmatrix}+\begin{bmatrix}4x_2\\5x_2\\6x_2\end{bmatrix}+\begin{bmatrix}7x_3\\8x_3\\9x_3\end{bmatrix}=\begin{bmatrix}x_1+4x_2+7x_3\\2x_1+5x_2+8x_3\\3x_1+6x_2+9x_3\end{bmatrix}$$

The results are of the form Ax, with the matrix A being respectively

$$\begin{bmatrix}1 & 3 & 5\\2 & 4 & 6\end{bmatrix}=[u\ v\ w]$$

$$\begin{bmatrix}1 & 4 & 7\\2 & 5 & 8\\3 & 6 & 9\end{bmatrix}=[a_1\ a_2\ a_3]$$

In general, we have

$x_1a_1 + x_2a_2 + ... + x_na_n = Ax$

where $A = [a_1\ a_2 ... a_n]$ is a matrix with the vectors $a_1, a_2, ..., a_n$ as columns.

Exercise: Express the vectors $b = (4, 6)$, $c = (1, 1)$, $e_1 = (1, 0)$, $e_2 = (0, 1)$ as linear combinations of the vectors $u = (1, 2)$, $v = (3, 4)$, $w = (5, 6)$. Can you express as linear combinations of the vectors u, v only?

Answer: We would like to write b, c, e_1, e_2 in the form $x_1(1, 2) + x_2(3, 4) + x_3(5, 6)$. From the previous exercise, we have

$$x_1\begin{bmatrix}1\\2\end{bmatrix} + x_2\begin{bmatrix}3\\4\end{bmatrix} + x_3\begin{bmatrix}5\\6\end{bmatrix} = \begin{bmatrix}x_1 + 3x_2 + 5x_3\\2x_1 + 4x_2 + 6x_3\end{bmatrix}$$

Therefore the problem is to solve the system of equations

$$x_1 + 3x_2 + 5x_3 = b_1$$
$$2x_1 + 4x_2 + 6x_3 = b_2$$

For $b_1 = 4$, $b_2 = 6$ on the right side, we find $x_1 = 1$, $x_2 = 1$ is one solution. Therefore $b = u + v$. In fact, the solution is not unique. For example, $x_1 = 2$, $x_2 = -1$, $x_3 = 1$ is another solution, from which we get $b = 2u - v + w$.

Incidentally, we have already expressed b as a linear combination of u and v: $b = u + v$. Such expressions are equivalent to solutions of

$$x_1 + 3x_2 = b_1$$
$$2x_1 + 4x_2 = b_2$$

We do the similar computation for the other vectors and find $c = -1/2\ u + 1/2\ v$, $e_1 = -2u + v$, $e_2 = 3/2\ u - 1/2\ v$.

Exercise: Express the vectors $b = (1, 1, 1)$, $c = (-1, 0, 1)$, $e_1 = (1, 0, 0)$ as linear combinations of the vectors $a_1 = (1, 2, 3)$, $a_2 = (4, 5, 6)$, $a_3 = (7, 8, 9)$. Is the expression (i.e., the coefficients) unique?

Answer: We would like to write b, c, e_1 in the form $x_1(1, 2, 3) + x_2(4, 5, 6) + x_3(7, 8, 9)$. The computation is similar to the previous exercise. From this exercise, we have

$$x_1\begin{bmatrix}1\\2\\3\end{bmatrix} + x_2\begin{bmatrix}4\\5\\6\end{bmatrix} + x_3\begin{bmatrix}7\\8\\9\end{bmatrix} = \begin{bmatrix}x_1 + 4x_2 + 7x_3\\2x_1 + 5x_2 + 8x_3\\3x_1 + 6x_2 + 9x_3\end{bmatrix}$$

Therefore the problem is to solve the system.

$$x_1 + 4x_2 + 7x_3 = b_1$$

$$2x_1 + 5x_2 + 8x_3 = b_2$$
$$3x_1 + 6x_2 + 9x_3 = b_3$$

For $b_1 = b_2 = b_3 = 1$ on the right side, we find $x_1 = -1/3, x_2 = 1/3, x_3 = 0$ is one solution. Therefore $b = -1/3\ a_1 + 1/3\ a_2$. Since the solution is not unique, the expression is not unique. Another possibility is $b = -1/6\ a_1 + 1/6\ a_3$.

For $b_1 = -1, b_2 = 0, b_3 = 1$ on the right side, we find $x_1 = 4/3, x_2 = 0, x_3 = -1/3$ is one solution. Therefore $c = 4/3\ a_1 - 1/3\ a_3$. Again the expression is not unique.

For $b_1 = 1, b_2 = 0, b_3 = 0$ on the right side, there is no solution. Therefore e_1 is not a linear combination of a_1, a_2, a_3.

Exercise: Express the matrix

$$X = \begin{bmatrix} 3 & 4 \\ 2 & 3 \end{bmatrix}$$

as linear combinations of the matrices.

$$A = \begin{bmatrix} 1 & 2 \\ 0 & 1 \end{bmatrix}, B = \begin{bmatrix} 1 & 0 \\ 2 & 1 \end{bmatrix}, C = \begin{bmatrix} 0 & 1 \\ -1 & 0 \end{bmatrix}$$

Is the expression (i.e., the coefficients) unique? Is it true that any 2 by 2 matrix is a linear combination of A, B and C?

Answer: By comparing the four entries on the both side of the equation $x_1A + x_2B + x_3C = X$,

$$x_1 \begin{bmatrix} 1 & 2 \\ 0 & 1 \end{bmatrix} + x_2 \begin{bmatrix} 1 & 0 \\ 2 & 1 \end{bmatrix} + x_3 \begin{bmatrix} 0 & 1 \\ -1 & 0 \end{bmatrix} = \begin{bmatrix} 3 & 4 \\ 2 & 3 \end{bmatrix}$$

we have the following system of equations.

$$x_1 + x_2 = 3$$
$$2x_1 + x_3 = 4$$
$$2x_2 - x_3 = 2$$
$$x_1 + x_2 = 3$$

From the row echelon form of the augmented matrix,

$$\begin{bmatrix} 1 & 1 & 0 & 3 \\ 0 & 2 & -1 & 2 \\ 0 & 0 & 0 & 0 \\ 0 & 0 & 0 & 0 \end{bmatrix}$$

we see that the system has non–unique solutions. One solution is $x_1 = 2$, $x_2 = 1$, $x_3 = 0$. Therefore $X = 2A + 1B + 0C$. Moreover, the expression is not unique.

For a general 2 by 2 matrix $X = (b_{ij})$ (i.e., b_{ij} is the entry of X at the i–th row and j–th column), the corresponding system is

$$x_1 + x_2 = b_{11}$$
$$2x_1 + x_3 = b_{12}$$
$$2x_2 - x_3 = b_{21}$$
$$x_1 + x_2 = b_{22}$$

Then from the row echelon form above (or by the fact that there are more equations than variables), we see that solutions do not always exist. In other words, not any X can be expressed as a linear combination of A, B and C.

Remark: In an earlier exercise, we have seen how the problem of expressing a vector as a linear combination of some other vectors is equivalent to solving a system of linear equations. This exercise shows that we have similar relation for matrices.

LINEAR TRANSFORMATION

TRANSFORMATION

Transformations are rules that convert one data into another data. We may denote a transformation by

$T: X \to Y$,

where T is the transformation (also called map, function), X is the source set (or domain), and Y is the target set (or codomain). Taking any element $x \in X$ as input, the transformation produces an element $y = T(x) \in Y$ as output. We also call y the image (or value) of x under the transformation T. We may also include more specific information about the transformation by writing $T: X \to Y, x \to$

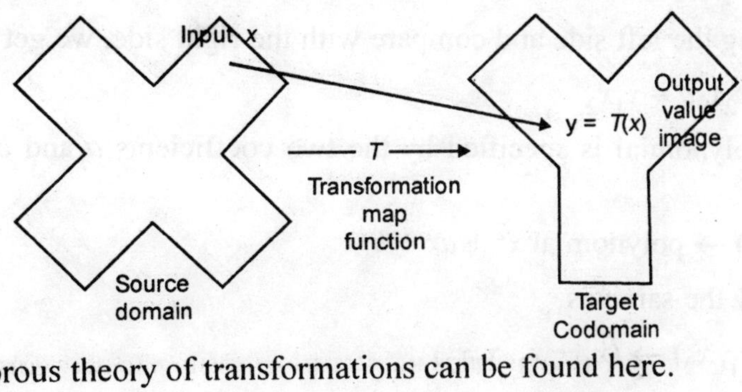

A more rigorous theory of transformations can be found here.

Example: The following are some transformations in everyday life.

Age: people $\to$ number

ID_s: student $\to$ number

ID_p: professor $\to$ number

Instructor: course $\to$ professor

Maker: product $\to$ manufacturer

Capital City: country $\to$ city

Population: city $\to$ number

Example In mathematics, transformations are often given by formulae.

Square: $R \to R, x \to x^2$

Reflection in x–axis: $R^2 \to R^2, (x, y) \to (x, -y)$

Polar–Cartesian: $R^2 \to R^2, (r, \theta) \to (r\cos\theta, r\sin\theta)$

Root–Polynomial: $R^2 \to R^2, (x_1, x_2) \to (x_1 + x_2, x_1x_2)$

$T: R^3 \to R^2, (x_1, x_2, x_3) \to (x_1 + 2x_2 + 3x_3, 4x_1 + 5x_2 + 6x_3)$

Here are two examples of mathematical transformations given by descriptions instead of formulae.

Sign: $R \to R$, positive number $\to 1$, negative number $\to -1$, $0 \to 0$

Dirichlet function: $R \to R$, rational number $\to 1$, irrational number $\to 0$

ROOT-POLYNOMIAL TRANSFORMATION

If x_1, x_2 are the roots of a quadratic polynomial $x^2 + ax + b$, then we have

$(x - x_1)(x - x_2) = x^2 + ax + b$.

By expanding the left side and compare with the right side, we get

$a = -(x_1 + x_2), b = x_1x_2.$

Since the polynomial is specified by the two coefficients a and b, we see that the transformation

roots $(x_1, x_2) \to$ polynomial $x^2 + ax + b$

is essentially the same as

$R^2 \to R^2, (x_1, x_2) \to (x_1 + x_2, x_1x_2)$

As an exercise for yourself, how would you define the transformation *Root–Polynomial*: $R^3 \to R^3$ for three roots of polynomials of degree 3?

Finally, you may try the general case *Root–Polynomial*: $R^n \to R^n$.

MATRIX TRANSFORMATION

We may think of the left side of a system of linear equations as a transformation. For example, the left side of the system

$$x_1 + 2x_2 + 3x_2 = b_1$$
$$4x_1 + 5x_2 + 6x_4 = b_2$$

can be considered as the linear transformation T in the example above. By expressing vectors in vertical form, we also see that

$$T\begin{bmatrix} x_1 \\ x_2 \\ x_3 \end{bmatrix} = \begin{bmatrix} x_1 + 2x_2 + 3x_3 \\ 4x_1 + 5x_2 + 6x_3 \end{bmatrix} = \begin{bmatrix} 1 & 2 & 3 \\ 4 & 5 & 6 \end{bmatrix}\begin{bmatrix} x_1 \\ x_2 \\ x_3 \end{bmatrix}$$

In particular, the formula for the transformation T is given by a matrix A:

$$T(x) = Ax.$$

In general, for any m by n matrix A, we may construct a transformation T: $R^n \to R^m$ by the above formula. Such transformations are called matrix transformations.

Example: The following matrices

$$\begin{bmatrix} 1 & -1 & 1 \\ 3 & 1 & -1 \\ 2 & 2 & 1 \end{bmatrix}$$

$$\begin{bmatrix} 1 & 3 & 2 & 0 & 1 \\ -1 & -1 & -1 & 1 & 0 \\ 0 & 4 & 2 & 4 & 3 \\ 1 & 3 & 2 & -2 & 0 \end{bmatrix}$$

give us matrix transformations,

$$T_1\begin{bmatrix} x_1 \\ x_2 \\ x_3 \end{bmatrix} = \begin{bmatrix} x_1 - x_2 + x_3 \\ 3x_1 + x_2 - x_3 \\ 2x_1 + x_2 + x_3 \end{bmatrix} : R^3 \to R^3$$

$$T_1\begin{bmatrix} x_1 \\ x_2 \\ x_3 \\ x_4 \\ x_5 \\ x_6 \end{bmatrix} = \begin{bmatrix} x_1 + 3x_2 + 2x_3 + x_5 \\ -x_1 - x_2 - x_3 + x_4 + x_6 \\ 4x_2 + 2x_3 + 4x_4 + 3x_5 + 3x_6 \\ x_1 + 3x_2 + 2x_3 - 2x_4 \end{bmatrix} : \mathrm{R}^6 \to \mathrm{R}^4$$

or written horizontally,

T_1: $\mathrm{R}^3 \to \mathrm{R}^3$, $(x_1, x_2, x_3) \to (x_1 - x_2 + x_3, 3x_1 + x_2 - x_3, 2x_1 + 2x_2 + x_3)$

T_2: $\mathrm{R}^6 \to \mathrm{R}^4$, $(x_1, x_2, x_3, x_4, x_5) \to (x_1 + 3x_2 + 2x_3 + x_5, -x_1 - x_2 - x_3 + x_4 + x_6, 4x_2 + 2x_3 + 4x_4 + 3x_5 + 3x_6, x_1 + 3x_2 + 2x_3 - 2x_4)$

Example The transformations

S: $\mathrm{R}^3 \to \mathrm{R}^3$, $(x_1, x_2, x_3) \to (x_1 - x_2 + x_3, 3x_1 + x_2 - x_3 - 2, 2x_1 + 2x_2 + x_3)$

T: $\mathrm{R}^2 \to \mathrm{R}^2$, $(x, y) \to (xy, x^2)$

are not matrix transformations because they cannot be expressed as Ax for a *constant* matrix A.

LINEAR TRANSFORMATION

The following example should convince you that the expression Ax is linear:

$A(x + y) = Ax + Ay$, $A(cx) = cAx$.

A more general proof can be found here.

Example: Let

$$A = \begin{bmatrix} a & b \\ c & d \end{bmatrix}$$

The corresponding linear transformation is

$$T\begin{bmatrix} x_1 \\ x_2 \end{bmatrix} = \begin{bmatrix} ax_1 + bx_2 \\ cx_1 + dx_2 \end{bmatrix} : \mathrm{R}^2 \to \mathrm{R}^2$$

We have

$$T\begin{bmatrix} x_1 + y_1 \\ x_2 + y_2 \end{bmatrix} = \begin{bmatrix} a(x_1 + y_1) + b(x_2 + y_2) \\ c(x_1 + y_1) + d(x_2 + y_2) \end{bmatrix} = \begin{bmatrix} ax_1 + bx_2 \\ cx_1 + dx_2 \end{bmatrix} + \begin{bmatrix} ay_1 + by_2 \\ cy_1 + dy_2 \end{bmatrix}$$

$$= T\begin{bmatrix} x_1 \\ x_2 \end{bmatrix} + T\begin{bmatrix} y_1 \\ y_2 \end{bmatrix}$$

and (we use k instead of c as the scalar to avoid confusion in notations)

$$T\begin{bmatrix} kx_1 \\ kx_2 \end{bmatrix} = \begin{bmatrix} akx_1 + bkx_2 \\ ckx_2 + dkx_2 \end{bmatrix} = \begin{bmatrix} k(ax_1 + bkx_2) \\ k(cx_1 + dx_2) \end{bmatrix} = k\begin{bmatrix} ay_1 + by_2 \\ cy_1 + dy_2 \end{bmatrix} kT\begin{bmatrix} x_1 \\ x_2 \end{bmatrix}$$

You are strongly encouraged to verify the linearity for matrices of other sizes.

The discussion suggests us to introduce the following concept.

A transformation T: $R^n \to R^m$ is a linear transformation if it satisfies

$$T(u + v) = T(u) + T(v),\ T(cu) = cT(u).$$

By taking $c = 0$ in the definition, we have

$$T(0) = 0.$$

Moreover, by combining the two properties in the definition, we see that linear transformations preserve linear combinations.

$$T(c_1u_1 + c_2u_2 + ... + c_ku_k) = c_1T(u_1) + c_2T(u_2) + ... + c_kT(u_k).$$

Example Suppose T: $R^2 \to R^2$ is a linear transformation. Suppose we also know (recall $e_1 = (1, 0)$, $e_2 = (0, 1)$)

$$T(e_1) = (1, 2),\ T(e_2) = (3, 4)$$

Then by

$$(5, -3) = 5\,(1, 0) - 3(0,1) = 5e_1 - 3e_2,$$
$$(-2, 4) = -2e_1 + 4e_2,$$

we have

$$T(5, -3) = 5T(e_1) - 3T(e_2) = 5(1, 2) - 3(3, 4) = (-4, -2),$$
$$T(-2, 4) = -2T(e_1) + 4T(e_2) = -2(1, 2) + 4(3, 4) = (10, 12).$$

In general, by

$$(x_1, x_2) = x_1e_1 + x_2e_2,$$

we have

$$T(x_1, x_2) = x_1T(e_1) + x_2T(e_2) = x_1(1, 2) + x_2(3, 4) = (x_1 + 3x_2, 2x_1 + 4x_2).$$

We remark that T is a matrix transformation given by

$$\begin{bmatrix} 1 & 3 \\ 2 & 4 \end{bmatrix} = [T(e_1)\ T(e_2)]$$

Example: The transformation

$$S(x_1, x_2, x_3) = (2x_1 + 3x_2 - 7x_3 + 1, 10x_1 - 4x_2 - 8x_3, -4x_1 + x_2 - 5)$$

is not a linear transformation because

$$S(0) = S(0, 0, 0) = (1, 0, -5) \text{ ``` } 0$$

The transformation

$$T(x, y) = (x + y, xy)$$

does satisfy $T(0) = 0$. However, we have

$T(1, 0) + T(0, 1) = (1 + 0, 1X0) + (0 + 1, 0X1) = (1, 0) + (1, 0) = (2, 0)$.

This is different from

$T((1, 0) + (0, 1)) = T(1, 1) = (1 + 1, 1X1) = (2, 1)$.

Therefore T is not linear.

Since the transformations are not linear, they are not matrix transformations.

The linearity of Ax helps us to construct solutions of systems of linear equations. The key idea is that, if x and x' are respectively solutions of $Ax = b$ and $Ax = b'$, then $cx + c'x'$ is a solution of $Ax = cb + c'b'$.

Example: The vectors $\boldsymbol{x} = (1, 1, 2)$ and $x' = (0, 1, -1)$ are the respective solutions of the following systems.

$$x_1 - x_2 + x_3 = 2$$
$$3x_1 + x_2 - x_3 = 2$$
$$2x_1 + 2x_2 + x_3 = 6$$
$$x_1 - x_2 + x_3 = -2$$
$$3x_1 + x_2 - x_3 = 2$$
$$2x_1 + 2x_2 + x_3 = 1$$

Note that the two systems have the same left side. The following systems

$$x_1 - x_2 + x_3 = 0$$
$$3x_1 + x_2 - x_3 = 4$$
$$2x_1 + 2x_2 + x_3 = 7$$
$$x_1 - x_2 + x_3 = 1$$
$$3x_1 + x_2 - x_3 = 1$$

$$2x_1 + 2x_2 + x_3 = 3$$
$$x_1 - x_2 + x_3 = 8$$
$$3x_1 + x_2 - x_3 = -4$$
$$2x_1 + 2x_2 + x_3 = 3$$

still have the same left side but different right sides. By

$$(0, 4, 7) = (2, 2, 6) + (-2, 2\ 1),$$
$$(1, 1, 3) = 1/2\ (2, 2, 6),$$
$$(8, -4, 3) = (2, 2, 6) - 3(-2, 2\ 1),$$

we see that

$$x + x' = (1, 1, 2) + (0, 1, -1) = (1, 2, 1),$$
$$1/2\ x = 1/2\ (1, 1, 2) = (1/2, 1/2, 1),$$
$$x - 3x' = (1, 1, 2) - 3(0, 1, -1) = (1, -2, 5),$$

are the respective solutions of the three new systems.

MATRIX TRANSFORMATIONS ARE LINEAR

Theorem $A(x + y) = Ax + Ay$, $A(cx) = cAx$.

Proof By an earlier exercise, we have

$x_1a_1 + x_2a_2 + ... + x_na_n = Ax$

where $A = [a_1\ a_2 ...\ a_n]$ is a matrix with the vectors $a_1, a_2, ..., a_n$ as columns, and $x_1, x_2, ..., x_n$ are the coordinates of x. Thus

$$\begin{aligned} Ax + Ay &= x_1a_1 + x_2a_2 + ... + x_na_n + y_1a_1 + y_2a_2 + ... + y_na_n \\ &= (x_1a_1 + y_1a_1) + (x_2a_2 + y_2a_2) + ... + (x_na_n + y_na_n) \\ &= (x_1 + y_1)a_1 + (x_2 + y_2)a_2 + ... + (x_n + y_n)a_n \end{aligned}$$

Since $x_1 + y_1, x_2 + y_2, ..., x_n + y_n$ are the coordinates of $x + y$, we see that the summation is $A(x + y)$.

The proof for $A(cx) = cAx$ is similar.

LINEARITY IN PICTURE

We can often visualize transformations between euclidean spaces. For example, the name of transformation *Reflection in x–axis*: $R^2 \to R^2$, $(x, y) \to (x, -y)$ is inspired by the picture.

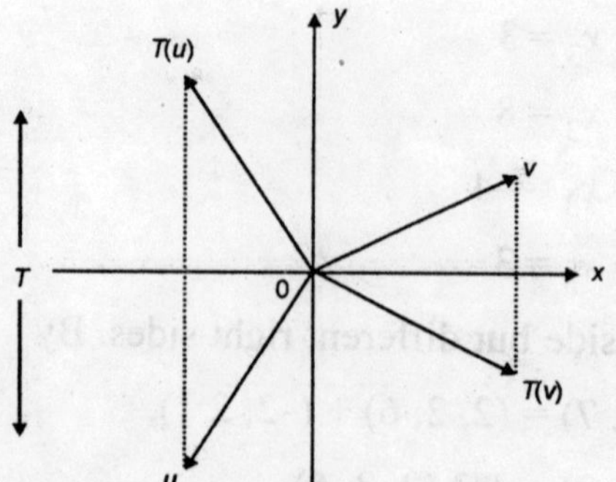

Example: The transformation *Reflection in y–axis*: $R^2 \to R^2$, $(x, y) \to (-x, y)$

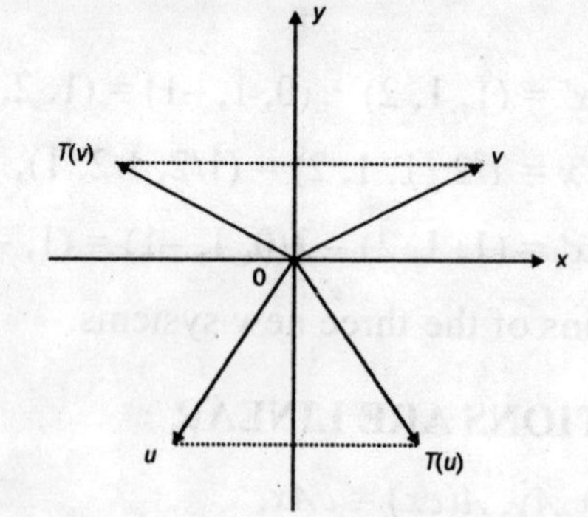

is given by the matrix

$$\begin{bmatrix} -1 & 0 \\ 0 & 1 \end{bmatrix}$$

The transformation *Rotation by &\theta;*: $R^2 \to R^2$, $(x, y) \to (x\cos\theta + y\sin\theta, -x\sin\theta + y\cos\theta)$

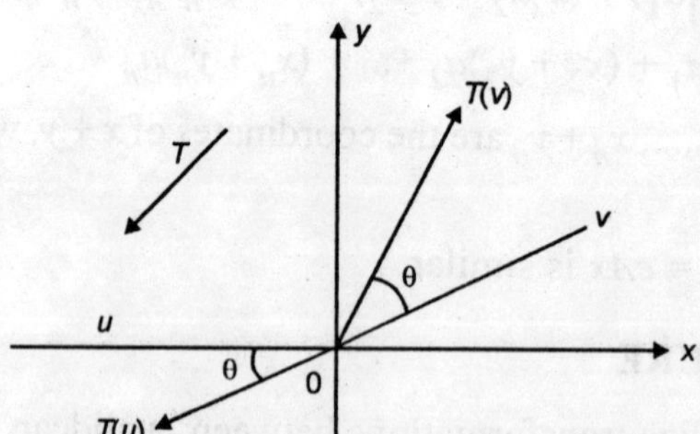

is given by the matrix (remember writing vectors vertically)

$$\begin{bmatrix} \cos\theta & -\sin\theta \\ \sin\theta & \cos\theta \end{bmatrix}$$

The transformation *Projection to* (x_1, x_3)*–plane*: $R^3 \to R^2$, $(x_1, x_2, x_3) \to (x_1, x_3)$

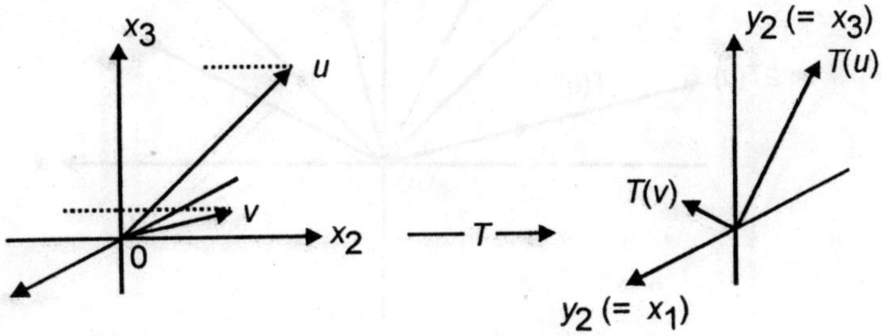

is given by the matrix

$$\begin{bmatrix} 1 & 0 & 0 \\ 0 & 0 & 1 \end{bmatrix}$$

Linear transformations are characterized by the property that addition and scalar multiplication are preserved. Geometrically, this means that parallelograms and stretching/ shrinking are preserved.

Example: The transformation *Reflection in x–axis* clearly preserves parallelograms and stretching/shrinking.

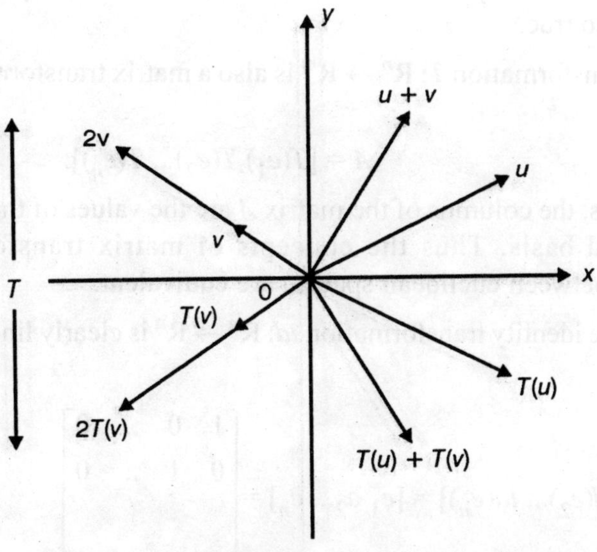

The rotation transformation also preserves the two operations.

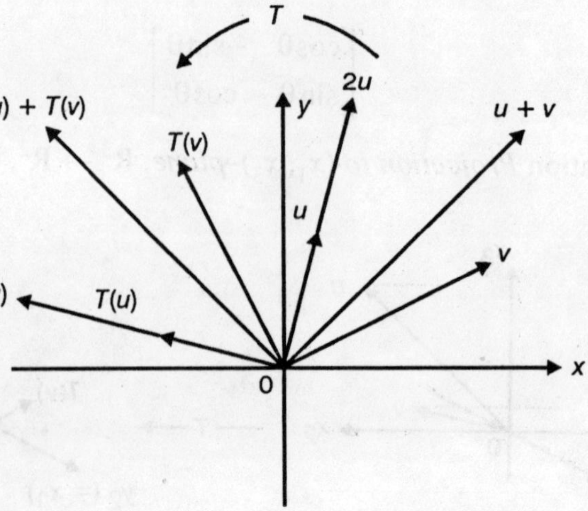

The projection transformation is also a linear transformation.

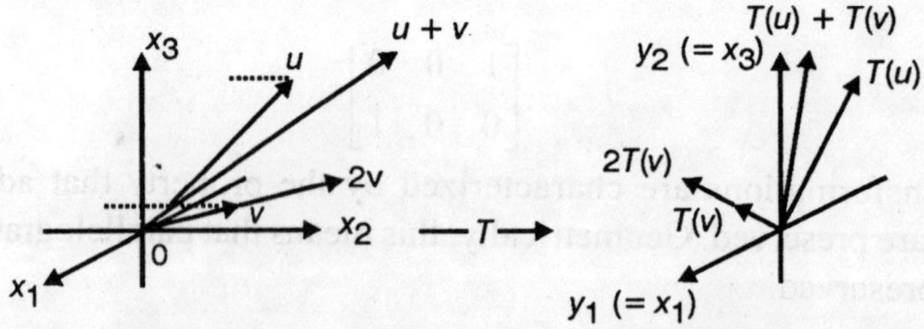

LINEAR TRANSFORMATION = MATRIX TRANSFORMATION

We have seen matrix transformations are linear. Conversely, this example suggests the converse is also true.

Any linear transformation $T: \mathrm{R}^n \rightarrow \mathrm{R}^m$ is also a matrix transformation, with the matrix given by

$$A = [T(e_1)\ T(e_2) \ldots\ T(e_n)].$$

In other words, the columns of the matrix A are the values of the linear transformation T at the standard basis. Thus the concepts of matrix transformations and linear transformations (between euclidean spaces) are equivalent.

Example: The identity transformation $id: \mathrm{R}^n \rightarrow \mathrm{R}^n$ is clearly linear. The corresponding matrix is

$$[id(e_1)\ id(e_2) \ldots\ id(e_n)] = [e_1\ e_2 \ldots\ e_n] = \begin{bmatrix} 1 & 0 & \ldots & 0 \\ 0 & 1 & \ldots & 0 \\ \vdots & \vdots & & \vdots \\ 0 & 0 & \ldots & 1 \end{bmatrix}$$

called the identity matrix and denoted by I_n.

Example: In this example, we started from a matrix

$$A = \begin{bmatrix} 1 & -1 & 1 \\ 3 & 1 & -1 \\ 2 & 2 & 1 \end{bmatrix}$$

and produced a matrix transformation (which is also a linear transformation).

$$T\begin{bmatrix} x_1 \\ x_2 \\ x_3 \end{bmatrix} = \begin{bmatrix} x_1 - x_2 + x_3 \\ 3x_1 + x_2 - x_3 \\ 2x_1 + 2x_2 + x_3 \end{bmatrix} \; R^3 \to R^3$$

Conversely, we may recover the columns of A as follows

$$[\text{col } 1] \text{ of } A = T\begin{bmatrix} 1 \\ 0 \\ 0 \end{bmatrix} = \begin{bmatrix} 1-0+0 \\ 3+0-0 \\ 0-1+0 \end{bmatrix} = \begin{bmatrix} 1 \\ 3 \\ 2 \end{bmatrix}$$

$$[\text{col } 2] \text{ of } A = T\begin{bmatrix} 0 \\ 0 \\ 1 \end{bmatrix} = \begin{bmatrix} 0-0-0 \\ 0+1-0 \\ 0-2+0 \end{bmatrix} = \begin{bmatrix} -1 \\ 1 \\ 2 \end{bmatrix}$$

$$[\text{col } 3] \text{ of } A = T\begin{bmatrix} 0 \\ 0 \\ 1 \end{bmatrix} = \begin{bmatrix} 0-0+1 \\ 0+0-1 \\ 0-0+1 \end{bmatrix}$$

Example: The reflections and rotations on R^2 are linear transformations because they preserve parallelogram and stretching/shrinking. To find their matrices, we simply need to apply reflections and rotations to the standard basis.

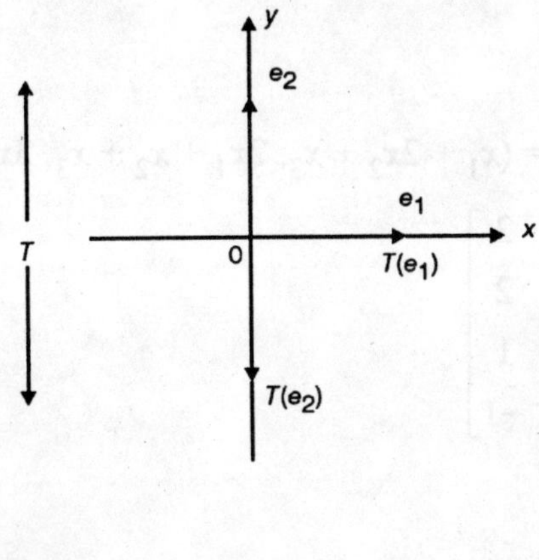

The picture shows that the reflections of e_1, e_2 in x–axis are $e_1 = (1, 0), -e_2 = (0, -1)$. Therefore the matrix for the reflection is

$$\begin{bmatrix} 1 & 0 \\ 0 & -1 \end{bmatrix}$$

Moreover, we also see from picture that for the rotation by angle θ, $T(e_1) = (\cos\theta, \sin\theta)$, $T(e_2) = (-\sin\theta, \cos\theta)$.

Therefore the matrix for the rotation is

$$\begin{bmatrix} \cos\theta & -\sin\theta \\ \sin\theta & \cos\theta \end{bmatrix}$$

Exercise: Find the corresponding transformations, and indicate the source and target euclidean spaces.

1. [1 2]

Answer: $T(x_1, x_2) = x_1 + 2x_2: R^2 \to R^1$

2. $\begin{bmatrix} 1 \\ 2 \end{bmatrix}$

Answer: $T(x_1) = (x_1, 2x_1): R^1 \to R^2$

3. $\begin{bmatrix} 2 & 0 & 2 & 1 \\ 1 & 1 & -1 & 0 \\ 0 & 1 & -2 & 1 \end{bmatrix}$

Answer: $T(x_1, x_2, x_3, x_4) = (x_1 + 2x_3 + x_4, x_1 + x_2 - x_3, x2 - 2x_3 + x_4): R^4 \to R^3$

4. $\begin{bmatrix} 1 & 2 & 1 \\ 2 & 1 & 1 \\ 0 & 3 & 1 \\ 3 & 0 & 1 \end{bmatrix}$

Answer: $T(x_1, x_2, x_3) = (x_1 + 2x_2 + x_3, 2x_1 + x_2 + x_3, 3x2 + x_3, 3x_1 + x_3): R^3 \to R^4$

5. $\begin{bmatrix} 1 & -1 & 1 & 0 & 2 \\ 2 & -2 & 0 & 2 & 2 \\ -1 & 1 & 2 & -3 & 1 \\ -2 & 2 & 1 & -3 & -1 \end{bmatrix}$

Answer: $T(x_1, x_2, x_3, x_4, x_5) = (x_1 - x_2 + x_3 + 2x_5, 2x_1 - 2x_2 + 2x_4 + 2x_5, - x_1 + x_2 + 2x_3 - 3x_4 + x5, - 2x_1 + 2x_2 + 2x_3 - 3x_4 - x5): R^5 \to R^4$

Exercise: Which of the following are matrix transformations? For the ones that are, find the corresponding matrices. For the ones that are not, provide the reason.

1. $T_1(x_1, x_2, x_3) = (2x_1 + 3x_2 - 7x_3, 10x_1 - 4x_2 - 8x_3, - 4x_1 + x_2)$

Answer: This is a matrix transformation.

$$\begin{bmatrix} 2 & 3 & -7 \\ 10 & -4 & -8 \\ -4 & 1 & 0 \end{bmatrix}$$

2. $T_2(x_1, x_2, x_3) = (2x_1 + 3x_2 - 7x_3 + 1, 10x_1 - 4x_2 - 8x_3, - 4x_1 + x_2 - 5)$

Answer: This is not a matrix/linear transformation because $T_2(0, 0, 0) = (1, 0, - 5)$, whereas any linear transformation must send zero vector to zero vector.

3. $T_3(x_1, x_2, x_3) = 2x_1 + 3x_2 - 7x_3$

Answer: This is a matrix transformation.

$$\begin{bmatrix} 2 & 3 & -7 \end{bmatrix}$$

4. $T_4(x_1, x_2, x_3) = 2x_1 + 3x_2 - 7x_3 + 1$

Answer: This is not a matrix/linear transformation because $T_4(0, 0, 0) = 1$, which is nonzero.

5. $T_5(x_1, x_2, x_3) = (2x_1, 3x_2, - 7x_3)$

Answer: This is a matrix transformation.

$$\begin{bmatrix} 2 & 0 & 0 \\ 0 & 3 & 0 \\ 0 & 0 & -7 \end{bmatrix}$$

6. $T_6(x_1, x_2, x_3) = (2x_2, 3x_3, - 7x_1)$

Answer: This is a matrix transformation.

$$\begin{bmatrix} 2 & 0 & 0 \\ 0 & 0 & 3 \\ -7 & 0 & 0 \end{bmatrix}$$

7. $T_7(x_1, x_2, x_3) = (3x_2, -7x_3, 2x_1)$

Answer: This is a matrix transformation.

$$\begin{bmatrix} 0 & 3 & 0 \\ 0 & 0 & -7 \\ 2 & 0 & 0 \end{bmatrix}$$

8. $T_8(x) = (2x, 3x, -7x)$

Answer: This is a matrix transformation.

$$\begin{bmatrix} 2 \\ 3 \\ -7 \end{bmatrix}$$

9. $T_9(x, y, z) = (x - y, y - z, z - x)$

Answer: This is a matrix transformation.

$$\begin{bmatrix} 1 & -1 & 0 \\ 0 & 1 & -1 \\ -1 & 0 & 1 \end{bmatrix}$$

10. $\boldsymbol{T}_{10}(x, y) = (x + y, xy)$

Answer: This is not a matrix/linear transformation because $T_{10}(1, 0) = T_{10}(0, 1) = (1,0)$, $T_{10}(1, 1) = (2, 1) \rightarrow T_{10}(1, 0) + T_{10}(0, 1)$.

11. $T_{11}(x, y) = (x + y, 2^{1/2}x + 3^{1/2}y)$

Answer: This is a matrix transformation.

$$\begin{bmatrix} 1 & 1 \\ 2^{1/2} & 3^{1/2} \end{bmatrix}$$

12. $\boldsymbol{T}_{12}(u, v) = (u + v, u^2 + v^2, u^3 + v^3)$

Answer: This is not a matrix/linear transformation because $T_{12}(1, 1) = (2, 2, 2)$, $T_{12}(2, 2) = (4, 8, 16)$, so that $T_{12}(2, 2)$ ‘‘‘ $2T_{12}(1, 1)$.

Exercise: Convince yourself that the following transformations are linear by drawing the picture and indicating how the parallelogram and scaling by 1/2 are preserved by the transformations. Then find the corresponding matrices.

1. *Reflection in diagonal line*: $R^2 \rightarrow R^2$

Answer: The following is the picture for the transformation.

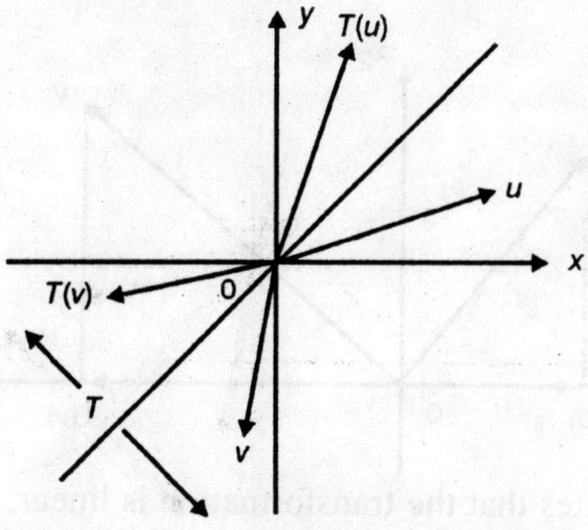

The following illustrates that the transformation is linear.

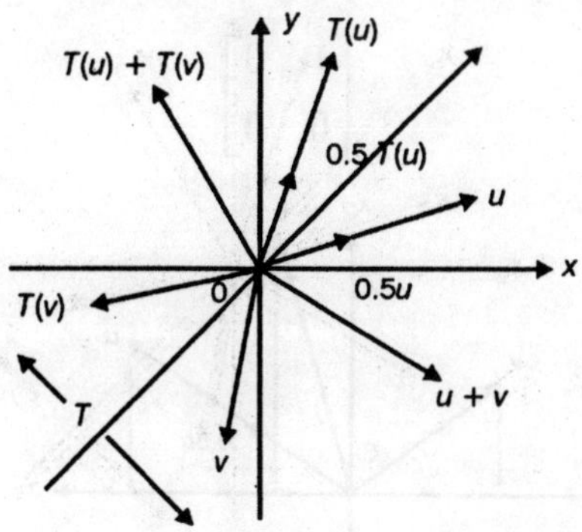

From this picture,

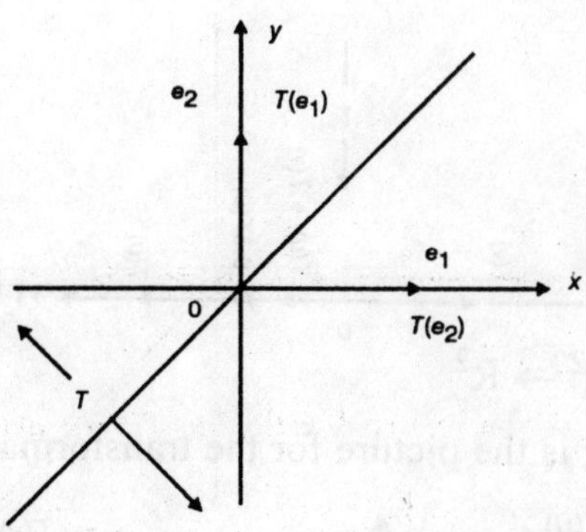

we see $T(e_1) = e_2 = (0, 1)$, $T(e_2) = e_1 = (1, 0)$. Therefore the matrix is

$$\begin{bmatrix} 0 & 1 \\ 1 & 0 \end{bmatrix}$$

2. *Projection to x_1–line*: $R^2 \to R^1$

Answer: The following is the picture for the transformation.

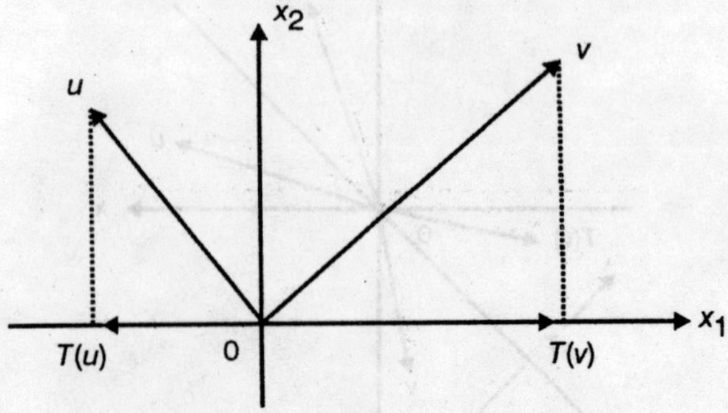

The following illustrates that the transformation is linear.

We clearly have $T(e_1) = e_1 = (1, 0)$, $T(e_2) = (0, 0)$. Therefore the matrix is

$$\begin{bmatrix} 1 & 0 \\ 0 & 0 \end{bmatrix}$$

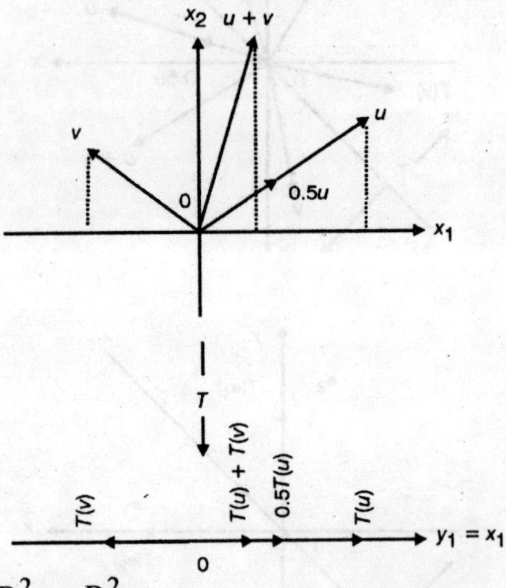

3. *Multiplying by* 2: $R^2 \to R^2$

Answer: The following is the picture for the transformation.

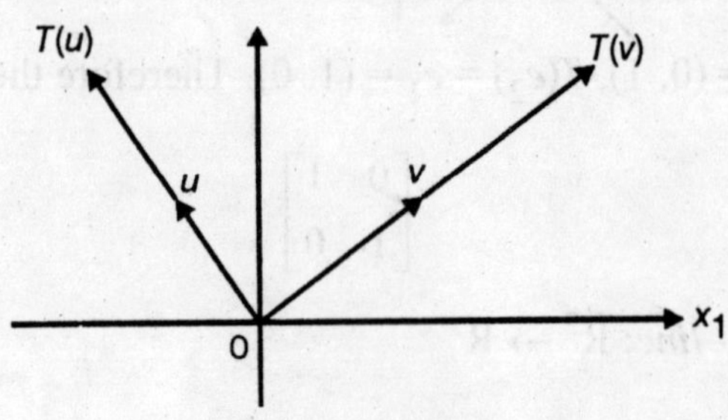

The following illustrates that the transformation is linear.

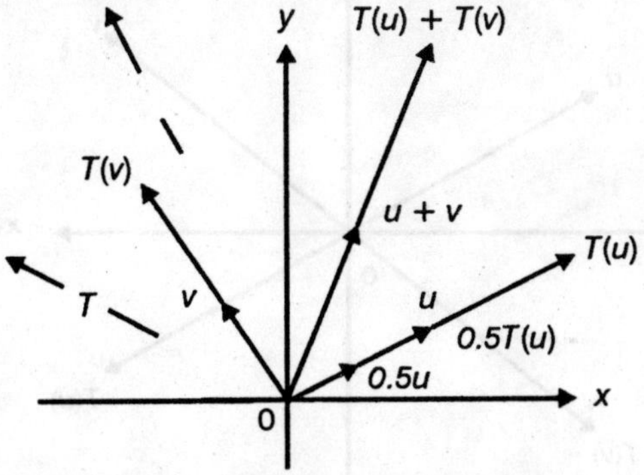

Since $T(e_1) = 2e_1 = (2, 0)$, $T(e_2) = 2e_2 = (0, 2)$, the matrix is

$$\begin{bmatrix} 2 & 0 \\ 0 & 2 \end{bmatrix}$$

4. *Embedding as diagonal*: $R^1 \to R^2$

Answer The following is the picture for the transformation.

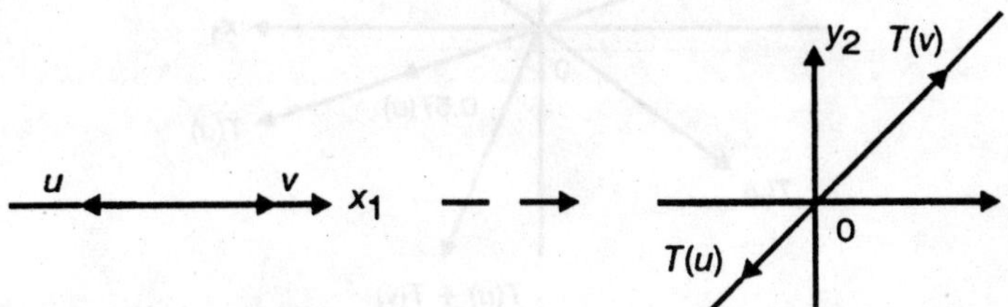

The following illustrates that the transformation is linear.

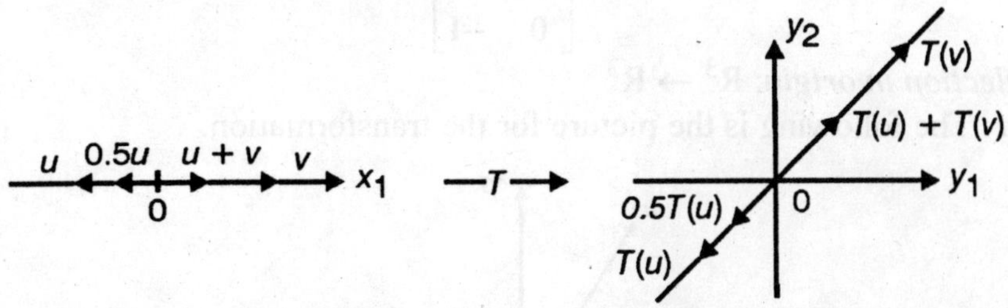

Since $T(e_1) = (1, 1)$ (note that $e_1 = 1$ is a 1–dimensional vector), the matrix is

$$\begin{bmatrix} 1 \\ 1 \end{bmatrix}$$

5. *Reflection in origin*: $R^2 \to R^2$

Answer: The following is the picture for the transformation.

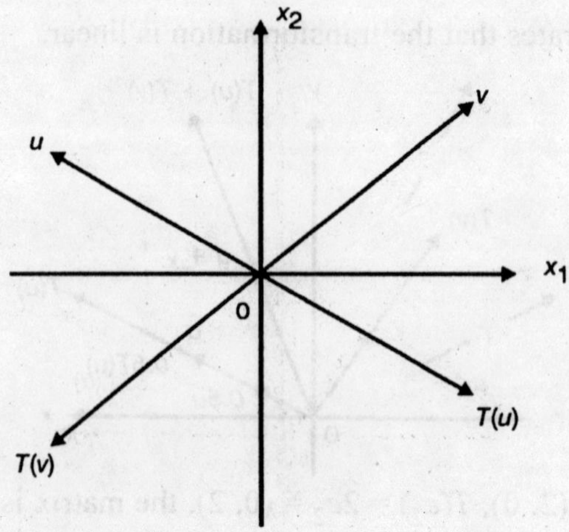

The following illustrates that the transformation is linear.

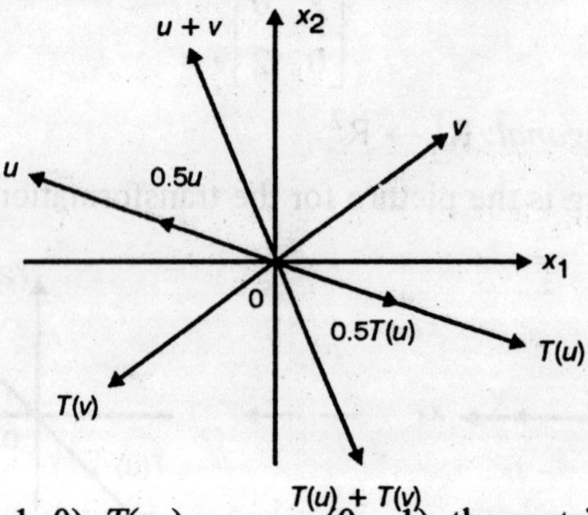

Since $T(e_1) = -e_1 = (-1, 0)$, $T(e_2) = -e_2 = (0, -1)$, the matrix is

$$\begin{bmatrix} -1 & 0 \\ 0 & -1 \end{bmatrix}$$

6. *Reflection in origin*: $R^3 \rightarrow R^3$
Answer: The following is the picture for the transformation.

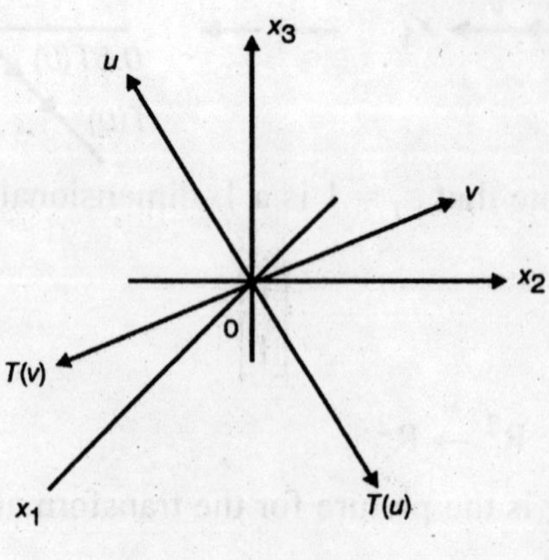

The following illustrates that the transformation is linear.

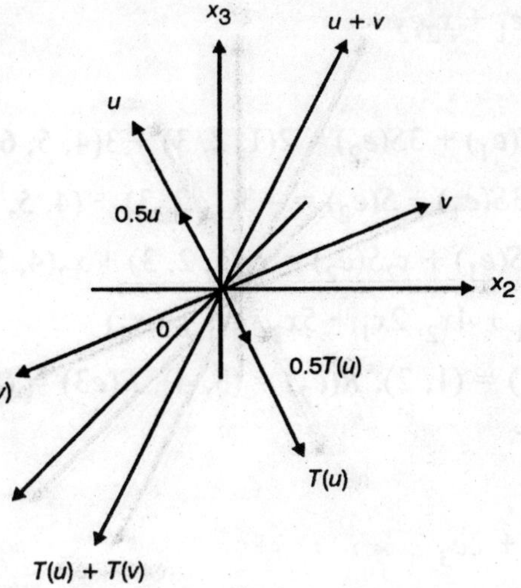

Since $T(e_1) = -e_1 = (-1, 0, 0)$, $T(e_2) = -e_2 = (0, -1, 0)$, $T(e_3) = -e_3 = (0, 0, -1)$, the matrix is

$$\begin{bmatrix} -1 & 0 & 0 \\ 0 & -1 & 0 \\ 0 & 0 & -1 \end{bmatrix}$$

Exercise: The values of the following linear transformations at the standard basis have been given. Find the value at the other vectors.

1. T: $R^2 \to R^2$, $T(e_1) = (5, -3)$, $T(e_2) = (-4, 2)$, find $T(2, 3)$, $T(-3, 1)$, $T(x_1, x_2)$.

Answer: By

$$(2, 3) = 2e_1 + 3e_2$$
$$(-3, 1) = -3e_1 + e_2$$
$$(x_1, x_2) = x_1e_1 + x_2e_2$$

We have

$$T(2, 3) = 2T(e_1) + 3T(e_2) = 2(5, -3) + 3(-4, 2) = (-2, 0)$$
$$T(-3, 1) = -3T(e_1) + T(e_2) = -3(5, -3) + (-4, 2) = (-19, 9)$$
$$T(x_1, x_2) = x_1T(e_1) + x_2T(e_2) = x_1(5, -3) + x_2(-4, 2) = (5x_1 - 4x_2, -3x_1 + 2x_2)$$

2. S: $R^2 \to R^3$, $S(e_1) = (1, 2, 3)$, $S(e_2) = (4, 5, 6)$, find $S(2, 3)$, $S(-3, 1)$, $S(x_1, x_2)$.

Answer: By

$$(2, 3) = 2e_1 + 3e_2$$

$(-3, 1) = -3e_1 + e_2$

$(x_1, x_2) = x_1e_1 + x_2e_2$

We have

$$S(2, 3) = 2S(e_1) + 3S(e_2) = 2(1, 2, 3) + 3(4, 5, 6) = (14, 19, 24)$$

$$S(-3, 1) = -3S(e_1) + S(e_2) = -3(1, 2, 3) + (4, 5, 6) = (1, -1, -3)$$

$$S(x_1, x_2) = x_1S(e_1) + x_2S(e_2) = x_1(1, 2, 3) + x_2(4, 5, 6)$$
$$= (x_1 + 4x_2, 2x_1 + 5x_2, 3x_1 + 6x_2)$$

3. R: $R^3 \to R^2$, $R(e_1) = (1, 2)$, $R(e_2) = (3, 4)$, $R(e3) = (5, 6)$, find $R(1, 0, 2)$, $R(-2, 1, 3)$, $R(x_1, x_2, x_3)$.

Answer: By

$(1, 0, 2) = e_1 + 2e_3$

$(-2, 1, 3) = -2e_1 + e_2 + 3e_3$

$(x_1, x_2, x_3) = x_1e_1 + x_2e_2 + x_3e_3$

We have

$$R(1, 0, 2) = R(e_1) + 2S(e_3) = (1, 2) + 2(5, 6) = (11, 14)$$

$$R(-2, 1, 3) = -2R(e_1) + R(e_2) + 3R(e_3) = -2(1, 2) + (3, 4) + 3(5, 6) = (16, 18)$$

$$R(x_1, x_2, x_3) = x_1R(e_1) + x_2R(e_2) + x_3R(e3) = x_1(1, 2) + x_2(3, 4) + x_3(5, 6)$$
$$= (x_1 + 3x_2 + 5x_3, 2x_1 + 4x_2 + 6x_3)$$

Exercise: Let T: $R^2 \to R^3$ be a linear transformation. Assume we already know $T(1, 2) = (1, 2, 3)$, $T(3, 4) = (4, 5, 6)$.

- Use this exercise and the linearity of T to compute $T(4, 6)$, $T(1, 1)$, $T(1, 0)$, $T(0, 1)$.
- Use $T(1, 0)$, $T(0, 1)$ to compute $T(x_1, x_2)$ in general.
- Use $3(1, 1, 1) = (4, 5, 6) - (1, 2, 3)$ to find x_1, x_2 such that $T(x_1, x_2) = (1, 1, 1)$

Answer: From the exercise, we have

$(4, 6) = (1, 2) + (3, 4)$

$(1, 1) = -1/2\ (1, 2) + 1/2\ (3, 4)$

$(1, 0) = -2(1, 2) + (3, 4)$

$(0, 1) = 3/2\ (1,2) - 1/2\ (3, 4)$

Thus

$\boldsymbol{T}(4, 6) = \boldsymbol{T}(1, 2) + \boldsymbol{T}(3, 4) = (1, 2, 3) + (4, 5, 6) = (5, 7, 9)$

$\boldsymbol{T}(1, 1) = -1/2\ \boldsymbol{T}(1, 2) + 1/2\ \boldsymbol{T}(3, 4) = -1/2\ (1, 2, 3) + 1/2\ (4, 5, 6) = (3/2, 3/2, 3/2)$

$\boldsymbol{T}(1, 0) = -2\boldsymbol{T}(1, 2) + \boldsymbol{T}(3, 4) = -2(1, 2, 3) + (4, 5, 6) = (2, 1, 0)$

$\boldsymbol{T}(0, 1) = 3/2\ \boldsymbol{T}(1,2) - 1/2\ \boldsymbol{T}(3, 4) = 3/2\ (1, 2, 3) - 1/2\ (4, 5, 6) = (-1/2, 1/2, 3/2)$

Moreover,

$T(x_1, x_2) = x_1 T(1, 0) + x_2 T(0, 1) = x_1(2, 1, 0) + x_2(-1/2, 1/2, 3/2) = (2x_1 - 1/2x_2, x_1 + 1/2x_2, 3/2x_2)$.

Finally, from the equality $3(1, 1, 1) = (4, 5, 6) - (1, 2, 3) = T(3, 4) - T(1, 2) = T(2, 2)$ we have $(1, 1, 1) = T(2/3, 2/3)$. Therefore for the last question, we may take $x_1 = x_2 = 2/3$.

Exercise: Let T: R^3 '! R^2 be a linear transformation. Assume we already know $T(1, 1, 1) = (1, 2)$, $T(1, 1, 0) = (3, 4)$, $T(1, 0, 0) = (5, 6)$. Find the general formula (or the matrix) for T.

Answer: We need to find $T(x_1, x_2, x_3)$. We use the same idea as in the previous exercise. First we solve the three systems $c_1(1, 1, 1) + c_2(1, 1, 0) + c3(1, 0, 0) = e_i$ for $i = 1, 2, 3$ and find

$$e_1 = (1, 0, 0)$$
$$e_2 = (1, 1, 0) - (1, 0, 0)$$
$$e_3 = (1, 1, 1) - (1, 1, 0)$$

Then we use the linearity of $\boldsymbol{T}$ to get

$$T(e_1) = T(1, 0, 0) = (5, 6)$$
$$T(e_2) = T(1, 1, 0) - T(1, 0, 0) = (3, 4) - (5, 6) = (-2, -2)$$
$$T(e_3) = T(1, 1, 1) - T(1, 1, 0) = (1, 2) - (3, 4) = (-2, -2)$$

Finally we use $(x_1, x_2, x_3) = x_1 e_1 + x_2 e_2 + x_3 e_3$ and the linearity of T to get

$T(x_1, x_2, x_3) = x_1 T(e_1) + x_2 T(e_2) + x_3 T(e_3) = x_1(5, 6) + x_2(-2, -2) + x_3(-2, -2) = (5x_1 - 2x_2 - 2x_3, 6x_1 - 2x_2 - 2x_3)$.

Exercise: Solve the following three systems of linear equations.

$$x_1 + 2x_2 = 1 \quad x_1 + 2x_2 = 0$$
$$2x_1 + 3x_2 = 0 \quad 2x_1 + 3x_2 = 1$$

Then use your results and the linearity to find solutions of the following systems

$$x_1 + 2x_2 = 2 \quad x_1 + 2x_2 = 13$$
$$2x_1 + 3x_2 = -3 \quad 2x_1 + 3x_2 = 9$$

Finally use the same technique to find the solution of the system

$$x_1 + 2x_2 = b_1$$
$$2x_1 + 3x_2 = b_2$$

Answer: For the two systems $Ax = e_1$ and $Ax = e_2$, we combine the augmented matrices together.

$$[A\ e_1\ e_2] = \begin{bmatrix} 1 & 2 & 1 & 0 \\ 2 & 3 & 0 & 1 \end{bmatrix}$$

Row operations give us

$$\begin{bmatrix} 1 & 0 & -3 & 2 \\ 0 & 1 & 2 & -1 \end{bmatrix}$$

Thus the solution for the system with $e_1 = (1, 0)$ on the right is $x = (-3, 2)$, and the solution with $e2 = (0, 1)$ on the right is $x' = (2, -1)$.

By $(2, 3) = 2(1, 0) + 3(0, 1)$, the solution of the system with $(2, 3)$ on the right is $2x + 3x' = 2(-3, 2) + 3(2, -1) = (0, 1)$.

By $(13, 9) = 13(1, 0) + 9(0, 1)$, the solution of the system with $(13, 9)$ on the right is $13x + 9x' = 13(-3, 2) + 9(2, -1) = (-21, 17)$.

By $(b_1, b_2) = b_1(1, 0) + b_2(0, 1)$, the solution of the system with (b_1, b_2) on the right is $b_1x + b_2x' = b_1(-3, 2) + b_2(2, -1) = (-3b_1 + 2b_2, 2b_1 - b_2)$.

Exercise: Solve the following three systems of linear equations.

$$\begin{array}{lll} x_1 + 2x_3 = 1 & x_1 + 2x_3 = 0 & x_1 + 2x_3 = 0 \\ 2x_1 - x_2 + 3x_2 = 0 & 2x_1 - x_2 + 3x_2 = 1 & 2x_1 - x_2 + 3x_2 = 0 \\ 4x_1 - x_2 + 8x_2 = 0 & 4x_1 - x_2 + 8x_2 = 0 & 4x_1 - x_2 + 8x_2 = 1 \end{array}$$

Then use your results and the linearity to find a solution of the system

$$x_1 + 2x_3 = b_1$$
$$2x_1 - x_2 + 3x_2 = b_2$$
$$4x_1 - x_2 + 8x_2 = b_3$$

Answer: For the three systems $Ax = e_1$, $Ax = e_2$, $Ax = e_3$, we combine the augmented matrices together.

$$[A\ e_1\ e_2\ e_3] = \begin{bmatrix} 1 & 0 & 2 & 1 & 0 & 0 \\ 2 & -1 & 3 & 0 & 1 & 0 \\ 4 & -1 & 8 & 0 & 0 & 1 \end{bmatrix}$$

Row operations give us

$$\begin{bmatrix} 1 & 0 & 0 & 5 & 2 & -2 \\ 0 & 1 & 0 & 4 & 0 & -1 \\ 0 & 0 & 1 & -2 & -1 & 1 \end{bmatrix}$$

Thus we conclude

$Ax = e_1$ has solution $x = (5, 4, -2)$

$Ax = e_2$ has solution $x' = (2, 0, -1)$

$Ax = e_3$ has solution $x'' = (-2, -1, 1)$

By $(b_1, b_2, b_3) = b_1e_1 + b_2e_2 + b_3e_3$, the combination

$b_1x + b_2x' + b_3x'' = b_1(5, 4, -2) + b_2(2, 0, -1) + b_3(-2, -1, 1) = (5b_1 + 2b_2 - 2b_3, 4b_1 - b_3, -2b_1 - b_2 + b_3)$

is a solution of the system $Ax = b$.

MATRIX OF A LINEAR TRANSFORMATION

Theorem Any linear transformation $T: \mathbb{R}^n \to \mathbb{R}^m$ is also a matrix transformation, with the matrix given by

$A = [T(e_1)\ T(e_2) ...\ T(e_n)]$.

Proof By an earlier exercise, any $x \in \mathbb{R}^n$ can be expressed as

$x = x_1e_1 + x_2e_2 + ... + x_ne_n$.

Then by the linearity of T, we have

$T(x) = T(x_1e_1 + x_2e_2 + ... + x_ne_n) = x_1T(e_1) + x_2T(e_2) + ... + x_nT(e_n)$.

In another earlier exercise, we already noticed the following equality

$x_1a_1 + x_2a_2 + ... + x_na_n = Ax$,

where $a_1\ a_2 ...\ a_n$ are the columns of A and $x_1, x_2, ..., x_n$ are the coordinates of x. (You are recommended to try to verify the equality for a general 3 by 3 matrix). Thus we conclude that

$T(x) = Ax$ for $A = [T(e_1)\ T(e_2) ...\ T(e_n)]$.

Chapter 4

Composition and Matrix Multiplication

COMPOSITION OF TRANSFORMATIONS

If the target of one transformation $S: X \to Y$ is the same as the source of another transformation $T: Y \to Z$, then we can combine the two transformations and form the composition $TS: X \to Z$, $TS(x) = T(S(x))$.

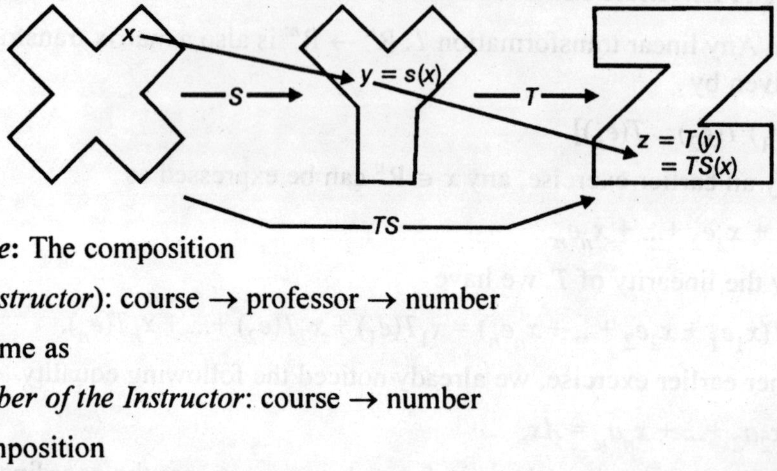

Example: The composition

$(ID_p)(Instructor)$: course → professor → number

is the same as

ID Number of the Instructor: course → number

The composition

(*Population*)(*Capital City*): country → city → number

is the same as

Population of the Capital City: country → number

Example: We are familiar with the following exercise: Suppose $y = x + 1$ and $z = y^2$. Then $z = (x + 1)^2 = x^2 + 2x + 1$. What we have done is actually composing two transformations

Add One: $R \to R, x \to x + 1$

Square: $R \to R, y \to y^2$

to form a transformation

Add One and then Take square: $R \to R$, $x \to (x + 1)^2$

For a multi–variable example, consider the transformations

T: $R^2 \to R^3$, $(y_1, y_2) \to (z_1, z_2, z_3) = (y_1^2 - 2y_2, y_1^3 - 3y_1y_2, y_1^4 - 4y_1^2y_2 - 2y_2^2)$

S: $R^2 \to R^2$, $(x_1, x_2) \to (y_1, y_2) = (x_1 + x_2, x_1x_2)$

The composition TS is computed by substituting the formulae of y_* (in terms of x_*) into the formulae of z_* (in terms of y_*).

$$z_1 = (x_1 + x_2)^2 - 2x_1x_2 = x_1^2 + x_2^2$$

$$z_2 = (x_1 + x_2)^3 - 3(x_1 + x_2)x_1x_2 = x_1^3 + x_2^3$$

$$z_3 = (x_1 + x_2)^4 - 4(x_1 + x_2)^2x_1x_2 - 2(x_1x_2)^2 = x_1^4 + x_2^4$$

Thus we have

$$TS(x_1, x_2) = (x_1^2 + x_2^2, x_1^3 + x_2^3, x_1^4 + x_2^4).$$

If transformations are given by formulae, then the composition is simply the substitution.

Example: Let T and S be the reflection in x–axis and the rotation by 90 degrees on the plane R^2. The following picture illustrates the composition TS.

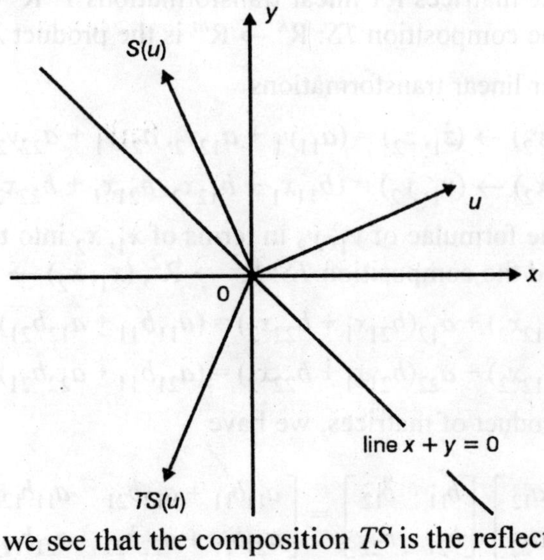

From the picture, we see that the composition TS is the reflection with respect to the line $x + y = 0$.

Similarly, the following picture illustrates that the composition ST is the reflection with respect to the diagonal line $x = y$.

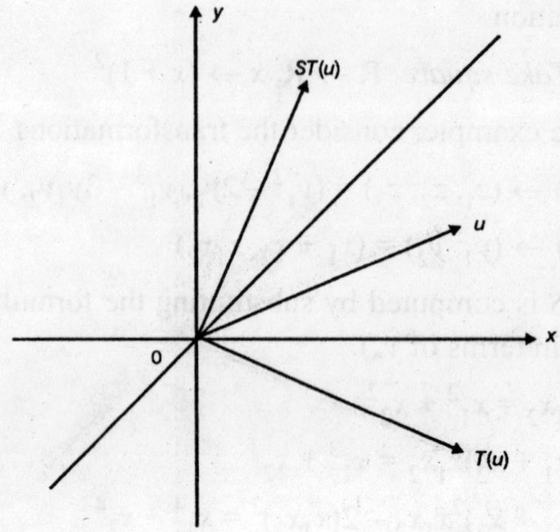

MULTIPLICATION OF MATRICES

The composition of two linear transformations is also linear. The geometrical reason is clear: if parallelograms and scalings are preserved by each transformation, then they are preserved by the combined transformation.

Since linear transformations between euclidean spaces are equivalent to matrices, the composition of linear transformations should have corresponding operation on matrices.

Let A and B be the matrices for linear transformations $T: \mathrm{R}^n \to \mathrm{R}^m$ and $S: \mathrm{R}^k \to \mathrm{R}^n$. Then the matrix for the composition $TS: \mathrm{R}^k \to \mathrm{R}^m$ is the product AB of A and B

Example Consider linear transformations

$T: \mathrm{R}^2 \to \mathrm{R}^2, (y_1, y_2) \to (z_1, z_2) = (a_{11}y_1 + a_{12}y_2, a_{21}y_1 + a_{22}y_2)$

$S: \mathrm{R}^2 \to \mathrm{R}^2, (x_1, x_2) \to (y_1, y_2) = (b_{11}x_1 + b_{12}x_2, b_{21}x_1 + b_{22}x_2)$.

By substituting the formulae of y_1, y_2 in terms of x_1, x_2 into the formulae of z_1, z_2 in terms of y_1, y_2, we find the composition $TS: \mathrm{R}^2 \to \mathrm{R}^2, (x_1, x_2) \to (z_1, z_2)$ is given by

$$z_1 = a_{11}(b_{11}x_1 + b_{12}x_2) + a_{12}(b_{21}x_1 + b_{22}x_2) = (a_{11}b_{11} + a_{12}b_{21})x_1 + (a_{11}b_{12} + a_{12}b_{22})x_2$$

$$z_2 = a_{21}(b_{11}x_1 + b_{12}x_2) + a_{22}(b_{21}x_1 + b_{22}x_2) = (a_{21}b_{11} + a_{22}b_{21})x_1 + (a_{21}b_{12} + a_{22}b_{22})x_2$$

In terms of the product of matrices, we have

$$\begin{bmatrix} a_{11} & a_{12} \\ a_{21} & a_{22} \end{bmatrix} \begin{bmatrix} b_{11} & b_{12} \\ b_{21} & b_{22} \end{bmatrix} = \begin{bmatrix} a_{11}b_{11} + a_{12}b_{21} & a_{11}b_{12} + a_{12}b_{22} \\ a_{21}b_{11} + a_{22}b_{21} & a_{21}b_{12} + a_{22}b_{22} \end{bmatrix}$$

Note that in the example above, the first entry $a_{11}b_{11} + a_{12}b_{21}$ of the matrix AB is obtained by "multiplying" the first row $(a_{11}\ a_{12})$ of A with the first column $(b_{11}\ b_{21})$ of B. Similar observation can be made to the other entries. In fact, we have the following general rule for multiplying an m by n matrix A and an m by k matrix B together.

(i, j)-entry of AB = (i–th row of A) • (j-th row of B)

The result AB is an m by k matrix. The rigorous proof of the rule can be found in the next section.

Example: Let

$$A = \begin{bmatrix} 1 & 2 \\ 3 & 4 \end{bmatrix}$$

$$B = \begin{bmatrix} -1 & 0 & 2 \\ 0 & 1 & 3 \end{bmatrix}$$

$$C = \begin{bmatrix} 0 & 1 & -1 \\ 2 & 0 & 3 \\ -1 & 1 & 2 \end{bmatrix}$$

$$D = \begin{bmatrix} 1 & 2 \\ 2 & 3 \\ 3 & 4 \end{bmatrix}$$

Then

$$AB = \begin{bmatrix} 1\times(-1)+2\times 0 & 1\times 0+2\times 1 & 1\times 2+2\times 3 \\ 3\times(-1)+4\times 0 & 3\times 0+4\times 1 & 3\times 2+4\times 3 \end{bmatrix} = \begin{bmatrix} -1 & 2 & 8 \\ -3 & 4 & 18 \end{bmatrix}$$

$$A^2 = AA = \begin{bmatrix} 1\times 1+2\times 3 & 1\times 2+2\times 4 \\ 3\times 1+4\times 3 & 3\times 2+4\times 4 \end{bmatrix} = \begin{bmatrix} 7 & 10 \\ 15 & 22 \end{bmatrix}$$

$$DA = \begin{bmatrix} 1\times 1+2\times 3 & 1\times 2+2\times 4 \\ 2\times 1+3\times 3 & 2\times 2+3\times 4 \\ 3\times 1+4\times 3 & 3\times 2+4\times 4 \end{bmatrix} = \begin{bmatrix} 7 & 10 \\ 11 & 16 \\ 15 & 22 \end{bmatrix}$$

$$BC = \begin{bmatrix} (-1)\times 0+0\times 2+2\times(-1) & (-1)\times 1+0\times 0+2\times 1 & (-1)\times(-1)+0\times 3+2\times 2 \\ 0\times 0+1\times 2+3\times(-1) & 0\times 1+1\times 0+3\times 1 & 0\times(-1)+1\times 3+3\times 2 \end{bmatrix}$$

$$= \begin{bmatrix} -2 & 1 & 5 \\ -1 & 3 & 9 \end{bmatrix}$$

$$(AB)C = \begin{bmatrix} (-1)\times 0+2\times 2+8\times(-1) & (-1)\times 1+2\times 0+8\times 1 & (-1)\times(-1)+2\times 3+8\times 2 \\ (-3)\times 0+4\times 2+18\times(-1) & (-3)\times 1+4\times 0+18\times 1 & (-3)\times(-1)+4\times 3+18\times 2 \end{bmatrix}$$

$$= \begin{bmatrix} -4 & 7 & 23 \\ -10 & 15 & 51 \end{bmatrix}$$

$$A(BC) = \begin{bmatrix} 1\times(-2)+2\times(-1) & 1\times 1+2\times 3 & 1\times 5+2\times 9 \\ 3\times(-2)+4\times(-1) & 3\times 1+4\times 3 & 3\times 5+4\times 9 \end{bmatrix} = \begin{bmatrix} -4 & 7 & 23 \\ -10 & 15 & 51 \end{bmatrix}$$

On the other hand, the products such as AD, AC, BA, CB, B^2 are meaningless.

Example: According to an earlier example, the matrix for the rotation by angle α is

$$R_a = \begin{bmatrix} \cos a & -\sin a \\ \sin a & \cos a \end{bmatrix}$$

Since rotation by β and then rotation by α is the same as rotation by $\alpha + \beta$, we must have $R_\alpha R_\beta = R_{\alpha+\beta}$, i.e.,

$$\begin{bmatrix} \cos a & -\sin a \\ \sin a & \cos a \end{bmatrix}\begin{bmatrix} \cos\beta & -\sin\beta \\ \sin\beta & \cos\beta \end{bmatrix} = \begin{bmatrix} \cos(a+\beta) & -\sin(a+\beta) \\ \sin(a+\beta) & \cos(a+\beta) \end{bmatrix}$$

If we compute matrix multiplication on the left side in the usual way, we get the following familiar formulae in trigonometry.

$\cos\alpha\cos\beta - \sin\alpha\sin\beta = \cos(\alpha + \beta)$

$\sin\alpha\cos\beta + \cos\alpha\sin\beta = \sin(\alpha + \beta)$

COMPOSITION OF LINEAR TRANSFORMATIONS IS LINEAR

Let T: $R^m \to R^n$ and S: $R^k \to R^n$ be linear transformations. Then by the definition, we have

- $T(u + v) = T(u) + T(v)$,
- $T(cu) = cT(u)$,
- $S(u + v) = S(u) + S(v)$,
- $S(cu) = cS(u)$.

To show the composition TS: $R^k \to R^m$ is also linear means the verification of the following equalities:

$TS(u + v) = TS(u) + TS(v)$,

$TS(cu) = cTS(u)$.

The first equality is verified as follows:

$$TS(u + v) = T(S(u + v))$$
$$= T(S(u) + S(v))$$
$$= T(S(u)) + T(S(v))$$
$$= TS(u) + TS(v)$$

The second equality is verified as follows:

$$TS(cu) = T(S(cu))$$
$$= T(Cs(u))$$
$$= cT(S(u))$$
$$= CTS(u)$$

This completes the proof that TS is a linear transformation.

PROPERTIES OF MATRIX MULTIPLICATION

The matrix multiplication is consistent with our notation for the left side of a system of linear equations. Specifically, by thinking of a vertical vector x as a 1 by n matrix, the matrix product

$$Ax = \begin{bmatrix} a_{11} & a_{12} & \cdots & a_{1n} \\ a_{21} & a_{22} & \cdots & a_{2n} \\ \vdots & \vdots & & \vdots \\ a_{m1} & a_{m2} & \cdots & a_{mn} \end{bmatrix}$$

is exactly the left of the system $Ax = b$. The remark also shows that the rule for matrix multiplication is equivalent to

$$B = [b_1\ b_2 \dots\ b_k] \Rightarrow AB = [Ab_1\ Ab_2 \dots\ Ab_k].$$

Now we give a rigorous proof of the rule for matrix multiplication by showing the above is true.

Proof Let A and B be matrices for linear transformations T: $R^m \to R^n$ and S: $R^k \to R^n$. Then by this formula and the additional reason provided below, we have

$AB = [TS(e_1)\ TS(e_2) \dots\ TS(e_k)]$ (definition: AB is matrix for TS)

$= [T(S(e_1))\ T(S(e_2)) \dots\ T(S(e_k))]$ (definition of composition TS)

$= [T(b_1)\ T(b_2) \dots\ T(b_k)]$ (B is matrix for S: $B = [S(e_1)\ S(e_2) \dots\ S(e_k)]$)

$= [Ab_1\ Ab_2 \dots\ Ab_k]$. ($A$ is matrix for T: $T(x) = Ax$)

We have seen the *associativity* $(AB)C = A(BC)$ for the matrix multiplication in this example. In general, this is the consequence of the corresponding associativity $(TS)R = T(SR)$ for the compositions of transformations (which is clearly true). Moreover, composing with the identity does not change the transformations: $id\,T = T = T\,id$. Since the identity transformation corresponds to the identity matrix I, we conclude that the product with the identity matrix does not change the matrix.

associative

$$(AB)C = A(BC)$$

identity

$$IA = A = AI$$

Thus the matrix multiplication is similar to number multiplication. However, the following example shows that one should be careful in stretching the analogue.

Example: In an earlier example, we have shown

(*Reflection in x–axis*)(*Rotation by 90 degrees*) = (*Reflection in the line x + y = 0*)

(*Rotation by 90 degrees*)(*Reflection in x–axis*) = (*Reflection in the line x = y*)

In particular, we see that the compositions TS and ST are different in general. In terms of the matrix product, we have

$$\begin{bmatrix} 1 & 0 \\ 0 & -1 \end{bmatrix}\begin{bmatrix} 0 & -1 \\ 1 & 0 \end{bmatrix} = \begin{bmatrix} 0 & -1 \\ -1 & 0 \end{bmatrix}$$

$$\begin{bmatrix} 0 & -1 \\ 1 & 0 \end{bmatrix}\begin{bmatrix} 1 & 0 \\ 0 & -1 \end{bmatrix} = \begin{bmatrix} 0 & 1 \\ 1 & 0 \end{bmatrix}$$

which shows $AB \neq BA$ in general.

Example: Consider two transformations T = *Projection to x–axis* and S = *Projection to y–axis* from R^2 to itself.

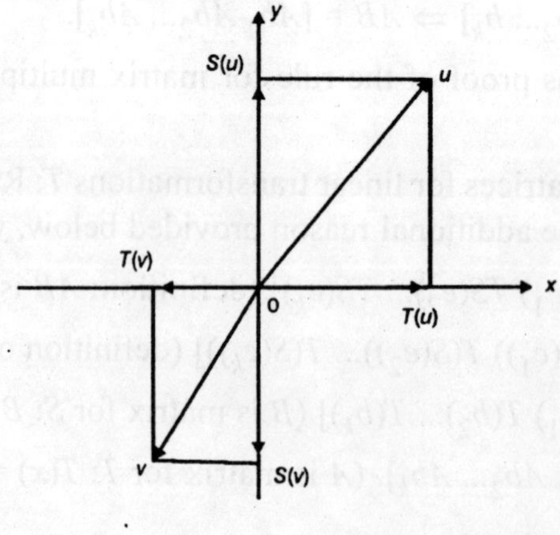

The composition *TS* is the zero transformation – taking everything to the zero vector 0.

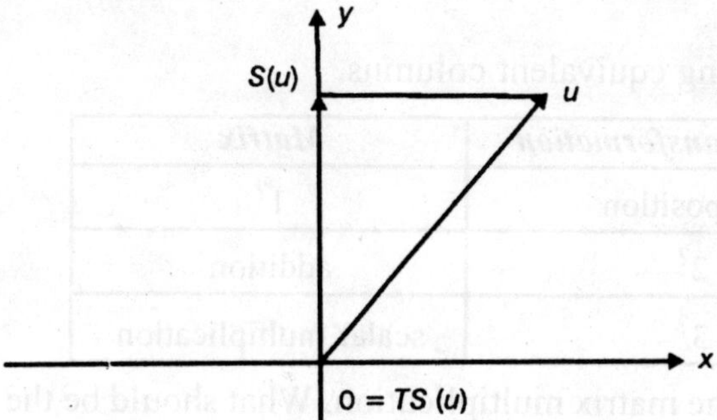

Correspondingly, we have two nonzero matrices with product being zero.

$$\begin{bmatrix} 1 & 0 \\ 0 & 0 \end{bmatrix}\begin{bmatrix} 0 & 0 \\ 0 & 1 \end{bmatrix} = \begin{bmatrix} 0 & 0 \\ 0 & 0 \end{bmatrix}$$

Finally, we denote

$$A^0 = I$$
$$A^1 = A$$
$$A^2 = AA$$
$$A^3 = AAA$$
....

In general, for positive *k* and square matrix *A*, we use *Ak* to denote *AA...A*, the product of *k* copies of *A*. The same notation is used for a transformation with the same same domain and codomain.

Example: For a diagonal matrix,

$$A = \begin{bmatrix} a_1 & 0 & .. & 0 \\ 0 & a_2 & .. & 0 \\ \vdots & \vdots & & \vdots \\ 0 & 0 & .. & a_n \end{bmatrix}$$

we have

$$AK = \begin{bmatrix} a_1^k & 0 & .. & 0 \\ 0 & a_2^k & .. & 0 \\ \vdots & \vdots & & \vdots \\ 0 & 0 & .. & a_n \end{bmatrix}$$

COMPOSITION/MULTIPLICATION MIXED WITH ADDITION AND SCALAR MULTIPLICATION

Consider the following equivalent columns.

Linear Transformation	*Matrix*
composition	1?
2?	addition
3?	scalar multiplication

Our answer to 1? is the matrix multiplication. What should be the answer to 2? and 3??

From the explicit formula for Ax, it is easy to see that

$(A + B)x = Ax + Bx$, $(cA)x = c(Ax)$.

Therefore the answer to 2? is the following definition,

For linear transformations $T, S: \mathrm{R}^n \to \mathrm{R}^m$, their addition $T + S$ is defined by the formula $(T + S)(x) = T(x) + S(x)$

and the answer to 3? is the following definition.

For a linear transformation $T: \mathrm{R}^n \to \mathrm{R}^m$, and a number r, the scalar multiplication rT is defined by the formula $(rT)(x) = r(T(x))$

Example: Let T and S be the reflection in x–axis and the rotation by 90 degrees on the plane R^2. The following picture illustrates the addition $T + S$.

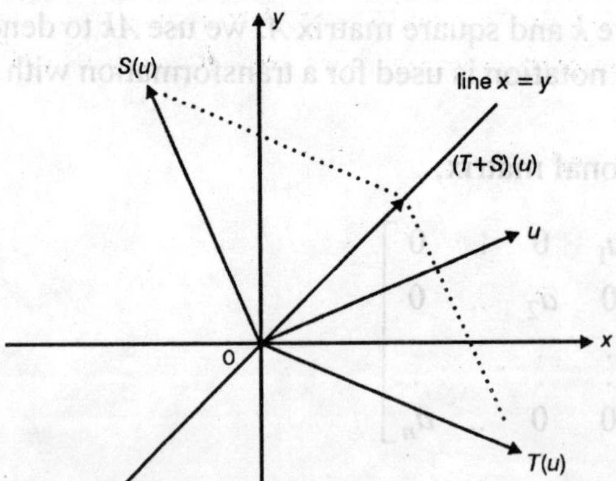

The formula for the addition is $(T + S)(x, y) = T(x, y) + S(x, y) = (x, -y) + (-y, x) = (x - y, x - y)$, and the corresponding matrix is

$$\begin{bmatrix} 1 & 0 \\ 0 & -1 \end{bmatrix} + \begin{bmatrix} 0 & -1 \\ 1 & 0 \end{bmatrix} = \begin{bmatrix} 1 & -1 \\ 1 & -1 \end{bmatrix}$$

For another example, consider two transformations T = *Projection to x–axis* and S = *Projection to y–axis* from R^2 to itself. The following picture shows their addition is the identity.

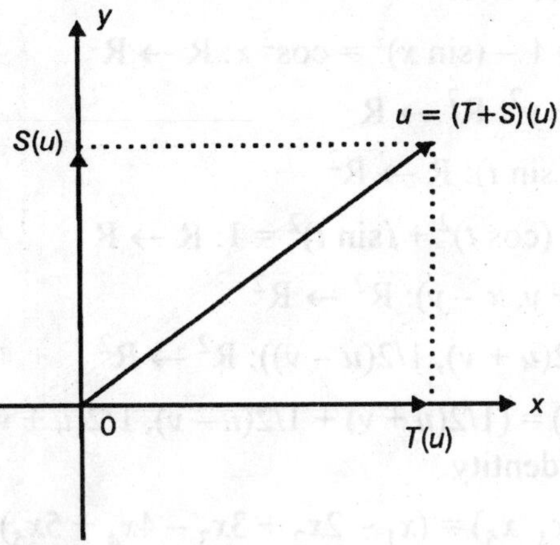

In terms of the corresponding matrix, we have

$$\begin{bmatrix} 1 & 0 \\ 0 & 0 \end{bmatrix} + \begin{bmatrix} 0 & 0 \\ 0 & 1 \end{bmatrix} = \begin{bmatrix} 1 & 0 \\ 0 & 1 \end{bmatrix}$$

When the multiplication is mixed with addition and scalar multiplication, we have the following properties.

distributive

$$(A + B)C = AC + BC$$

distributive

$$A(B + C) = AB + AC$$

associative

$$A(rB) = r(AB) = (rA)B$$

The properties are easy to prove in terms of linear transformations. For example, the following argument proves $(T + S)R = TR + SR$.

$$\begin{aligned} ((T + S)R)(x) &= (T + S)(R(x)) \text{ (definition of composition)} \\ &= T(R(x)) + S(R(x)) \text{ (definition of addition)} \\ &= (TR)(x) + (SR)(x) \text{ (definition of composition)} \\ &= (TR + SR)(x). \end{aligned}$$

The other properties can be proved in similar way.

Exercise: **Compute the composition** ***TS***

1. $T(y) = 1 - y^2: R \to R$

   $S(x) = \sin x: R \to R$

Answer: $TS(x) = 1 - (\sin x)^2 = \cos^2 x: R \to R$

2. $T(x, y) = x^2 + y^2: R^2 \to R$

   $S(t) = (\cos t, \sin t): R \to R^2$

Answer: $TS(t) = (\cos t)^2 + (\sin t)^2 = 1: R \to R$

3. $T(x, y) = (x + y, x - y): R^2 \to R^2$

   $S(u, v) = (1/2(u + v), 1/2(u - v)): R^2 \to R^2$

Answer: $TS(u, v) = (1/2(u + v) + 1/2(u - v), 1/2(u + v) - 1/2(u - v)) = (u, v): R^2 \to R^2$. The composition is identity.

4. $T(x_1, x_2, x_3, x_4, x_5) = (x_1 - 2x_2 + 3x_3 - 4x_4 + 5x_5): R^5 \to R$

   $S(t) = (8t, 7t, 6t, 5t, 4t): R \to R^5$

Answer: $TS(t) = (8t - 2X7t + 3X6t - 4X5t + 5X4t) = 12t$: R '! R

5. $T(t) = (8t, 7t, 6t, 5t, 4t): R \to R^5$

   $S(x_1, x_2, x_3, x_4, x_5) = (x_1 - 2x_2 + 3x_3 - 4x_4 + 5x_5): R^5 \to R$

Answer: $TS(x_1, x_2, x_3, x_4, x_5)$

$= (8(x_1 - 2x_2 + 3x_3 - 4x_4 + 5x_5), 7(x_1 - 2x_2 + 3x_3 - 4x_4 + 5x_5), 6(x_1 - 2x_2 + 3x_3 - 4x_4 + 5x_5), 5(x_1 - 2x_2 + 3x_3 - 4x_4 + 5x_5), 4(x_1 - 2x_2 + 3x_3 - 4x_4 + 5x_5))$

$= (8x_1 - 16x_2 + 24x_3 - 32x_4 + 40x_5, 7x_1 - 14x_2 + 21x_3 - 28x_4 + 35x_5, 6x_1 - 12x_2 + 18x_3 - 24x_4 + 30x_5, 5x_1 - 10x_2 + 15x_3 - 20x_4 + 25x_5, 4x_1 - 8x_2 + 12x_3 - 16x_4 + 20x_5)$: R^5 '! R^5

6. $T(x_1, x_2, x_3) = (x_1 - 2x_2 + 3x_3, - 2x_1 + 3x_2 - 4x_3)$: R^3 '! R^2

   $S(y_1, y_2) = (- 2y_1 - y_2, y_2, y_1 + y_2): R^2 \to R^3$

Answer: $TS(y_1, y_2) = ((- 2y_1 - y_2) - 2y_2 + 3(y_1 + y_2), - 2(- 2y_1 - y_2) + 3y_2 - 4(y_1 + y_2)) = (y_1, y_2): R^2 \to R^2$. The composition is identity.

7. $T(y_1, y_2) = (- 2y_1 - y_2, y_2, y_1 + y_2): R^2 \to R^3$

   $S(x_1, x_2, x_3) = (x_1 - 2x_2 + 3x_3, - 2x_1 + 3x_2 - 4x_3): R^3 \to R^2$

Answer: $TS(x_1, x_2, x_3)$

$= (- 2(x_1 - 2x_2 + 3x_3) - (- 2x_1 + 3x_2 - 4x_3), - 2x_1 + 3x_2 - 4x_3, (x_1 - 2x_2 + 3x_3) + (- 2x_1 + 3x_2 - 4x_3))$

$= (x_2 - 2x_3, - 2x_1 + 3x_2 - 4x_3, - x_1 + x_2 - x_3): R^3 \to R^3$

Exercise: Draw the picture and identify the composition *TS*.

1. *T* = *Reflection in x–axis*: $R^2 \to R^2$

   S = *Reflection in y–axis*: $R^2 \to R^2$

Answer: The composition *TS* is *Reflection in origin*: $R^2 \to R^2$

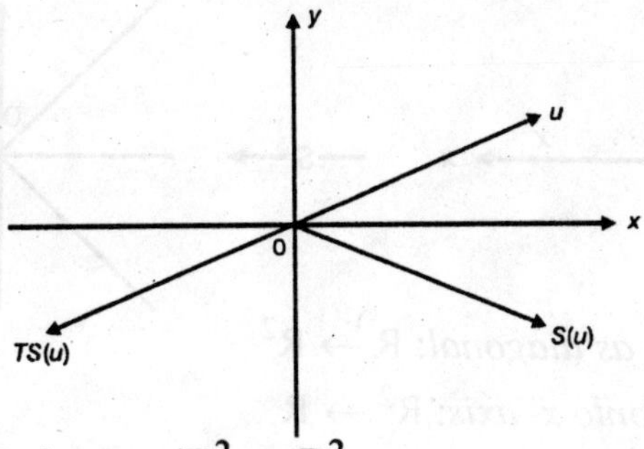

2. *T* = *Rotation by 90 degrees*: $R^2 \to R^2$

   S = *Reflection in origin*: $R^2 \to R^2$

Answer: The composition *TS* is *Rotation by –90 degrees*: $R^2 \to R^2$

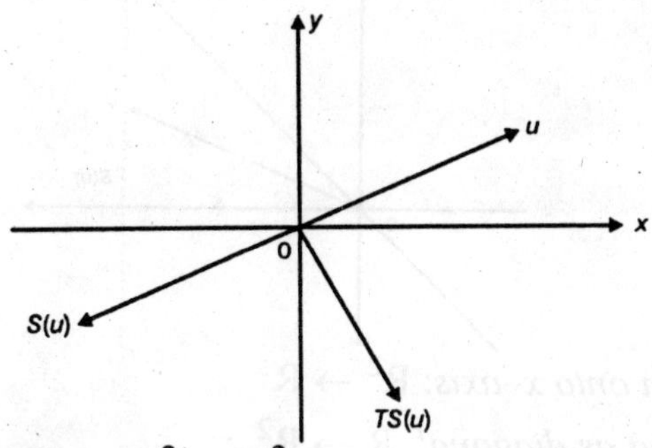

3. *T* = *Reflection in origin*: $R^2 \to R^2$

   S = *Rotation by 90 degrees: $R^2 \to R^2$*

Answer: The composition *TS* is *Rotation by –90 degrees*: $R^2 \to R^2$

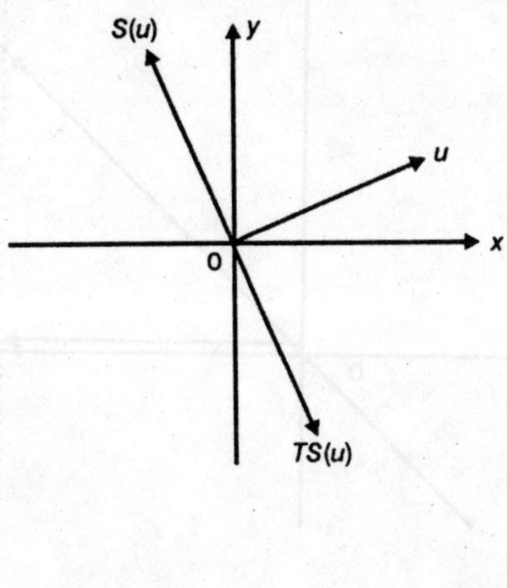

4. T = *Reflection in x–axis*: $R^2 \to R^2$

   S = *Embedding as diagonal*: $R \to R^2$

Answer: The composition TS is *Embedding as skew–diagonal*: $R^2 \to R^2$

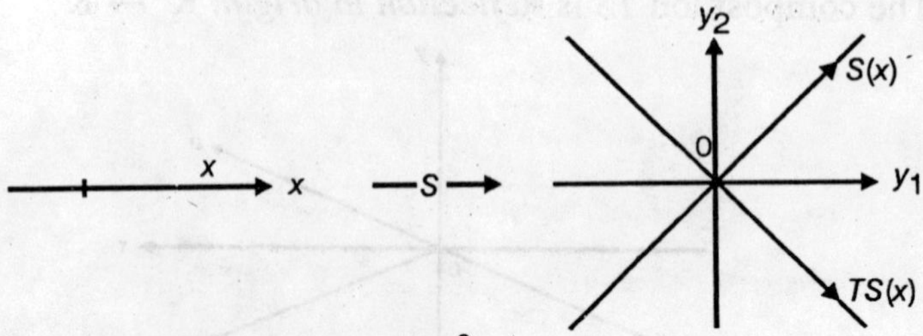

5. T = *Embedding as diagonal*: $R \to R^2$

   S = *Projection onto x–axis*: $R^2 \to R$

Answer: The composition TS is *Vertical intersection with diagonal*: $R^2 \to R^2$

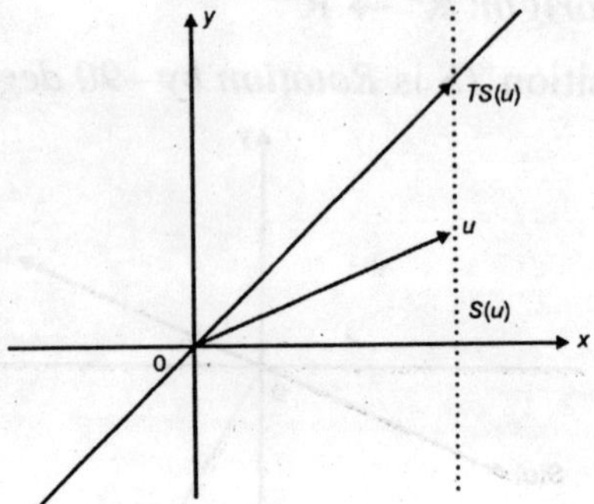

6. T = *Projection onto x–axis*: $R^2 \to R$

   S = *Embedding as diagonal*: $R \to R^2$

Answer: The composition TS is the identity.

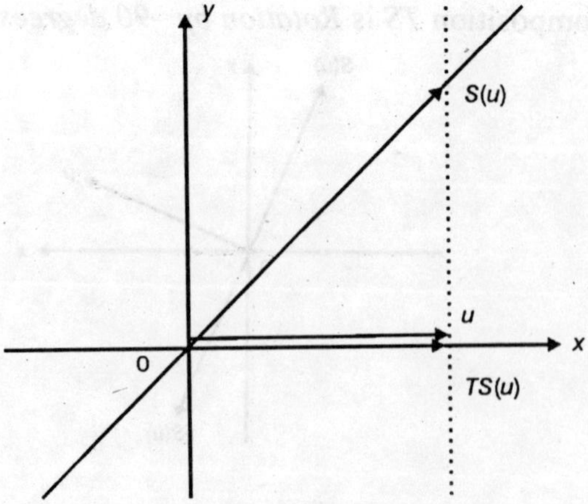

Exercise: Compute matrix products.

Answer: $$[1 \quad 1][1/2 \quad 1/2] = [1 \quad 0]$$

1 – 1 1/2 – 1/2 01

$$[1 \quad -2 \quad 3 \quad -4 \quad 5]\begin{bmatrix} 8 \\ 7 \\ 6 \\ 5 \\ 4 \end{bmatrix} = [12]$$

$$\begin{bmatrix} 8 \\ 7 \\ 6 \\ 5 \\ 4 \end{bmatrix}[1 \quad -2 \quad 3 \quad -4 \quad 5] = \begin{bmatrix} 8 & -16 & 24 & -32 & 40 \\ 7 & -14 & 21 & -28 & 35 \\ 6 & -12 & 18 & -24 & 30 \\ 5 & -10 & -15 & -20 & 25 \\ 4 & -8 & 12 & -16 & 20 \end{bmatrix}$$

$$\begin{bmatrix} 1 & -2 & 3 \\ -2 & 3 & -4 \end{bmatrix}\begin{bmatrix} -2 & -1 \\ 0 & 1 \\ 1 & 1 \end{bmatrix} = \begin{bmatrix} 1 & 0 \\ 0 & 1 \end{bmatrix}$$

$$\begin{bmatrix} -2 & -1 \\ 0 & 1 \\ 1 & 1 \end{bmatrix}\begin{bmatrix} 1 & -2 & 3 \\ -2 & 3 & -4 \end{bmatrix} = \begin{bmatrix} 0 & 1 & -2 \\ -2 & 3 & -4 \\ -1 & 1 & -1 \end{bmatrix}$$

$$\begin{bmatrix} -2 & -1 \\ 0 & 1 \end{bmatrix}\begin{bmatrix} 1 & -2 & 3 \\ -2 & 3 & -4 \end{bmatrix} = \begin{bmatrix} 0 & 1 & -2 \\ -2 & 3 & -4 \end{bmatrix}$$

$$\begin{bmatrix} 1 & 2 & 3 & 4 \\ 5 & 6 & 7 & 8 \\ 9 & 10 & 11 & 12 \end{bmatrix}\begin{bmatrix} 1 & 1 \\ -1 & -1 \\ 1 & -1 \\ -1 & 1 \end{bmatrix} = \begin{bmatrix} -2 & 0 \\ -2 & 0 \\ -2 & 0 \end{bmatrix}$$

$$\begin{bmatrix} 1 & -1 & 1 \\ -1 & 2 & -1 \\ 2 & 0 & -1 \end{bmatrix}\begin{bmatrix} 1 & 2 & 3 & 4 \\ 5 & 6 & 7 & 8 \\ 9 & 10 & 11 & 12 \end{bmatrix} = \begin{bmatrix} 5 & 6 & 7 & 8 \\ 0 & 0 & 0 & 0 \\ -7 & -6 & -5 & -4 \end{bmatrix}$$

The following exercises provide more examples that $AB = BA$ is generally not true.

Exercise: For each of the following matrix A, find all the matrices X satisfying $AX = XA$.

1. $A = [1 \quad 2]$

Answer: No X will satisfy $AX = XA$ because AX and XA have different sizes. Specifically, in order for AX to be meaningful, X must have two rows. In order for XA to be meaningful X must have one column. Then AX is a 1 by 1 matrix, and XA is a 2 by 2 matrix.

2. $A = \begin{bmatrix} 1 & 1 \\ 1 & -1 \end{bmatrix}$

Answer: Similar argument as in the first question shows that $\boldsymbol{X}$ must be a 2 by 2 matrix. Let

$$X = \begin{bmatrix} x & y \\ z & w \end{bmatrix}$$

Then

$$AX = \begin{bmatrix} x+z & y+w \\ x-z & y-w \end{bmatrix}, XA = \begin{bmatrix} x+y & x-y \\ z-w & z-w \end{bmatrix}$$

so that $AX = XA$ means

$x + z = x + y, y + w = x - y, x - z = z + w, y - w = z - w.$

The solution is $x = w + 2z$, $y = z$, and w, z are arbitrary. In summary, $\boldsymbol{AX = XA}$ if and only if

$$X = \begin{bmatrix} x+2z & z \\ z & w \end{bmatrix} = w\begin{bmatrix} 1 & 0 \\ 0 & 1 \end{bmatrix} + z\begin{bmatrix} 2 & 1 \\ 1 & 0 \end{bmatrix}$$

3. $A = \begin{bmatrix} 1 & 0 \\ 0 & 2 \end{bmatrix}$

Answer: Again we assume

$$X = \begin{bmatrix} x & y \\ z & w \end{bmatrix}$$

and find the condition for $AX = XA$ to be $x = x$, $y = 2y$, $2z = z$, $2w = 2w$. The solution is $y = z = 0$, and x, w arbitrary. In other words, X is a diagonal matrix.

4. $A = \begin{bmatrix} 0 & 0 \\ 0 & 0 \end{bmatrix}$

Answer: For any 2 by 2 matrix X, we have $AX = O = XA$ (O is the 2 by 2 zero matrix). Therefore $AX = XA$ for any X.

5. $\begin{bmatrix} 2 & 0 \\ 0 & 2 \end{bmatrix}$

Answer: For any 2 by 2 matrix X, we have $AX = 2X = XA$. Therefore $AX = XA$ for any X.

6. $\begin{bmatrix} 2 & 0 & 0 \\ 0 & 2 & 0 \end{bmatrix}$

Answer For AX and XA to be meaningful, X has to be a 3 by 2 matrix. Then AX is a 2 by 2 matrix, XA is a 3 by 3 matrix, and they cannot be equal.

Remark: Here are some further problems for you to think about.

1. If A is not a square matrix, then we can never have $AX = XA$ due to the contradiction in size.
2. In the 2nd and the 3rd questions, we have seen that two entries in X can be arbitrary. In the 4th and the 5th questions, all four entries in X can be arbitrary. Can you find examples of 2 by 2 matrix A, so that only one entry can be arbitrary? How about three arbitrary entries, or no arbitrary entries, in X?
3. What matrix commutes with all the other matrices? In other words, what A has the property that $AX = XA$ for all X? First try the 2 by 2 case, and then guess the general case.
4. What matrix commutes with all the diagonal matrices? First try the 2 by 2 diagonal matrices

$$X = \begin{bmatrix} x & 0 \\ 0 & w \end{bmatrix}$$

5. And then the general n by n matrices.
6. What matrix commutes with all the matrices of the form

$$X = \begin{bmatrix} w+2z & z \\ z & w \end{bmatrix} = w\begin{bmatrix} 1 & 0 \\ 0 & 1 \end{bmatrix} + z\begin{bmatrix} 2 & 1 \\ 1 & 0 \end{bmatrix}$$

7. We know the matrix A in the third problem is one such. The problem is to find all the others.

Remark: The following exercises show that, for the matrix product, $AB = O$ does not necessarily imply $A = O$ or $B = O$.

Exercise: For each of the following matrix A, find all the two column matrices X satisfying $AX = O$ (the zero matrix).

1. $A = [1 \;\; 2]$

Answer: The matrix X has to be 2 by 2. Let

$$X = \begin{bmatrix} x & y \\ z & w \end{bmatrix}$$

Then

$$AX = [x + 2z \quad y + 2w]$$

so that $AX = O$ means

$x + 2z = 0,$

$y + 2w = 0.$

This further means $x = -2z$, $y = -2w$, with z and w arbitrary. We conclude that

$$X = \begin{bmatrix} -2z & -2w \\ z & w \end{bmatrix} = \begin{bmatrix} -2 & 0 \\ 1 & 0 \end{bmatrix} + \begin{bmatrix} 0 & -2 \\ 0 & 1 \end{bmatrix}$$

2. $A = \begin{bmatrix} 1 \\ 2 \end{bmatrix}$

Answer: The matrix X has to be 1 by 2. Let $X = [x \; y]$. Then

$$AX = \begin{bmatrix} x & y \\ 2x & 2y \end{bmatrix}$$

so that $AX = O$ means

$x = 0, 2x = 0,$

$y = 0, 2y = 0.$

We conclude that $X = O = [0 \;\; 0]$.

3. $A = \begin{bmatrix} 1 & 1 \\ 1 & -1 \end{bmatrix}$

Answer: The matrix X has to be 2 by 2. Let

$$X = \begin{bmatrix} x & y \\ z & w \end{bmatrix}$$

Then

$$AX = \begin{bmatrix} x+z & y+w \\ x-z & y-w \end{bmatrix}$$

so that $AX = O$ means

$x + z = 0, x - z = 0,$

$y + w = 0, y - w = 0.$

The first two equations tell us $x = z = 0$. The last two equations tell us $y = w = 0$. We conclude that $AX = O$ implies $X = O$.

4. $A = \begin{bmatrix} 2 & 4 \\ -1 & -2 \end{bmatrix}$

Answer: The matrix X has to be 2 by 2. Let X be as before. Then

$$AX = \begin{bmatrix} 2x+4z & 2y+4w \\ -x-2z & -y-2w \end{bmatrix}$$

so that $AX = O$ means

$2x + 4z = 0, - x - 2z = 0,$

$2y + 4w = 0, - y - 2w = 0.$

The first two equations tell us $x = - 2z$, z arbitrary. The last two equations tell us $y = - 2w$, w arbitrary. We conclude that

$$X = \begin{bmatrix} -2z & -2w \\ z & w \end{bmatrix} = z\begin{bmatrix} -2 & 0 \\ 1 & 0 \end{bmatrix} + w\begin{bmatrix} 0 & -2 \\ 0 & 1 \end{bmatrix}$$

5. $A = \begin{bmatrix} 0 & 0 \\ 0 & 0 \end{bmatrix}$

Answer: For any 2 by 2 matrix X, we always have $AX = O$.

6. $A = \begin{bmatrix} 1 & -2 & 3 \\ -2 & 3 & -4 \end{bmatrix}$

Answer: The matrix X has to be 3 by 2. Let

$$X = \begin{bmatrix} x & y \\ z & w \\ u & v \end{bmatrix}$$

Then $AX = O$ means

$$x - 2z + 3u = 0, -2x + 3z - 4u = 0,$$

$$y - 2w + 3v = 0, -2y + 3w - 4v = 0.$$

The first two equations tell us $x = u$, $z = 2u$, u arbitrary. The last two equations tell us $y = v$, $w = 2v$, v arbitrary. We conclude that

$$X = \begin{bmatrix} u & v \\ 2u & 2v \\ u & v \end{bmatrix} = u\begin{bmatrix} 1 & 0 \\ 2 & 0 \\ 1 & 0 \end{bmatrix} + v\begin{bmatrix} 0 & 1 \\ 0 & 2 \\ 0 & 1 \end{bmatrix}$$

7. $A = \begin{bmatrix} 1 & -2 \\ -2 & 3 \\ 3 & -4 \end{bmatrix}$

Answer: The matrix X has to be 2 by 2. Let

$$X = \begin{bmatrix} x & y \\ z & w \end{bmatrix}$$

Then $AX = O$ means

$$x - 2z = 0, -2x + 3z = 0, 3x - 4z = 0,$$

$$y - 2w = 0, -2y + 3w = 0, 3y - 4w = 0.$$

From the equations, we conclude that X has to be the zero matrix.

Remark: Here are some further problems for you to think about.

1. Let $X = [x\ y]$. Then $AX = [Ax\ Ay]$ by this formula. Thus $AX = O$ means solving $Ax = 0$ and $Ay = 0$. More precisely, we solve the homogeneous system $Ax = 0$ and put two independent solutions as the two columns of X. Check out the discussion with the concrete examples we worked out above.
2. When can we find nonzero X so that $AX = 0$? Answer the question in terms of properties of the systems $Ax = b$.
3. What if X has 3 columns? four columns?
4. For a given matrix A, find X such that $XA = O$. First try the problem with the matrices given in the exercise.

Exercise: For each of the following matrix A, find all the matrices X satisfying $AX = I$ (the identity matrix).

1. $A = [1\ 2]$

Answer: The matrix X has to be 2 by 1. Let

$$X = \begin{bmatrix} x \\ z \end{bmatrix}$$

Then $AX = [x + 2z]$, and $AX = I$ means $x + 2z = 1$. Thus we conclude that

$$X = \begin{bmatrix} 1-2z \\ z \end{bmatrix} = \begin{bmatrix} 1 \\ 0 \end{bmatrix} + z\begin{bmatrix} -2 \\ 1 \end{bmatrix}$$

where z is arbitrary.

2. $A = \begin{bmatrix} 1 \\ 2 \end{bmatrix}$

Answer: The matrix X has to be 1 by 2. Let $X = [x\ y]$. Then

$$AX = \begin{bmatrix} x & y \\ 2x & 2y \end{bmatrix}$$

so that $AX = I$ means $x = 1$, $y = 0$, $2x = 0$, $2y = 1$. The conditions are contradictory. Therefore there is no X such that $AX = I$.

3. $A = \begin{bmatrix} 1 & 1 \\ 1 & -1 \end{bmatrix}$

Answer: The matrix X has to be 2 by 2. Let

$$X = \begin{bmatrix} x & y \\ z & w \end{bmatrix}$$

Then

$$AX = \begin{bmatrix} x+z & y+w \\ x-z & y-w \end{bmatrix}$$

so that $AX = I$ means

$$x + z = 1, x - z = 0,$$

$$y + w = 0, y - w = 1.$$

The first two equations tell us $x = z = 1/2$. The last two equations tell us $y = 1/2$, $w = -1/2$. We conclude that $AX = I$ implies

$$X = \begin{bmatrix} 1/2 & 1/2 \\ 1/2 & -1/2 \end{bmatrix}$$

4. $A = \begin{bmatrix} 2 & 4 \\ -1 & -2 \end{bmatrix}$

Answer: The matrix X has to be 2 by 2. Let X be as before. Then

$$AX = \begin{bmatrix} 2x+4z & 2y+4w \\ -x-2z & -y-2w \end{bmatrix}$$

so that $AX = I$ means

$2x + 4z = 1, - x - 2z = 0,$

$2y + 4w = 0, - y - 2w = 1.$

The first two equations are contradictory. The last two equations are also contradictory. We conclude that there is no X such that $AX = I$.

5. $A = \begin{bmatrix} 0 & 0 \\ 0 & 0 \end{bmatrix}$

Answer: We have $AX = O$ for all X. Therefore AX can never be the identity matrix.

6. $A = \begin{bmatrix} 1 & -2 & 3 \\ -2 & 3 & -4 \end{bmatrix}$

Answer The matrix X has to be 3 by 2. Let

$$X = \begin{bmatrix} x & y \\ z & w \\ u & v \end{bmatrix}$$

Then $AX = I$ means

$x - 2z + 3u = 1, - 2x + 3z - 4u = 0,$

$y - 2w + 3v = 0, - 2y + 3w - 4v = 1.$

The first two equations tell us $x = 1 + u$, $z = 2u$, u arbitrary. The last two equations tell us $y = - 2 + v$, $w = -1 + 2v$, v arbitrary. We conclude that

$$X = \begin{bmatrix} 1+u & -2+v \\ 2u & 2v \\ u & v \end{bmatrix} =$$

where u and v are arbitrary.

7. $A = \begin{bmatrix} 1 & -2 \\ -2 & 3 \\ 3 & -4 \end{bmatrix}$

Answer: The matrix X has to be 2 by 3. Let

$$X = \begin{bmatrix} x & y & z \\ u & v & w \end{bmatrix}$$

Then $AX = I$ means

$x - 2u = 1,\ -2x + 3u = 0,\ 3x - 4u = 0,$

$y - 2v = 0,\ -2y + 3v = 1,\ 3y - 4v = 0,$

$z - 2w = 0,\ -2z + 3w = 0,\ 3z - 4w = 1.$

Note that we actually have three systems of linear equations, with different variables ((x, u) for the first system, (y, v) for the second, (z, w) for the third), but with the same matrix A as the coefficient matrix. Since the system in (x, u) has no solutions, there is no X such that $AX = I$.

Exercise: Let

$$A = \begin{bmatrix} 2 & 0 & 2 & 1 \\ 1 & 1 & -1 & 0 \\ 0 & 1 & -2 & 1 \end{bmatrix}$$

For each of the following matrix B, find a matrix X such that $AX = B$.

1. $B = \begin{bmatrix} 1 \\ -1 \\ -2 \end{bmatrix}$

Answer: X must be a vertical vector of dimension 4. The equation $AX = B$ is the usual system of 3 linear equations in 4 variables. It is in fact this system. By the earlier computation, the augmented matrix [A B] has the following row echelon form.

$$\begin{bmatrix} 1 & 1 & -1 & 0 & -1 \\ 0 & 1 & -2 & 1 & -2 \\ 0 & 0 & 0 & 3 & -1 \end{bmatrix}$$

From this we find one solution

$$X = \begin{bmatrix} -1 \\ -2 \\ 0 \\ -1/3 \end{bmatrix}$$

2. $B = \begin{bmatrix} 1 & 1 \\ -1 & 0 \\ -2 & 0 \end{bmatrix}$

Answer: Let

$$b_1 = \begin{bmatrix} 1 \\ -1 \\ -2 \end{bmatrix}, \quad b_2 = \begin{bmatrix} 1 \\ 0 \\ 1 \end{bmatrix}$$

be the two columns of B. Let x, y be the two columns of X. By $AX = A[x\ y] = [Ax\ Ay]$, $AX = B$ is equivalent to two separate systems $Ax = b_1$, $Ay = b_2$. The first system has been studied in the problem above and a solution has been obtained. The second system has an obvious solution

$$y = \begin{bmatrix} 0 \\ 0 \\ 0 \\ 1 \end{bmatrix}$$

Combining the solutions together, we get a solution

$$X = \begin{bmatrix} 1 & 0 \\ -2 & 0 \\ 0 & 0 \\ -1/3 & 1 \end{bmatrix}$$

of $AX = B$.

3. $B = \begin{bmatrix} 1 & 0 & 0 \\ 0 & 1 & 0 \\ 0 & 0 & 1 \end{bmatrix}$

Answer: The columns of B are the standard basis vectors e_1, e_2, $e3$. By a similar argument as before, the columns of X are the solutions of the systems $Ax = e_i$. Combining the augmented matrices of the three systems, we get

$$[A\ e_1\ e_2\ e_3] = [A\ I] = \begin{bmatrix} 2 & 0 & 2 & 1 & 1 & 0 & 0 \\ 1 & 1 & -1 & 0 & 0 & 1 & 0 \\ 0 & 1 & -2 & 1 & 0 & 0 & 1 \end{bmatrix}$$

Row operations (e.g., as in this exercise) change the matrix into

$$\begin{bmatrix} 1 & 1 & -1 & 0 & 0 & 1 & 0 \\ 0 & 1 & -2 & 1 & 0 & 0 & 1 \\ 0 & 0 & 0 & 3 & 1 & -2 & 2 \end{bmatrix}$$

By taking [col 1–5], we get a solution (1/3, –1/3, 0, 1/3) of $Ax = e_1$. By taking [col 1–4] and [col 6], we get a solution (1/3, 2/3, 0, –2/3) of $Ax = e_2$. By taking [col 1–4] and [col 7], we get a solution (–1/3, 1/3, 0, 2/3) of $Ax = e_3$. Putting the three columns together, we get a solution

$$X = \begin{bmatrix} 1/3 & 1/3 & -1/3 \\ -1/3 & 2/3 & 1/3 \\ 0 & 0 & 0 \\ 1/3 & -2/3 & 2/3 \end{bmatrix}$$

of $AX = B$.

Remark: From this sequence of exercises, you can see the relation between solving equations like $AX = B$ and our usual systems $Ax = b$ of linear equations. Apply your new insight to this exercise and this exercise. Moreover, study the existence/uniqueness/general solution problem for $AX = B$, just as we have done for systems of linear equations.

Exercise: Let

$$A = \begin{bmatrix} 2 & -2 & 0 & 2 & 2 \\ -1 & 1 & 2 & -3 & 1 \\ -2 & 2 & 1 & -3 & -1 \end{bmatrix}, B = \begin{bmatrix} 0 & 2 & 0 \\ 0 & 3 & 2 \\ 0 & 0 & 1 \end{bmatrix}$$

Can you find a matrix X such that $AX = B$? In general, what is the criterion for the existence of X such that $AX = B$? What about the uniqueness of X?

Answer: As we have seen in the previous exercise, the equation $AX = B$ is equivalent to solving three separate systems $Ax = b_1$, $Ax = b_2$, $Ax = b_3$, where b_1, b_2, b_3 are the three columns of b. In an earlier exercise, we have argued that the system $Ax = b_1$ has no solution. Therefore the equation $AX = B$ has no solution.

In general, $AX = B$ has solutions if and only if no column of B is a pivot column of the "augmented matrix" $[A\ B]$. The solution is unique if and only if all columns of A are pivot.

Exercise: Describe the first two rows and the last three columns of AB in terms of row and columns of A and B.

Answer: The first two rows of AB is the product of the first two rows of A with B. The last three columns of AB is the product of A with the last three columns of B.

Exercise: The trace trA of a square matrix is the sum of its diagonal entries. For example,

$$tr\begin{bmatrix} 1 & 2 \\ 3 & 4 \end{bmatrix} = 1 + 4 = 5$$

$$tr\begin{bmatrix} 0 & 3 & -4 \\ 2 & 5 & 7 \\ 0 & 4 & -3 \end{bmatrix} = 0 + 5 - 3 = 2$$

Prove the trace has the following properties

$$tr(A + B) = trA + trB,\ tr\ cA = c\ trA,\ trA^{T} = trA,\ trAB = trBA.$$

Answer: The first two properties mean that the trace is linear. The third property means the trace is not changed by the transpose. All are quite obvious.

In the property $trAB = trBA$, we do not need A and B to be square matrices. We will only assume A is an m by n matrix, and B is an n by m matrix.

The i-th diagonal entry of AB is $\Sigma_{1 \le j \le n} a_{ij} b_{ji} = a_{i1}b_{1i} + a_{i2}b_{2i} + \ldots + a_{in}b_{ni}$. Thus

$trAB = \Sigma_{1 \le i \le m}(a_{i1}b_{1i} + a_{i2}b_{2i} + \ldots + a_{in}b_{ni}) = \Sigma_{1 \le i \le m,\ 1 \le j \le n} a_{ij}b_{ji}$.

Exchanging B and A, we have

$trBA = \Sigma_{1 \le j \le n}(b_{j1}a_{1j} + b_{j2}a_{2j} + \ldots + b_{jm}a_{mj}) = \Sigma_{1 \le j \le n, 1 \le i \le m} b_{ij}a_{ji} = \Sigma_{1 \le i \le m, 1 \le j \le n} a_{ji}b_{ij} = trAB.$

Exercise: What does the multiplication of two upper triangular matrices look like? What about lower triangular matrices? diagonal matrices?

Answer: Given two upper triangular matrices,

$$A=\begin{bmatrix} a_1 & * & .. & * \\ 0 & a_2 & .. & * \\ \vdots & \vdots & & \vdots \\ 0 & 0 & .. & a_n \end{bmatrix}, B=\begin{bmatrix} b_1 & * & .. & * \\ 0 & b_2 & .. & * \\ \vdots & \vdots & & \vdots \\ 0 & 0 & .. & b_n \end{bmatrix}$$

their multiplication is still upper triangular, with the diagonal entries simply multiplied together (the other entries are more complicated).

$$AB=\begin{bmatrix} a_1b_1 & * & .. & * \\ 0 & a_2b_2 & .. & * \\ \vdots & \vdots & & \vdots \\ 0 & 0 & .. & a_nb_n \end{bmatrix}$$

The similar statement is true for the multiplication of two lower triangular matrices.

For two diagonal matrices,

$$A=\begin{bmatrix} a_1 & 0 & .. & 0 \\ 0 & a_2 & .. & 0 \\ \vdots & \vdots & & \vdots \\ 0 & 0 & .. & a_n \end{bmatrix}$$

their multiplication is still diagonal, with the diagonal entries simply multiplied together.

$$AB=\begin{bmatrix} a_1b_1 & 0 & .. & 0 \\ 0 & a_2b_2 & .. & 0 \\ \vdots & \vdots & & \vdots \\ 0 & 0 & .. & a_nb_n \end{bmatrix}$$

Exercise: What are the matrices A satisfying $AX = XA$ for any X?

Answer If A is an m by n matrix, then in order for both AX and XA to make sense, X has to be an n by m matrix. Then AX is an m by m matrix, and XA is an n by n matrix. Now for the two matrix to be equal, we must have $m = n$. In other words, A has to be a square matrix.

Now assume A is an n by n matrix. Let E_{ij} be the n by n matrix with 1 as the ij–entry and 0 as all the other entries. For example, in case $n = 3$ we have

$$E_{11} = \begin{bmatrix} 1 & 0 & 0 \\ 0 & 0 & 0 \\ 0 & 0 & 0 \end{bmatrix}, E_{23} = \begin{bmatrix} 0 & 0 & 0 \\ 0 & 0 & 1 \\ 0 & 0 & 0 \end{bmatrix}, E_{31} = \begin{bmatrix} 0 & 0 & 0 \\ 0 & 0 & 0 \\ 1 & 0 & 0 \end{bmatrix}$$

Then AE_{ij} is obtained by moving the i-th column of A to j-th column, and leaving all the other columns zero. For example,

$$AE_{23} = \begin{bmatrix} a_{11} & a_{12} & a_{13} \\ a_{21} & a_{22} & a_{23} \\ a_{31} & a_{32} & a_{33} \end{bmatrix} \begin{bmatrix} 0 & 0 & 0 \\ 0 & 0 & 1 \\ 0 & 0 & 0 \end{bmatrix} = \begin{bmatrix} 0 & 0 & a_{12} \\ 0 & 0 & a_{22} \\ 0 & 0 & a_{32} \end{bmatrix}$$

And $E_{ij}A$ is obtained by moving the j-th row of A to i-th row, and leaving all the other rows zero. For example,

$$E_{23}A = \begin{bmatrix} 0 & 0 & 0 \\ 0 & 0 & 1 \\ 0 & 0 & 0 \end{bmatrix} \begin{bmatrix} a_{11} & a_{12} & a_{13} \\ a_{21} & a_{22} & a_{23} \\ a_{31} & a_{32} & a_{33} \end{bmatrix} = \begin{bmatrix} 0 & 0 & 0 \\ a_{31} & a_{32} & a_{23} \\ 0 & 0 & 0 \end{bmatrix}$$

Thus from $AE_{ij} = E_{ij}A$, we have

- Any entry of AE_{ij} not in the i-th row should vanish. This means in the i-th column of A, all but a_{ii} should vanish. In the E_{23} example above, we mean $a_{12} = a_{32} = 0$.
- Any entry of $E_{ij}A$ not in the j-th column should vanish. This means in the j-th row of A, all but a_{jj} should vanish. In the E_{23} example above, we mean $a_{31} = a_{32} = 0$.
- The ij–entry of AE_{ij} and $E_{ij}A$ are equal. This means that $a_{ii} = a_{jj}$.

Since the above should hold for any i, j, we conclude that $a_{ij} = 0$ for $i \neq j$, and $a_{ii} = a_{jj}$ for any i, j. This is the same as $A = aI$, which is a constant multiple of the identity.

Exercise:

1. Prove that transformations T and S are onto $\Rightarrow$ The composition TS is onto.

Answer: For any b, by T onto, we can find y, such that $T(y) = b$. For this y, by S onto, we can find x, such that $S(x) = y$. Then we have

$$TS(x) = T(S(x)) = T(y) = b.$$

Thus any b is the image of some element under the transformation TS. This shows TS is onto.

2. What about the converse? In other words, if the composition TS is onto, are the transformations T and S onto?

Answer: If TS is onto, then for any b, we have x, such that $T(S(x)) = b$. In particular, b is the image of $S(x)$ under the transformation T. The observation tells us T must be onto.

On the other hand, S is not necessarily onto. An example is given by $S = ID_s$: student → integer, T = *Last–Digit*: integer → {0, 1, 2, 3, 4, 5, 6, 7, 8, 9}.

3. Rephrase your conclusions in the first and second part in terms of matrix products.

Answer: If $Ay = b$ and $Bx = c$ has solutions for any b and c, then $ABx = b$ has solution for any b.

Conversely, if $ABx = b$ has solutions for any b, then $Ay = b$ has solutions for any b. However, $Bx = c$ may not have solution for any c.

Alternatively: If all rows of A and B are pivot, then all rows of AB are pivot. Conversely, if all rows of AB are pivot, then all rows of A (but not necessarily B) are also pivot.

Exercise:

1. Prove that transformations T and S are ono–to–one ή The composition TS is one–to–one.

Answer: Let T and S be one–to–one. Then

$$TS(x) = TS(x') \Rightarrow S(x) = S(x') \Rightarrow x = x'$$

where the first ⇒ follows from T being one–to–one, and the second ⇒ follows from S being one–to–one. The whole implication means TS is one–to–one.

2. What about the converse? In other words, if the composition TS is ono–to–one, are the transformations T and S ono–to–one?

Answer: Suppose TS is ono–to–one. Then we have

$$S(x) = S(x') \Rightarrow TS(x) = TS(x') \Rightarrow x = x'$$

where the first ⇒ is obtained by applying T to both sides, and the second ⇒ follows from TS being one–to–one. The whole implication means S is one–to–one.

On the other hand, T is not necessarily one–to–one. An example is given by $S = ID_s$: student → integer, T = *Absolute–Value*: integer → non–negative integer. Note that student ID numbers are always positive.

3. Rephrase your conclusions in the first and second part in terms of matrix products.

Answer: If solutions of $Ay = b$ and $Bx = c$ are unique, then the solution of $ABx = b$ is unique.

Conversely, if the solution of $ABx = b$ is unique, then solution of $Bx = c$ is unique. However, solution of $Ay = b$ may not be unique.

Alternatively: If all columns of A and B are pivot, then all columns of AB are pivot. Conversely, if all columns of AB are pivot, then all columns of B (but not necessarily A) are also pivot.

Chapter 5

Inverse Transformation

INVERSE

A transformation $T: X \to Y$ is invertible if there is a transformation $S: Y \to X$ satisfying

$ST = id_X$, and $TS = id_Y$.

In other words, we have

$ST(x) = x$ for any $x \in X$, and $TS(y) = y$ for any $y \in Y$.

Intuitively, the first equality means that S reverses what T does. The first equality means that T reverses what S does. We call S the inverse of T and denote $S = T^{-1}$.

Clearly, T is also the inverse of S: $T = S^{-1}$.

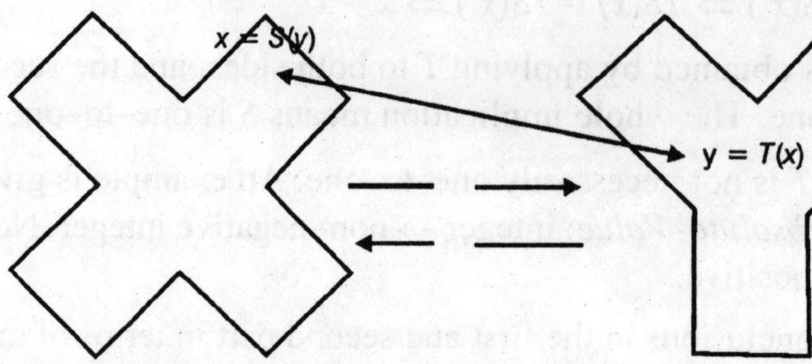

More discussions on inverse transformations can be found here.

Example: The transformation

Capitalize: $\{a, b, c, \ldots, z\} \to \{A, B, C, \ldots, Z\}$

has

de–Capitalize: $\{A, B, C, \ldots, Z\} \to \{a, b, c, \ldots, z\}$

as the inverse transformation. The transformation

Name: {+, –, >, <. = 0, 1} → {plus, minus, bigger, smaller, equal, zero, one}

has

Symbol: {plus, minus, bigger, smaller, equal, zero, one} → {+, –, >, <, =, 0, 1}

as the inverse transformation.

For the transformations

Capital City: country → city,

Which Country: city → country,

we have

(*Which Country*)(*Capital City*) = *id*

(*Capital City*)(*Which Country*) ≠ *id*.

Thus the two transformations are not inverse to each other. In fact, neither transformation is invertible.

Example: For T = *Reflection in x–axis*, we have $T^2 = TT = id$ (reflecting twice goes back to itself).

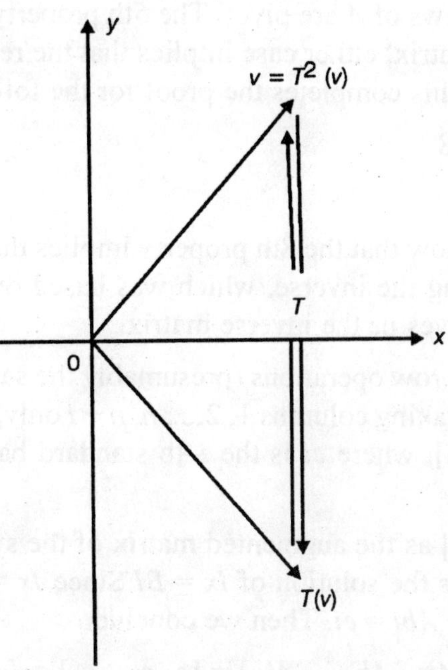

Therefore we have $T = T^{-1}$. The rotation R_θ by angle θ, is clearly reversed by the rotation $R_{-\theta}$ by angle –θ. Therefore we have $R_{-\theta} = R_\theta^{-1}$.

For a non–invertible linear transformation, let us consider the zero transformation O (from R^2 to R^2, for example), that takes any vector to the zero vector.

Clearly for any other linear transformation X, we have $XO = O$ and $OX = O$. Therefore we cannot find an inverse for O.

One general way of computing the inverse is based on the following: If S is the inverse of T, then

$T(x) = y \Leftrightarrow S(y) = x.$

Exercise: To find the inverse of the transformation $T(x) = 3x - 2$: R → R, we solve $3x - 2 = y$ and get $x = y/3 + 2/3$. Therefore $T^{-1}(y) = y/3 + 2/3$.

We also recall that one way to introduce the log function *ln*: $(0, \infty) \to$ R is to define it as the inverse of the exponential function e^t: R $\to (0, \infty)$. The exponential function can be sensibly defined as the function satisfying $f' = f$ and $f(0) = 1$.

You may also recall how the inverse trigonometric functions are defined.

CRITERION FOR INVERTIBILITY OF A TRANSFORMATION

Theorem A transformation is invertible ⇔ It is onto and one–to–one.

Proof. By definition of invertibility, the 1st property implies the 2nd and the 3rd. Then by this result, the 2nd property implies the 4th, and the 3rd implies the 5th. The 4th property means that all rows of A are pivot. The 5th property means that all columns of A are pivot. For a square matrix, either case implies that the reduced row echelon form of A is the identity matrix I. This completes the proof for the following implications.

$1 \Rightarrow 2 \Rightarrow 4 \Rightarrow 6 \Rightarrow 8$

$1 \Rightarrow 3 \Rightarrow 5 \Rightarrow 7\ 8$

Thus it remains to show that the 8th property implies the 1st. The idea is to show that our method for computing the inverse, which was based on the assumption that the 8th property holds, indeed gives us the inverse matrix.

The method is to use row operations (presumably the same ones that change A to I) to change $[A\ I]$ to $[I\ B]$. By taking columns $1, 2, \ldots, n, n + i$ only, we see that $[A\ ei]$ can be row operated to become $[I\ bi]$, where ei is the i–th standard basis vector, and BI is the i–th column of B.

By thinking of $[A\ ei]$ as the augmented matrix of the system $Ax = ei$, the solution of the system is the same as the solution of $Ix = BI$ Since $Ix = BI$ clearly has $x = BI$ as the unique solution, we have $Abi = ei$. Then we conclude

$$AB = [Ab_1, Ab_2, \ldots, Ab_n] = [e_1\ e_2 \ldots\ e_n] = I.$$

To show B is really the inverse of A, we still need to establish $BA = I$. The following argument shows that this is a consequence of $AB = I$ and A is a square matrix.

$AB = I$

$\Rightarrow Ax = b$ has solution for any b

$\Rightarrow$ Solution of $Ax = b$ is unique

$AB = I$

$\Rightarrow A(BA) = (AB)A = IA = A = AI$

$\Rightarrow A(i\text{-th column of } BA) = Ae_i$

$\Rightarrow i\text{-th column of } BA = e_i$

$\Rightarrow BA = [e_1\ e_2 \ldots e_n] = I.$

INVERSE MATRIX

By the equivalence between the composition of linear transformations and the multiplication of matrices, we define a matrix B to be the inverse of a matrix A if

$$BA = I, \text{ and } AB = I,$$

where I denotes the identity matrix. We also denote $B = A^{-1}$, similar to the notation for inverse transformations.

Our study of the inverse matrix is based on the following result. A generalization of the result can be found in this earlier exercise and this earlier exercise. The converse of the result can be found in this exercise.

$AB = I \Rightarrow Ax = b$ has solution for any b, and solution of $By = c$ is unique.

Proof For any b, the following verifies that $x = Bb$ is a solution of $Ax = b$:

$$Ax = A(Bb) = (AB)b = Ib = b.$$

As for the uniqueness for $By = c$, let y and y' be two solutions. Then

$$By = c,\ By' = c \Rightarrow ABy = Ac,\ ABy' = Ac \ \square\ y = Ac = y',$$

where $AB = I$ is used in the last step.

Thus the two conditions in the definition of inverse matrices imply that $Ax = b$ must have a unique solution for any b. In fact from the proof we also know what the solution is.

A is invertible $\Rightarrow$ For any b, $Ax = b$ has $x = A^{-1}b$ as the unique solution.

Moreover, by the numerical consequences of always existence and uniqueness, we have

An invertible matrix must be a square matrix.

Translated to linear transformations, we see that there is no invertible linear transformations between euclidean spaces of different dimensions.

If there is an invertible linear transformation from R^n to R^m, then $m = n$.

Example: We gave the zero transformation as an example of non–invertible linear transformation. Now by the size consideration, we know that the linear transformation

$T(x_1, x_2, x_3) = (x_1 + 2x_2 + 3x_3, 4x_1 + 5x_2 + 6x_3)$: $R^3 \rightarrow R^2$

is not invertible. Correspondingly, the matrix

is not invertible.

For the square matrix

$$\begin{bmatrix} 1 & 3 & 5 \\ 2 & 4 & 6 \end{bmatrix}$$

because the corresponding system does not always have solutions, the matrix is not invertible.

$$\begin{bmatrix} 3 & 1 & -1 \\ 1 & -1 & 1 \\ 2 & -1 & 1 \end{bmatrix}$$

INVERSE OF 2 BY 2 MATRIX

A 1 by 1 matrix $A = (a)$ is invertible if and only if a "' 0. Moreover, the inverse $A^{-1} = (a^{-1})$.

Assume the 2 by 2 matrix

$$A = \begin{bmatrix} a & b \\ c & d \end{bmatrix}$$

has

$$B = \begin{bmatrix} x & y \\ y & w \end{bmatrix}$$

as the inverse. Then from $AB = I$ we have

$ax + by = 1$, $cx + dy = 0$, $az + bw = 0$, $cz + dw = 1$.

The first two is a system of linear equations with A as the coefficient matrix and x, y as variables. Then d[equation 1] – b[equation 2] and – c[equation 1] + a[equation 2] give us

$$(ad - bc)x = d, (ad - bc)y = - c.$$

Similarly, from the 3rd and the 4th equations we get

$$(ad - bc)z = - b, (ad - bc)w = a.$$

Thus we conclude the ά direction of the following result.

A 2 by 2 matrix $\begin{bmatrix} a & b \\ c & d \end{bmatrix}$ is invertible $\Leftrightarrow ad - bc \neq 0$. Moreover, the inverse is

$$\frac{1}{ad-bc}\begin{bmatrix} d & -b \\ -c & a \end{bmatrix}$$

For the $\Rightarrow$ direction, see this exercise.

Example: We have

$$\begin{bmatrix} 1 & 3 \\ 2 & 4 \end{bmatrix}^{-1} = (1\times 4 - 3\times 2)^{-1}\begin{bmatrix} 4 & -3 \\ -2 & 1 \end{bmatrix}^{-1}$$

We verify our computation as follows.

$$\begin{bmatrix} 1 & 3 \\ 2 & 4 \end{bmatrix}\begin{bmatrix} -2 & 3/2 \\ 1 & -1/2 \end{bmatrix} = \begin{bmatrix} 1\times(-2)+3\times 1 & 1\times 3/2 + 3\times(-1/2) \\ 2\times(-2)+4\times 1 & 2\times 3/2 + 4\times(-1/2) \end{bmatrix} = \begin{bmatrix} 1 & 0 \\ 0 & 1 \end{bmatrix}$$

$$\begin{bmatrix} -2 & 3/2 \\ 1 & -1/2 \end{bmatrix}\begin{bmatrix} 1 & 3 \\ 2 & 4 \end{bmatrix} = \begin{bmatrix} (-2)\times 1+(3/2)\times 2 & (-2)\times 3+(3/2)\times 4 \\ 1\times 1+(-1/2)\times 2 & 1\times 3+(-1/2)\times 4 \end{bmatrix} = \begin{bmatrix} 1 & 0 \\ 0 & 1 \end{bmatrix}$$

The inverse also tells us that the system

$$\begin{aligned} x_1 + 3x_2 &= 1 \\ 2x_1 + 4x_2 &= -4 \end{aligned}$$

has

$$\begin{bmatrix} x_1 \\ x_2 \end{bmatrix} = \begin{bmatrix} 1 & 3 \\ 2 & 4 \end{bmatrix}^{-1}\begin{bmatrix} 1 \\ -4 \end{bmatrix} = \begin{bmatrix} 4 \\ -1 \end{bmatrix}$$

as the unique solution.

Example: The following matrices

$$\begin{bmatrix} 2 & 4 \\ -1 & -2 \end{bmatrix}$$

$$\begin{bmatrix} 1 & 5 \\ 3 & 15 \end{bmatrix}$$

are not invertible because $2\times(-2) - 4\times(-1) = 0$, $1\times 15 - 5\times 3 = 0$.

The formula for the inverse of 2 by 2 matrices can be generalized to bigger matrices. However, the amount of computation involved in such a formula will become too big to be practical.

COMPUTATION OF INVERSE MATRIX

Recall that an invertible matrix A must be a square matrix such that $Ax = b$ has a unique solution for any b. In particular, the reduced row echelon form of A must be the identity matrix. In other words, we can use row operations to change A to I. Conversely, we may use the observation to compute the inverse matrix.

Row operations change $[A\ I]$ to $[I\ B] \Rightarrow x$ is the inverse of A

The justification of the method can be found later in the proof of criterion for invertibility.

Example: To find the inverse of

$$A = \begin{bmatrix} 3 & 1 & -1 \\ 1 & -1 & 1 \\ 2 & 2 & 1 \end{bmatrix}$$

we form

$$[A\ I] = \begin{bmatrix} 3 & 1 & -1 & 1 & 0 & 0 \\ 1 & -1 & 1 & 0 & 1 & 0 \\ 2 & 2 & 1 & 0 & 0 & 1 \end{bmatrix}$$

The row operations corresponding to what we have done to A in this example and this example give us

$$\begin{bmatrix} 1 & -1 & 1 & 0 & 1 & 0 \\ 0 & 4 & -4 & 1 & -3 & 0 \\ 0 & 0 & 3 & -1 & 1 & 1 \end{bmatrix}$$

Further row operations give us

$$\begin{bmatrix} 1 & 0 & 0 & 1/4 & 1/4 & 0 \\ 0 & 1 & 0 & -1/2 & -5/12 & 1/3 \\ 0 & 0 & 1 & -1/3 & 1/3 & 1/3 \end{bmatrix}$$

Thus we conclude

$$A^{-1} = \begin{bmatrix} 1/4 & 1/4 & 0 \\ -1/12 & -5/12 & 1/3 \\ -1/3 & 1/3 & 1/3 \end{bmatrix}$$

By the way, the vector

$$\begin{bmatrix} 1/4 & 1/4 & 0 \\ -1/12 & -5/12 & 1/3 \\ -1/3 & 1/3 & 1/3 \end{bmatrix}\begin{bmatrix} 2 \\ 2 \\ 6 \end{bmatrix} = \begin{bmatrix} 1 \\ 1 \\ 2 \end{bmatrix}$$

is exactly the solution of the system in this example.

Exercise: To find the inverse of

$$A = \begin{bmatrix} 1 & 0 & 1 & 0 \\ 1 & 1 & 0 & 1 \\ 0 & 0 & 1 & 1 \\ 1 & 1 & 1 & 1 \end{bmatrix}$$

we form

$$[A\ I] = \begin{bmatrix} 1 & 0 & 1 & 0 & 1 & 0 & 0 & 0 \\ 1 & 1 & 0 & 1 & 0 & 1 & 0 & 0 \\ 0 & 0 & 1 & 1 & 0 & 0 & 1 & 0 \\ 1 & 1 & 1 & 1 & 0 & 0 & 0 & 1 \end{bmatrix}$$

Row operations on the matrix give us

$$\begin{bmatrix} 1 & 0 & 0 & 0 & 1 & 1 & 0 & -1 \\ 0 & 1 & 0 & 0 & -1 & -1 & -1 & 2 \\ 0 & 0 & 1 & 0 & 0 & -1 & 0 & 1 \\ 0 & 0 & 0 & 1 & 0 & 1 & 1 & -1 \end{bmatrix}$$

Thus we have

$$A^{-1} = \begin{bmatrix} 1 & 0 & 0 & 0 & 1 & 1 & 0 & -1 \\ 0 & 1 & 0 & 0 & -1 & -1 & -1 & 2 \\ 0 & 0 & 1 & 0 & 0 & -1 & 0 & 1 \\ 0 & 0 & 0 & 1 & 0 & 1 & 1 & -1 \end{bmatrix}$$

In terms of linear transformations, the inverse of

$$T(x_1, x_2, x_3, x_4) = (x_1 + x_3, x_1 + x_2 + x_4, x_3 + x_4, x_1 + x_2 + x_3 + x_4): \mathrm{R}^4 \to \mathrm{R}^4$$

is

$$T^{-1}(x_1, x_2, x_3, x_4) = (x_1 + x_2 - x_4, -x_1 - x_2 - x_3 + 2x_4, -x_2 + x_4, x_2 + x_3 - x_4): \mathrm{R}^4 \to \mathrm{R}^4$$

PROPERTIES OF INVERSE

We have seen the relation between invertibility and the existence of unique solution, which has immediate implication on the size of the matrix. In addition to this, we have more properties for inverse transformations.

We start with a simple property.

A is invertible $\Rightarrow A^{-1}$ is invertible, and $((A)^{-1})^{-1} = A$.

This is the consequence of the symmetry of the definition of inverse: B is inverse of A $\Leftrightarrow A$ is inverse of B. Here is another simple property.

A and B are invertible $\Rightarrow AB$ is invertible, and $(AB)^{-1} = B^{-1}A^{-1}$.

It is important to note the order of multiplication.

A consequence of the two properties is that $(A^k)^{-1} = (A^{-1})^k$ for any positive k and invertible A. We denote $(AK)^{-1}$ or $(A^{-1})^k$ by AK Combined with earlier notation, we have defined AK for any invertible A and any integer k.

Proof $(AB)(B^{-1}A^{-1}) = A(BB^{-1})A^{-1} = A(I)A^{-1} = AA^{-1} = I$. Similarly, $(B^{-1}A^{-1})(AB) = I$.

The inverse of a matrix (or a linear transformation) is unique.

Proof Suppose B and C are inverses of A. Then $B = IB = (CA)B = C(AB) = CI = C$.

Finally, we assemble equivalent *invertibility criterion for square matrices.*

For a square matrix A, the following are equivalent

- A is invertible.
- $AB = I$ for some matrix B.
- $BA = I$ for some matrix B.
- $Ax = b$ has solutionss for any b.
- Solution of $Ax = b$ is unique.
- All rows of A are pivot.
- All columns of A are pivot.
- A can be row operated to become I.

Note that in the original definition of inverse matrix, we need both the 2nd and the 3rd properties hold. Now we see that *for a square matrix*, they are equivalent. The following are the same criterion stated for linear transformations.

For a linear transformation T: $R^n \to R^n$, the following are equivalent.

- T is invertible.
- $TS = id$ for some transformation S.

- $ST = id$ for some transformation S.
- T is onto.
- T is one–to–one.

If we do not know apriori the dimensions on both sides of T, we may use the following criterion for invertibility.

For a linear transformation $T: \mathbb{R}^n \to \mathbb{R}^m$, the following are equivalent.

- T is invertible.
- T is onto and one–to–one.
- *Range* $T = \mathbb{R}^m$, *kernel*$T = \{0\}$.

Proof. By considering the matrices associated to linear transformations, this result then implies that T is invertible $\Leftrightarrow$ T is onto and one–to–one. Conversely, suppose T is onto and one–to–one, then this result shows $m \leq n$ and this result shows $m \geq n$.

Thus $m = n$, and we may use the previous criterion to conclude that T is invertible. This completes the proof of the equivalence between the first two statements.

The equivalence between the second and the third statements follows from this definition and this result.

CRITERION FOR INVERTIBILITY OF A MATRIX

Theorem For a square matrix A, the following are equivalent

- A is invertible
- $AB = I$ for some matrix B
- $BA = I$ for some matrix B
- $Ax = b$ has solution for any b
- Solution of $Ax = b$ is unique
- All rows of A are pivot
- All columns of A are pivot
- A can be row operated to become I

Proof. By definition of invertibility, the 1st property implies the 2nd and the 3rd. Then by this result, the 2nd property implies the 4th, and the 3rd implies the 5th. The 4th property means that all rows of A are pivot.

The 5th property means that all columns of A are pivot. For a square matrix, either case implies that the reduced row echelon form of A is the identity matrix I. This completes the proof for the following implications.

$1 \Rightarrow 2 \Rightarrow 4 \Rightarrow 6 \Rightarrow 8$

$1 \Rightarrow 3 \Rightarrow 5 \Rightarrow 7\ 8$

Thus it remains to show that the 8th property implies the 1st. The idea is to show that our method for computing the inverse, which was based on the assumption that the 8th property holds, indeed gives us the inverse matrix.

The method is to use row operations (presumably the same ones that change A to I) to change $[A\ I]$ to $[I\ B]$. By taking columns 1, 2,..., n, $n+i$ only, we see that $[A\ ei]$ can be row operated to become $[I\ bi]$, where ei is the i–th standard basis vector, and BI is the i–th column of B. By thinking of $[A\ ei]$ as the augmented matrix of the system $Ax = ei$, the solution of the system is the same as the solution of $Ix = BI$ Since $Ix = BI$ clearly has $x = BI$ as the unique solution, we have $Abi = ei$. Then we conclude

$$AB = [Ab_1, Ab_2,..., Ab_n] = [e_1\ e_2...\ e_n] = I.$$

To show B is really the inverse of A, we still need to establish $BA = I$. The following argument shows that this is a consequence of $AB = I$ and A is a square matrix.

$AB = I$

$\Rightarrow Ax = b$ has solution for any b

$\Rightarrow$ Solution of $Ax = b$ is unique

$AB = I$

$\Rightarrow A(BA) = (AB)A = IA = A = AI$

$\Rightarrow A(i\text{-th column of } BA) = Ae_i$

$\Rightarrow i\text{-th column of } BA = e_i$

$\Rightarrow BA = [e_1\ e_2...\ e_n] = I.$

Exercise: Find invertible matrices and compute the inverse.

1. $\begin{bmatrix} 3 & -2 \\ 5 & -3 \end{bmatrix}$

Answer: By the conclusion about the inverse of 2 by 2 matrices, since $3 \times (-3) - 5 \times (-2) \neq 0$, the matrix is invertible. The inverse is

$$(3\times(-3)-5\times(-2))^{-1}\begin{bmatrix} 3 & -2 \\ -5 & 3 \end{bmatrix} = \begin{bmatrix} -3 & 2 \\ -5 & 3 \end{bmatrix}$$

2. $\begin{bmatrix} 3 & 5 \\ -2 & -3 \end{bmatrix}$

Answer: Since $3\times(-3) - (-2) \times 5 \neq 0$, the matrix is invertible. The inverse is

$$(3\times(-3)-5\times(-2))^{-1}\begin{bmatrix} -3 & 5 \\ 2 & 3 \end{bmatrix} = \begin{bmatrix} -3 & -5 \\ 2 & 3 \end{bmatrix}$$

3. $\begin{bmatrix} 6 & 4 \\ 9 & 6 \end{bmatrix}$

Answer: Since $6 \times 6 - 4 \times 9 = 0$, the matrix is not invertible.

4. $\begin{bmatrix} 6 & 4 \\ 9 & 7 \end{bmatrix}$

Answer: Since $6 \times 7 - 4 \times 9 \neq 0$, the matrix is invertible. The inverse is

$$(6\times 7-4\times 9)^{-1}\begin{bmatrix} 7 & -4 \\ -9 & 6 \end{bmatrix} = \begin{bmatrix} 7/6 & -2/3 \\ -3/2 & 1 \end{bmatrix}$$

5 – 8)

$$\begin{bmatrix} 2 & 0 & 1 \\ 1 & 1 & 0 \\ 0 & 1 & 1 \end{bmatrix}$$

$$\begin{bmatrix} 2 & 2 & 1 \\ 1 & -1 & 0 \\ 0 & -2 & 1 \end{bmatrix}$$

$$\begin{bmatrix} 2 & 0 & 2 \\ 1 & 1 & -1 \\ 0 & 1 & -2 \end{bmatrix}$$

$$\begin{bmatrix} 2 & 0 & 2 & 1 & 1 \\ 1 & 1 & -1 & 0 & -1 \\ 0 & 1 & -2 & 1 & -2 \end{bmatrix}$$

Answer: The first three matrices are obtained by selecting columns from the fourth matrix. In an earlier exercise, we have computed the row echelon form

$$\begin{bmatrix} 1 & 1 & -1 & 0 & -1 \\ 0 & 1 & -2 & 1 & -2 \\ 0 & 0 & 0 & 3 & -1 \end{bmatrix}$$

for the fourth matrix. The first matrix, being [col 1], [col 2], [col 4], can be row changed to

$$\begin{bmatrix} 1 & 1 & 0 \\ 0 & 1 & 1 \\ 0 & 0 & 3 \end{bmatrix}$$

which shows that the first matrix is invertible. To actually compute the inverse, we use this method. So we append the first matrix by the identity matrix.

$$\begin{bmatrix} 2 & 0 & 1 & 1 & 0 & 0 \\ 1 & 1 & 0 & 0 & 1 & 0 \\ 0 & 1 & 1 & 0 & 0 & 1 \end{bmatrix}$$

The same row operations as in the earlier exercise give us

$$\begin{bmatrix} 1 & 1 & 0 & 0 & 1 & 0 \\ 0 & 1 & 1 & 0 & 0 & 1 \\ 0 & 0 & 3 & 1 & -2 & 2 \end{bmatrix}$$

Then the operations 1/3[row 3], – [row 3] + [row 2], – [row 2] + [row 1] gives us

$$\begin{bmatrix} 1 & 0 & 0 & 1/3 & 1/3 & -1/3 \\ 0 & 1 & 0 & -1/3 & 2/3 & 1/3 \\ 0 & 0 & 1 & 1/3 & -2/3 & 2/3 \end{bmatrix}$$

From this we conclude

$$\begin{bmatrix} 2 & 0 & 1 \\ 1 & 1 & 0 \\ 0 & 1 & 1 \end{bmatrix}^{-1} = \begin{bmatrix} 1/3 & 1/3 & -1/3 \\ -1/3 & 2/3 & 1/3 \\ 1/3 & -2/3 & 2/3 \end{bmatrix}$$

Similarly, the second matrix, being [col 1], [col 3], [col 4], has the following row echelon form

$$\begin{bmatrix} 1 & -1 & 0 \\ 0 & -2 & 1 \\ 0 & 0 & 3 \end{bmatrix}$$

which shows that the first matrix is invertible. To compute the inverse, we again append the identity matrix.

$$\begin{bmatrix} 2 & 2 & 1 & 1 & 0 & 0 \\ 1 & -1 & 0 & 0 & 1 & 0 \\ 0 & -2 & 1 & 0 & 0 & 1 \end{bmatrix}$$

The same row operations give us

$$\begin{bmatrix} 1 & -1 & 0 & 0 & 1 & 0 \\ 0 & -2 & 1 & 0 & 0 & 1 \\ 0 & 0 & 3 & 1 & -2 & 2 \end{bmatrix}$$

Then the operations 1/3[row 3], – [row 3] + [row 2], (–1/2)[row 2], [row 2] + [row 1] gives us

$$\begin{bmatrix} 1 & 0 & 0 & 1/6 & 2/3 & -1/6 \\ 0 & 1 & 0 & 1/6 & -1/3 & -1/6 \\ 0 & 0 & 1 & 1/3 & -2/3 & 2/3 \end{bmatrix}$$

From this we conclude

$$\begin{bmatrix} 2 & 0 & 1 \\ 1 & 1 & 0 \\ 0 & 1 & 1 \end{bmatrix}^{-1} = \begin{bmatrix} 1/6 & 2/3 & -1/6 \\ 1/6 & -1/3 & -1/6 \\ 1/3 & -2/3 & 2/3 \end{bmatrix}$$

The third matrix has the following row echelon form

$$\begin{bmatrix} 1 & 1 & -1 \\ 0 & 1 & -2 \\ 0 & 0 & 0 \end{bmatrix}$$

which shows that the matrix is not invertible. Finally, the fourth matrix is not invertible because it is not a square matrix.

9. $\begin{bmatrix} 1 & 2 & 1 & 1 \\ 2 & 1 & 1 & 0 \\ 0 & 3 & 1 & 0 \\ 3 & 0 & 1 & 0 \end{bmatrix}$

Answer: In an earlier exercise, we computed the row echelon form.

$$\begin{bmatrix} 1 & 2 & 1 & 1 \\ 0 & 3 & 1 & 0 \\ 0 & 0 & 0 & 1 \\ 0 & 0 & 0 & 0 \end{bmatrix}$$

Since [col 3] (and [row 4]) are not pivot, by the criterion for the invertibility, the matrix is not invertible.

10. $\begin{bmatrix} 1 & 2 & 3 & 4 \\ 0 & 1 & 2 & 3 \\ 0 & 0 & 1 & 2 \\ 0 & 0 & 0 & 1 \end{bmatrix}$

Answer: Since this is a square matrix with all rows (and columns) pivot, the matrix is invertible. To compute the inverse, we form

$$\begin{bmatrix} 1 & 2 & 3 & 4 & 1 & 0 & 0 & 0 \\ 0 & 1 & 2 & 3 & 0 & 1 & 0 & 0 \\ 0 & 0 & 1 & 2 & 0 & 0 & 1 & 0 \\ 0 & 0 & 1 & 0 & 0 & 0 & 0 & 1 \end{bmatrix}$$

Then –2[row 4] + [row 3], –3[row 4] + [row 2], –4[row 4] + [row 1] give us

$$\begin{bmatrix} 1 & 2 & 3 & 0 & 1 & 0 & 0 & -4 \\ 0 & 1 & 2 & 0 & 0 & 1 & 0 & -3 \\ 0 & 0 & 1 & 0 & 0 & 0 & 1 & -2 \\ 0 & 0 & 0 & 1 & 0 & 0 & 0 & 1 \end{bmatrix}$$

The operations –2[row 3] + [row 2], –3[row 3] + [row 1] further give us

$$\begin{bmatrix} 1 & 2 & 0 & 0 & 1 & 0 & -3 & 2 \\ 0 & 1 & 0 & 0 & 0 & 1 & -2 & 1 \\ 0 & 0 & 1 & 0 & 0 & 0 & 1 & -2 \\ 0 & 0 & 0 & 1 & 0 & 0 & 0 & 1 \end{bmatrix}$$

The operations –2[row 2] + [row 1] finally give us

$$\begin{bmatrix} 1 & 0 & 0 & 0 & 1 & -2 & 4 & 0 \\ 0 & 1 & 0 & 0 & 0 & 1 & -2 & 1 \\ 0 & 0 & 1 & 0 & 0 & 0 & 1 & -2 \\ 0 & 0 & 0 & 1 & 0 & 0 & 0 & 1 \end{bmatrix}$$

We conclude

$$\begin{bmatrix} 1 & 2 & 3 & 4 \\ 0 & 1 & 2 & 3 \\ 0 & 0 & 1 & 2 \\ 0 & 0 & 0 & 1 \end{bmatrix}^{-1} = \begin{bmatrix} 1 & -2 & 4 & 0 \\ 0 & 1 & -2 & 1 \\ 0 & 0 & 1 & -2 \\ 0 & 0 & 0 & 1 \end{bmatrix}$$

11. $\begin{bmatrix} 10.2 & 0.3 & -5.4 \\ -7.9 & 8.2 & 12.4 \\ 0.8 & -0.9 & 1.6 \\ 33.5 & -44.9 & 38.4 \\ -5.9 & -7.3 & 0.1 \end{bmatrix}$

Answer: Since the matrix is not square, it is not invertible.

12. $\begin{bmatrix} 1 & 1 & 1 & 1 & 1 \\ 1 & 0 & 1 & 1 & 1 \\ 1 & 1 & 0 & 1 & 1 \\ 1 & 1 & 1 & 0 & 1 \\ 1 & 1 & 1 & 1 & 0 \end{bmatrix}$

Answer: We append the identity matrix

$$\begin{bmatrix} 1 & 1 & 1 & 1 & 1 & 1 & 0 & 0 & 0 & 0 \\ 1 & 0 & 1 & 1 & 1 & 0 & 1 & 0 & 0 & 0 \\ 1 & 1 & 0 & 1 & 1 & 0 & 0 & 1 & 0 & 0 \\ 1 & 1 & 1 & 0 & 1 & 0 & 0 & 0 & 1 & 0 \\ 1 & 1 & 1 & 1 & 0 & 0 & 0 & 0 & 0 & 1 \end{bmatrix}$$

By –[row 1] + [row 2], –[row 1] + [row 3], –[row 1] + [row 4], –[row 1] + [row 5], we get

$$\left[\begin{array}{ccccc|ccccc} 1 & 1 & 1 & 1 & 1 & 1 & 0 & 0 & 0 & 0 \\ 0 & -1 & 0 & 0 & 0 & -1 & 1 & 0 & 0 & 0 \\ 0 & 0 & -1 & 0 & 0 & -1 & 0 & 1 & 0 & 0 \\ 0 & 0 & 0 & -1 & 0 & -1 & 0 & 0 & 1 & 0 \\ 0 & 0 & 0 & 0 & -1 & -1 & 0 & 0 & 0 & 1 \end{array}\right]$$

By adding [row 2] through [row 5] to [row 1], we get

$$\left[\begin{array}{ccccc|ccccc} 1 & 0 & 0 & 0 & 0 & -3 & 1 & 1 & 1 & 1 \\ 0 & -1 & 0 & 0 & 0 & -1 & 1 & 0 & 0 & 0 \\ 0 & 0 & -1 & 0 & 0 & -1 & 0 & 1 & 0 & 0 \\ 0 & 0 & 0 & -1 & 0 & -1 & 0 & 0 & 1 & 0 \\ 0 & 0 & 0 & 0 & -1 & -1 & 0 & 0 & 0 & 1 \end{array}\right]$$

By multiplying –1 to [row 2] through [row 5], we get

$$\left[\begin{array}{ccccc|ccccc} 1 & 0 & 0 & 0 & 0 & -3 & 1 & 1 & 1 & 1 \\ 0 & 1 & 0 & 0 & 0 & 1 & -1 & 0 & 0 & 0 \\ 0 & 0 & 1 & 0 & 0 & 1 & 0 & -1 & 0 & 0 \\ 0 & 0 & 0 & 1 & 0 & 1 & 0 & 0 & -1 & 0 \\ 0 & 0 & 0 & 0 & 1 & 1 & 0 & 0 & 0 & -1 \end{array}\right]$$

Thus we conclude

$$\begin{bmatrix} 1 & 1 & 1 & 1 & 1 \\ 1 & 0 & 1 & 1 & 1 \\ 1 & 1 & 0 & 1 & 1 \\ 1 & 1 & 1 & 0 & 1 \\ 1 & 1 & 1 & 1 & 0 \end{bmatrix} = \begin{bmatrix} -3 & 1 & 1 & 1 & 1 \\ 1 & -1 & 0 & 0 & 0 \\ 1 & 0 & -1 & 0 & 0 \\ 1 & 0 & 0 & -1 & 0 \\ 1 & 0 & 0 & 0 & -1 \end{bmatrix}$$

Remark: One can see the general pattern and conclude

$$\begin{bmatrix} 1 & 1 & 1 & .. & 1 \\ 1 & 0 & 1 & .. & 1 \\ 1 & 1 & 0 & .. & 1 \\ : & : & : & & : \\ 1 & 1 & 1 & .. & 0 \end{bmatrix}^{-1} = \begin{bmatrix} 2-n & 1 & 1 & .. & 1 \\ 1 & -1 & 0 & .. & 0 \\ 1 & 0 & -1 & .. & 0 \\ : & : & : & & : \\ 1 & 0 & 0 & .. & -1 \end{bmatrix}^{-1}$$

Exercise: Find invertible linear transformations and compute the inverse.

1. $T_1(x_1, x_2) = (2x_1 + 3x_2, 5x_1 + 7x_2)$: $R^2 \to R^2$

Answer: The matrix for T_1 is

$$\begin{bmatrix} 2 & 3 \\ 5 & 7 \end{bmatrix}$$

Since $2 \times 7 - 3X5 \neq 0$, the matrix is invertible. The inverse is

$$(2\times 7 - 3\times 5)^{-1}\begin{bmatrix} 7 & -3 \\ -5 & 2 \end{bmatrix} = \begin{bmatrix} -7 & 3 \\ 5 & -2 \end{bmatrix}$$

Thus the inverse transformation is $T_1^{-1}(x_1, x_2) = (-7x_1 + 3x_2, 5x_1 - 2x_2)$.

2. $T_2(x, y) = (2x - 6y, -5x + 15y)$: $R^2 \to R^2$

Answer: The matrix for T_2 is

$$\begin{bmatrix} 2 & -6 \\ -5 & 15 \end{bmatrix}$$

Since $2 \times 15 - (-5) \times (-6) = 0$, the matrix is not invertible. Correspondingly, the linear transformation is not invertible.

3. $T_3(x_1, x_2, x_3) = (-2x_2 + x_3, x_1 - x_3, 3x_1 + x_2 - 3x_3)$: $R^3 \to R^3$

Answer: The matrix of the transformation is

$$\begin{bmatrix} 0 & -2 & 1 \\ 1 & 0 & -1 \\ 3 & 1 & -3 \end{bmatrix}$$

We append the identity matrix.

$$\begin{bmatrix} 0 & -2 & 1 & 1 & 0 & 0 \\ 1 & 0 & -1 & 0 & 1 & 0 \\ 3 & 1 & -3 & 0 & 0 & 1 \end{bmatrix}$$

Row operations change this to

$$\begin{bmatrix} 1 & 0 & 0 & 1 & -5 & 2 \\ 0 & 1 & 0 & 0 & -3 & 1 \\ 0 & 0 & 1 & 1 & -6 & 2 \end{bmatrix}$$

Thus we have

$$\begin{bmatrix} 0 & -2 & 1 \\ 1 & 0 & -1 \\ 3 & 1 & -3 \end{bmatrix}^{-1} = \begin{bmatrix} 1 & -5 & 2 \\ 0 & -3 & 1 \\ 1 & -6 & 2 \end{bmatrix}$$

and the inverse transformation is $T_3^{-1}(x_1, x_2, x_3) = (x_1 - 5x_2 + 2x_3, -3x_2 + x_3, x_1 - 6x_2 + 2x_3)$.

4. $T_4(x_1, x_2, x_3) = (-2x_2 + x_3, 3x_1 + x_2 - 3x_3)$: $R^3 \to R^2$

Answer: The transformation is not invertible because the domain and codomain have different dimensions.

5. $T_5(x_1, x_2, x_3) = (-2x_2 + x_3, x_1 - x_3, 3x_1 + 2x_2 - 4x_3)$: $R^3 \to R^3$

Answer: The matrix of the transformation is

$$\begin{bmatrix} 0 & -2 & 1 \\ 1 & 0 & -1 \\ 3 & 2 & -4 \end{bmatrix}$$

Row operations change this to

$$\begin{bmatrix} 1 & 0 & -1 \\ 0 & -2 & 1 \\ 0 & 0 & 0 \end{bmatrix}$$

Thus the matrix, as well as the linear transformation, is not invertible.

6. $T_6(x_1, x_2, x_3, x_4) = (x_2 + 2x_3 + 4x_3, x_1 + x_2 + 2x_3, x_1 + x_2 + 2x_3 - 2x_4, x_1 + x_2 + x_3 + 2x_4)$: $R^4 \to R^4$

Answer: The corresponding matrix is

$$\begin{bmatrix} 0 & 1 & 2 & 4 \\ 1 & 1 & 2 & 0 \\ 1 & 1 & 2 & -2 \\ 1 & 1 & 1 & 2 \end{bmatrix}$$

The matrix is invertible, with

$$\begin{bmatrix} -1 & 3 & -2 & 0 \\ 1 & -6 & 4 & 2 \\ 0 & 2 & -1 & -1 \\ 0 & 1/2 & 1/2 & 0 \end{bmatrix}$$

as the inverse matrix. Thus $T_6^{-1}(x_1, x_2, x_3, x_4) = (-x_1 + 3x_2 - 2x_3, x_1 - 6x_2 + 4x_3 + 2x_4, 2x_2 - x_3 - x_4, 1/2\, x_2 + 1/2\, x_3)$.

7. $T_7(x_1, x_2, x_3, x_4, x_5) = (x_1 + x_2, x_2 + x_3, x_3 + x_4, x_4 + x_5, x_5 + x_1)$: $R^5 \to R^5$

Answer: The corresponding matrix is

$$\begin{bmatrix} 1 & 1 & 0 & 0 & 0 \\ 0 & 1 & 1 & 0 & 0 \\ 0 & 0 & 1 & 1 & 0 \\ 0 & 0 & 0 & 1 & 1 \\ 1 & 0 & 0 & 0 & 1 \end{bmatrix}$$

The matrix is invertible, with

$$1/2\begin{bmatrix} 1 & -1 & 1 & -1 & 1 \\ 1 & 1 & -1 & 1 & -1 \\ -1 & 1 & 1 & -1 & 1 \\ 1 & -1 & 1 & 1 & -1 \\ -1 & 1 & -1 & 1 & 1 \end{bmatrix}$$

as the inverse matrix. Thus $T_7^{-1}(x_1, x_2, x_3, x_4, x_5) = 1/2(x_1 - x_2 + x_3 - x_4 + x_5, x_1 + x_2 - x_3 + x_4 - x_5, -x_1 + x_2 + x_3 - x_4 + x_5, x_1 - x_2 + x_3 + x_4 - x_5, -x_1 + x_2 - x_3 + x_4 + x_5)$.

Remark: For $T(x_1, x_2,..., x_n) = (x_1 + x_2, x_2 + x_3, x_3 + x_4,..., x_{n-1} + x_n, x_n + x_1)$: $R^n \to R^n$, we have two cases: If n is odd, then T is invertible, with

$T_{-1}(x_1, x_2,..., x_n) = 1/2(x_1 - x_2 + x_3 -... + x_{n-2} - x_{n-1} + x_n, x_1 + x_2 - x_3 +... - x_{n-2} + x_{n-1} - x_n, -x_1 + x_2 + x_3 -... + x_{n-2} - x_{n-1} + x_n,..., x_1 - x_2 + x_3 -... + x_{n-2} + x_{n-1} - x_n, -x_1 + x_2 - x_3 +... - x_{n-2} + x_{n-1} + x_n)$.

If n is even, then for the nonzero vector $(1, -1, 1, -1,..., 1, -1)$, we have $T(1, -1, 1, -1,..., 1, -1) = 0$. Thus T is not one–to–one and is not invertible.

Exercise: For what choices of the parameters, the following matrices are invertible?

1. $\begin{bmatrix} a+1 & a-1 \\ a & a+2 \end{bmatrix}$

Answer The condition for the matrix to be invertible is $(a + 1)(a + 2) - a(a - 1) \neq 0$. Since $(a + 1)(a + 2) - a(a - 1) = 4a + 2$, we see that the condition is $a \neq -1/2$.

2. $\begin{bmatrix} 1 & 1 & 1 & 1 \\ 1 & 0 & 0 & 0 \\ 2 & 2 & 0 & a \\ 0 & 1 & 3 & a \end{bmatrix}$

Answer: The matrix appears as [col 1], [col 2], [col 4], [col 5] of the matrix in an earlier exercise. According to the computation there, row operations change the matrix into

$$\begin{bmatrix} 1 & 1 & 1 & 1 \\ 0 & -1 & 0 & -1 \\ 0 & 0 & -2 & a-2 \\ 0 & 0 & 0 & 2a-3 \end{bmatrix}$$

By the criterion, we need all columns (and rows) to be pivot in order for the matrix to be invertible. Thus the condition is $a \neq 3/2$.

3. $\begin{bmatrix} 1 & 1 & -1 \\ 1 & 0 & -a \\ 1 & a & 0 \end{bmatrix}$

Answer: The matrix appears as the coefficient matrix of the system in an earlier exercise. According to the computation there, row operations change the matrix into

$$\begin{bmatrix} 1 & 1 & -1 \\ 0 & -1 & 1-a \\ 0 & 0 & 2a-a^2 \end{bmatrix}$$

The condition for the matrix to be invertible is that all columns (and rows) are pivot, which means $2a - a^2 \neq 0$, i.e., $a \neq 0$ or 2.

4. $\begin{bmatrix} 0 & 1 & 1 & a \\ 1 & 0 & a & 1 \\ 1 & 1 & 0 & 0 \\ 0 & a & -1 & 1 \end{bmatrix}$

Answer: According to the computation in an earlier exercise, row operations change the matrix into

$$\begin{bmatrix} 1 & 0 & a & 1 \\ 0 & 1 & 1 & a \\ 0 & 0 & -a-1 & -a-1 \\ 0 & 0 & 0 & -a^2+a+2 \end{bmatrix}$$

The condition for the matrix to be invertible is $-a-1 \neq 0$ and $-a^2+a+2 \neq 0$, which means $a \neq -1$ or 2.

5. $\begin{bmatrix} 1 & 0 & 0 & 0 \\ a & 1 & 0 & 0 \\ 0 & a & 1 & 0 \\ 0 & 0 & a & 1 \end{bmatrix}$

Answer: By $(-a)$[row 1] + [row 2], $(-a)$[row 2] + [row 3], $(-a)$[row 3] + [row 4], the matrix becomes the identity. Therefore the matrix is invertible for any a.

6. $\begin{bmatrix} 1 & 1 & 1 & 1 \\ a & 1 & 1 & 1 \\ a & a & 1 & 1 \\ a & a & a & 1 \end{bmatrix}$

Answer: By $(-a)$[row 1] + [row 2], $(-a)$[row 1] + [row 3], $(-a)$[row 1] + [row 4], the matrix becomes

$$\begin{bmatrix} 1 & 1 & 1 & 1 \\ 0 & 1-a & 1-a & 1-a \\ 0 & 0 & 1-a & 1-a \\ 0 & 0 & 0 & 1-a \end{bmatrix}$$

Therefore the matrix is invertible for $a \neq 1$.

7. $\begin{bmatrix} a & b & b & b \\ a & a & b & b \\ a & a & a & b \\ a & a & a & a \end{bmatrix}$

Answer: By – [row 3] + [row 4], – [row 2] + [row 3], – [row 1] + [row 2], the matrix becomes

$$\begin{bmatrix} a & b & b & b \\ 0 & a-b & 0 & 0 \\ 0 & 0 & a-b & 0 \\ 0 & 0 & 0 & a-b \end{bmatrix}$$

Therefore the matrix is invertible for $a \neq 0, b$.

Exercise: Compute the inverse of the following matrices. Then use your result to solve the following systems.

1. $\begin{bmatrix} 2 & 3 \\ 7 & 9 \end{bmatrix}$

$2x_1 + 3x_2 = 2, \quad 2x_1 + 3x_2 = 1$

$7x_1 + 9x_2 = -1 \quad 7x_1 + 9x_2 = 0$

Answer: The inverse of the matrix is

$$(2 \times 9 - 7 \times 3)^{-1} \begin{bmatrix} 9 & -3 \\ -7 & 2 \end{bmatrix} = \begin{bmatrix} -3 & 1 \\ 7/3 & -2/3 \end{bmatrix}$$

Since

$$\begin{bmatrix} 2 & 3 \\ 7 & 9 \end{bmatrix}^{-1} \begin{bmatrix} 2 \\ -1 \end{bmatrix} = \begin{bmatrix} -3 & 1 \\ 7/3 & -2/3 \end{bmatrix} \begin{bmatrix} 2 \\ -1 \end{bmatrix} = \begin{bmatrix} -7 \\ 16/3 \end{bmatrix}$$

the solution of the first system is $x_1 = -7, x_2 = 4$. Similarly, the solution of the second system is

$$\begin{bmatrix} x_1 \\ x_2 \end{bmatrix} = \begin{bmatrix} 7 & 3 \\ 7 & 9 \end{bmatrix}^{-1} \begin{bmatrix} 1 \\ 0 \end{bmatrix} = \begin{bmatrix} -3 \\ 7/3 \end{bmatrix}$$

2. $\begin{bmatrix} 1 & 0 & 2 \\ 2 & -1 & 3 \\ 4 & -1 & 8 \end{bmatrix}$

$x_1 + 2x_3 = 1, \qquad x_1 + 2x_3 = 0$

$2x_1 - x_2 + 3x_3 = 0 \qquad 2x_1 - x_2 + 3x_3 = 2$

$4x_1 - x_2 + 8x_3 = -1 \qquad 4x_1 - x_2 + 8x_3 = 1$

Answer: Row operations change

$$\begin{bmatrix} 1 & 0 & 2 & 1 & 0 & 0 \\ 2 & -1 & 3 & 0 & 1 & 0 \\ 4 & -1 & 8 & 0 & 0 & 1 \end{bmatrix}$$

into

$$\begin{bmatrix} 1 & 0 & 0 & 5 & 2 & -2 \\ 0 & 1 & 0 & 4 & 0 & -1 \\ 0 & 0 & 1 & -2 & -1 & 1 \end{bmatrix}$$

Thus the inverse of the matrix is

$$\begin{bmatrix} 5 & 2 & -2 \\ 4 & 0 & -1 \\ -2 & -1 & 1 \end{bmatrix}$$

and the solution to the two systems are respectively

$$\begin{bmatrix} x_1 \\ x_2 \\ x_3 \end{bmatrix} = \begin{bmatrix} 5 & 2 & -2 \\ 4 & 0 & -1 \\ -2 & -1 & 1 \end{bmatrix} \begin{bmatrix} 1 \\ 0 \\ -1 \end{bmatrix} = \begin{bmatrix} 7 \\ 5 \\ -3 \end{bmatrix}$$

$$\begin{bmatrix} x_1 \\ x_2 \\ x_3 \end{bmatrix} = \begin{bmatrix} 5 & 2 & -2 \\ 4 & 0 & -1 \\ -2 & -1 & 1 \end{bmatrix} \begin{bmatrix} 0 \\ 2 \\ 1 \end{bmatrix} = \begin{bmatrix} 2 \\ -1 \\ -1 \end{bmatrix}$$

3. $\begin{bmatrix} 2 & 0 & 2 & 1 \\ 1 & 1 & -1 & 0 \\ 0 & 1 & -2 & -1 \\ 1 & -1 & 2 & 1 \end{bmatrix}$

$2x_1 + 2x_3 + x_4 = 1$
$x_1 + x_2 - x_3 = 0$
$x_2 - 2x_3 - x_4 = 1$
$x_1 - x_2 + 2x_3 + x_4 = 0$

Answer: The inverse of the matrix is

$$\begin{bmatrix} 0 & 0 & 1 & 1 \\ 1 & 0 & -1 & -2 \\ 1 & -1 & 0 & -1 \\ -1 & 2 & -2 & 0 \end{bmatrix}$$

The solution of the system is

$$\begin{bmatrix} x_1 \\ x_2 \\ x_3 \\ x_4 \end{bmatrix} = \begin{bmatrix} 0 & 0 & 1 & 1 \\ 1 & 0 & -1 & -2 \\ 1 & -1 & 0 & -1 \\ -1 & 2 & -2 & 0 \end{bmatrix} \begin{bmatrix} 1 \\ 0 \\ 1 \\ 0 \end{bmatrix} = \begin{bmatrix} 1 \\ 0 \\ 1 \\ -3 \end{bmatrix}$$

$$\begin{bmatrix} x_1 \\ x_2 \\ x_3 \end{bmatrix} = \begin{bmatrix} 5 & 2 & -2 \\ 4 & 0 & -1 \\ -2 & -1 & 1 \end{bmatrix} \begin{bmatrix} 0 \\ 2 \\ 1 \end{bmatrix} = \begin{bmatrix} 2 \\ -1 \\ -1 \end{bmatrix}$$

Exercise: Show that

$$A = \begin{bmatrix} 1 & 0 & -3 \\ 0 & 1 & 0 \\ 2 & 1 & -3 \end{bmatrix}, B = \begin{bmatrix} -3 & -2 & 2 \\ 0 & 1 & 0 \\ -2 & -1 & 1 \end{bmatrix}$$

are inverse to each other. Then compute

$$C = A \begin{bmatrix} 1 & 0 & 0 \\ 0 & -1 & 0 \\ 0 & 0 & 0 \end{bmatrix} B$$

and C^{2001}, $(AC)^{-1}$, A^{-2}.

Answer: By this criterion, to verify A and B are inverse to each other, it suffices to show $AB = I$. A straightforward computation confirms this. Another straightforward computation gives us

$$C = \begin{bmatrix} -3 & -2 & 2 \\ 0 & -1 & 0 \\ -6 & -5 & 4 \end{bmatrix}$$

Then by $C = ADB$, $C^{2001} = (ADB)(ADB)...(ADB) = AD(BA)D(BA)D...D(BA)DB = ADIDID...DIDB = AD^{2001}B$, and the obvious equality $D^{2001} = D$, we conclude that $C^{2001} = C$.

By this equality, we have

$$(AC)^{-1} = C^{-1}A^{-1} = CA^{-1} = \begin{bmatrix} 1 & 0 & 0 \\ 0 & -1 & 0 \\ 2 & -1 & 0 \end{bmatrix}$$

where we used $C^{-1} = (ADB)^{-1} = B^{-1}D^{-1}A^{-1} = ADB = C$.

Finally, we have

$$A^{-2} = B^2 = \begin{bmatrix} 5 & 2 & -4 \\ 0 & 1 & 0 \\ 1 & 2 & -3 \end{bmatrix}$$

Exercise: What is the condition for a diagonal matrix to be invertible? What is the inverse? What about an upper (lower) triangular matrix?

Answer: Based on the discussion in an earlier exercise, it is easy to see that the condition for a diagonal matrix

$$\begin{bmatrix} a_1 & 0 & .. & 0 \\ 0 & a_2 & .. & 0 \\ \vdots & \vdots & & \vdots \\ 0 & 0 & .. & a_n \end{bmatrix}$$

to be invertible is that the diagonal numbers $a_1, a_2, ..., a_n$ are nonzero. Moreover, the inverse is

$$\begin{bmatrix} a_1^{-1} & 0 & .. & 0 \\ 0 & a_2^{-1} & .. & 0 \\ \vdots & \vdots & & \vdots \\ 0 & 0 & .. & a_n^{-1} \end{bmatrix}$$

Similar statement holds for upper and lower triangular matrices.

Exercise: Prove that

$$\begin{bmatrix} a & b \\ c & d \end{bmatrix}$$

is invertible $\Leftrightarrow ad - bc \neq 0$.

Answer Denote the matrix by A.

In the lecture, we have already shown that $ad - bc \neq 0$ is necessary for A to be invertible.

Conversely, assume $ad - bc \neq 0$. Then either $a \neq 0$ or $c \neq 0$. If $a \neq 0$, then the operation $-a^{-1}c$[row 1] + [row 2] on A gives us

$$R = \begin{bmatrix} a & b \\ 0 & d - a^{-1}cb \end{bmatrix}$$

Since $a \neq 0$ and $d - a^{-1}cb = a^{-1}(ad - bc) \neq 0$, all rows and columns of R are pivot. By this criterion, A is invertible. If $c \neq 0$, then the operations [row 1] $\to$ [row 2], $-c^{-1}a$[row 1] + [row 2] on A give us

$$R = \begin{bmatrix} c & d \\ 0 & b - c^{-1}ad \end{bmatrix}$$

Since $c \neq 0$ and $b - c^{-1}ad = -c^{-1}(ad - BC) \neq 0$, all rows and columns of R are pivot. By this criterion, A is invertible.

Exercise: Prove that if row operations change $[A\ B]$ to $[I\ X]$, then $X = A^{-1}B$.

Answer: Since the same row operations change A to I, by this criterion, we see that A is invertible.

Let the i-th columns of B and X be b_i and x_i. Then the same operations change $[A\ b_i]$ to $[I\ x_i]$. Equivalently, $Ax = b_i$ and $Ix = x_i$ have the same solutions. This implies $Ax_i = b_i$. Then by this formula, $AX = [Ax_1\ Ax_2 \ldots Ax_n] = [b_1\ b_2 \ldots b_n] = B$. Since A is invertible, we get $X = A^{-1}B$.

Exercise: Let A and B be square matrices. Prove that A and B are invertible Ÿ AB is invertible. Please also explain why the square condition is necessary.

Answer: In the lecture, we have already shown the direction $\Rightarrow$.

Now we assume AB is invertible and try to prove A and B are invertible. Let C be the inverse of AB. Then we have $ABC = I$ and $CAB = I$.

By this criterion, the fact that $A(BC) = I$ and A is square implies A is invertible. Similarly, the fact that $(CA)B = I$ and B is square implies B is invertible. This completes the proof for the direction $\Leftarrow$.

The square condition is necessary because nonsquare matrices are not invertible. For a concrete example, consider

$$A = [1 \quad -2],\ B = \begin{bmatrix} 3 \\ 1 \end{bmatrix}$$

We have $AB = I$, but A and B are not invertible.

Exercise:

1. Prove that $Ax = b$ has solutions for any $b \Leftrightarrow$ There is a matrix X, such that $AX = I$.

Answer: Assume $Ax = b$ has solutions for any b. By taking b to be vectors in the standard basis $e_1, e_2,..., e_n$, we get vectors $x_1, x_2,..., x_n$, satisfying $Ax_i = e_i$. Thus if we take $X = [x_1\ x_2...\ x_n]$, then by this formula, $AX = [Ax_1\ Ax_2...\ Ax_n] = [e_1\ e_2...\ e_n] = I$.

Conversely, assume $AX = I$. Then for any b, we have $A(Xb) = Ib = b$. In other words, $x = Xb$ is a solution of $Ax = b$.

2. Prove that solution of $Ax = b$ is unique $\Leftrightarrow$ There is a matrix X, such that $XA = I$.

Answer: Assume A is an m by n matrix such that the solution of $Ax = b$ is unique. Then all columns of A are pivot, which means that the reduced row echelon form of A is

$$J = [e_1\ e_2 ... e_n] = \begin{bmatrix} I_n \\ O \end{bmatrix}$$

where $e_1, e_2,..., e_n$ are the first n vectors in the standard basis of R^m (recall that the uniqueness implies $n \le m$). In other words, we can use row operations to change A to become J. The same row operations will change $[I_n\ A]$ to $[Y\ J]$, where Y is some n by n matrix. Then the reversed row operations change $[Y\ J] = [Y\ e_1\ e_2...\ e_n]$ to $[I_n\ A] = [I_n\ a_1\ a_2...\ a_n]$. Consequently, $[Y\ e_i]$ can be row operated to become $[I_n\ a_i]$. In other words, the systems $Yx = e_i$ and $I_n x = a_i$ have the same solutions. Since $x = AI$ is the unique solution of $I_n x = AI$, we conclude that $Ya_i = e_i$. Thus

$$YA = [Ya_1\ Ya_2...\ Ya_n] = [e_1\ e_2...\ e_n] = J$$

and if we take $X = [I_n\ O]Y$ (i.e., X is the first n rows of Y), then

$$XA = [I_n\ O]YA = [I_n\ O]J = I_n.$$

Conversely, assume $XA = I$. Then $Ax = 0$ implies $XAx = X0 = 0$. Since $XA = I$, we conclude that $x = 0$. The whole implication is that the homogeneous system $Ax = 0$ has only the trivial solution.

3. Rephrase the first and second parts in terms of linear transformations.

Answer: A linear transformation T is onto $\Leftrightarrow$ There is a linear transformation S, such that $TS = id$.

A linear transformation T is one–to–one $\Leftrightarrow$ There is a linear transformation S, such that $ST = id$.

We remark that the statements also hold for general transformations.

TRANSPOSE

The transpose A^T of a matrix A is the exchange of rows and columns of the matrix.

Example: Let

$$A=\begin{bmatrix}1 & 1 & -2\\ 3 & 0 & 1\end{bmatrix}, B=\begin{bmatrix}-1 & 0 & 1\\ -2 & 1 & 1\end{bmatrix}, C=\begin{bmatrix}1 & 2\\ 3 & 4\end{bmatrix}, D=\begin{bmatrix}0 & 1\\ 1 & 2\end{bmatrix}, X=\begin{bmatrix}a & b & c\\ x & y & z\end{bmatrix}$$

Then

$$A^{\mathrm{T}}=\begin{bmatrix}1 & 3\\ 1 & 0\\ -2 & 1\end{bmatrix}, B^{\mathrm{T}}=\begin{bmatrix}-1 & -2\\ 0 & 1\\ 1 & 1\end{bmatrix}, C^{\mathrm{T}}=\begin{bmatrix}1 & 3\\ 2 & 4\end{bmatrix}, D^{\mathrm{T}}=\begin{bmatrix}0 & 1\\ 1 & 2\end{bmatrix}, X^{\mathrm{T}}=\begin{bmatrix}a & x\\ b & y\\ c & z\end{bmatrix}$$

In general, if we write a matrix in a compact form $A = (a_{ij})$, then $A^{\mathrm{T}} = (a_{ji})$.

$$A=\begin{bmatrix}a_{11} & a_{12} & \dots & a_{1n}\\ a_{21} & a_{22} & \dots & a_{2n}\\ \vdots & \vdots & & \vdots\\ a_{m1} & a_{m2} & \dots & a_{mn}\end{bmatrix} \Rightarrow A^{\mathrm{T}}=\begin{bmatrix}a_{11} & a_{21} & \dots & a_{m1}\\ a_{12} & a_{22} & \dots & a_{m2}\\ \vdots & \vdots & & \vdots\\ a_{1n} & a_{2n} & \dots & a_{mn}\end{bmatrix}$$

Moreover, if A is an m by n matrix, then A^{T} is an n by m matrix. The rows of A are the columns of A^{T}, and the columns of A are the rows of A^{T}.

A matrix A is called symmetric if $A^{\mathrm{T}} = A$. It is called skew–symmetric if $A^{\mathrm{T}} = -A$. The matrix D in the example above is symmetric. Clearly, a symmetric (or skew–symmetric) matrix must be a square matrix.

Example: The following are the general forms of 2 by 2 and 3 by 3 symmetric matrices.

$$\begin{bmatrix}a & b\\ b & c\end{bmatrix}$$

$$\begin{bmatrix}a & b & c\\ b & d & e\\ c & e & f\end{bmatrix}$$

The following are the general forms of 2 by 2 and 3 by 3 skew–symmetric matrices.

$$\begin{bmatrix}0 & a\\ -a & 0\end{bmatrix}$$

$$\begin{bmatrix}0 & a & b\\ -a & 0 & c\\ -b & -c & 0\end{bmatrix}$$

The transpose has the following properties. The linearity and involutive property are quite obvious. The product property is not so obvious.

linearity

$$(A + B)^T = A^T + B^T, (cA)^T = cA^T$$

product

$$(AB)^T = B^T A^T$$

involutive

$$(A^T)^T = A$$

inverse

$$(A^{-1})^T = (A^T)^{-1}$$

The properties can also be explained by the interpretation of the transpose in terms of the linear transformations. In the equivalence between matrices and linear transformations, we would like to find the answer to $4^?$ in the following table.

Linear Transformation	Matrix
Composition	Multiplication
addition	addition
Scalar multiplication	Scalar multiplication
inverse	inverse
$4^?$	transpose

The answer is the dual of linear transformations. The linearity and the product properties are quite easy to see from the transformation viewpoint. The interpretation of the involutive property is somewhat complicated and can be found here.

Exercise: Prove that for any square matrix A, there are unique symmetric matrix X and skew–symmetric matrix Y such that $A = X + Y$.

Answer: Given $A = X + Y$ as described, we have $A^T = X^T + Y^T = X - Y$. Thus we have

$$X = (A + A^T)/2, Y = (A - A^T)/2$$

and X and Y are uniquely determined by A through the above formulae – hence the uniqueness. On the other hand, for the matrices X and Y given as above, we certainly have $A = X + Y$, with X symmetric:

$$X^T = (A + A^T)^T/2 = (A^T + (A^T)^T)/2 = (A^T + A)/2 = X,$$

and Y skew–symmetric:

$$Y^T = (A - A^T)^T/2 = (A^T - (A^T)^T)/2 = (A^T - A)/2 = - Y.$$

Exercise: **Prove that if A is invertible, then A^T is also invertible, and $(A^T)^{-1} = (A^{-1})^T$.**

***Answer*:** **We use the properties $(AB)^T = B^TA^T$ and $(A^T)^T = A$ to get**

$$A^T(A^{-1})^T = (A^{-1}A)^T = I^T = I.$$

and

$$(A^{-1})^TA^T = (AA^{-1})^T = I^T = I.$$

Then by the definition of inverse matrix, $(A^{-1})^T$ is the inverse of A^T.

Exercise: **By making use of the computation in this exercise, solve the following systems.**

$$2x_1 + 7x_2 = 2$$
$$3x_1 + 9x_2 = -1$$

$$x_1 + 2x_2 + 4x_3 = 1$$
$$-x_2 - x_3 = 0$$
$$2x_1 + 3x_2 + 8x_3 = -1$$

$$2x_1 + x_2 + x_4 = 1$$
$$x_2 + x_3 - x_4 = 0$$
$$2x_1 - x_2 - 2x_3 + 2x_4 = 1$$
$$x_1 - x_3 + x_4 = 0$$

***Answer*:** **The coefficient matrices of the systems are the transposes of the coefficient matrices in the earlier exercise. By taking the transpose, we get the inverse of the transposed matrix. Thus the solution of the first system is**

$$\begin{bmatrix} x_1 \\ x_2 \end{bmatrix} = \begin{bmatrix} 2 & 7 \\ 3 & 9 \end{bmatrix}^{-1} \begin{bmatrix} 2 \\ -1 \end{bmatrix} = \begin{bmatrix} -25/3 \\ 8/3 \end{bmatrix}$$

The solution of the second system is

$$\begin{bmatrix} x_1 \\ x_2 \\ x_3 \end{bmatrix} = \begin{bmatrix} 5 & 2 & -2 \\ 4 & 0 & -1 \\ -2 & -1 & 1 \end{bmatrix} T \begin{bmatrix} 1 \\ 0 \\ -1 \end{bmatrix} = \begin{bmatrix} 7 \\ 3 \\ -3 \end{bmatrix}$$

The solution of the third system is

$$\begin{bmatrix} x_1 \\ x_2 \\ x_3 \\ x_4 \end{bmatrix} = \begin{bmatrix} 0 & 0 & 1 & 1 \\ 1 & 0 & -1 & -2 \\ 1 & -1 & 0 & -1 \\ -1 & 2 & -2 & 0 \end{bmatrix} T \begin{bmatrix} 1 \\ 0 \\ 1 \\ 0 \end{bmatrix} = \begin{bmatrix} 1 \\ -1 \\ 1 \\ 0 \end{bmatrix}$$

PARTITIONED MATRIX

A 4 by 5 matrix can be considered as a 2 by 2 matrix

$$\begin{bmatrix} 1 & 2 & 3 & 4 & 5 \\ 6 & 7 & 8 & 9 & 10 \\ 11 & 12 & 13 & 14 & 15 \\ 16 & 17 & 18 & 19 & 20 \end{bmatrix} = \begin{bmatrix} A & B \\ C & D \end{bmatrix}$$

with the matrices

$$A = \begin{bmatrix} 1 & 2 & 3 \\ 6 & 7 & 8 \\ 11 & 12 & 13 \end{bmatrix}, B = \begin{bmatrix} 4 & 5 \\ 9 & 10 \\ 14 & 15 \end{bmatrix}, C =, D = [19 \quad 20]$$

as the entries. This way of dividing a matrix into compatible smaller matrices is a partition of the matrix.

We have partitioned matrices before. For example, for a system of m equations in n variables, the augmented matrix $[A\ b]$ is partitioned into an m by n matrix A and an m by 1 matrix b. Moreover, when we think of the columns, a matrix $A = [a_1\ a_2 ... a_n]$ is partitioned into n m by 1 matrices.

Example: The $(m + n)$ by $(m+n)$ identity matrix may be partitioned as follows

$$I_{m+n} = \begin{bmatrix} IM & \cdot\ O_{n\times m} \\ O_{m\times n} & I_n \end{bmatrix}$$

One may perform the usual operations with partitioned matrices as usual, as long as we keep the following in mind:

- Sizes must match,
- Matrices must be multiplied in the correct order.

Example: For any two m by n matrices X and Y, we have

$$\begin{bmatrix} IM & X \\ O & I_n \end{bmatrix} \begin{bmatrix} IM & X \\ O & I_n \end{bmatrix} = \begin{bmatrix} II + XO & IY + XI \\ OI + IO & OY + II \end{bmatrix} = \begin{bmatrix} I & X + Y \\ O & I \end{bmatrix}$$

In particular, we conclude that

$$\begin{bmatrix} I & X \\ O & I \end{bmatrix}^{-1} = \begin{bmatrix} I & -X \\ O & I \end{bmatrix}$$

Example: Let A and B be square matrices, we would like to find out when

$$M = \begin{bmatrix} A & C \\ O & B \end{bmatrix}$$

is invertible. Assume

$$M^{-1} = \begin{bmatrix} X & Y \\ Z & W \end{bmatrix}$$

Then $MM^{-1} = I$ means

$$AX + CZ = I, AY + CW = O, BZ = O, BW = I.$$

Since B is a square matrix, the equation $BW = I$ means that B is invertible, and $W = B^{-1}$. Then by multiplying B^{-1} to the left of $BZ = O$, we find $Z = O$. Substituting $Z = O$ into $AX + CZ = I$, we get $AX = I$. Since A is a square matrix, this means that A is invertible, and $X = A^{-1}$. Then substituting $W = B^{-1}$ into $AY + CW = O$ and solve the equation, we get $Y = -A^{-1}CB^{-1}$. Thus we conclude that

M is invertible T A and B are invertible

Moreover, the inverse is given by

$$M^{-1} = \begin{bmatrix} A^{-1} & -A^{-1}CB^{-1} \\ O & B^{-1} \end{bmatrix}$$

The partitioned matrix also has interpretation as the linear transformation between direct sums of euclidean spaces.

Exercise: Let A be an invertible matrix. Find X, Y, S, such that

$$\begin{bmatrix} A & B \\ C & D \end{bmatrix} = \begin{bmatrix} I & O \\ X & I \end{bmatrix} \begin{bmatrix} A & O \\ O & S \end{bmatrix} \begin{bmatrix} I & Y \\ O & I \end{bmatrix}$$

Then find out the condition for the matrix to be invertible.

Answer: The multiplication on the right side is

$$\begin{bmatrix} A & AY \\ XA & XAY+S \end{bmatrix}$$

Therefore the problem is reduced to solving the following equations.

$$B = AY, C = XA, D = XAY + S$$

Since A is invertible, we get from the three equations that

$$Y = A^{-1}B, X = CA^{-1}, S = D - XAY = D - CA^{-1}B.$$

The answer is then

$$\begin{bmatrix} A & B \\ C & D \end{bmatrix} = \begin{bmatrix} I & O \\ CA^{-1} & I \end{bmatrix} \begin{bmatrix} A & O \\ O & D-CA^{-1} \end{bmatrix} \begin{bmatrix} I & A^{-1}B \\ O & I \end{bmatrix}$$

Based on the discussion in this example (or this example), it is easy to see that among the three matrices on the right, the left and the right ones are invertible. Therefore the whole product is invertible if and only if the middle one is invertible. Then by the discussion in this example, the middle one is invertible if and only if $D - CA^{-1}B$ is invertible.

Remark: The matrix $D - CA^{-1}B$ is called the Schur complement of A. Similarly, if D is invertible, then $A - BD^{-1}C$ is the Schur complement of D. By the way, the inverse is

$$\begin{bmatrix} A & B \\ C & D \end{bmatrix}^{-1} = \begin{bmatrix} I & A^{-1}B \\ O & I \end{bmatrix}^{-1} \begin{bmatrix} A & O \\ O & S \end{bmatrix}^{-1} \begin{bmatrix} I & O \\ CA^{-1} & I \end{bmatrix}^{-1}$$

$$= \begin{bmatrix} I & -A^{-1}B \\ O & I \end{bmatrix}^{-1} \begin{bmatrix} A^{-1} & O \\ O & S^{-1} \end{bmatrix} \begin{bmatrix} I & O \\ -CA1 & I \end{bmatrix}$$

$$= \begin{bmatrix} A^{-1} - A^{-1}BS^{-1}CA^{-1} & -A^{-1}BS^{-1} \\ -S^{-1}CA & \end{bmatrix}$$

Exercise: A block upper triangular matrix is a partitioned matrix similar to the usual upper triangular matrix, except the entries are matrices instead of numbers, and the diagonal blocks are square matrices.

$$\begin{bmatrix} A_1 & \# & .. & \# \\ O & A_2 & .. & \# \\ : & : & & : \\ O & O & .. & A_k \end{bmatrix}, A_1, A_2, ..., A_k \text{ are square matrices.}$$

What does the multiplication of two block upper triangular matrices look like? What about the inverse of a block upper triangular matrix? What about the similar problems for block lower triangular matrix and block diagonal matrix.

Answer: Given two block upper triangular matrices partitioned in the same way,

$$M = \begin{bmatrix} A_1 & \# & .. & \# \\ O & A_2 & .. & \# \\ \vdots & \vdots & & \vdots \\ O & O & .. & A_k \end{bmatrix} \quad N = \begin{bmatrix} B_1 & \# & .. & \# \\ O & B_2 & .. & \# \\ \vdots & \vdots & & \vdots \\ O & O & .. & B_k \end{bmatrix}$$

their multiplication is still a block upper triangular matrix, with the diagonal entries simply multiplied together (the other entries are more complicated).

$$MN = \begin{bmatrix} A_1B_1 & \# & .. & \# \\ O & A_2B_2 & .. & \# \\ \vdots & \vdots & & \vdots \\ O & O & .. & A_kB_k \end{bmatrix}$$

Moreover, M is invertible if and only if all the diagonal matrices $A_1, A_2, \ldots, AK$ are invertible. The inverse is of the form

$$M^{-1} = \begin{bmatrix} A_1^{-1} & \# & .. & \# \\ O & A_2^{-1} & .. & \# \\ \vdots & \vdots & & \vdots \\ O & O & .. & AK-1 \end{bmatrix}$$

One way to see the reason behind the claim is to study the 2 by 2 blocked case first, as was done in this example. The general case follows by taking

$$M = \begin{bmatrix} A & C \\ O & B \end{bmatrix}$$

with

$$A = A_1, B = \begin{bmatrix} A_2 & .. & \# \\ \vdots & & \vdots \\ O & .. & AK \end{bmatrix}, C = [\# \;\; .. \;\; \#]$$

and then induct.

The similar statement is true for block lower triangular matrix and block diagonal matrix.

NUT SHELL OF TRANSFORMATION

Definition of Transformation

We have introduced some concepts for transformations in the process of developing lienar algebra. Now we discuss general transformations in more detail.

Transformations have been defined as rules that convert one data into another data. In fact, this is not quite rigorous as a definition. The following is more precise.

A transformation T from a set X (called source or domain) to another set Y (called target or codomain) is a rule that assigns for $x \in X$ an element (called image or value) $T(x) \in Y$. The rule should be well–defined in the following sense.

applicability

The rule applies to any input $x \in X$ and produces some output $T(x)$.

unambiguity

For any input $x \in X$, the output $T(x)$ is unique.

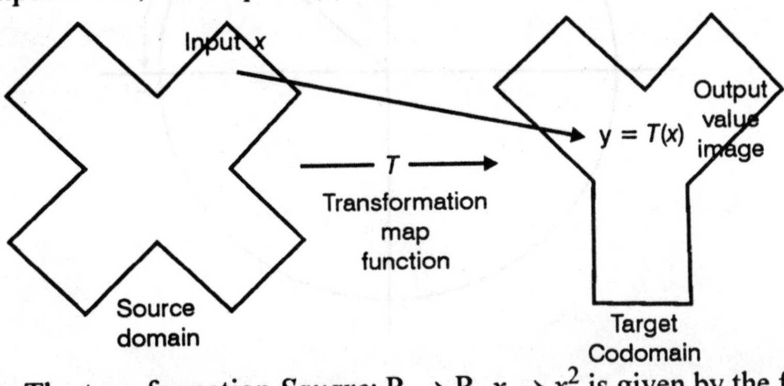

Example: The transformation *Square*: R $\to$ R, $x \to x^2$ is given by the following rule:

For any real number $x \in$ R, multiply x to itself.

Note that multiplying to itself is a process that can be applied to any real number. Moreover, the result is unambiguous, such as the square of –2 is 4, not any other number. The transformation is well–defined.

Example: Consider the rule *Square Root*: R $\to$ R given by the following:

For any $x \in$ R, find a number $y \in$ R, such that $y^2 = x$.

Unfortunately, the rule cannot be applied to the negative numbers. Moreover, even if we can apply the rule to non–negative numbers, such as 4, the result is ambiguous because we have $2^2 = (-2)^2 = 4$, for example. Thus both conditions for the transformation fail, and *Square Root*: R → R is not a transformation.

To make the rule well–defined, we first need to restrict ourself to non–negative numbers and consider ∈: [0, ∞) → R. Moreover, we make a specific choice among the two positive candidates for *y* with the following modified rule:

For any *x* Ξ [0, “), find a number $y \in$ R, such that $y^2 = x$ and $y \geq 0$.

Then √ is well-defined and is the usual square root function.

Example Let $b = (1, 0)$ be a base point of the unit circle S^1 in R^2. The transformation *Sin*: R → R is given by the following rule:

For any real number $x \in$ R, start from the *b* and move along S^1 in counterclowise direction by distance *x*. Then take the *y*–coordinate of the point we arrive at.

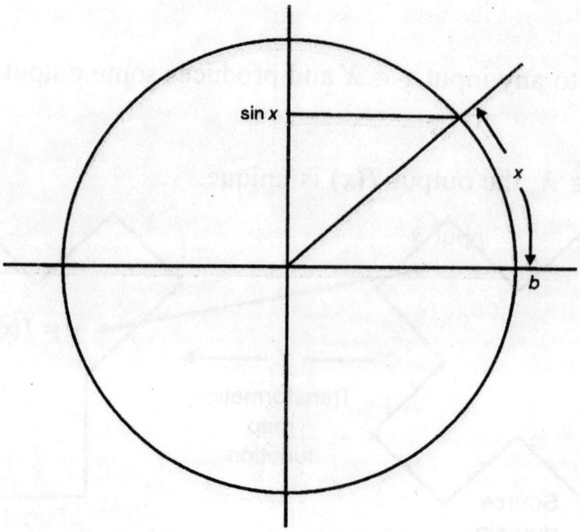

Onto

Let T: $X \to Y$ be a transformation. Recall that *T* is onto (surjective) if its range is the whole target set. More specifically, this means

For any $y \in Y$, there is $x \in X$, such that $T(x) = y$.

The following characterizes onto transformations. More properties can be found in this exercise.

A transformation T: $X \to Y$ is onto ⇔ There is a transformation S: $Y \to X$, such that $TS = id_Y$.

Proof. If $TS = id_Y$, then for any $y \in Y$, the element $x = S(y)$ satisfies $T(x) = TS(y) = id_Y(y) = y$. This proves that any element of *Y* is in the range.

Conversely, suppose T is onto. Then for any $y \in Y$, there is $x \in X$, such that $(x, y) \in T$. We choose $S(y)$ to be one of such x. After $S(y)$ is chosen for all $y \in Y$, we have a transformation $S: Y \to X$. By the way S is constructed, we clearly have $TS = id_Y$.

One-to-one

Let $T: X \to Y$ be a transformation. Recall that T is one–to–one (injective) if any target element can be the image of at most one source element. More specifically, by the usual definition, this means

$T(x) = T(x')$ implies $x = x'$.

The following characterizes one–to–one transformations. More properties can be found in this exercise.

A transformation $T: X \to Y$ is one–to–one $\Leftrightarrow$ There is a transformation $S: Y \to X$, such that $ST = id_X$.

Proof. $T(x) = T(x')$ implies $ST(x) = ST(x')$. If $ST = id_X$, then this further impliesx $= x'$. This proves the direction.

Conversely, suppose T is one–to–one. Then for any $y \in Y$, we choose $S(y) \in X$ according to to cases

- If y is in the range of T, then $S(y)$ = the unique x satisfying $T(x) = y$. The uniqueness of x follows from T being one–to–one.
- If y is not in the range of T, then choose $S(y)$ to be any element of X.

Now starting from $x \in X$, we apply T to get $y = T(x)$. Since T is one–to–one, this x is the unique element with y as the image. Thus we are in the first case above and get $ST(x) = S(y) = x$. Since this is true for any x, we conclude $ST = id_X$.

Inverse

Recall a transformation $T: X \to Y$ is invertible if there is $S: Y \to X$ satisfying $TS = id_Y$ and $ST = id_X$. Because of the following characterization, an invertible transformation is also called a one–to–one correspondence.

For a transformation, invertible $\Leftrightarrow$ onto and one–to–one.

Proof By this characterization, $T: X \to Y$ is onto means there is $S: Y \to X$ satisfying $TS = id_Y$. By this characterization, $T: X \to Y$ is one–to–one means there is $S': Y \to X$ satisfying $S'T = id_X$. Thus

$T: X \to Y$ is onto and one–to–one $\Leftrightarrow$ There are $S, S': Y \to X$ satisfying $TS = id_Y$ and $S'T = id_X$.

The proof is then complete by showing that S and S' must be equal:

$TS = id_Y$ and $S'T = id_X S = id_X S = S'TS = S' id_Y = S'$.

Graph of Transformation

Our more precise definition of transformations is in fact still not up to the rigorous standard of mathematics. The formal mathematical definition of transformations relies on set theory. We need to use the basic concepts such as set, elements of a set, subset, and the cartesian product of two sets

$$XXY = \{(x, y): x \in X, y \in Y\}.$$

Now we can present the formal definition of transformations.

A transformation from a set X to another set Y is a subset T of XXY satisfying the following property.

For any $x \in X$, there is a unique $y \in Y$ such that $(x, y) \in T$.

The unique y in the definition is completely determined by x. It is the value of y under the transformation T and is denoted as $T(x)$.

What we have done is basically define a transformation T by its graph $\{(x, T(x)): \text{all } x \in X\}$.

Example: The identity transformation $id(x) = x: X \to X$ is defined as the diagonal subset $\Delta_X = \{(x, x): \text{all } x \in X\}$.

For a fixed $b \in Y$, the constant transformation $C(x) = b: X \to Y$ is defined as the subset $XXb = \{(x, b): \text{all } x \in X\}$.

Example: The transformation $Square(x) = x^2$: R → R is defined as the parabola $\{(x, x^2): \text{all } x \in R\}$ in the plane.

Two transformations T and S are equal if they have the same source X and the same target Y, and the subsets T and S of $X \times Y$ are equal. The following result shows that this definition is consistent with our intuition.

For transformations T and S from X to Y, the following are equivalent.

- $T = S$ (as subsets of XXY).
- T M S (as subsets of XXY).
- $T(x) = S(x)$ for all $x \in X$.

Proof. The first property trivially implies the second.

Assume T M S. For any $x \in X$, since T is a transformation, we have $y = T(x) \in Y$ such that $(x, y) \in T$. Then applying T M S, we also have $(x, y) \in S$. This means $y = S(x)$. Thus we have proved $T(x) = y = S(x)$.

Finally, the third property implies the first because the subsets, being the graphs of the transformations, are completely determined by the values of the transformations.

Formal Theory of Transformation

Using the formal definition, for a transformation $T: X \to Y$, onto means

For any $y \in Y$, there is $x \in X$, such that $(x, y) \in T$,

and one–to–one means

(x, y) and $(x', y) \in T$ implies $x = x'$.

The composition of $S: X \to Y$ and $T: Y \to Z$ is the subset

$\{(x, z)$: There is $y \in Y$ such that $(x, y) \in S$ and $(y, z) \in T\}$.

A transformation $S: Y \to X$ is the inverse of a transformation $T: X \to Y$ if the composition TS is the diagonal of Y and ST is the diagonal of X.

Technically, if the logical rigor is your only concern, then proofs about transformations should use the formal definition. However, this is not recommended (and should even be discouraged) because such proofs are usually very dry and not intuitive. For an example:

onto and one–to–one $\Rightarrow$ invertible.

Proof. Let T M $X \times Y$ be an onto and one–to–one transformation. Consider the subset

$S = \{(y, x): (x, y) \in T\}$ M $Y \times X$.

We verify that the subset satisfies the condition in the formal definition:

For any $y \in Y$, there is a unique $x \in X$ such that $(y, x) \in S$.

By the construction of S, this is the same as

For any $y \in Y$, there is a unique $x \in X$ such that $(x, y) \in T$.

In the statement above, the existence of x means exactly T is onto, and the uniqueness of x means exactly T is one–to–one. Therefore we have

S is a transformation $\Leftrightarrow$ T is onto and one-to-one.

Now starting with any $x \in X$, since T is a transformation, we have $(x, y) \in T$ for some (unique) $y \in Y$. Then by the construction of S, we have $(y, x) \in S$. Combining $(x, y) \in T$ and $(y, x) \in S$, we deduce that (x, x) is in the composition subset $ST \in X \times X$. Thus we have proved that the diagonal Δ_X is a subset of ST. Since the diagonal Δ_X is the identity transformation $id_X: X \to X$, by this result we conclude that $ST = id_X$. By the same reason, we also have $TS = id_Y$. Therefore S is the inverse of T.

Chapter 6

Linear Independence

DEFINITION AND RELATION TO UNIQUENESS

The concept of span is derived from the existence problem for the vector equation

$$c_1v_1 + c_2v_2 + \dots + c_kv_k = b,$$

where $c_1, c_2, \dots, c_k$ are the unknown variables. Since the left side of the vector equation is linear in the variable $c = (c_1, c_2, \dots, c_k)$, the solution is unique if and only if the corresponding homogeneous equation

$$c_1v_1 + c_2v_2 + \dots + c_kv_k = 0,$$

has only the trivial solution. This leads to the following definition.

Vectors $v_1, v_2, \dots, v_k$ are linearly independent if

$$c_1v_1 + c_2v_2 + \dots + c_kv_k = 0 \Rightarrow c_1 = c_2 = \dots = c_k = 0.$$

In other words, if a linear combination is zero, then the coefficients must be zero.

Example: The equality

$$x = (x_1, x_2, x_3) = x_1e_1 + x_2e_2 + x_3e_3$$

has been used in deriving the fact that the standard basis e_1, e_2, e_3 span R^3. The same equality also gives us the following implication

$$x_1e_1 + x_2e_2 + x_3e_3 = 0 \Rightarrow x = 0 \Rightarrow \text{all coordinates } x_1, x_2, x_3 \text{ of } x = 0.$$

The implication means exactly that e_1, e_2, e_3 are linearly independent. The same reason also shows that the standard basis $e_1, e_2, \dots, e_n$ of R^n is linearly independent.

Example: The trivial statement

$$a_0 + a_1t + a_2t^2 + \dots + a_nt^n = 0 \Rightarrow a_0 = a_1 = a_2 = \dots = a_n = 0$$

means exactly that $1, t, t^2, \dots, t^n$ are linearly independent.

The opposite of linear independence is

Vectors $v_1, v_2, ..., v_k$ are linearly dependent if there are $c_1, c_2, ..., c_k$, not all zero, such that

$$c_1v_1 + c_2v_2 + ... + c_kv_k = 0.$$

By "not all zero", we mean at least one c_i `" 0. Note that this is different from "all not zero", which would mean all $c_i$ `" 0.

Example: To find out whether the vectors (1, 3, 2) and (1, –1, 1) are linearly independent, we check whether the following implication holds

$$c_1(1, 3, 2) + c_2(1, -1, 1) = (0, 0, 0) \Rightarrow c_1 = c_2 = 0.$$

By considering the three coordinates, this is the same as whether a homogeneous system of linear equations has only the the trivial solution

$$c_1 + c_2 = 0, 3c_1 - c_2 = 0, 2c_1 + c_2 = 0 \Rightarrow c_1 = c_2 = 0.$$

It is easy to see that the implication indeed holds. Therefore we conclude that (1, 3, 2) and (1, –1, 1) are linearly independent.

On the other hand, since

$$(-1)(1, 3, 1) + 2(-1, 1, 1) + 1(3, 1, -1) = (0, 0, 0),$$

$$(-2)(1, 3, 1) + 0(-1, 1, 1) + 1(2, 6, 2) = (0, 0, 0),$$

the vectors (1, 3, 1), (–1, 1, 1), (3, 1, –1) are linearly dependent, and the vectors (1, 3, 1), (–1, 1, 1), (2, 6, 2) are also linearly dependent.

Example: To find out whether cos$t$ and sin$t$ are linearly independent, we start with the equation

$$a \cos t + b \sin t = 0$$

and try to deduce that (the constants) $a = b = 0$. By taking $t = 0$, we have $a = 0$. Substituting into the equation, we get $b \sin t = 0$. Since $\sin t = 0$ is a nonzero function, we have $b = 0$. Therefore the implication

$$a \cos t + b \sin t = 0 \Rightarrow a = b = 0$$

does hold, and cos$t$ and sin$t$ are linearly independent.

On the other hand, the equalities

$$1 \cos^2 t + 1 \sin^2 t + (-1)\,1 = 0,$$

$$\sin 2t - 2 \sin t \cos t + 0\,t = 0,$$

show that the functions $\cos^2 t$, $\sin^2 t$, 1 are linearly dependent, and the functions sin2t, sin$t$cos$t$, t are also linearly dependent.

Example: We want to determine whether polynomials $p_1 = 1 - t + t^3$, $p_2 = 3 - t + 4t^2$

$+ 3t^3, p_3 = 2 - t + 2t^2 + 2t^3, p_4 = t + 4t^2 - 2t^3, p_5 = 1 + 3t^2$ are linearly independent. Thus we study the equation

$$c_1(1 - t + t^3) + c_2(3 - t + 4t^2 + 3t^3) + c_3(2 - t + 2t^2 + 2t^3) + c_4(t + 4t^2 - 2t^3) + c_5(1 + 3t^2) = 0$$

and see whether we can deduce $c_1 = c_2 = c_3 = c_4 = c_5 = 0$ from the equation.

By comparing the coefficients of 1, t, t^2, and t^3, the equation is the same as a system of four linear homogeneous equations.

$$\begin{aligned} x_1 + 3x_2 + 2x_3 + x_5 &= 0 \\ -x_1 - x_2 - x_3 + x_4 &= 0 \\ 4x_2 + 2x_3 + 4x_4 + 3x_5 &= 0 \\ x_1 + 3x_2 + 2x_3 - 2x_4 &= 0 \end{aligned}$$

The question of deducing $c_1 = c_2 = c_3 = c_4 = c_5 = 0$ is the same as the uniqueness of the solution. By this example, a row echelon form of the coefficient matrix

$$[\,[p_1]\,[p_2]\,[p_3]\,[p_4]\,[p_5]\,] = \begin{bmatrix} 1 & 3 & 2 & 0 & 1 \\ -1 & -1 & -1 & 1 & 0 \\ 0 & 4 & 2 & 4 & 3 \\ 1 & 3 & 2 & -2 & 0 \end{bmatrix}$$

is

$$\begin{bmatrix} 1 & 3 & 2 & 0 & 1 \\ 0 & 2 & 1 & -1 & 0 \\ 0 & 0 & 0 & 2 & 1 \\ 0 & 0 & 0 & 0 & 0 \end{bmatrix}.$$

Since not all columns are pivot, the solution is not unique. Therefore the five polynomials are linearly dependent.

Similar to the earlier example, what we have really done was to use

$$p = a + bt + ct^2 + dt^3 \in P_3 \leftrightarrow [p] = (a, b, c, d) \in \mathrm{R}^4$$

to translate a linear independence problem about polynomials to the a linear independence problem about euclidean vectors.

The later problem is the same as the uniqueness of a system of linear equations and can be solved by looking at the pivot columns of the coefficient matrix. For example, the computation of the row echelon form also tells us that

$$[\,[p_1]\;[p_2]\;[p_4]\,] = \begin{bmatrix} 1 & 3 & 0 \\ -1 & -1 & 1 \\ 0 & 4 & 4 \\ 1 & 3 & -2 \end{bmatrix}$$

has

$$\begin{bmatrix} 1 & 3 & 0 \\ 0 & 2 & -1 \\ 0 & 0 & 2 \\ 0 & 0 & 0 \end{bmatrix}$$

as a row echelon form. Since all three columns are pivot, we have the uniqueness for the solution of the corresponding linear equation. In particular, this translates back to polynomials and tells us that p_1, p_2, p_4 are linearly independent.

GEOMETRIC INTUITION OF LINEAR INDEPENDENCE

One vector u is linearly independent if

$cu = 0 \Rightarrow c = 0.$

This implication holds if and only if $u \neq 0$. Therefore we conclude

u is linearly independent $\Leftrightarrow u \neq 0$.

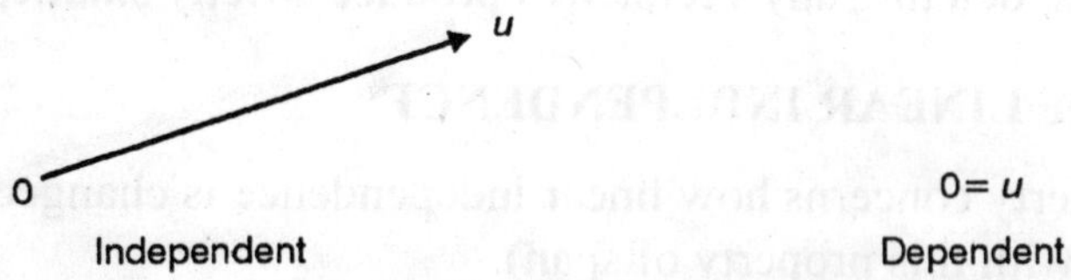

Two vectors u, v are linearly *dependent* if there are a and b, not all zero, such that $au + bv = 0$. If $a \neq 0$, then we have $u = (-b/a)v$. If $b \neq 0$, then we have $v = (-a/b)u$. Thus we conclude

Two vectors are linearly dependent $\Leftrightarrow$ One is a scalar multiple of another.

In other words, the two vectors are parallel to each other. The opposite statement is

Two vectors are linearly dependent $\Leftrightarrow$ The vectors are not parallel to each other.

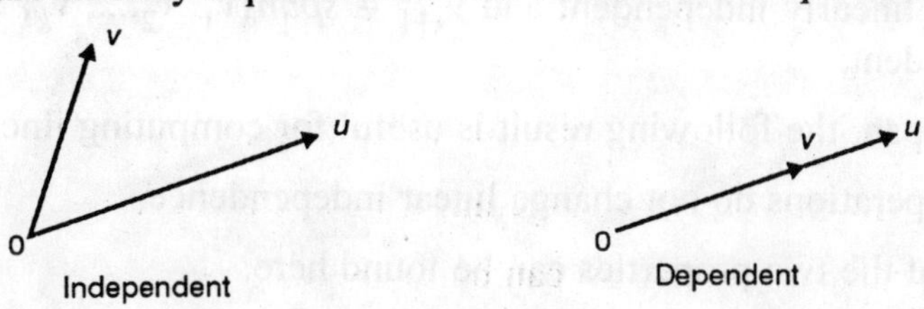

For three vectors u, v, w to be linearly dependent, we have a, b, c, not all zero, such that $au + bv + cw = 0$. If $a \neq 0$, then $u = (-b/a)v + (-c/a)w$ is in the plane (or line, or even a point) spanned by v and w. In particular, the three vectors lie on the same plane (called coplanar). Similar arguments can be made in case other coefficients are nonzero. Thus we conclude

Three vectors are linearly independent $\Leftrightarrow$ The vectors are on the same plane.

The linear independence of three vectors can be then understood by the opposite of the statement.

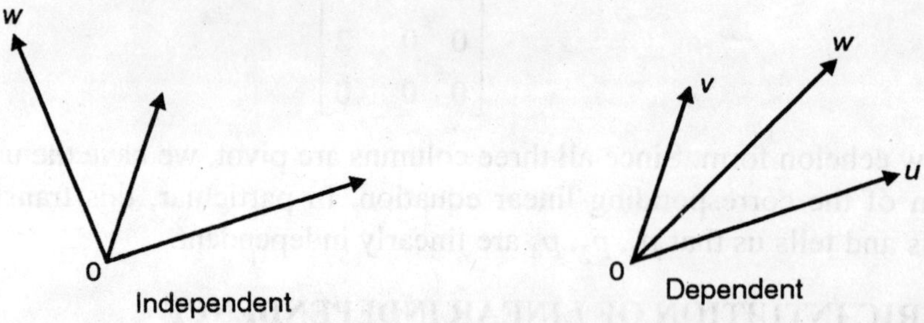

The discussion on the linear dependence can be extended to many vectors.

Vectors are linearly dependent $\Leftrightarrow$ One is a linear combination of others.

Compared with this property of span, we conclude with the slogan

linear independence $\Leftrightarrow$ no waste (direction in the span) $\Leftrightarrow$ most efficient span.

In other words, deleting any vector will produce strictly smaller span.

PROPERTIES OF LINEAR INDEPENDENCE

The first property concerns how linear independence is changed when we add/delete vectors (compare with this property of span).

$v_1, v_2, \ldots, v_k, v_{k+1}$ linearly independent $\Rightarrow v_1, v_2, \ldots, v_k$ linearly independent.

It is easy to understand the property from the intuition that linear independence means no waste: If all five people are necessary for the job, then surely any four of the five (or any three, any two, etc) are also necessary.

The following is the converse of the property above.

$v_1, v_2, \ldots, v_k$ linearly independent and $v_{k+1} \notin span\{v_1, v_2, \ldots, v_k\} \Rightarrow v_1, v_2, \ldots, v_k, v_{k+1}$ linearly independent.

Similar to span, the following result is useful for computing linear independence.

The three operations do not change linear independence.

The proof of the two properties can be found here.

Example: By the method similar to this example, we know $\cos^2 t$ and $\sin^2 t$ are linearly independent. Then we have

$\cos^2 t$, $\sin^2 t$ are linearly independent

$\Rightarrow \cos^2 t$, 1 are linearly independent

$\Rightarrow -2\cos^2 t$, 1 are linearly independent

$\Rightarrow \cos^2 t$, 1 are linearly independent

$\Rightarrow \cos^2 t$, $2\sin^2 t$ are linearly independent

$\Rightarrow \cos^2 t$, $\sin^2 t$ are linearly independent.

Example: In this example, we effectively showed that the vectors $v_1 = (1, -1, 0, 1)$, $v_2 = (3, -1, 4, 3)$, $v_3 = (2, -1, 2, 2)$, $v_4 = (0, 1, 4, -2)$, $v_5 = (1, 0, 3, 0)$ are linearly independent by *row operations* on the matrix

$$A = [v_1\ v_2\ v_3\ v_4\ v_5] = \begin{bmatrix} 1 & 3 & 2 & 0 & 1 \\ -1 & -1 & -1 & 1 & 0 \\ 0 & 4 & 2 & 4 & 3 \\ 1 & 3 & 2 & -2 & 0 \end{bmatrix}$$

Alternatively, we may use the *column operations* on A to simplify the vectors and get the *Answer*:. By the computation in the earlier example, the column echelon form is

$$\begin{bmatrix} 1 & 0 & 0 & 0 & 0 \\ -1 & 1 & 0 & 0 & 0 \\ 0 & 2 & 1 & 0 & 0 \\ 1 & 0 & -1 & 0 & 0 \end{bmatrix}.$$

The five column vectors in the column echelon form are clearly linearly dependent (for example, the linear combination with coefficients 0, 0, 0, 1, 0 is the zero vector). Since linear independence is not changed by the three operations, we conclude that v_1, v_2, v_3, v_4, v_5 are linearly dependent.

The example above leads to some interesting observations. First for euclidean vectors, the preservation of the linear independence under the three operations can be rephrased in terms of column operations.

If $A \to B$ by column operations, then columns of A are linearly independent $\Leftrightarrow$ columns of B are linearly independent.

Second is the numerical consequence of linear independence.

$v_1, v_2, \ldots, v_k \in \mathbb{R}^n$ are linearly independent $\Rightarrow k \le n$.

The equivalent statement in terms of the null space appeared in the first part of this exercise.

PROPERTIES OF LINEAR INDEPENDENCE

Theorem: $v_1, v_2, \ldots, v_k, v_{k+1}$ are linearly independent $\Rightarrow v_1, v_2, \ldots, v_k$ are linearly independent.

Proof. In case $k = 2$, we assume v_1, v_2, v_3 are linearly independent. Then we have the following implication

$$c_1v_1 + c_2v_2 = 0 \Rightarrow c_1v_1 + c_2v_2 + 0v_3 = 0 \Rightarrow c_1 = c_2 = 0,$$

where the second $\Rightarrow$ follows from the linear independence of v_1, v_2, v_3 (take $c_3 = 0$ in the definition). The implication means exactly that v_1, v_2 are linearly independent.

The proof for general k is similar.

Theorem: $v_1, v_2, \ldots, v_k$ are linearly independent and v_{k+1} O $span\{v_1, v_2, \ldots, v_k\} \Rightarrow v_1, v_2, \ldots, v_k, v_{k+1}$ are linearly independent.

Proof. In case $k = 3$, we assume v_1, v_2, v_3 are linearly independent and $v_4 \notin span\{v_1, v_2, v_3\}$.

Assume v_1, v_2, v_3, v_4 are linearly dependent, then we have

$$c_1v_1 + c_2v_2 + c_3v_3 + c_4v_4 = 0,$$

with at least one of the coefficients c_1, c_2, c_3, c_4 being nonzero. If $c_4 = 0$, then at least one of the coefficients c_1, c_2, c_3 is nonzero, and the equality above becomes $c_1v_1 + c_2v_2 + c_3v_3 = 0$. This means that v_1, v_2, v_3 are linearly dependent. The contradiction shows that we must have c_4 ‘‘‘ 0. Thus we may divide c_4 and get

$$v_4 = (-c_1/c_4)v_1 + (-c_2/c_4)v_2 + (-c_3/c_4)v_3 \in span\{v_1, v_2, v_3\},$$

again contradicting with the assumption.

Therefore we conclude that v_1, v_2, v_3, v_4 must be linearly independent. The proof for general k is similar.

Theorem: The three operations do not change linear independence.

Proof. We only consider special case of the operations on three vectors. The general case is similar.

For the first operation, we will prove

v_1, v_2, v_3 linearly independent $\Leftrightarrow v_1, v_2 + rv_1, v_3$ linearly independent.

Assuming left side holds, we have

$$c_1v_1 + c_2(v_2 + rv_1) + c_3v_3 = 0$$

$\Rightarrow (c_1 + rc_2)v_1 + c_2v_2 + c_3v_3 = 0$

$\Rightarrow c_1 + rc_2 = 0, c_2 = 0, c_3 = 0$ (v_1, v_2, v_3 linearly independent)

$\Rightarrow c_1 = 0, c_2 = 0, c_3 = 0.$

The whole implication means exactly the right side. Assume the right side holds. Then

$c_1v_1 + c_2v_2 + c_3v_3 = 0$

$\Rightarrow (c_1 - rc_2)v_1 + c_2(v_2 + rv_1) + c_3v_3 = 0$

$\Rightarrow c_1 - rc_2 = 0, c_2 = 0, c_3 = 0$ ($v_1, v_2 + rv_1, v_3$ linearly independent)

$\Rightarrow c_1 = 0, c_2 = 0, c_3 = 0.$

The whole implication means exactly the left side.

For the second operation, we will prove

v_1, v_2, v_3 linearly independent $\Leftrightarrow$ v_2, v_1, v_3 linearly independent.

Assume left side holds. Then

$c_1v_2 + c_2v_1 + c_3v_3 = 0$

$\Rightarrow c_2v_1 + c_1v_2 + c_3v_3 = 0$

$\Rightarrow c_2 = 0, c_1 = 0, c_3 = 0$ (v_1, v_2, v_3 linearly independent)

$\Rightarrow c_1 = 0, c_2 = 0, c_3 = 0.$

The whole implication means exactly the right side. The implication in the other direction can be proved similarly.

For the third operation, we will prove that for $d \neq 0$,

v_1, v_2, v_3 linearly independent $\Leftrightarrow$ dv_1, v_2, v_3 linearly independent.

Assume left side holds. Then

$c_1(dv_1) + c_2v_2 + c_3v_3 = 0$

$\Rightarrow c_1d = 0, c_2 = 0, c_3 = 0$ (v_1, v_2, v_3 linearly independent)

$\Rightarrow c_1 = 0, c_2 = 0, c_3 = 0$ (d “‘ 0)

The whole implication means exactly the right side. The implication in the other direction can be proved similarly.

Exercise: Determine whether the vectors are linearly independent.

1. (1, –1, 0, 1)

Answer: The vector is linearly independent because it is nonzero.

2. (1, –1, 0, 1), (–2, 3, 1, 2)

Answer: The vectors are linearly independent because they are not scalar multiples of each other.

3. (2, –2, 0, 2), (–5, 5, 0, –5)

Answer: The vectors are linearly dependent because the second vector is –2.5 multiple of the first.

4. (1, –1, 0, 1), (–2, 3, 1, 2), (1, 0, 1, 5)

Answer: The linear independence of the three vectors is the same as the uniqueness for the system with (the given vectors are the columns)

$$A = \begin{bmatrix} 1 & -2 & 1 \\ -1 & 3 & 0 \\ 0 & 1 & 1 \\ 1 & 2 & 5 \end{bmatrix}$$

as the coefficient matrix. From the row echelon form computed in an earlier exercise, we see that [col 3] is not pivot. Therefore the 3 vectors are linearly dependent.

5. $1 - t + t^3, -2 + 3t + t^2 + 2t^3, 1 + t^2 + 5t^3$

Answer: The linear independence means

$c_1(1 - t + t^3) + c_2(-2 + 3t + t^2 + 2t^3) + c_3(1 + t^2 + 5t^3) = 0 \Rightarrow c_1 = c_2 = c_3.$

Using the translation $p = a + bt + ct^2 + dt^3 \in P_3 \leftrightarrow [p] = (a, b, c, d) \in \mathbb{R}^4$, the left side is equivalent to

$c_1[1 - t + t^3] + c_2[-2 + 3t + t^2 + 2t^3] + c_3[1 + t^2 + 5t^3] = c_1(1, -1, 0, 1) + c_2(-2, 3, 1, 2) + c_3(1, 0, 1, 5) = 0,$

so that the linear independence of the polynomials is equivalent to the linear independence of the euclidean vectors (1, –1, 0, 1), (–2, 3, 1, 2), (1, 0, 1, 5). By the fourth problem, the 3 euclidean vectors are linearly dependent. Therefore the 3 polynomials are linearly dependent.

6. $\begin{bmatrix} 1 & -1 \\ 0 & 1 \end{bmatrix}, \begin{bmatrix} -2 & 3 \\ 1 & 2 \end{bmatrix}, \begin{bmatrix} 1 & 0 \\ 1 & 5 \end{bmatrix}$

Answer: Using the translation

$$A = \begin{bmatrix} a & b \\ c & d \end{bmatrix}, \leftrightarrow [A] = (a, b, c, d)$$

the linear independence of the three matrices is equivalent to the linear independence of the euclidean vectors (1, –1, 0, 1), (–2, 3, 1, 2), (1, 0, 1, 5). By the fourth problem, the 3 euclidean vectors are linearly dependent. Therefore the 3 matrices are linearly dependent.

7. $\cos^2 t$, $\sin^2 t$

Answer: Suppose $a \cos^2 t + b \sin^2 t = 0$ for constants a, b and all $t \in$ R. By taking $t = 0$, we get $a = 0$. Thus we have $b \sin^2 t = 0$ for all $t \in$ R, which further implies $b = 0$. We have proved that a linear combination of $\cos^2 t$, $\sin^2 t$ is zero implies the coefficients are zero. Therefore $\cos^2 t$, $\sin^2 t$ are linearly independent.

Exercise: From the given vectors, find as many linearly independent ones as possible (i.e., maximal independent subset).

1. (1, –1, 0, 1), (–2, 3, 1, 2), (1, 0, 1, 5)

Answer: With the 3 given vectors as the columns, we form a matrix.

$$\begin{bmatrix} 1 & -2 & 1 \\ -1 & 3 & 0 \\ 0 & 1 & 1 \\ 1 & 2 & 5 \end{bmatrix}$$

In an earlier exercise, a row echelon form has been computed.

$$\begin{bmatrix} 1 & -2 & 1 \\ 0 & 1 & 1 \\ 0 & 0 & 0 \\ 0 & 0 & 0 \end{bmatrix}$$

Thus if we take [col 1] and [col 2] only (which are all the pivot columns), then the 4 by 2 matrix has all columns pivot. This implies the first 2 vectors (1, –1, 0, 1), (–2, 3, 1, 2) are linearly independent. Moreover, since the 3rd column is not pivot, adding the 3rd vector to the two will give us 3 linearly dependent vectors. Therefore the two vectors is a maximal linearly independent subset.

2. (1, –1, 0, 1), (–2, 3, 1, 2), (1, 0, 1, 5), (1, 2, 3, 13)

Answer: With the 4 given vectors as the columns, we form a matrix.

$$\begin{bmatrix} 1 & -2 & 1 & 1 \\ -1 & 3 & 0 & 2 \\ 0 & 1 & 1 & 3 \\ 1 & 2 & 5 & 13 \end{bmatrix}$$

In an earlier exercise, a row echelon form has been computed.

$$\begin{bmatrix} 1 & -2 & 1 & 1 \\ 0 & 1 & 1 & 3 \\ 0 & 0 & 0 & 0 \\ 0 & 0 & 0 & 0 \end{bmatrix}$$

Since first two columns are pivot, the first 2 vectors (1, –1, 0, 1), (–2, 3, 1, 2) are linearly independent. Since [col 1], [col 2], [col 3] do not form a matrix with all columns pivot, adding the 3rd vector to (1, –1, 0, 1), (–2, 3, 1, 2) will give us 3 linearly dependent vectors. Similarly, adding the 4th vector to (1, –1, 0, 1), (–2, 3, 1, 2) will also give us 3 linearly dependent vectors. Therefore the two vectors is a maximal linearly independent subset.

Remark: The argument made in the last two problems is a general one. The idea is based on the relation

independence ⇔ uniqueness ⇔ all columns pivot

which suggests that in order to get independent ones from any (not necessarily independent) vectors, we should pick the "pivot" vectors only. Indeed, all the pivot vectors are not only linearly independent, they are actually maximal linearly independent – adding any more vector (which are necessarily nonpivot) will give us a linear dependent collection.

We note that the process is only one way of finding maximal linear independence. In fact, in the second problem above, the 1st vector (1, –1, 0, 1) and the 3rd vector (1, 0, 1, 5) also form a maximal independent set.

The non–uniqueness of the maximal independent set can also be seen from this picture, among the three vectors "front", "left", "diagonal", any two form a maximal independent set.

3. (1, –1, 0, 1), (–2, 3, 1, 2), (1, 0, 1, 5), (2, –2, 4, 5), (0, 0, 0, 1)

Answer: With the 5 given vectors as the columns, we form a matrix.

$$\begin{bmatrix} 1 & -2 & 1 & 2 & 0 \\ -1 & 3 & 0 & -2 & 0 \\ 0 & 1 & 1 & 4 & 0 \\ 1 & 2 & 5 & 5 & 1 \end{bmatrix}$$

In an earlier exercise, a row echelon form has been computed.

$$\begin{bmatrix} 1 & -2 & 1 & 2 & 0 \\ 0 & 1 & 1 & 0 & 0 \\ 0 & 0 & 0 & 4 & 0 \\ 0 & 0 & 0 & 0 & 1 \end{bmatrix}$$

Since all columns except [col 3] are pivot, all vectors (1, –1, 0, 1), (–2, 3, 1, 2), (2, –2, 4, 5), (0, 0, 0, 1) except the third one form a maximal linearly independent sbset.

4. $1 - t + t^3, -2 + 3t + t^2 + 2t^3, 1 + t^2 + 5t^3$

Answer: By $p = a + bt + ct^2 + dt^3 \in P_3 \leftrightarrow [p] = (a, b, c, d) \in R^4$, the problem can be converted to the first problem. Therefore $1 - t + t^3, -2 + 3t + t^2 + 2t^3$ is maximal linearly independent.

Exercise: Find the condition on the parameters so that the vectors are linearly independent.

1. $(1, -1, 1), (1, 0, a), (-1, -a, 0)$

Answer: The problem is the same as whether the solution to the system

$$\begin{aligned} x_1 + x_2 - x_3 &= b_1 \\ x_1 - ax_3 &= b_2 \\ x_1 + ax_2 &= b_3 \end{aligned}$$

is unique. By the computation in an earlier exercise, row operations change the coefficient matrix to

$$\begin{bmatrix} 1 & 1 & -1 \\ 0 & -1 & 1-a \\ 0 & 0 & 2a-a^2 \end{bmatrix}$$

Linear independence is then equivalent to all columns pivot, which means $a \neq 0$ and 2. We remark that the condition is the same as the one for spanning the whole space.

2. $t + t^2, 1 + t^2 + at^3, 1 + at - t^3$

Answer: We may use

$p = a + bt + ct^2 + dt^3 \in P_3 \leftrightarrow [p] = (a, b, c, d) \in R^4$

to translate the polynomial problem into whether the vectors

$[t + t^2] = (0, 1, 1, 0), [1 + t^2 + at^3] = (1, 0, 1, a), [1 + at - t^3] = (1, a, 0, -1)$

are linearly independent. We use the vectors as columns and form a matrix

$$\begin{bmatrix} 0 & 1 & 1 \\ 1 & 0 & a \\ 1 & 1 & 0 \\ 0 & a & -1 \end{bmatrix},$$

which can be changed by row operations to

$$\begin{bmatrix} 1 & 0 & a \\ 0 & 1 & 1 \\ 0 & 0 & 1-a \\ 0 & 0 & 0 \end{bmatrix}$$

The condition for linear independence is all columns are pivot, which means a “‘ -1.

3. $\begin{bmatrix} a & 1 \\ 1 & 0 \end{bmatrix}, \begin{bmatrix} 1 & a \\ 0 & 1 \end{bmatrix}, \begin{bmatrix} 1 & 0 \\ 1 & a \end{bmatrix}$

Answer: We may use

$$A = \begin{bmatrix} a & b \\ c & d \end{bmatrix}, \leftrightarrow [A] = (a, b, c, d)$$

to translate the problem into the linear independence of the euclidean vectors $(a, 1, 1, 0), (1, a, 0, 1), (1, 0, a, 1), (0, 1, 1, a)$.

This is the same as all columns of the following matrix (with the four vectors as columns) are pivot.

$$\begin{bmatrix} a & 1 & 1 & 0 \\ 1 & a & 0 & 1 \\ 1 & 0 & a & 1 \\ 0 & 1 & 1 & a \end{bmatrix}$$

By row operations, we get

$$\begin{bmatrix} 1 & 0 & a & 1 \\ 0 & 1 & 1 & a \\ 0 & 0 & 2a & a^2 \\ 0 & 0 & 0 & 4a - a^3 \end{bmatrix}$$

The condition for the matrix to have all columns pivot is $a \neq 0, 2, -2$. This is the condition for the vectors to be linearly independent.

We remark that the condition is the same as the one for spanning the whole space.

Exercise: Prove that for a subspace H spanned by any give vectors, H is also spanned by the maximal linearly independent vectors .

Answer: Let u, v, w be maximal linearly independent among u, v, w, x, y. Then the maximality means that by adding x to the collection, the four vectors u, v, w, x are linearly dependent. Thus we have a, b, c, d, not all zero, such that

$$au + bv + cw + dx = 0.$$

If $d = 0$, then $au + bv + cw = 0$ and a, b, c are not all zero. This contradicts with the assumption that u, v, w are linearly independent. Therefore we have $d \neq 0$, and

$$x = (-a/d)u + (-b/d)v + (-c/d)w \in span\{u, v, w\}.$$

Similarly, we have $y \in span\{u, v, w\}$. Then by this property, we have

$$span\{u, v, w, x, y\} = span\{u, v, w\}.$$

This finishes the proof in case 3 out of 5 vectors are maximal linearly independent. The general case is similar.

Exercise: We already know that the linear independence among the columns of a matrix is preserved by column operations. Is the linear independence among the columns preserved by row operations?

Answer: Suppose a matrix A is changed to a matrix B by row operations. Then the same operations change $[A\ 0]$ to $[B\ 0]$. Therefore the solutions of the homogeneous systems $Ax = 0$ and $Bx = 0$ are the same. In particular, we have

$Ax = 0$ has only the trivial solution $\Leftrightarrow$ $Bx = 0$ has only the trivial solution.

In other words, the uniqueness with coefficient matrix A is equivalent to the uniqueness with coefficient matrix B. Since the uniqueness is equivalent to the linear independence of the columns, we conclude that column of A linearly independent $\Leftrightarrow$ column of B linearly independent.

Remark: Since the null space is defined as the solutions of the homogeneous system, it is also preserved by row operations. In the special case the null space is $\{0\}$, we conclude that linear independence is preserved by row operations.

Remark: Why do row operations have different effects on span (column space) and on linear independence? The reason is that from the viewpoint of systems of linear equations, the span is explicitly related to the right side of the system, while the linear independence is not affected by the right side. Since row operations do change *nonzero* right sides, the span is changed, and the indpendence is not.

Exercise: Let u, v, w be three vectors in a vector space V. Then we have a linear transformation $T(a, b, c) = au + bv + cw\colon R^3 \to V$. Prove that u, v, w are linearly independent $\Leftrightarrow T$ is one-to-one.

Answer: By this characterization of one–to–one in terms of the kernel, we have

T is one–to–one $\Leftrightarrow kernelT = \{0\}$.

However, $kernelT$ consists of (a, b, c) such that $T(a, b, c) = au + bv + cw = 0$. Therefore $kernelT = \{0\} \Leftrightarrow au + bv + cw = 0$ implies $(a, b, c) = 0$.

Since the right side is exactly the linear independence of u, v, w, by combining the two equivalences, we get

T is one–to–one $\Leftrightarrow u$, v, w are linearly independent.

Exercise: How are the linear independences of the three collections $\{u, v, w\}$, $\{x, y\}$, $\{u, v, w, x, y\}$ related? How are linear independences of $\{u, v, w\}$, $\{v, w, x\}$, $\{v, w\}$ related?

Answer: By this property, we have

u, v, w, x, y linearly independent $\Rightarrow u$, v, w linearly independent.

u, v, w, x, y linearly independent $\Rightarrow x$, y linearly independent.

On the other hand, we cannot combine the linear independence of u, v, w and the linear independence of x, y to conclude the linear independence of u, v, w, x, y.

Similarly, we have

u, v, w linearly independent $\Rightarrow v$, w linearly independent.

v, w, x linearly independent $\Rightarrow v$, w linearly independent.

Again the converse does not hold.

Exercise: Let $v_1, v_2,..., v_k$ be linearly independent. Prove that $v_1, v_2,..., v_k, v_{k+1}$ are also linearly independent $\Leftrightarrow v_{k+1}$ O $span\{v_1, v_2,..., v_k\}$.

Answer: We will prove the equivalent statement.

$v_1, v_2,..., v_k, v_{k+1}$ are linearly dependent $\Leftrightarrow v_{k+1} \in span\{v_1, v_2,..., v_k\}$.

The left side means that we have

$c_1v_1 + c_2v_2 +... + c_kv_k + c_{k+1}v_{k+1} = 0$, where not all $c_1, c_2,..., c_k, c_{k+1}$ are zero.

If $c_{k+1} = 0$, then the above becomes

$c_1v_1 + c_2v_2 +... + c_kv_k = 0$, where not all $c_1, c_2,..., c_k$ are zero.

This is contradicts the assumption that $v_1, v_2,..., v_k$ are linearly independent. Thus $c_{k+1} \neq 0$, and $v_1, v_2,..., v_k, v_{k+1}$ being linearly dependent means

$c_1v_1 + c_2v_2 +... + c_kv_k + c_{k+1}v_{k+1} = 0,\ c_{k+1} \neq 0.$

This is equivalent to $v_{k+1} \in span\{v_1, v_2, ..., v_k\}$.

Exercise: Let $v_1, v_2, ..., v_k$ be vectors in V. Let $T: V \to W$ be a linear transformation.

1. Prove that $T(v_1), T(v_2), ..., T(v_k)$ are linearly independent $\Rightarrow v_1, v_2, ..., v_k$ are linearly independent.

Answer: By the assumptions, we have

$$c_1v_1 + c_2v_2 + ... + c_kv_k = 0$$
$$\Rightarrow c_1T(v_1) + c_2T(v_2) + ... + c_kT(v_k) = 0 \ (T \text{ is linear})$$
$$\Rightarrow c_1 = c_2 = ... = c_k = 0. \ (T(v_1), T(v_2), ..., T(v_k) \text{ are linearly independent})$$

2. Prove that T is one–to–one and $v_1, v_2, ..., v_k$ are linearly independent $\Leftrightarrow T(v_1), T(v_2), ..., T(v_k)$ are linearly independent.

Answer: The á direction has been proved in the first part. For the $\Rightarrow$ direction, we have

$$c_1T(v_1) + c_2T(v_2) + ... + c_kT(v_k) = 0$$
$$\Rightarrow T(c_1v_1 + c_2v_2 + ... + c_kv_k) = 0 \ (T \text{ is linear})$$
$$\Rightarrow c_1v_1 + c_2v_2 + ... + c_kv_k = 0 \ (T \text{ is one–to–one})$$
$$\Rightarrow c_1 = c_2 = ... = c_k = 0. \ (v_1, v_2, ..., v_k \text{ are linearly independent})$$

3. What if T is not assumed to be one–to–one in the second part?

Answer: If T is not one–to–one, then we have a nonzero $v \in V$ such that $T(v) = 0$. In particular, $\{v\}$ is linearly independent and $\{T(v)\}$ is linearly dependent. Therefore the statement is no longer true.

Chapter 7

Algebra of Complex Numbers

Complex analysis is the study of complex numbers together with their derivatives, manipulation, and other properties. Complex analysis is an extremely powerful tool with an unexpectedly large number of practical applications to the solution of physical problems. They also have applications in a variety of areas, ranging from fluid flow, to electric circuits, to the mysterious quantum world. Generally, they are described as belonging to the area of mathematics known as complex analysis.

The key result in complex analysis is the Cauchy integral theorem, which is the reason that single-variable complex analysis has so many nice results. A fundamental result of complex analysis is the Cauchy-Riemann equations, which give the conditions a function must satisfy in order for a complex generalization of the derivative, the so-called complex derivative, to exist. When the complex derivative is defined "everywhere," the function is said to be analytic.

The complex numbers are the field $\mathbb{C}$ of numbers of the form $x + iy$, where x and y are real numbers and i is the imaginary unit equal to the square root of -1.

COMPLEX NUMBERS

Complex numbers are useful abstract quantities that can be used in calculations and result in physically meaningful solutions. However, recognition of this fact is one that took a long time for mathematicians to accept.

THE ORIGIN OF COMPLEX NUMBERS CS

Complex analysis can roughly be thought of as the subject that applies the theory of calculus to imaginary numbers. But what exactly are imaginary numbers? Usually, students learn about them in high school with introductory remarks from their teachers along the following lines: "We can't take the square root of a negative number. But let's pretend we can and begin by using the symbol $i = \sqrt{-1}$."

If $i = \sqrt{-1}$, it stands to reason that $i^2 = -1$.

ALGEBRA OF COMPLEX NUMBERS

We have seen that complex numbers came to be viewed as ordered pairs of real numbers. That is, a complex number z is defined to be

$$z = (x, y),$$

where x and y are both real numbers.

The reason we say ordered pair is because we are thinking of a point in the plane. The point (2, 3), for example, is not the same as (3, 2).

The order in which we write x and y in the equation makes a difference. Clearly, then, two complex numbers are equal if and only if their x coordinates are and their y coordinates are equal. In other words,

$$(x, y) = (u, v) \text{ iff } x = u \text{ and } y = v.$$

A meaningful number system requires a method for combining ordered pairs. The definition of algebraic operations must be consistent so that the sum, difference, product, and quotient of any two ordered pairs will again be an ordered pair.

The key to defining how these numbers should be manipulated is to follow Gauss's lead and equate (x, y) with $x + iy$. Then, if $z_1 = (x_1, y_1)$ and $z_2 = (x_2, y_2)$ are arbitrary complex numbers, we have

$$\begin{aligned} z_1 + z_2 &= (x_1, y_1) + (x_2, y_2) \\ &= (x_1 + iy_1) + (x_2 + iy_2) \\ &= (x_1 + x_2) + i\,(y_1 + y_2) \\ &= (x_1 + x_2, y_1 + y_2) \end{aligned}$$

Thus, the following definitions should make sense.

ADDITION OF COMPLEX NUMBERS

Two complex numbers $z = x + iy$ and $z' = x' + iy'$ are added together component-wise,

$$z + z' = (x + x') + i\,(y + y').$$

In component form,

$$(x, y) + (x', y') = (x + x', y + y')$$

$$z_1 + z_2 = (x_1 + iy_1) + (x_2 + iy_2) = (x_1 + x_2, y_1 + y_2).$$

$$z_1 = x_1 + iy_1,\ z_2 = x_2 + iy_2$$

$$z_1 + z_2 = x_1 + x_2 + i\,y_1 + i\,y_2)$$

$$z_1 + z_2 = x_1 + x_2 + i\,(y_1 + y_2)$$

SUBTRACTION OF COMPLEX NUMBERS

Formula (1-9) $z_1 - z_2 = (x_1 + i\,y_1) - (x_2 + i\,y_2) = (x_1 - x_2, y_1 - y_2)$.

$$z_1 = x_1 + i\,y_1, \quad z_2 = x_2 + i\,y_2$$

$$z_1 - z_2 = x_1 - x_2 + i\,y_1 - i\,y_2$$

$$z_1 - z_2 = x_1 - x_2 + i\,(y_1 - y_2)$$

Example: Given $z_1 = 3 + 7\,i$ and $z_2 = 5 - 6\,i$.

(a) Find $z_1 + z_2$ and (b) $z_1 - z_2$.

$z_1 + z_2 = (3, 7) + (5, -6) = (3 + 5, 7 - 6) = (8, 1)$ and

$z_1 + z_2 = (3, 7) - (5, -6) = (3 - 5, 7 + 6) = (-2, 13)$.

We can also use the notation $z_1 = 3 + 7\,i$ and $z_2 = 5 - 6\,i$:

$z_1 + z_2 = 3 + 7\,i + 5 - 6\,i = 8 + i$ and $z_1 + z_2 = 3 + 7\,i - (5 - 6\,i) = -2 + 13\,i$.

COMPLEX MULTIPLICATION

Two complex numbers $x = a + i\,b$ and $c + i\,d$ are multiplied as follows:

$$x\,y = (a + i\,b)\,(c + i\,d) = ac + i\,b\,c + i\,a\,d - bd$$

$$= (ac - bd) + i\,(ad + bc).$$

In component form,

$$(x, y)\,(x', y') = (x\,x' - y\,y', x\,y' + y\,x')$$

The special case of a complex number multiplied by a scalar is then given by

$$a\,(x, y) = (a, 0)\,(x, y) = (a\,x, a\,y).$$

Surprisingly, complex multiplication can be carried out using only three real multiplications, ac, bd, and $(a + b)\,(c + d)$ as

$$R\,[(a + i\,b)\,(c + i\,d)] = a\,c - b\,d$$

$$\text{I}\,[(a + i\,b)\,(c + i\,d)] = (a + b)\,(c + d) - ac - bd.$$

Complex multiplication has a special meaning for elliptic curves.

$$z_1 z_2 = (x_1, y_2)\,(x_2, y_2) = (x_1 x_2 - y_1 y_2,\ x_1 y_2 + x_2 y_1).$$

Example: Given $z_1 = 3 + 7\,i$ and $z_2 = 5 - 6\,i$. Find $z_1\,z_2$.

Solution.

Ist Method

$$z_1\,z_2 = (3, 7)\,(5, -6)$$

$= (3 * 5 - 7(-6), 3(-6) + 5 * 7)$
$= (15 + 42, -18 + 35)$
$= (57, 17)$

IInd Method

$z_1 = 3 + 7i$ and $z_2 = 5 - 6i$:

$z_1 z_2 = (3, 7)(5, -6) = (3 + 7i)(5 - 6i)$

$= 15 - 18i + 35i - 42i^2$

$= 15 - 42(-1) + (-18 + 35)i$

$= 57 + 17i$

$= (57, 17)$

Of course, it makes sense that the answer came out as we expected because we used the notation $x + iy$ as motivation for our definition in the first place.

To motivate our definition for division, we will proceed along the same lines as we did for multiplication, assuming $z_2 \neq 0$:

Complex Division

$$\frac{z_1}{z_2} = \frac{(x_1, y_1)}{(x_2, y_2)} = \frac{x_1 + i\, y_1}{x_2 + i\, y_2}$$

$$\frac{z_1}{z_2} = \frac{(x_1, i\, y_1)(x_2 - i\, y_2)}{(x_2, i\, y_2)(x_2 - i\, y_2)}$$

$$= \frac{x_1, x_2 + y_1 y_2 + i(-x_1 y_2 + x_2 y_1)}{x_2^2 + y_2^2}$$

$$= \frac{x_1 x_2 + y_1 y_2}{x_2^2 + y_2^2} + i\frac{-x_1 y_2 + x_2 y_1}{x_2^2 + y_2^2}$$

$$= \left(\frac{x_1 x_2 + y_1 y_2}{x_2^2 + y_2^2}, \frac{-x_1 y_2 + x_2 y_1}{x_2^2 + y_2^2}\right)$$

COMPLEX DIVISION

The division of two complex numbers can be accomplished by multiplying the numerator and denominator by the complex conjugate of the denominator, for example, with $z_1 = a + b\,i$ and $z_2 = c + d\,i$, $z = z_1/z_2$ is given by

$$z = \frac{a + bi}{c + di} = \frac{(a + bi)\overline{c + di}}{(c + di)\,\overline{c + di}}$$

$$= \frac{(a+bi)(c-di)}{(c+di)(c-di)} = \frac{(ac+bd)+i(bc-ad)}{c^2+d^2},$$

where $\bar{z}$ denotes the complex conjugate. In component notation with $(x, y) = x + i\,y$,

$$\frac{(a+b)}{(c+d)} = \left(\frac{ac+bd}{c^2+d^2}, \frac{bc-ad}{c^2+d^2}\right).$$

$$\frac{z_1}{z_2} = \frac{(x_1, y_1)}{(x_2, y_2)} = \left(\frac{x_1x_2+y_1y_2}{x_2^2+y_2^2}, \frac{-x_1d_2+x_2y_1}{x_2^2+y_2^2}\right), \text{ for } z_2 \neq 0.$$

Example: Given $z_1 = 3 + 7$ II and $z_2 = 5 - 6$ II. Find $\frac{z_1}{z_2}$.

$$\frac{z_1}{z_2} = \frac{(3, 7)}{(5, -6)} = \left(\frac{15-42}{25+36}, \frac{18+35}{25+36}\right) = \left(-\frac{27}{61}, \frac{53}{61}\right)$$

As with the example for multiplication, we also get this answer if we use the notation x + IIY:

$$\frac{z_1}{z_2} = \frac{(3, 7)}{(5, -6)} = \left(\frac{3+7\text{II}}{5-6\text{II}}\right)$$

$$= \frac{(3+7\pi)\,(5+6\pi)}{(5-6\pi)\,(5+6\pi)} = \frac{15+18\pi+35\pi+42\pi^2}{25+30\pi-30\pi-36\pi^2}$$

$$= \frac{15-42+(18+35)\pi}{(25+3)} = -\frac{27}{61}+\frac{53}{61}\pi$$

$$= \left(-\frac{27}{61}, \frac{53}{61}\right)$$

To perform operations on complex numbers, most mathematicians would use the notation $x + i\,y$ and.engage in algebraic manipulations, as we did here, rather than apply the complicated-looking definitions we gave for those operations on ordered pairs.

It turns out that our algebraic definitions give complex numbers all the properties we normally ascribe to the real number system. Taken together, they describe what algebraists call a field. In formal terms, a field is a set (in this case, the complex numbers) together with two binary operations (in this case, addition and multiplication) having the following properties.

LAWS ASSOCIATED WITH ADDITION WITH COMPLEX NUMBER

1. *Commutative Law*

$$z_1 + z_2 = z_2 + z_1.$$

2. *Associative Law*

$$z_1 + (z_2 + z_3) = (z_1 + z_2) + z_3.$$

3. *Additive Identity.*

There is a complex number ω such that

$z + \omega = z$ for all complex numbers z.

The number ω is obviously the ordered pair (0, 0).

4. *Additive Inverses.*

Given any complex number z, there is a complex number n (depending on z) with the property that

$$z + n = (0, 0).$$

Obviously, if $z = (x, y) = x + i\,y$, the number n will be $n = (-x, -y) = -x - i\,y = -z$.

LAWS ASSOCIATED WITH MULTIPLICATION WITH COMPLEX NUMBER

1. *Commutative Law*

$$z_1 z_2 = z_2 z_1.$$

2. *Associative Law*

$$z_1\,(z_2 z_3) = (z_1 z_2)\,z_3.$$

3. *Multiplicative Identity*

There is a complex number ξ such that

$\xi z = z$ for all complex numbers z.

As one might expect, it turns out that (1, 0) is the unique complex number ξ with this property.

4. *Multiplicative Inverses*

Given any number z other than the number (0, 0), there is a complex number (depending on z) which we shall denote by z^{-1} with the property that

$z\,z^{-1} = (1, 0) = 1.$

Based on our definition for division, it seems reasonable that the number z^{-1} would be

$$z^{-1} = \frac{(1,0)}{z} = \frac{1}{x+iy}$$

$$= \frac{x-iy}{(x+iy)(x-iy)} = \frac{x-iy}{x^2+y^2}$$

$$= \frac{x}{x^2 + y^2} + i\frac{-y}{x^2 + y^2}$$

$$= \left(\frac{x}{x^2 + y^2}, \frac{-y}{x^2 + y^2}\right)$$

5. The Distributive Law

$$z_1(z_2 + z_3) = z_1(z_2 + z_3).$$

•**Real Part of z:** The real part of z = x + π y, denoted by Re(z), is the real number x.

•**Imaginary Part of z:** The imaginary part of z = x + π y, denoted by Im(z), is the real number y.

Conjugate of z

The complex conjugate of a complex number $z \equiv a + bi$ is defined to be

$$\bar{z} \equiv a - bi.$$

The conjugate matrix of a matrix A = (a_{ij}) is the matrix obtained by replacing each element a_{ij} with its complex conjugate, $\bar{A} = (\bar{a}_{ij})$.

By definition, the complex conjugate satisfies

$$\bar{\bar{z}} = z.$$

The complex conjugate is distributive under complex addition,

$$\overline{z_1 + z_2} = \bar{z}_1 + \bar{z}_2,$$

since

$$\overline{(a_1 + ib_1) + (a_2 + ib_2} = \overline{(a_1 + a_2) + i(b_1 + b_2)}$$
$$= (a_1 + a_2) - i(b_1 + b_2)$$
$$= (a_1 - ib_1) + (a_2 - ia_2)$$
$$= \overline{a_1 + ib_1} + \overline{a_1 + ib_2},$$

and distributive over complex multiplication,

$$\overline{z_1 z_2} = \bar{z}_1\, \bar{z}_2,$$

since

$$\overline{(a_1 + b_1 i)(a_2 + b_2 i)} = \overline{(a_1 a_2 - b_1 b_2 + i(a_1 b_2 + a_2 b_1)}$$

$$= (a_1a_2 - b_1b_2) - i(a_1b_2 + a_2b_1)$$
$$= (a_1 - ib_1)(a_2 - ib_2)$$
$$= \overline{a_1 + ib_1}\ \overline{a_2 + ib_2}.$$

The conjugate of $z = x + i\,y$, denoted by $\bar{z}$, is the complex number (x, –y) = x – $\bar{z}$ y.

The following theorem gives some important facts relating to these operations. You will be asked for a proof in the exercises

Theorem. **Suppose z_1, z_2 and z_3 are arbitrary complex numbers. Then if $z_2 \neq 0$**

$$\bar{\bar{z}} = z$$

$$\overline{z_1 + z_2} = \bar{z}_1 + \bar{z}_2$$

$$\overline{z_1 z_2} = \bar{z}_1\,\bar{z}_2$$

$$\overline{\frac{z_1}{z_2}} = \frac{\bar{z}_1}{\bar{z}_2} \text{ if } z_2 \neq 0$$

$$\text{Im}[z] = \frac{z - \bar{z}}{2i}$$

$$\text{Re}[i\,z] = -\,\text{Im}[z]$$

$$\text{Im}\,[i\,z] = \text{Re}[z]$$

Exerecise:

1. **Find each of the following:**
   (*a*) (1 + 2i)(3 + 4i) (*b*) $(1 - 2i)^2/(1 - i)$
   (*c*) {1/(1 + i)+ 1/(1 – 2i)}
2. **Find all the roots of the equation $(z^8 - 1)(z^3 + 8) = 0$.**
3. **Suppose that z = x + iy, where x, y ∈ R. Express each of the following in terms of x and y:**

   (*a*) $\left|\dfrac{z+1}{z-1}\right|$ (*b*) $\left|\dfrac{z+i}{1-iz}\right|$
4. **For each of the following, compute all the values and plot them on the plane:**

   (*a*) $(1+i)^{\frac{1}{2}}$ (*b*) $(-4)^{\frac{3}{4}}$ (*c*) $(1-i)^{\frac{3}{8}}$
5. **Solve the equation $z^2 + z + 1 = 0$ for $z = (x, y)$ by writing (*x*, *y*) (*x*, *y*) + (*x*, *y*) + (1, 0) = (0, 0) and then solving a pair of simultaneous equations in *x* and *y*.**

COMPLEX FUNCTIONS AND LINEAR MAPPINGS

A complex-valued function f of the complex variable z is a rule that assigns to each complex number z in a set D one and only one complex number w. We write $w = f(z)$ and call w the image of z under f. A simple example of a complex-valued function is given by the formula $w = f(z) = z^4$. The set D is called the domain of f, and the set of all images $\{w = f(z): z \in D\}$ is called the range of f. When the context is obvious, we omit the phrase complex-valued, and simply refer to a function f, or to a complex function f.

We can define the domain to be any set that makes sense for a given rule, so for $w = f(z) = z^4$, we could have the entire complex plane for the domain D, or we might artificially restrict the domain to some set such as $D = D_1(0) = \{z: |z| < 1\}$. Determining the range for a function defined by a formula is not always easy, but we will see plenty of examples later on. In some contexts functions are referred to as mappings or transformations.

In Section 1.6, we used the term domain to indicate a connected open set. When speaking about the domain of a function, however, we mean only the set of points on which the function is defined. This distinction is worth noting, and context will make clear the use intended.

Just as z can be expressed by its real and imaginary parts, $z = x + iy$, we write $f(z) = w = v + iv$, where u and v are the real and imaginary parts of w, respectively. Doing so gives us the representation.

$$w = f(z) = f(x, y) = f(x + iy) = u + iv.$$

Because u and v depend on x and y, they can be considered to be real-valued functions of the real variables x and y; that is,

$$u = u(x, y) \text{ and } v = v(x, y).$$

Combining these ideas, we often write a complex function f in the form

$$f(z) = f(x + iy) = u(x, y) + iv(x, y).$$

Figure illustrates the notion of a function (mapping) using these symbols.

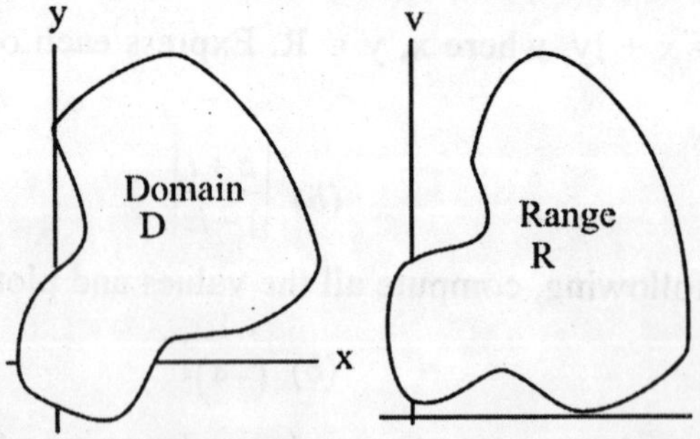

Fig. The mapping $w = f(z) = u(x, y) + iv(x, y)$.

Example: Write $f(z) = z^4$ in the for $f(z) = u(x, y) + iv(x, y)$.

Solution. Using the binomial formula, we obtain

$$\begin{aligned} f(z) &= f(x+iy) = (x+iy)4 \\ &= x^4 + 4x^3(iy) + 6x^2(iy)^2 + 4x(iy)^3 + (iy)^4 \\ &= x^4 + 4ix^3y - 6x^2y^2 - 4ixy^3 + y^4 \\ &= x^4 - 6x^2y^2 + y^4 + i(4x^3y - 4xy^3) \\ &= u(x,y) + iv(x,y) \end{aligned}$$

so that $u(x, y) = x^4 - 6x^2y^2 + y^4$ and $v(x + y) = 4x^3y - 4xy^3$.

Example: Express the function $f(z) = \bar{z}\ \text{Re}[z] + z^2\ \text{Im}[z]$ in the form $f(z) = u(x,y) + v(x,y)$.

Solution. Using the elementary properties of complex numbers, it follows that

$$f[z] = (x - iy)x + (x + iy)^2 + y = (2x^2 - y^2 + y) + i(xy)$$

so that $u(x, y) = 2x^2 - y^2 + y$ and $(x, y) = xy$.

Examples 1 and 2 show how to find $u(x, y)$ and $v(x, y)$ when a rule for computing f is given. Conversely, if $u(x, y)$ and $v(x, y)$ are two real-valued functions of the real variables x and y, they determine a complex-valued function

$f(z) = u(x, y) + iv(x, y)$, and we can use the formulas

$$x = \frac{z+\bar{z}}{2} \quad \text{and} \quad y = \frac{z-\bar{z}}{2i}$$

to find a formula for f involving the variables z and $\bar{z}$.

Example: Express $f(z) = 4x^2 + i4y^2$ by a formula involving the variables z and $\bar{z}$.

Solution. Calculation reveals that

$$\begin{aligned} f(z) &= 4\left(\frac{z+\bar{z}}{2}\right)^2 + i4\left(\frac{z-\bar{z}}{2i}\right)^2 \\ &= z^2 + 2z\bar{z} + \bar{z}^2 - i(z^2 - 2z\bar{z} + \bar{z}^2) \\ &= (1-i)z^2 + (2+2i)z\bar{z} + (1-i)\bar{z}^2 \end{aligned}$$

Using $z = re^{i\theta}$ in the expression of a complex function f may be convenient. It gives us the polar representation

$$f(z) = f(re^{i\theta}) = u(r,\theta) + iv(r,\theta),$$

where U and V are real functions of the real variables r and θ.

Remark. For a given function *f*, the functions *u* and *v* defined above are different from those used previously in $f(z) = f(x + iy) = u(x, y) + i\,v(x, y)$ which used Cartesian coordinates instead of polar coordinates.

***Example*:** Express $f(z) = z^2$ in both Cartesian and polar form.

Solution. For the Cartesian form, a simple calculation gives

$$f(z) = f(x + iy) = (x + iy)^2$$

$$x^2 - y^2 + 2\,i\,x\,y$$

$$u(x, y) + i\,v(x, y)$$

so that $u(x, y) = x^2 - y^2$ and $v(x, y) = x\,y$.

For the polar form, we get v

$$f(z) = f(re^{i\theta}) = (re^{i\theta})^2 = r^2 e^{i2\theta}$$

$$= r^2(\cos 2\theta + i \sin 2\theta)$$

$$= r^2 \cos 2\theta + ir^2 \sin 2\theta$$

$$= U(r,\theta) + i\,V(r,\theta)$$

so that $= u(r,\theta) + r^2 \cos 2\theta$ and $v(r,\theta) = r^2 \sin 2\theta$.

Remark: Once we have defined *u* and *v* for a function *f* in Cartesian form, we must use different symbols if we want to express *f* in polar form.

As is clear here, the functions *u* and *U* are quite different, as are *v* and *V*. Of course, if we are working only in one context, we can use any symbols we choose.

For a given function *f*, the functions *u* and *v* defined here are different from those defined by equation (1), because equation (2) involves Cartesian coordinates and equation (3) involves polar coordinates.

***Example*:** Express $f(z) = z^5 + 4z^2 - 6$ in polar form.

Solution. We obtain

$$f(z) = f(re^{i\theta}) = (re^{i\theta})^5 + 4(re^{i\theta})^2 - 6$$

$$= r^5 e^{i5\theta} + 4r^2 e^{i2\theta} - 6$$

$$= r^5 \cos 5\theta + 4\,r^2 \cos 2\theta - 6 + i\,(r^5 \sin 5\theta + 4\,r^2 \sin 2\theta)$$

$$= U(r,\theta) + i\,V(r,\theta)$$

so that $U(r,\theta) = r^2 \cos 5\theta + 4r^2 \cos 2\theta - 6$ and $V(r,\theta) = r^2 \sin 5\theta + 4r^2 \sin 2\theta$.

We now look at the geometric interpretation of a complex function. If D is the domain of real-valued functions $u(x, y)$ and $v(x, y)$, the equations

$$u = u(x, y) \text{ and } v = v(x, y)$$

describe a transformation (or mapping) from D in the xy plane into the uv plane, also called the w plane. Therefore, we can also consider the function

$$w = f(z)\ u(x,y) + i\,v(x,y)$$

to be a transformation (or mapping) from the set D in the z plane onto the range R in the w plane. This idea was illustrated in Figure 1. In the following paragraphs we present some additional key ideas. They are staples for any kind of function, and you should memorize all the terms in bold.

If A is a subset of the domain D of f, the set $B = \{w = f(z): z \in A\}$ is called the image of the set A, and f is said to map A onto B. The image of a single point is a single point, and the image of the entire domain, D, is the range, R. The mapping $w = f(z)$ is said to be from A into S if the image of A is contained in S. Mathematicians use the notation $f: A \to S$ to indicate that a function maps A into S. Figure illustrates a function f whose domain is D and whose range is R. The shaded areas depict that the function maps A onto B. The function also maps A into R, and, of course, it maps D onto R.

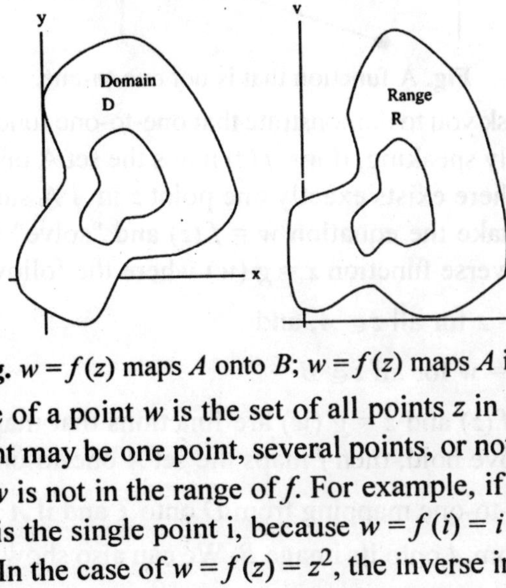

Fig. $w = f(z)$ maps A onto B; $w = f(z)$ maps A into R.

The inverse image of a point w is the set of all points z in D such that $w = f(z)$. The inverse image of a point may be one point, several points, or nothing at all. If the last case occurs then the point w is not in the range of f. For example, if $w = f(z) = i\,z$, the inverse image of the point -1 is the single point i, because $w = f(i) = i\,(i) = -1$, and i is the only point that maps to -1. In the case of $w = f(z) = z^2$, the inverse image of the point -1 is the set $\{i, -i\}$. You will learn that, if $w = f(z) = e^z$, the inverse image of the point 0 is the empty set---there is no complex number z such that $e^z = 0$.

The inverse image of a set of points, S, is the collection of all points in the domain that map into S. If f maps D onto R it is possible for the inverse image of R to be function as well, but the original function must have a special property: a function f is said to be one-to-one if it maps distinct points $z_1 \neq z_2$, onto distinct points $f(z_1) \neq f(z_2)$. Many times an easy way to prove that a function f is one-to-one is to suppose $f(z_1) = f(z_2)$, and from this assumption deduce that z_1 must equal z_2. Thus, $f(z) = iz$ is one-to-one because if $f(z_1) = f(z_2)$, then $iz_1 = iz_2$. Dividing both sides of the last equation by i gives $z_1 = z_2$. Figure illustrates the idea of a one-to-one function: distinct points get mapped to distinct points.

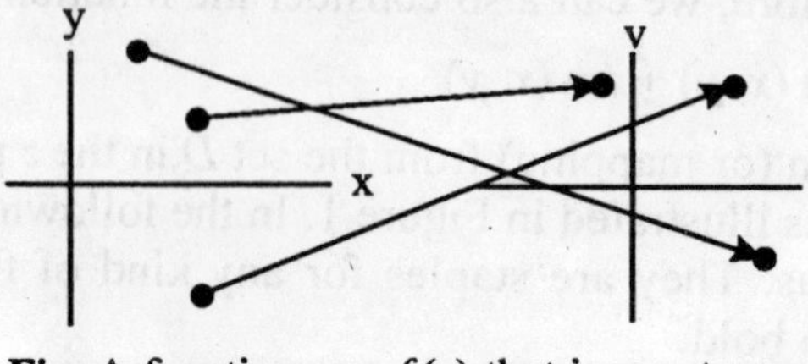

Fig. A function $w = f(z)$ that is one-to-one.

The function $f(z) = z^2$ is not one-to-one because $-i \neq i$, but $f(i) = f(-i) = -1$. Figure depicts this situation: at least two different points get mapped to the same point.

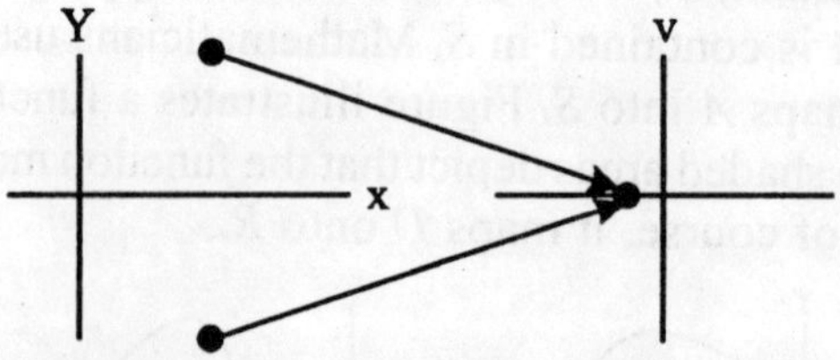

Fig. A function that is not one-to-one.

In the exercises we ask you to demonstrate that one-to-one functions give rise to inverses that are functions. Loosely speaking, if $w = f(z)$ maps the set A one-to-one and onto the set B, then for each w in B there exists exactly one point z in A A such that $w = f(z)$. For any such value of z we can take the equation $w = f(z)$ and "solve" for z as a function of w. Doing so produces an inverse function $z = g(w)$ where the following equations hold:

$g(f(z)) = z$ for all $z \in A$, and

$f(g(w)) = w$ for all $z \in B$

Conversely, if $w = f(z)$ and $z = g(w)$ are functions that map A into B and B into A, respectively, and the above hold, then f maps the set A one-to-one and onto the set B.

Further, if f is a one-to-one mapping from D onto T and if A is a subset of D, then f is a one-to-one mapping from A onto its image B. We can also show that, if $\zeta = f(z)$ is a one-to-one mapping from A onto B and $w = g(\zeta)$ is a one-to-one mapping from B onto S, then the composite mapping $w = g(f(z))$ is a one-to-one mapping from A onto S.

We usually indicate the inverse of f by the symbol f^{-1}. If the domains of f and f^{-1} are A and B respectively, then we write

$f^{-1}(f(z)) = z$ for all $z \in A$, and $f(f^{-1}(w)) = w$ for all $w \in B$.

Also, for $z_0 \in A$ and $w_0 \in B$.

$$w_0 = f(z_0) \text{ iff } f^{-1}(w_0) = z_0, \text{ and}$$
$$z_0 = f^{-1}(w_0) \text{ iff } f(w_0) = z_0.$$

Example: If $w = f(z) = i\,z$ for any complex number z, find $f^{-1}(w)$. Also find $f(f^{-1}(w))$

Solution. We can easily show f is one-to-one and onto the entire complex plane. We solve for z, given $w = f(z) = i$, to get $z = \frac{w}{i} = -i\,w$. This result implies that $f^{-1}(w) = -i\,w$ for all complex numbers w.

Since,

$$w = f[z] = i\,z$$
$$\{\{z \to i(-4x^2 - 4\,i\,y^2)\}\}$$
$$z = f^{-1}[w] = i(-4x^2 - 4\,i\,y^2)$$
$$f^{-1}[f[z]] = -i\,(i\,z)$$
$$f^{-1}[f[z]] = i(-4x^2 - 4\,i\,y^2)$$
$$f[f^{-1}[w]] = i\,(i\,(-4\,x^2 - 4\,i\,y^2))$$
$$f[f^{-1}[w]] = 4\,x^2 + 4\,i\,y^2$$

Remark: Once we have specified $f^{-1}(w) = -i\,w$ for all complex numbers w, we note that there is nothing magical about the symbol w. We could just as easily write $f^{-1}(z) = -i\,z$ for all complex numbers z.

We now show how to find the image B of a specified set A under a given mapping $u + iv = w = f(z)$.

The set A is usually described with an equation or inequality involving x and y. Using inverse functions, we can construct a chain of equivalent statements leading to a description of the set B in terms of an equation or an inequality involving u and v.

Example: Show that the function $f[z] = i\,z$ maps the line y = x + 1 in the xy plane onto the line $v = -u - 1$ in the w plane.

Solution.

First Method: With $A = \{(x, y)\} : y = x + 1$, we want to describe $B = f(A)$. We let $z = x + i\,y \in A$ and get

$$u + i\,v = w = f(z) \in B$$
$$\Leftrightarrow f^{-1}(w) = z = x + i\,y \in A$$
$$\Leftrightarrow i\,w \in A$$
$$\Leftrightarrow v - i\,u \in A$$
$$\Leftrightarrow (v, -u) \in A$$
$$\Leftrightarrow u = v + 1$$
$$\Leftrightarrow v = -u - 1$$

where $\Leftrightarrow$ is the notation for "if and only if." Note what this result says: $u + i\,v = w + B \Leftrightarrow v = -u - 1$. The image of A under f, therefore, is the set $B = \{(u, v): v = -u - 1\}$.

Second Method: We write $u + i\,v = w = f(z) = i\,(x + i\,y) = -y + i\,x$ and note that the transformation can be given by the equations $u = -y$ and $v = x$. Because A is described by $A = \{x + i\,y: y = x + 1\}$, we can substitute $u = -y$ and $v = x$ into the equation $y = x + 1$ to obtain $-u = v + 1$, which we can rewrite as $v = -u - 1$. If you use this method, be sure to pay careful attention to domains and ranges. We now look at some elementary mappings. If we let $B = a + ib$ denote a fixed complex constant, the transformation $w = T(z) = z + B = x + a + i\,(y + b)$ is a one-to-one mapping of the z-plane onto the w-plane and is called a translation.

TRANSFORMATION

A transformation consisting of a constant offset with no rotation or distortion. In n-dimensional Euclidean space, a translation may be specified simply as a vector giving the offset in each of the $n$coordinates. This transformation can be visualized as a rigid translation whereby the point z is displaced through the vector $B = a + ib$ to its new position $w = T(z)$. The inverse mapping is given by

$$z = T^{-1}(w) = w - B = u - a + i\,(v - b)$$

and shows that T is a one-to-one mapping from the z-plane onto the w-plane. The effect of a translation is depicted in Figure.

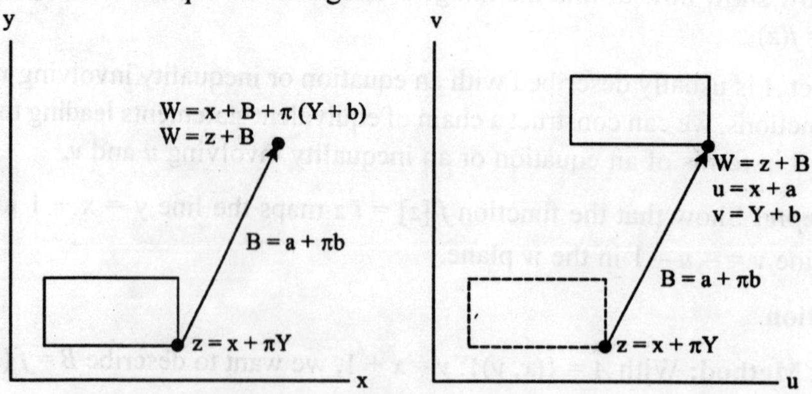

Fig. The Translation $w = T(z) = z + B = x + a + i\,(y + b)$.

If we let α be a fixed real number, then for $z = re^{i\theta}$, the transformation

$$w = R(z) = ze^{i\alpha} = re^{i\theta}e^{i\alpha} = re^{i(\theta+\alpha)}$$

is a one-to-one mapping of the z-plane onto the w-plane and is called a rotation.

ROTATION TRANSFORMATION

The symmetry operation corresponding to a rotation followed by an inversion operation, also called a roto-inversion.

This operation is denoted $\bar{n}$ for an improper rotation by 360°/n, so the crystallography restriction gives only $\bar{1}$, $\bar{2}$, $\bar{3}$, $\bar{4}$, $\bar{6}$ for crystals. The mirror plane symmetry operation is $(x, y, z) \to (x, y, -z)$, etc., which is equivalent to $\bar{2}$.

The turning of an object or coordinate system by an angle about a fixed point. A rotation is an orientation-preserving orthogonal transformation.

Euler's rotation theorem states that an arbitrary rotation can be parameterized using three parameters. These parameters are commonly taken as the Euler angles. Rotations can be implemented using rotation matrices.

Rotation in the plane can be concisely described in the complex plane using multiplication of complex numbers with unit modulus such that the resulting angle is given by arg (z). For example, multiplication by i represents a rotation to the right by 90° and by $-i$ represents rotation to the left by 90°.

So starting with $z = 1$ and rotating left twice gives $z = 1(-i)(-i) = i^2 = -1$, which is the same as rotating right twice, $z = 1(i)(i) = i^2 = -1$, and $\arg(-1) = \pi = 180°$. For multiplication by multiples of 90°, the possible positions are then concisely represented by $z = 1, i, -1$, and $-i$.

The rotation symmetry operation for rotation by 360°/n is denoted "n." For periodic arrangements of points ("crystals"), the crystallography restriction gives the only allowable rotations as 1, 2, 3, 4, and 6.

It can be visualized as a rigid rotation whereby the point z is rotated about the origin through an angle α to its new position $w = R(z)$. If we use polar coordinates and designate $w = \rho e^{i\phi}$ in the w-plane, then the inverse mapping is

$$z = R^{-1}(w) = we^{-i\alpha} = \rho e^{i\phi} e^{-i\alpha} \rho e^{i(\phi-\alpha)}.$$

This analysis shows that R is a one-to-one mapping from the z-plane onto the w-plane. The effect of rotation is depicted in Figure.

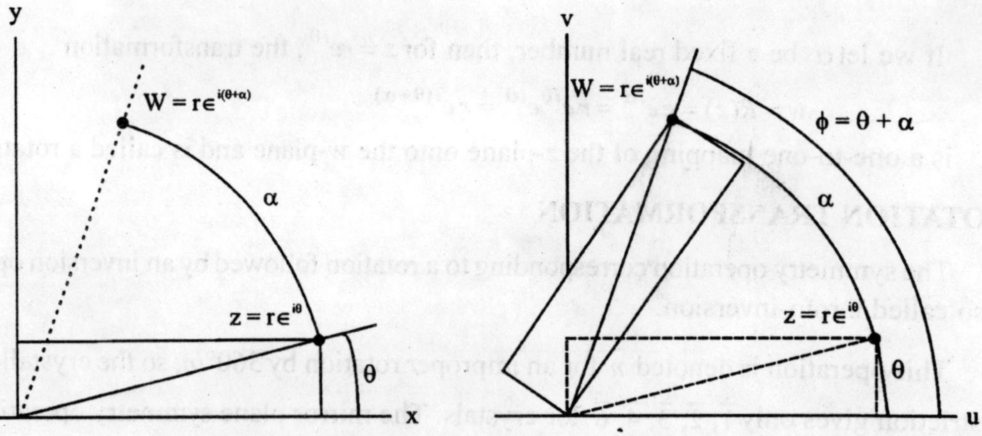

Fig. The Rotation $w = R(z) = re^{i(\theta+\alpha)}$.

Example: The ellipse centered at the origin with a horizontal major axis of 4 units and vertical minor axis of 2 units can be represented by the parametric equation

$$s(t) = 2\cos t + i\sin t = (2\cos t, \sin t), \text{ for } 0 \le t \le 2\pi.$$

Solution:

Suppose we wanted to rotate the ellipse by an angle of $\frac{\pi}{6}$ radians and shift the center of the ellipse 2 units to the right and 1 unit up. Using complex arithmetic, we can easily generate a parametric equation $r(t)$ that does so:

$$r[t] = s[t]e^{-i\pi/6} + (2+i)$$

$$= (2\cos t + i\sin t)\left(\cos\frac{\pi}{6} + i\sin\frac{\pi}{6}\right) + (2+i)$$

$$= \left(2\cos t\cos\frac{\pi}{6} + -\sin t\sin\frac{\pi}{6}\right) + i\left(2\cos t\sin\frac{\pi}{6} + \sin t\cos\frac{\pi}{6}\right) + (2+i)$$

$$= \left(\sqrt{3}\cos t - \frac{1}{2}\sin t + 2\right) + i\left(\cos t + \frac{\sqrt{3}}{2}\sin t + 1\right)$$

$$\sqrt{3}\cos t - \frac{1}{2}\sin t + 2,\ \cos t + \frac{\sqrt{3}}{2}\sin t + 1$$

for $0 \le t \le 2\pi$. Figure shows parametric plots of these ellipses.

$$s(t) = 2\cos t + i\sin t\ r(t) = s(t)e^{i\pi/6} + (2+i)$$

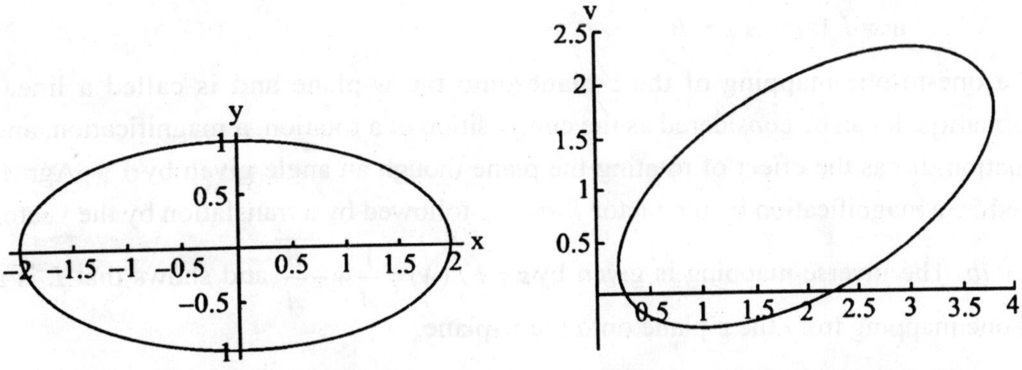

Fig. (*a*) Plot of the Original Ellipse (*b*) Plot of the Rotated Ellipse

If we let $K > 0$ be a fixed positive real number, then the transformation

$$w = S(z) = Kz = Kx + iKy$$

is a one-to-one mapping of the *z*-plane onto the *w*-plane and is called a magnification. If $K > 1$, it has the effect of stretching the distance between points by the factor *K*. If $K >$ 1, then it reduces the distance between points by the factor *K*. The inverse transformation is given by

$$z = S^{-1}(w) = \frac{1}{K}w = \frac{1}{K}u + i\frac{1}{k}v$$

and shows that *S* is a one-to-one mapping from the *z*-plane onto the *w*-plane. The effect of magnification is shown in Figure.

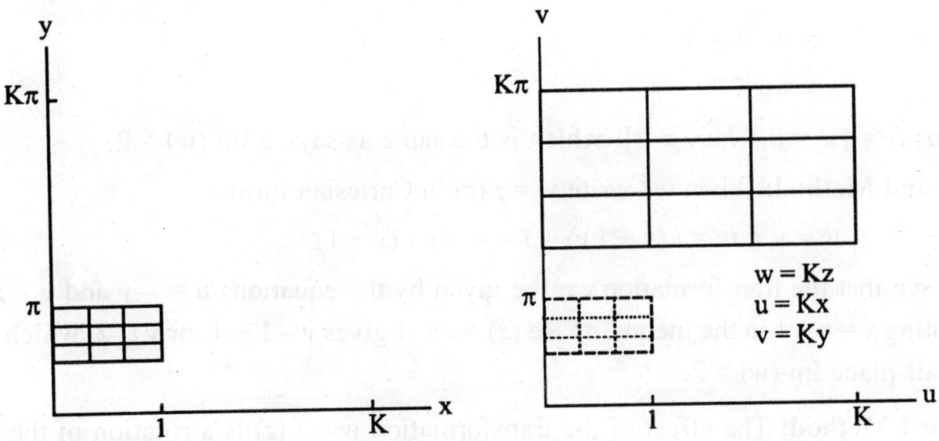

Fig. The Magnification $w = S(z) = Kz = Kx + i\,Ky$.

Finally, if we let $A = Ke^{i\alpha}$ and $B = a + i\,b$, where $K > 0$ is a positive real number, then the transformation

$$w = L(z) = Az + B$$

is a one-to-one mapping of the z-plane onto the w-plane and is called a linear transformation. It can be considered as the composition of a rotation, a magnification, and a translation. It has the effect of rotating the plane though an angle given by α = Agr A, followed by a magnification by the factor $K = |A|$, followed by a translation by the vector $B = a + ib$. The inverse mapping is given by $z = L^{-1}(w) = \frac{1}{A}w - \frac{B}{A}$ and shows that L is a one-to-one mapping from the z-plane onto the w-plane.

***Example*:** Show that the linear transformation $w = f(z) = iz + i$ maps the right half plane Re $(z) > 1$ onto the upper half plane Im $(w) > 2$.

Solution.

First Method: Let A = {(x, y): x≥ 1}. To describe $B = F(A)$, we solve $w = iz + i$ for z to get $z = \frac{w-i}{i} = -iw - 1 = f^{-1}(w)$. We have the following

$$u + iv = w = f(z) \in B$$

$$\Leftrightarrow f^{-1}(w) = z \in A$$

$$\Leftrightarrow -iw - 1 \in A$$

$$\Leftrightarrow v - 1 - iu \in A$$

$$\Leftrightarrow (v - 1, -u) \in A$$

$$\Leftrightarrow v - 1 \geq 1$$

$$\Leftrightarrow v \geq 2$$

Thus $B = \{w = u + iv: v \geq 2\}$, which is the same as saying Im $(w) > 2$.

Second Method: When we write $w = f(z)$ in Cartesian form as

$$w = u + iv = i(x + iy) + i = -y + i(x + 1),$$

we see that the transformation can be given by the equations $u = -y$ and $v = x + 1$. Substituting $x = v - 1$ in the inequality Re $(z) = x \geq 1$ gives $v - 1 \geq 1$, or $v \geq 2$, which is the upper half-plane Im $(w) > 2$.

Third Method: The effect of the transformation $w = f(z)$ is a rotation of the plane through the angle $\alpha = \frac{\pi}{2}$ (when z is multiplied by i) followed by a translation by the vector $B = i$. The first operation yields the set Im $(w) \geq 1$. The second shifts this set up 1 unit, resulting in the set Im $(w) > 2$. We illustrate this result in Figure.

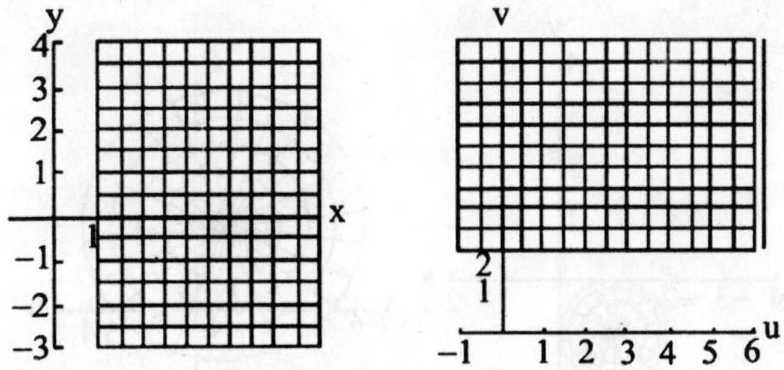

Fig. The Linear Transformation $w = f(z) = i\,z + i$.

Translations and rotations preserve angles. First, magnifications rescale distance by a factor *K*, so it follows that triangles are mapped onto similar triangles, preserving angles.

Then, because a linear transformation can be considered to be a composition of a rotation, a magnification, and a translation, it follows that linear transformations preserve angles.

Consequently, any geometric object is mapped onto an object that is similar to the original object; hence linear transformations can be called similarity mappings.

Example: Show that the image of the open disk $D_1(-1-i) = \{z: |z+1+i| < 1\}$ under the linear transformation $w = f(z) = (3-4i)z + 6 + 2i$ is the open disk $D_5(-1+3i) = \{w: |w+1-3i| < 5\}$.

Solution. The inverse transformation is $z = \dfrac{w-6-2i}{3-4i}$, so if we designate the range of *f* as *B*, then

$$w = f(z) \in B \Leftrightarrow f^{-1}(w) = z \in D_1(-1-i)$$

$$\Leftrightarrow \frac{w-6-2i}{3-4i} \in D_1(-1-i)$$

$$\Leftrightarrow \left|\frac{w-6-2i}{3-4i} + 1 + i\right| < 1$$

$$\Leftrightarrow \left|\frac{w-6-2i}{3-4i} + 1 + i\right| |3-4i| < 1\,|3-4i|$$

$$\Leftrightarrow |w-6-2i+(1+i)(3-4i)| < 5$$

$$\Leftrightarrow |w+1-3i| < 5$$

Hence the disk with center $-1-i$ and radius 1 is mapped one-to-one and onto the disk with center $-1+3i$ and radius 5 as shown in Figure.

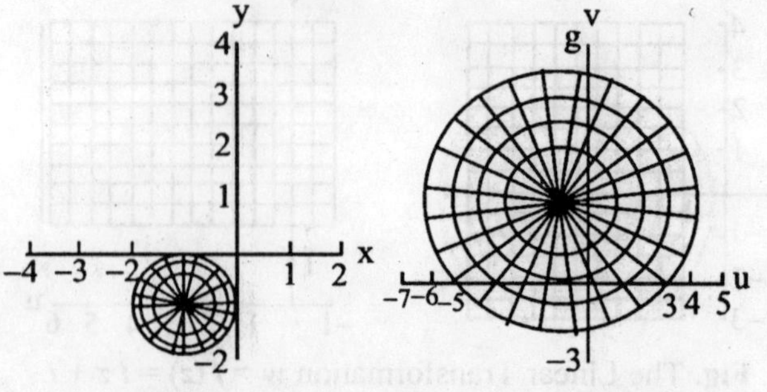

Fig. The Mapping $w = f(z) = (3 - 4i)z + 6 + 2i$.

Example: **Show that the image of the right half plane Re $(z) = x > 1$ under the linear transformation w = $(-1 + i)z - 2 + 3i$ is the half plane $v > u + 7$.**

Solution. The inverse transformation is given by

$$z = \frac{w+2-3i}{-1+i} = \frac{u+2+i(v-3)}{-1+i},$$

which we write as

$$x + iy = \frac{-u+v-5}{2} + i\frac{-u-v+1}{2}$$

Substituting $x = \dfrac{-u+v-5}{2}$ into e $\text{Re}(z) = x > 1$ gives $\dfrac{-u+v-5}{2} > 1$, which simplifies $v > u + 7$. Figure illustrates the mapping.

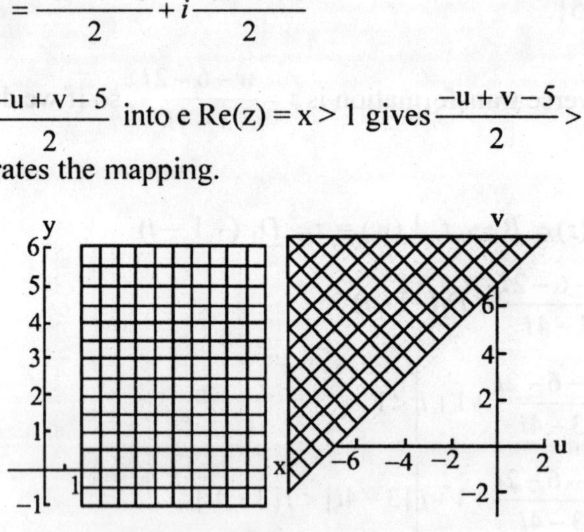

Fig. The the Linear Transformation

THE MAPPINGS $W = Z^N$ AND $W = Z^{1/N}$

In this section we turn our attention to power functions. For $z = re^{i\theta}$, we can express the function $w = f(z) = z^2$ in polar coordinates by

$$w = f(z) = z^2 r^2 e^{i2\theta}$$

If we also use polar coordinates for $w = \rho e^{i\theta}$ in the w –plane, we can express this mapping by the system of equations

$$\rho = r^2 \text{ and } \phi = 2\theta.$$

Because an argument of the product (z)(z) is twice an argument of z, we say that f doubles angles at the origin. Points that lie on the ray $r > 0, \theta = \alpha$ are mapped onto points that lie on the ray $\rho = 0, \phi = 2\alpha$. If we now restrict the domain of $w = f(z) = z^2$ to the region

$$A = \left\{ re^{i\theta} : r > 0 \text{ and } \frac{-\pi}{2} < \theta \le \frac{\pi}{2} \right\},$$

then the image of A under the mapping $w = z^2$ can be described by the set

$$B = \left\{ \rho e^{i\theta} : \rho > 0 \text{ and } -\pi < \phi \le \pi \right\},$$

which consists of all points in the w-plane except the point $w = 0$.

The inverse mapping of f, which we shall denote by g, is then

$$z = g(w) = w^{1/2} \rho^{1/2} e^{i\phi/2}$$

where $w \in B$. That is

$$z = g(w) = w^{1/2} = |w|^{1/2} e^{i\mathrm{Arg}(w)/2},$$

where $w \neq 0$. The function g is so important that we call special attention to it with a formal definition.

SQUARE ROOT FUNCTION

A square root, also called a radical or surd, of *x* is a number *r* such that $r^2 = x$. The function $\sqrt{x}$ is therefore the inverse function of $f(x) = x^2$ for $x \ge 0$.

There are a number of square root algorithms that can be used to approximate the square root of a given number. These include the Bhaskara-Brouckner algorithm and Wolfram's iteration. The simplest algorithm for $\sqrt{n}$ is Newton's iteration

$$x_{k+1} = \frac{1}{2}\left(x_k + \frac{n}{x_k} \right)$$

with $x_0 = 1$.

Note that any positive real number has two square roots, one positive and one negative.

For example, the square roots of 9 are -3 and $+3$, since $(-3)^2 = (+3)^2 = 9$. Any nonnegative real number x has a unique nonnegative square root r; this is called the principal square root and is written $r = x^{1/2}$ or $r = \sqrt{x}$. For example, the principal square root of 9 is $\sqrt{9} = +3$, while the other square root of 9 is $-\sqrt{9} = -3$.

The square roots of a complex number $z = x + iy$ are given by

$$\sqrt{x+iy} = \pm(x^2+y^2)^{1/4}\left\{\cos\left[\frac{1}{2}\tan^{-1}(x,y)\right] + i\sin\left[\frac{1}{2}\tan^{-1}(x,y)\right]\right\}$$

In addition,

$$\sqrt{x+iy} = \frac{1}{2}\sqrt{2}\left[\sqrt{\sqrt{x^2+y^2}+x} + x + i\sin(y)\sqrt{\sqrt{x^2+y^2}-x}\right]$$

As can be seen in the above figure, the imaginary part of the complex square root function has a branch cut along the negative real axis.

A nested radical of the form $\sqrt{a \pm b\sqrt{c}}$ can sometimes be simplified into a simple square root by equating

$$\sqrt{a \pm b\sqrt{c}} = \sqrt{d} \pm \sqrt{e}$$

Squaring gives

$$a \pm b\sqrt{c} = d + e \pm 2\sqrt{de}$$

so

$$a = d + e$$

$$b^2 c = 4de.$$

Solving for d and e gives

$$d,\, e = \frac{a \pm \sqrt{a^2 - b^2 c}}{2}.$$

For example,

$$\sqrt{5+2\sqrt{6}} = \sqrt{2} + \sqrt{3}$$

$$\sqrt{3-2\sqrt{2}} = \sqrt{2} - 1$$

A counterintuitive property of inverse functions is that

$$\sqrt{z}\sqrt{\frac{1}{z}} = \begin{cases} -1 & \text{for } J[z] = 0 \text{ and } R[z] < 0 \\ \text{undefined} & \text{for } z = 0 \\ 1 & \text{otherwise,} \end{cases}$$

so the expected identity (i.e., canceling of the $\sqrt{z}$ s) does not hold along the negative real axis.

PRINCIPAL SQUARE ROOT FUNCTION

The function

$$z = g(w) = w^{1/2} = |w|^{1/2} e^{iArg(w)/2}, \text{ for } w \neq 0.$$

is called the principal square root function.

It is left as an exercise to show that f and g satisfy equations $g(f(z)) = z$ and $f(g(w)) = w$, and thus are inverses of each other that map the set A one-to-one and onto the set B and the set B one-to-one and onto the set A, respectively. Figure illustrates this relationship.

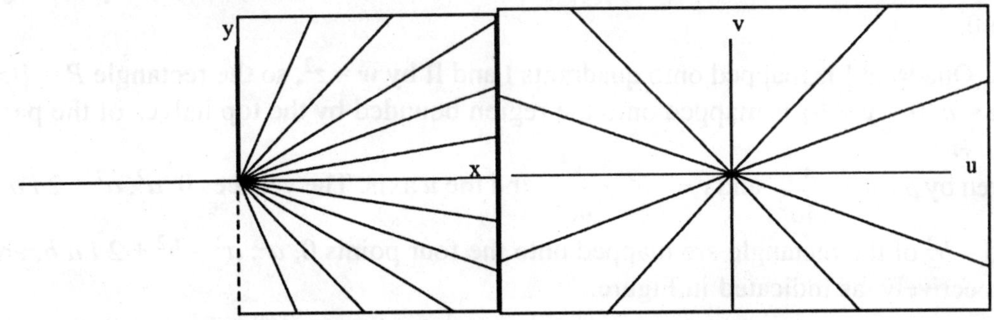

Fig. The Mapping $w = f(z) = z^2$ (and $z = g(w) = w^{1/2}$).

What are the images of rectangles under the mapping $w = z^2$? To find out, we make use of the Cartesian form

$$w = u + i\,v = f(z) = z^2 = x^2 - y^2 + i\,2\,x\,y = (x^2 - y^2, 2\,x\,y) = (u, v)$$

and the resulting system of equations

$$u = x^2 - y^2, \text{ and}$$

$$v = 2\,x\,y.$$

Example: Show that the transformation $w = f(z) = z^2$, usually maps vertical and horizontal lines onto parabolas and use this fact to find the image of the rectangle $R = \{(x, y)\}: 0 < x < a, 0 < x < a, 0 < y < b\}$. (a) Find the image of the vertical line $x = a$. (b) Find the image of the horizontal line y = b.

Solution. Using the above equations, we determine that the vertical line $x = a$ is mapped onto the set of points given by the equations $u = a^2 - y^2$ and $v = 2\,a\,y$. If a $\neq$ 0, then $y = \dfrac{v}{2a}$

and $u = a^2 - \dfrac{v^2}{2a^2}$,

this equation represents a parabola with vertex at $(a^2, 0)$, oriented horizontally, and opening to the left. If $a > 0$, the set $\{(u, v): u = a^2 - y^2, v = 2\,a\,y\}$ has $v > 0$ precisely when $y > 0$, so the part of the line $x = a$ lying above the x axis is mapped to the top half of the parabola.

The horizontal line $y = b$ is mapped onto the parabola given by the equations $u = x^2 - b^2$ and $v = 2\,x\,b$. If $b \neq 0$, then as before we get

$$v = -b^2 + \frac{v^2}{4b^2},$$

this equation represents a parabola with vertex at $-b^2$, oriented horizontally and opening to the right. If $b > 0$, the part of the line $y = b$ to the right of the y axis is mapped to the top half of the parabola because the set $\{(u, v): u = x^2 - b^2, v = 2\,b\,x\}$ has $v > 0$ precisely when $x > 0$.

Quadrant I is mapped onto quadrants I and II by $w = z^2$, so the rectangle $R = \{(x, y): 0 < x < a, 0 < y < b\}$ is mapped onto the region bounded by the top halves of the parabolas given by $u = a^2 - \frac{v^2}{4a^2}$ and $v = -b^2 + \frac{v^2}{4b^2}$ and the u axis. The vertices $0, a^2, a^2 + 2\,i\,a\,b - b^2$, and $-b^2$ of the rectangle are mapped onto the four points $0, a^2, a^2 - b^2 + 2\,i\,a\,b$, and $-b^2$ respectively, as indicated in Figure.

Finally, we can easily verify that the vertical line $\{(0, y): -\infty < y < \infty$ is mapped to the set $\{(-y^2, 0): -\infty < y < \infty\}$. This is simply the set of non-positive real numbers. Similarly, the horizontal line $\{(x, 0): -\infty < x < \infty\}$ is mapped to the set $\{(x^2, 0): -\infty < x < \infty\}$, which is the set of non-negative real numbers.

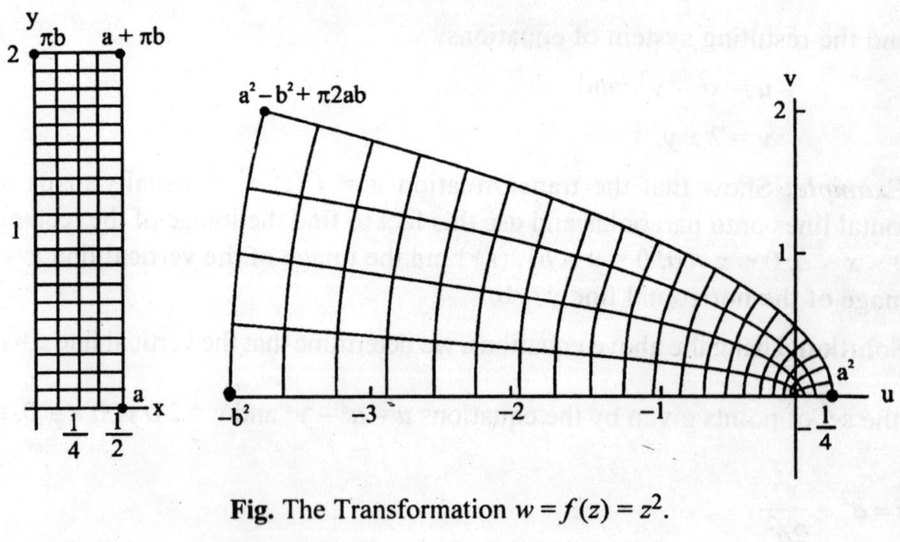

Fig. The Transformation $w = f(z) = z^2$.

Example: Consider the mapping $w = z^2$. Find the image of the ray $r > 0, \theta = \alpha$ and circle $r = c$.

Solution

Here we find the image of the ray $r > 0, \theta = \alpha$ and circle $r = c$.

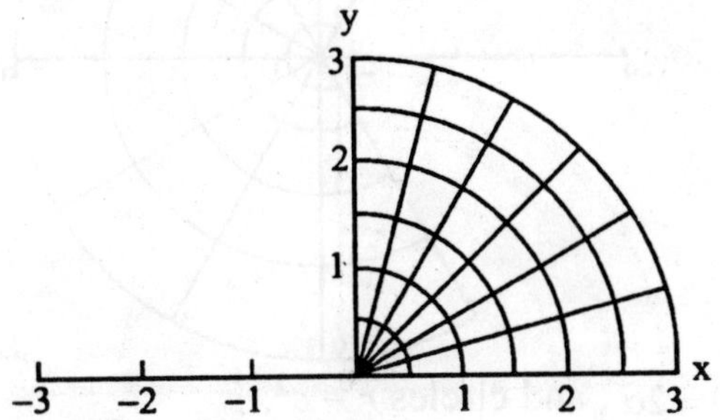

The rays $r > 0, \theta = \alpha$ and circles $r = c$.

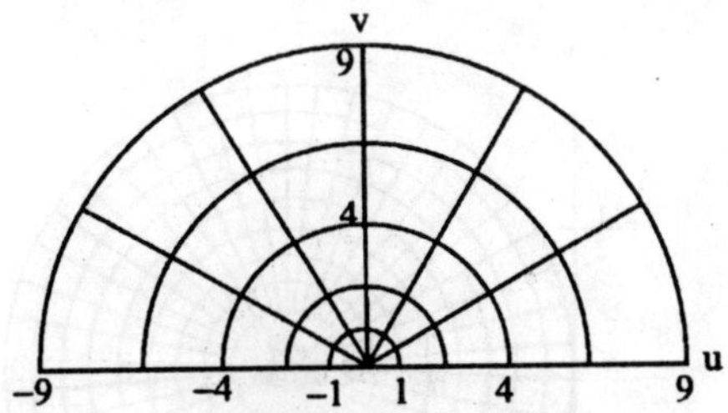

The image rays $r > 0, \theta = 2\alpha$, and circles $r = c^2$, under the mapping $w = z^2$

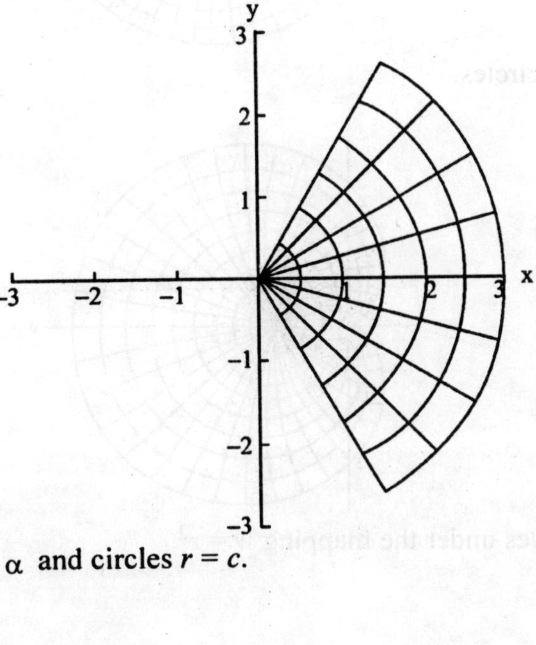

The rays $r > 0, \theta = \alpha$ and circles $r = c$.

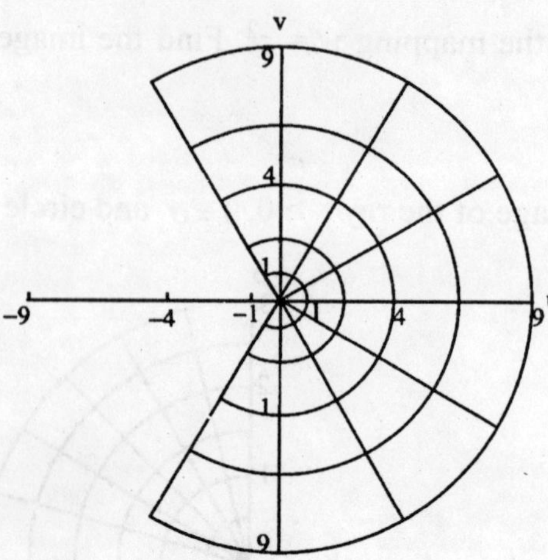

The image $r > 0, \theta = 2\alpha$, and circles $r = c^2$,

under the mapping $w = z^2$

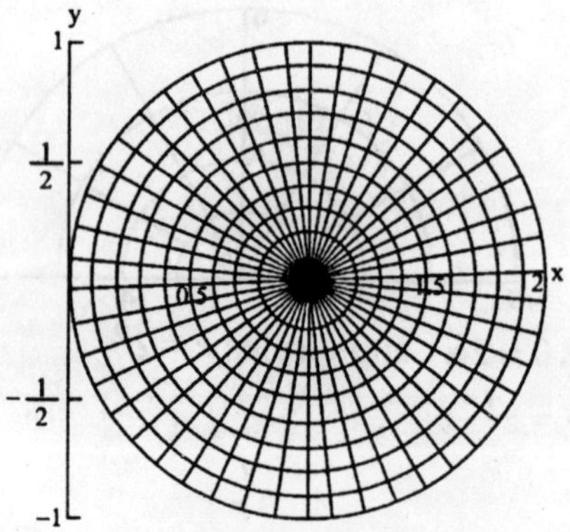

The radii, and circles.

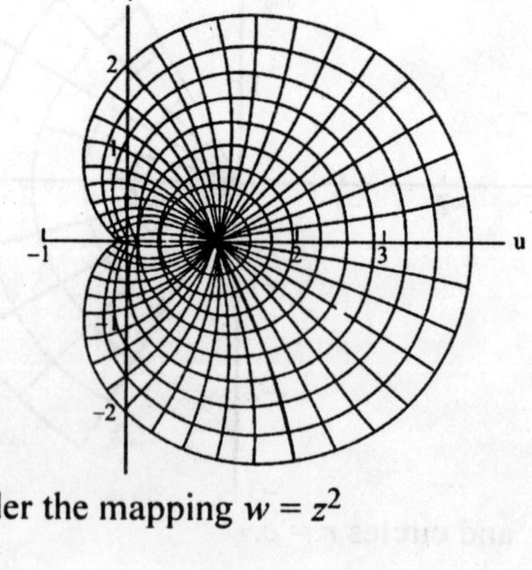

The image curves under the mapping $w = z^2$

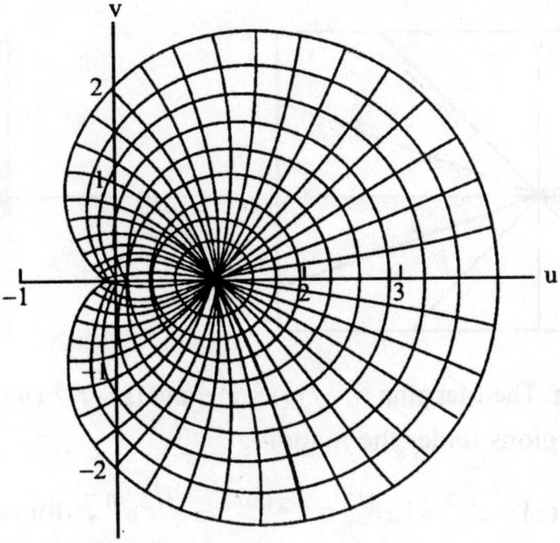

The outer boundary curve is a cardioid.

$(u^2 + v^2 - 2u)^2 = 4(u^2 + v^2)$

(u, v) = (2 cos [t] (1 + cos [t]), 2 sin [t] (1 + cos [t]))

We can use knowledge of the inverse mapping $z = w^2$ to get further insight into how the mapping $w = z^{1/2}$ acts on rectangles. If we let $z = x + iy \neq 0$, then

$$z = x + iy = w^2 = u^2 - v^2 + i\,2uv,$$

and we note that the point $z = x + iy$ in the z plane is related to the point $w = u + iv = z^{1/2}$ in the w-plane by the system of equations

$$x = u^2 - v^2, \text{ and } y = 2uv.$$

Example: The transformation $w = f(z) = z^{1/2}$ usually maps vertical and horizontal lines onto portions of hyperbolas.

- Find the image of the vertical line $x = a$.
- Find the image of the horizontal line $y = b$.

Solution. Let $a > 0$. The above equations map the right half-plane given by Re $(z) > a$ (i.e., $x > a$) onto the region in the right half-plane satisfying $u^2 - v^2 > a$ and lying to the right of the hyperbola $u^2 - v^2 = a$.

If $b > 0$, the equations $x = u^2 - v^2$ and $y = 2uv$ map the upper half-plane Im $(z) > b$ (i.e., $y > b$) onto the region in quadrant I satisfying $2uv > b$ and lying above the hyperbola $2uv = b$. This situation is illustrated in Figure.

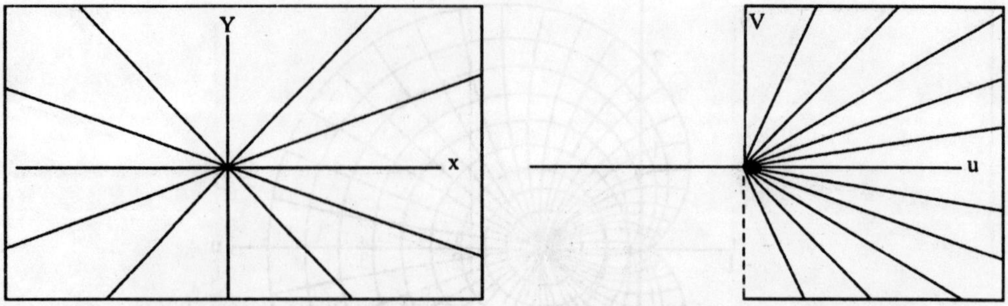

Fig. The Mapping $w = f(z) = z^{1/2}$ and $(z = f^{-1}(w) = w^2)$.

The images of regions under the mapping

$$w = f(z) = z^{1/2} = |z|^{1/2} e^{i\,Arg(z)/2} = r^{1/2} e^{i\theta/2} \text{ for } re^{i\theta} \neq 0,$$

where $-\pi < \theta \leq \pi$. By the use of polar coordinates for $w = \rho e^{i\phi}$ in the w plane, we can represent this mapping by the system

$$\rho = r^{1/2} \text{ and } \phi = \frac{\theta}{2}.$$

This Equation indicate that the argument of $f(z)$ is half the argument of z and that the modulus of $f(z)$ is the square root of the modulus of z. Points that lie on the ray $r > 0, \theta = \alpha$ are mapped onto the ray $\rho > 0, \phi = \frac{\alpha}{2}$. The image of the z plane (with the point $z = 0$ deleted) consists of the right half-plane Re $(w) > 0$ together with the positive v axis, which is shown in Figure 15.

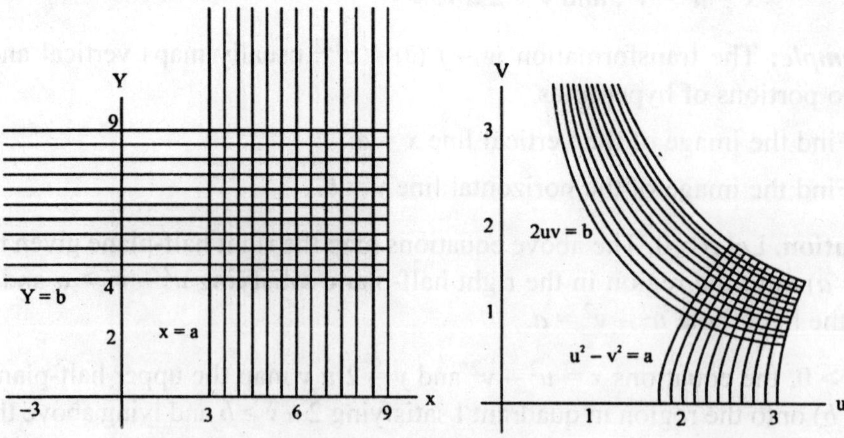

Fig. The mapping $w = f(z) = z^{1/2}$ (and $z = f^{-1}(w) = w^2$).

PRINCIPAL ROOT FUNCTION

The function

$$z = g(w) = w^{1/n} = |w|^{1/n} e^{i\,Arg(\omega)/n}, \text{ for } w \neq 0$$

is called the principal n^{th} root function.

We leave as an exercise to show that f and g are inverses of each other that map the set E one-to-one and onto the set F and the set F one-to-one and onto the set E, respectively. Figure illustrates this relationship.

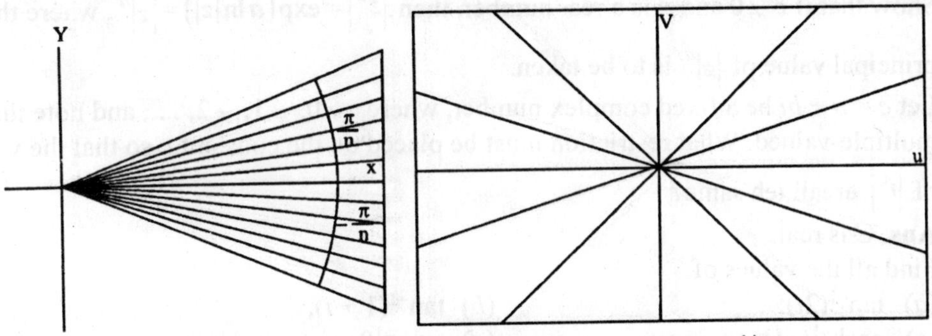

Fig. The Mapping $w = f(z) = z^n$ (and $z = g(w) = w^{1/n}$).

EXERCISE

1. Suppose z and w are complex numbers. What is the curve described by $\gamma(t) = (1-t)w + tz$ $0 \leq t \leq 1$?
2. Find a function γ that describes that part of the curve $y = 4x^3 + 1$ between $x = 0$ and $x = 10$.
3. Find the image of the strip {z: |Rez| < 1} and of the disc {z: |z| < 1} under each of the following mappings:

   (*a*) $w = (1+i)z + 1$ (*b*) $w = 2z^2$

   (*c*) $w = z^{-1}$ (*d*) $w = \left|\dfrac{z+1}{z-1}\right|$

4. A function f(z) is said to be an isometry if $|f(z_1) - f(z_2)| = |z_1 - z_2|$ for every $z_1, z_2 \in C$; in other words, if it preserves distance.

   (*a*) Suppose that f(z) is an isometry. Show that for every a, b $\in$ C with |a| = 1, the function g(z) = af(z) + b is also an isometry.

   (b) Show that the function h(z) = $\dfrac{f(z) - f(0)}{f(1) - f(0)}$ is an isometry with h(0) = 0 and h(1) = 1.

5. Show that when n = 0, ± 1, ± 2, ...

(*a*) $(1+\mathrm{i})\mathrm{i} = \exp\left(-\frac{\pi}{4}+2n\pi\right)\exp\left(\frac{i}{2}ln2\right)$;

(*b*) $(-1)^{1/\pi} = e^{(2n+1)i}$

6. Find the principal value of

(*a*) 1^{i} (*b*) $\left[\frac{e}{2}\left(-1-\sqrt{3i}\right)\right]^{3\pi i}$

(*c*) $(1-\mathrm{i})^{4\mathrm{i}}$

7. Show that if $z \neq 0$ and a is a rea. number, then $\left|z^a\right| = \exp\left(a\ln|z|\right) = |z|^a$, where the principal value of $|z|^a$ is to be taken.

8. Let $c = a + bi$ be a fixed complex number, where $c \neq 0, \neq 1, \neq 2, \ldots$, and note that i^c is multiple-valued. What restriction must be placed on the constant c so that the values of $\left|i^c\right|$ areall teh same?

Ans. C is real.

9. Find all the values of

(*a*) $\tan^{-1}(2i)$; (*b*) $\tan^{-1}(1+i)$;

(*c*) $\cosh^{-1}(-1)$; (*d*) $\tanh^{-1}0$.

Ans. (*a*) $\left(n+\frac{1}{2}\right)\pi + \frac{i}{2}\ln 3\left(n = 0, \pm 1, \pm 2 \ldots\right)$;

(*b*) $n\pi i\ (n = 0, \pm 1, \pm 2\ldots)$.

10. Solve the equation $\sin z = 2$ for z by equating real parts and imaginary parts in that equation;

11. Solve the equation $\cos z = \sqrt{2}$ for z.

12. $f(x) = (z+2)/z$ and C is

(*a*) The semicircle $z = 2e^{i\theta}$ $(0 \leq \theta \leq \pi)$;

(*b*) The semicircle $z = 2e^{i\theta}$ $(\pi \leq \theta \leq 2\pi)$;

(*c*) The semicircle $z = 2e^{i\theta}$ $(0 \leq \theta \leq 2\pi)$; s

13. $f(z) = z - 1$ and C is the arc from $z = 0$ to $z = 2$ consisting of

(*a*) The semicircle $z = 1 + e^{i\theta}$ $(\pi \leq \theta \leq 2\pi)$;

(*b*) The segment $0 \leq x \leq 2$ of the real axis.

14. $f(z)$ is defined by the equations

$$f(x) = \begin{cases} 1 & \text{wheny } < 0, \\ 4 & \text{ywheny} > 0, \end{cases}$$

Chapter 8

Functions of a Complex Variable

ANALYTIC FUNCTIONS

Functions of (x, y) i.e., $f(x,y)$ that depend only on the combination $(x + i\,y)$ are called functions of a complex variable. The combination $(x + i\,y)$ is generally called z, and we can define such functions as z^n, exp(z), sin z and all the standard functions of z as well as of x.A symbol, such as z, which can stand for any one of a set of complex numbers is called a *complex variable*.

They are defined in exactly the same way the only difference being that they are actually complex valued functions, that is, they are vectors in this two dimensional complex number space, each with a real and an imaginary part (or component).

Since we can multiply z by itself and by any other complex number, we can form any polynomial in z and any power series as well.

We define the exponential and sine functions of z by their power series expansions which converge everywhere in the complex plane.

We can produce all the standard functions of a complex variable by the same steps as go to producing standard functions of real variables.

If to each value which a complex variable z can assume there corresponds one or more values of a complex variable w, say that w is a function of z and write $w = f(z) = f(x+iy)$ $= u(x,y)+iv(x,y) = u+iv$.

Single-Valued Function

A single-valued complex function of a complex variable is a complex function $f: C \to C$ that has the same value (unique value) at every point z_0 independent of the path along which it is reached by analytic continuation

Example: If $w = f(z) = z$, then to each value of z there is only one value of w. Hence $w = f(z)$ single-valued function.

Multiple-Valued Function

A multiple-valued function, also known as a multiple-valued function, is a "function" that assumes two or more distinct values in its range for at least one point in its domain. While these "functions" are, not functions in the normal sense of being one-to-one or many-to-one.

The trigonometric, hyperbolic, exponential, and integer power functions are all single-valued functions, their inverses are multiple-valued.

Example: If $w = z^{\frac{1}{2}}$, then to each value of z there are two values of w. Hence $w = f(z) = z^{\frac{1}{2}}$ is a multiple-valued function.

The discontinuities of multiple-valued functions in the complex plane are commonly handled through the adoption of branch cuts, but use of Riemann surfaces is another possibility.

Inverse Functions

If $w = f(z)$, then there exist the function f^{-1} is often called the inverse function corresponding to f. Thus $w = f(z) =$ and $w = f^{-1}(z)$ are inverse function of each other.

MAPPING

If $w = u + i\,v$ (where u and v are real) is a single-valued function of $z = x + i\,y$, we can write $w = f(z) = u + i\,v = f(x + i\,y)$. By equating real and imaginary parts we get,

$u = u\,(x, y), v = v\,(x, y).$

Thus, for given a point P(x, y) in the z plane such as P in Figure there corresponds a point P'(u, v) in the z plane. We say that z is mapped or transformed into (u, v) by means of the mapping (transformation) and call P' the image of P.

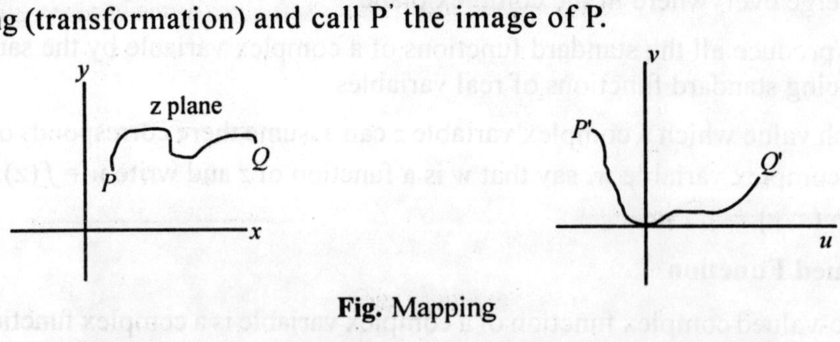

Fig. Mapping

LIMIT AND CONTINUITY OF COMPLEX NUMBER

COMPLEX NUMBER SYSTEM

The complex plane with metric $d(x,y) := |x - y|$ is also a metric space. There are two different types of limits when we consider complex-valued functions.

LIMIT OF A FUNCTION AT A POINT

Suppose f is a complex-valued function, then we write

$$\lim_{x \to p} f(x) = L$$

if and only if

for every $\varepsilon > 0$ there exists a $\delta > 0$ such that for all real numbers x with $0<|x-p|<\delta$, we have $|f(x)-L| < \varepsilon$

It is just a particular case of functions over metric spaces with both M and N are the complex plane.

LIMIT OF A FUNCTION OF MORE THAN ONE VARIABLE

By noting that $|x-p|$ represents a distance, the definition of a limit can be extended to functions of more than one variable.

$$\lim_{(x,y)\to(p,q)} f(x,y) = L$$

if and only if

for every $\varepsilon > 0$ there exists a $\delta > 0$ such that for all real numbers x with $0 < \|(x, y)-(p,q)\| < \delta$, we have $|f(x, y)-L| < \varepsilon$

where $\|(x, y)-(p, q)\|$ represents the Euclidean distance. This can be extended to any number of variables.

FUNCTIONS OF A COMPLEX VARIABLE

The real excitement begins when we consider function $f: D \to C$ in which the domain D is a subset of the complex numbers. In some sense, these too are familiar to us from elementary calculus—they are simply functions from a subset of the plane into the plane:

$$f(z) = f(x,y) = u(x,y) + iv(x,y) = (u(x,y), y(x,y))$$

Thus $f(z) = z^2$ looks like $f(z) = z^2 = (x + iy)^2 = x^2 - y^2 + 2xyi$. In other words, $u(x,y) = x^2 - y^2$ and $v(x,y) = 2xy$. The complex perspective, as we shall see, generally provides richer and more profitable insights into these function.

The definition of the limit of a function f at a point $z = z_0$ is essentially the same as that which we learned in elementary calculus:

$$\lim_{z \to z0} f(z) = L$$

means that given and $\varepsilon > 0$, there is a δ so that $|f(z) - L| < e$ whenever $0 < |z - z_0| < \delta$. As you could guess, we say that is continuous at z_0 if it is true $\lim_{z \to z0} f(z) = f(z_0)$. If f is continuous at each point of its domain, we say simply that f is continuous.

PROPERTIES OF LIMITS

Suppose both $\lim_{z \to z0} f(z)$ and $\lim_{z \to z_0} g(z)$ exist. Then the following properties are easy to establish:

$$\lim_{z\to z_0}[f(z)\pm g(z)] = \lim_{z\to z_0} f(z)\pm \lim_{z\to z_0} g(z)$$

$$\lim_{z\to z_0} f(z)g(z) = \lim_{z\to z_0} f(z)\lim_{z\to z_0} g(z)$$

and, $$\lim_{z\to z_0}\frac{f(z)}{g(z)} = \frac{\lim_{z\to z_0} f(z)}{\lim_{z\to z_0} g(z)}$$

provided, of course, that $\lim_{z\to z_0} g(z) \neq 0$.

It now follows at once from these properties that the sum, difference, product, and quotient of two functions are continuous at z_0 are z_0. (We must, as usual, except the dereaded 0 in the denominator.)

It should not be too difficult to convince yourself that if $z = (x,y)$, $z_0 = (x_0,y_0)$, and $f(z) = u(x,y) + iv(x,y)$, then

$$\lim_{z\to z_0} f(z) = \lim_{(x,y)\to(x_0,y_0)} u(x,y) + i \lim_{(x,y)\to(x_0,y_0)} v(x,y)$$

Thus f is continuous at $z_0 = (x_0,y_0)$ precisely when u and v are.

Let $u = u(x,y)$ be a real-valued function of the two real variables x and y. Recall that u has the limit u_0 as (x,y) approaches (x_0, y_0) provided that the value of u (x,y) can be made to get as close as we please to the value u_0 by taking (x, y) to be sufficiently close to (x_0, y_0). When this happens we write

$$\lim_{(x,y)\to(x_0,y_0)} u(x, y) = u_0.$$

In more technical language, u has the limit u_0 as (x,y) approaches (x_0, y_0) iff $|u(x, y) - u_0|$ can be made arbitrarily small by making both $|x-x_0|$ and $|y-y_0|$ small. This condition is like the definition of a limit for functions of one variable. The point (x, y) is in the x-y plane, and the distance between (x, y) and (x_0, y_0) is $\sqrt{(x-x_0)^2+(y-y_0)^2}$. With this perspective we can now give a precise definition of a limit.

LIMIT OF U(*X*, *Y*)

The expression $\lim_{(x,y)\to(x_0,y_0)} u(x, y)$ means that for each number $\in > 0$, there corresponds a number $\delta > 0$ such that

$|u(x, y) - u_0| < \in$ whenever

$0 < \sqrt{(x-x_0)^2+(y-y_0)^2} < \delta$.

Example: Show, if $u(x, y) = \dfrac{2x^2}{x^2+y^2}$,

then $\lim_{(x,y)\to(0,0)} u(x, y) = 0$.

Solution. If $x = r\cos\theta$ and $y = r\sin\theta$, and $z = x + iy \neq 0$ then

$$u(x, y) = \frac{2r^3 \cos\theta^3\theta}{r^2 \cos^2\theta + r^2 \sin^2\theta} = 2r\cos^3\theta.$$

Because $\sqrt{(x-0)^2 + (y-0)^2} = r$ and

because $|\cos^3\theta| < 1$, we have

$|u(r, \theta)| = 2r|\cos^3\theta < \in$ whenever

$$\sqrt{(x-0)^2 + (y-0)^2} = r < \frac{\in}{2}.$$

Hence, for any $\in > 0$, Inequality is satisfied for $\delta = \frac{\in}{2}$; that is, u (x, y) has the limit $u_0 = 0$ as (x, y) approaches (0, 0).

Example: Show that the function u (*x*, *y*) $\frac{xy}{x^2 + y^2}$ does not have a limit as (*x*, *y*) approaches (0, 0).

Solution. If we let (x,y) approach (0,0) along the x axis, then

$$\lim_{(x,0)\to(0,0)} u(x, y) = \lim_{(x,0)\to(0,0)} \frac{(x)(o)}{x^2 + o^2} = 0$$

But if we let (x,y) approach (0, 0) along the line y = x, then

$$\lim_{(x,y)\to(0,0)} u(x, y) = \lim_{(x,x)\to(0,0)} \frac{(x)(x)}{x^2 + x^2} = \frac{1}{2}$$

Because the value of the limit differs depending on how (x,y) approaches (0, 0), we conclude that u (x, y) does not have a limit as (*x*, *y*)approaches (0, 0).

Let f(z) be a complex function of the complex variable z that is defined for all values of z in some neighborhood of z_0, except perhaps at the point z_0. We say that f has the limit w_0 as z approaches z_0 provided that the value f(z) can be made as close as we please to the value w_0 by taking z to be sufficiently close to z_0. When this happens we write

$\lim_{x\to z_0} f(z) = w_0$

The distance between the points z and z_0 can be expressed by $|z - z_0|$, so we can give a precise definition similar to the one for a function of two variables.

LIMIT OF *F(Z)*

The expression $\lim_{z\to z_0} f(z) = w_0$ means that for each number $\in > 0$, there exists a real number $\delta > 0$ such that

$|f(z) - w_0| < \in$ whenever $O < |z - z_0| < \delta$.

We can also express the above relationship as

$f(z) \in D_\in(w_0)$ whenever $z \in D^*_\delta(z_0)$.

The formulation of limits in terms of open disks provides a good context for looking at this definition. It says that for each disk of radius $\in > 0$ about the point w_0 (represented by $D_\in(w_0)$ there is a punctured disk of radius $\delta > 0$ about the point z_0 (represented by $D^*_\delta(z_0)$ such that the image of each point in the punctured δ-disk lies in the $\in$-disk. The image of the δ-disk does not have to fill up the entire $\in$-disk; but if z approaches z_0 along a curve that ends at z_0, then w = f(z) approaches w_0.

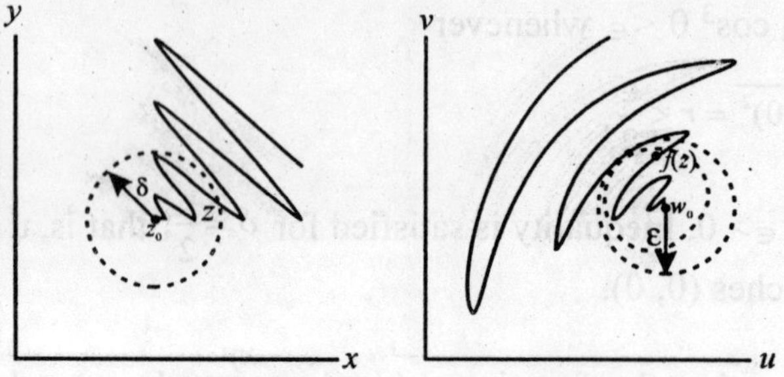

Fig. The limit f (z)→ w_0 as $z \to z_0$

Example: Show that if $f(z) = \bar{z}$, then

$\lim_{z \to z_0} f(z) = \bar{z}_0$, where z_0 is any complex number.

Solution. As f merely reflects points about the y axis, we suspect that any $\in$-disk about the point $\bar{z}_0$ would contain the image of the punctured δ-disk about z_0 if $\delta = \in$. To confirm thisconjecture, we let $\in$ be any positive number and set $\delta = \in$. Then we suppose that $z \in D^*_\delta(z_0) = D^*_\in(z_0)$, which means that $0 <| \overline{z - z_0} < \in$. The modulus of a conjugate is the same as the modulus of the number itself, so the last inequality implies that $0 <| \overline{z - z_0} | < \in$. This is the same as $0 <| \bar{z} - \bar{z}_0 | < \in$. Since $f(z) = \bar{z}$ and $w_0 = \bar{z}_0$, this is the same as $0 <| f(z) - w_0 | < \in$, which in turn is the same as $f(z) \in D_\in(w_0)$, which is what we needed to show.

Theorem: Let f (z) = u (x, y) + ṅ v(x, y) be a complex function that is defined in some neighborhood of z_0, except perhaps at $z_0 = x_0 + ṅy_0$. Then

$\lim_{z \to z_0} f(z) = w_0 = u_0 + ṅv_0$

iff both

$\lim_{(x,y) \to (x_0,y_0)} u(x,y) = u_0$ and $\lim_{(x,y) \to (x_0,y_0)} v(x,y) = v_0$

Example: Show that $\lim_{z \to 1+i} (z^2 - 2z + 1) = -1$

Solution. We have

$$f(z) = 1 - 2z + z^2$$

$$f(x+ṅy) = (x^2 - y^2 - 2x + 1) + ṅ(2xy - 2y)$$

$$u(x, y) = x^2 - y^2 - 2x + 1$$

$$v(x, y) = 2xy - 2y$$

Computing the limits for u and v, we obtain

$\lim_{(x,y)\to(1,1)} u(x,y) = \lim_{(x,y)\to(1,1)}(x^2 - y^2 - 2x + 1) = -1$

and

$\lim_{(x,y)®(1,1)} v(x,y) = \lim_{(x,y)\to(1,1)}(2xy) - 2y = 0$,

so our previous theorem implies that $\lim_{z\to 1+i} f(z) = -1$.

Limits of complex functions are formally the same as those of real functions, and the sum, difference, product, and quotient of functions have limits given by the sum, difference, product, and quotient of the respective limits. We state this result as a theorem and leave the proof as an exercise.

Theorem: Suppose that $\lim_{z\to z_0}$ f(z) = A and $\lim_{z\to z_0} g(z) = B$Then

- $\lim_{z\to z_0} f(z)+g(z) = A + B$, *and*
- $\lim_{z\to z_0} f(z) - g(z) = A - B$,
- $\lim_{z\to z_0} f(z)\, g(z) = AB$,
- $\lim_{z\to z_0} \dfrac{f(z)}{g(z)} = \dfrac{A}{B}$ *where* $B \neq 0$.

CONTINUITY

The graph of *f* is unbroken at *a* i.e., it's continuous there. It has a hole at *b* i.e., it's broken, or discontinuous (ie, not contnuous) there; we notice that *f*(*b*) doesn't exit. The graph jumps abruptly at *c*; it's discontinuous there; we notice that *f*(*c*) exists but $\lim_{x\to c} f(x)$ doesn't exist (both $\lim_{x\to c+} f(x)$ and $\lim_{x\to c-} f(x)$) exist but aren't equal). The graph has a hole at *d*; it's discontinuous there; we notice that *f*(*d*) exists and $\lim_{x\to d} f(x)$ exists, but they aren't equal.

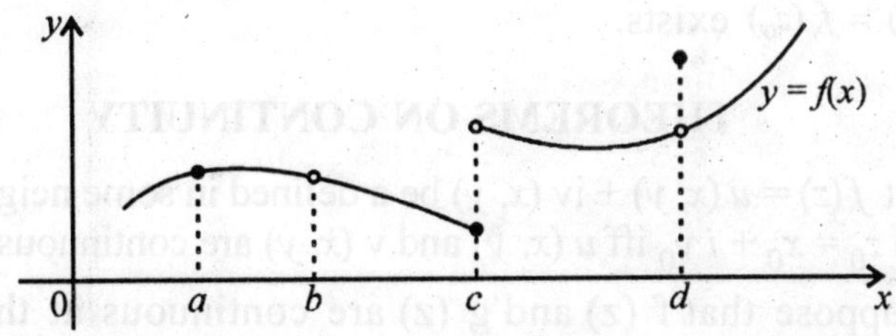

Fig. *f* is Continuous at *a*, discontinuous at *b*, *c*, and *d*.

This means that:

$\lim_{x\to a} f(x)$, exists and equals $f(a)$.

All that's just been examined forms the basis for the definition of continuity.

Definition of Continuity: Three conditions for continuity of *f*(*x*) at a point $x = c$:

- The function must be defined at *c*, or *f*(*c*) must be defined.

- $\lim_{x \to c} f(x)$ must exist.
- $\lim_{x \to c} f(x) = f(c)$

CONTINUITY OF *U(X, Y)*

Let u(x,y) be a real-valued function of the two real variables x and y. We say that u is continuous at the point (x_0, y_0) if the three conditions are satisfied:

- $\lim_{(x,y)\to(x_0,y_0)} u(x,y)$ exists.
- $u(x_0, y_0)$ exists,
- $\lim_{(x,y)\to(x_0,y_0)} u(x,y) = u(x_0 y_0)$

Condition (iii) actually implies Conditions (i) and (ii) because the existence of the quantity on each side of Equation (iii) is implicitly understood to exist. For example, if u $(x, y) = \frac{x^2}{x^2 + y^2}$ when $(x, y) \neq (0, 0)$ and if u (0, 0)= 0, then $\lim_{(x, y)\to(0,0)} u(x, y) = 0$ so that Conditions (i), (ii), and (iii) are satisfied. Hence u (x, y) is continuous at (0, 0).

There is a similar definition for complex valued functions.

CONTINUITY OF COMPLEX VALUED FUNCTION

Definition: (Continuity of *f*(*z*)): Let f(z) be a complex function of the complex variable z that is defined for all values of z in some neighborhood of z_0. We say that f is continuous at z_0 if three conditions are satisfied:

- $\lim_{z \to z_0} f(z)$ exists,
- $f(z_0)$ exists,
- $\lim_{z \to z_0} f(z) = f(z_0)$ exists.

THEOREMS ON CONTINUITY

Theorem: Let $f(z) = u(x, y) + iv(x, y)$ be a defined in some neighborhood of z_0. Then $f(z)$ is continuous at $z_0 = x_0 + i y_0$ iff $u(x, y)$ and $v(x, y)$ are continuous at (x_0, y_0).

Theorem: Suppose that f (z) and g (z) are continuous at the point z_0. Then the following functions are continuous at z_0.

- The sum $f(z) + g(z)$,
- The difference $f(z) - g(z)$,
- The product $f(z)\, g(z)$,
- The quotient $\frac{f(z)}{g(z)}$, provided that $g(z_0) \neq 0$.

· The composition f (g(z)), provided that f (z) is continuous in a neighborhood of the point g (z_0).

Example: Show that the polynomial p (z) = $1 - z - z^2 + z^3 - z^4 + z^5$ is continuous at the point $z_0 = 1$ in the complex plane.

Solution. We have p(z) = $1 - z - z^2 + z^3 - z^4 + z^5$ and find the limit.

$$p(z) = 1 - z - z^2 + z^3 - z^4 + z^5$$

$$p(1) = 0$$

$$\lim_{z\to 1} p(z) = \lim_{x\to 1} (1 - z - z^2 + z^3 - z^4 + z^5)$$

$$\lim_{z\to 1} p(z) = 0$$

Therefore, f(z) is continous at z0 = 1.

Example: Show that the polynomial p (z) = $1 - z - z^2 + z^3 - z^4 + z^5$ is continuous at the point $z_0 = \frac{1}{2}$ in the complex plane.

Solution. We have, p (z) = $1 - z - z^2 + z^3 - z^4 + z^5$ and find the limit.

$$p(z) = 1 - z - z^2 + z^3 - z^4 + z^5$$

$$p\left[\frac{1}{2}\right] = \frac{11}{32}$$

$$\lim_{z\to \frac{1}{2}} p(z) = \lim_{z\to \frac{1}{2}} (1 - z - z^2 + z^3 - z^4 + z^5)$$

$$\lim_{z\to \frac{1}{2}} p(z) = \frac{11}{32}$$

Therefore, f(z) is continous $z_0 = \frac{1}{2}$

If we let P and Q be polynomials and if $Q(z_0) \neq 0$, then

$$\lim_{z\to z_0} \frac{P(z)}{Q(z)} = \frac{p(z_0)}{Q(z_0)}.$$

Another technique, involves factoring polynomials. If both p (z_0) = 0 and Q (z_0) = 0, then P and Q can be factored as

$$p(z) = (z - z_0)\, p_1(z)$$

and Q (z) = (z − z_0) Q_1 (z). If Q (z_0) ≠ 0

Q (z) = (z − z_0) Q_1 (z). If Q (z_0) ≠ 0, then the limit is

$$\lim_{z\to z_0} \frac{P(z)}{Q(z)} = \frac{(z - z_0)P_1(z)}{(z - z_0)Q_1(z)} = \frac{P_1(z_0)}{Q_1(z_0)}.$$

Example: Show that $\lim_{z\to 1+i} \frac{z^2 - 2i}{z^2 - 2z + 2} = 1 - i$

Solution. Here P and Q can be factored in the form

$$P(z) = (z-1-i)(z+1+i),$$

and

$$Q(z) = (z-1-i)(z-1+i),$$

so that the limit is obtained by the calculation

$$\lim_{x\to 1+i} \frac{z^2-2i}{z^2-2z+2} = \lim_{z\to 1+i} \frac{(z-1-i)(z+1+i)}{(z-1-i)(z-1+i)}$$

$$= \lim_{z\to 1+i} \frac{z+1+i}{z-1+ia}$$

$$= \frac{(1+i)+1+i}{(1+i)-1+i}$$

$$= \frac{2+2i}{2i} = 1-i$$

DIFFERENTIATION OF COMPLEX NUMBER

Earlier we have learnt about the derivative of real valued functions, here we study the derivative of complex-valued functions.A complex function is a function in which the independent variable and the dependent variable are both complex numbers. More precisely, a complex function is a function whose domain Ω is a subset of the complex plane and whose range is also a subset of the complex plane.

For any complex function, both the independent variable and the dependent variable may be separated into real and imaginary parts:

$z = x + iy$ and $w = f(z) = f(x+iy) = u(x,y) + iv(x,y) = u + iv$

where $x,\ y \in R$ and $u(z), v(z)$ are real-valued functions.

In other words, the components of the function *f*(z), $u(x,y)$ and $v(x,y)$ can be interpreted as real valued functions of the two real variables, *x* and *y*.

THE DERIVATIVE OF A FUNCTION OF A COMPLEX VARIABLE

Let *f* be a function whose domain of definition contains a neigbourhood of a point z_0.

The derivative of *f* at z_0 written $f'(z_0)$ is defined by the equation

$$f'(z_0) = \lim_{z\to z_0} \frac{f(z)-f(z_0)}{z-z_0},$$

provided this limits exists. The function *f* is said to be differentiable at z_0 when its derivable at z_0 exists.

To begin with, the usual calculus definition of a derivative can be applied to complex functions just as well as to real functions. It is just that the norm $|\Delta z|$ of an increment must be sent to zero, rather than the absolute value of a real difference, $|\Delta x|$.

$$\frac{d}{dz}f(z)=c\lim_{|\Delta z|\to 0}\frac{f(z+\Delta z)-f(z)}{\Delta z},$$

supposing that the limit exists.

From this, the usual rules governing the derivatives of sums, products, differences and quotients can be deduced. Derivatives of powers can be obtained from the binomial theorem

$$(z+\Delta z)^n = \sum_{i=0}^{n}\frac{n!}{i!(n-i)!}z^{n-i}\Delta z^i$$

in the usual way with the result

$$\frac{d}{dz}z^n = n\, z^{n-1}$$

The chain rule for composite functions is applicable, so that the derivatives of inverse functions can be obtained. For the exponenial, the derivative of products and the definition holding that

$$e^z = \lim_{x\to\infty}\left(1+\frac{z}{n}\right)^n$$

SUFFICIENT CONDITIONS FOR DIFFRENTIABILITY

Theorem: Let the function

$f(z)=u(x,y)+iv(x,y)$

be defined throughout some ε neighborhood of a point $z_0 = x_0 + iy_0$. Suppose that the first-order partial derivatives of the functions u and v with respect to x and y exist everywhere in that neighborhood and that they are continuous at (x_0, y_0). Then if those partial derivatives satisfy the Cauchy-Riemann equations

$u_x = v_y,\ u_y = -v_x$

at (x_0, y_0), the derivative $f'(z_0)$ exists.

To check that the derivative is well-defined, separate the complex function into the sum of two real functions, just as z can be written as the sum of a real and an imaginary part:

$$z = x + i\,y\, f = u(x,y) + i\, v(x,y).$$

By first taking Δz real,

$$\frac{df}{dz}=\lim_{\Delta z\to 0}\frac{f(z+\Delta x)}{\Delta x}$$

$$=\lim_{\Delta x\to 0}\frac{u(x+\Delta x,y)+iv(x+\Delta x,y)-(u(x,y)+iv(x,y))}{\Delta x}$$

$$= \underset{\Delta x \to 0}{lt} \ \underset{\Delta x \to 0}{lt} \frac{u(x+\Delta xy)-u(x,y)}{\Delta x + iu}$$

$$\underset{\Delta x \to 0}{lt} \frac{(x+\Delta x, y)-u(x,y)}{\Delta x}$$

$$f'(t) = \frac{\partial u}{\partial x} + i\frac{\partial v}{dx},$$

and then imaginary,

$$\frac{df}{dz} = \lim_{\Delta z \to 0} \frac{f(z+i\Delta y)}{i\Delta y}.$$

$$= \lim_{\Delta y \to 0} \frac{u(x,y+\Delta y)+iv(x,y+\Delta y)-(u(x,y)+iv(x,y))}{i\Delta y}$$

$$\underset{\Delta y \to 0}{lt} \frac{u(x,y+\Delta y)-u(x,y)}{i\Delta y} + \underset{\Delta y \to 0}{lt} \frac{u(x,y+\Delta y)-u(x,y)}{\Delta y}$$

$$f'(t) = -i\frac{\partial u}{\partial y} + \frac{dv}{dy}.$$

CAUCHY RIEMAN EQUATIONS

Supposing that the derivative is independent of phase, at least in the context taken here, comparison shows

$$\frac{\partial u}{\partial x} = \frac{\partial v}{\partial y}$$

$$\frac{\partial v}{\partial x} = \frac{\partial u}{\partial y}$$

The members of this pair of equations are known as the Cauchy - Riemann equations, relating the real and imaginary parts of a complex function which is supposed to have a derivative.

Consider the Jacobian matrix for the real functions $u(x,y)$ and $v(x,y)$ of the real variables x and y,

$$\frac{d(u,v)}{d(x,y)} = \begin{bmatrix} \frac{\partial u}{\partial x} & \frac{\partial v}{\partial x} \\ \frac{\partial u}{\partial y} & \frac{\partial v}{\partial y} \end{bmatrix} = \begin{bmatrix} \frac{\partial u}{\partial x} & \frac{\partial v}{\partial x} \\ -\frac{\partial v}{\partial x} & \frac{\partial v}{\partial y} \end{bmatrix}$$

$$= \frac{\partial u}{\partial x}1 + \frac{\partial v}{\partial x}i$$

The impact of the Cauchy-Riemann equations is to give the Jacobian matrix the form of a complex number in quaternion.

By knowing one of the pair, one effectively knows the other. That is,

$$v(x,y) - v(a,y) = \int_a^x \frac{\partial v(\sigma,y)}{\partial \sigma} d\sigma$$
$$= \int_a^x \frac{\partial u(\sigma,y)}{\partial y} d\sigma$$

Knowing u, take its derivative and integrate to get v. For example, according to Euler's formula,

$$e^z = e^x(\cos y + i \sin y)$$

Then

$$u = e^x \cos y$$

$$\frac{\partial u}{dy} = -e^x \sin y$$

$$-\int_a^b e^\sigma \sin y d\sigma = e^\sigma \sin y\Big|_a^x$$

$$v(x,y) - v(a,y) = e^x \sin y - e^a \sin a$$

which is consistent with

$$v(x,y) = e^x \sin y$$

So trying to think of something like $e^x(\cos y + i \tan y)$, as an analytic function just wouldn't work.

HARMONIC FUNCTIONS OF REAL VARIABLES

Statement: If D is region in the plane, a real-valued function $u(x,y)$ having continuous second partial derivatives is said to be harmonic on D if it satisfies Laplace's equation on D:

$$\frac{\partial^2 u}{\partial x^2} + \frac{\partial^2 u}{\partial y^2} = 0.$$

There is an intimate relationship between harmonic functions and analytic functions. Suppose f is analytic on D, and let $f(z) = u(x,y) + iv(x,y)$. Since, we know

$$\frac{\partial u}{\partial x} = \frac{\partial v}{\partial y}$$

and , $$\frac{\partial u}{\partial y} = \frac{\partial v}{\partial x}.$$

If we differentiate the first and the second with respect to x and respectively y, and then add the two results, we have

$$\frac{\partial^2 u}{\partial x^2}+\frac{\partial^2 u}{\partial y^2}=\frac{\partial^2 y}{\partial x dy}-\frac{\partial^2 y}{\partial y \partial x}=0.$$

Thus the real part of any analytic function is harmonic! Next, if we differentiate the first of the Cauchy-Riemann equations with respect to y and the second with respect to x, and then substract the second from the first, we have

$$\frac{\partial^2 v}{\partial x^2}+\frac{\partial^2 v}{\partial y^2}=0,$$

and we see that the imaginary part of an analytic function is also harmonic.

There is even more excitement. Suppose we are given a function φ harmonic in a simply connected region D. Then there is a function f analytic on D which is such that Ref $=\varphi$. First, define g by

$$g(z)=\frac{\partial \varphi}{\partial x}-i\frac{\partial \varphi}{\partial y}.$$

We'll show that g is analytic by verifying that the real and imaginary parts satisfy the Cauchy-Riemann equations:

$$\frac{\partial}{\partial x}\left(\frac{\partial \varphi}{\partial x}\right)=\frac{\partial^2 \varphi}{\partial x^2}=-\frac{\partial^2 \varphi}{\partial y^2}=\frac{\partial}{\partial y}\left(-\frac{\partial \varphi}{\partial y}\right),$$

since φ is harmonic. Next,

$$\frac{\partial}{\partial y}\left(\frac{\partial \varphi}{\partial x}\right)=\frac{\partial^2 \varphi}{\partial y \partial x}=\frac{\partial^2 \varphi}{\partial x \partial y}=-\frac{\partial}{\partial x}\left(-\frac{\partial \varphi}{\partial y}\right).$$

since g is analytic on the simply connected region D, we know that the intgral of g around any closed curve is zero, and so it has an antiderivative $G(z) = u + iv$. This antiderivative is, of course, analytic on D, and we know that

$$G'(z)=\frac{\partial u}{\partial x}-i\frac{\partial u}{\partial y}=\frac{\partial \varphi}{\partial x}-i\frac{\partial \varphi}{\partial y}.$$

Thus, $u(x, y) = \varphi(x, y)\, h(y)$. From this,

$$\frac{\partial u}{\partial y}=\frac{\partial \varphi}{\partial y}+h'(y),$$

and so $h'(y) = 0$, or h = constant, from which it follows that $u(x, y) = \varphi(x, y) + c$. In other words, Re $G = u$, as we promised to show.

An important practical consequence is that such functions never have maxima nor minima. By writing the second derivatives in finite difference form

$$\frac{d^2 f}{dx^2} = \frac{f(x-\Delta x) - 2f(x) + f(x+\Delta x)}{\Delta x^2}$$

so that

$$u(x,y) = \frac{1}{4}\{u(x+\Delta x,y) + u(x-\Delta x,y) + u(x,y+\Delta y) + u(x,y+\Delta y)\}$$

or, in other words, each central value is the average of its neighbors. No average can exceed all the values it is averaging, nor fall below all of them, either.

Example: The Function $\varphi(x,y) = x^3 - 3xy^2$ is harmonic everywhere.

Solution. We shall find function G that Re $G = \varphi$. We know that $G(z) = x^3 - 3xy^2 + iv$, and so from the Cauchy-Riemann equations:

$$\frac{\partial v}{\partial x} = \frac{\partial u}{\partial y} = 6xy$$

Hence,

$$v(x,y) = 3x^2 y + k(y).$$

To find $k(y)$ differentiate with respect to y:

$$\frac{\partial v}{\partial y} = 3x^2 + k'(y) = \frac{\partial u}{\partial x} = 3x^2 - 3y^2,$$

and so,

$$k'(y) = -3y^2, \text{ or}$$

$$k(y) = -y^3 + \text{any constant.}$$

If we choose the constant to be zero, this gives us

$$v = 3x^2 y + k(y) = 3x^2 y - y^3,$$

and finally,

$$G(z) = u + iv = (x^3 - 3xy^2) + i(3x^2 y - y^3).$$

THE SCHWARTZ DERIVATIVE

Consider the basic linear fractional transformation

$$w = \frac{az+b}{cz+d}$$

which is equivalent to the vanishing of the implicit expression

$$cwz + dw - az - b = 0.$$

To get a relation between w and z which does not depend on a, b, c and d, take three derivatives with respect to a still unspecified parameter:

$$c(wz)' + dw' - az' = 0$$

$$c(wz)'' + dw'' - az'' = 0$$

$$c(wz)''' + dw''' - az''' = 0.$$

In equivalent matrix form,

$$\begin{bmatrix} (wz)' & w' & -z' \\ (wz)'' & w'' & -z'' \\ (wz)''' & w''' & -z''' \end{bmatrix} \begin{bmatrix} c \\ d \\ a \end{bmatrix} = \begin{bmatrix} 0 \\ 0 \\ 0 \end{bmatrix}.$$

The determinant of this matrix, which is a Wronskian, must vanish to get a nontrivial solution. But

$$\begin{vmatrix} (wz)' & w' & -z' \\ (wz)'' & w'' & -z' \\ (wz)''' & w''' & -z''' \end{vmatrix}$$

$$= \begin{vmatrix} w'z + wz' & w' & -z' \\ w''z + 2w'z' + wz'' & w'' & -z'' \\ w'''z + 3w''z' + 3w'z'' + wz''' & w''' & -z''' \end{vmatrix}$$

Operate $C_1, (-z)z + w3$, we get

$$= \begin{vmatrix} 0 & w' & -z' \\ 2w'z' & w'' & -z'' \\ 3w''z' + 3w'z'' & w''' & -z''' \end{vmatrix}$$

Expowding along first raw, we get

$$= -w'z' \begin{vmatrix} 0 & w' & -z' \\ 2 & w'' & -z'' \\ 3\dfrac{w''}{w'} + 3\dfrac{z''}{z'z} & w''' & -z''' \end{vmatrix}$$

$$= -w'z' \left\{ -w' \begin{vmatrix} 2 & z'' \\ 3\left(\dfrac{w''}{w'} + \dfrac{z''}{z'}\right) & z''' \end{vmatrix} - z' \begin{vmatrix} 2 & w'' \\ 3\left(\dfrac{w''}{w'} + \dfrac{z''}{z'}\right) & w''' \end{vmatrix} \right\}$$

$$= -w'z' \left\{ -w' \left(2z''' - 3z'' \left(\frac{w''}{w'} + \frac{z''}{z'} \right) \right) + z' \left(2w''' - 3w'' \left(\frac{w''}{w'} + \frac{z''}{z'} \right) \right) \right\}$$

$$= -w'z' \left\{ -2w'z''' + 3w'z'' \left(\frac{w''}{w'} + \frac{z''}{z'} \right) + 2z'w''' - 3w'' \left(\frac{w''}{w'} + \frac{z''}{z'} \right) \right\}$$

$$= -(w'z')^2 \left\{ -2\frac{z'''}{z'} + 3\frac{z''}{z'} \left(\frac{w''}{w'} + \frac{z''}{z'} \right) + 2\frac{w'''}{w'} - 3w'' \left(\frac{w''}{w'} + \frac{z''}{z'} \right) \right\}$$

Since this expression vanishes, the factor $(w'z')^2$ can be discarded (unless it happens to vanish everywhere) to leave

$$2\frac{w'''}{w'}-3\left(\frac{w''}{w'}\right)^2=2\frac{z'''}{z'}-3\left(\frac{z''}{z'}\right)^2,$$

which is a relationship which is expected to hold after applying a linear fractional transformation to a variable, no matter what the linear fractional transformation. Besides, if z is the variable, the right hand is zero, which leaves the left hand side as a constraint on any function that supports invariance. Some rearrangement can make the invariant read

$$\frac{2w'w'''-3(w'')^2}{(w')^2}$$

half of which is called a *Schwartz derivative*. The symbol for the Schwartz derivative of w with respect to z is $\{w,z\}$, some variants of whose definition are:

$$\{w,z\}=\frac{w'''}{w'}-\frac{3}{2}\left(\frac{w''}{w'}\right)^2,$$

$$=\left\{\frac{w''}{w'}\right\}'-\frac{1}{2}\left(\frac{w''}{w'}\right)^2,$$

$$=\frac{d^2}{dz^2}\left[\ln\frac{dw}{dz}\right]-\frac{1}{2}\left\{\frac{d}{dz}\left[\ln\frac{dw}{dz}\right]\right\}^2.$$

The Schwartz derivative is not really a derivative (it is a differential invariant) because it doesn't follow the product rule when applied to products, nor does it even work linearly on sums, and it is furthermore immune to non-zero multiplicative constants. As the derivative was derived, w and z depended on a parameter with respect to which two differential expressions were identical. When z itself was the parameter, one result was obviously zero, forcing the other to assume the same value; which of the two was chosen creates an asymmetry by which $\{w,z\}\neq\{z,w\}$.

Nevertheless, the invariant can be calculated for various w's all depending on the same z, for which the following rules and particular results can be verified through calculation:

$$\left\{\frac{aw+b}{cw+d},z\right\}=\{w,z\},$$

$$\left\{w,\frac{az+b}{cz+d}\right\}\frac{(ad-bc)}{(cz+d)^4}=(w,z),$$

$$\{w,z\}=\{w,u\}\left(\frac{du}{dz}\right)^2+\{u,z\},$$

$$\{w,z\} = -\{z,w\}\left(\frac{dw}{dz}\right)^2$$

$$\{w,z\} = 0 \Rightarrow w = \frac{az+b}{cz+d'}$$

$$\{z^n, z\} = \frac{1-n^2}{2z^2},$$

$$\{e^{\lambda z}, z\} = -\frac{1}{2}\lambda^2$$

The last three lines suggest the task of inverting the Schwartz derivative; that is, of finding $w(z)$ for which $\{w,z\} = Q(z)$, for some prescribed function Q. The answer is that one should solve the differential equation

$$y''(z) + \frac{1}{2}Q(z)y(z) = 0,$$

for which one can expect to obtain two linearly independent solutions $y_1(z)$ and $y_2(z)$. Their quotient

$$w(z) = \frac{y_2(z)}{y_1(z)}$$

will then solve the Schwartzian equation.

HIGHER ORDER DERIVATIVES

Note that a partial derivative is itself a function of two variables, and so further partial derivatives can be calculated. We write

$$\frac{\partial}{\partial x}\frac{\partial f}{\partial x} = \frac{\partial^2 f}{\partial x^2}, \frac{\partial}{\partial x}\frac{\partial f}{\partial y} = \frac{\partial^2 f}{\partial x \partial y}, \frac{\partial}{\partial y}\frac{\partial f}{\partial x} = \frac{\partial^2 f}{\partial y \partial x} \text{ and } \frac{\partial}{\partial y}\frac{\partial f}{\partial y} = \frac{\partial^2 f}{\partial y^2}.$$

This notation generalises to more than two variables, and to more than two derivatives in the way you would expect. There is a complication that does not occur when dealing with functions of a single variable; there are *four* derivatives of second order, as follows:

$$\frac{\partial^2 f}{\partial x^2}, \frac{\partial^2 f}{\partial x \partial y}, \frac{\partial^2 f}{\partial y \partial x} \text{ and } \frac{\partial^2 f}{\partial y^2}.$$

Fortunately, when f has mild restrictions, the order in which the differentiation is done doesn't matter.

Proposition: Assume that all second order derivatives of f exist and are continuous. Then the mixed second order partial derivatives of f are equal. i.e.

$$\frac{\partial^2 f}{\partial x \partial y} = \frac{\partial^2 f}{\partial y \partial x}.$$

Example: Suppose that $f(x, y)$ is written in terms of u and v where $x = u + \log v$ and $y = u - \log v$. Show that, with the usual convention,

$$\frac{\partial^2 f}{\partial u^2} = \frac{\partial^2 f}{\partial x^2} + 2\frac{\partial^2 f}{\partial x \partial y} + \frac{\partial^2 f}{\partial y^2}$$

and

$$v^2 \frac{\partial^2 f}{\partial u^2} = \frac{\partial f}{\partial y} - \frac{\partial f}{\partial x} + \frac{\partial^2 f}{\partial x^2} + 2\frac{\partial^2 f}{\partial x \partial y} + \frac{\partial^2 f}{\partial y^2}$$

You may assume that all second order derivatives of f exist and are continuous.

Solution. Using the chain rule, we have

$$\frac{\partial f}{\partial u} = \frac{\partial f}{\partial x}\frac{\partial x}{\partial u} + \frac{\partial f}{\partial y}\frac{\partial y}{\partial u} = \frac{\partial f}{\partial x} + \frac{\partial f}{\partial y}$$

$$\text{and} \frac{\partial f}{\partial v} = \frac{\partial f}{\partial x}\frac{\partial x}{\partial v} + \frac{\partial f}{\partial y}\frac{\partial y}{\partial v} = \frac{1}{v}\frac{\partial f}{\partial x} - \frac{1}{v}\frac{\partial f}{\partial y}$$

Thus using both these and their operator form, we have

$$\frac{\partial^2 f}{\partial v^2} = \frac{\partial}{\partial u}\left(\frac{\partial f}{\partial u}\right) = \frac{\partial}{\partial u}\left(\frac{\partial f}{\partial x} + \frac{\partial f}{\partial y}\right)$$

$$= \frac{\partial}{\partial x}\left(\frac{2f}{\partial x} + \frac{\partial f}{\partial y}\right)\frac{\partial x}{\partial u} + \frac{\partial}{\partial y}\left(\frac{\partial f}{\partial x} + \frac{\partial f}{\partial y}\right)\frac{\partial y}{\partial u}$$

$$= \left(\frac{\partial^2 f}{\partial x^2} + \frac{\partial^2 f}{\partial y}\right)^1 + \left(\frac{\partial^2 f}{\partial y \partial x} + \frac{\partial^2 f}{\partial y^2}\right)^1$$

$$\frac{\partial^2 f}{\partial x^2} + \frac{\partial^2 f}{\partial x \partial y} + \frac{\partial^2 f}{\partial y \partial x} + \frac{\partial^2 f}{\partial y^2}$$

While differentiating with respect to u, we have

$$\frac{\partial^2 f}{\partial v^2} = \frac{\partial}{\partial u}\left(\frac{\partial f}{\partial u}\right) = \frac{\partial}{\partial v}\left(\frac{1}{v}\frac{\partial f}{\partial x} - \frac{1}{v}\frac{\partial f}{\partial y}\right)$$

$$= \frac{-1}{v^2}\frac{\partial f}{\partial x} + \frac{1}{v}\frac{\partial}{\partial v}\left(\frac{\partial f}{\partial x}\right) + \frac{1}{v^2}\frac{\partial f}{\partial y} - \frac{1}{v}\frac{\partial}{\partial v}\left(\frac{\partial f}{\partial y}\right)$$

$$= \frac{-1}{v^2}\frac{2f}{2x} + \frac{1}{v}\left(\frac{\partial}{2x}\left(\frac{\partial f}{\partial x}\right)\frac{\partial x}{\partial v} + \frac{\partial}{\partial y}\left(\frac{\partial f}{\partial y}\right)\frac{\partial y}{\partial v}\right)$$

$$+\frac{1}{v^2}\frac{\partial f}{\partial y}\frac{-1}{v}\left(\frac{\partial}{\partial x}\left(\frac{\partial f}{\partial y}\right)\frac{\partial x}{\partial v}\right)$$

$$=\frac{-1}{v^2}\frac{2f}{2x}+\frac{1}{v}\left(\frac{\partial}{2x}\left(\frac{\partial f}{\partial x}\right)\frac{\partial x}{\partial v}+\frac{\partial}{\partial y}\left(\frac{\partial f}{\partial y}\right)\frac{\partial y}{\partial v}\right)$$

$$=\frac{1}{v^2}\left(\frac{\partial f}{\partial y}-\frac{\partial f}{\partial x}+\frac{\partial^2 f}{\partial x^2}-2\frac{\partial^2 f}{\partial x\partial y}+\frac{\partial^2 f}{\partial y^2}\right)$$

Theorem: If $f(z)$ is analytic in a region $\Re$, so also are $f'(z), f''(z), \ldots$ analytic in $\Re$, i.e., all higher derivatives exists in $\Re$.

L' HOSPITAL'S RULE

Let $f(z)$ and $g(z)$ be an analytic in a region containing the point z_0 and suppose that $f(z_0)=g(z_0)=0$, but $g'(z_0)\neq 0$. Then L' Hospital's rule state that

$$\lim_{z\to z_0}\frac{f(z)}{g(z)}=\frac{f'(z_0)}{g'(z_0)}$$

In case $f'(z_0)=g'(z_0)=0$, the rule may be extended.

Limits represented by so-called indeterminate forms $\frac{\infty}{\infty}, 0\circ\infty, \infty^0, 0^0, 1^\infty$ and $\infty-\infty$ can be evaluated by appropriate modifications of L' Hospital's rule.

EXERCISE

1. Evaluate each of the following limits:

(*a*) $\lim_{z\to i}\left(4z^3-6z+3\right)$ (*b*) $\lim_{w\to 3i}\left(\frac{w^2+9}{w-3i}\right)$ (*c*) $\lim_{z\to i}\left(\frac{z^4-1}{z^2+1}\right)$

2. For each of the following, find $\lim_{n\to\infty} z_n$. Also, plot $z_1, z_2, z_3, \ldots, z_{15}$ in the complex plane:

(*a*) $z_n=\left(\frac{3-n}{n}\right)+i\left(\frac{n+1}{2n+3}\right)$ (*b*) $z_n=\left(\frac{2-n}{n^2}\right)-i\left(4+\frac{6}{n}\right)$

(c) $z_n=3e^{i\frac{\pi}{n}}$ (*d*) $z_n=e^{i\frac{\pi(n-1)}{n}}$

3. Suppose $f(z)=3xy+i\left(x-y^2\right)$. Find $\lim_{z\to 3+2i} f(z)$ or explain carefully why it does not exist.

4. Prove that if f has a derivative at z, then f is continuous at z.

5. Suppose that $f(z) = \dfrac{a_0 + a_1 z + a_2 z^2}{b_0 + b_1 z + b_2 z^2}$

   where $a_0, a_1, a_2, b_0, b_1, b_2 \in C$. Examine the behaviour of f(z) at z = 0 and at $z = \infty$.

6. Each of the following functions is not defined at z = z0. What value must f(z0) take to ensure continuity at z = z0?

   (*a*) $f(z) = \dfrac{z - z_0}{z - z_0}$ (*b*) $f(z) = \dfrac{z^3 - z_0{}^3}{z - z_0}$ (c) $f(z) = \dfrac{1}{z - z_0}\left(\dfrac{1}{z} - \dfrac{1}{z_0}\right)$

7. Prove that polynomial is continuous everywhere in the finite plane.

8. Find all points at which the valued function *f* defined by $f(z) = \bar{z}$ has a derivative.

9. Find all points at which the valued function *f* defined by $f(z) = (2+i)z^3 - iz^2 + 4z - (1+7i)$ has a derivative.

10. At what points is the function *f* given by $f(z) = x^3 + i(1-y)^3$ analytic? Explain.

11. Do the real and imaginary parts of the function *f* satisfy the Cauchy-Riemann equations at $z = 0$? What do you make of your answer?

12. Find all points at which $f(z) = \dfrac{x}{x^2 + y^2} - i\dfrac{y}{x^2 + y^2}$ is differentiable. At what points is *f* analytic? Explain.

13. (*a*) Show that if f = u + iv is a complex function, where u and v are the real part and the imaginary part of f respectively, then the Cauchy-Riemann equation $\dfrac{\partial f}{\partial \bar{z}} = 0$ is equivalent to $u_x = v_y$, $u_y =$ "v_x. (1.11)

    (*b*) Use (1.11) to check that the function $f(x, y) = e^x \cos y + ie^x \sin y$ is analytic.

    (c) Use the previous exercise or other methods to show that in polar coordinates is $u_r = \dfrac{1}{r}v_\theta v_r = -\dfrac{1}{r}u_\theta$

14. For each of the following functions f(z), determine whether the Cauchy-Riemann equations are satisfied:

    (a) $f(z) = x^3 - 3y^2 + 2x + i(3x^2y - y^3 + 2y)$

    (b) $f(z) = \log(x2 - y2) + 2i \tan^{-1}(y/x)$

15. Show that a real valued analytic function is constant.

ELEMENTARY FUNCTIONS OF COMPLEX NUMBERS

Once we have mastered using *i* as an exponent, we can advance to functions of complex numbers. Exponential functions, of course, are simply functions which use *i* as an exponent.

POLYNOMIAL FUNCTIONS

It is defined by $w = a_0 z^n + a_1 z^{n-1} + \dots + a_{n-2} z^2 + a_{n-1} z + a_n = P(z)$

where the coefficients $a_0 \neq 0, a_1, \dots, a_{n-2}, a_{n-1}$ and a_n are all constants and n is a positive integer called degree of the polynomial $P(z)$.

The transformation $w = az + b$ is called a linear transformation.

RATIONAL ALGEBRAIC FUNCTION

It is defined by $w = \dfrac{P(z)}{Q(z)}$

Where $P(z)$ and $Q(z)$ are polynomials. We sometimes call as a rational transformations

The special case $w = \dfrac{az+b}{cz+d}$, where $ad - bc \neq 0$ is often called a bilinear transformation.

EXPONENTIAL FUNCTIONS

It is defined by

$$z = x + iy = r(\cos\theta + i\sin\theta) = re^{i\theta} = re^{i(\theta + 2n\pi)}$$

where n is an integer.

Common Operations of Complex Numbers

$$z_1 z_2 = r_1 e^{i\theta_1} r_2 e^{i\theta_2} = r_1 r_2 e^{i(\theta_1 + \theta_2)}$$

$$z_1 / z_2 = r_1 e^{i\theta_1} / r_2 e^{i\theta_2} = \frac{r_1}{r_2} e^{i(\theta_1 - \theta_2)}$$

$$z^p = \left(re^{i\theta}\right)^p = r^p e^{ip\theta}$$

$$\sqrt[p]{z} = \sqrt[p]{re^{i\theta}} = \left(re^{i\theta}\right)^{\frac{1}{p}} = \left[re^{i(\theta+2n\pi)}\right]^{\frac{1}{p}} = \sqrt[p]{r}e^{\frac{i(\theta+2n\pi)}{p}}$$

where n is an integer.

LOGARITHM FUNCTIONS

Since a complex number can be written in exponential form as re^{ip}, a logarithm is almost immediate. We get

$$\ln z = \ln\left(re^{i\theta}\right) = \ln\left(re^{i(\theta+2n\pi)}\right) = \ln r + i\theta + 2n\pi i$$

which gives the principal value of the natural logarithm. But the angle P is not unique, as it can vary by multiples of 2π. Therefore, the formula for the natural logarithm of a complex number is

$$\ln z = \ln\left(re^{i\theta}\right) = \ln\left(re^{i(\theta+2n\pi)}\right) = \ln r + i\theta + 2n\pi i$$

For example, we can find the natural logarithm of i. Its principal value is

$$\ln i = \ln\left(e^{i\pi/2}\right) = \frac{\pi}{2}i$$

Another interesting example is the natural logarithm of negative one. Its principal value is $\ln(-1) = \ln(1e^{\pi}) = \pi i$. (And you thought you couldn't take logarithms of negative numbers! You can, but the answers are not real numbers.)

We can use the change-of-base formula to find other base logarithms (including complex bases). For example,

$$\log_i e = \frac{\ln e}{\ln i} = \frac{1}{\frac{\pi}{2}i} = -\frac{2i}{\pi}.$$

TRIGONOMETRIC FUNCTIONS

From formula $e^{ix} = \cos x + : \sin x$, we can immediately obtain some formulas for the sine and cosine functions, in terms of complex exponentials. These formulas are often used as definitions of the trigonometric functions for complex numbers.

$$\cos x = \frac{e^{ix} + e^{-ix}}{2}, \sin x = \frac{e^{ix} - e^{-ix}}{2i}$$

Substituting the quantity ix in place of the variable x, we can produce formulas for pure imaginary numbers. Notice that the results are related to the hyperbolic functions.

$$\cos ix = \frac{e^{x} + e^{-x}}{2} = \cosh x,\ \sin ix = \frac{e^{-x} - e^{x}}{2i} i \sinh x$$

Then, using the formulas for the sine and cosine of the sum of two angles, we can obtain formulas for the sine and cosine of complex numbers.

$$\cos(a + bi) = \cos a \cosh b - i \sin a \sinh b$$

$$\sin(a + bi) = \sin a \cosh b + i \cos a \text{ singh } b$$

As an example, consider the cosine of i.

We get

$$\cos i = \cos 0 \cosh 1 - i \sin 0 \sinh 1 = \cosh 1 = \frac{e^{1} + e^{-1}}{2} \approx 1.5431$$

HYPERBOLIC FUNCTIONS

Using the connection between hyperbolic functions and trigonometric functions, the results for hyperbolic functions are almost immediate.

$\cosh(a + bi) = \cosh a \cos b + i \sinh a \sin b$

$\sinh(a + bi) = \sinh a \cos b + i \cosh a \sin b$

As an example, consider the hyperbolic cosine of i. We get

$\cosh i = \cosh 0 \cos 1 + i \sinh 0 \sin 1 = \cos 1 \approx 0.5403$

INVERSE TRIGONOMETRIC FUNCTIONS

Since $\cos x = \dfrac{e^{ix} + e^{-ix}}{2}$, we can find the inverse by replacing the function with the variable y, swapping x and y, then solving for y. After the variable swap, we have $x = \dfrac{e^{iy} + e^{-iy}}{2}$. Multiplying both sides by $2e^{iy}$ gives the equation $0 = e^{2iy} - 2xe^{iy} + 1$. This can be solved using the quadratic formula, to get $e^{iy} = \dfrac{2x \pm \sqrt{4x^2 - 4}}{2}$. After simplifying, we can take the natural logarithm of both sides and divide by i. Then, the result is $\cos^{-1}x = -i \ln\left[x \pm i\sqrt{1 - x^2}\right]$. (A factor of i typically appears as the coefficient of the square root so that the similarities between the inverse cosine and inverse sine formulas are more apparent.)

By similar arguments, the inverse sine formula is found to be $\sin^{-1} x = -i \ln\left[ix \pm \sqrt{1 - x^2}\right]$, and the inverse tangent formula is $\tan^{-1} x = \dfrac{i}{2} \ln \dfrac{i + x}{i - x}$. As an example, consider the inverse cosine of i. We shall find only one of its many values, by using the plus sign in the formula. We get: $\cos^{-1} i = -i \ln\left[i + i\sqrt{1 - i^2}\right] = -i \ln\left[\left(1 + \sqrt{2}\right)i\right]$ Then we use the logarithm formula to continue:

$\cos^{-1} i = -i \ln\left[\left(1 + \sqrt{2}\right)e^{i\pi/2}\right] = -i\left[\ln\left(1 + \sqrt{2}\right)\right] + \dfrac{\pi}{2}i$ This can be simplified, and we get $\cos^{-1} i = \dfrac{\pi}{2} - i \ln\left(1 + \sqrt{2}\right) \approx 1.57 - 0.88$

INVERSE HYPERBOLIC FUNCTIONS

The formulas are identical to those previously found for inverse hyperbolic functions:

$$\cosh^{-1} x = \ln\left[x \pm \sqrt{x^2 - 1}\right]$$

$$\sinh^{-1} x = \ln\left[x \pm \sqrt{x^2 + 1}\right]$$

For example, one of the values of the inverse hyperbolic cosine of i (using the plus sign in the formula) is:

$\cosh^{-1} i = \ln\left[i + \sqrt{i^2 - 1}\right] = \ln\left[\left(1 + \sqrt{2}\right)i\right]$. Then using the logarithm formula, we get:

$\cosh^{-1} i = \ln\left(1 + \sqrt{2}\right) + \frac{\pi}{2} i \approx 0.88 + 1.57i$

THE FUNCTION

z^{α}, α may be complex, is defined as $e^{\alpha \ln z}$. Similarly if $f(z)$ and $g(z)$ are two given functions of z, we can define $f(z)^{g(z)} = e^{g(z)\ln f(z)}$. In general such functions are multiple-valued.

Algebraic and Transcendental Functions

If w is a solution of the polynomial equation

$$P_0(z)w^n + P_1(z)w^{n-1} + P_2(z)w^{n-2} + P_3(z)w^{n-3} + \ldots + P_{n-1}(z)w^n + P_n(z) = 0 \ldots(2)$$

Where $P_0 \neq 0$, $P_1(z)$, $P_2(z)$, ..., $P_n(z)$ are polynomials in z and n is a positive integer then $w = f(z)$ is called an algebraic function of z.

Example: $w = z^{\frac{1}{2}}$ is a solution of the equation $w^2 - z = 0$ and so is an algebraic function of z.

Any function which cannot be expressed as a solution of (6) is called a transcendental function. The logarithmic, trigonometric and hyperbolic function and their corresponding inverses are example of transcendental functions.

BRANCH POINT AND BRANCH CUT

In complex analysis, a branch is a portion of the range of a multiple-valued function over which the function is single-valued. Combining all the branches gives the full structure of the function. It is often convenient to choose a particular branch of a function to work with, and this choice is often designated the principal branch.

In complex analysis, a branch point may be thought of informally as a point z_0 at which a "multiple-valued function" changes values when one winds once around z_0.

A branch point of an analytic function is a point in the complex plane whose complex argument can be mapped from a single point in the domain to multiple points in the range. For example, consider the behavior of the point z = 0 under the power function

$$f(z) = z^a$$

Examples:

- 0 is a branch point of the square root function. Suppose $w = \Phi 2$ and z starts at 4 and moves along a circle of radius 4 centered at 0. The dependent variable w changes while depending on z in a continuous manner. When z has made one full circle, going from 4 back to 4 again, w will have made one half-circle, going from the positive square root of 4, i.e., from 2, to the negative square root of 4, i.e., –2.

- 0 is also a branch point of the natural logarithm. Since e^0 is the same as $e^{\pi i}$, both 0 and $2\pi i$ are among the multiple values of Log (1). As z moves along a circle of radius 1 centered at 0, w = Log (z) goes from 0 to $2\pi i$.
- In trigonometry, since tan ($\pi/4$) and tan ($5\pi/4$) is both equal to 1, the two numbers $\pi/4$ and $5\pi/4$ are among the multiple values of arc-tan (1). The imaginary units i and $-i$ are branch points of the arctangent function. That this is so may be seen by observing that the derivative $\frac{d}{dz}\{\text{arc-tan}(z)\} = 1/(1 + z^2)$ has simple poles at these two points i.e., at z = i, $-i$ $\frac{d}{dz}\left(\tan^{-1} z\right) = 0$ since the denominator is zero at those points.
- If the derivative f' of a function f has a simple pole at a point a, then f has a branch point at a.

The branch cut device may appear arbitrary (it is); but it is very useful, for example in the theory of special functions. An invariant explanation of the branch phenomenon is developed in Riemann surface theory (of which it is historically the origin), and more generally in the ramification and monodromy theory of algebraic functions and differential equations.

BRANCH CUT

A branch cut is a curve (with ends possibly open, closed, or half-open) in the complex plane across which an analytic multiple-valued function is discontinuous. For convenience, branch cuts are often taken as lines or line segments. Branch cuts (even those consisting of curves) are also known as cut lines, slits, or branch lines.

For example, consider the function z^2 which maps each complex number z to a well-defined number z^2. Its inverse function $\sqrt{z}$, on the other hand, maps, for example, the value $z = 1$ to $\sqrt{1} = \pm 1$. While a unique principal value can be chosen for such functions, the choices cannot be made continuous over the whole complex plane. Instead, lines of discontinuity must occur. The most common approach for dealing with these discontinuities is the adoption of so-called branch cuts. In general, branch cuts are not unique, but are instead chosen by convention to give simple analytic properties.

Some functions have a relatively simple branch cut structure, while branch cuts for other functions are extremely complicated. An alternative to branch cuts for representing multiple-valued functions is the use of Riemann surfaces.

In addition to branch cuts, singularities known as branch points also exist. It should be noted, however, that the endpoints of branch cuts are not necessarily branch points.

Branch cuts do not arise for the single-valued trigonometric, hyperbolic, integer power, and exponential functions. However, their multiple-valued inverses do require branch cuts.

Function Name	Function	Branch cut(s)
inverse cosecant	$\csc^{-1} z$	$(-1, 1)$
inverse cosine	$\cos^{-1} z$	$(-\infty, -1)$ and $(1, \infty)$
inverse cotangent	$\cot^{-1} z$	$(-i, i)$
inverse hyperbolic cosecant	csch^{-1}	$(-i, i)$
inverse hyperbolic cosine	$\cosh^{-1}$	$(-\infty, 1)$
inverse hyperbolic cotangent	$\coth^{-1}$	$(-1, 1)$
inverse hyperbolic secant	sech^{-1}	$(-\infty\ 0)$ and $(1, \infty)$
inverse hyperbolic sine	$\sinh^{-1}$	$(-i\infty, -i)$ and $(i, i\infty)$
inverse hyperbolic tangent	$\tanh^{-1}$	$(-\infty, -1)$ and $(1, \infty)$
inverse secant	$\sec^{-1} z$	$(-1, 1)$
inverse sine	$\sin^{-1} z$	$(-\infty, -1)$ and $(1, \infty)$
inverse tangent	$\tan^{-1} z$	$(-i\infty, -i)$ and $(i, i\infty)$
natural logarithm	$\ln z$	$(-\infty, 0)$
power	$z^n, n \notin \mathbb{Z}$	$(-\infty, 0)$ for $R[n] \leq 0$; $(-\infty, 0)$ for $R[n] > 0$
square root	$\sqrt{z}$	$(-\infty, 0)$

Next we discuss about 'Riemann surface'.

RIEMANN SURFACE

A Riemann surface is a surface-like configuration that covers the complex plane with several, and in general infinitely many, "sheets." These sheets can have very complicated structures and interconnections. Riemann surfaces are one way of representing multiple-valued functions; another is branch cuts.

For example: they can look like a sphere or a torus or a couple of sheets glued together.

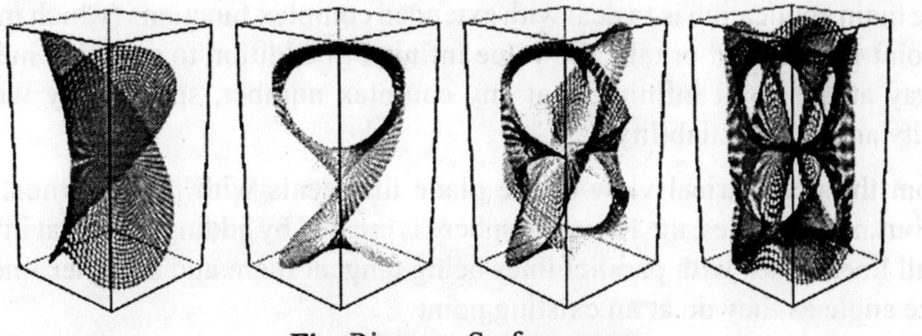

Fig. Riemann Surface.

The above plot shows Riemann surfaces for solutions of the equation

$[w(z)]^d + w(z) + z^{d-1} = 0$

with $d = 2, 3, 4$, and 5, where $w(z)$ is the Lambert W-function.

The main point of Riemann surfaces is that holomorphic functions may be defined between them. Riemann surfaces are nowadays considered the natural setting for studying the global behavior of these functions, especially multi-valued functions such as the square root or the logarithm.

Examples:

- The complex plane C is perhaps the most trivial Riemann surface. The map $f(z) = z$ (the identity map) defines a chart for C, and $\{f\}$ is an atlas for C. The map $g(z) = \bar{z}$ (the conjugate map) also defines a chart on C and $\{g\}$ is an atlas for C. The charts f and g are not compatible, so this endows C with two distinct Riemann surface structures. In fact, given a Riemann surface X and its atlas A, the conjugate atlas $B = \{f^* : f \to A\}$ is never compatible with A, and endows X with a distinct, incompatible Riemann structure.
- Every open subset of the complex plane can be viewed as a Riemann surface in a natural way. More generally, every open subset of a Riemann surface is a Riemann surface.

Every non-compact Riemann surface admits non-constant holomorphic functions (with values in). In fact, every non-compact Riemann surface is a Stein manifold.

In constrast, on a compact Riemann surface every holomorphic function with value in $\mathbb{C}$ is constant due to the maximum principle. However, there always exists non-constant meromorphic functions.

RIEMANN SPHERE

The Riemann sphere, is the unique way of viewing the extended complex plane (the complex plane plus a point at infinity) so that it looks exactly the same at the point infinity as at any complex number.

The main application is to deal with extended complex functions (which may be defined at the point infinity and/or take the value infinity, in addition to complex numbers) in the same way at the point infinity as at any complex number, specifically with respect to continuity and differentiability.

From the geometrical view of the plane that deals with points, lines, circles and angles but not distances, the Riemann sphere is created by adding a point at infinity through which all lines cross, with parallel lines being tangent there and all other lines crossing at the same angle as they do at an existing point.

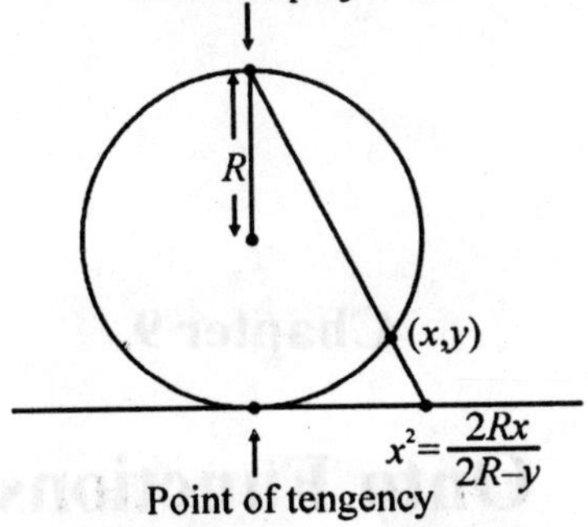

Fig. Riemann Sphere

Topologically, the Riemann sphere is the one-point compactification of the complex plane. In all of these viewpoints, the point at infinity acquires an identical role to any point in the complex plane.

EXERCISE

1. Evaluate each of the following limits:

(*a*) $\lim_{z \to i}(4z^3 - 6z + 3)$ (*b*) $\lim_{z \to 1-i}(z^2 - 3z)$

(c) $\lim_{w \to 3i}\left(\frac{w^2+9}{w-3i}\right)$ (*d*) $\lim_{z \to i}\left(\frac{z^4-1}{z^2+1}\right)$

2. Suppose $f(z) = 3xy + i(x - y^2)$. Find $\lim_{z \to 3+2i} f(z)$ or explain carefully why it does not exist.

3. Prove that if f has a derivative at z, then f is continuous at z.

4. Suppose that $f(z) = \frac{a_0 + a_1 z + a_2 z^2}{b_0 + b_1 z + b_2 z^2}$ where $a_0, a_1, a_2, b_0, b_1, b_2 \in$ C. Examine the behaviour of f(z) at z = 0 and at $z = \infty$.

5. Each of the following functions is not defined at z = z0. What value must f(z0) take to ensure continuity at z = z0?

(*a*) $f(z) = \frac{z - z_0}{z - z_0}$ (*b*) $f(z) = \frac{z^3 - z_0^3}{z - z_0}$

(c) $f(z) = \frac{1}{z - z_0}\left(\frac{1}{z} - \frac{1}{z_0}\right)$ (*d*) $f(z) = \frac{1}{z - z_0}\left(\frac{1}{z^3} - \frac{1}{z_0^3}\right)$

6. Prove that polynomial is continuous everywhere in the finite plane.

Chapter 9

Onto Functions

RANGE AND EXISTENCE

The left side of a system of linear equations has been considered as a transformation. Now we raise some natural questions about transformations.

Example: For the transformation T = *Capital City*: X = country $\to$ Y = city, we ask:

- What is the capital city of China?
- What countries have Beijing as the capital city ?
- Is there a country with Beijing as the capital? Is any city the capital of some country?

In mathematical terminology, the questions can be rephrased as:

- Find the value of x = China $\in X$ under T.
- What $x \in X$ has b = Beijing $\in Y$ as the value under T? Equivalently, find x satisfying $T(x) = b$.
- Is b = Beijing $\in Y$ the value of some $x \in X$ under T? Equivalently, do the equation $T(x) = b$ have solutions for b = Beijing? What about any city b?

Motivated by the last question in the example, we define the range of a transformation $T: X \to Y$ to be all the images under the transformation.

$$\textit{range}\ T = \{T(x)\text{: any } x \in X\}$$

$$= \{b \in Y\text{: There is } x \in X\text{, such that } T(x) = b\}$$

$$= \{b \in Y\text{: } T(x) = b \text{ has solutions}\}$$

Example: The range of the transformation

Capital City: country $\to$ city

is all the capital cities in the world. For example, Beijing is an element of *range* (*Capital City*), while Hong Kong is not an element of *range*(*Capital City*).

The range of the transformation

Instructor: course → professor

is all the professors who teach at least one course. For example, I belong to the set *range(Instructor)*, while the president of the university (who is so busy) is unlikely an element of *range(Instructor)*.

As for the transformation

Age: people → number

we know 20 is in *range(Age)* because there are millions of people at age 20. We are also pretty sure 100 is in *range(Age)*, although we are less sure about 110. On the other hand, we are absolutely certain that 200 and –10 are not in *range(Age)*.

Example: The range of *Square*: R → R is all the non–negative numbers. The range of *Reflection in x–axis*: $R^2 \to R^2$ is R^2.

Since the system in this example has solutions, the vector (2, 2, 6) (which is the right side of the system) is in the range of

$$T\begin{bmatrix} x_1 \\ x_2 \\ x_3 \end{bmatrix} = \begin{bmatrix} 3x_1 + x_2 - x_3 \\ x_1 - x_2 + x_3 \\ 2x_1 + 2x_2 + x_3 \end{bmatrix} R^3 \to R^3$$

In fact, further discussion in this example shows that the range is the whole space R^3.

In contrast, this example shows that (2, 2, 6) is not in the range of

$$T\begin{bmatrix} x_1 \\ x_2 \\ x_3 \end{bmatrix} = \begin{bmatrix} 3x_1 + x_2 - x_3 \\ x_1 - x_2 + x_3 \\ 2x_1 - 2x_2 + x_3 \end{bmatrix} : R^3 \to R^3$$

In particular, the range of the transformation is not R^3 (strictly smaller than R^3).

ONTO AND ALWAYS EXISTENCE

Since the equation $T(x) = b$ has solutions $\Leftrightarrow b \in rangeT$, the equation $T(x) = b$ has solutions for *any* b is equivalent to the following definition.

A transformation $T: X \to Y$ is onto (surjective) if its range is the whole target set Y

Intuitively, we may think of a transformation as a way of "shooting" from source to target. The transformation is onto if any element of the target set is "hit" by some element of the source.

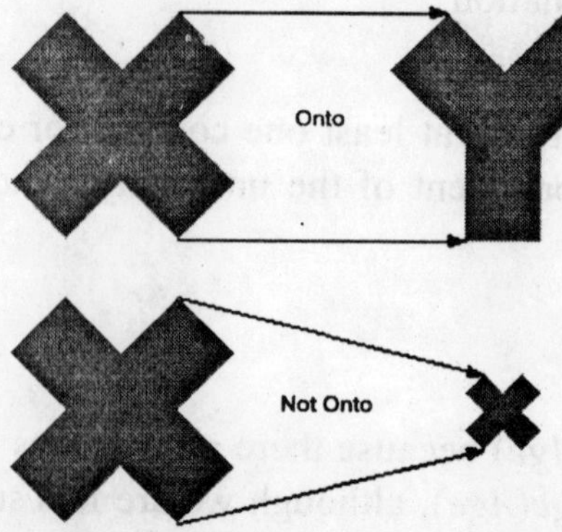

More discussion on onto transformations can befound here.

Example: As explained in an earlier example, the *Capital City* and *Age* transformations are definitely not onto. For the *Instructor* transformation to be onto, we need every professor to teach some courses. This is unlikely to happen, so that the *Instructor* transformation is unlikely to be onto.

From the discussion in this example, the *Square* transformation is not onto because the range does not include negative numbers. If we change the transformation to *Square'*: $R \to [0, \infty)$ (due to different target set, we consider this transformation to be *different* from *Square*: $R \to R$), then *Square'* is onto.

Since the range of *Reflection in x–axis* is the whole target set R^2, the transformation is onto. Similarly, the *Rotation* transformation is also onto.

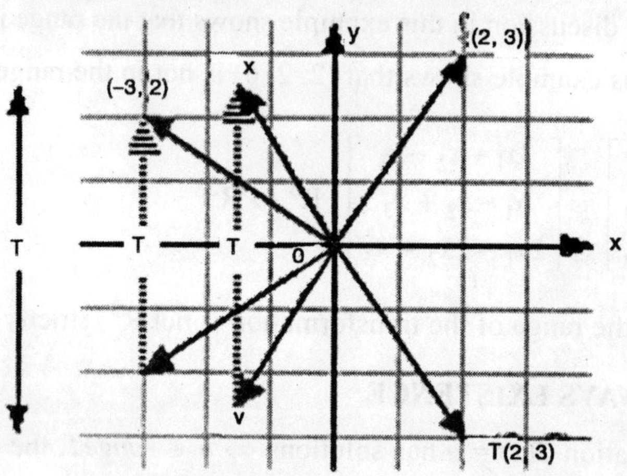

We also see that

$$T\begin{bmatrix} x_1 \\ x_2 \\ x_3 \end{bmatrix} = \begin{bmatrix} 3x_1 + x_2 - x_3 \\ x_1 - x_2 + x_3 \\ 2x_1 + 2x_2 + x_3 \end{bmatrix} : R^3 \to R^3$$

is onto, while

$$T\begin{bmatrix} x_1 \\ x_2 \\ x_3 \end{bmatrix} = \begin{bmatrix} 3x_1 + x_2 - x_3 \\ x_1 - x_2 + x_3 \\ 2x_1 + 2x_2 + x_3 \end{bmatrix} : \mathrm{R}^3 \to \mathrm{R}^3$$

is not onto.

If a linear transformation T: $\mathrm{R}^n \to \mathrm{R}^m$ is given by a matrix A, then T is onto $\Leftrightarrow Ax = b$ has solutions for any right side b. By this criterion, this is equivalent to all rows of A are pivot.

Example: To determine whether the linear transformation

$$T\begin{bmatrix} x_1 \\ x_2 \\ x_3 \\ x_4 \\ x_5 \\ x_6 \end{bmatrix} = \begin{bmatrix} x_1 + 3x_2 + 2x_3 + x_5 \\ -x_1 - x_2 - x_3 + x_4 + x_6 \\ 4x_2 + 2x_3 + 4x_4 + 3x_5 + 3x_6 \\ x_1 + 3x_2 + 2x_3 - 2x_4 \end{bmatrix} : \mathrm{R}^6 \to \mathrm{R}^4$$

is onto or not, we find the row echelon form of the corresponding matrix. In this example, the matrix corresponding to T

$$\begin{bmatrix} 1 & 3 & 2 & 0 & 1 & 0 \\ -1 & -1 & -1 & 1 & 0 & 1 \\ 0 & 4 & 2 & 4 & 3 & 3 \\ 1 & 3 & 2 & -2 & 0 & 0 \end{bmatrix}$$

has been simplified, by row operations, to

$$\begin{bmatrix} 1 & 3 & 2 & 0 & 1 & 0 \\ 0 & 2 & 1 & -1 & 0 & 1 \\ 0 & 0 & 0 & 2 & 1 & 0 \\ 0 & 0 & 0 & 0 & 0 & 1 \end{bmatrix}$$

Since all rows are pivot, the transformation T is onto.

To determine whether the linear transformation

$$S\begin{bmatrix} x_1 \\ x_2 \\ x_3 \end{bmatrix} = \begin{bmatrix} x_1 + 3x_2 + 2x_3 \\ -x_1 - x_2 - x_3 \\ 4x_2 + 2x_3 \\ x_1 + 3x_2 + 2x_3 \end{bmatrix} : \mathrm{R}^3 \to \mathrm{R}^4$$

is onto or not, we could certainly appeal to the same computation. However, we do not even need any computation. Since the corresponding 4 equations in 3 variables cannot always have solutions, the transformation is not onto.

We expand on the remark made at the end of the last example and get

A linear transformation T: $R^n \to R^m$ is onto $\Rightarrow m \le n$

Exercise: Which transformations are onto?

1. ID_s: student → number

Answer: ID_s is onto means that any number is the ID of some student. This is certainly not true because no student has –1 or 0.02 as the ID number.

2. ID_p: professor '→ number

Answer: ID_p is not onto for the same reason as ID_s is not onto.

3. *Maker*: product → manufacturer

Answer: *Maker* should be onto because any manufacturer, by the very definition of the word, produce something. (Of course you may also argue that somebody may set up some bogus manufacturing company for dubious purpose. But such a company should not be recognized as a manufacturer.)

4. *Population*: city → number

Answer: *Population* is not onto because negative numbers (such as –10) or really large numbers (such as 1 trillion) cannot be the population of any city.

5. *First–Alphabet*: word → character = $\{a, b, \dots, z\}$

Answer: *First–Alphabet* is onto because every alphabet is the first alphabet of some word. You may verify this one by one, like: *a* is the first alphabet of *alligator*, *b* is the first alphabet of *bee*, *c* is the first alphabet of *cat*, etc.

What if you change the domain from words to names of animals?

6. *Sex*: people → {male, female}

Answer To show *Sex* is onto, all we need to do is to find one male person (such as *Jacky Chan*) and one female person (such as *Celine Dion*).

7. *Sign*: number → {+, –, 0}, positive number → 1, negative number → –1, 0 → 0

Answer: *Sign* is onto because + is the sign of 2, – is the sign of –0.1, 0 is the sign of 0.

8. *Sin*: number → number

Answer: *Sin* is not onto because the value of sine function must be between –1 and 1. For example, 2 is not the sine of any angle.

9. *Sin'*: number → [–1, 1]

Answer: *Sin'* is onto because any number between –1 and 1 is the sine of some angle.

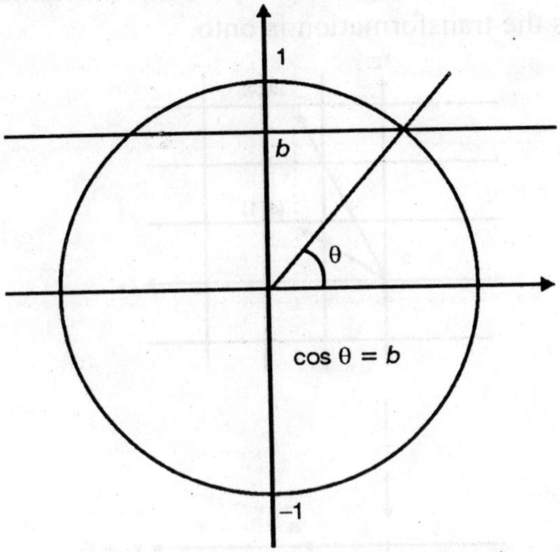

Geometrically, for the given –1 d" *b* d" 1, we draw the picture above to find the angle θ so that $\sin\theta = b$.

10. *Polar–Cartesian*: $R^2 \to R^2$, $(r, \theta) \to (r\cos\theta, r\sin\theta)$

Answer: The transformation is onto because any vector (x, y) has the polar coordinate given by the length $r = (x^2 + y^2)^{1/2}$ and the angle.

11. *Root–Polynomial*: $R^2 \to R^2$, $(x_1, x_2) \to (x_1 + x_2, x_1x_2)$

Answer: *Root–Polynomial* is not onto because the roots of a real polynomial may be complex. For example, the polynomial $x^2 + x + 1 = 0$ (equivalent to $x_1 + x_2 = -1$, $x_1x_2 = 1$) has no real solution.

12. *Reflection in diagonal line*: $R^2 \to R^2$

Answer: Any vector (a, b) is the image of (b, a). Thus the transformation is onto.

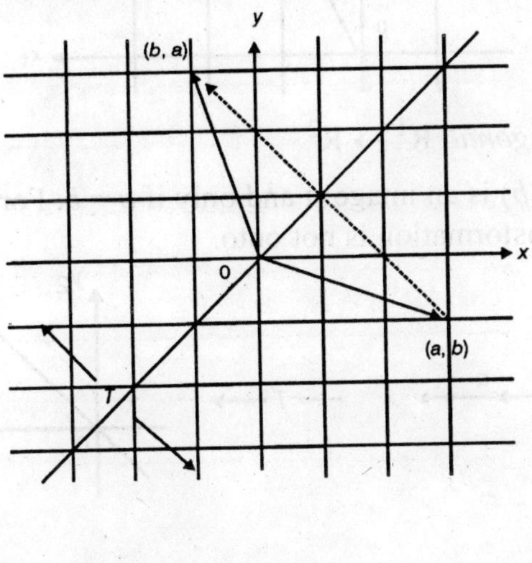

13. *Projection to x_1–line*: $R^2 \to R^1$

Answer: Any number a is the image of $(a, 1)$ (in fact the number 1 can be replaced by any other number). Thus the transformation is onto.

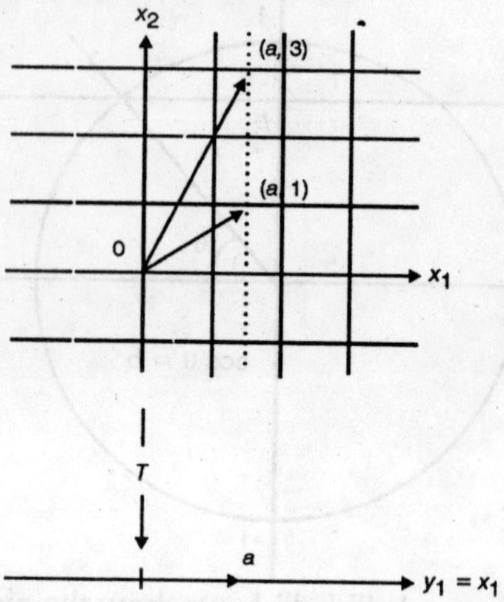

14. *Multiplying by* 2: $R^2 \to R^2$

Answer: Any vector (a, b) is the image of $(1/2\ a, 1/2\ b)$. Thus the transformation is onto.

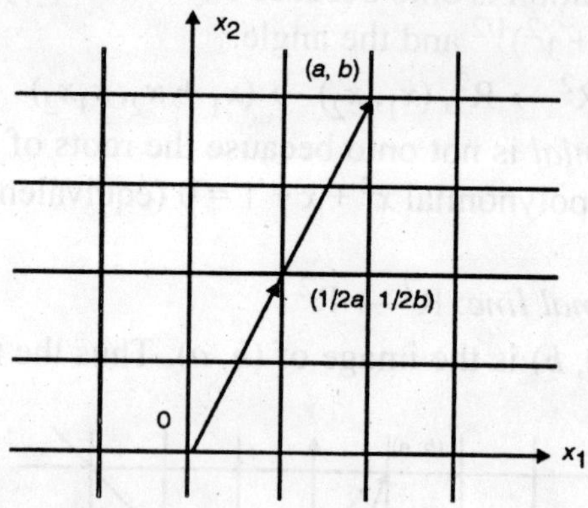

15. *Embedding as diagonal*: $R^1 \to R^2$

Answer: A vector (a, b) is an image if and only if $a = b$. For example, $(1, 2)$ is not an image. Therefore, the transformation is not onto.

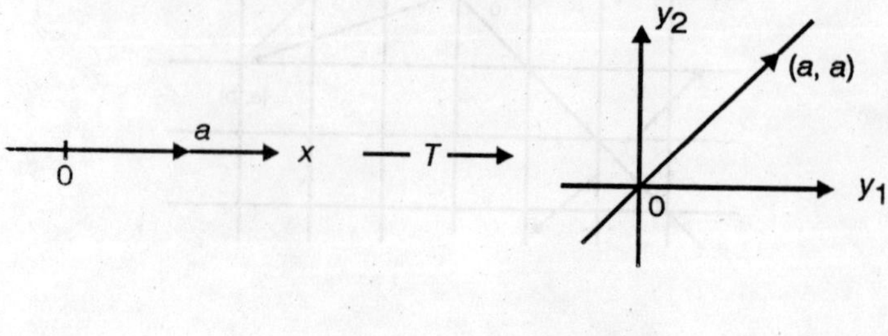

16. *Reflection in origin*: $R^2 \to R^2$

Answer: Any vector (a, b) is the image of $(-a, -b)$. Thus the transformation is onto.

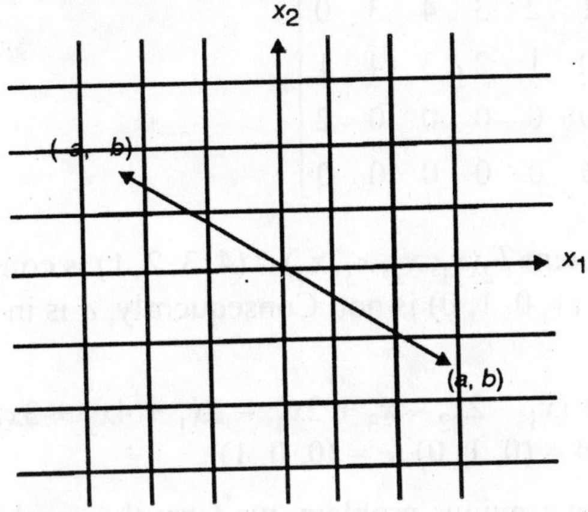

17. *Reflection in origin*: $R^3 \to R^3$

Answer: Similar to the last transformation, this is onto.

Exercise: Determine whether the vectors are in the ranges of the transformations.

1. $T_1(x_1, x_2, x_3) = (x_1 + 2x_2 + x_3, 2x_1 + x_2 + x_3, 3x_2 + x_3, 3x_1 + x_3)$, $a = (1, 0, 0, 0)$, $b = (1, 1, 1, 1)$

Answer: The problem is the same as whether the systems $T_1(x_1, x_2, x_3) = (1, 0, 0, 0)$ and $T_1(x_1, x_2, x_3) = (1, 1, 1, 1)$ have solutions.

The system $T_1(x_1, x_2, x_3) = (1, 0, 0, 0)$ appeared in this exercise. The conclusion was that the system has no solution. Therefore a is not in the range of T_1.

The system $T_1(x_1, x_2, x_3) = (1, 1, 1, 1)$ has an obvious solution $x_1 = x_2 = 0$, $x_3 = 1$. Therefore b is the image of $(0, 0, 1)$ and is in the range of T_1.

2. $T_2(x_1, x_2, x_3, x_4) = (x_2 + 2x3 + 3x_4, x_1 + 2x_2 + 3x_3 + 4x_4, 2x_1 + 3x_2 + 4x_3 + 5x_4, 3x_1 + 4x_2 + 5x_3 + 6x_4)$, $a = (4, 3, 2, 1)$, $b = (1, 0, 1, 0)$

Answer: We need to determine whether the systems $T_2(x_1, x_2, x_3, x_4) = (4, 3, 2, 1)$ and $T_2(x_1, x_2, x_3, x_4) = (1, 0, 1, 0)$ are consistent. The combined augmented matrix

$$[A\ a\ b] = \begin{bmatrix} 0 & 1 & 2 & 3 & 4 & 1 \\ 1 & 2 & 3 & 4 & 3 & 0 \\ 2 & 3 & 4 & 5 & 2 & 1 \\ 3 & 4 & 5 & 6 & 1 & 0 \end{bmatrix}$$

Row operations change this into

$$\begin{bmatrix} 1 & 2 & 3 & 4 & 3 & 0 \\ 0 & 1 & 2 & 3 & 4 & 1 \\ 0 & 0 & 0 & 0 & 0 & 2 \\ 0 & 0 & 0 & 0 & 0 & 0 \end{bmatrix}$$

Therefore the first system $T_2(x_1, x_2, x_3, x_4) = (4, 3, 2, 1)$ is consistent, while the second system $T_2(x_1, x_2, x_3, x_4) = (1, 0, 1, 0)$ is not. Consequently, a is in the range of T_2, and b is not in the range.

3. $T_3(x_1, x_2, x_3, x_4) = (x_1 - 2x_2 - x_3 + 2x_4, -2x_1 + 4x_2 + 3x_3 - x_4, -x_1 + 2x_2 + x_3 - x_4)$, $a = (1, 0, 0)$, $b = (0, 1, 0)$, $c = (0, 0, 1)$

Answer: Similar to the previous problem, we form the combined augmented matrix

$$\begin{bmatrix} 1 & -2 & -1 & 2 & 1 & 0 & 0 \\ -2 & 4 & 3 & -1 & 0 & 1 & 0 \\ -1 & 2 & 1 & -1 & 0 & 0 & 1 \end{bmatrix}$$

Row operations change this into

$$\begin{bmatrix} 1 & -2 & 0 & 0 & -2 & 1 & -5 \\ 0 & 0 & 1 & 0 & -1 & 1 & -3 \\ 0 & 0 & 0 & 1 & 1 & 0 & 1 \end{bmatrix}$$

Therefore the all three systems are consistent. Consequently, all three vectors a, b, c are in the range of T_3.

Exercise: Determine the conditions on the parameters so that the vectors are in the ranges of the transformations.

1. $T_1(x, y, u, v) = (u + 2v, -2x + 2y - u + 4v, x - y + u - v)$, $b = (1, h, 1)$

Answer: The system $T_1(x, y, u, v) = (1, h, 1)$ appeared in this exercise. The conclusion was that we need $h = -1$ in order for the system to be consistent. Therefore b is in the range of T_1 if and only if $b = -1$.

2. $T_2(x_1, x_2, x_3, x_4) = (x_1 + x_2 + x_3 + x_4, x_1 + ax_4, x_2 + ax_3)$, $b = (1, 1, 1)$

Answer: The system $T_2(x_1, x_2, x_3, x_4) = (1, 1, 1)$ appeared in this exercise. The conclusion was that we need a “‘ 1 in order for the system to be consistent. Therefore $(1, 1, 1)$ is in the range of T_2 if and only if a “‘ 1.

3. $T_3(u, v, w) = (v + w, u + aw, u + v, av - w)$, $b = (a, 1, 0, 1)$

Answer: The system $T_3(u, v, w) = (a, 1, 0, 1)$ appeared in this exercise. The conclusion was that we need $a = -1$ or 2 in order for the system to be consistent. Therefore $(a, 1, 0, 1)$ is in the range of T_3 if and only if $a = -1$ or 2.

Exercise: Which linear transformations are onto?

1. $T_1(x_1, x_2, x_3) = (x_1 + 2x_2 + x_3, 2x_1 + x_2 + x_3, 3x_2 + x_3, 3x_1 + x_3)$: $R^3 \to R^4$

Answer: By the numerical consequence about onto linear transformations, T_1 cannot be onto.

2. $T_2(x_1, x_2, x_3, x_4) = (x_2 + 2x3 + 3x_4, x_1 + 2x_2 + 3x_3 + 4x_4, 2x_1 + 3x_2 + 4x_3 + 5x_4, 3x_1 + 4x_2 + 5x_3 + 6x_4)$: $R^4 \to R^4$

Answer: The transformation is onto only if all rows of the matrix corresponding to T_2

$$\begin{bmatrix} 0 & 1 & 2 & 3 \\ 1 & 2 & 3 & 4 \\ 2 & 3 & 4 & 5 \\ 3 & 4 & 5 & 6 \end{bmatrix}$$

are pivot. In this exercise, a row echelon form of the matrix was computed.

$$\begin{bmatrix} 1 & 2 & 3 & 4 \\ 0 & 1 & 2 & 3 \\ 0 & 0 & 0 & 0 \\ 0 & 0 & 0 & 0 \end{bmatrix}$$

Thus we conclude that T_2 is not onto.

3. $T_3(x_1, x_2, x_3, x_4) = (x_1 - 2x_2 - x_3 + 2x_4, -2x_1 + 4x_2 + 3x_3 - x_4, -x_1 + 2x_2 + x_3 - x_4)$: $R^4 \to R^3$

Answer: Again we need to compute the row echelon form of the matrix

$$\begin{bmatrix} 1 & -2 & -1 & 2 \\ -2 & 4 & 3 & -1 \\ -1 & 2 & 1 & -1 \end{bmatrix}$$

of the linear transformation T_3. In this exercise, the row echelon form was computed.

$$\begin{bmatrix} 1 & -2 & 0 & 0 \\ 0 & 0 & 1 & 0 \\ 0 & 0 & 0 & 1 \end{bmatrix}$$

Since all rows are pivot, T_3 is onto.

4. $T_4(x_1, x_2, x_3, x_4) = (3.547x_1 + 5.782x_2 + 0.003x_3, -2.358x_1 - 2.683x_2 - 3.907x_3, -11.291x_1 + 4.902x_2 + 2.013x_3, 7.912x_1 + 3.091x_2 + 7.892x_3)$: $R^3 \to R^4$

Answer: By the numerical consequence about onto linear transformations, T_4 cannot be onto.

Exercise: Determine the conditions on the parameters so that the linear transformations are onto.

1. $T_1(x, y, z) = (x + y + az, x + ay + AZ, ax + ay + AZ)$

Answer: The matrix of the transformation is the following.

$$\begin{bmatrix} 1 & 1 & a \\ 1 & a & a \\ a & a & a \end{bmatrix}$$

Row operations change this into

$$\begin{bmatrix} 1 & 1 & a \\ 0 & a-1 & 0 \\ 0 & 0 & a-a^2 \end{bmatrix}$$

For all the rows to be pivot, we need $a - 1$ and $a - a^2$ nonzero. Thus T_1 is onto if and only if $a \neq 0, 1$.

2. $T_2(x_1, x_2, x_3, x_4) = (x_1 + x_2 + x_3 + x_4, x_1 + ax_4, x_2 + ax_3)$

Answer: By this exercise, the matrix of T_2

$$\begin{bmatrix} 1 & 1 & a \\ 0 & a-1 & 0 \\ 0 & 0 & a-a^2 \end{bmatrix}$$

can be changed by row operations into

$$\begin{bmatrix} 1 & 1 & 1 & 1 \\ 0 & 1 & 1 & 1-a \\ 0 & 0 & a-1 & a-1 \end{bmatrix}$$

For all the rows to be pivot, we need $a - 1$ nonzero. Thus T_2 is onto if and only if $a \neq 1$.

3. $T_3(u, v, w) = (v + w, u + aw, u + v, av - w)$

Answer: Since T_3 is a linear transformation from R^3 to R^4, according to this numerical consequence about onto linear transformations, T_3 cannot be onto.

ONE–TO–ONE

ONE–TO–ONE AND UNIQUENESS

After the discussion about the relation between onto and existence, we turn to uniqueness from the transformation viewpoint. Again the relation can be seen through everyday life questions.

Example: For the transformation

Capital City: country → city

we already know the city of Beijing is in the range. Now we ask

How many cities has Beijing as the capital city?

The answer is that Beijing is the capital city of a *unique* contry – China. In other words,

For $\boldsymbol{b}$ =Beijing, the solution to the equation *Capital City*(x) = $\boldsymbol{b}$ is unique.

In fact, we know that the uniqueness for any capital city $\boldsymbol{b}$ in the world. This means that for any city $\boldsymbol{b}$, we have

Capital City(country$_1$) = *Capital City*(country$_2$) ⇒ country$_1$ = country$_2$.

In other words, countries with the same capital must be the same country.

The discussion leads to the following definition.

A transformation T: $X \rightarrow Y$ is one–to–one (injective) if $T(x) = T(x')$ implies $x = x'$.

We note that the condition is also equivalent to

$x \neq x' \Rightarrow T(x) \neq T(x')$

(different elements are transformed to different elements)

If a transformation is not one–to–one, then it is several–to–one: Different elements of X may have the same image.

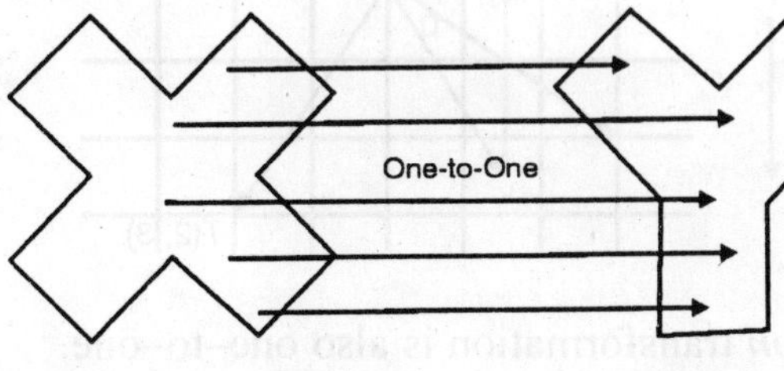

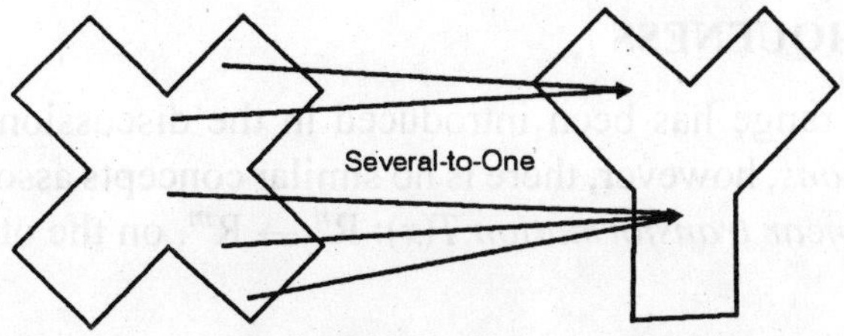

More discussion on one–to–one transformations can be found here.

Example: For the transformation

Instructor: course → professor

the following are equivalent statements

- *Instructor* is one–to–one
- *Instructor*(course_1) = *Instructor*(course_2) ⇒ $\text{course}_1 = \text{course}_2$ (courses taught by the same professor are the same course)
- $\text{Course}_1 \neq \text{course}_2$ ή *Instructor*(course_1) ≠ *Instructor*(course_2) (different courses are taught by different professors)
- No professor teaches more than one courses

The answer is (almost) definitely no.

Because there are millions of people at age 20, the transformation *Age*: people → number is not one–to–one.

Example: The transformation *Square*: R → R is two–to–one at nonzero numbers (and is one–to–one at 0). Overall, the transformation is not one–to–one.

For *Reflection in x–axis*, note that two points with the same reflections must be the same point. Therefore the transformation is one–to–one.

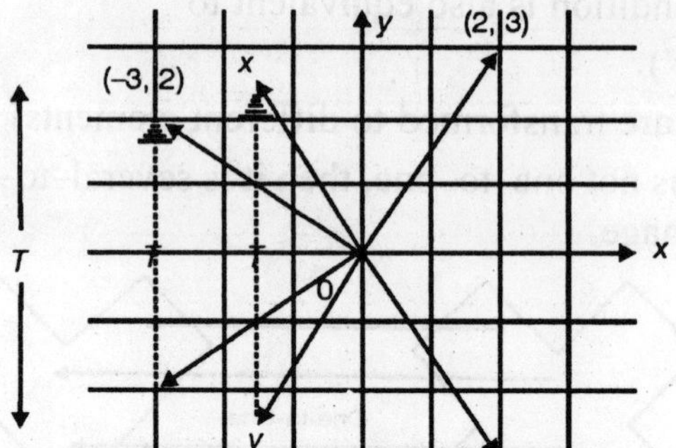

Similarly, the *Rotation* transformation is also one–to–one.

KERNEL AND UNIQUENESS

The concept of range has been introduced in the discussion of existence/onto. For *general transformations*, however, there is no similar concepts associated with uniqueness/one–to–one. For a *linear transformation* $T(x)$: $R^n \to R^m$, on the other hand, the definition of uniqueness

$T(x) = T(x') \Rightarrow x = x'$.

is equivalent to

$T(x - x') = 0 \Rightarrow x - x' = 0.$

This leads to the following definition.

kernel $T = \{x \in \mathrm{R}^n : T(x) = 0\}$

The discussion before the definition leads to the following conclusion.

A *linear* transformation T is one–to–one $\Leftrightarrow$ *kernelT* $= \{0\}$

We remark that the conclusion is comparable to the earlier observation that the following are equivalent

- Solution of $Ax = b$ is unique
- The homogeneous system $Ax = 0$ has only the trivial solution.

We also note that if T is given by a matrix A, then *kernelT* is all solutions of $Ax = 0$.

Example: The system in this example has the vector (2, 2, 6) as the unique solution. Since the uniqueness is independent of the right side, we conclude that the linear transformation

$$T\begin{bmatrix} x_1 \\ x_2 \\ x_3 \end{bmatrix} = \begin{bmatrix} 3x_1 + x_2 - x_3 \\ x_1 - x_2 + x_3 \\ 2x_1 + 2x_2 + x_3 \end{bmatrix} : \mathrm{R}^3 \to \mathrm{R}^3$$

is one–to–one.

In contrast, this system has infinitely many solutions. The corresponding linear transformation

$$T\begin{bmatrix} x_1 \\ x_2 \\ x_3 \end{bmatrix} = \begin{bmatrix} 3x_1 - x_2 + x_3 \\ x_1 + x_2 + x_3 \\ 2x_1 + x_2 - x_3 \end{bmatrix} : \mathrm{R}^3 \to \mathrm{R}^3$$

is not one–to–one.

Finally, in this example we have a system with no solution. In general, the inconsistency of the system alone would not tell us whether the corresponding linear transformation

$$T\begin{bmatrix} x_1 \\ x_2 \\ x_3 \end{bmatrix} = \begin{bmatrix} 3x_1 + x_2 - x_3 \\ x_1 - x_2 + x_3 \\ 2x_1 - x_2 + x_3 \end{bmatrix} : \mathrm{R}^3 \to \mathrm{R}^3$$

is one–to–one or not. More computation or argument is needed. In this case, fortunately, we do not need to do any computation. By applying the basic principle of linear algebra, since 3 (number of variables/columns) = 3 (number of equations/rows), the system cannot have unique solution for any right side. Therefore T is not one–to–one.

Example: To determine whether the linear transformation

$$T_1\begin{bmatrix} x_1 \\ x_2 \\ x_3 \\ x_4 \\ x_5 \\ x_6 \end{bmatrix} =: \begin{bmatrix} x_1 + 3x_2 \\ -x_1 - x_2 + x_4 + x_6 \\ 4x_2 + 4x_4 + 3x_6 \\ x_1 + 3x_2 - 2x_4 \end{bmatrix} R^6 \to R^4$$

is one–to–one or not, we find the row echelon forms of the corresponding matrix. In this example, the matrix corresponding to T_1

$$\begin{bmatrix} 1 & 3 & 2 & 0 & 1 & 0 \\ -1 & -1 & -1 & 1 & 0 & 1 \\ 0 & 4 & 2 & 4 & 3 & 3 \\ 1 & 3 & 2 & -2 & 0 & 0 \end{bmatrix}$$

has been simplified, by row operations, to

$$\begin{bmatrix} 1 & 3 & 2 & 0 & 1 & 0 \\ 0 & 2 & 1 & -1 & 0 & 1 \\ 0 & 0 & 0 & 2 & 1 & 0 \\ 0 & 0 & 0 & 0 & 0 & 1 \end{bmatrix}$$

Since not all columns are pivot, we do not have uniqueness, and the transformation T is not one–to–one.

In fact, we do not even need to make any computation in order to show T is not one–to–one, since the solutions to the corresponding 4 equations in 6 variables cannot be unique.

In order to modify the transformation to become one–to–one, we need to make sure to have all rows pivot. Since the pivots columns are 1, 2, 4, 6, we delete x_3 and x_5 to get

$$T_2\begin{bmatrix} x_1 \\ x_2 \\ x_3 \\ x_4 \end{bmatrix} = \begin{bmatrix} x_1 + 3x_2 \\ -x_1 - x_2 + x_4 + x_6 \\ 4x_2 + 4x_4 + 3x_6 \\ x_1 + 3x_2 - 2x_4 \end{bmatrix} : R^4 \to R^4$$

The row echelon form of the corresponding matrix is

$$\begin{bmatrix} 1 & 3 & 0 & 0 \\ 0 & 2 & -1 & 1 \\ 0 & 0 & 2 & 0 \\ 0 & 0 & 0 & 1 \end{bmatrix}$$

with all columns pivot, the transformation T_2 is one–to–one.

As in the case of onto, one–to–one linear transformations also has numerical consequences. Expanding on the argument in the last example, we have

A linear transformation T: $\mathbf{R}^n \to \mathbf{R}^m$ is one–to–one $\Rightarrow m \geq n$

Exercise: Which transformations are one–to–one?

1. $T_1(x_1, x_2, x_3) = (x_1 + 2x_2 + x_3, 2x_1 + x_2 + x_3, 3x_2 + x_3, 3x_1 + x_3)$

Answer: One–to–one is the same as the solutions to $T_1(x_1, x_2, x_3) = b$ is unique. According to the answer to the 2nd problem in this exercise, the solution is not unique for $b = (1, 1, 1, 1)$. Therefore T_1 is not one–to–one.

2. $T_2(x_1, x_2, x_3, x_4) = (x_2 + 2x3 + 3x_4, x_1 + 2x_2 + 3x_3 + 4x_4, 2x_1 + 3x_2 + 4x_3 + 5x_4, 3x_1 + 4x_2 + 5x_3 + 6x_4)$

Answer: The matrix for T_2 is

$$\begin{bmatrix} 0 & 1 & 2 & 3 \\ 1 & 2 & 3 & 4 \\ 2 & 3 & 4 & 5 \\ 3 & 4 & 5 & 6 \end{bmatrix}$$

By the computation in this exercise, row operations change this into

$$\begin{bmatrix} 1 & 2 & 3 & 4 \\ 0 & 1 & 2 & 3 \\ 0 & 0 & 0 & 0 \\ 0 & 0 & 0 & 0 \end{bmatrix}$$

Since not all columns are pivot, T_2 is not one–to–one.

3. $T_3(x_1, x_2, x_3, x_4) = (x_1 - 2x_2 - x_3 + 2x_4, -2x_1 + 4x_2 + 3x_3 - x_4, -x_1 + 2x_2 + x_3 - x_4)$

Answer: By the numerical consequence for one–to–one linear transformations, T_3 cannot be one–to–one.

4. $T_4(x_1, x_2, x_3, x_4) = (0.75x_1 + 3.55x_2 + 2.73x_3 + 0.39x_4, -3.24x_1 - 0.01x_2 - 6.01x_3 + 0.03x_4, 5.32x_1 + 4.22x_2 + 0.36x_3 - 7.56x_4)$

Answer: By the numerical consequence for one–to–one linear transformations, T_4 cannot be one–to–one.

5. $T_5(x_1, x_2) = (x_1 + 2x_2, 2x_1 + 3x_2, 3x_1 + 4x_2)$

Answer: The matrix for T_5

$$\begin{bmatrix} 1 & 2 \\ 2 & 3 \\ 3 & 4 \end{bmatrix}$$

can be changed by row operation to

$$\begin{bmatrix} 1 & 2 \\ 0 & -1 \\ 0 & 0 \end{bmatrix}$$

Since all columns are pivot, T_5 is one–to–one.

Exercise: Determine the conditions on the parameters so that the linear transformations are one–to–one.

1. $T_1(x, y, z) = (x + y + az, x + ay + AZ, ax + ay + AZ)$

Answer: By this exercise, row operations change the matrix of the transformation to

$$\begin{bmatrix} 1 & 1 & a \\ 0 & a-1 & 0 \\ 0 & 0 & a-a^2 \end{bmatrix}$$

T_1 is one–to–one if and only if all the columns are pivot, which means a """ 0, 1.

We also remark that it is no accident that the condition for T_1 to be one–to–one is the same as the the condition for T_1 to be onto.

2. $T_2(x_1, x_2, x_3, x_4) = (x_1 + x_2 + x_3 + x_4, x_1 + ax_4, x_2 + ax_3)$

Answer: By this exercise, row operations change the matrix of the transformation to

$$\begin{bmatrix} 1 & 1 & 1 & 1 \\ 0 & 1 & 1 & 1-a \\ 0 & 0 & a-1 & a-1 \end{bmatrix}$$

Since for any a, there is always some nonpivot column, we conclude that $\boldsymbol{T_2}$ cannot be one–to–one.

Alternatively, by the numerical consequence for one–to–one linear transformations, T_2 cannot be one–to–one.

3. $T_3(u, v, w) = (v + w, u + aw, u + v, av - w)$

Answer: In this exercise, row operations change the matrix of the transformation to

$$\begin{bmatrix} 1 & 0 & a \\ 0 & 1 & 1 \\ 0 & 0 & -a-1 \\ 0 & 0 & 0 \end{bmatrix}$$

For T_3 to be one–to–one, we need all columns pivot, which means $a \neq -1$.

Exercise: Prove that if T: $R^n \to R^m$ is a one–to–one and onto linear transformation, then $m = n$.

Answer: According this remark, T being onto implies $m \leq n$. According to this remark, T being one–to–one implies $m \geq n$. Combining the two conclusions, we get $m = n$.

Exercise: Using the equivalence between matrices and linear transformations, rephrase the basic principle of linear algebra in terms of linear transformations.

Answer: In terms of systems of linear equations, the basic principle is: For a square matrix A, the following are equivalent

- $Ax = b$ has a unique solution for any b
- $Ax = b$ has solutions for any b
- The solution of a consistent system $Ax = b$ is unique

Now the left side Ax is a linear transformation T: $R^n \to R^m$. The condition for the matrix to be square is $m = n$. The existence of solution for any b is the same as onto. The uniqueness of solution means one–to–one. Thus we may translate the basic principle to the following: For a linear transformation T: $R^n \to R^n$ between euclidean spaces of the *same dimension*, the following are equivalent

- The transformation is one–to–one and onto
- The transformation is onto
- The transformation is one–to–one

Exercise: Show that for nonlinear transformations, the uniqueness for the solution of $T(x) = b$ may depend on the right side b.

Answer: A simple example is *Square*: R → R. The equation *Square*(t) = 0 has $t = 0$ as the unique solution. However, the equation *Square*(t) = 4 has $t = 2$ or -2 as two solutions.

For another example, the *Root–Polynomial* transformation is two–to–one in case the polynomial has no double root. It is one–to–one when the polynomial has double root.

Chapter 10

Polynomial

ROOT AND MULTIPLICITY

We have used polynomials as examples of vector spaces. We also use polynomials in the computation of eigenvalue. Here we briefly sketch the basic theory of polynomial.

Polynomials can be defined over any field. A polynomial over a field F is

$p(t) = a_0 + a_1 t + a_2 t^2 + ... + a_n t^n$,

where the coefficients $a_i \in F$. Although we will only discuss polynomials with real or complex numbers as coefficients, our discussion will be valid for general fields. Whenever a claim requires complex number, the claim will still be valid for algebraically closed fields.

A root of the polynomial $p(t)$ is a number r satisfying $p(r) = 0$. We will show that

$p(r) = 0 \Leftrightarrow p(t) = (t - r)q(t)$ for another polynomial q.

By repeatedly using this, we have

$p(r) = 0 \Leftrightarrow p(t) = (t - r)^k s(t)$ for an integer $k > 0$ and another polynomial s satisfying $s(r)$ “‘‘ 0.

The natural number k is the multiplicity of the root r. Thus we have the following terminologies

$k = 1$: r is a simple root

$k = 2$: r is a double root

$k = 3$: r is a triple root

$k > 1$: r is a multiple root

For example, $-t^2 + t^4 = t^2(t + 1)(t - 1)$ has 1 and -1 as simple roots, and has 0 as a double root.

Define the derivative of a polynomial by

$p'(t) = a_1 + 2a_2t + ... + na_nt^{n-1}$.

Then we have $(p + q)' = p' + q'$, $(cp)' = cp'$, $(pq)' = p'q + pq'$. Thus

r is a simple root

$\Rightarrow p(t) = (t - r)s(t), s(r) \neq 0$

$\Rightarrow p'(t) = s(t) + (t - r)s'(t), s(r) \neq 0$

$\Rightarrow p'(r) = s(r) + (r - r)s'(r) = s(r) \neq 0.$

On the other hand,

r is a multiple root

$\Rightarrow p(t) = (t - r)^k s(t), k > 1$

$\Rightarrow p'(t) = k(t - r)^{k-1}s(t) + (t - r)^k s'(t), k > 1$

$\Rightarrow p'(r) = k(r - r)^{k-1}s(r) + (r - r)^k s'(r) = 0.$

Thus we have the following criterion for multiple roots.

r is a multiple root of $p(t) \Leftrightarrow p(r) = 0, p'(r) = 0.$

REPRESENTING POLYNOMIAL BY ROOTS

The real polynomial $1 + t^2$ does not have any real root. In order for it to have roots, we must add a special number $i = \sqrt{-1}$ to the real numbers R. This gives rise to the complex numbers C. It turns out that this extension is sufficient to enable any polynomial (not just the special $1 + t^2$) of positive degree to have roots. This striking result is the fundamental theorem of algebra.

Now consider a complex polynomial $p(t)$ of degree n. By the fundamental theorem, p has a root r_1. This implies $p(t) = (t - r_1)p_1(t)$ for another polynomial $p_1(t)$, which must have degree $n - 1$. Then by aplying the fundamental theorem to p_1, we have $p(t) = (t - r_1)p_1(t) = (t - r_1)(t - r_2)p_2(t)$ for another polynomial $p_2(t)$ of degree $n - 2$. Continuing like this, we eventually prove the following.

Any complex polynomial of degree n can be decomposed as follows

$p(t) = a(t - r_1)(t - r_2)...(t - r_n), a \neq 0.$

Assume the highest order term in $p(t)$ is t^n. Then $a = a_n = 1$. Expanding the decomposition, we get

$p(t) = (t - r_1)(t - r_2)...(t - r_n) = t^n - \sigma_1 t^{n-1} + \sigma_2 t^{n-2} + ... + (-1)^{n-1}\sigma_{n-1}t + (-1)^n\sigma_n,$

where

$\sigma_1 = \Sigma_i r_i = r_1 + r_2 + r_3 + ... + r_n,$

$\sigma_2 = \Sigma_{i<j} r_i r_j = r_1 r_2 + r_1 r_3 + r_2 r_3 + ... + r_{n-1} r_n,$

$\sigma_3 = \Sigma_{i<j<k} r_i r_j r_k = r_1 r_2 r_3 + r_1 r_3 r_4 + r_2 r_3 r_4 + ... + r_{n-2} r_{n-1} r_n,$

......

$\sigma_n = \Pi_i r_i = r_1 r_2 r_3 ... r_n,$

are the symmetric polynomials of the roots $r_1, r_2, ..., r_n$. This leads to the so called Viete Theorem.

Let $r_1, r_2, ..., r_n$ be all the roots of $p(t) = a_0 + a_1 t + a_2 t^2 + ... + a_{n-1} t^{n-1} + t^n$. Then

$a_i = (-1)^{n-i} \sigma_n(r_1, r_2, ..., r_n).$

In the special case $p(t) = a_0 + a_1 t + t^2$, $p(t)$ has two roots r_1, r_2, and we have

$a_0 = r_1 r_2,\ a_1 = -(r_1 + r_2).$

DIVISION OF POLYNOMIALS

Let a and b be integers, $b \neq 0$. Then we can always find unique integers q, r, such that

$a = qb + r,\ 0 \leq r < |b|.$

The number q is the quotient of a divided by b, and r is the remainder of the division. In case $r = 0$, we say b divides a, or b is a divisor of a, or a is divisible by b.

We have similar division for polynomials.

Let $a(t)$ and $b(t)$ be polynomials, $b(t)$ ‘‘‘ 0. Then there are unique polynomials $q(t)$, $r(t)$, such that

$a(t) = q(t)b(t) + r(t),\ 0 \leq \deg r < \deg b.$

Example: Let $a(t) = t^4 + t^3 + 3t^2 + t + 1$, $b(t) = t^2 - t + 1$. We may carry out the following divisions.

$t^2 + 2t + 4$ quotient

$t^2 - t + 1 \mid t^4 + t^3 + 3t^2 + t + 1$

$t^4 - t^3 + t^2$

$2t^3 + 2t^2 + t$

$2t^3 - 2t^2 + 2t$

$4t^2 - t + 1$

$4t^2 - 4t + 4$

remainder $3t - 3$

Each time, by trying to match the highest degree term, the degree of the remaining term is decreased by at least one. The algorithm will stop after the degree of remaining term is < *deg b*. The result of the division is

$$t^4 + t^3 + 3t^2 + t + 1 = (t^2 + 2t + 4)(t^2 - t + 1) + (3t - 3).$$

If the remainder $r(t) = 0$, then we have $a(t) = q(t)b(t)$ for a polynomial $q(t)$. Similar to the case of numbers, we say $b(t)$ divides $a(t)$, or $b(t)$ is a divisor of $a(t)$, or $a(t)$ is divisible by $b(t)$. Note that the equality $a(t) = q(t)b(t)$ has the following implication

$$r \text{ is a root of } b(t) \Rightarrow b(r) = 0 \Rightarrow a(r) = q(r)b(r) = 0.$$

This proves the following result.

$b(t)$ divides $a(t) \Rightarrow$ roots of $b(t)$ are roots of $a(t)$.

Dividing any polynomial $p(t)$ by a degree 1 polynomial $b(t) = t - r$ gives us

$$p(t) = q(t)(t - r) + c,$$

where c is a number. Thus $p(r) = c$, and

$$r \text{ is a root of } p(t) \Leftrightarrow c = 0 \Leftrightarrow p(t) = q(t)(t - r).$$

We just proved the following result.

r is a root of $p(t) \Leftrightarrow p(t) = (t - r)q(t)$ for another polynomial q.

GREATEST COMMON DIVISOR

For integers a and b, denote by $Div(a, b)$ the collection of common divisors, which are the natural numbers that divide both a and b. For example, 3 $Div(102, 45)$ because $102 = 34 \times 3$ and $45 = 15 \times 3$.

Starting from $a = 102$ and $b = 45$, we carry out the following series of divisions (called euclidean algorithm)

$$102 = 2 \times 45 + 12,$$
$$45 = 3 \times 12 + 9,$$
$$12 = 1 \times 9 + 3,$$
$$9 = 3 \times 3.$$

The first equality implies that

102 and 45 are divisible by $c \Leftrightarrow$ 45 and 12 are divisible c.

Thus $Div(102, 45) = Div(45, 12)$. In general, it is easy to see that if $a = qb + r$, then $Div(a, b) = Div(b, r)$. Then the series of divisions tells us

$$Div(102, 45) = Div(45, 12) = Div(12, 9) = Div(9, 3) = \{1, 3\},$$

where the last equality is due to the fact that 9 is a multiple of 3, so that $Div(9, 3)$ consists of divisors of 3.

The biggest number in $Div(a, b)$ is denoted $gcd(a, b)$ and is called the greatest common divisor between a and b. The euclidean algorithm above shows that $gcd(102, 45) = 3$. Since the algorithm produces smaller and smaller natural numbers, it always stops (say at (a', b')), when one number is an integer multiple of the other (say $a' = qb'$). Then $gcd(a, b) = gcd(a', b') = b'$. In particular, we conclude that for integers, the greatest common divisor always exists.

We may introduce similar concepts for polynomials.

The greatest common divisor between two polynomials $p(t)$ and $q(t)$ is a polynomial $d(t) = gcd(p(t), q(t))$ with the following properties:

- $d(t)$ divides $p(t)$ and $q(t)$.
- $c(t)$ divides $p(t)$ and $q(t) \Rightarrow c(t)$ divides $d(t)$.

Since polynomials can also be divided, the euclidean algorithm can also be carried out for polynomials. By the same reason, the algorithm produces the greatest common divisor between two polynomials. In particular, this proves the existence of the *gcd* for polynomials.

Example: From this example, we already know

$$t^4 + t^3 + 3t^2 + t + 1 = (t^2 + 2t + 4)(t^2 - t + 1) + (3t - 3).$$

Further computations give us

$$t^2 - t + 1 = (t/3)(3t - 3) + 1,$$

$$3t - 3 = (3t - 3)\ 1.$$

Thus $gcd(t^4 + t^3 + 3t^2 + t + 1, t^2 - t + 1) = 1$.

If both $d_1(t)$ and $d_2(t)$ satisfy the conditions for *gcd*, then $d_1(t) = u(t)d_2(t)$ and $d_2(t) = v(t)d_1(t)$ for some polynomials $u(t)$ and $v(t)$. This implies $u(t)v(t) = 1$, so that both $u(t)$ and $v(t)$ are actually nonzero numbers. Thus we conclude that *gcd* for polynomials is unique up to a nonzero number multiple.

By tracing the euclidean algorithm between 102 and 45 backwards, we have

$$\begin{aligned} 3 &= 12 - 1\times9 \\ &= 12 - 1\times(45 - 3\times12) = (-1)\times45 + (1 + 3)\text{times;}12 \\ &= (-1)\times45 + 4\times(102 - 2\times45) = 4\times102 + (-1 - 8)\times45 \\ &= 4\times102 - 9\times45. \end{aligned}$$

Similarly, from the example above, we have

$$1 = (t^2 - t + 1) - (t/3)(3t - 3)$$
$$= (t^2 - t + 1) - (t/3)[(t^4 + t^3 + 3t^2 + t + 1) - (t^2 + 2t + 4)(t^2 - t + 1)]$$
$$= (-t/3)(t^4 + t^3 + 3t^2 + t + 1) + [1 + (t/3)(t^2 + 2t + 4)](t^2 - t + 1)$$
$$= [(-1/3)t](t^4 + t^3 + 3t^2 + t + 1) + [(1/3)t^3 + (2/3)t^2 + (4/3)t + 1](t^2 - t + 1).$$

Such computations can always be carried out in general, and we the following conclusion.

$d(t) = gcd(p(t), q(t)) \Rightarrow d(t) = u(t)p(t) + v(t)q(t)$ for some polynomials $u(t)$ and $v(t)$.

COPRIME POLYNOMIALS

Two integers are coprime if their *gcd* is 1. Similarly, two polynomials $p(t)$ and $q(t)$ are coprime if $gcd(p(t), q(t)) = 1$ (or any nonzero number). Specifically this means that

$c(t)$ divides $p(t)$ and $q(t) \Rightarrow c(t)$ is a nonzero constant.

From this example, we find that the polynomials $t^4 + t^3 + 3t^2 + t + 1$ and $t^2 - t + 1$ are coprime.

For two complex polynomials $p(t)$ and $q(t)$, the following are equivalent.

- $p(t)$ and $q(t)$ are coprime.
- There are polynomials $u(t)$ and $v(t)$, such that $u(t)p(t) + v(t)q(t) = 1$.
- $p(t)$ and $q(t)$ do not have common roots.

Proof. Applying this result to the special case that $d(t) = 1$, we find that the first property implies the second.

Now assume $u(t)p(t) + v(t)q(t) = 1$ for some polynomials $u(t)$ and $v(t)$. If r is a of $p(t)$, then we have $p(r) = 0$, so that $v(r)q(r) = u(r)p(r) + v(r)q(r) = 1$. In particular, this implies $q(r) \neq 0$. This proves that any root of $p(t)$ cannot be a root of $q(t)$. This completes the proof that the second property implies the third.

Finally, if $p(t)$ and $q(t)$ are not coprime, then $gcd(p(t), q(t)) = d(t)$ is a polynomial of positive degree. By the fundamental theorem of algebra, $d(t)$ must have a complex root r. Since $d(t)$ divides $p(t)$ and $q(t)$, by this result, r is also a common root of $p(t)$ and $q(t)$. This completes the proof that the third property implies the first.

How can we determine whether two given polynomials have common roots? Let

$$p(t) = a_0 + a_1t + a_2t^2 + ... + a_{m-1}t^{m-1} + t^m = (t - r_1)(t - r_2)...(t - r_m),$$
$$q(t) = b_0 + b_1t + b_2t^2 + ... + b_{n-1}t^{n-1} + t^n = (t - s_1)(t - s_2)...(t - s_n).$$

Then for $p(t)$ and $q(t)$ to have a common root means that the following expression, called the resultant, is zero.

$$Res(p, q) = \Pi_{ij}(r_i - s_j) = q(r_1)q(r_2)\ldots q(r_m) = (-1)^{mn}p(s_1)p(s_2)\ldots p(s_n)$$
$$= (r_1 - s_1)(r_1 - s_2)\ldots(r_1 - s_n)\,(r_2 - s_1)(r_2 - s_2)\ldots(r_2 - s_n)\ldots\ldots(r_m - s_1)(r_m - s_2)\ldots(r_m - s_n).$$

Note that the expression is symmetric in the roots $r_1, r_2, \ldots r_m$ of $p(t)$, and is also symmetric in the roots $s_1, s_2, \ldots, s_n$ of $q(t)$. On the other hand, we have seen in Viete theorem that the coefficients of a polynomial are also the symmetric polynomials of the roots. This suggests that $Res(p, q)$ may be expressed in terms of the coefficients in $p(t)$ and $q(t)$.

Example: Let $p(t) = a_0 + a_1 t + t^2$ and $q(t) = b_0 + b_1 t + t^2$ be of second order. Then by $r_1 + r_2 = -a_1$ and $r_1 r_2 = a_0$, we have

$$Res(p, q) = (b_0 + b_1 r_1 + r_1^2)(b_0 + b_1 r_2 + r_2^2)$$
$$= b_0^2 + b_0 b_1(r_1 + r_2) + b_0(r_1^2 + r_2^2) + b_1^2 r_1 r_2 + b_1 r_1 r_2(r_1 + r_2) + r_1^2 r_2^2$$
$$= b_0^2 + b_0 b_1(-a_1) + b_0((-a_1)^2 - 2a_0) + b_1^2 a_0 + b_1 a_0(-a_1) + a_0^2$$
$$= a_0^2 + b_0^2 - 2a_0 b_0 + a_1^2 b_0 + a_0 b_1^2 - a_1 b_0 b_1 - a_0 a_1 b_1.$$

Chapter 11

Complex Diagonalization

COMPLEX EIGENVALUE AND EIGENVECTOR

By the Fundamental Theorem of Algebra, any (real or complex) polynomial have roots, as long as complex numbers are allowed. In fact, if we count the repeated roots, a polynomial of degree *n* has *n* roots.

Since the eigenvalues were computed as the roots of the characteristic polynomial, we need to consider the possibility of complex eigenvalues, even for real matrices.

Example: Consider the matrix

$$A = \begin{bmatrix} 1 & 5 \\ -2 & 3 \end{bmatrix}.$$

The characteristic polynomial is $det(A - \lambda I) = (1 - \lambda)(3 - \lambda) - (-2)5 = 13 - 4\lambda + \lambda^2$, and A has two eigenvalues $\lambda_1 = 2 + 3i$, $\lambda_2 = 2 - 3i$.

For $\lambda_1 = 2 + 3i$, we have

$$A - (2 + 3i)I = \begin{bmatrix} -1 - 3i & 5 \\ -2 & 1 - 3i \end{bmatrix}$$

The two equations $(-1 - 3i)x_1 + 5x_2 = 0$ and $-2x_1 + (1 - 3i)x_2 = 0$ are equivalent (you may verify that $(1 - 3i)$[equation 1] = 2[equation 2]). The solutions of $(A - (2 + 3i)I)x = 0$ are of the form $x = c(1 - 3i, 2)$, c arbitrary (complex number). We get an eigenvector $v_1 = (1 - 3i, 2)$, which is really a basis of the *complex* eigenspace $nul(A - (2 + 3i)I)$ (a subspace of C^2).

For $\lambda_1 = 2 - 3i$, we have

$$A - (2 - 3i)I = \begin{bmatrix} -1 + 3i & 5 \\ -2 & 1 + 3i \end{bmatrix}$$

The solutions of $(A-(2-3i)I)x=0$ are of the form $x=c(1+3i, 2)$, c arbitrary. We get an eigenvector $v_2=(1+3i, 2)$, which is really a basis of the eigenspace $nul(A-(2-3i)I)$.

We conclude that A is diagonalizable, with diagonalization

$$A=\begin{bmatrix}1+3i & 1-3i\\ 2 & 2\end{bmatrix}\begin{bmatrix}2+3i & 0\\ 0 & 2-3i\end{bmatrix}\begin{bmatrix}1+3i & 1-3i\\ 2 & 2\end{bmatrix}^{-1}.$$

We can also try to diagonalize complex matrices.

Example: The matrix

$$A=\begin{bmatrix}i & -i & 1\\ -1 & 1 & i\\ 0 & 0 & 1+1\end{bmatrix}$$

has the characteristic polynomial is $det(A-\lambda I)=[(i-\lambda)(1-\lambda)-i](1+i-\lambda)=-\lambda(1+i-\lambda)^2$. We get two eigenvalues $\lambda_1=0$, $\lambda_2=1+i$.

For $\lambda_1=0$, we solve $Ax=0$ and find solutions $x=x_2(1, 1, 0)$, x_2 arbitrary. Thus $v_1=(1, 1, 0)$ form a basis of the eigenspace $nulA$.

For $\lambda_2=1+i$, we have

$$A-(1+i)I=\begin{bmatrix}-1 & -i & 1\\ -1 & -i & i\\ 0 & 0 & 0\end{bmatrix}.$$

The solutions of $(A-(1+i)I)x=0$ are of the form $x=x_2(-i, 1, 0)$, x_2 arbitrary. Thus $v_2=(-i, 1, 0)$ form a basis of the eigenspace $nul(A-(1+i)I)$.

The matrix is not diagonalizable.

COMPLEX EIGENVALUE AND EIGENVECTOR FOR REAL MATRIX

The complex conjugation A of a matrix A is obtained by taking the complex conjugations of the entries of A. For example, the complex conjugation of

$$A=\begin{bmatrix}i & -1 & 1\\ -1 & 1 & i\\ 0 & 0 & 1+i\end{bmatrix},$$

is

$$A=\begin{bmatrix}-i & i & 1\\ -1 & 1 & -i\\ 0 & 0 & 1-i\end{bmatrix}.$$

It is easy to see that the conjugation operation has the following properties.

$$\overline{A+B}+\overline{A}+\overline{B},\quad \overline{cA}=\overline{c}\overline{A},\quad \overline{AB}=\overline{A}\overline{B},\quad \overline{(A^1)}=(\overline{A}),(\overline{A})^{T},\quad \overline{A}=B\Leftrightarrow \overline{B}=A$$

As a consequence, we have $Av = \lambda v \Leftrightarrow Av = \lambda v$, where v is obtained by taking complex conjugates of the coordinates of v. Combined with the following fact

A is a real matrix $\Leftrightarrow A = A$,

we have

For a *real* matrix, v is an eigenvector with eigenvalue $\lambda \Leftrightarrow v$ is an eigenvector with eigenvalue λ.

Example: Consider the linear transformation $T(x_1, x_2) = (x_1 + 5x_2, -2x_1 + 3x_2)$: $R^2 \to R^2$ given by the matrix

$$A = \begin{bmatrix} 1 & 5 \\ -2 & 3 \end{bmatrix}$$

In an earlier example, we found eigenvectors $v = (1 - 3i, 2)$, $v = (1 + 3i, 2)$ with eigenvalues $\lambda = 2 + 3i$, $\lambda = 2 - 3i$.

To understand the transformation T on the real euclidean space R^2, we write

$$v = u + iw, \text{ for } u = (1, 2) \text{ and } w = (-3, 0).$$

By comparing the real and the imaginery parts of

$$T(v) = T(u) + iT(w),$$

$$\lambda v = (2 + 3i)(u + iw) = (2u - 3w) + i(3u + 2w),$$

we find

$$T(u) = 2u - 3w,$$

$$T(w) = 3u + 2w.$$

In other words, the matrix for T with respect to the basis $\{u, w\}$ of R^2 is

$$B = \begin{bmatrix} 2 & 3 \\ -3 & 2 \end{bmatrix}.$$

If we further make use of the polar expression of the eigenvalue:

$2 + 3i = \sqrt{13}e^{i\theta} = \sqrt{13}\cos\theta + i\sqrt{13}\sin\theta$, where $\theta = tan^{-1}(2/3) \approx 33.69°$,

we find

$$B = \sqrt{13}\begin{bmatrix} \cos(-\theta) & -\sin(-\theta) \\ \sin(-\theta) & \cos(-\theta) \end{bmatrix}.$$

Thus if we use the coordinate system with respect to the basis $\{(1, 2), (-3, 0)\}$, then T appears to be a rotation by angle $-33.69°$ followed by multiplying $\sqrt{13}$.

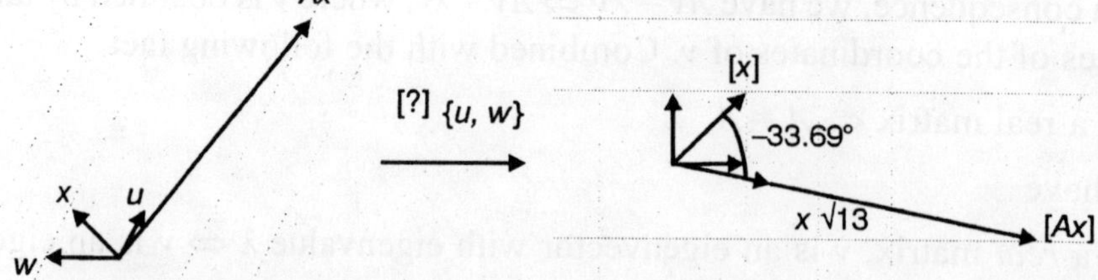

In general, suppose a *real* linear transformation $T: R^2 \to R^2$ has a conjugate pair of *truly complex* eigenvalues $\lambda = \mu + iv$ and $\overline{\lambda} = \mu - iv$, where $v \neq 0$. Let $v = u + iw$ and $\overline{v} = u - iw$ be the corresponding eigenvectors. Then $\{u, v\}$ form a basis of R^2, and the matrix of T with respect to the basis is

$$\Lambda = \begin{bmatrix} \mu & v \\ -v & \mu \end{bmatrix}$$

Geometrically, if we use the polar expression: $\lambda = re^{i\theta} = r(\cos\theta + i\sin\theta)$, then T is a rotation by angle $-\theta$ and scaling by r with respect to the $\{u, v\}$–coordinates.

COMPLEX EIGENVALUE IN REAL DIAGONALIZATION

In the previous part, we studied the geometric meaning of one conjugate pair of complex eigenvalues and eigenvectors. In this part, we consider several such pairs.

We illustrate the situation by considering a real linear transformation $T: R^5 \to R^5$. We assume T has a *complex* basis (for C^5) of eigenvectors

$$v_1, v_2 = u_2 + iw_2, \overline{v_2} = u_2 - iw_2, v_3 = u_3 + iw_3, \overline{v_3} = u_3 - iw_3,$$

with corresponding eigenvalues

$$\lambda_1, \lambda_2 = \mu_2 + iv_2, \overline{\lambda}_2 = \mu_2 - iv_2, \overline{\lambda}_3 = \mu_3 + iv_3, \lambda_3 = \mu_3 - iv_3.$$

First we prove that the *real* vectors v_1, u_2, w_2, u_3, w_3 form a basis of R^5. In fact, for *real* numbers c_1, a_2, b_2, a_3, b_3, we have

$$c_1v_1 + a_2u_2 + b_2w_2 + a_3u_3 + b_3w_3 = c_1v_1 + c_2v_2 + \overline{c_2}\overline{v_2} + c_3v_3 + \overline{c_3}\overline{v_3},$$

where $c_2 = (a_2 - ib_2)/2$, $c_3 = (a_3 - ib_3)/2$. Then we have

$$c_1v_1 + a_2u_2 + b_2w_2 + a_3u_3 + b_3w_3 = 0$$

$$\Rightarrow c_1v_1 + c_2v_2 + \overline{c_2}\overline{v_2} + c_3v_3 + \overline{c_3}\overline{v_3} = 0$$

$\Rightarrow c_1 = c_2 = c_3 = 0$

$\Rightarrow c_1 = a_2 = b_2 = a_3 = b_3 = 0$

This means that v_1, u_2, w_2, u_3, w_3 are linear independent vectors. By this result, they must form a basis of R^5.

Similar to the computation in the example in the previous part, we have

$$T(v_1) = \lambda_1 v_1,$$
$$T(u_2) = \mu_2 u_2 - v_2 w_2,$$
$$T(w_2) = v_2 u_2 + \mu_2 w_2,$$
$$T(u_3) = \mu_3 u_3 - v_3 w_3,$$
$$T(w_3) = v_3 u_3 + \mu_3 w_3.$$

Note that the basis $\{v_1, u_2, w_2, u_3, w_3\}$ gives us a direct sum $span\{v_1\} \oplus span\{u_2, w_2\} E span\{u_3, w_3\} = R^5$. This gives the geometric interpretation below.

Let V be a finite dimensional *real* vector space. Let T: $V \rightarrow V$ be *complex* diagonalizable. Then V is decomposed as a direct sum of 1-dimensional subspaces $span\{v_i\}$ and 2-dimensional subspaces $span\{u_j, w_j\}$, such that

- v_i is a real eigenvector of T, and T restricts to a scalar multiplication on $span\{v_i\}$.
- $u_j + iw_j$ and $u_j - iw_j$ is a pair of complex eigenvectors of T, and T restricts to a rotation and scalar multiplication on $span\{u_j, w_j\}$.

The scalar multiplication on $span\{v_i\}$ is by the real eigenvalue associated to v_i. The rotation and scalar multiplication on $span\{u_j, w_j\}$ is determined by the polar expressions of the complex eigenvalue associated to $u_j + iw_j$.

Finally let us consider the discussion from the matrix viewpoint. Let the 5 by 5 matrix A be the usual matrix of T: $R^5 \rightarrow R^5$ (with respect to the standard basis). Then we have $A = PDP^{-1}$, with

$$P = [v_1\ u_2\ w_2\ u_3\ w_3],\ D = \begin{bmatrix} \lambda_1 & 0 & 0 & 0 & 0 \\ 0 & \mu_2 & v_2 & 0 & 0 \\ 0 & -v_2 & \mu_2 & 0 & 0 \\ 0 & 0 & 0 & \mu_3 & v_3 \\ 0 & 0 & 0 & -v_3 & \mu_3 \end{bmatrix}.$$

Note that D is a block diagonal matrix, obtained by replacing the conjugate pairs of complex eigenvalues in the usual complex diagonal matrix by special 2 by 2 matrices.

Exercise: Find the eigenvalues and eigenvectors. Then find diagonalization if possible.

1. $\begin{bmatrix} i & 1 \\ 1 & -i \end{bmatrix}$

Answer:

$$det\begin{bmatrix} i-\lambda & 1 \\ 1 & -i-\lambda \end{bmatrix} = (i-\lambda)(-i-\lambda) - 1 = \lambda^2.$$

The only eigenvalue is $\lambda = 0$. The general solution of the system (same as $Ax = 0$).

$$ix_1 + x_2 = 0$$
$$x_1 - ix_2 = 0$$

is $x = (ix_2, x_2)$, with x_2 arbitrary (complex number). Thus $v = (i, 1)$ is a basis of the eigenspace *nulA*. The matrix is not diagonalizable.

2. $\begin{bmatrix} i & 1 \\ 1 & i \end{bmatrix}$

Answer:

$$det\begin{bmatrix} i-\lambda & 1 \\ 1 & i-\lambda \end{bmatrix} = (i-\lambda)^2 - 1 = (1+i-\lambda)(-1+i-\lambda).$$

For $\lambda_1 = 1 + i$, we have

$$A - (1+i)I = \begin{bmatrix} -1 & 1 \\ 1 & -1 \end{bmatrix},$$

and $v_1 = (1, 1)$ is a basis of the eigenspace $nul(A - (1 + i)I)$. For $\lambda_2 = -1 + i$, we have

$$A - (-1+i)I = \begin{bmatrix} 1 & 1 \\ 1 & 1 \end{bmatrix},$$

and $v_2 = (1, -1)$ is a basis of the eigenspace $nul(A - (-1 + i)I)$. A diagonalization of A is

$$A = \begin{bmatrix} 1 & 1 \\ 1 & -1 \end{bmatrix}\begin{bmatrix} 1+i & 0 \\ 0 & -1+i \end{bmatrix}\begin{bmatrix} 1 & 1 \\ 1 & -1 \end{bmatrix}^{-1}.$$

3. $\begin{bmatrix} 1 & -i \\ i & 1 \end{bmatrix}$

Answer:

$$det\begin{bmatrix} i-\lambda & -i \\ i & i-\lambda \end{bmatrix} = (1-\lambda)^2 - 1 = (-\lambda)(2-\lambda).$$

For $\lambda_1 = 0$, $v_1 = (i, 1)$ is a basis of the eigenspace *nulA*. For $\lambda_2 = 2$, we have

$$A - 2I = \begin{bmatrix} -1 & -i \\ i & -1 \end{bmatrix},$$

and $v_2 = (i, -1)$ is a basis of the eigenspace *nul*$(A - 2I)$. A diagonalization of A is

$$A = \begin{bmatrix} i & i \\ 1 & -1 \end{bmatrix} \begin{bmatrix} 0 & 0 \\ 0 & 2 \end{bmatrix} \begin{bmatrix} i & i \\ 1 & -1 \end{bmatrix}^{-1}.$$

4. $\begin{bmatrix} 1+2i & 0 & 0 \\ 1+2i & 2+3i & 0 \\ 1+2i & 2+3i & 3+4i \end{bmatrix}$

Answer: The lower triangular matrix has the diagonal entries $\lambda_1 = 1 + 2i$, $\lambda_2 = 2 + 3i$, $\lambda_3 = 3 + 4i$ as the eigenvalues. For $\lambda_1 = 1 + 2i$, we have

$$A - (1 + 2i)I = \begin{bmatrix} 0 & 0 & 0 \\ 1+2i & 1+i & 0 \\ 1+2i & 2+3i & 2+2i \end{bmatrix}$$

and $v_1 = (4/5 + 2/5i, -1 - i, 1/2 + i)$ is a basis of the eigenspace. For $\lambda_2 = 2 + 3i$, we have

$$A - (2 + 3i)I = \begin{bmatrix} -1-i & 0 & 0 \\ 1+2i & 1+i & 0 \\ 1+2i & 2+3i & 1+i \end{bmatrix}$$

and $v_2 = (0, -1 - i, 2 + 3i)$ is a basis of the eigenspace. For $\lambda_3 = 3 + 4i$, we have

$$A - (3 + 4i)I = \begin{bmatrix} -2-2i & 0 & 0 \\ 1+2i & -1-i & 0 \\ 1+2i & 2+3i & 0 \end{bmatrix}$$

and $v_3 = (0, 0, 1)$ is a basis of the eigenspace. A diagonalization is

$$A = \begin{bmatrix} 4/5+2/5i & 0 & 0 \\ -1-i & -1-i & 0 \\ 1/2+i & 2+3i & 1 \end{bmatrix} \begin{bmatrix} 1+2i & 0 & 0 \\ 0 & 2+3i & 0 \\ 0 & 0 & 3+4i \end{bmatrix} \begin{bmatrix} 4/5+2/5i & 0 & 0 \\ -1-i & -1-i & 0 \\ 1/2+i & 2+3i & 1 \end{bmatrix}^{-1}$$

5. $\begin{bmatrix} 1+2i & 1 & 1 \\ 0 & 1+i & 1 \\ 0 & -1 & 1+i \end{bmatrix}$

Answer:

$$det\begin{bmatrix} 1+2i-\lambda & 1 & 1 \\ 0 & 1+i-1 & 1 \\ 0 & -1 & 1+i-\lambda \end{bmatrix} = (1 + 2i - \lambda)[(1 + i - \lambda)^2 + 1] = (1 + 2i - \lambda)^2(1 - \lambda).$$

For $\lambda_1 = 1 + 2i$, we have

$$A - (1 + 2i)I = \begin{bmatrix} 0 & 1 & 1 \\ 0 & -i & 1 \\ 0 & 1 & -i \end{bmatrix}$$

and $v_1 = (1, 0, 0)$ is a basis of the eigenspace. For $\lambda_2 = 1$, we have

$$A - I = \begin{bmatrix} 2i & 1 & 1 \\ 0 & i & 1 \\ 0 & -1 & i \end{bmatrix}$$

and $v_2 = (1 - i, 2i, 2)$ is a basis of the eigenspace. The matrix is not diagonalizable.

Exercise: Find the eigenvalues and eigenvectors. Then find diagonalization if possible.

1. $\begin{bmatrix} 4 & 2 \\ -1 & 2 \end{bmatrix}$

Answer:

$$det\begin{bmatrix} 4-\lambda & 2 \\ -1 & 2-\lambda \end{bmatrix} = 10 - 6\lambda + \lambda^2 = (3 + i - \lambda)(3 - i - \lambda).$$

For $\lambda = 3 + i$. we have

$$A - (3 + i)I = \begin{bmatrix} 1-i & 2 \\ -1 & -1-i \end{bmatrix},$$

and $v = (1 + i, -1)$ is a basis of the eigenspace $nul(A - (3 + i)I)$. Since the matrix is real, by taking conjugate, we also have the basis $v = (1 - i, -1)$ of the eigenspace $nul(A - (3 - i)I)$. A diagonalization is

$$A=\begin{bmatrix}1+i & 1-i\\ -1 & -1\end{bmatrix}\begin{bmatrix}3+i & 0\\ 0 & 3-i\end{bmatrix}\begin{bmatrix}1+i & 1-i\\ -1 & -1\end{bmatrix}^{-1}.$$

On the other hand, we may separate real and imaginary parts for the eigenvector of $\lambda = 3 + i$

$v = (1 + i, -1) = (1, -1) + i(1, 0) = u + iw.$

Then we have the real block (there is only one block) diagonalization

$$A=\begin{bmatrix}1 & 1\\ -1 & 0\end{bmatrix}\begin{bmatrix}3 & 1\\ -1 & 3\end{bmatrix}\begin{bmatrix}1 & 1\\ -1 & 0\end{bmatrix}^{-1}.$$

2. $\begin{bmatrix}1 & 0 & 4\\ -1 & 2 & 1\\ -3 & 3 & 2\end{bmatrix}$

Answer: By [col 2] + [col 1] and [row 1] + [row 2], we have

$$det\begin{bmatrix}1-\lambda & 0 & 4\\ -1 & 2-\lambda & 1\\ -3 & 3 & 2-\lambda\end{bmatrix} = det\begin{bmatrix}1-\lambda & 0 & 4\\ 1-\lambda & 2-\lambda & 1\\ 0 & 3 & 2-\lambda\end{bmatrix} = det\begin{bmatrix}1-\lambda & 0 & 4\\ 0 & 2-\lambda & -3\\ 0 & 3 & 2-\lambda\end{bmatrix}$$

$$= (1-\lambda)det\begin{bmatrix}2-\lambda & -3\\ 3 & 2-\lambda\end{bmatrix} = (1-\lambda)(2+3i-\lambda)(2-3i-\lambda).$$

For $\lambda_1 = 1$, we have

$$A-I=\begin{bmatrix}0 & 0 & 4\\ -1 & 1 & 1\\ -3 & 3 & 1\end{bmatrix}$$

and $v_1 = (1, 1, 0)$ is a basis of the eigenspace. For $\lambda_2 = 2 + 3i$, we have

$$A-(2+3i)I=\begin{bmatrix}-1-3i & 0 & 4\\ -1 & -3i & 1\\ -3 & 3 & -3i\end{bmatrix}$$

and $v_2 = (2 - 2i, 1, 2 + i)$ is a basis of the eigenspace. By taking complex conjugate, for $\lambda_2 = 2 - 3i$, we have a basis $v_2 = (2 + 2i, 1, 2 - i)$ of the eigenspace. A diagonalization is

$$A = \begin{bmatrix} 1 & 2-2i & 2+2i \\ 1 & 1 & 1 \\ 0 & 2+i & 2-i \end{bmatrix} \begin{bmatrix} 1 & 0 & 0 \\ 0 & 2+3i & 0 \\ 0 & 0 & 2-3i \end{bmatrix} \begin{bmatrix} 1 & 2-2i & 2+2i \\ 1 & 1 & 1 \\ 0 & 2+i & 2-i \end{bmatrix}^{-1}$$

On the other hand, the decomposition

$v_2 = (2 - 2i, 1, 2 + i) = (2, 1, 2) + i(-2, 0, 1)$

gives us the real block (there is one 1 by 1 block and one 2 by 2 block) diagonalization

$$A = \begin{bmatrix} 1 & 2 & -2 \\ 1 & 1 & 0 \\ 0 & 2 & 1 \end{bmatrix} \begin{bmatrix} 1 & 0 & 0 \\ 0 & 2 & 3 \\ 0 & -3 & 2 \end{bmatrix} \begin{bmatrix} 1 & 2 & -2 \\ 1 & 1 & 0 \\ 0 & 2 & 1 \end{bmatrix}^{-1}$$

3. $\begin{bmatrix} 4 & 2 & 1 & 3 \\ -1 & 2 & 3 & 4 \\ 0 & 0 & 1 & 5 \\ 0 & 0 & -1 & 5 \end{bmatrix}$

Answer:

$$det\begin{bmatrix} 4-\lambda & 2 & 1 & 3 \\ -1 & 2-\lambda & 3 & 4 \\ 0 & 0 & 1-\lambda & 5 \\ 0 & 0 & -1 & 5-\lambda \end{bmatrix}$$

$$det\begin{bmatrix} 4-\lambda & 2 \\ -1 & 2-\lambda \end{bmatrix} = det\begin{bmatrix} 1-\lambda & 5 \\ -1 & 5-\lambda \end{bmatrix} = (3 + i - \lambda)^2 (3 - i - \lambda)^2.$$

For $\lambda = 3 + i$, we have

$$A - (3+i)I = \begin{bmatrix} 1-i & 2 & 1 & 2 \\ -1 & -1-i & 3 & 4 \\ 0 & 0 & -2-i & 5 \\ 0 & 0 & -1 & 2-i \end{bmatrix}$$

By row operations, we get

$$\begin{bmatrix} -1 & -1-i & 3 & 4 \\ 0 & 0 & -1 & 2-i \\ 0 & 0 & 0 & 11-14i \\ 0 & 0 & 0 & 0 \end{bmatrix}.$$

Thus $v = (1 + i, -1, 0, 0)$ is a basis of the eigenspace. By taking complex conjugate, $\bar{v} = (1 - i, -1, 0, 0)$ is a basis of the eigenspace for $\bar{\lambda} = 3 - i$. The matrix is not diagonalizable.

4. $\begin{bmatrix} 4 & 2 & -1 & 1 \\ -1 & 2 & 0 & 1 \\ 0 & 0 & 1 & 5 \\ 0 & 0 & -1 & 5 \end{bmatrix}$

Answer: By the same computation as in the previous example, we find the eigenvalues $\lambda = 3 + i$ and $\lambda = 3 - i$. For $\lambda = 3 + i$, we have

$$A - (3 + i)I = \begin{bmatrix} 1-i & 2 & -1 & 1 \\ -1 & -1-i & 0 & 1 \\ 0 & 0 & -2-i & 5 \\ 0 & 0 & -1 & 2-i \end{bmatrix}$$

By row operations, we get

$$\begin{bmatrix} -1 & -1-i & 0 & 1 \\ 0 & 0 & -1 & 2-i \\ 0 & 0 & 0 & 0 \\ 0 & 0 & 0 & 0 \end{bmatrix}$$

Thus $v_1 = (1 + i, -1, 0, 0)$ and $v_2 = (1, 0, 2 - i, 1)$ form a basis of the eigenspace. By taking complex conjugate, $\bar{v}_1 = (1 - i, -1, 0, 0)$ and $\bar{v}_2 = (1, 0, 2 + i, 1)$ form a basis of the eigenspace for $\bar{\lambda} = 3 - i$. A diagonalization of the matrix is

$$A = \begin{bmatrix} 1+i & 1-i & 1 & 1 \\ -1 & -1 & 0 & 0 \\ 0 & 0 & 2-i & 2+i \\ 0 & 0 & 1 & 1 \end{bmatrix} \begin{bmatrix} 3+i & 0 & 0 & 0 \\ 0 & 3-i & 0 & 0 \\ 0 & 0 & 3+i & 0 \\ 0 & 0 & 0 & 3-i \end{bmatrix} \begin{bmatrix} 1+i & 1-i & 1 & 1 \\ -1 & -1 & 0 & 0 \\ 0 & 0 & 2-i & 2+i \\ 0 & 0 & 1 & 1 \end{bmatrix}^{-1}$$

On the other hand, the decompositions

$v_1 = (1 + i, -1, 0, 0) = (1, -1, 0, 0) + i(1, 0, 0, 0)$,

$v_2 = (1, 0, 2 + i, 1) = (1, 0, 2, 1) + i(0, 0, 1, 0)$,

give us a real block (there are two 2 by 2 blocks) diagonalization

$$A = \begin{bmatrix} 1 & 1 & 1 & 0 \\ -1 & 0 & 0 & 0 \\ 0 & 0 & 2 & 1 \\ 0 & 0 & 1 & 0 \end{bmatrix} \begin{bmatrix} 3 & 1 & 0 & 0 \\ -1 & 0 & 0 & 0 \\ 0 & 0 & 2 & 1 \\ 0 & 0 & 1 & 0 \end{bmatrix} \begin{bmatrix} 1 & 1 & 1 & 0 \\ -1 & 0 & 0 & 0 \\ 0 & 0 & 2 & 1 \\ 0 & 0 & 1 & 0 \end{bmatrix}^{-1}.$$

Another way of expressing this is

$$A = \begin{bmatrix} 1 & 1 & 1 & 0 \\ -1 & 0 & 0 & 0 \\ 0 & 2 & 0 & 1 \\ 0 & 1 & 0 & 0 \end{bmatrix} \begin{bmatrix} 3 & 0 & 1 & 0 \\ 0 & 3 & 0 & 1 \\ -1 & 0 & 3 & 0 \\ 0 & -1 & 0 & 3 \end{bmatrix} \begin{bmatrix} 1 & 1 & 1 & 0 \\ -1 & 0 & 0 & 0 \\ 0 & 2 & 0 & 1 \\ 0 & 1 & 0 & 0 \end{bmatrix}^{-1}.$$

APPLICATION

DISCRETE DYNAMICAL SYSTEM

We have already seen the use of eigenvalue and eigenvector in the computation of powers of matrix such as

$$\begin{bmatrix} 13 & -4 \\ -4 & 7 \end{bmatrix}^k.$$

In a general *linear* discrete dynamical system, we are given an initial vector $x_0 \Xi V$ and a linear transformation $T: V \to V$. Then we have a sequence

$$x_0,$$
$$x_1 = T(x_0),$$
$$x_2 = T(x_1) = T^2(x_0),$$
$$x_3 = T(x_2) = T^3(x_0),$$
$$\vdots$$
$$x_k = T(x_{k-1}) = T^k(x_0),$$
$$\vdots$$

Suppose T is *diagonalizable*, with a basis $v_1, v_2,..., v_n$ of eigenvectors and the corresponding eigenvalues $\lambda_1, \lambda_2,..., \lambda_n$. Then we may express x_0 as a linear combination of the basis vectors

$$x_0 = c_1v_1 + c_2v_2 + ... + c_nv_n,$$

and obtain the formula for x_k:

$$x_k = c_1T^k(v_1) + c_2T^k(v_2) + ... + c_nT^k(v_n) = c_1\lambda_1{}^kv_1 + c_2\lambda_1{}^kv_2 + ... + c_n\lambda_n{}^kv_n.$$

Example: Suppose the three sequences

$x_1, x_2, \ldots, x_k, \ldots$

$y_1, y_2, \ldots, y_k, \ldots$

$z_1, z_2, \ldots, z_k, \ldots$

start with $x_1 = y_1 = z_1 = 1$ and are related by the recursive relation

$$x_{k+1} = x_k + 3y_k - 3z_k$$

$$y_{k+1} = -3x_k + 7y_k - 3z_k$$

$$z_{k+1} = -6x_k + 6y_k - 2z_k$$

Then we denote

$$\mathbf{x}_k = \begin{bmatrix} x_k \\ y_k \\ z_k \end{bmatrix}, A = \begin{bmatrix} 1 & 3 & -3 \\ -3 & 7 & -3 \\ -6 & 6 & -2 \end{bmatrix}$$

and get $\mathbf{x}_k = A\mathbf{x}_{k-1} = A^{k-1}\mathbf{x}_1$. In an earlier example, we found a basis of eigenvectors (1, 1, 0), (1, 0, –1), (1, 1, 2) with eigenvalues 4, 4, –2. Since

$$\mathbf{x}_1 = \begin{bmatrix} 1 \\ 1 \\ 1 \end{bmatrix} = (1/2)\begin{bmatrix} 1 \\ 1 \\ 0 \end{bmatrix} + 0\begin{bmatrix} 1 \\ 0 \\ -1 \end{bmatrix} + (1/2)\begin{bmatrix} 1 \\ 1 \\ 2 \end{bmatrix},$$

we conclude that

$$\mathbf{x}_k = A^{k-1}\begin{bmatrix} 1 \\ 1 \\ 1 \end{bmatrix} = (1/2)4^{k-1}\begin{bmatrix} 1 \\ 1 \\ 0 \end{bmatrix} + (1/2)(-2)^{k-1}\begin{bmatrix} 1 \\ 1 \\ 2 \end{bmatrix} = \begin{bmatrix} 2\times 4^{k-2} + (-1)^{k-1}2^{k-2} \\ 2\times 4^{k-2} + (-1)^{k-1}2^{k-2} \\ 2\times 4^{k-2} + (-1)^{k-1}2^{k-1} \end{bmatrix}$$

In other words, the three sequences are given by

$x_k = 2 \times 4^{k-2} + (-1)^{k-1}2^{k-2}$, $y_k = 2 \times 4^{k-2} + (-1)^{k-1}2^{k-2}$, $z_k = 2 \times 4^{k-2} + (-1)^{k-1}2^{k-1}$.

Some problems may be converted to discrete dynamical systems.

Example: The Fibonacci series

1, 1, 2, 3, 5, 8, 13, 21, 34, 55, 89, 144,...

is determined by the initial data $a_0 = a_1 = 1$ and the recursive relation $a_{k+2} = a_{k+1} + a_k$. We would like to find the general formula for a_k.

Denote

$$x_k = \begin{bmatrix} a_{k+1} \\ a_k \end{bmatrix}.$$

Then we have

$$x_0 = \begin{bmatrix} 1 \\ 1 \end{bmatrix}, x_{k+1} = \begin{bmatrix} a_{k+2} \\ a_{k+1} \end{bmatrix} = \begin{bmatrix} a_{k+1} + a_k \\ a_{k+1} \end{bmatrix} = \begin{bmatrix} 1 & 1 \\ 1 & 0 \end{bmatrix} \begin{bmatrix} a_{k+1} \\ a_k \end{bmatrix} = Ax_k.$$

The characteristic polynomial of the matrix is $det(A - \lambda I) = (1 - \lambda)(-\lambda) - 1 = -1 - \lambda + \lambda^2$, and A has two eigenvalues

$$\lambda_1 = \frac{1+\sqrt{5}}{2}, \lambda_2 = \frac{1-\sqrt{5}}{2}.$$

We also find the corresponding eigenvectors

$$v_1 = \left(\frac{1+\sqrt{5}}{2}, 1\right) v_2 = \left(\frac{1-\sqrt{5}}{2}, 1\right).$$

From

$$x_0 = (1, 1) = \frac{1+\sqrt{5}v_1 - (1-\sqrt{5})v_2}{2\sqrt{5}},$$

we then get

$$x_k = A^k x_0 = \frac{(1+\sqrt{5})\lambda_1 k v_1 - (1-\sqrt{5})\lambda_2 k v_2}{2\sqrt{5}}.$$

Since a_k is the second coordinate of x_k, we only need to calculate the second coordinate of the combination on the right:

$$a_k = \frac{(1+\sqrt{5})\lambda_1^k - (1-\sqrt{5})\lambda_2^k}{2\sqrt{5}} = \frac{1}{\sqrt{5}}\left[\left(\frac{1+\sqrt{5}}{2}\right)^{k+1} - \left(\frac{1-\sqrt{5}}{2}\right)^{k+1}\right].$$

FUNCTIONS OF MATRIX

In a discrete dynamical system, we compute the power of linear transformation (or matrix). From the matrix viewpoint, what we have done is the following. Suppose $A = PDP^{-1}$ for an invertible P and diagonal

$$D = \begin{bmatrix} \lambda_1 & 0 & .. & 0 \\ 0 & \lambda_2 & .. & 0 \\ \vdots & \vdots & & \vdots \\ 0 & 0 & .. & \lambda_n \end{bmatrix}$$

The power of the diagonal matrix is easy to compute (try the case $n = k = 3$ to convince yourself)

$$D^k = \begin{bmatrix} \lambda_1^k & 0 & .. & 0 \\ 0 & \lambda_2^k & .. & 0 \\ : & : & & : \\ 0 & 0 & .. & \lambda_n^k \end{bmatrix}$$

Thus by $P^{-1}P = I$ we have

$$A^k = (PDP^{-1})^k = PDP^{-1}PDP^{-1}...PDP^{-1} = PD^kP^{-1} = P\begin{bmatrix} \lambda_1^k & 0 & .. & 0 \\ 0 & \lambda_2^k & .. & 0 \\ : & : & & : \\ 0 & 0 & .. & \lambda_n^k \end{bmatrix}P^{-1}$$

We see it is quite easy to compute the powers of a matrix from its diagonalization.

Example: From an earlier example, we have

$$\begin{bmatrix} 13 & -4 \\ -4 & 7 \end{bmatrix} = \begin{bmatrix} 1 & 2 \\ -2 & 1 \end{bmatrix}\begin{bmatrix} 5 & 0 \\ 0 & 15 \end{bmatrix}\begin{bmatrix} 1 & 2 \\ -2 & 1 \end{bmatrix}^{-1}.$$

Thus

$$\begin{bmatrix} 13 & -4 \\ -4 & 7 \end{bmatrix} = \begin{bmatrix} 1 & 2 \\ -2 & 1 \end{bmatrix}\begin{bmatrix} 5^k & 0 \\ 0 & 15^k \end{bmatrix}\begin{bmatrix} 1 & 2 \\ -2 & 1 \end{bmatrix}^{-1}.$$

$$= (1/5)\begin{bmatrix} 5^k + 4\cdot 15^k & -2\cdot 5^k + 2\cdot 15^k \\ -2\cdot 5^k + 2\cdot 15^k & 4\cdot 5^k + 15^k \end{bmatrix} = 5^{k-1}\begin{bmatrix} 1 + 4\cdot 3^k & -2 + 2\cdot 3^k \\ -2 + 2\cdot 3^k & 4 + 3^k \end{bmatrix}$$

The computation of the power of a diagonalizable matrix $A = PDP^{-1}$ can be extended to other functions. For the polynomial $p(t) = 1 + 2t + 3t^2$, we have

$$p(A) = I + 2A + 3A^2 = I + 2PDP^{-1} + 3(PDP^{-1})^2 = P(I + 2D + 3D^2)P^{-1}$$

$$= P\begin{bmatrix} 1 + 2\lambda_1 + 3\lambda_1^2 & 0 & .. & 0 \\ 0 & 1 + 2\lambda_1 + 3\lambda_1^2 & .. & 0 \\ : & : & & : \\ 0 & 0 & .. & 1 + 2\lambda_n + 3\lambda_1^2 \end{bmatrix}P^{-1}.$$

For the exponential function

$$e^t = 1 + t + \frac{1}{2!}t^2 + \frac{1}{3!}t^3 + ... + \frac{1}{k!}t^k + ...,$$

we find

$$e^D = I + D + \frac{1}{2!}D^2 + \frac{1}{3!}D^3 + ... + \frac{1}{k!}D^k + ...$$

to be a diagonal matrix with the exponential of the eigenvalues

$$e^\lambda = 1 + \lambda + \frac{1}{2!}\lambda^2 + \frac{1}{3!}\lambda^3 + ... + \frac{1}{k!}\lambda^k + ...$$

as the diagonal entries. Then we have

$$e^A = P(I + D + \frac{1}{2!}D^2 + \frac{1}{3!}D^3 + ... + \frac{1}{k!}D^k + ...)\, P^{-1} = Pe^D P^{-1}.$$

In general, for a function $f(t)$ and a diagonalizable matrix $A = PDP^{-1}$, we would expect

$$f(A) = \begin{bmatrix} f(\lambda_1) & 0 & .. & 0 \\ 0 & f(\lambda_2) & .. & 0 \\ : & : & & : \\ 0 & 0 & .. & f(\lambda_n) \end{bmatrix} P^{-1}.$$

Strictly speaking, for the similar argument to work, we need $f(t)$ to have power series expansion at $t = 0$ and have to worry about the convergence issue. The existence of power series means that the function $f(t)$ should be analytic at 0. The convergence means that the "size" of A (called the norm) should be less than the radius of convergence for the power series.

By a more advanced theory, it is possible to define the continuous functions of symmetric matrices (or self–adjoint linear transformations).

LINEAR DIFFERENTIAL SYSTEM

Continuous dynamical systems are described by systems of ordinary differential equations:

$$x' = f(t, x),$$

where t is the independent variable (usually considered as the time for evolution), and

$$x = x(t) = (x_1(t), x_2(t),..., x_n(t)),$$

$$f(t, x) = (f_1(t, x_1, x_2,..., x_n), f_2(t, x_1, x_2,..., x_n),..., f_n(t, x_1, x_2,..., x_n)).$$

The system is linear if $f(t, x)$ is a linear transformation of x:

$$x'_1 = a_{11}(t)x_1 + a_{12}(t)x_2 + \ldots + a_{1n}(t)x_n$$
$$x'_2 = a_{21}(t)x_1 + a_{22}(t)x_2 + \ldots + a_{2n}(t)x_n$$
$$\vdots$$
$$x'_n = a_{n1}(t)x_1 + a_{n2}(t)x_2 + \ldots + a_{nn}(t)x_n$$

and may be abbreviated as $x' = A(t)x$. The following result tells us that the key to solving the system is to find n linearly independent solutions.

The general solution of a system $x' = A(t)x$ of n first order linear differential equations in n dependent variables $x = (x_1(t), x_2(t), \ldots, x_n(t))$ is

$$x(t) = c_1v_1(t) + c_2v_2(t) + \ldots + c_nv_n(t),$$

where $v_1(t), v_2(t), \ldots, v_n(t)$ are n special linearly independent solutions of the system and $c_1, c_2, \ldots, c_n$ are arbitrary constants.

Since $x \to x' - A(t)x$ is a linear transformation, the system $x' = A(t)x$ may be considered as homogeneous. The result above can be compared with the structure of the solutions of $Ax = 0$.

In the special case the coefficient functions are constants: $a_{ij}(t) = a_{ij}$, we have a system $x' = Ax$ of linear differential equations with constant coefficients. Suppose v is an eigenvector of A with eigenvalue λ, then $x(t) = e^{\lambda t}v$ satisfies the system:

$$\begin{aligned}(e^{\lambda t}v)' &= (e^{\lambda t})'v \\ &= e^{\lambda t}\lambda v \\ &= e^{\lambda t}Av \\ &= A(e^{\lambda t}v).\end{aligned}$$

Thus we conclude the following.

Suppose $v_1, v_2, \ldots, v_n$ is a basis of eigenvectors of A with eigenvalues $\lambda_1, \lambda_2, \ldots, \lambda_n$. Then the general solution of the system $x' = Ax$ of first order linear differential equations is

$$x(t) = c_1e^{\lambda_1 t}v_1 + c_2e^{\lambda_2 t}v_2 + \ldots + c_ne^{\lambda_n t}v_n.$$

Example: From the eigenvalues 5, 15 and the eigenvectors (1, –2), (2, 1) found in this example for the matrix

$$\begin{bmatrix} 13 & -4 \\ -4 & 7 \end{bmatrix},$$

we conclude that the system

$$x' = 13x - 4y$$

$$y' = -4x + 7y$$

has two (linearly independent) solutions

$$\begin{bmatrix} x_1(t) \\ y_1(t) \end{bmatrix} = e^{5t}\begin{bmatrix} 1 \\ -2 \end{bmatrix}, \begin{bmatrix} x_2(t) \\ y_2(t) \end{bmatrix} = e^{15t}\begin{bmatrix} 2 \\ 1 \end{bmatrix},$$

and the general solution for the system is

$$\begin{bmatrix} x(t) \\ y(t) \end{bmatrix} = c_1 e^{5t}\begin{bmatrix} 1 \\ -2 \end{bmatrix} + c_2 e^{15t}\begin{bmatrix} 2 \\ 1 \end{bmatrix} = \begin{bmatrix} c_1 e^{5t} + 2c_2 e^{15t} \\ -2c_1 e^{5t} + 2c_2 e^{15t} \end{bmatrix}.$$

Example We try to find the solution of the system

$$x' = x + 5y$$
$$y' = -2x + 3y$$

satisfying the initial condition

$$x(0) = -1, y(0) = 14.$$

Based on this example, we find the general solution

$x(t) = (1 - 3i)c_1 e^{(2+3i)t} + (1 + 3i)c_2 e^{(2-3i)t}$,

$y(t) = 2c_1 e^{(2+3i)t} + 2c_2 e^{(2-3i)t}$

of the system. By

$e^{(2+3i)t} = e^{2t}(\cos 3t + i\sin 3t)$

and the similar formula for $e^{(2-3i)t}$, we may rewrite the general solution as

$x(t) = ae^{2t}(\cos 3t + 3\sin 3t) + be^{2t}(-3\cos 3t + \sin 3t)$,

$y(t) = ae^{2t}(2\cos 3t) + be^{2t}(2\sin 3t)$,

where $a = c_1 + c_2$, $b = i(c_1 - c_2)$.

To find the solution satisfying the initial condition, we substitute the initial condition to the general solution and get

$a - 3b = -1$,

$2a + 2b = 12$.

Then we get $a = 5$ and $b = 2$ and the special solution

$x(t) = e^{2t}(-\cos 3t + 17\sin 3t)$,

$y(t) = e^{2t}(10\cos 3t + 4\sin 3t)$,

satisfying the initial condition.

In the example above, we saw that if we take a_1 and a_2 to be real numbers, then we get the real–valued general solution in case of complex eigenvalues. In general, suppose a *real* matrix A has a conjugate pair of *truly complex* eigenvalues $\lambda = \mu + iv$ and $\lambda = \mu - iv$, where v ‘‘‘ 0. Let $v = u + iw$ and $v = u - iw$ be the corresponding eigenvectors. Then we get a linear combination $(a = c_1 + c_2,\ b = i(c_1 - c_2))$

$$c_1 e^{(\mu + iv)t}(u + iw) + c_2 e^{(\mu - iv)t}(u - iw)$$

$$= ae^{\mu t}((\cos vt)u - (\sin vt)w) + be^{\mu t}((\sin vt)u + (\cos vt)w)$$

as part of the general solution of the system $x' = Ax$. If we take a and b to be real numbers for all the conjugate pairs of eigenvalues, then we get the general *real–valued* solutions.

Exercise: For each recursive relation and initial data, find the general formula.

1. $x_{k+1} = 3x_k + y_k,\ x_0 = 3$

$y_{k+1} = 2x_k + 4y_k,\ y_0 = 0$

Answer: The recursive relation is

$$\begin{bmatrix} x_{k+1} \\ y_{k+1} \end{bmatrix} = A\begin{bmatrix} x_k \\ y_k \end{bmatrix}, \text{ where } A = \begin{bmatrix} 3 & 1 \\ 2 & 4 \end{bmatrix}.$$

From an earlier exercise, the matrix A is diagonalizable, with eigenvalues 2, 5 and corresponding eigenvectors $(1, -1)$, $(1, 2)$. Since

$$\begin{bmatrix} x_0 \\ y_0 \end{bmatrix} = \begin{bmatrix} 3 \\ 0 \end{bmatrix} = 2\begin{bmatrix} 1 \\ -1 \end{bmatrix} + \begin{bmatrix} 1 \\ 2 \end{bmatrix},$$

we conclude that

$$\begin{bmatrix} x_k \\ y_k \end{bmatrix} = A^k\begin{bmatrix} 3 \\ 0 \end{bmatrix} = 2 \times 2^k\begin{bmatrix} 1 \\ -1 \end{bmatrix} + 5^k\begin{bmatrix} 1 \\ 2 \end{bmatrix} = \begin{bmatrix} 2^{k+1} + 5^k \\ -2^{k+1} + 2 \times 5^k \end{bmatrix}.$$

2. $x_{k+1} = 3x_k + y_k + 1,\ x_0 = 3$

$y_{k+1} = 2x_k + 4y_k + 7, \qquad y_0 = 0$

Answer: The recursive relation is

$$\begin{bmatrix} x_{k+1} \\ y_{k+1} \end{bmatrix} = A\begin{bmatrix} x_k \\ y_k \end{bmatrix} + \begin{bmatrix} 1 \\ 7 \end{bmatrix}, \text{ where } A = \begin{bmatrix} 3 & 1 \\ 2 & 4 \end{bmatrix}.$$

To get rid of the extra vector $(1, 7)$, we solve the system

$$\begin{bmatrix} a \\ b \end{bmatrix} = A \begin{bmatrix} a \\ b \end{bmatrix} + \begin{bmatrix} 1 \\ 7 \end{bmatrix}.$$

and get a solution $a = 1$, $b = -3$. Then for $x'_k = x_k - 1$ and $y'_k = y_k + 3$, we have

$$\begin{bmatrix} x'_{k+1} \\ y'_{k+1} \end{bmatrix} = A \begin{bmatrix} x'_k \\ y'_k \end{bmatrix}, \text{ and } \begin{bmatrix} x'_0 \\ y'_0 \end{bmatrix} = \begin{bmatrix} 2 \\ 3 \end{bmatrix}.$$

Recall A has eigenvalues 2, 5 and corresponding eigenvectors (1, –1), (1, 2). Then by

$$\begin{bmatrix} x'_0 \\ y'_0 \end{bmatrix} = (1/3) \begin{bmatrix} 1 \\ -1 \end{bmatrix} + (5/3) \begin{bmatrix} 1 \\ 2 \end{bmatrix},$$

we conclude that

$$\begin{bmatrix} x'_k \\ y'_k \end{bmatrix} = (1/3) \times 2^k \begin{bmatrix} 1 \\ -1 \end{bmatrix} + (5/3) \times 5^k \begin{bmatrix} 1 \\ 2 \end{bmatrix} = (1/3) \begin{bmatrix} 2^k + 5^{k+1} \\ -2^k + 2 \times 5^{k+1} \end{bmatrix},$$

and

$$x_k = x'_k + 1 = (1/3)(2^k + 5^{k+1}) + 1, \; y_k = y'_k - 3 = (1/3)(-2^k + 2X5^{k+1}) - 3.$$

Remark: If $x_{k+1} = Ax_k + u$, then $x_k = A^k x_0 + A^{k-1}u + A^{k-2}u + ... + Au + u$. In case $u = (I - A)v$ (this is the same as $v = Av + u$. A concrete example is $v = (a, b)$, $u = (1, 7)$ above), we have

$$x_k = A^k x_0 + (A^{k-1} + A^{k-2} + ... + A + I)(I - A)v = A^k x_0 + (I - A^k)v.$$

In other words, $x'_k = x_k + v$ satisfies the recursive relation $x'_{k+1} = Ax'_k$. This is exactly what we have done in the exercise above.

However, if we cannot find v satisfying $u = (I - A)v$, then we have to find another way of computing $A^{k-1}u + A^{k-2}u + ... + Au + u$, and the general formula may become more complicated.

Remark: Another way is to convert the recursive relation of two variables into the following recursive relation of three variables.

$$\begin{aligned} x_{k+1} &= 3x_k + y_k + z_k, \; x_1 = 3 \\ y_{k+1} &= 2x_k + 4y_k + 7z_k, \; y_1 = 0 \\ z_{k+1} &= z_k, \; z_1 = 1 \end{aligned}$$

3. $a_{k+1} = 2a_k + b_k$, $a_0 = 3$

   $b_{k+1} = -a_k + 4b_k$, $b_0 = 3$

Answer: The recursive relation is

$$\begin{bmatrix} a_{k+1} \\ b_{k+1} \end{bmatrix} = A \begin{bmatrix} a_k \\ b_k \end{bmatrix}, \text{ where } A = \begin{bmatrix} 2 & 1 \\ -1 & 4 \end{bmatrix}.$$

From an earlier exercise, the matrix A has eigenvalue 3 and corresponding eigenvector $v = (1, 1)$. Since the initial data $(x_0, y_0) = 3v$ is a multiple of the eigenvector, we have

$$\begin{bmatrix} a_k \\ b_k \end{bmatrix} = A^k(3v) = 3 \times 3^k v = \begin{bmatrix} 3^{k+1} \\ 3^{k+1} \end{bmatrix}.$$

Remark: We are lucky that the initial data is a multiple of an eigenvector. Otherwise, say $(a_0, b_0) = (1, 0)$, then the general formula is not easy to compute.

4. $u_{k+1} = -5u_k + 2v_k - w_k$, $u_1 = 0$

   $v_{k+1} = 2u_k - 2v_k - 2w_k$, $v_1 = 0$

   $w_{k+1} = -u_k - 2v_k \quad - 5w_k$, $w_1 = 1$

Answer: The recursive relation is

$$\begin{bmatrix} u_{k+1} \\ v_{k+1} \\ w_{k+1} \end{bmatrix} = A \begin{bmatrix} u_k \\ v_k \\ w_k \end{bmatrix}, \text{ where } A = \begin{bmatrix} -5 & 2 & -1 \\ 2 & -2 & -2 \\ -1 & -2 & -5 \end{bmatrix}.$$

From an earlier exercise, the matrix A is diagonalizable, with bases {(–1,–2, 1)} and {(1, 0, 1), (–1, 1, 1)} for the eigenspaces with eigenvalues 0 and –6. Since

$$\begin{bmatrix} u_1 \\ v_1 \\ w_1 \end{bmatrix} = \begin{bmatrix} 0 \\ 0 \\ 1 \end{bmatrix} = (1/6) \begin{bmatrix} 1 \\ -2 \\ 1 \end{bmatrix} + (1/2) \begin{bmatrix} 1 \\ 0 \\ 1 \end{bmatrix} + (1/3 \begin{bmatrix} -1 \\ 1 \\ 1 \end{bmatrix},$$

we conclude that for $k > 1$,

$$\begin{bmatrix} u_k \\ v_k \\ w_k \end{bmatrix} = A^{k-1} \begin{bmatrix} 0 \\ 0 \\ 1 \end{bmatrix} = (1/2)(-6)^{k-1} \begin{bmatrix} 1 \\ 0 \\ 1 \end{bmatrix} + (1/3)(-6)^{k-1} \begin{bmatrix} -1 \\ 1 \\ 1 \end{bmatrix} = \begin{bmatrix} -(-6)^{k-2} \\ -2(-6)^{k-2} \\ -5(-6)^{k-2} \end{bmatrix}$$

5. $p_{k+1} = 2p_k + 3q_k - 3r_k$, $p_1 = 0$

$q_{k+1} = 2q_k - 3r_{k,}\ q_1 = 3$

$r_{k+1} = r_{k,}\ r_1 = 1$

Answer: The recursive relation is

$$\begin{bmatrix} p_{k+1} \\ q_{k+1} \\ r_{k+1} \end{bmatrix} = A \begin{bmatrix} p_k \\ q_k \\ r_k \end{bmatrix}, \text{ where } A = \begin{bmatrix} 2 & 3 & -3 \\ 0 & 2 & -3 \\ 0 & 0 & 1 \end{bmatrix}.$$

From an earlier example, the matrix A has eigenvalues 1 and 2, with corresponding eigenvectors (–6, 3, 1) and (1, 0, 0). Although A is not diagonalizable, we still have

$$\begin{bmatrix} p_1 \\ q_1 \\ r_1 \end{bmatrix} = \begin{bmatrix} 0 \\ 3 \\ 1 \end{bmatrix} = \begin{bmatrix} -6 \\ 3 \\ 1 \end{bmatrix} + 6 \begin{bmatrix} 1 \\ 0 \\ 0 \end{bmatrix},$$

and

$$\begin{bmatrix} p_k \\ q_k \\ r_k \end{bmatrix} = A^{k-1} \begin{bmatrix} 0 \\ 3 \\ 1 \end{bmatrix} = 1^{k-1} \begin{bmatrix} -6 \\ 3 \\ 1 \end{bmatrix} + 6 \times 2^{k-1} \begin{bmatrix} 1 \\ 0 \\ 0 \end{bmatrix} = \begin{bmatrix} -6 + 3 \times 2^k \\ 3 \\ 1 \end{bmatrix}.$$

Remark: As long as the initial data is a linear combination of eigenvectors, we may easily compute the general term.

Exercise: Find the solution of the linear differential system with the initial condition.

1. $x' = 3x + y,\ x(0) = 3$

   $y' = 2x + 4y,\ y(0) = 0$

Answer: From an earlier exercise, the coefficient matrix

$$\begin{bmatrix} 3 & 1 \\ 2 & 4 \end{bmatrix}$$

is diagonalizable, with eigenvalues 2, 5 and corresponding eigenvectors (1, –1), (1, 2). Thus the general solution of the system is

$$\begin{bmatrix} x(t) \\ y(t) \end{bmatrix} = c_1 e^{2t} \begin{bmatrix} 1 \\ -1 \end{bmatrix} + c_2 e^{5t} \begin{bmatrix} 1 \\ 2 \end{bmatrix} = \begin{bmatrix} c_1 e^{2t} + c_2 e^{5t} \\ -c_1 e^{2t} + 2c_2 e^{5t} \end{bmatrix}.$$

Substituting in the initial condition, we have

$$\begin{bmatrix} x(0) \\ y(0) \end{bmatrix} = \begin{bmatrix} 3 \\ 0 \end{bmatrix} = c_1 \begin{bmatrix} 1 \\ -1 \end{bmatrix} + c_2 \begin{bmatrix} 1 \\ 2 \end{bmatrix}$$

from which we get $c_1 = 2$ and $c_2 = 1$. Thus the solution

$$x(t) = 2e^{2t} + e^{5t},$$
$$y(t) = -2e^{2t} + 2e^{5t},$$

satisfies the initial condition.

2. $x' = 3x + y + e^t, x(0) = 3$
   $y' = 2x + 4y + 7e^t, y(0) = 0$

Answer: The linear differential system is $x' = Ax + e^t u$, where

$$x = \begin{bmatrix} x \\ y \end{bmatrix}, u = \begin{bmatrix} 1 \\ 7 \end{bmatrix}, A = \begin{bmatrix} 3 & 1 \\ 2 & 4 \end{bmatrix}.$$

To get rid of the extra term $e^t u$, we try to find one special solution $x_0 = x_0(t)$ of the system. Then the new variable $z = x - x_0$ satisfies the system $z' = Az$. We set our special solution to be

$$x_0(t) = e^t a, \text{ where } a = \begin{bmatrix} a_1 \\ a_2 \end{bmatrix} \text{ is a constant vector.}$$

Subsitituting into the system $x' = Ax + e^t u$, we get $e^t a = e^t Aa + e^t u$. In other words, we have

$$\begin{bmatrix} a_1 \\ a_2 \end{bmatrix} = \begin{bmatrix} 3 & 1 \\ 2 & 4 \end{bmatrix} \begin{bmatrix} a_1 \\ a_2 \end{bmatrix} + \begin{bmatrix} 1 \\ 7 \end{bmatrix}.$$

The solution of the system of linear equations is $a_1 = 1, a_2 = -3$. Then the new variable

$$z = \begin{bmatrix} z \\ w \end{bmatrix} = x - x_0 \begin{bmatrix} x - e^t \\ y + 3e^t \end{bmatrix}.$$

satisfies

$$z' = 3z + w, z(0) = 3 - e^0 = 2$$
$$w' = 2z + 4w, w(0) = 0 + 3e^0 = 3$$

Recall A has eigenvalues 2, 5 and corresponding eigenvectors (1, –1), (1, 2). Thus

$$\begin{bmatrix} z(t) \\ w(t) \end{bmatrix} = c_1 e^{2t} \begin{bmatrix} 1 \\ -1 \end{bmatrix} + c_2 e^{5t} \begin{bmatrix} 1 \\ 2 \end{bmatrix} = \begin{bmatrix} c_{1e^{2t}} + c_{2e^{5t}} \\ -c_{1e^{2t}} + 2c_{2e^{5t}} \end{bmatrix},$$

with the constants determined by substituting in the initial condition

$$\begin{bmatrix} z(0) \\ w(0) \end{bmatrix} = \begin{bmatrix} c_1 + c_2 \\ -c_1 + c_2 \end{bmatrix} = \begin{bmatrix} 2 \\ 3 \end{bmatrix}.$$

We find $c_1 = 1/3$, $c_2 = 5/3$, and

$$x(t) = z(t) + e^t = (1/3)e^{2t} + (5/3)e^{5t} + e^t,$$

$$y(t) = w(t) - 3e^t = -(1/3)e^{2t} + (10/3)e^{5t} - 3e^t,$$

is the solution that satisfies the initial condition.

Remark: The system in the first problem is of the form $x' = Ax$, which is a homogeneous linear system because it can be rewritten as $T(x) = x' - Ax = 0$, and T is a linear transformation. The system in the second problem is of the form $T(x) = b$ ($= e^t u$ in the specific case). According to this general result, the solution of $T(x) = b$ is of the form $x = x_0 + x_h$, where

- x_0 is one particular solution of $T(x) = b$.
- x_h is any solution of $T(x) = 0$ (i.e., solution of the first problem).

The solution of the second problem follows exactly this result.

Remark We tried the form $x_0(t) = e^t a$ for a special solution of $x' = Ax + e^t u$ because we know the derivative of $e^t a$ is still of the form $e^t a$. Similarly, we should try $x_0(t) = e^{ct} a$ for a special solution of $x' = Ax + e^{ct} u$, and the problem is reduced to solving $ca = Aa + u$. Noting that if c is not an eigenvalue of A, then we always have $a = (A - cI)^{-1} u$ as the solution.

3. $x' = 3x + y + e^{2t,}\ x(0) = 3$

   $y' = 2x + 4y + 7e^{2t,}\ y(0) = 0$

Answer: The linear differential system is $x' = Ax + e^{2t}$, where

$$x = \begin{bmatrix} x \\ y \end{bmatrix}, u = \begin{bmatrix} 1 \\ 7 \end{bmatrix}, A = \begin{bmatrix} 3 & 1 \\ 2 & 4 \end{bmatrix}.$$

Using the idea of the previous problem, we try a special solution

$$x_0(t) = e^{2t} a, \text{ where } a = \begin{bmatrix} a_1 \\ a_2 \end{bmatrix} \text{ is a constant vector.}$$

Subsitituting into the system $x' = Ax + e^{2t} u$, we get

$$\begin{bmatrix} 1 & 1 \\ 2 & 2 \end{bmatrix}\begin{bmatrix} a_1 \\ a_2 \end{bmatrix} = \begin{bmatrix} -1 \\ -7 \end{bmatrix},$$

which unfortunately has no solution (The theoretical reason is that 2 is an eigenvalue of A). Thus we try a more sophisticated special solution

$$x_0(t) = e^{2t}(a + tb), \text{ where } a = \begin{bmatrix} a_1 \\ a_2 \end{bmatrix} \text{ and } \begin{bmatrix} b_1 \\ b_2 \end{bmatrix} \text{ are constant vectors.}$$

Subsitituting into the system $x' = Ax + e^{2t}u$, we get $2e^{2t}(a + tb) + e^{2t}b = e^{2t}(Aa + tAb) + e^{2t}u$. Dividing e^{2t} and then comparing the coefficients, we get $2a + b = Aa + u$, $2b = Ab$. More specifically, we have

$$\begin{bmatrix} 1 & 1 \\ 2 & 2 \end{bmatrix}\begin{bmatrix} a_1 \\ a_2 \end{bmatrix} = \begin{bmatrix} b_1 - 1 \\ b_2 - 7 \end{bmatrix},$$

$$\begin{bmatrix} 1 & 1 \\ 2 & 2 \end{bmatrix}\begin{bmatrix} b_1 \\ b_2 \end{bmatrix} = \begin{bmatrix} 0 \\ 0 \end{bmatrix},$$

For the first system to have solutions, we must have $2(b_1 - 1) = b_2 - 7$. From the second system, we also have $b_1 + b_2 = 0$. Thus we get $b_1 = -5/3$, $b_2 = 5/3$. Substituting into the first system, we find one solution $a_1 = -8/3$, $a_2 = 0$. In conclusion, we find a special solution

$$x_0(t) = e^{2t}\begin{bmatrix} -8/3 - (5/3)t \\ (5/3)t \end{bmatrix}.$$

Combined with the general solution of the homogeneous system in the first problem, we have the general solution of the current problem

$$\begin{bmatrix} x(t) \\ y(t) \end{bmatrix} = \begin{bmatrix} [c_1 - 8/3 - (5/3)t]e^{2t} + c_{2e^{5t}} \\ [-c_1 + (5/3)t]e^{2t} + 2c_{2e^{5t}} \end{bmatrix}.$$

Substituting in the initial condition, we get

$$\begin{bmatrix} x(0) \\ y(0) \end{bmatrix} = \begin{bmatrix} c_1 - 8/3 + c_2 \\ -c_1 + 2c_2 \end{bmatrix} = \begin{bmatrix} 3 \\ 0 \end{bmatrix}.$$

We find $c_1 = 34/9$, $c_2 = 17/9$, and

$$x(t) = [10/9 - (5/3)t]e^{2t} + (17/9)e^{5t},$$

$$y(t) = [-\,34/9 + (5/3)t]e^{2t} + (34/9)e^{5t},$$

is the solution that satisfies the initial condition.

4. $x' = 3x + y + 3 + 3t + t^2,\ x(0) = 3$

$y' = 2x + 4y + 6 + 4t^2,\ y(0) = 0$

Answer: The linear differential system is $x' = Ax + u + tv + t^2w$, where

$$x = \begin{bmatrix} x \\ y \end{bmatrix},\ u = \begin{bmatrix} 3 \\ 6 \end{bmatrix},\ v = \begin{bmatrix} 3 \\ 0 \end{bmatrix},\ w = \begin{bmatrix} 1 \\ 4 \end{bmatrix},\ A = \begin{bmatrix} 3 & 1 \\ 2 & 4 \end{bmatrix}.$$

We try to find one special solution of the form $x_0(t) = a + tb + t^2c$, where a, b, c are constant vectors. Subsitituting into the system $x' = Ax + u + tv + t^2w$, we get $b + 2tc = Aa + tAb + t^2Ac + u + tv + t^2w$. In other words, we have

$b = Aa + u,$

$c = Ab + v,$

$0 = Ac + w.$

From the third system, we find $c = (0, -1)$. Substituting the value of c into the second system, we find $b = (-1, 0)$. Substituting the value of c into the first system, we find $a = (-1, -1)$. Thus we find a special solution

$$x_0(t) = \begin{bmatrix} -1 & -t \\ -1 & -t^2 \end{bmatrix}.$$

Combined with the general solution of the homogeneous system in the first problem, we have the general solution of the current problem

$$\begin{bmatrix} x(t) \\ y(t) \end{bmatrix} = \begin{bmatrix} c_{1e^{2t}} + c_{2e^{5t}} - 1 - t \\ -c_{1e^{2t}} + 2c_{2e^{5t}} - 1 - t^2 \end{bmatrix}.$$

Substituting in the initial condition, we get

$$\begin{bmatrix} x(0) \\ y(0) \end{bmatrix} = \begin{bmatrix} c_1 - c_2 - 1 \\ -c_1 + 2c_2 - 1 \end{bmatrix} = \begin{bmatrix} 3 \\ 0 \end{bmatrix}.$$

We find $c_1 = 7/3$, $c_2 = 5/3$, and

$x(t) = (7/3)e^{2t} + (5/3)e^{5t} - 1 - t,$

$y(t) = -(7/3)e^{2t} + (10/3)e^{5t} - 1 - t^2,$

is the solution that satisfies the initial condition.

5. $x' = 2x + y,\ x(0) = 3$

$y' = -x + 4y,\ y(0) = 3$

Answer: From an earlier exercise, the coefficient matrix

$$\begin{bmatrix} 2 & 1 \\ -1 & 4 \end{bmatrix}$$

has eigenvalue 3 and corresponding eigenvector $v = (1, 1)$. This implies that

$$\begin{bmatrix} x \\ y \end{bmatrix} = ce^t \begin{bmatrix} 1 \\ 1 \end{bmatrix}$$

is a solution of the system. Since the initial data $(x(0), y(0)) = 3v$ happens to be a multiple of the eigenvector, we find that, by taking $c = 3$,

$$x = 3e^t,\ y = 3e^t,$$

is a solution satisfying the initial condition.

Remark: We are lucky that the initial condition is a multiple of an eigenvector. Also note that if the initial condition is changed to $x(-2) = 5$, $y(-2) = 5$, which has different location at -2 but is still a multiple of the eigenvector, the solution satisfying the initial condition is $x = 5e^{t+2}$, $y = 5e^{t+2}$.

6. $x' = 4x + 2y,\ x(0) = 1$

   $y' = -x + 2y,\ y(0) = 2$

Answer: From an earlier exercise, the coefficient matrix

$$\begin{bmatrix} 4 & 2 \\ -1 & 2 \end{bmatrix}$$

is complex diagonalizable, with conjugate eigenvalues $3 \pm i$ and corresponding conjugate eigenvectors $(1 \pm i, -1)$. Thus the general real solution of the system is

$$\begin{bmatrix} x(t) \\ y(t) \end{bmatrix} = ae^{3t}(\cos t \begin{bmatrix} 1 \\ -1 \end{bmatrix} - \sin t \begin{bmatrix} 1 \\ 0 \end{bmatrix}) + be^{3t}(\sin t \begin{bmatrix} 1 \\ -1 \end{bmatrix} + \cos t \begin{bmatrix} 1 \\ 0 \end{bmatrix})$$

Substituting in the initial condition, we have

$$\begin{bmatrix} x(0) \\ y(0) \end{bmatrix} = \begin{bmatrix} 1 \\ 2 \end{bmatrix} = a \begin{bmatrix} 1 \\ -1 \end{bmatrix} + b \begin{bmatrix} 1 \\ 0 \end{bmatrix},$$

from which we get $a = -2$ and $b = 3$. Thus the solution

$$x(t) = -2e^{3t}(\cos t - \sin t) + 3e^{3t}(\sin t + \cos t) = e^{3t}(\cos t + 5\sin t),$$

$$y(t) = -2e^{3t}(-\cos t) + 3e^{3t}(-\sin t) = e^{3t}(2\cos t - 3\sin t),$$

satisfies the initial condition.

7. $u' = -5u + 2v - w,\ u(1) = 0$

$v' = 2u - 2v - 2w,\ v(1) = 0$

$w' = -u - 2v - 5w,\ w(1) = 1$

Answer: From an earlier exercise, the coefficient matrix

$$\begin{bmatrix} -5 & 2 & -1 \\ 2 & -2 & -2 \\ -1 & -2 & -5 \end{bmatrix}$$

is diagonalizable, with eigenvalues 0, –6 and bases {(–1, –2, 1)}, {(1, 0, 1), (–1, 1, 1)} for the corresponding eigenspaces. Thus the general solution of the system is

$$\begin{bmatrix} u(t) \\ v(t) \\ w(t) \end{bmatrix} = c_1 \begin{bmatrix} -1 \\ -2 \\ 1 \end{bmatrix} + c_2 e^{6t} \begin{bmatrix} 1 \\ 0 \\ 1 \end{bmatrix} + c_3 e^{6t} \begin{bmatrix} -1 \\ 1 \\ 1 \end{bmatrix}.$$

Substituting in the initial condition, we have

$$\begin{bmatrix} 0 \\ 0 \\ 1 \end{bmatrix} = c_1 \begin{bmatrix} -1 \\ -2 \\ 1 \end{bmatrix} + c_2 e^{6} \begin{bmatrix} 1 \\ 0 \\ 1 \end{bmatrix} + c_3 e^{6} \begin{bmatrix} -1 \\ 1 \\ 1 \end{bmatrix},$$

from which we get $c_1 = 1/6$, $c_2 = (1/2)e^{-6}$, $c_3 = (1/3)e^{-6}$. Thus the solution

$x(t) = -1/6 + (1/2)e^{6(t-1)} - (1/3)e^{6(t-1)} = -1/6 + (1/6)e^{6(t-1)}$,

$y(t) = -1/3 + (1/3)e^{6(t-1)}$,

$z(t) = 1/6 + (1/2)e^{6(t-1)} + (1/3)e^{6(t-1)} = 1/6 + (5/6)e^{6(t-1)}$,

satisfies the initial condition.

8. $p' = 2p + 3q - 3r,\ p(0) = 0$

$q' = 2q - 3r,\ q(0) = 3$

$r' = r,\ r(0) = 1$

Answer: From an earlier example, although the coefficient matrix

$$\begin{bmatrix} 2 & 3 & -3 \\ 0 & 2 & -3 \\ 0 & 0 & 1 \end{bmatrix}$$

is not diagonalizable, ww still have eigenvalues 1, 2 and the corresponding eigenvectors (–6, 3, 1), (1, 0, 0). Thus implies that for any constants c_1, c_2,

$$\begin{bmatrix} p(t) \\ q(t) \\ r(t) \end{bmatrix} = c_1 e^t \begin{bmatrix} -6 \\ 3 \\ 1 \end{bmatrix} + c_2 e^{2t} \begin{bmatrix} 1 \\ 0 \\ 0 \end{bmatrix}$$

is a solution of the system. Substituting in the initial condition, we have

$$\begin{bmatrix} 0 \\ 3 \\ 1 \end{bmatrix} = c_1 \begin{bmatrix} -6 \\ 3 \\ 1 \end{bmatrix} + c_2 \begin{bmatrix} 1 \\ 0 \\ 0 \end{bmatrix}$$

from which we get $c_1 = 1, c_2 = 6$. Thus the solution

$p(t) = -6e^t + 6e^{2t}$,

$q(t) = 3e^t$,

$r(t) = e^t$,

satisfies the initial condition.

Remark: As long as the initial condition is a linear combination of eigenvectors, we still get the solution.

Exercise: Diagonalize the matrix

$$A = \begin{bmatrix} 2 & -3 & 1 \\ 1 & -2 & 1 \\ 1 & -3 & 2 \end{bmatrix}.$$

Then use the diagonalization to compute A^{10}, $A^5 - 2A^4 + A^3$, e^{A-I}, $\cos\pi A$, $\sin\pi A$.

Answer: $det(A - \lambda I) = -\lambda(1 - \lambda)^2$. The eigenvaues are 0 and 1. For the eigenvalue 0, we get a basis {(1, 1, 1)} of the eigenspace. For the eigenvalue 1, we get a basis {(3, 1, 0), (–1, 1, 0)} of the eigenspace. Then we have the diagonalization

$$A = \begin{bmatrix} 1 & 3 & -1 \\ 1 & 1 & 1 \\ 1 & 0 & 0 \end{bmatrix} \begin{bmatrix} 0 & 0 & 0 \\ 0 & 1 & 0 \\ 0 & 0 & 1 \end{bmatrix} \begin{bmatrix} 1 & 3 & -1 \\ 1 & 1 & 1 \\ 1 & 0 & 0 \end{bmatrix}^{-1}.$$

From this we have

$$A^{10} = \begin{bmatrix} 1 & 3 & -1 \\ 1 & 1 & 1 \\ 1 & 0 & 0 \end{bmatrix} \begin{bmatrix} 0 & 1^{10} & 0 \\ 0 & 0 & 1^{10} \\ 1 & 3 & -1 \end{bmatrix} \begin{bmatrix} 1 & 3 & -1 \\ 1 & 1 & 1 \\ 1 & 0 & 0 \end{bmatrix}^{-1} = A.$$

$$A^5 - 2A^4 + A^3 = \begin{bmatrix} 1 & 3 & -1 \\ 1 & 1 & 1 \\ 1 & 0 & 0 \end{bmatrix} \begin{bmatrix} 0^5 - 2\times 0^4 + 1^3 & 0 & 0 \\ 0 & 1^5 - 2\times 0^4 + 1^3 & 0 \\ 0 & 0 & 1^5 - 2\times 0^4 + 1^3 \end{bmatrix}$$

$$\begin{bmatrix} 1 & 3 & -1 \\ 1 & 1 & 1 \\ 1 & 0 & 0 \end{bmatrix}^{-1} = O.$$

$$e^{A-I} = \begin{bmatrix} 1 & 3 & -1 \\ 1 & 1 & 1 \\ 1 & 0 & 0 \end{bmatrix} \begin{bmatrix} e^{0-1} & 0 & 0 \\ 0 & e^{1-1} & 0 \\ 0 & 0 & e^{1-1} \end{bmatrix} \begin{bmatrix} 1 & 3 & -1 \\ 1 & 1 & 1 \\ 1 & 0 & 0 \end{bmatrix}^{-1} = \begin{bmatrix} 1 & 0 & e^{-1}-1 \\ 1 & 1 & e^{-1}-1 \\ 0 & 0 & e^{-1} \end{bmatrix}.$$

$$\cos\pi A = \begin{bmatrix} 1 & 3 & -1 \\ 1 & 1 & 1 \\ 1 & 0 & 0 \end{bmatrix} \begin{bmatrix} \cos 0\pi & 0 & 0 \\ 0 & \cos 1\pi & 0 \\ 0 & 0 & \cos 1\pi \end{bmatrix} \begin{bmatrix} 1 & 3 & -1 \\ 1 & 1 & 1 \\ 1 & 0 & 0 \end{bmatrix}^{-1}$$

$$= \begin{bmatrix} 1 & 3 & -1 \\ 1 & 1 & 1 \\ 1 & 0 & 0 \end{bmatrix} \left(\begin{bmatrix} 1 & 0 & 0 \\ 0 & 1 & 0 \\ 0 & 0 & 1 \end{bmatrix} - 2 \begin{bmatrix} 0 & 0 & 0 \\ 0 & 1 & 0 \\ 0 & 0 & 1 \end{bmatrix} \right) \begin{bmatrix} 1 & 3 & -1 \\ 1 & 1 & 1 \\ 1 & 0 & 0 \end{bmatrix}^{-1} =$$

$$I - 2A = \begin{bmatrix} -3 & 6 & -2 \\ -2 & 5 & -2 \\ -2 & 6 & -3 \end{bmatrix}.$$

$$\sin p A = \begin{bmatrix} 1 & 3 & -1 \\ 1 & 1 & 1 \\ 1 & 0 & 0 \end{bmatrix} \begin{bmatrix} \sin 0\pi & 0 & 0 \\ 0 & \sin 1\pi & 0 \\ 0 & 0 & \sin 1\pi \end{bmatrix} \begin{bmatrix} 1 & 3 & -1 \\ 1 & 1 & 1 \\ 1 & 0 & 0 \end{bmatrix}^{-1} = O.$$

Exercise: The city of Linear Algebra maintains a constant population of 30,000. A political science study estimated that there are 15,000 independents (a name of political party, not a concept of linear algebra), 9,000 spans, and 6,000 bases. It was also estimated that each year 20% of spans and 10% of bases become independents, 20% of independents and 10% of bases become spans, 10% of independents and 10% of spans become bases.

1. Let $x_k = (I_k, S_k, B_k)$ be the "political distribution vector" k years from now. Find the transition from x_k to x_{k+1}, which is given by a matrix.

Answer: In the year $k + 1$, $0.2S_k + 0.1B_k$ number of people move from outside of independents to become indpendents. Moreover, among the existing independents, $0.2I_k +$

$0.1I_k$ are lost to the spans and bases. Therefore the total number of independents in the year $k + 1$ is

$$I_{k+1} = I_k + (0.2S_k + 0.1B_k) - (0.2I_k + 0.1I_k) = 0.7I_k + 0.2S_k + 0.1B_k.$$

Similarly, we have

$$S_{k+1} = S_k + (0.2I_k + 0.1B_k) - (0.2S_k + 0.1S_k) = 0.2I_k + 0.7S_k + 0.1B_k,$$

and

$$B_{k+1} = B_k + (0.1I_k + 0.1S_k) - (0.1B_k + 0.1B_k) = 0.1I_k + 0.1S_k + 0.8B_k.$$

In other words, we have

$$x_{k+1} = \begin{bmatrix} 0.7 & 0.2 & 0.1 \\ 0.2 & 0.7 & 0.1 \\ 0.1 & 0.1 & 0.8 \end{bmatrix} x_k.$$

2. Find the eigenvalue and eigenvector of matrix of the linear transformation.

Answer: In the first equality, we use [row 2] + [row 1] and [row 3] + [row 1]. In the second equality, we use (–1)[col 1] + [col 2] and (–1)[col 1] + [col 3].

$$det\begin{bmatrix} 0.7-\lambda & 0.2 & 0.1 \\ 0.2 & 0.7-\lambda & 0.1 \\ 0.1 & 0.1 & 0.8-\lambda \end{bmatrix} = det\begin{bmatrix} 1-\lambda & 1-\lambda & 1-\lambda \\ 0.2 & 0.7-\lambda & 0.1 \\ 0.1 & 0.1 & 0.8-\lambda \end{bmatrix} =$$

$$det\begin{bmatrix} 1-\lambda & 0 & 0 \\ 0.2 & 0.5-\lambda & -0.1 \\ 0.1 & 0 & 0.7-\lambda \end{bmatrix}.$$

Thus the eigenvalues of the transition matrix are 0.5, 0.7, 1. For the eigenvalue 0.5, the null space of

$$\begin{bmatrix} 0.7-0.7 & 0.2 & 0.1 \\ 0.2 & 0.7-0.7 & 0.1 \\ 0.1 & 0.1 & 0.8-0.7 \end{bmatrix} = \begin{bmatrix} 0 & 0.2 & 0.1 \\ 0.2 & 0 & 0.1 \\ 0.1 & 0.1 & 0.1 \end{bmatrix}$$

has basis $\{(1, -1, 0)\}$. For the eigenvalue 0.7, the null space of

$$\begin{bmatrix} 0.7-1 & 0.2 & 0.1 \\ 0.2 & 0.7-1 & 0.1 \\ 0.1 & 0.1 & 0.8-0.7 \end{bmatrix} = \begin{bmatrix} -0.3 & 0.2 & 0.1 \\ 0.2 & -0.3 & 0.1 \\ 0.1 & 0.1 & -0.2 \end{bmatrix}$$

has basis $\{(1, 1, -2)\}$. For the eigenvalue 1, the null space of

$$\begin{bmatrix} 0.7-1 & 0.2 & 0.1 \\ 0.2 & 0.7-1 & 0.1 \\ 0.1 & 0.1 & 0.8-1 \end{bmatrix} = \begin{bmatrix} -0.3 & 0.2 & 0.1 \\ 0.2 & -0.3 & 0.1 \\ 0.1 & 0.1 & -0.2 \end{bmatrix}$$

has basis $\{(1, 1, 1)\}$.

3. Use the diagonalization to find the formula for x_k.

Answer: We have

$$\begin{bmatrix} 0.7 & 0.2 & 0.1 \\ 0.2 & 0.7 & 0.1 \\ 0.1 & 0.1 & 0.8 \end{bmatrix} = \begin{bmatrix} 1 & 1 & 1 \\ -1 & 1 & 1 \\ 0 & -2 & 1 \end{bmatrix} \begin{bmatrix} 0.5 & 0 & 0 \\ 0 & 0.7 & 0 \\ 0 & 0 & 1 \end{bmatrix} \begin{bmatrix} 1 & 1 & 1 \\ -1 & 1 & 1 \\ 0 & -2 & 1 \end{bmatrix}^{-1}.$$

Therefore

$$x_k = PD^kP^{-1}x_0 = \begin{bmatrix} 1 & 1 & 1 \\ -1 & 1 & 1 \\ 0 & -2 & 1 \end{bmatrix} \begin{bmatrix} 0.5^k & 0 & 0 \\ 0 & 0.7^k & 0 \\ 0 & 0 & 1^k \end{bmatrix} \begin{bmatrix} 1 & 1 & 1 \\ -1 & 1 & 1 \\ 0 & -2 & 1 \end{bmatrix}^{-1} \begin{bmatrix} 15,000 \\ 9,000 \\ 6,000 \end{bmatrix}$$

$$= \begin{bmatrix} 0.5^k & 0.7^k & 1 \\ -0.5^k & 0.7^k & 1 \\ 0 & -2\times 0.7^k & 1 \end{bmatrix} \begin{bmatrix} 3,000 \\ 2,000 \\ 10,000 \end{bmatrix} = \begin{bmatrix} 3000\times 0.5^k + 2000\times 0.7^k + 10000 \\ -3000\times 0.5^k + 2000\times 0.7^k + 10000 \\ -4,000\times 0.5^k + 10000 \end{bmatrix}$$

and we get

$I_k = 3000\text{X}0.5^k + 2000\text{X}0.7^k + 10000,$

$S_k = -3000\text{X}0.5^k + 2000\text{X}0.7^k + 10000,$

$B_k = -4000\text{X}0.5^k + 10000.$

4. Use the formula for x_k to predict the long term trend of the political distribution.

Answer: In the long run (as k gets very big), all three parties approach 10000 each.

Exercise: Consider the recursive relation $x_{k+2} = ax_{k+1} + bx_k$.

1. Prove that if the equation $\lambda^2 = a\lambda + b$ has two distinct solutions λ_1 and λ_2, then the general formula for x_k is $x_k = c_1\lambda_1{}^k + c_2\lambda_2{}^k$.

Answer: We may derive the formula for x_k by using the method similar to the computaiton of the Fibonacci series.

Denote

$$y_k = \begin{bmatrix} x_{k+1} \\ x_k \end{bmatrix}.$$

Then we have

$$y_{k+1} = \begin{bmatrix} x_{k+2} \\ x_{k+1} \end{bmatrix} = \begin{bmatrix} x_{k+1} + bx_k \\ x_{k+1} \end{bmatrix} = \begin{bmatrix} a & b \\ 1 & 0 \end{bmatrix} \begin{bmatrix} x_{k+1} \\ x_k \end{bmatrix} = Ay_k$$

The characteristic polynomial of the matrix is $det(A - \lambda I) = (a - \lambda)(-\lambda) - b = -b - a\lambda + \lambda^2$. Thus the eigenvalues of A are the solutions of the equation $\lambda^2 = a\lambda + b$.

For any eigenvalue λ of A, we have

$$A - \lambda I = \begin{bmatrix} a-\lambda & b \\ 1 & -\lambda \end{bmatrix}.$$

Since the system $(A - \lambda I)v = 0$ must have nontrivial solutions, the system must be equivalent to the equation given by the (nontrivial) second row. In other words, the corresponding eigenvector is $(\lambda, 1)$.

Now assume the equation $\lambda^2 = a\lambda + b$ has two distinct solutions λ_1 and λ_2. Then the corresponding eigenvectors $(\lambda_1, 1)$ and $(\lambda_2, 1)$ clearly form a basis of R^2. Thus we can always find (unique) c_1 and c_2 such that

$$y_0 = (x_0, x_1) = c_1(\lambda_1, 1) + c_2(\lambda_2, 1).$$

Then we get

$$y_k = (x_{k+1}, x_k) = A^k y_0 = c_1\lambda_1{}^k(\lambda_1, 1) + c_2\lambda_2{}^k(\lambda_2, 1).$$

By taking the second coordinates, we get the general formula

$$x_k = c_1\lambda_1{}^k + c_2\lambda_2{}^k.$$

2. Derive the general formula for the Fibonacci series by using the first part.

Answer: For the Fibonacci series, we have $a = b = 1$. The solutions of the equation $\lambda^2 = \lambda + 1$ are

$$\lambda_1 = \frac{1+\sqrt{5}}{2}, \lambda_2 \frac{1-\sqrt{5}}{2}.$$

Then from the equation

$$(x_0, x_1) = (1, 1) = c_1\left(\frac{1+\sqrt{5}}{2}, 1\right) + c_2\left(\frac{1-\sqrt{5}}{2}, 1\right),$$

we find

$$c_1 = \frac{1+\sqrt{5}}{2},\ c_2 = \frac{1-\sqrt{5}}{2}.$$

Therefore

$$x_k = \frac{1+\sqrt{5}}{2}\left(\frac{1+\sqrt{5}}{2}\right)^k + \frac{1-\sqrt{5}}{2}\left(\frac{1-\sqrt{5}}{2}\right)^k = \frac{1}{\sqrt{5}}\left[\left(\frac{1+\sqrt{5}}{2}\right)^{k+1} - \left(\frac{1-\sqrt{5}}{2}\right)^{k+1}\right]$$

3. Prove that if the equation $\lambda^2 = a\lambda + b$ has a nonzero double root λ, then the general formula for x_k is $x_k = (c_1 + c_2 k)\lambda^k$.

Answer: The fact that λ is a double root means that $a = 2\lambda$, $b = -\lambda^2$. Then the recursive relation is $x_{k+2} = 2\lambda x_{k+1} - \lambda^2 x_k$, which can be rewritten as

$$y_{k+1} = \lambda y_k, \text{ where } y_k = x_{k+1} - \lambda x_k.$$

Then it is easy to see that

$$y_k = \lambda^k y_0, \text{ where } y_0 = x_1 - \lambda x_0.$$

Using $y_k = x_{k+1} - \lambda x_k$, we have

$$x_k - \lambda x_{k-1} = \lambda^{k-1} y_0,$$
$$x_{k-1} - \lambda x_{k-2} = \lambda^{k-2} y_0,$$
$$x_{k-2} - \lambda x_{k-3} = \lambda^{k-3} y_0,$$
$$\vdots$$
$$x_1 - \lambda x_0 = y_0.$$

Multiplying suitable powers of λ, we have

$$x_k - \lambda x_{k-1} = \lambda^{k-1} y_0,$$
$$\lambda x_{k-1} - \lambda^2 x_{k-2} = \lambda^{k-1} y_0,$$
$$\lambda^2 x_{k-2} - \lambda^3 x_{k-3} = \lambda^{k-1} y_0,$$
$$\vdots$$
$$\lambda^{k-1} x_1 - \lambda^k x_0 = \lambda^{k-1} y_0.$$

Adding all the equalities together, we get

$$x_k - \lambda^k x_0 = k\lambda^{k-1} y_0.$$

Letting $c_1 = x_0$, $c_2 = \lambda^{-1} y_0 = \lambda^{-1} x_1 - x_0$, we conclude that $x_k = (c_1 + c_2 k)\lambda^k$.

4. What about the general formula for the series given by the more general recursive relation $x_{k+n} = a_{n-1} x_{k+n-1} + a_{n-2} x_{k+n-2} + ... + a_0 x_k$?

Answer: By changing x_{k+i} to λ^i, the recursive relation becomes

$\lambda^n = a_{n-1}\lambda^{n-1} + a_{n-2}\lambda^{n-2} + ... + a_0\lambda + a_0$.

If the polynomial equation has no multiple roots, then we have n distinct roots λ_1, $\lambda_2, \ldots, \lambda_n$. The general formula for x_k is then

$$x_k = c_1\lambda_1{}^k + c_2\lambda_2{}^k \ldots + c_n\lambda_n{}^k.$$

If the equation has multiple roots, then some terms similar to the one in third part appears. For example, if $n = 5$, and the polynomial equation has a double root λ_1 and a triple root λ_2, then

$$x_k = (c_{11} + c_{12}k)\lambda_1{}^k + (c_{21} + c_{22}k + c_{23}k^2)\lambda_2{}^k.$$

PROPERTIES

ABOUT SINGLE MATRIX

In this part, we compare the eigenvalues and eigenvectors of a matrix and its modification.

- A and A^T have the same characteristic polynomial and eigenvalues.
- A is diagonalizable $\Leftrightarrow A^T$ is diagonalizable.

Proof. Since the determinant is not changed by transpose, we have

$det(A^T - \lambda I) = det((A - \lambda I)^T) = det(A - \lambda I)$.

By the definition of diagonalization, we have

$A = PDP^{-1} \Leftrightarrow A^T = (P^{-1})^T\, D^T\, P^T = QD^TQ^{-1}$,

where $Q = (P^{-1})^T$ is still invertible and D^T is still diagonal.

The relation between eigenvectors of A and A^T is rather complicated in general: The eigenvectors of A are columns of P. The eigenvectors of A^T are columns of $Q = (P^{-1})^T$.

- A and PAP^{-1} have the same characteristic polynomial and eigenvalues.
- v is an eigenvector of A with eigenvalue $\lambda \Leftrightarrow Pv$ is an eigenvector of PAP^{-1} with eigenvalue λ.
- A is diagonalizable Ÿ PAP^{-1} is diagonalizable.

Proof. The first part follows from

$det(PAP^{-1} - \lambda I) = det(P(A - \lambda I)P^{-1}) = det(P^{-1}P(A - \lambda I)) = det(A - \lambda I)$.

The second part follows from

$Av = \lambda v \Leftrightarrow PAP^{-1}(Pv) = \lambda(Pv)$.

For the third part, we note that since P is invertible, we have

$\{v_1, v_2, \ldots, v_n\}$ is a basis of eigenvectors for $A \Leftrightarrow \{Pv_1, Pv_2, \ldots, Pv_n\}$ is a basis of eigenvectors for PA.

Then we apply the definition of diagonalizability in terms of bases of eigenvectors.

- A square matrix A is invertible $\Leftrightarrow$ 0 is not an eigenvalue.
- v is an eigenvector of A with eigenvalue $\lambda \Leftrightarrow v$ is an eigenvector of A^{-1} with eigenvalue λ^{-1}.
- A is diagonalizable $\Leftrightarrow A^{-1}$ is diagonalizable.

Proof. In the discussion leading to the computation of eigenvalues, we showed

λ is an eigenvalue of $A \Leftrightarrow A - \lambda I$ is not invertible.

In particular, taking $\lambda = 0$ gives us

0 is an eigenvalue of $A \Leftrightarrow A$ is not invertible.

This is the same as the first part.

Now assume A is invertible and v is an eigenvector of A with eigenvalue λ. Then $\lambda \neq 0$ and $Av = \lambda v$. Applying $\lambda^{-1}A^{-1}$ to both sides, we get $\lambda^{-1}v = A^{-1}v$. This is the second part.

The third part can be proved by using either definition of diagonalizability, similar to the proof of the previous two problems.

The properties were stated in terms of matrices. They also hold for linear transformations. For example, $T: V \to V$ and its dual $T^*: V^* \to V^*$ have the same eigenvectors and T is diagonalizable if and only if T^* is diagonalizable.

ABOUT SEVERAL MATRICES

In this part, we consider the eigenvalues and eigenvectors of several matrices.

Let v be an eigenvector of A and B with eigenvalues λ and μ. Then v is an eigenvector of $aA + bB$ with eigenvalue $a\lambda + b\mu$, and an eigenvector AB with eigenvalues $\lambda\mu$.

Proof From $Av = \lambda v$, $Bv = \mu v$, we have

$(aA + bB)v = aAv + bBv = a\lambda v + b\mu v = (a\lambda + b\mu)v,$

$(AB)v = A(Bv) = A(\mu v) = \lambda\mu v.$

Example: Consider the matrices

$$A = \begin{bmatrix} 13 & -4 \\ -4 & 7 \end{bmatrix}, B = \begin{bmatrix} 6 & 2 \\ 2 & 9 \end{bmatrix}.$$

From an earlier example, we know $v_1 = (1, 2)$, $v_2 = (-2, 1)$ are eigenvectors of A with eigenvalues 5 and 15. It is also easy to verify that they are eigenvectors of B with eigenvalues 10 and 5. We may then conclude that $v_1 = (1, 2)$, $v_2 = (-2, 1)$ are eigenvectors of the following matrices with the following eigenvalues:

$$A + B = \begin{bmatrix} 19 & -2 \\ -2 & 16 \end{bmatrix}, \lambda_1 = 5 + 10 = 15, \lambda_2 = 15 + 5 = 20.$$

$$A + 2B = \begin{bmatrix} 25 & 0 \\ 0 & 25 \end{bmatrix}, \lambda_1 = 5 + 20 = 15, \lambda_2 = 15 + 10 = 25.$$

$$AB = \begin{bmatrix} 70 & -10 \\ -40 & -30 \end{bmatrix}, \lambda_1 = 5 \times 10 = 50, \lambda_2 = 15 \times 5 = 75.$$

$$AB - B^2 = \begin{bmatrix} 30 & -40 \\ -40 & -30 \end{bmatrix}, \lambda_1 = 50 - 100 = -50, \lambda_2 = 75 - 25 = 50.$$

Let $AB = BA$. Let v be an eigenvector of A with eigenvalue λ and $Bv \neq 0$. Then Bv is again an eigenvector of A with eigenvalue λ.

Proof. $A(Bv) = (AB)v = (BA)v = B(Av) = B(\lambda v) = \lambda Bv$.

Example: Let A be a 3 by 3 matrix commuting with

$$B = \begin{bmatrix} 0 & 2 & 0 \\ 1 & -1 & 1 \\ 3 & -5 & 3 \end{bmatrix}$$

Suppose $v_1 = (1, 1, 0)$, $v_2 = (1, 1, 2)$ are eigenvectors of A with eigenvalues 4 and –2. Then $Bv_1 = (2, 0, -2)$ and $Bv_2 = (2, 2, 4)$ are also eigenvectors of A with eigenvalues 4 and –2. Note that v_1, v_2, Bv_1 form a basis of eigenvectors of A with eigenvalues 4, –2, 4. Therefore for

$$P = [v_1\ v_2\ Bv_1] = \begin{bmatrix} 1 & 1 & 2 \\ 1 & 1 & 0 \\ 0 & 2 & -2 \end{bmatrix}, D = \begin{bmatrix} 4 & 0 & 0 \\ 0 & -2 & 0 \\ 0 & 0 & 4 \end{bmatrix},$$

we have

$$A = PDP^{-1} = \begin{bmatrix} 1 & 3 & -3 \\ -3 & 7 & -3 \\ -6 & 6 & -2 \end{bmatrix}.$$

The matrix appeared in this example. You may also directly verify that $AB = BA$.

Finally, we remark that the properties also holds for linear transformations $T, S: V \to V$. For example, if v is an engenvector for both T and S, then v is an engenvector for $aT + bS$ and TS.

ABOUT POLYNOMIALS OF MATRIX

Polynomials can be applied to square matrices. For example,

$$p(t) = -2 - t + 2t^2 + t^3 \to p(A) = -2I - A + 2A^2 + A^3.$$

Moreover, factorizations are preserved,

$$-2-t+2t^2+t^3=(t-1)(t+1)(t+2) \to -2I-A+2A^2+A^3=(A-I)(A+I)(A+2I).$$

The following summarizes what we know about eigenvalues and eigenvectors of $p(A)$.

Let $p(t)$ be a polynomial. Let A be a square matrix.

- v is an eigenvector of A with eigenvalue $\lambda \Rightarrow v$ is an eigenvector of $p(A)$ with eigenvalue $p(\lambda)$.
- A is diagonalizable $\Leftrightarrow p(A)$ is diagonalizable.
- Any eigenvalue of $p(A)$ is of the form $p(\lambda)$ for some eigenvalue λ of A.

Note that the third part is the "partial converse" of the first part. See the second example below and the subsequent discussion.

Also note that the third part implicitly allows the use of complex numbers as eigenvalues.

Proof. $p(A)$ can be constructed from A and I by repeatedly making use of matrix product and linear combination. The first then follow from this result. Together with basis of eigenvectors interpretation, the second part also follows from the result.

To prove the third part, let μ be an eigenvalue of $p(A)$. We may decompose the polynomial $p(t)-\mu$

$$p(t)-\mu=c(t-\lambda_1)(t-\lambda_2)...(t-\lambda_k),\ c \neq 0$$

(this requires us to allow complex numbers) and get

$$p(A)-\mu I=c(A-\lambda_1 I)(A-\lambda_2 I)...(A-\lambda_k I).$$

Then by the property $detAB = detA\ detB$, we have

$$det(p(A)-\mu I)=c^n\ det(A-\lambda_1 I)det(A-\lambda_2 I)...det(A-\lambda_k I).$$

Since μ is assumed to be an eigenvalue of $p(A)$, we have $det(p(A)-\mu I)=0$. Consequently, $det(A-\lambda_i I)=0$ for some λ_i. Substituting $t=\lambda_i$ into the decomposition of $p(t)-\mu$, we find μ is of the form $p(\lambda_i)$ for some eigenvalue λ_i of A.

Example: Let A be a square matrix such that the eigenvalues of $I+2A+A^2$ are 0, 4, 16. Then any eigenvalue λ of A must be the solution of one of the following equations.

$$1+2\lambda+\lambda^2=0 \text{ (solution: } \lambda=-1)$$

$$1+2\lambda+\lambda^2=4 \text{ (solution: } \lambda=1,-3)$$

$$1+2\lambda+\lambda^2=16 \text{ (solution: } \lambda=3,-5)$$

Thus λ must be one of the following five numbers: –5, –3, –1, 1, 3.

Moreover, the eigenvalues of A must include at least one from each of the three groups above. For example, the eigenvalues of A cannot be –3, –1, 1 only, because either 3 or –5 must also be an eigenvalue of A.

Example: The matrices

$$A = \begin{bmatrix} 0 & 0 \\ 0 & 0 \end{bmatrix}, B = \begin{bmatrix} 0 & 1 \\ 0 & 0 \end{bmatrix}$$

satisfy $A^2 = B^2 = O$. Note that although eigenvalues (= 0) of A and B can be determined from the eigenvalues (= 0) of O, the eigenvectors cannot be determined.

In general, all we can say is that the inclusion of the eigenspace $nul(A - \lambda I)$ M $nul(p(A) - p(\lambda)I)$ (equivalent to the first property above). The equality does not necessarily hold.

Example: In an earlier example, we found the matrix

$$A = \begin{bmatrix} 13 & -4 \\ -4 & 7 \end{bmatrix}$$

has a basis {(1, 2), (–2, 1)} of eigenvectors, with eigenvalues 5, 15. For the characteristic polynomial $p(t) = 75 - 20t + t^2 = (5 - t)(15 - t)$, we know that {(1, 2), (–2, 1)} is a basis of eigenvectors for $p(A) = 75I - 20A + A^2$, with $p(5) = 0, p(15) = 0$ as eigenvalues. Then it is easy to conclude that $p(A) = O$.

In another earlier example, we found the matrix

$$B = \begin{bmatrix} 1 & 3 & -3 \\ -3 & 7 & -3 \\ -6 & 6 & -2 \end{bmatrix}$$

has a basis {(1, 1, 0), (1, 0, –1), (1, 1, 2)} of eigenvectors, with eigenvalues 4, 4, –2. For the polynomial $p(t) = (4 - t)(-2 - t) = -8 - 2t + t^2$, we know that {(1, 1, 0), (1, 0, –1), (1, 1, 2)} is a basis of eigenvectors for $p(B) = -8I - 2B + B^2$, with $p(4) = 0, p(4) = 0, p(-2) = 0$ as eigenvalues. Then it is easy to conclude that $p(B) = O$.

Note that the equality $-8I - 2B + B^2 = O$ is the same as $B(-2I + B) = 8I$. Therefore

$$B^{-1} = (1/8)(-2I + B) = (1/8)\begin{bmatrix} 1 & 3 & -3 \\ -3 & 5 & -3 \\ -6 & 6 & -4 \end{bmatrix}.$$

TRACE AND DETERMINANT

The eigenvalues are the roots of the characteristic polynomial $det(A - \lambda I)$. To see why $det(A - \lambda I)$ is a polynomial of λ, we consider a general 3 by 3 matrix $A = [a_1\ a_2\ a_3]$. Then

$$det(A - \lambda I) = det(a_1 - \lambda e_1, a_2 - \lambda e_2, a_3 - \lambda e_3).$$

By multilinearity, we have

$$det(A - \lambda I) = det(a_1, a_2, a_3)$$

$+ [det(e_1, a_2, a_3) + det(a_1, e_2, a_3) + det(a_1, a_2, e_3)] (-\lambda)$
$+ [det(e_1, e_2, a_3) + det(e_1, a_2, e_3) + det(a_1, e_2, e_3)] (-\lambda)^2$
$+ det(e_1, e_2, e_3) (-\lambda)^3.$

Noting that the terms are obtained by making choices among a_i and $-\lambda e_i$ in each variable, and the power of $-\lambda$ is the number of choices of $-\lambda e_i$. The observation holds in general, and we have the following conclusion.

For an n by n matrix A, $det(A - \lambda I)$ is a polynomial of degree n.

Looking at $det(A - \lambda I)$ in more detail, we see the constant term $det(a_1, a_2, a_3) = detA$, and the top degree term $det(e_1, e_2, e_3) (-\lambda)^3 = (-1)^3\lambda^3$. The observation also holds in general.

Let us consider the second highest degree term. By the column operations $(-a_{31})$[col 1] + [col 3], $(-a_{32})$[col 2] + [col 3], we have

$det(e_1, e_2, a_3) = det(e_1, e_2, a_{31}e_1 + a_{32}e_2 + a_{33}e_3) = det(e_1, e_2, a_{33}e_3) = a_{33}.$

Similarly, we have $det(e_1, a_2, e_3) = a_{22}$, $det(a_1, e_2, e_3) = a_{11}$. Thus the second order term is $(a_{11} + a_{22} + a_{33}) (-\lambda)^2$.

In general, for an n by n matrix A, the trace of A is the sum of its diagonal entries

$trA = a_{11} + a_{22} + ... + a_{nn}$

and the characteristic polynomial is of the form

$det(A - \lambda I) = detA + ... + (-1)^{n-1} trA\, \lambda^{n-1} + (-1)^n \lambda^n.$

Let $\lambda_1, \lambda_2, ..., \lambda_n$ be all the roots of $det(A - \lambda I)$. Since the top degree term of $det(A - \lambda I)$ is $(-1)^n \lambda^n$, we have

$det(A - \lambda I) = (\lambda_1 - \lambda)(\lambda_2 - \lambda)...(\lambda_n - \lambda)$
$= \lambda_1\lambda_2...\lambda_n + ... + (-1)^{n-1}(\lambda_1 + \lambda_2 + ... + \lambda_n) \lambda^{n-1} + (-1)^n \lambda^n.$

Comparing the two formulae for $det(A - \lambda I)$, we conclude the following.

Let $\lambda_1, \lambda_2, ..., \lambda_n$ be all the eigenvalues of an n by n matrix A. Then

$\lambda_1\lambda_2...\lambda_n = detA$
$\lambda_1 + \lambda_2 + ... + \lambda_n = trA.$

We note that decomposing a polynomial into a product of $\lambda_i - \lambda$ is generally possible only if we allow *complex* numbers (or *algebraically closed fields*). Thus some of the eigenvalues in the statement above may be complex.

Example: From this example, we have

$$det\begin{bmatrix} 13 & -4 \\ -4 & 7 \end{bmatrix} = 5 \times 15 = 75,\ tr\begin{bmatrix} 13 & -4 \\ -4 & 7 \end{bmatrix} = 5 + 15 = 20.$$

From this example, we have

$$det\begin{bmatrix}1 & 3 & -3\\ -3 & 7 & -3\\ -6 & 6 & -2\end{bmatrix} = 4 \times 4 \times (-2) = -32,\ tr\begin{bmatrix}1 & 3 & -3\\ -3 & 7 & -3\\ -6 & 6 & -2\end{bmatrix} = 4 + 4 - 2 = 6.$$

Exercise: The following are some statements about eigenvalue and eigenvectors of a single matrix A. Prove the true statement and give counterexample to the wrong statement.

1. u and v are eigenvectors of A with eigenvalue λ ή $u + v$ is an eigenvector of A with eigenvalue λ.

Answer: The statement is true provided the vector $u + v$ is nonzero. Suppose $Au = \lambda u$ and $Av = \lambda v$. Then

$$A(u + v) = Au + Av = \lambda u + \lambda v = \lambda(u + v).$$

2. v is an eigenvector of A with eigenvalue $\lambda \Rightarrow cv$ is an eigenvector of A with eigenvalue λ.

Answer: The statement is true provided the number c is nonzero. Suppose $Av = \lambda v$. Then

$$A(cv) = c(Av) = c(\lambda u) = \lambda(cv).$$

3. u and v are eigenvectors of $A \Rightarrow u + v$ is an eigenvector of A.

Answer: The statement is wrong. The matrix

$$A = \begin{bmatrix}1 & 0\\ 0 & 2\end{bmatrix}$$

has eigenvalues 1, 2 and corresponding eigenvectors (1, 0), (0, 1). But (1, 0) + (0, 1) = (1, 1) is not an eigenvector

$$\begin{bmatrix}1 & 0\\ 0 & 2\end{bmatrix}\begin{bmatrix}1\\ 1\end{bmatrix} = \begin{bmatrix}1\\ 2\end{bmatrix} \neq \lambda\begin{bmatrix}1\\ 1\end{bmatrix}.$$

Remark: The difference between the first and the third statements is the specified eigenvalues. A later exercise contains more discussion.

4. v is an eigenvector of $A \Rightarrow cv$ is an eigenvector of A.

Answer: By the second statement, this statement is also true.

Exercise: Suppose $AB = BA$, and v is an eigenvector of A with eigenvalue λ. Prove that if $Bv \neq 0$, then Bv is an eigenvector of A with eigenvalue λ.

Answer: $A(Bv) = (AB)v = (BA)v = B(Av) = B(\lambda v) = \lambda Bv$.

Exercise: Use the $A = PDP^{-1}$ definition of diagonalizability to prove that A is diagonalizable T! The polynomials and the inverse (if invertible) of A are diagonalizable.

Answer: If $A = PDP^{-1}$, then

$A^2 = PDP^{-1}PDP^{-1} = PD^2P^{-1}$,

$A^3 = PDP^{-1}PDP^{-1}PDP^{-1} = PD^3P^{-1}$,

...

By taking the linear combinations of above, for any polynomial $p(t)$ we get

$p(A) = P\,p(D)\,P^{-1}$.

It is also easy to verify

$$D = \begin{bmatrix} \lambda_1 & 0 & .. & 0 \\ 0 & \lambda_2 & .. & 0 \\ : & : & & : \\ 0 & 0 & .. & \lambda_n \end{bmatrix} \Rightarrow p(D) = \begin{bmatrix} p(\lambda_1) & 0 & .. & 0 \\ 0 & p(\lambda_2) & .. & 0 \\ : & : & & : \\ 0 & 0 & .. & p(\lambda_n) \end{bmatrix},$$

so that $p(D)$ is a diagonal matrix. Thus $p(A)$ is diagonalizable.

As for the inverse, we have

$A = PDP^{-1} \Rightarrow A^{-1} = PD^{-1}P^{-1}$,

where

$$D^{-1} = \begin{bmatrix} \lambda_1^{-1} & 0 & .. & 0 \\ 0 & \lambda_2^{-1} & .. & 0 \\ : & : & & : \\ 0 & 0 & .. & \lambda_n^{-1} \end{bmatrix}$$

is still diagonal.

Exercise: (complex number required) Prove that for a 2 by 2 matrix A, the following are all the posibilities regarding diagonalizability

- A has two distinct eigenvalues and is diagonalizable,
- A has only one eigenvalue and is not diagonalizable,
- $A = cI$ for a constant c.

Answer: The characteristic polynomial of A has degree 2. Thus A has either one eigenvalue or two distinct eigenvalues.

Suppose A has two distinct eigenvalues λ_1 and λ_2. Then for each eigenvalue, we can find at least one eigenvector. Let $v_1, v_2 \in R^2$ be eigenvectors with λ_1, λ_2 as eigenvalues. Then we claim v_1, v_2 are linearly independent. Assume $c_1v_1 + c_2v_2 = 0$. Then applying A and multiplying λ_2 give us

$$c_1Av_1 + c_2Av_2 = c_1\lambda_1v_1 + c_2\lambda_2v_2 = 0, c_1\lambda_2v_1 + c_2\lambda_2v_2 = 0.$$

The difference between the two equalities is

$$c_1(\lambda_1 - \lambda_2)v_1 = 0.$$

Since $\lambda_1 \neq \lambda_2$ and $v_1 \neq 0$, we conclude that $c_1 = 0$. Similarly, we also have $c_2 = 0$. Thus v_1, v_2 are two linearly independent vectors in R^2. Then by this result, the two eigenvectors form a basis. In other words, A is diagonalizable.

If A has one eigenvalue λ and is diagonalizable, then $A = PDP^{-1}$. Because the two diagonal entries of D are the same eigenvalue λ, we have $D = \lambda I$ and $A = PDP^{-1} = \lambda I$.

The remaining case is that A has one eigenvalue and is not diagonalizable. This completes the discussion of the three cases as long as the exercise is concerned.

We continue with a more detailed study of the case that A has one eigenvalue λ and is not diagonalizable. Let v be the one eigenvector we can find for λ. Choose another vector w, such that $\{w, v\}$ form a basis of R^2. Then

$$Aw = aw + bv$$

for some numbers a, b. Since A has no basis of eigenvectors, w is not an eigenvector. This means b “‘ 0. Let $u = b^{-1}w$. Then $\{u, v\}$ is still a basis, and by dividing the equality above by b, we get

$$Au = au + v$$

Combined with

$$Av = \lambda v,$$

we see the matrix for the linear transformation $x \to Ax$ with respect to the basis $\{u, v\}$ is

$$\begin{bmatrix} a & 1 \\ 0 & \lambda \end{bmatrix}$$

In particular, a is an eigenvalue of A. Since λ is assumed to be the only eigenvalue, we have $a = \lambda$ and

$$Au = \lambda u + v, Av = \lambda v.$$

Thus the matrix for the linear transformation with respect to the basis $\{u, v\}$ is

$$\begin{bmatrix} \lambda & 1 \\ 0 & \lambda \end{bmatrix}.$$

This is equivalent to

$$A = P\begin{bmatrix} \lambda & 1 \\ 0 & \lambda \end{bmatrix}P^{-1}.$$

Remark Any 2 by 2 matrix is similar to one of the following two matrices

$$\begin{bmatrix} \lambda_1 & 0 \\ 0 & \lambda_2 \end{bmatrix}, \begin{bmatrix} \lambda & 1 \\ 0 & \lambda \end{bmatrix}.$$

These are the possible Jordan canonical forms of A. The vector u we found in the discussion is a generalized eigenvector.

Exercise: Find 2 by 2 diagonalizable matrices A and B, such that $A + B$ and AB are not diagonalizable.

Answer: The idea is to find 2 by 2 matrices A and B with two distinct eigenvalues (which are necessarily diagonalizable by this exercise), such that $A + B$ or AB is the following typical non–diagonalizable matrix

$$\begin{bmatrix} 1 & 0 \\ 0 & 2 \end{bmatrix}.$$

One possible choice for the $A + B$ counterexample is

$A = \begin{bmatrix} 1 & 1 \\ 0 & 0 \end{bmatrix}$ diagonalizable, $B = \begin{bmatrix} -1 & 0 \\ 0 & 0 \end{bmatrix}$ diagonalizable, $A + B = \begin{bmatrix} 0 & 1 \\ 0 & 0 \end{bmatrix}$ non-diagonalizable.

One possible choice for the AB counterexample is

$A = \begin{bmatrix} 1 & 1 \\ 0 & 0 \end{bmatrix}$ diagonalizable, $B = \begin{bmatrix} 0 & 0 \\ 0 & 1 \end{bmatrix}$ diagonalizable, $AB = \begin{bmatrix} 0 & 1 \\ 0 & 0 \end{bmatrix}$ non-diagonalizable.

Exercise: For each of the following three equations, what can you say about the solution A?

1. $A^2 = \begin{bmatrix} 13 & -6 \\ 18 & -8 \end{bmatrix}$

Answer: A^2 has a basis $\{(1, 2), (2, 3)\}$ of eigenvectors, with eigenvalues 1, 4:

$$A^2 = \begin{bmatrix} 1 & 3 \\ 2 & 3 \end{bmatrix}\begin{bmatrix} 1 & 0 \\ 0 & 4 \end{bmatrix}\begin{bmatrix} 1 & 3 \\ 2 & 3 \end{bmatrix}^{-1}.$$

By this result, A must have a solution 1 or –1 of $\lambda^2 = 1$ as one eigenvalue. Moreover, if v is an eigenvector of A with 1 or -1 as the eigenvalue, then v is an eigenvector of A^2 with $(\pm 1)^2 = 1$ as the eigenvalue. From the diagonalization of A^2, we find v must be a nonzero multiple of $(1, 2)$.

Similarly, A also have a solution 2 or -2 as an eigenvalue. By the similar argument as above, the corresponding eigenvector must be a nonzero multiple of $(2, 3)$.

Thus we conclude the following four possible choices of A.

$$A = \begin{bmatrix} 1 & 2 \\ 2 & 3 \end{bmatrix} \begin{bmatrix} 1 & 0 \\ 0 & 2 \end{bmatrix} \begin{bmatrix} 1 & 3 \\ 2 & 3 \end{bmatrix}^{-1} = \begin{bmatrix} 5 & -2 \\ 6 & -2 \end{bmatrix},$$

$$A = \begin{bmatrix} 1 & 2 \\ 2 & 3 \end{bmatrix} \begin{bmatrix} -1 & 0 \\ 0 & -2 \end{bmatrix} \begin{bmatrix} 1 & 2 \\ 2 & 3 \end{bmatrix}^{-1} = \begin{bmatrix} -5 & 2 \\ -6 & 2 \end{bmatrix},$$

$$A = \begin{bmatrix} 1 & 2 \\ 2 & 3 \end{bmatrix} \begin{bmatrix} -1 & 0 \\ 0 & 2 \end{bmatrix} \begin{bmatrix} i & 2 \\ 2 & 3 \end{bmatrix}^{-1} = \begin{bmatrix} 11 & -6 \\ -18 & -10 \end{bmatrix},$$

$$A = \begin{bmatrix} 1 & 2 \\ 2 & 3 \end{bmatrix} \begin{bmatrix} 1 & 0 \\ 0 & -2 \end{bmatrix} \begin{bmatrix} 1 & 2 \\ 2 & 3 \end{bmatrix}^{-1} = \begin{bmatrix} 11 & 6 \\ -18 & -10 \end{bmatrix}$$

2. $A^2 = \begin{bmatrix} 0 & 1 \\ 0 & 0 \end{bmatrix}$.

Answer: A^2 is has one eigenvector $(1, 0)$ with eigenvalue 0, and is not diagonalizable. By this result and an argument similar to the one in the first problem, A must also have $(1, 0)$ as an eigevector, with the solution 0 of $\lambda^2 = 0$ as eigenvalue, and is not diagonalizable. By the discussion in this exercise, we must have

$$A = P \begin{bmatrix} 0 & 1 \\ 0 & 0 \end{bmatrix} P^{-1}.$$

However, this A satisfies

$$A^2 = P \begin{bmatrix} 0 & 1 \\ 0 & 0 \end{bmatrix}^2 P^{-1} = O,$$

which contradicts the original equation. We conclude that the equation has no solution.

3. $A^2 = \begin{bmatrix} 1 & 0 \\ 0 & 1 \end{bmatrix}$

Answer: By this result, the eigenvalues of A must satisfy $\lambda^2 = 1$. Thus we have the following posibilities.

First, A has two distinct eigenvalues 1, –1. Then A is diagonalizable and the general formula is

$$A = P\begin{bmatrix} 1 & 0 \\ 0 & -1 \end{bmatrix} P^{-1}.$$

Second, A has one eigenvalue 1 or –1, and is diagonalizable. By this exercise, we conclude $A = I$ or $-I$.

Third, A has one eigenvalue 1 or –1, and is not diagonalizable. By this exercise again, the general formula for A is one of the following

$$A = P\begin{bmatrix} 1 & 1 \\ 0 & 1 \end{bmatrix} P^{-1},$$

$$A = P\begin{bmatrix} -1 & 1 \\ 0 & -1 \end{bmatrix} P^{-1}.$$

However, the first formula would imply

$$A^2 = P\begin{bmatrix} 1 & 2 \\ 0 & 1 \end{bmatrix} P^{-1},$$

which would never be the identity. Similarly, the second formula also does not satisfy the original equation.

Thus the solutions are one of the following.

$$A = P\begin{bmatrix} 1 & 0 \\ 0 & -1 \end{bmatrix} P^{-1}, A = I, A = -I.$$